Stochastic Point Processes:

Statistical Analysis, Theory, and Applications

Applied Probability and Statistics (Continued)

BROWNLEE · Statistical Theory and Methodology in Science and Engineering, *Second Edition*
CHAKRAVARTI, LAHA and ROY · Handbook of Methods of Applied Statistics, Vol. I
CHAKRAVARTI, LAHA and ROY · Handbook of Methods of Applied Statistics, Vol. II
CHERNOFF and MOSES · Elementary Decision Theory
CHIANG · Introduction to Stochastic Processes in Biostatistics
CLELLAND, deCANI, BROWN, BURSK, and MURRAY · Basic Statistics with Business Applications
COCHRAN · Sampling Techniques, *Second Edition*
COCHRAN and COX · Experimental Designs, *Second Edition*
COX · Planning of Experiments
COX and MILLER · The Theory of Stochastic Processes
DANIEL and WOOD · Fitting Equations to Data
DAVID · Order Statistics
DEMING · Sample Design in Business Research
DODGE and ROMIG · Sampling Inspection Tables, *Second Edition*
DRAPER and SMITH · Applied Regression Analysis
ELANDT-JOHNSON · Probability Models and Statistical Methods in Genetics
GOLDBERGER · Econometric Theory
GUTTMAN, WILKS and HUNTER · Introductory Engineering Statistics, *Second Edition*
HAHN and SHAPIRO · Statistical Models in Engineering
HALD · Statistical Tables and Formulas
HALD · Statistical Theory with Engineering Applications
HOEL · Elementary Statistics, *Third Edition*
HUANG · Regression and Econometric Methods
JOHNSON and KOTZ · Distributions in Statistics
Discrete Distributions
Continuous Univariate Distributions-1
Continuous Univariate Distributions-2
Continuous Multivariate Distributions
JOHNSON and LEONE · Statistics and Experimental Design: In Engineering and the Physical Sciences, Volumes I and II
LANCASTER · The Chi Squared Distribution
LEWIS · Stochastic Point Processes
MILTON · Rank Order Probabilities: Two-Sample Normal Shift Alternatives
RAO and MITRA · Generalized Inverse of Matrices and Its Applications
SARD and WEINTRAUB · A Book of Splines
SARHAN and GREENBERG · Contributions to Order Statistics
SEAL · Stochastic Theory of a Risk Business
SEARLE · Linear Models
THOMAS · An Introduction to Applied Probability and Random Processes
WHITTLE · Optimization under Constraints
WILLIAMS · Regression Analysis

continued on back

Stochastic Point Processes:

Statistical Analysis, Theory, and Applications

PETER A. W. LEWIS, Editor

Naval Postgraduate School

Monterey, California

WILEY - INTERSCIENCE

a division of John Wiley & Sons, Inc.

New York London Sydney Toronto

Library of Congress Cataloging in Publication Data:

Main entry under title:

Stochastic point processes.

(Wiley series in probability and mathematical statistics)

"Contains the bulk of the papers presented at a conference held at the IBM Center, Yorktown Heights, New York, on August 2-7, 1971."

1. Stochastic processes--Congresses. 2. Analysis of variance--Congresses. I. Lewis, Peter A. W., 1932 ed.

QA274.A1S76 519.2 72-3954

ISBN 0-471-53350-5

Printed in the United States of America

10 9 8 7 6 5 4 3 2 1

To Lydia

For patience and understanding in
the face of many series of events

Preface

This volume contains the bulk of the papers presented at a conference held at the IBM Research Center, Yorktown Heights, New York on August 2-6, 1971. The title of the meeting was "Stochastic Point Processes: Statistical Analysis, Theory and Applications."

The idea of holding the meeting grew out of a seminar series and some special lectures on stochastic point processes given in the Mathematics Department at Imperial College, University of London, during the academic year 1969-70, and I wish to express my thanks to the National Institutes of Health for providing me with the fellowship which made my presence at the seminar possible. At the end of that seminar series, it became evident to Professor D. R. Cox and myself that it would be timely to hold a fairly comprehensive conference to pull together the many aspects of point processes, all rapidly developing, and to bring together the applied and theoretical people in the field.

Since the International Statistical Institute (ISI) was to hold its biennial meeting in Washington, D. C., from August 10-20, 1971, a date was set for August 2-6, 1971, for the point process conference and an organizing committee was formed. The committee members were P.A.W. Lewis, Chairman, M. S. Bartlett, D. R. Cox, J. Gani, K. Matthes, P. A. P. Moran, E. Parzen, R. Pyke, W. L. Smith, and D. Vere-Jones. Subsequently the conference became one of the satellite symposia organized by Professor J. Neyman and the International Association for Statistics in the Physical Sciences (IASPS) in conjunction with the ISI meeting in Washington, D. C.. Other satelite symposia were held on statistical problems in the fields of turbulence, metallurgy, pollution, and hydrology.

The conference was started with a small grant from the IBM Mathematical Sciences Department through Drs. R. E. Gomory and S. Winograd. To them and the IBM Research Division, who provided physical facilities and numerous services, I am deeply grateful. The main financial support for the conference was provided by the Office of Naval Research (ONR) and the National Science Foundation (NSF) and thanks are due to Dr. Bruce McDonald of ONR and Dr. William Pell of NSF for all of their help.

On the technical side much help was provided, in addition to the committee members, by Dr. D. J. Daley, Dr. A. J. Lawrance and Professor M. R. Leadbetter. The splendid organization of the conference was due primarily to Mrs. M. Casey of IBM, Dr. A. J. Lawrance (temporarily of IBM) and Dr. W. Turner of IBM. To Mrs. Casey, too, go thanks for constant good cheer and gentle direction under often trying circumstances.

Responsibility for editing this volume is my own. Enormous help was provided, especially with the initial editing at the time of the conference, by Professors Cox and Gani. Subsequent editorial help was also provided by Dr. D. J. Daley, Dr. A. J. Lawrance, Professor D. Vere-Jones and Dr. M. Westcott. The primary aim in editing the volume was to produce the volume as soon as possible after the conference; to do this some liberty had to be taken with editorial niceties. Thus while notational conventions, etc., are substantially the same, certain concessions were made to the very different conventions prevailing in different fields. Also no effort was made to provide absolute consistency in reference to obscure journals and articles.

Thanks for typing of the volume are due to Mrs. Ruth Guthrie and Miss Rosemarie Stampfel.

Finally, the editing of the volume was done at the Naval Postgraduate School, Monterey, California, to whom I am grateful for support.

P. A. W. Lewis
Naval Postgraduate School
March, 1972

Contents

INTRODUCTION xii
by *P. A. W. Lewis*

STATISTICAL ANALYSIS

Recent results in the statistical analysis of univariate point processes 1
by *P. A. W. Lewis*

The statistical analysis of dependencies in point processes 55
by *D. R. Cox*

Statistical analysis of non-homogeneous Poisson processes 67
by *M. Brown*

Statistical inference for doubly stochastic Poisson processes 90
by *J. Grandell*

The use of distance methods for the analysis of spatial distribution of points 122
by *P. Holgate*

Some applications of multivariate point processes 136
by *M. S. Bartlett*

Estimation of traffic densities at the Lincoln tunnel from time series of flow and speed data 151
by *D. C. Gazis* and *M. W. Szeto*

Statistical properties of traffic counts 166
by *G. F. Newell* and *G. A. Sparks*

Time information in nuclear physics experiments 175
by *M. W. Sachs*

MODELS

Some models for stationary series of univariate events 199
by *A. J. Lawrance*

Summary of 'Further results for Gauss-Poisson processes' 257
by *R. K. Milne* and *M. Westcott*

Spectra of some mutually exciting point processes with associated variables 261
by *A. G. Hawkes*

A generalization of the two-state two-interval semi-Markov model 272
by *A. Ekholm*

Certain chance mechanisms related to birth and death processes which have classical discrete distributions as special cases 285
by *M. T. Boswell* and *G. P. Patil*

THEORY

A summary of the theory of point processes 299
by *D. J. Daley* and *D. Vere-Jones*

Infinitely divisible point processes 384
by *K. Matthes*

The prediction of stationary point processes 405
by *J. Jowett* and *D. Vere-Jones*

Point processes generated by level crossings 436
by *M. R. Leadbetter*

A survey of the mathematical theory of multidimensional point processes 468
by *L. Fisher*

Theory of hyperplane processes 514
by *K. Krickeberg*

Summary of some results on point and line processes 522
by *F. Papangelou*

Stochastic theory of evolutionary processes (1937-71) 533
by *A. Ramakrishnan*

Superposition of point processes — 549
by *E. Cinlar*

On the convergence of sums of point processes with integer marks — 607
by *D. Szasz*

On weak convergence of the sums of multivariate stochastic point processes — 616
by *B. Grigelionis*

Palm distributions and superpositions of independent point processes in R^n — 626
by *R. V. Ambartzumian*

APPLICATIONS

Processes of clustering and applications — 646
by *J. Neyman* and *E. L. Scott*

Some point process models of lunar and planatary surfaces — 682
by *A. Marcus*

The stochastic properties of spike trains recorded from nerve cells — 700
by *R. B. Stein*

Some point processes in motor and sensory neurophysiology — 732
by *B. W. Knight*

Point processes in epidemiology — 757
by *J. Gani*

Point process problems in reliability — 775
by *D. P. Gaver*

Point processes in forestry — 801
by *W. G. Warren*

Some spatial and temporal point processes in photographic science — 818
by *J. F. Hamilton, W. H. Lawton* and *E. A. Trabka*

Potential contributions of the theory of stochastic point processes to the social science study of map distributions — 869
by *M. F. Dacey*

AUTHOR INDEX — 887

INTRODUCTION

P. A. W. Lewis
Naval Postgraduate School, Monterey, California

The aim of this introduction is to provide a brief summary of the papers in this volume and to put them in the context of the whole field of point process methodology as it is developing today. The papers do represent a good cross-section of the field, but time and space limitations induced some truncation. Thus, for example, interest in and development of techniques for handling doubly stochastic Poisson process has burgeoned very recently, but it was only possible to devote one paper (Grandell, p. 90) to it completely, although some aspects of the problems involved are touched on in Lewis (p. 1), Lawrance (p. 199), Daley and Vere-Jones (p. 299) and Krickeberg (p. 514). Some truncation was also produced by the limitations imposed by my own interests; as the organization of the conference proceeded, and more particularly during the discussion at the conference, my own knowledge of the field and its interrelationship with other aspects of stochastic process theory grew. However, by that time the program was already set!

It is perhaps worthwhile to start by reproducing the prospectus from the original call for papers for the conference.

"The aim of the conference is to bring together mathematicians and statisticians working in the field of point processes and workers in applied fields such as ecology, neurophysiology, traffic studies, reliability, geography, forestry, epidemiology, physics and geophysics. Consequently, there will be three categories of papers presented at the conference;

i) Invited survey papers on the mathematical theory, statistical analysis and models of univariate point processes, multivariate point processes, multidimensional point processes and line processes;

ii) Invited review papers on the types of problems involving point processes encountered in fields of application such as ecology, neurophysiology, physics, forestry, reliability, traffic, geography, etc.;

iii) A very limited number of contributed and invited papers on new work in the field."

Two problems arose from this call.

One was that the distinction between survey and review papers, never really too clear, became blurred. Moreover many invited survey and/or review papers missed some of the coverage which was asked for when they were solicited, and also overlapped into other areas. This of course is inevitable and desirable in light of the professed aim of the conference to join together some of the disparate and overlapping areas of endeavour.

A second problem was that of nomenclature; this was discussed at some length at the conference. The general name "point process" appears to go back to Wold (1948) and is now fairly widely used; other names are "event processes," "series of events," "streams of events" and "flows of events."

On the other hand, the taxonomy of point processes is not so easy to resolve. The usage "univariate point processes" in this volume refers to events occurring in a one dimensional continuum, usually time, the events being distinguishable only by where they occur, i.e., having no qualitative or quantitative information attached to them. The usage "multidimensional point processes" refers to the same situation where the events occur in a multidimensional continuum. However, "multidimensional point processes" is also used to refer to the situation where different types of events occur (events distinguished by qualitative information), especially in the Russian literature. The tenor of the discussion at the conference was that "multivariate point processes" (Cox and Lewis, 1972) was preferable in this latter situation, since events of two types occur in higher dimensions, e.g., different types of trees in a forest. This latter terminology has been adopted here. It is also perhaps worthwhile to point out that there are two categories of multivariate point processes; in the one the typing arises because of qualitative information and should perhaps be called multitype point processes; in the other the typing is in fact because of physical (time or space) separation of events (e.g., arrivals and departures from a queue; this spring's flowers and last spring's flowers) and should perhaps be called multiple point processes, following Hannan (1970).

Another fairly confused usage is that of "line processes." Bartlett (1967) used this to describe processes in which the events were accompanied by quantitative information. For example, earthquakes not only occur at "points" in time, but also have space and energy information attached to them. This is a special case of a general situation defined by Matthes (1963) and called "marked point processes" (see Daley and Vere-Jones, p. 299). This designation of the situation where events have quantitative information attached to them seems best referred to as marked point processes, even though Matthes' definition is broad enough to cover multitype processes too.

True "line processes" do occur. For example, planes flying in the air and particles in space or bubble chambers, the "lines" usually being of finite length, although it is mathematically convenient to think of infinite lines. These "line processes" when sampled at several instants of time produce multiple, multivariate point processes, the space being usually three dimensional. A survey of line process theory is given by Krickeberg (p. 514), who calls them "hyperplane processes." This perhaps is a better name than "line processes."

A related situation which occurs frequently in physics (Ramakrishnan, p. 533) is that of events occurring in time and space. This is distinguished from multidimensional point processes because the time and space dimensions usually do not occur on an equal footing. Sampling of these spatial point processes by pooling events occurring in a time period produces ordinary point processes, e.g., the records of places of occurrence of traffic accidents on successive days.

Analysis and discussion of these "line processes" or "hyperplane processes" and "spatial point processes" by mathematicians and statisticians has barely begun and they are clearly a broad enough category that a detailed attempt at naming would be futile. Moreover taxonomies are at best leaky.

To conclude the discussion it should be pointed out that these line processes occur in metallurgy, and the interested reader might also refer to the papers presented at the IASPS Satellite Symposium on Metallurgy which will appear in the journal "Advances in Applied Probability," in 1972.

We now describe briefly the papers in the volume and their interrelationships. The volume contains all of the papers presented at the conference except three. The paper by Milne and Westcott (p. 257) is a summary of a long paper which will appear in "Advances in Applied Probability" in 1972, as will the paper presented at the conference by J. E. Moyal entitled, "Number operators and population processes in quantum field theories." Dr. Moyol graciously offered to submit his paper to "Advances in Applied Probability" when it became evident that this present volume was far too long to be practicable. The paper by W. L. Smith entitled "Queues; a broad view" which was presented at the conference will appear elsewhere. The papers by Grigelionis and Ambartzumian were scheduled for the conference, but the authors were unable to present them in person.

The organization of the papers in the volume is not exactly the same as the session organization at the conference. This has been done partly because of editorial convenience and partly because the overlaps and abutments of the papers were clearer in hindsight. Thus there is an implied classification of the papers by their appearance in Statistical Analysis, Models, Theory or Applications and the order in which they appear in those four categories. Again, the classification is leaky and only partially valid. The paper by Neyman and Scott, for instance, describes a long term effort to apply statistics and probability to the analysis of the dispersion of stars in space. However, it also describes some important results on cluster models and the statistical analysis of cluster point processes.

Starting with the section entitled Statistical Analysis, the summary paper by Lewis (p. 1) attempts to survey results obtained on the statistical analysis of univariate point processes since the publication of the monograph by Cox and Lewis (1966). There is particular emphasis on new results, both theoretical and tabular, on tests for renewal processes and on (time) trend analysis in non-homogeneous Poisson processes. Both cyclical and evolutionary trends are considered, and the methodology is illustrated by results obtained on the analysis of a very long series of arrivals at an intensive care unit. Cox (p. 55) gives a general model for the analysis of trends in Poisson processes and extends it to time-dependent (modulated) renewal processes. The methodology given for handling these situations is then used to derive formally a test for dependence in a bivariate point process which was given in Cox and Lewis (1972). The paper by Brown (p. 67) supplies much of the theory needed to rigorize the models and methodology for non-homogeneous Poisson processes given in the papers by Lewis (p. 1) and by Cox (p. 55). Other aspects of the analysis of these processes are also considered.

Methods based on least squares theory for making inferences in the important class of doubly stochastic point processes are given by Grandell (p. 90). Analysis of doubly stochastic Poisson processes is presently a very active field, with interest in applications in physics, engineering and medicine. References to the recent work by, in particular, Snyder, Karp and Clark is given in Lewis (Section 4.3). Other recent work is by Macchi (1971). Krickeberg (p. 514) refers to these doubly stochastic Poisson processes as "Cox processes;" they were first discussed in Cox (1955).

The fifth paper in this section is a survey paper by Holgate (p. 122) on the use of distance methods for analyzing spatial patterns; the fields of application are very numerous.

Bartlett (p. 136) takes up two important problems in multivariate and multidimensional point processes which are only just beginning to receive attention. One problem is the analysis of dependencies between point processes and continuous stochastic processes; in Bartlett's paper these are the times at which vehicles pass a point on a road and their velocities. In addition, typing by which lane the vehicle is in is added and spectral methods are suggested for the analysis of data. A similar problem in multivariate two-dimensional point processes is also discussed. The papers by Gazis and Szeto (p. 151) and Newell and Sparks

(p. 166) give analyses of two specific situations involving marked multivariate point processes occurring in traffic studies.

Finally the paper by Sachs (p. 175) in this section on Statistical Analysis is an attempt to inform statisticians of the very complex and large scale point process problems which are faced and analyzed by physicists quite routinely. Both the types of problems and the techniques of analysis are described.

The section on Models starts with a survey paper by Lawrance (p. 199) of models for stationary series of events. There is a table of contents on pages 199 and 200. The coverage is primarily of processes (such as semi-Markov processes) with some kind of Markovian dependency, generally defined on the intervals between events; doubly stochastic Poisson processes of several kinds, and cluster processes. The emphasis is on aspects, such as arbitrary event and arbitrary time initial conditions and marginal distributions of interevent intervals, which are useful in modeling and statistical analysis. The paper by Milne and Westcott (p. 257) summarizes results the authors have obtained on a process, defined by Newman, through its characteristic functional and called a Gauss-Poisson process. The authors have shown it to be a special case of the Bartlett-Lewis cluster process (see the paper by Lawrance, p. 237 for a description of the Bartlett-Lewis process). Hawkes' paper (p. 261) extends results on the mutually exciting point processes (which he introduced recently) by attaching auxiliary variables to the points. Count spectra and cross spectra are given.

The generalization of the (univariate) two state semi-Markov process considered by Ekholm (p. 272) allows for other than geometrically distributed runs of intervals of each type. Since the process is considered to be univariate, the type of event or interval is not observable; however this ingenious generalization of the physical mechanism of the model is intuitively appealing. Ekholm has applied the model in neurophysiological contexts. Finally Boswell and Patil (p. 285) discuss several processes which can be considered to be multivariate point processes of finite duration.

The section entitled Theory starts on page 299 with a comprehensive survey paper by Daley and Vere-Jones. This was originally a survey paper on the theory of univariate point processes but its scope is much wider, as can be seen from the detailed table of contents on pages 299 and 300. It overlaps into models, particularly in discussing some of the operations such as clustering which generates one of the most important theoretical and practical point process models. The paper by Matthes (p. 384) compliments that of Daley and Vere-Jones, going into much greater detail on results of the author and his coworkers on infinitely divisible point processes and the important sub-class of cluster point processes. Many of these results will appear in a forthcoming book by Kerstan, Matthes and Mecke (1972). Prediction theory for univariate point processes, both for the interval and counting processes, is considered by Jowett and Vere-Jones (p. 405). Results of prediction theory for generalized random functions are used, although (as the authors point out) the solutions are not entirely satisfactory. Much of this is because of the distinctive role of intervals between events in univariate point processes. Ad hoc methods are considered for certain cluster processes. To conclude this section on univariate point processes, Leadbetter (p. 436) discusses the important class of point processes consisting of times at which a stochastic process crosses a level or curve. This updates results given by Cramer and Leadbetter (1967).

The theory of multidimensional point processes is considered by Fisher (p. 468). Much of this theory developed separately from the theory of univariate point processes, even though there were points of contact, as in the early work of Bartlett. Fisher's concluding remarks (p. 505) are of great interest. He notes that there is an essential difference between multivariate and univariate point processes, this being the presence or lack of ordering, i.e. the fact that in univariate point processes intervals between events are so important. Moreover, because of the wide use of multivariate point processes in fields such as ecology, forestry, etc., he notes that there has been more development of statistical analysis techniques than of theory. This is the reverse of the situation for univariate point processes. I would add that much remains to be done in statistical theory for multidimensional point processes, especially since in most fields of application, stationarity would appear to me to be an unjustifiable artifice.

The theory of line process is considered by Krickeberg (p. 514), who calls them hyperplane processes, and in summary by Papangelou (p. 522) while Ramakrishnan (p. 533) discusses the space-time point process which were considered at an early stage in physics and attacked, very often using the idea of product densities, by Bhaba, Bartlett, and Moyal. It seems certain that there will be further developments in this area, and migration of some of the ideas current in univariate point process theory and (static) multivariate point process theory to this more difficult and complex area.

The last group of papers in this section on Theory concerns the superposition of point processes. The survey paper by Çinlar covers finite superpositions of univariate point processes (an area of great importance and very active at present) and the known and fairly complete results on convergence to Poisson limits. In Sections 6 and 7 important work on rates of convergence to the Poisson limit and on the "error" involved in approximating a finite superposition by a Poisson process is covered. The discussion is mainly in terms of counting distributions; for results using spectra and intensity functions see Cox and Lewis (1966, Chapter 8).

Generalizations of superposition results to processes with marks and multivariate point processes are given by Szasz (p. 607) and Grigelionis (p. 616) respectively, while Ambartzumian (p. 626) considers superpositions of multidimensional point processes and new techniques for analyzing such superpositions using Palm distributions.

The Applications section starts with a paper by Neyman and Scott (p. 646), mentioned earlier, on cluster processes, generally multidimensional, and the applications which these processes have had in several fields. Of particular interest is the description of the authors own work, carried out over a long period of time, on the spatial distribution of galaxies and the interrelationship of their results with a developing cosmological theory. A similar application is reported by Marcus (p. 682).

Perhaps of all the fields of application to which point processes (and other stochastic processes) are central, none will be more important and as rapidly developing in the future as the spike train analyses of neurophysiology. This is the subject of two fascinating papers, one by Knight (p. 732) and the other by Stein (p. 700). Knight describes some specific and recent analyses while Stein's paper is a survey. An interesting development reported there from Stein's co-workers is a technique to allow the Fast Fourier Transform (Cooley-Tukey algorithm) to be used for rapid computation of the Bartlett (count) spectrum of a point process.

Stein's presentation featured a display on an oscilloscope of spike trains originating in the speakers hand; this unfortunately could not be presented in this volume.

Epidemiology is an important and old field concerned with point processes. There should be more contact between the work reported by Gani (p. 757) and other work in point processes. In particular, Gani discusses research attempting to analyze the interaction between the time of occurrence and place of occurrence. This problem of space-time interaction, a problem in the analysis of marked point processes, is also important (perhaps crucial), in the analysis of earthquakes. (This application of point processes was not presented at the conference, but has recently been surveyed by Vere-Jones (1970).)

Gaver (p. 775) details point process problems in reliability, a field with a lore of its own and somewhat separate from the mainstream of point process theory. Of particular interest in Gaver's paper is his descriptions of dependency structures in reliability situations and the description of models for correlated positive random variables. This has points of contact with the tests for renewal processes discussed in Lewis (p. 1).

Two important sources of spatial point processes, forestry and photographic science, are the subject of the papers by Warren (p. 802) and Hamilton, Lawton and Trabka (p. 818).

Forestry problems generated much of the statistical work reported on by Holgate (p. 122) and Warren's paper gives an ample description of these problems. Reading of both of these papers will reinforce the conclusions of Fisher (p. 468), and of Warren in his Summation (p. 811) that there is a large gap to be filled between the available theory and the needs of model builders and data analysts. In addition Warren points out the need for interpreters of the theory to practioners.

The paper by Hamilton, Lawton and Trabka is highly recommended as a source of new and important problems; of interest too are the descriptions of problems solved or partially solved by photographic scientists and well known in other fields such as queueing theory.

The last paper by Dacey (p. 869) describes another, perhaps tenuous field of application of point processes, the study of map distributions. Perhaps modesty detered Dacey from referring to some of his own contributions, some of which are referenced in the book by King (1969) on statistical geography.

REFERENCES

Bartlett, M. S. (1967). The spectral analysis of line processes. Proc. 5th Berkeley Symposium 3, 133-54.

Cox, D. R. (1955). Some statistical methods connected with series of events. J. R. Statist. Soc. B 17, 129-64.

Cox, D. R. & Lewis, P. A. W. (1972). *The Statistical Analysis of Series of Events.* London: Methuen and New York: Barnes & Noble.

Cox, D. R. & Lewis, P. A. W. (1972). Multivariate point processes. To appear in Proc. 6th Berkeley Symposium.

Cramer, H. & Leadbetter, M. R. (1967). *Stationary and Related Stochastic Processes.* New York: Wiley.

Hannan, E. J. (1970). *Multiple Time Series.* New York: Wiley.

Kerstan, J., Matthes, K. & Mecke, J. (1972). *Unbegrenzt teilbare Punktprozesse.* Berlin: Akademie-Verlag (in preparation).

King, L. J. (1969). *Statistical Analysis in Geography.* Englewood Cliffs, N.J.: Prentice-Hall.

Macchi, O. (1971). Stochastic point processes and multicoincidences. IEEE Trans. Information Theory IT-17, 2-6.

Matthes, K. (1963). Stationare zufallige Punktfolgen, I. Jber. Deutsch. Math-Verein 66, 66-79.

Vere-Jones, D. (1970). Stochastic models for earthquake occurrence. J. R. Statist. Soc. B 32, 1-62.

Wold, H. (1948). Sur les processus stationnaires ponctuels. Colloques Int. Cent. Natn. Rech. Scient. 13, 75-86.

RECENT RESULTS IN THE STATISTICAL ANALYSIS OF UNIVARIATE POINT PROCESSES

P. A. W. Lewis
Naval Postgraduate School, Monterey, California

SUMMARY

Results on the statistical analysis of series of events published subsequent to the monograph by Cox and Lewis on this subject are surveyed. Special emphasis is given to tests for renewal processes, a considerable amount now being known about the distributions of some of the test statistics involved, and to testing the functional form of a trend in a nonhomogeneous Poisson process, as well as the point process model itself. A survey of work in special processes such as cluster processes and doubly stochastic Poisson processes is also given.

1. INTRODUCTION

The object of this paper is to survey recent results in the statistical analysis of univariate point processes (series of events). The survey is personal, reflecting my own present interests, and is not totally comprehensive. For convenience, I have taken "recent" to mean anything published since Cox and Lewis (1966). I have also mentioned some topics and results which were omitted, for reasons of emphasis or ignorance, from that monograph. Finally, I point out areas where further work is required.

A survey of work in this field is not so difficult as a survey of, say, the theory of point processes because the advances in the statistical analysis of point processes are, by comparison with the theory, few. This may reflect the relative difficulties, but that thought may be a personal bias.

Some of the shortcomings of the Cox and Lewis monograph were, in hindsight, the following:

(i) Not enough consideration of grouped data. In some cases where point events occur, such as epidemiology, recording limitations or the volume of data force one to work with numbers of events in fixed intervals whose width may or may not be controllable in advance. I have touched on this problem recently (Lewis, 1970); separate spectral analyses of the intervals and of the counting process (Bartlett, 1963) are, for example, not possible. One has to use a spectral analysis of the grouped counts (i.e., number of events in a fixed time interval), which is very nonstandard in its distribution theory, but in which well-known problems of aliasing arise. Results of Cox (1970) may be useful here, and Cox (1955) has considered some other aspects of the analysis of grouped events.

Interesting examples of this type of problem are found in physics and optics. For example (Helstrom, 1964; Karp and Clark, 1970) photon or other particle emissions are known from physical considerations to be generated by a doubly stochastic Poisson process and it is required to determine parameter values of the driving process from counts of the number of photons emitted in successive periods. Here it is prohibitively costly to record exact times of occurrence. However, the recording interval can be determined in advance by the experimenter. Problems of grouped Poisson counts (McNeill, 1971) are also common.

(ii) Very little emphasis was given to sequential methods. While this was to some extent deliberate to save space it is also true that most data I have come across is presented for analysis in toto. This may change as better recording methods are introduced and, hopefully, as statisticians are called in before the fact. There is still a problem in that simple sequential methods are known only for Poisson processes (homogeneous or non-homogeneous) and also that rather unsmooth inhomogeneities occur in practice which make the application of formal sequential methods based on very definite assumptions quite hazardous.

Less formal sequential methods are useful; in particular, an analysis of the data in successive sections is very useful, both to cut down on computation time in, say, spectral analyses (Lewis, 1970) and to examine the time evolution of the process.

As an example, consider a series of arrivals at an intensive care unit in a hospital. This data will be used for illustration throughout the paper; it was supplied by Dr. R. Barr, of the Oxford Regional Hospital Board, England. The times-of-arrival in the data were exceptionally well recorded. Of the 1468 arrivals in the 1420-day period from 4 February 1963, to 6 February 1968, only one time-of-day was not recorded. Nine tied arrivals occurred. Generally, recording times seemed to be at the five-minute intervals of the hour, although other times do occur.

The first section of the data, consisting of n = 251 arrivals in t_0 = 409 days (4 February 1963 - 18 March 1964), was analyzed in Chapter 1 of Cox and Lewis (1966) and recorded in the Appendix of that volume. Later on, the arrivals up to 6 February 1968 were received and are recorded in Lewis (1971). Three successive sections of length t_0 = 409 days were taken from these later arrivals for comparison and their statistics, as well as those for the total record, are shown in Table 1.

Table 1. *Arrivals at an Intensive Care Unit (No Ties)*

	Period 1	Period 2	Period 3	Period 4	Total
	4 Feb '63- 18 Mar '64	19 Mar '64- 2 May '65	3 May '65- 15 June '66	16 June '66- 29 July '67	4 Feb '63- 6 Feb '68
n	251	350	372	316	1458
t_0	409	409	409	409	1829
$\underline{U}$(days)	218.47	218.53	183.69	197.20	954.25
$\overline{U}$	1.874	2.222	-3.399	-1.099	2.874
$\tilde{m}=n/t_0$	0.614	0.856	0.910	0.773	0.797

The statistic $U = \sum_{i=1}^{n} t_i/n$, where the t_i are the times to events and n the number of events in the period of observation of length t_0, is the centroid of the t_i's and is used to test (Cox and Lewis, 1966, p.47) for the trend parameter β in a nonhomogeneous Poisson process with rate

$$\lambda(t) = \exp(\alpha + \beta t) = \lambda\exp(\beta t). \tag{1.1}$$

In testing for $\beta = 0$ against $\beta \neq 0$, α is a nuisance parameter with sufficient statistic n and the test is based on the distribution of U given n. The normalized statistic $\bar{U}$ in line three of Table 1 should be distributed approximately as a N(0,1) variable. There was fairly strong evidence of a monotone trend at the end of the first observation period ($\simeq$7 percent level), but later on in the series, as in Section 3, there is a definite decreasing trend. However, the total arrival process gives fairly strong evidence of an increase in $\lambda(t)$, the value $\bar{U} = 2.874$ being significant at about a 2 percent level. For Period 1 and Period 2 combined, $\bar{U}$ was found to be 4.492 which is significant at a level much smaller than 1 percent. An actual sequential test of $\beta = 0$ against $\beta \neq 0$ rejects the null hypothesis at a 1 percent level after 550 days. The sectioned analysis of four periods suggests a long cycle or quadratic term in the exponential trend (1.1). The nonhomogeneity is confirmed very strongly by a dispersion test (Cox and Lewis, 1966, p.232) applied to the numbers of events in the four sections. This has a value of 26.74; its distribution is that of a chi-squared variable with 3 degrees of freedom which has an 0.9999 point equal to 21.11.

A plot of the times-to-events t_i against serial number i is given in Fig. 1, and a plot of the average values of successive sets of twenty intervals between arrivals is shown in Fig. 2. Again the long cycle or quadratic trend is quite evident graphically. We return to this example later in the paper.

(iii) Sectioning brings up in some ways the analysis of replicated point processes, which again was not considered in detail by Cox and Lewis (1966). This replication can occur quite commonly in experimental situations, for instance in neurophysiology where experiments can be repeated many times. Here the signals are trains of very narrow spikes of apparently fixed height, so it is appropriate to analyze the times of occurrence of the spikes as a point process. Observation over too long a period can be deceiving, because of physiological deterioration, and replication is sometimes preferable. Of course, care must still be taken that physiological or experimental conditions have not changed.

Comparison of rates and trends, mainly in Poisson processes, was

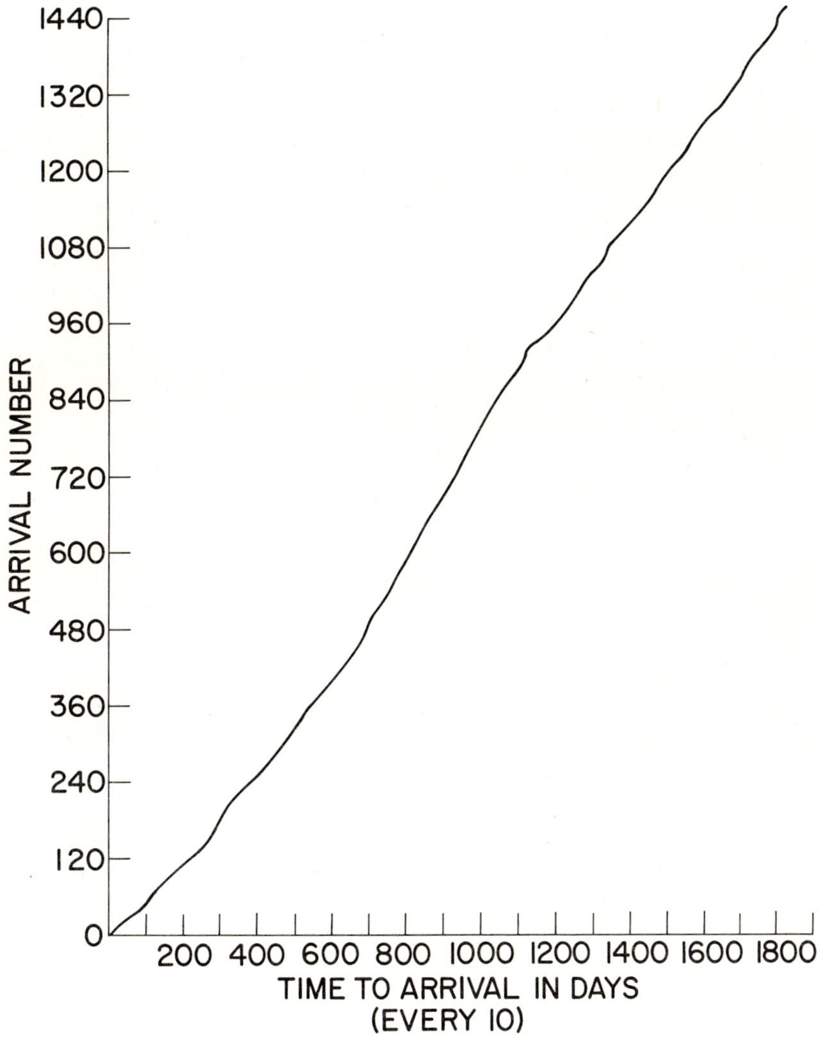

Figure 1. *Arrival of patients at an intensive care unit.* Complete record, 4 February 1963 to 8 February 1968. Number of arrivals $n = 1458$ in a period of $t_0 = 1829$ days. Arrival number vs. time to arrival. $\tilde{m} = n/t_0 = 0.7972$.

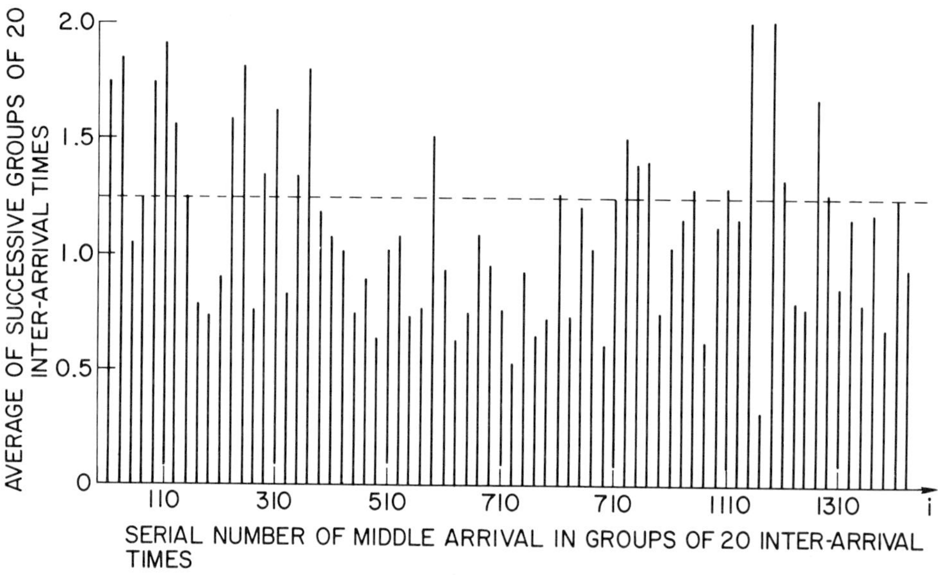

Figure 2. *Arrival of patients in an intensive care unit.* Averages of successive groups of 20 inter-arrival times. Number of arrivals n =1458 in period of t = 1829 days. Overall average is 1.254 days. Homogeneity of variance statistic = 318.000 (χ^2_{71}: mean 71, σ = 11.92).

discussed in Cox and Lewis (1966). See also Qureishi (1964) for the comparison of rates in two Weibull processes. General problems of multiple point processes (multivariate point processes) are discussed in Cox and Lewis (1972) and Perkel, Gerstein and Moore (1967b), but are beyond the scope of this paper. Lewis (1970) discussed estimates of the spectrum of counts from sectioned data, as well as pooling and comparison of the spectra.

(iv) Although trend analysis was discussed in Chapter 3 of Cox and Lewis (1966), I don't believe it received enough emphasis. Also, problems such as the analysis of cyclic trends were barely touched upon. These and other problems in trend analysis are discussed in more detail later in the paper.

(v) Finally, perhaps more emphasis could have been given to

graphical methods. These were discussed in Chapter 1, but no reference was made to probability plotting, for example, in later chapters. See, for example, Wilk and Gnadadesikan (1968).

The field where most use has been made of techniques for the analysis of point processes is in neurophysiological work on the signals occurring on nerve fibers. A summary of techniques for this type of analysis given in Perkel, Gerstein and Moore (1967a, 1967b) largely parallels Cox and Lewis (1966), with more information on the special problems of neurophysiology. For later work and applications see, for example, Moore, Segundo, Perkel and Levitan (1970) and the references given in that paper. It would be impossible to try to summarize all of this work here; some of it will be touched on later, but there is virtually no statistical methodology given in these papers which is not given in Cox and Lewis (1966).

Another interesting field of application is to the study of the occurrence of earthquakes. While strictly not a univariate point process, an approximation to the earthquake process as a univariate point process yields useful insights. For such analyses, see Vere-Jones, Turnovsky and Eiby (1964), Vere-Jones and Davies (1966), Vere-Jones (1970), and Shlien and Toksöz (1970). The discussion in Vere-Jones (1970) contains extensive comments on the problem of analyzing earthquake occurrence data.

Computational problems in the spectral analysis of point processes have been discussed by Lewis (1970). By spectral analysis here and in the rest of the paper we mean Bartlett's spectral analysis of the counting process N(t) of a point process (Bartlett, 1963; Cox and Lewis, 1966, Chapter 5). The spectral analysis of the intervals between events in a point process has been greatly facilitated by the availability of the fast Fourier transform algorithm; see Cooley and Tukey (1965) and Cooley, Lewis and Welch (1970) for details.

Most of the statistical methodology for the analysis of univariate series of events given in Cox and Lewis (1966) has been implemented in a computer program called SASE IV. For details, see Lewis (1966) and Lewis, Katcher and Weiss (1969). SASE IV is a large FORTRAN program with graphical output which is available from the IBM Program Information Department, 40 Saw Mill River Road, Hawthorne, New York 10532, as Program No. 360 L 130001.

In subsequent sections of this paper we discuss first two central problems in the statistical analysis of point processes, namely tests for renewal processes and tests for Poisson processes. Then techniques for specific non-renewal processes, such as doubly stochastic and clustering Poisson processes are discussed. The inability to write down a likelihood function makes the formal analysis of non-renewal point processes difficult; it is only for the important doubly stochastic Poisson process that techniques are beginning to appear.

The final sections deal with trend analyses, mostly of non-homogeous Poisson processes. First we consider the case of monotone trends then the case of cyclic trends and their relationship to the spectral analysis of point processes. Following this, some general problems of trend analysis are considered; these include tests for particular rate functions and the nonhomogeneous Poisson process model per se, and the definition of "residuals" in the analysis of nonhomogeneous Poisson models.

2. TESTS FOR RENEWAL PROCESSES

2.1. Markov Interval Processes and Serial Correlation Coefficients.

A natural extension of the renewal process model is to processes with first order dependence of intervals (Wold, 1948; Cox, 1955). The naturalness may be only mathematical since point processes of this type have not been commonly found in applications, although they have recently been postulated in neurophysiological contexts (Lampard, 1968; Walloe et al, 1969). In the process of Walloe et al (1969), the first order Markov interval property depends on the input to a neuron being a Poisson process. However, since the input is the superposition of an unknown number of fairly regular neuron signals, the hypothesis is tenuous. A drawback in the use of the first order Markov interval process as an approximate model for serial dependence in series of events is the difficulty of obtaining analytical results on, for example, the spectrum of counts or the variance-time curve. This difficulty is closely related to the fact that there are no regeneration points in the process.

Another problem with the model is the dearth (until recently) of useful bivariate distribution models. Cox (1955) used a bivariate exponential of the form

$$f_{X_{i+1}}(x;x_i = y) = \lambda(y)\exp\{-\lambda(y)x\}, \tag{2.1}$$

where

$$\lambda(y) = \lambda_0(1+\lambda_1 y) > 0,$$

which had the drawback of nonlinear restrictions on the parameter. Another widely used model for bivariate gamma distributions which arises quite naturally has been discussed by Moran (1967a), Vere-Jones (1967), Lampard (1968), and Gaver (1970). (See also Griffiths, 1969, for general properties of bivariate gamma distributions.) These bivariate distributions all have positive correlations. Bivariate exponentials with negative correlation have been derived by Gaver (1972) and bivarite, negatively correlated intervals have been observed in a neurophysiological context by Walloe et al (1969).

For statistical analysis, one of the most important theoretical results on bivariate exponential distributions is that of Moran (1967a), who showed from more general results that the serial correlation coefficient (of order 1) for bivariate exponential distributions is always greater than -0.6649;

$$\rho_1 = \frac{E[\{X_i - E(X)\}\{(X_{i+1} - E(X)\}]}{\text{var }(X_i)} \geq 1 - \frac{1}{6}\pi^2 = -0.6649. \tag{2.2}$$

Estimates of the serial correlation and tests of the renewal hypothesis against a first order Markov interval process are based on the sample serial correlation coefficient $\tilde{\rho}_1$, defined as follows:

For simplicity, assume observation starts with an event and ends with an event, there being n observed intervals between events $x_1, x_2, \ldots, x_n$ with mean $\bar{x} = n^{-1}\Sigma x_i$. Let $z_i = x_i - \bar{x}$. Then

$$\tilde{\rho}_1 = \frac{n}{n-1} \frac{\sum_{i=1}^{n-1} z_i z_{i+1}}{\sum_{i=1}^{n} z_i^2} . \tag{2.3}$$

Before discussing theoretical results on tests for renewal processes based on $\tilde{\rho}_1$, consider its distribution for renewal processes, a great deal about which is now known.

The expected value of $\tilde{\rho}_1$ under the renewal hypothesis, for any distribution of the intervals x_i, is (Moran, 1967b)

$$E(\tilde{\rho}_1) = -\frac{1}{n-1} , \tag{2.4}$$

and while its variance is known to be asymptotically n^{-1} if $\rho = 0$, the exact variance depends on the distributions of the x_i (Moran, 1967b):

$$\mathrm{var}(\tilde{\rho}_1) = \frac{n^2-n+1}{n^2(n-1)} - \frac{n+1}{n(n-1)} E\{\frac{\Sigma z_i^4}{(\Sigma z_i^2)^2}\} \leq \frac{n-2}{n(n-1)} . \tag{2.5}$$

Moran (1967b) obtained an approximation by replacing the expectation by the ratio of the expectations of the numerator and denominator. This result is exact for normally distributed x_i's, giving

$$\mathrm{var}(\tilde{\rho}) = \frac{(n-2)^2}{n^2(n-1)} . \tag{2.6}$$

For exponentially distributed x_i, the Moran approximation gives

$$\mathrm{var}(\tilde{\rho}_1) \sim \frac{1}{n} - \frac{7}{n^2} + \frac{52}{n^3} - \frac{398}{n^4} + \dots . \tag{2.7}$$

Moran (1970) compared his approximation to sampling results obtained by Cox (1966) and a brief summary of their results on the variances follows.

(a) The variance of $\tilde{\rho}_1$ tends to be smaller for random variables with long tails than in the normal case.

(b) Moran's approximation works reasonably well for gamma distributions with shape parameter $k \geq 0$, $n \geq 50$, or $k \geq 8$, $n \geq 10$. Outside these limits the approximation underestimates the variance

($k = 0$ is the exponential distribution).

(c) For the exponential distribution, a partial reconstruction of Table 1 from Moran (1970) using sampling results from Goodman and Lewis (1972) follows.

Table 2. *Variance of $\tilde{\rho}_1$ in Exponentially Distributed Random Samples.*

Sample size n	Observed variance (simulation)	Moran's approximation	Difference	Difference/ Observed
20	0.0419	0.0371	0.0048	.115
50	0.0182	0.0176	0.0010	.055
100	0.00945	0.00935	0.00010	.011
450	0.00219	0.00221	0.000020	.009

(d) For very skewed distributions, a better approximation to $\mathrm{var}(\tilde{\rho}_1)$ is needed. This is apparent from Table 3 for random samples where the x_i have a Weibull distribution with shape parameter $\frac{1}{2}$($\frac{1}{2}$ Weibull). A random variable with this Weibull distribution is the square of a random variable with a unit exponential distribution.

Table 3. *Variance of $\tilde{\rho}_1$ in 1/2 Weibull Random Samples.*

Sample size n	Observed variance (simulation)	Moran's approximation	Difference	Difference/ Observed
50	0.01410	0.00058	0.01352	.96
100	0.00782	0.00054	0.00728	.93
450	0.00201	0.00047	0.00154	.77

Returning to the distribution of $\tilde{\rho}_1$, Moran(1967b) showed that $n^{1/2}\tilde{\rho}_1$ converges in distribution to a unit normal variate if the first four moments of the x_i exist. Cox (1966) examined the distribution by synthetic sampling for the case where the x_i had a normal distribution, and also distributions with the probability density function

$$\frac{1}{\Gamma(k+1)} e^{-x} x^k, \qquad (x \geq 0), \tag{2.8}$$

with k = 24, 8, 1, 0, -1/2, a rectangular distribution, a double exponential distribution and a Cauchy distribution. The first two moments and coefficients of skewness and kurtosis were examined. For the positive random variables, the distributions were generally positively skewed with convergence to normality apparently fairly rapid.

These results have been extended for the case of x_i exponentially distributed in large scale simulations reported in Goodman and Lewis (1972).

Figure 3 shows a computer plot of the estimated mean (*), standard deviation (x) and coefficients of skewness (+) and kurtosis (λ) of $\tilde{\rho}_1$, as a function of n(RHO(1,N)) for n = 11(1)120, 130, 140, 150. The simulations involved at least 3,000,000 replications for each n, so that the sampling variances of the estimates are small. The curves have not been smoothed. Note that the skewness is positive, with a maximum value of about 0.32 at n = 35, after which it starts back toward its asymptotic value of zero. The kurtosis is small, going from positive to negative at about n = 35 and then going back toward its asymptotic value of zero very slowly.

The departure from normality (positive skewness) is seen much more clearly in Fig. 4, where the computer plot gives estimated normalized values of, from top to bottom, the quantiles at levels 0.001, 0.002, 0.005, 0.010, 0.020, 0.025, 0.050, 0.100, the estimated mean plus $(n-1)^{-1}$, and the quantiles at levels 0.900, 0.950, 0.975, 0.980, 0.990, 0.995, 0.998, 0.999. These should converge to the corresponding quantiles of the unit normal distribution.

Note two things: the departures from normality are relatively small for the inner quantiles and convergence to the normal quantiles is very slow.

More detailed results from which the departure from normality and the slow rate of convergence can be assessed are shown in Table 4. The quantiles and moments of $\tilde{\rho}_1$ (called RHO(1,N) in the computer) for n = 450 are shown in the rows marked exponential. Of these rows, the first row is the actual estimated value of the moment or quantile, the bracketed quantities in the next row are the estimated sampling standard deviations of the estimates (5 degrees of freedom), and the third row is the quantile minus $\tilde{\mu}$, all divided by $\tilde{\sigma}$. This row illustrates the departure from normality, $\tilde{x}_{0.975}$ being 2.039 instead of 1.960.

Much more serious departures occur for the quantiles of $\tilde{\rho}_1$ from random samples with 1/2 Weibull intervals. These are given in the rows marked "1/2 Weibull" in Table 4. Thus, not only is Moran's approximation poor, as seen above, but the normal approximation is considerably off. The simulation results for the 1/2 Weibull case are discussed more fully in Goodman and Lewis (1972).

Moran (1967b) showed that tests for independence against first order serially correlated intervals with exponentially distributed marginals based on $\tilde{\rho}_1$ are asymptotically most powerful. He also conjectured that this is true for Gamma distributed intervals. More formal results have been reported by Yang (1970).

Moran (1967b) also proposed two modifications of the serial correlation coefficient. One is obtained (see also Cox, 1955) by replacing the estimate of the variance in the denominator of (2,3) by the square of the estimate of the mean (the mean and the standard deviation are equal for the exponential distribution). The other modification is due to Ogawara. Moran (1967b, 1970) gives the first two moments of these test statistics and shows that asymptotically they involve no loss of efficiency against the first-order Markov interval process. However, Moran's conjecture that the distribution of Ogawara's statistic converges rapidly to the normal is not correct (Goodman and Lewis, 1972); in fact, its distribution is very similar to that of $\tilde{\rho}_1$ in the null exponential case.

An interesting application of serial correlation statistics in examining neurophysiological models is given by Walloe et al (1969). There is very evident negative correlation between successive intervals; given the known structure of the neuronal process, first order Markov dependence would have been expected if the input to the neuron had been a Poisson process, and this was tested by checking if $(\tilde{\rho}_1)^2 \approx \rho_2$. Unfortunately (personal communication) ρ_1 was estimated by averaging together estimates obtained from successive sections of length 20, and

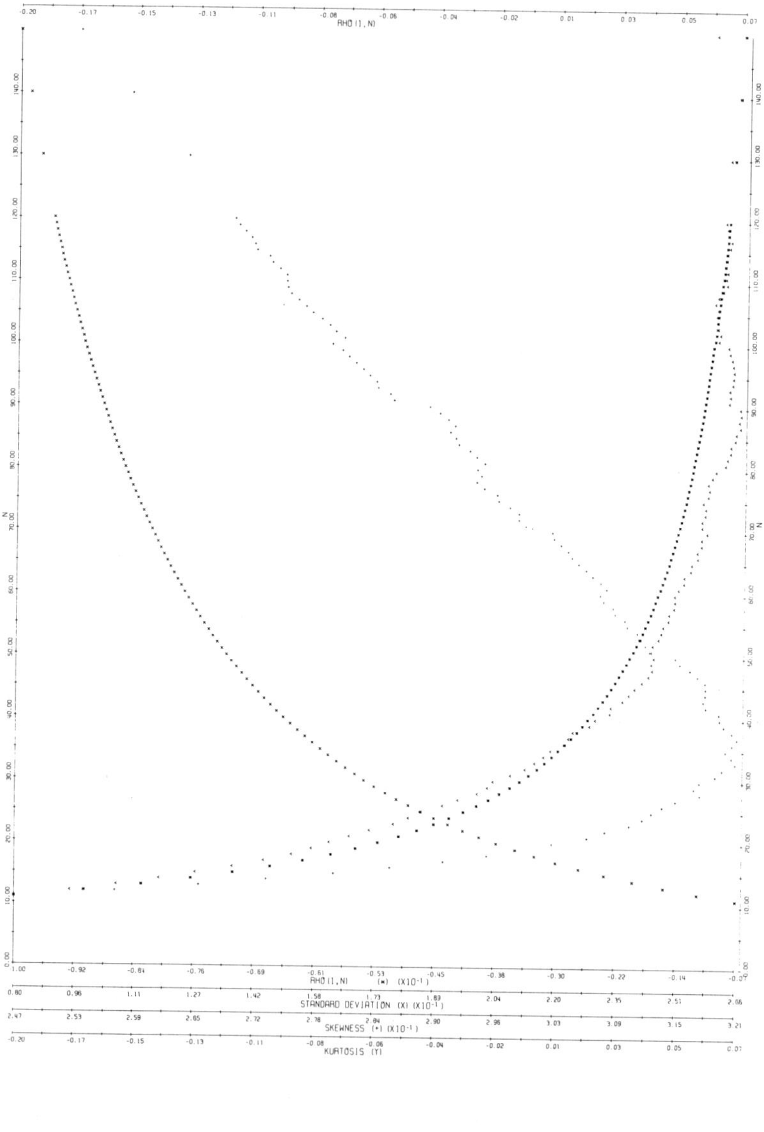

Figure 3. *First serial correlation coefficient.* Computer plot of the estimated mean (*), standard deviation (x) and coefficients of skewness (+) and kurtosis (λ) of the first serial correlation coefficients $\tilde{\rho}_1$(RHO(1,N)) as a function of sample size n. Plot is for values n = 11(1)120, 130, 140, 150. Each n involved 6 replications of sampling experiments with 750,000 samples. The data has not been smoothed.

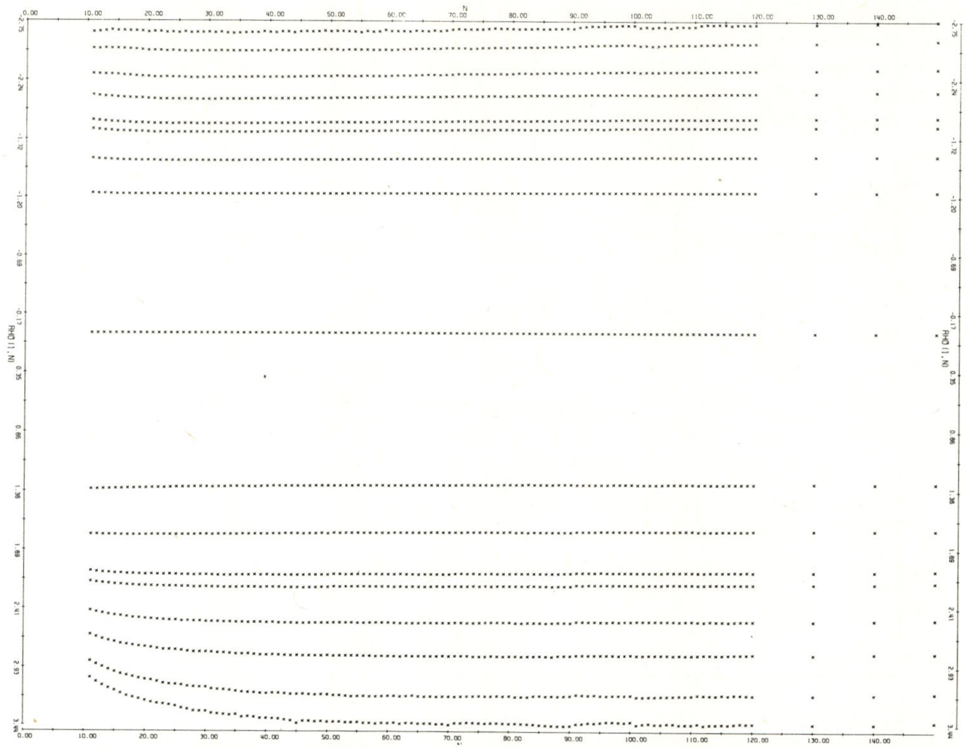

Figure 4. *First serial correlation coefficient.* Computer plot of the normalized sample serial correlation coefficient $\tilde{\rho}_1$(RHO(1,N)) for series of length n of unit exponential variables as a function of n. Values of n given are 11(1)120, 130, 140, 150. Excluding the center zero mean line, quantiles plotted are, from top to bottom, for levels 0.001, 0.002, 0.005, 0.010, 0.020, 0.025, 0.050, 0.100 and 0.900, 0.950, 0.975, 0.980, 0.990, 0.995, 0.998, 0.999. Each value of n involved 6 replications of sampling experiments with 750,000 samples. The data has not been smoothed.

	Mean	Stand. Dev.	Lower Quantiles							
	$\tilde{\mu}$	$\tilde{\sigma}$	$\tilde{x}_{0.001}$	$\tilde{x}_{0.002}$	$\tilde{x}_{0.005}$	$\tilde{x}_{0.010}$	$\tilde{x}_{0.020}$	$\tilde{x}_{0.025}$	$\tilde{x}_{0.050}$	$\tilde{x}_{0.100}$
Exponential	-0.00207	0.04683	-0.13604	-0.12776	-0.11541	-0.10522	-0.09390	-0.09002	-0.07673	-0.06113
	(0.00002)	(0.00001)	0.00008	(0.00007)	(0.00005)	(0.00006)	(0.00007)	(0.00006)	(0.00005)	(0.00004)
			-2.861	-2.684	-2.420	-2.203	-1.961	-1.878	-1.595	-1.261
1/2 Weibull	-0.00214	0.04483	-0.10169	-0.09609	-0.08798	-0.08123	-0.07377	-0.07120	-0.06238	-0.05187
	(0.00002)	(0.00001)	(0.00016)	(0.00014)	(0.00008)	(0.00007)	(0.00004)	(0.00005)	(0.00003)	(0.00003)
			-2.221	-2.095	-1.915	-1.764	-1.598	-1.540	-1.343	-1.109

	Skewness	Kurtosis	Upper Quantiles							
	$\tilde{\gamma}_1$	$\tilde{\gamma}_2$	$\tilde{x}_{0.900}$	$\tilde{x}_{0.950}$	$\tilde{x}_{0.975}$	$\tilde{x}_{0.980}$	$\tilde{x}_{0.990}$	$\tilde{x}_{0.995}$	$\tilde{x}_{0.998}$	$\tilde{x}_{0.999}$
Exponential	0.17041	0.04871	0.05872	0.07712	0.09343	0.09836	0.11281	0.12610	0.14252	0.15439
	(0.00082)	(0.00219)	(0.00005)	(0.00004)	(0.00007)	(0.00007)	(0.00007)	(0.00008)	(0.00018)	(0.00017)
			1.298	1.691	2.039	2.145	2.453	2.737	3.088	3.341
1/2 Weibull	1.02894	2.12262	0.05600	0.08064	0.10460	0.11228	0.13588	0.15964	0.19099	0.21524
	(0.00209)	(0.01146)	(0.00006)	(0.00003)	(0.00007)	(0.00007)	(0.00014)	(0.00026)	(0.00019)	(0.00032)
			1.297	1.847	2.381	2.552	3.079	3.609	4.308	4.849

Table 4. *First Serial Correlation Coefficient.* Estimated quantiles, means, standard deviations and coefficients of skewness and kurtosis of the sample serial correlation coefficient of order 1(RHO(1,N)). Series of n = 450 independent exponential and 1/2 Weibull variates. Averaged estimates from 6 synthetic samples each with 750,000 trials. Quantities in brackets are estimates of the standard deviation of the estimates (5 degrees of freedom). The third line in each group gives the estimated quantiles normalized by subtracting $\tilde{\mu}$ and dividing by $\tilde{\sigma}$.

the bias (2.4) accounts for a large part of the observation that for several experiments $(\tilde{\rho}_1)^2$ was systematically significantly larger than $\tilde{\rho}_2$. A jackknifed estimate (Quenouille, 1956) could have been used, or a correction for bias introduced. There probably still is real higher order dependence in this data accounted for by the fact that the input to the neuron is a superposition of a finite number of inputs and therefore not quite a Poisson process.

2.2. Product-Moment Score Statistics

An alternative to the serial correlation coefficient is obtained (Cox and Lewis, 1966, pp. 166-7) by replacing the actual interval values x_i by their ranks r_i or exponential scores $e(r_j;n)$ (Cox and Lewis, 1966, pp. 26-27), and computing a first order product moment statistic

$$R_1(n) = \sum_{i=1}^{n-1} e(r_i;n) \times e(r_{i+1};n). \tag{2.9}$$

A test for serial independence is then based on the null distribution of $R_1(n)$ under a permutation hypothesis.

An advantage of using $R_1(n)$ is that it controls outliers in the series of events; i.e., missing points. Its distribution has been tabulated by Lewis and Goodman (1969, 1970) for ranks, exponential scores, scores from gamma populations with parameter $k = -1/2$ and scores from Weibull populations with shape parameter 1/2 (1/2 Weibull).

The distributions for these product moment rank statistics are similar to those for the ordinary serial correlation coefficient $\tilde{\rho}_1$ from the equivalent parent population; positively skewed and very slow to converge to the asymptotic normal distribution. The distribution of the normalized exponential score product-moment statistic does in fact converge to the distribution of $\tilde{\rho}_1$ when the series length is $n = 10{,}000$; quantiles are shown for $n = 450$ in Table 5 and should be compared with the exponential values in Table 4.

Table 5. *Normalized Exponential Score Product-Moment Statistic; n = 450*

$\tilde{x}_{0.001}$	$\tilde{x}_{0.002}$	$\tilde{x}_{0.005}$	$\tilde{x}_{0.010}$	$\tilde{x}_{0.020}$	$\tilde{x}_{0.025}$	$\tilde{x}_{0.050}$	$\tilde{x}_{0.100}$
-2.799	-2.623	-2.370	-2.159	-1.929	-1.848	-1.567	-1.240
$\tilde{x}_{0.900}$	$\tilde{x}_{0.950}$	$\tilde{x}_{0.975}$	$\tilde{x}_{0.980}$	$\tilde{x}_{0.990}$	$\tilde{x}_{0.995}$	$\tilde{x}_{0.998}$	$\tilde{x}_{0.999}$
1.270	1.656	1.994	2.096	2.395	2.669	3.009	3.245

The sampling error in the values in Table 5 is approximately 0.001.

Note that the idea behind these tests is similar to that for the technique of random shuffling described in Perkel et al, (1967a) and apparently commonly used in neurophysiological work. The use of the product-moment statistic is more sophisticated and does not require a computer to do the shuffling.

2.3. Tests of Serial Independence Based on the Periodogram

The first serial correlation coefficient $\tilde{\rho}_1$ can be used as a test for serial independence for alternatives other than the first order Markov interval process, but has certain drawbacks. In particular, for the more common alternatives such as clustering processes, there is no imperative for looking at just the first serial correlation and tests combining serial correlations of several orders present difficulties because of the correlation between these statistics.

Tests based on the periodogram were advocated by Bartlett (1954) and described in Cox and Lewis (1966, pp. 168-170) and Durbin (1969). Denote the finite Fourier transform of the n intervals $x_1,\ldots,x_n$ by

$$H_n(\omega_p) = \frac{1}{(2\pi n)^{1/2}} \sum_{s=1}^{n} x_s e^{is\omega_p}, \qquad (\omega_p = 2\pi p/n), \quad (p = 1,2,\ldots,[1/2n] = \ell),$$

and the periodogram by

$$I_n(\omega_p) = |H_n(\omega_p)|^2.$$

The test is essentially for a "flat" spectrum using the distribution theory for the $I_n(\omega_p)$'s. The theory states that the $I_n(\omega_p)$'s are approximately independent, exponentially distributed random variables, the result being exact for $x_1,\ldots,x_n$ independent and identically normally distributed.

This hypothesis of a "flat" spectrum can be tested using a modified homogeneity of variance statistic, or the cumulated periodogram values

$$U_{(i)} = \sum_{p=1}^{i} I_n(\omega_p) / \sum_{p=1}^{\ell} I_n(\omega_p) \qquad (i = 1,2,\ldots,\ell-1),$$

can be tested as order statistics from a uniform distribution. Bartlett showed that the distribution theory is essentially independent of k_3, the third cumulant of the intervals between events, but it can be shown to be sensitive to k_4, the fourth cumulant.

Again, results have been obtained by synthetic sampling by Goodman and Lewis (1972). The distribution of the maximum values of the periodogram is shown in Table 6, the values for normally distributed x_i's being exact, those for exponential and half Weibull being taken from a large simulation. The deviations from the normal case for exponential x_i's is small; for 1/2 Weibull x_i's, which have a coefficient of kurtosis of 84.720, the departure is dramatic.

In Fig. 5, for exponential variates, we give a computer plot of the distribution of the modified Kolmogorov-Smirnov statistic

$$\text{KSTWO(N)} = n^{\frac{1}{2}} C_n = n^{\frac{1}{2}} \max_{1\le i\le \ell-1} |U_{(i)} - i/\ell|$$

in the form of computer printed plots of sixteen quantiles as a function of n, the length of the series. (n = N in the plot and the statistic is called KSCTWO(N)). The quantiles are a little smaller than for the normal case in which the upper (asymptotic) 5 percent and 1 percent points are 1.358 and 1.628, respectively. Note that for N = 140 the statistic is based on $\ell = 69$ $U_{(i)}$'s, so that the convergence of the quantiles is faster than it appears to be in the figure, although not as fast as in the normal case. Of course, it is not known that the distribution based on the spectrum for non-normal variates converges to the distribution for normal variates, but it is likely if the first four moments exist.

Table 6. *Maximum Value of the Periodogram.* Estimated quantiles, means, standard deviations and coefficients of skewness and kurtosis of the maximum value of the periodograms of series of N = 450 independent exponential and 1/2-Weibull variates. Averaged estimates from 6 synthetic samples each with 750,000 trials. Quantities in brackets are estimates of the standard deviations of the estimates (5 degrees of freedom). Also exact values for normally distributed time series for comparison.

	Mean	Stand. Dev.	Lower Quantiles							
	$\tilde{\mu}$	$\tilde{\sigma}$	$\tilde{x}_{0.001}$	$\tilde{x}_{0.002}$	$\tilde{x}_{0.005}$	$\tilde{x}_{0.010}$	$\tilde{x}_{0.020}$	$\tilde{x}_{0.025}$	$\tilde{x}_{0.050}$	$\tilde{x}_{0.100}$
Normal (Exact)	11.982	2.562	6.989	7.197	7.512	7.789	8.113	8.229	8.642	9.165
Exponential	11.824 (0.001)	2.827 (0.001)	6.133 (0.005)	6.378 (0.003)	6.746 (0.003)	7.068 (0.002)	7.445 (0.002)	7.580 (0.002)	8.063 (0.002)	8.668 (0.001)
1/2 Weibull	213.828 (0.011)	85.678 (0.033)	67.580 (0.067)	72.483 (0.080)	80.052 (0.082)	87.068 (0.059)	95.410 (0.032)	98.523 (0.030)	109.707 (0.036)	124.384 (0.032)

	Skewness	Kurtosis	Upper Quantiles							
	$\tilde{\gamma}_1$	$\tilde{\gamma}_2$	$\tilde{x}_{0.900}$	$\tilde{x}_{0.950}$	$\tilde{x}_{0.975}$	$\tilde{x}_{0.980}$	$\tilde{x}_{0.990}$	$\tilde{x}_{0.995}$	$\tilde{x}_{0.998}$	$\tilde{x}_{0.999}$
Normal (Exact)	1.442		15.324	16.764	18.176	18.627	20.024	21.415	23.251	24.638
Exponential	1.061 (0.003)	2.120 (0.023)	15.514 (0.004)	17.064 (0.005)	18.572 (0.005)	19.061 (0.006)	20.552 (0.009)	22.045 (0.009)	24.055 (0.024)	25.557 (0.033
1/2 Weibull	1.557 (0.003)	5.082 (0.035)	323.012 (0.084)	373.912 (0.059)	425.816 (0.136)	442.749 (0.172)	496.882 (0.182)	553.934 (0.328)	634.068 (0.472)	700.534 (1.696)

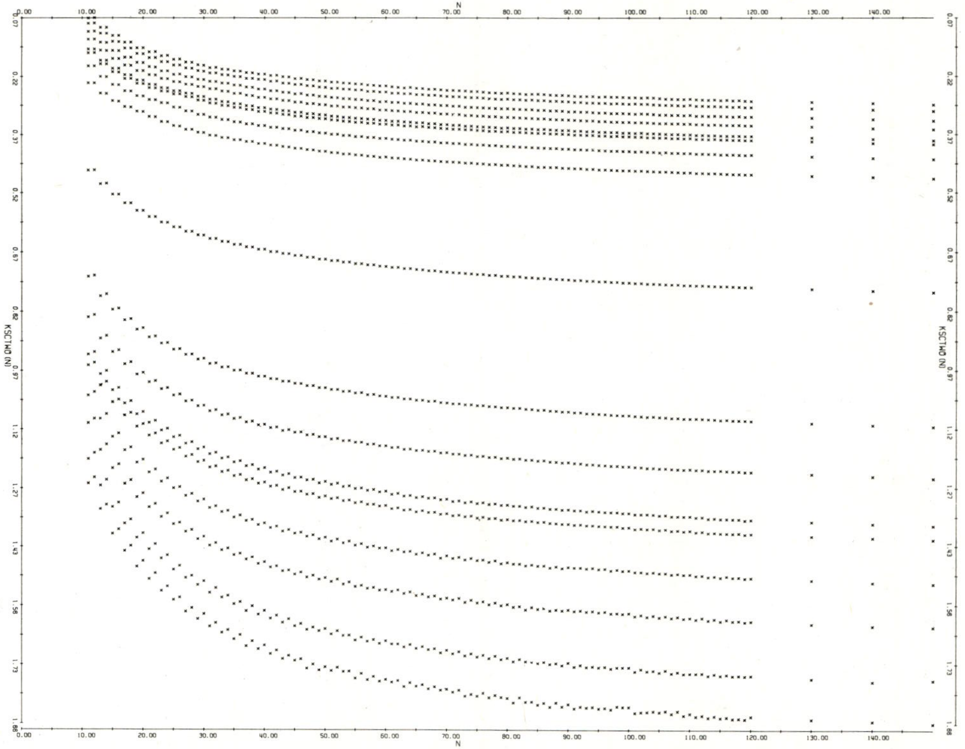

Figure 5. *Kolmogorov-Smirnov statistic.* Computer plot of the quantiles of the modified Kolmogorov-Smirnov statistic $n^{\frac{1}{2}}C_n$ for series of length n of unit exponential variables as a function of n. Values of n are 11(1)120, 130, 140, 150. Excluding the center zero-mean line, the quantiles plotted are, from top to bottom, for levels 0.001, 0.002, 0.005, 0.010, 0.020, 0.025, 0.050, 0.100 and 0.900, 0.950, 0.980, 0.990, 0.995, 0.998, 0.999. Each value of n involved 6 replications with 750,000 samples. The data has not been smoothed.

3. TESTS FOR POISSON PROCESSES

We will discuss here primarily tests for the Poisson hypothesis against stationary alternatives, trends being discussed in Section 5. We assume as before that n events are observed at times $t_1<t_2<\ldots<t_n$ in the fixed interval $(0,t_0]$. Intervals between events are denoted by $x_1, x_2, \ldots, x_n$, and it is convenient to denote the residual interval at the end of the period of observation by $x_{n+1} = t_0 - t_n$.

There are four major categories of tests.

a) Tests based on the t_i's, conditional upon $n = N(t_0)$, which is a sufficient statistic for the nuisance parameter λ, the rate in a homogeneous Poisson process. Since the t_i's are (conditionally) the order statistics from a uniform distribution, the empirical count function $N(t)$ (see Fig. 1) is proportional to the empirical cumulative distribution function for the uniformly distributed samples. The intervals x_i are then just the gap statistics for the sample.

Tests based on the maximum deviation between $N(t)$ and t/t_0 (Kolmogorov-Smirnov statistics and modifications) or other metrics (Anderson-Darling statistic) have well known distribution theories (see, for example, Durbin, 1967), but are sensitive mainly to trend departures from the homogeneous Poisson hypothesis. In fact, Lewis (1965) showed for the special case of gamma renewal alternatives, that the test is not consistent. These results can probably be extended by results on empirical processes for renewal processes reported in Pyke (1972).

The most important thing to note here is that while the distribution theory of these tests is well known because of the identity with the problem of testing, in a random sample of size n, a distribution $F_0(t)$ against an alternative $F_1(t)$, the distribution theory under alternatives is completely different. In the one case $F_0^{-1}(t_i)$ are order statistics from nonuniform random samples; for a point process other than the (homogeneous or nonhomogeneous) Poisson process, both the uniformity *and* the randomness (independence) disappear.

Note that the dispersion test for Poisson variates (Cox and Lewis, 1966, p. 158) is equivalent to testing the uniformity of the t_i's with a chi-squared goodness of fit statistic (Lewis, 1965).

b) The gap statistics for the x_i's, $i = 1,\ldots,(n+1)$, which, under the Poisson hypothesis, are a random number n of independent exponential variates, are formed from the order statistics of the x_i's as

$$D_{ni} = (x_{(i)} - x_{(i-1)})(n+2-i) \qquad (x_0=0,\ i=1,\ldots,n+1). \tag{3.1}$$

These are again a random number of independent exponentials (Cox and Lewis, 1966, p. 26) with sum t_0, and the statistics

$$t_i' = \sum_{j=1}^{i} D_{nj} \tag{3.2}$$

$$= \{x_{(1)} + x_{(2)} + \ldots + (n+2-i)x_{(i)}\}/t_0 \qquad (i=1,\ldots,n) \tag{3.3}$$

are, conditional upon n, order statistics from a uniform (0,1) distribution (see Cox and Lewis, 1966, p. 154 for a more complete derivation; the conditioning upon n and the fixed total t_0 is very subtle).

A great deal is now known about the null distributions of test statistics based on the gaps D_{ni}, or on the ordered x_i's, and the reader is referred to the excellent review articles by Pyke (1965, 1972). Also, against renewal alternatives, a good deal is known about asymptotic and, in some cases, small sample power. Alternatives are generally specific distributions or those in which the distribution is limited by specifying that the intervals have, say, distributions with monotone increasing or decreasing failure rate (Proschan and Pyke, 1967). Again see Pyke (1972) for a summary, and specifically Jackson (1967), Bickel (1969), and Bickel and Doksum (1969).

Empirically and from a common sense point of view it is quite clear that against stationary alternatives tests for Poisson processes based on the t_i''s are more useful than tests based on the t_i's. This has been observed in using the SASE program, in which the uniformity of the t_i''s is tested using both one-sided and two-sided Kolmogorov-Smirnov statistics and the Cramer-von Mises statistic (Cox and Lewis, 1966, p. 147).

More information about the power of these procedures, especially against cluster alternatives (Lewis, 1969; Vere-Jones, 1970) would be valuable.

c) There are also specific tests against renewal alternatives. The Moran test (Cox and Lewis, 1966, p. 161) is an example.

d) A useful test can be based on the empirical spectrum of counts, as pointed out by Bartlett (1963). Thus, let the finite Fourier-Stieltjes transform of the sample function N(t) be

$$H_{t_0}(\omega) = (\pi t_0)^{-1/2} \int_{t=0}^{t_0} e^{it\omega} dN(t) \tag{3.4}$$

$$= (\pi t_0)^{-1/2} \{ \sum_{j=1}^{n} \cos(t_j\omega) + i \sum_{j=1}^{n} \sin(t_j\omega) \} \tag{3.5}$$

$$= (\pi t_0)^{-1/2} \{A_{t_0}(\omega) + iB_{t_0}(\omega)\}, \tag{3.6}$$

and the periodogram

$$I_{t_0}(\omega) = (\pi t_0)^{-1} \{A^2_{t_0}(\omega) + B^2_{t_0}(\omega)\}.$$

We will refer again to the distribution theory of $I_{t_0}(\omega)$ later. Noting, however, that for $\omega t_0 = 2\pi p, I_{t_0}(\omega)$ is approximately exponentially distributed with mean $\lambda\pi$, and that for any two such frequencies the correlation between the periodogram points is $\sim 1/(1+n)$, we can see that the spectral test for independence (Section 2) can be applied here to test for a Poisson process.

The main drawback here is that while the spectrum of intervals is limited to approximately n/2 periodogram points, the count spectrum is not limited in such a way. Since there are roughly n/2 degrees of freedom available, it is a problem as to which n/2 points of the periodogram with frequency of the form $\omega t_0 = 2\pi p$ to use. Using more would invalidate, for example, use of standard distribution theory of Kolmogorov-Smirnov statistics to test the cumulated periodogram.

Despite this, the spectral tests are useful in an informal way, and the shape of the spectrum can suggest physical reasons for departures from a Poisson process. In neurophysiology (Perkel et al, 1967), the tendency has been to use the estimated intensity function (Cox and Lewis, 1966, p. 121) rather than the spectrum of counts to assess departures from the Poisson hypothesis. While this is useful in specific cases the distribution theory for the estimated intensity function is difficult; see Cox (1965) for the distribution theory for the null Poisson case.

4. STATISTICAL ANALYSIS OF NON-RENEWAL POINT PROCESSES

We consider here several non-renewal point processes which are of great practical importance. For the most part, however, the structure is too complicated to write down a likelihood function so that estimation and testing is ad hoc. Examples of such analyses are cited.

4.1. Cluster Point Processes

Cluster point processes (branching point processes) are important because they arise naturally in practice and have interesting mathematical properties (Lewis, 1969; Vere-Jones, 1970).

Briefly, a main process (usually a Poisson process) generates at each point a sequence of subsidiary events. In Vere-Jones (1970) the main processes are initial earthquake shocks, the subsidiary or cluster events are aftershocks. The processes of subsidiary events are independent and can, in general, be arbitrary point processes which terminate with probability one after a finite number of events occur. Two important special cases arise:

a) The subsidiary events are generated cumulatively as a finite renewal process. This is known as the Bartlett-Lewis process.

b) The subsidiary events are generated additively, each one of the random number of events being independently displaced from its (main) generating event. The resultant subsidiary process is generally non-stationary and the complete process is called a Neyman-Scott cluster process.

Computer failure patterns generated by this type of mechanism have been analyzed by Lewis (1964), and earthquakes by Vere-Jones and Davies

(1966), Vere-Jones (1970) and Shlien and Toksöz (1970). A fairly good ad hoc analysis can be given for the Bartlett-Lewis process since the marginal distribution of intervals is known (Lewis, 1964), as well as the spectrum of counts and in some cases the spectrum of intervals (Gilles and Lewis, 1967). The Neyman-Scott process does not yield a simple expression for the interval distribution and it is not yet known if the coefficient of variation of the intervals is greater than one, as it is for the Bartlett-Lewis process.

These cluster processes are overdispersed relative to a Poisson process; the asymptotic form of the variance-time curve (Lewis 1969) is independent of the fine structure of the subsidiary processes. This is a help in analyzing the data; basically for large time periods the counts of events $N(t)$ behave as though the subsidiary events were concentrated at the main, generating event, i.e., like a bulk Poisson process.

Despite these points, the situation with regard to cluster processes is unsatisfactory; questions such as the discrimination of Bartlett-Lewis and Neyman-Scott processes arise in practice and are not solved (see the discussion in Vere -Jones, 1970).

When the cluster process has clusters with only one event, the process is equivalent to an infinite server queue. For Poisson main events (parameter λ) the process of subsidiary events is, in equilibrium, a Poisson process also having parameter λ and the delay distribution cannot be determined.

An important special case occurs when the main process is regular and each event is independently delayed. Appointment processes are very often of this type and the determination of the delay distribution from observation of the subsidiary events (arrivals) has been considered in detail by Govier and Lewis (1967) for several delay distributions. Since the variance-time curve has a finite limit, they have called these processes with controlled variability.

4.2. Superposed Processes

Statistical analysis of processes which are superpositions of point processes is important, particularly in neurophysiological contexts, and usually hinges on questions of estimating the number of contributing processes or, when this is known, identifying the structure of the component processes.

These are difficult questions whose solution has not progressed much beyond the basic work of Cox and Smith reported in Cox and Lewis (1966, Chapter 8). Without specific assumptions, i.e., that the component processes are renewal processes with, say, gamma distributed intervals, very little can be done to estimate the number of processes. The problem of Walloe et al (1969) is of this kind, the nature of the input processes and the neural mechanism being well known, and the main question being how many inputs impinged on the neuron.

Identifying the component processes is, again, difficult, without specific assumptions. Work such as that of Ross (1970) on identifying the interval distribution in a known number k of superposed processes is rather technical. Note that the variance-time curve and spectra of counts are additive; e.g., the spectrum of counts of the superposed process is k times the spectrum of the individual processes. Thus, since the spectrum of counts of a renewal process is simply related to the Laplace transform of the probability density of the intervals, $f_X^*(s)$, i.e.,

$$g_x(\omega) = (\pi\mu)^{-1}\{1 + \frac{f_X^*(i\omega)}{1-f_X^*(i\omega)} + \frac{f_X^*(-i\omega)}{1-f_X^*(-i\omega)}\}, \qquad (4.1)$$

where μ is the mean of X, the spectrum of the superposed process is simply $\bar{g}_+(\omega) = kg_+(\omega)$ and it is natural, and probably efficient, to use the estimated spectrum or the estimated covariance density to estimate $f_X(x)$. Convergence follows from results of Brillinger (1972).

Nonhomogeneous superpositions also occur (Blumenthal, Greenwood and Herbach, 1971), but as in the homogeneous case, when the number of contributing processes is large, the resultant process over realistic periods of observation is almost indistinguishable from a (homogeneous or nonhomogeneous) Poisson process.

4.3. Doubly Stochastic Point Processes

Let $\Lambda(t)$ be a real valued, nondecreasing stochastic process. A doubly stochastic point process is a generalization of the nonhomogenous

Poisson process in which the integrated intensity function is replaced by $\Lambda(t)$. Thus, given a realization of $\Lambda(t)$, the point process is a non-homogeneous Poisson process. Generally $\Lambda(t)$ is differentiable and $\lambda(t) = \Lambda'(t)$ might be a stationary stochastic process, a deterministic trend or a combination of both. The process was introduced by Cox (1955); see Bartlett (1966, p. 325), Cox and Lewis (1966, p. 179), and Grandell (1970) for details. Gaver (1963) considered the case where $\lambda(t)$ changes level at random times and called the process a random hazard process.

The doubly stochastic mechanism in a point process is very realistic and probably quite common. Thus computer failure processes depend to some extent on temperature, humidity, etc. Unfortunately, analytic properties of the process when $\lambda(t)$ has a stochastic element are very difficult to derive; see Lawrance (1972). If $\lambda(t)$ is a stationary stochastic process with mean $\bar{\lambda}$, variance σ_λ^2 and autocorrelation function

$$\rho(\tau) = (1/2\pi) \int_{-\infty}^{\infty} \rho(\tau)\, e^{-i\omega\tau} dF(\omega), \tag{4.2}$$

where $F(\omega)$ is the integrated spectrum and $f(\omega) = F'(\omega)$ when it exists, then several useful results can be obtained (Cox and Lewis, 1966, pp. 179-183). In particular, the covariance density $\gamma_+(t)$ of the doubly stochastic Poisson process is

$$\gamma_+(t) = \sigma_\lambda^2\, \rho(t), \tag{4.3}$$

and the spectrum of counts is

$$g_+(\omega) = \bar{\lambda}/\pi + 2f(\omega). \tag{4.4}$$

Also, in general, the index of dispersion is

$$I(\infty) = 1 + 2\sigma_\lambda^2 \int_0^{\infty} \rho(u)\, du, \tag{4.5}$$

so that the process is overdispersed relative to the Poisson process.

There has been a great deal of interest in the doubly stochastic Poisson process in physics, optics, and engineering, and some work on the estimation problems.

a) In physics photon emission is a well understood physical process, and it is known that the process modulating the Poisson emission is, for instance, an Ornstein-Uhlenbeck process. The parameters of this process are physical parameters which it is of interest to estimate, and attempts have been made (e.g., Pusey, 1971; Jakeman, Pyke and Swain, 1971, and Koppel, 1971) to do this via the spectrum and the relationship (4.4). Again, laser light is deliberately focused on a physical system and the resultant intensity fluctuations in the scattered light reflects rates of molecular motions and interactions in the system.

In most cases the number of photon counts in these experiments is very large and it is convenient, and very often necessary, to cumulate counts. There are problems of determining the best sampling interval, a problem which is often made simpler by detailed knowledge of the modulating process.

Perhaps because of the large amount of data involved and the fact that the $\lambda(t)$ process is fairly well understood, physicists rarely bother about details such as the efficiency or optimality of estimation procedures. Moreover, the spectral estimation is very often automated and done digitally.

b) In engineering, two areas are involved. One is biomedical engineering in which, for example, radioactive substances are injected into the blood stream and the counts of the radioactive emission are used to estimate a decay function which is related to a physical process of interest (Snyder, 1971).

A second area is optics, where modulated light is used to transmit information (Reiffen and Sherman, 1963; Bar-David, 1969; Karp and Clark, 1970; Clark and Hoversten, 1970). Here $\lambda(t)$ is very often a signal changing levels, a possible problem being the discrimination of several $\lambda_i(t)$'s from photon counts of the noise. Very often a stationary "noise" element is also present; see Bédard (1966) for the physical considerations.

In all of this engineering literature, there is much concern with optimality, in some sense, and procedures are based on likelihood ratios, Bayesian posterior statistics and maximum likelihood detectors. It is difficult literature to penetrate and my overall impression is that there are many hidden assumptions involved, one being that samples are large, the other a normality assumption which may be quite incorrect. Moreover, most explicit results are for very simple situations such as mixed Poisson processes of one type or another, or rates (Reiffen and Sherman, 1963; Snyder, 1972) changing at known times. The latter problem is very simple and straightforward, especially when compared to the case of unknown change points which gives a true doubly stochastic Poison process and an inference problem similar to the difficult change point problem.

c) Grandell (1971, 1972) has considered inference in doubly stochastic processes from the viewpoint of mathematical statistics, his motivation apparently being primarily problems in actuarial work. His approach is quite different from that in the engineering literature, cited above, and he has considered, for example, optimum estimates of $E\{\lambda(t)\}$ based on linear combinations of the data. The problems are related to the general problem of curve estimation from random data and, while Grandell (1972) gives specific examples, much remains to be done.

5. TREND ANALYSIS IN NON-HOMOGENEOUS POISSON PROCESSES

We discuss now the analysis of trends in non homogeneous Poisson processes, extending the results of Cox and Lewis (1966, Chapter 3). Initially we discuss results based on specific parametric models for the rate function $\lambda(t)$ of the non-homogeneous Poisson process. These results are based on the fact that the likelihood function for n observations in the fixed period $(0,t_0]$ at times $t_1 < t_2 <,\dots,< t_n$ is

$$L(t_1,\dots,t_n,n;\underline{\theta}) = \prod_{i=1}^{n} \lambda(t_i;\underline{\theta}) \exp\{-\int_0^{t_0} \lambda(u;\underline{\theta})du\}, \tag{5.1}$$

where $\underline{\theta}$ denotes the vector of parameters in the model. Moreover, given that n events occur in $(0,t_0]$, the times to events t_i are the order statistics of a random sample from the probability density function

$$f(t;\underline{\Theta}) = \frac{\lambda(t;\underline{\Theta})}{\int_0^{t_0} \lambda(u;\underline{\Theta})\,du} = \frac{\lambda(t;\underline{\Theta})}{\Lambda(t_0;\underline{\Theta})} \qquad (0 \le t \le t_0) \tag{5.2}$$

and the conditional likelihood is

$$L(t_1,\ldots,t_n;n;\underline{\Theta}) = n! \prod_{i=1}^{n} \lambda(t_i;\underline{\Theta})/\{\Lambda(t_0;\underline{\Theta})\}^n. \tag{5.3}$$

Later in the section we discuss procedures for examining the adequacy of the model for $\lambda(t)$ and adequacy of the non-homogeneous Poisson process model itself.

5.1. Monotone and Evolutionary Trends

The estimate of $\lambda(t)$ when there is no trend present i.e., $\lambda(t) = \lambda$, is n/t_0.

In Cox and Lewis (1966, Chapter 3) the exponential linear trend

$$\lambda(t) = \exp\{\alpha+\beta t\} = \lambda \exp\{\beta t\} \tag{5.4}$$

was discussed and it was shown, using (5.1), that $(n,\Sigma t_i)$ were a set of sufficient statistics for (α,β). A test for $\beta = 0$ against $\beta \neq 0$, where α is a nuisance parameter, is based on the distribution of $\Sigma t_i/n$, given n. This is an optimum (conditional) test. The maximum likelihood estimate β was also given implicitly, but no small sample properties of the estimate are known.

The test for $\beta = 0$ was applied in Section 1 to analyze a sequence of arrivals at an intensive care unit (Table 1) and a sectioned analysis indicated that the trend was not monotone, although same overall increase in the rate was present.

An exponential quadratic trend uses the model

$$\lambda(t) = \exp\{\alpha+\beta t+\lambda t^2\} = \lambda\exp\{\beta t+\gamma t^2\}, \tag{5.5}$$

for which we get

$$\ln L(\alpha,\beta,\gamma) = n\ln\lambda + \beta\Sigma t_i + \gamma\Sigma t_i^2 - \lambda \int_0^{t_0} \exp(\beta u+\gamma u^2)\,du. \tag{5.6}$$

The maximum likelihood estimates of λ, α, β are given as the solution of the equations

$$\hat{\lambda} = n/\{ \int_0^{t_0} \exp(\hat{\beta}u + \hat{\gamma}u^2)\,du\}, \tag{5.7}$$

$$\frac{\Sigma t_i}{n} - \frac{\int_0^{t_0} u \exp(\hat{\beta}u+\hat{\gamma}u^2)\,du}{\int_0^{t_0} \exp(\hat{\beta}u+\hat{\gamma}u^2)\,du} = 0, \tag{5.8}$$

$$\frac{\Sigma t_i^2}{n} - \frac{\int_0^{t_0} u^2 \exp(\hat{\beta}u+\hat{\gamma}u^2)\,du}{\int_0^{t_0} \exp(\hat{\beta}u+\hat{\gamma}u^2)\,du} = 0, \tag{5.9}$$

and it is clear that $(n, \Sigma t_i, \Sigma t_i^2)$ are a set of sufficient statistics for (α, β, γ).

There are several interesting open question here.

What are the small sample properties of the estimators and what are their large sample properties? Note that the usual theory for maximum likelihood estimates is not directly applicable because of the random sample size.

What effect does the quadratic term (i.e., $\gamma \neq 0$) have on the estimates of β in the model (5.4)?

For the arrival data the following estimates were obtained.

Linear model; $\hat{\lambda} = 0.6342$; $\hat{\beta} = 0.001427$;
Quadratic model; $\hat{\hat{\lambda}} = 0.4792$; $\hat{\hat{\beta}} = 0.009788$; $\hat{\gamma} = -0.4481 \times 10^{-6}$

Note the large difference in $\hat{\beta}$ in the two cases.

Another problem of interest is to test for the quadratic term in the trend. It is clear from Table 1 that either a quadratic term or a long cycle is necessary to adequately model the arrival data from Section 1. A formal test can be based on the idea that, for any given γ, n and Σt_i are a set of sufficient statistics for α, β. Therefore a test for γ is based on the statistic Σt_i^2 and its conditional distribution, given n and Σt_i. This distribution is difficult to obtain analytically, Σt_i^2 being the square of the distance to the sample point, which is constrained to lie in the (n-1)-dimensional hyperplane defined by $\Sigma t_i = C$.

For large samples this distribution can be obtained from the fact that $\Sigma t_i/n$ and $\Sigma t_i^2/n$, conditioned on n, are jointly normally distributed for large n, and the following exact moment results:

$$\mu_1 = (\Sigma t_i/n) = t_0/2; \qquad \sigma_1^2 = \mathrm{var}(\Sigma t_i/n) = t_0^2(12n)^{-1};$$

$$\mu_2 = (\Sigma t_i^2/n) = t_0^2/3: \qquad \sigma_2^2 = \mathrm{var}(\Sigma t_i^2/n) = 4t_0^4(45n)^{-1};$$

$$\rho = \mathrm{corr}(\frac{\Sigma t_I}{n}, \frac{\Sigma t_I^2}{n} = \frac{\sqrt{15}}{4} = 0.968.$$

Thus, using normal theory results we test with $\Sigma t_i^2/n$ having a distribution with mean and standard deviation

$$\mu = \mu_2 + \rho\frac{\sigma_1}{\sigma_2}(\frac{\Sigma t_i}{n} - \mu_i)$$

$$= t_0^2/3 + t_0\,(\frac{\Sigma t_i}{n} - \frac{t_0}{2}) = t_0^2(\frac{\Sigma t_i}{nt_0} - \frac{1}{6}), \tag{5.10}$$

$$\sigma = \sigma_1(1-\rho^2)^{\frac{1}{2}} = \frac{1}{4}\frac{t_0}{(12n)^{\frac{1}{2}}} \tag{5.11}$$

For the complete set of arrival data $\Sigma t_i/n = 954.25$, giving $\mu = 1{,}187{,}783$ $\sigma = 6{,}530$ and $\Sigma t_i^2/n = 1{,}161{,}565$. These give $(\Sigma t_i^2/n-\mu)/\sigma = -4.015$ and this is clearly a highly significant rejection of the hypothesis that $\gamma = 0$.

Boswell (1966) and Boswell and Brunk (1969) have considered tests of the hypothesis that $\lambda(t)$ is not constant but is non-decreasing. Using a likelihood ratio criterion, conditional on the fixed value of $N(t_0) = n$, they found the test statistic

$$(\prod_{i=1}^{n} \bar{x}_i)^{-1}, \tag{5.12}$$

where

$$\bar{x}_i = \max_{1\le j\le i} \min_{i\le k\le n} (k-j+1)/(t_{k+1}-t_j), \tag{5.13}$$

the null hypothesis being rejected for large values of the statistic.

The statistic (5.12) is not simple to compute, but Boswell gave an iterative procedure for the computation, and some results on the limiting distribution of the statistic.

It would be of interest to compare the power of this test against the power of the test based on $\Sigma t_i/n$ for the exponential linear trend. Some results have been obtained by Mr. Ian White of the University of Edinburgh.

5.2. Cyclic Trends of Fixed Frequency

Cyclic time trends (as opposed to cycles on serial number) occur frequently in point processes but were treated only as an exercise in Cox & Lewis (1966). An example of such a series is given in Forrest (1950), who was investigating thunderstorm severity in Great Britain and its effect on power lines. He found a tendency for thunderstorms to occur in the morning (time-of-day effect) and of course a very strong time of year (seasonal) effect.

The arrival data discussed in Section 1 might be expected to have such effects, although a time-of-day effect is by no means evident since there is only about one arrival every day and a half. In Fig. 6 we show a collapsed plot of the numbers of arrivals in the successive hourly periods of all days of observation. The plot should be compared to Fig. 1.8a in Cox & Lewis (1966) for the arrivals from 4 February 1963 to 18 March 1964. The plot for the complete period (4 February 1963-6 February 1968) in Fig. 6 is similar in shape to the plot for the earlier period and no formal statistical test is needed to conclude that there is a real time-of-day effect.

As a model for the rate $\lambda(t)$ in a non-homogeneous Poisson process with fixed frequency $\omega_0 = 2\pi/T_0$ we take

$$\lambda(t) = \exp\{\alpha + k_s \sin \omega_0 t + k_c \cos \omega_0 t\}, \tag{5.14}$$

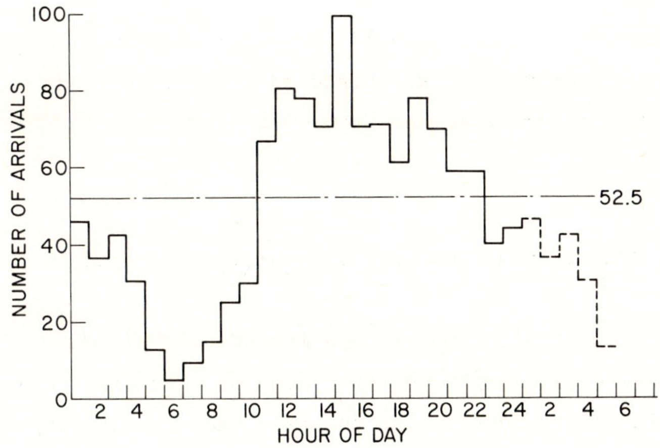

Figure 6. *Arrival of Patients at an Intensive Care Unit.* Collapsed hourly plot of number of arrivals to investigate "time-of-day" effect. Second period of observation: 19 March 1964 to 8 February 1968. Total number of arrivals (including ties): 1216. Observation time t_0 = 1420 days; $\tilde{m}$ = 0.856.

$$\lambda(t)=\frac{\lambda\exp\{k\ \sin(\omega_0 t+\theta)\}}{I_0(k)}, \tag{5.15}$$

where $k = \{k_s^2+k_c^2\}^{\frac{1}{2}}$, $\theta = \tan^{-1}(\frac{k_c}{k_s})$, $\lambda = I_0(k)\exp(\alpha)$

and $I_0(k)$ is a modified Bessel function of the first kind of zero order. This reparameterization is used because

$$\int_0^{T_0} \exp\{k\ \sin\ (\frac{2\pi t}{T_0} + \theta)\}\ dt = I_0(k), \tag{5.16}$$

so that if t_0, the total period of observation, is a multiple p of T_0 (here one day), we have

$$\Lambda(t_0) = \int_0^{t_0} \lambda(u)du = \lambda t_0 = \lambda p T_0. \tag{5.17}$$

In effect we have separated out multiplicatively a linear growth from a cyclic effect which takes place only within periods of length T_0.

The reason for using the form (5.14) rather than, say,

$$\lambda(t) = \alpha + k \sin(\omega_0 t + \theta) \tag{5.18}$$

is that $\lambda(t)$ must be positive, and to achieve this with (5.18) requires $k < \alpha$.

For $k << \alpha$ the two models are essentially the same. In addition, the rate (5.14) gives simple results based on sufficient statistics, this being closely connected with the fact that the density (5.2) belongs to the exponential family of density functions.

An additional reason for using (5.14) is that such nonsymmetrical cycles are probably more natural with series of events, possibly because of the positivity of the rate function. Professor J. W. Tukey has suggested a model which is (5.18) squared; this could be useful with data in which there are a large number of arrivals in each period, so that regression techniques using the square root of the numbers of arrivals in fixed intervals can be used (Cox and Lewis, 1966, Ch. 3). This model has also been used by Fisher (1953).

In Lewis (1970) the following results were given for the nonhomogeneous Poisson process with the rate $\lambda(t)$ given in (5.14).

Using (5.3) we find that the observations enter into the likelihood function only through

$$n, \quad A_{t_0}(\omega_0) = \sum_{i=1}^{n} \cos \omega_0 t_i, \quad B_{t_0}(\omega_0) = \sum_{i=1}^{n} \sin \omega_0 t_i, \tag{5.19}$$

and these are a set of sufficient statistics for the parameters α, k_s and k_c in (5.14). Note that $A_{t_0}(\omega_0)$ and $B_{t_0}(\omega_0)$ are the components of the periodogram (3.6).

Maximum likelihood estimates of λ, θ, k are

$$\hat{\theta} = \tan^{-1}\left\{\frac{A_{t_0}(\omega_0)}{B_{t_0}(\omega_0)}\right\}, \tag{5.20}$$

$$\hat{\lambda} = n/t_0, \tag{5.21}$$

and $\hat{k}$ is the solution of the equation

$$(1/n)\{A_{t_0}^2(\omega_0) + B_{t_0}^2(\omega_0)\} = \frac{I_1(\hat{k})}{I_0(\hat{k})} = \psi(\hat{k}), \tag{5.22}$$

where $I_1(k)$ is the derivative of $I_0(k)$ and $\psi(k)$ increases monotonically from 0 to 1 as k goes from 0 to ∞. This latter fact allows one to use the Neyman-Pearson lemma to show that, conditional on the observed value n of $N(t_0)$, the most powerful test of $k = 0$ against $k \neq 0$ is based on the statistic

$$(t_0/n)\left\{\frac{A_{t_0}^2(\omega_0)}{t_0} + \frac{B_{t_0}^2(\omega_0)}{t_0}\right\} = (\pi t_0/n)\ I_{t_0}(\omega_0). \tag{5.23}$$

Since, for $k = 0$, $2I_{t_0}(\omega_0)$ has asymptotically (Cox and Lewis, 1966, Chapter 5) an exponential distribution, the test for $k = 0$ reduces to accepting the hypothesis at, say, a 5% level if

$$2I_{t_0}(\omega_0) \leq (6/\pi)\times(t_0/n) = (6/\pi)\times\tilde{m}. \tag{5.24}$$

The factor 6 arises because prob $\{\chi_2^2 = 6) = .95$, where χ_2^2 is a random variable with a chi-square distribution on two degrees of freedom.

This test applied in Lewis (1970) to the first section of the arrival data gave a strong indication of the presence of the cycle at a period of one day.

For the complete record of arrivals at an intensive care unit discussed in Section 1 we get for the periodogram at $p = 1829$ or $\omega_0 = 2\pi$, which is the day frequency, since $t_0 = 1829$ and $\omega_0 = 2\pi p/t_0$,

$$2I_{t_0}(\omega_0) = 27.094 \text{ and } 6\tilde{m}/\pi = 1.52.$$

This is, not surprisingly, highly significant. For the first 409 arrivals the results are (Lewis, 1970)

$$2I_{t_0}(\omega_0) = 5.0 \text{ and } 6\tilde{m}/\pi = 1.10.$$

Thus the periodogram component is increasing roughly in proportion to n, as it should for a true fixed-cycle component.

When Lewis (1970) was written, the connection of this model for a cycle of fixed frequency with tests for directionality on the circle was not realized. In fact, the conditional test (5.24) is equivalent to testing that n observations on a circle (here a 24-hour clock) have the von Mises or circular normal distribution (Gumbel, Greenwood and Durand, 1953)

$$f(\ell) = \frac{\exp\{k \sin(\ell+\theta)\}}{2\pi I_0(k)} \qquad (0 \le \ell \le 2\pi). \tag{5.25}$$

For $k = 0$, this is a uniform distribution; otherwise, it has a mode at θ, the vector in the direction θ being called the modal vector. Greenwood and Durand (1955) obtained the distribution of the square of the resultant

$$R^2 = (\sum_{i=1}^{n} \cos \ell_i)^2 + (\sum_{i=1}^{n} \sin \ell_i)^2 \tag{5.26}$$

of n such vectors when $k = 0$, generalizing earlier work of Pearson on the problem of random walks on the circle. The formal analogy of (5.26) with the periodogram (3.5) should be clear. Watson and Williams (1956) generalized these distributional results, using results for sufficient statistics, and found the conditional p.d.f. of the quantity on the left of equation (5.22), which we denote by r, as

$$f(n^{1/2}r) = r\, I_0(kr) \int_0^{\infty} \{J_0(u)\}^n J_0(ru)du/\{I_0(k)\}^n, \tag{5.27}$$

where $J_0(\cdot)$ is the ordinary Bessel function of zero order.

It is not at all clear that (5.27) is more useful for computation than the generating function, which is given, for example, for $k = 0$ in Cox and Lewis (1966, Chapter 5) in the discussion of the distribution of the periodogram at one point ω_0, where ω_0 is of the form $t_0\omega_0 = 2\pi p$, for p integer. However, Stephens (1969a, 1969b) has used (5.27) to tabulate, among other things, the power of the test that $k = 0$ against alternatives $k > 0$. The most complete discussion of these tests for the von Mises distribution is given in Watson and Williams (1956). The function $\psi(k)$ is tabulated by Gumbel, Greenwood, and Durand (1953). Stephens (1969b) has also discussed tests for $\theta = \theta_0$, and joint tests for k and θ.

It is clear that problems involving cycles at two or three fixed frequences, e.g., ω_0, ω_1, ω_2, will arise in analyzing series of events. For example, in the problem of analyzing the arrivals at an intensive care unit, a time of week effect and a time-of-year effect are distinct possibilities. Surprisingly enough, there does not seem to be a strong time-of-week effect in that data, but there is not space here to go into this.

Formal tests for more than one cycle follow from results in Cox and Lewis (1966, Chapter 5) to the effect that at two different frequencies ω_0 and ω_1, both of which when multiplied by t_0 are 2π times an integer, the correlation between the periodogram values $I_{t_0}(\omega_0)$ and $I_{t_0}(\omega_1)$ is approximately $(1 + n)^{-1}$. Thus, a test for a time-of-week effect (ω_1), based on the conditional distribution of $I_{t_0}(\omega_1)$, given the values of $N(t_0)=n$ and $I_{t_0}(\omega_0)$, reduces in effect to an independent test at ω_1 based on (5.23).

For the arrival data, considered over a period of 1456 days, we get for $p = 208$ $(\omega_p = 2\pi/7)$, or a period of one week, the value

$$2I_{t_0}(\omega_p) = \frac{2}{\pi}\left\{\frac{A^2_{t_0}(\omega_p)}{t_0} + \frac{B^2_{t_0}(\omega_p)}{t_0}\right\} = 0.467 .$$

Thus there is no formal indication of a time-of-week effect. This is also evident from the collapsed plot of the numbers of arrivals on different days of the week (for all days of observation) given in Fig. 7.

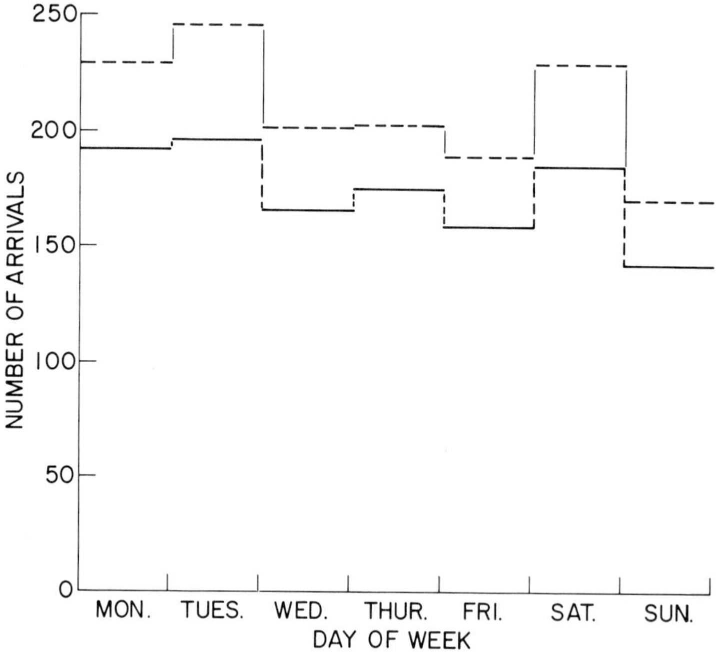

Figure 7. *Arrival of Patients at an Intensive Care Unit.* Collapsed daily plot of arrivals. All arrivals including ties. Solid line is 2nd period, 19 March 1964 to 8 February 1968 (1216 arrivals in 1420 days). Dashed line is complete record, 4 February 1963 to 8 February 1968 (1467 in 1829 days). $\sum_i (n_i - \bar{n})^2/(\bar{n}) = 12.93$. Note that $\chi^2_{6.095} \sim 12.6$.

5.3 The Spectrum of Counts and Cycles of Unknown Frequency

In the previous subsection, we showed the connection between tests for a cyclic component at a known frequency in a nonhomogeneous Poisson process and the periodogram, the periodogram being the basis for estimation of the (second order) spectrum of counts $g_+(\omega)$. Clearly, one might want to look for cycles at unknown frequency and this will, intuitively, be based on the spectrum of counts. More precisely, it will be based on the periodogram

$$I_{t_0}(\omega) = (1/\pi t_0)\{A^2_{t_0}(\omega) + B^2_{t_0}(\omega)\}. \tag{5.28}$$

The analogous problem in ordinary time-series analysis is the classical problem of hidden periodicities, discussed at length by Hannan (1970, p.463). This problem has not been tackled in point processes, and is more difficult for two reasons. First, the periodogram points are not quite uncorrelated, as we saw in the previous section, and also the spectrum is, in theory, not limited in the frequency domain. (There will always be some band limiting due to jitter; see Lewis, 1970.) Thus, it is a problem how to use the distribution theory for the spectrum given in Bartlett (1963) and Cox and Lewis (1966, Chapter 5) and how to pick relevent parts of the spectrum from which to estimate the unknown frequency.

As an example, we give in Fig. 8 the spectrum of the arrival

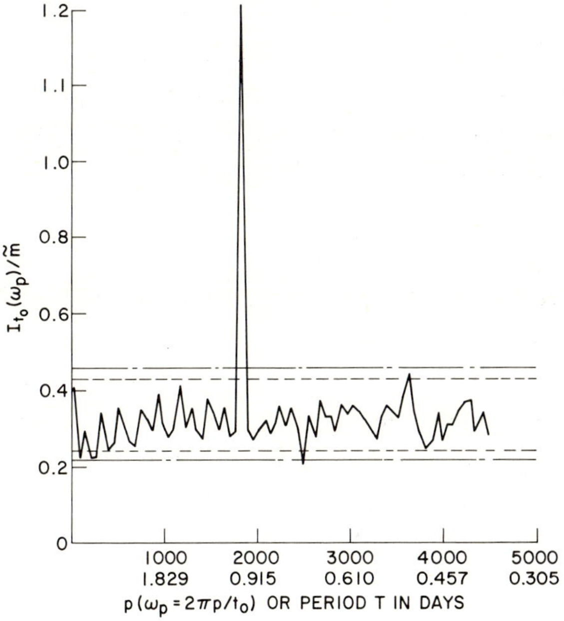

Figure 8. *Arrival of Patients at an Intensive Care Unit.* Spectrum of counts for the complete record: 4 February 1963 to 8 February 1968. n = 1458 arrivals in a period of t_0 = 1829 days. 60 point uniform smoothing. Bands are 0.95 and 0.99 confidence regions for individual values. $\tilde{m}$ = 0.7972.

data of Section 1. The peak for the day effect at p = 1829 ($\omega_p = 2\pi$) is evident, but it would be difficult to read into this spectrum anything but a conclusion that it is a nonhomogeneous Poisson process with a single fixed cycle (day-effect). There is possibly a harmonic at p =3658 and some inhibition at low frequencies, but all the points (except for p = 1829) are within 1 percent confidence bands for individual values, and would be within bands for the maximum of the periodogram values (Bartlett, 1963).

The spectrum of counts is generally a useful tool, and even more particularly so for non-Poisson processes, but has found very limited acceptance up to now amongst applied workers. This is partly due to confusion with the spectrum of intervals, with which neurophysiologists, for example, are much more familiar.

Lewis (1970) has given a heuristic justification for the spectrum of counts and discussed computation and smoothing. Bartlett (1967) has discussed finer points in this type of analysis.

Finally, Brillinger (1972) has given a general theory for spectral analysis of point processes, and has defined higher order spectra (cumulant spectra of order k) for point processes. These are generalizations of the cumulant spectra of order k of continuous time series. Brillinger's work is based on a general spectral representation for $N(t)$, the counting process of the point process, and provides an asymptotic theory for the spectral estimates of order 2 (periodogram) considered in this paper. Brillinger's paper is far too extensive to do justice to here. The problem of whether cumulant spectra of order higher than 2 will be useful in practice remains open and should be explored. Brillinger's spectral decomposition may provide an answer to whether spectral analysis can be useful as a representation in real systems; i.e., neurons.

5.4. Mixed Models

The arrival data considered in Section 1 has been shown to have a long term trend which can be represented by the exponential quadratic function (5.5) and a strong day cycle. A combined model for this data could be a nonhomogeneous Poisson process with rate $\lambda(t)$ of the form

$$\lambda(t) = \exp\{\alpha+\beta t+\gamma t^2 + k \sin(\omega_0 t+\theta)\}. \tag{5.29}$$

This is difficult to handle because $\Lambda(t) = \int_0^t \lambda(u)du$ is not tractable. However, in the context of the arrival data the evolutionary trend changes little within a cycle and an approximation in which the linear and quadratic terms are assumed constant over the period of each cycle yields a tractable model. The estimates of k and θ are then simply (5.20) and (5.22), γ and β are estimated from (5.8) and (5.9) and $\hat{\lambda}$ is given by (5.7) with $I_0(\hat{k})$ multiplying the term in the denominator. For the arrival data, we obtain the following values:

$\hat{\lambda} = 0.4792$;
$\hat{k} = 0.654$;
$\hat{\theta} = 3.519$;
$\hat{\beta} = 0.0009788$;
$\hat{\gamma} = -0.4481 \times 10^{-6}$.

Clearly, mixed models such as (5.29) will be needed in many cases for the analysis of real data. The model can be generalized, for example, to more general linear combinations of parameters and functions of time (Cox, 1972).

5.5. Residual Analysis of Point Processes

Many observed point processes have rate functions which are clearly not easily representable by simple, tractable models such as those discussed in the previous section. For example, it is common in observing arrivals of jobs in a computing center to find (roughly) the rate increasing sharply throughout the morning, dipping around lunch time, picking up slightly in the afternoon, and then dipping off sharply at night. The situation is fairly simple in this example, since many days observations are available, showing similar variations, and careful pooling gives a smoothed tabulated estimate of the rate. This rate is needed to model scheduling algorithms for the computer.

In general, the problem of smoothing the point process to obtain an estimate of $\lambda(t)$ is, in my opinion, open, despite the work on doubly stochastic Poisson processes cited in Section 4, which should be applicable. It is generally connected with the statistical problem of curve smoothing and density estimation and will not be considered further.

Both in the case where $\lambda(t)$ is obtained by smoothing or from a specific model, say exp $\{\alpha + \beta t\}$, it will often be necessary to test the adequacy of both the trend model and/or the assumption of an underlying nonhomogeneous Poisson process.

For example, an intensive care unit is a service facility, and to provide adequate service, one must know something about the arrival process; it might be very regular if all arrivals are from surgical operations, or exhibit clustering if all arrivals are from the emergency room. The form of the underlying process will also affect the test for cyclic trend given in (5.23), the attempt to estimate the spectrum and filter out the spike usually being a selfdefeating process (Bartlett, 1967; Lewis, 1970).

Both formal and informal methods, usually graphical, are required for this question, analogous to the similar problem in regression analysis which is usually approached by examining the residuals after estimating parameters for the mean value in the model.

While keeping in mind that there are many ways to approach this in different situations, the following proposal seems to offer a systematic approach to examining such questions (Lewis, 1970). It is well known that on the transformed time scale

$$\tau = \int_0^t \lambda(u)du, \tag{5.30}$$

a nonhomogeneous Poisson process with rate function $\lambda(t)$ becomes a homogeneous Poisson process of rate 1. Thus, we propose to examine the residual process $\{\tau_i\}$ where

$$\tau_i = \int_0^{t_i} \hat{\lambda}(u)du, \tag{5.31}$$

and $\hat{\lambda}(t)$ is some estimate of $\lambda(t)$.

There are a number of ways of looking at this transformation.

1) If $\lambda(t)$ is very smooth over periods compared to the mean time between intervals, we can write

$$\chi_i = \tau_i - \tau_{i-1} \simeq \lambda(t_{i-1})x_i. \tag{5.32}$$

Then a logarithmic transformation reduces this to standard regression models, and the exponential trend functions discussed above are appealing since we have (approximately) standard linear regression methods. These are considered in Cox and Lewis (1966, Chapter 3).

Of course, in data such as the arrival data of Section 1, this approximation is not possible, and the integral transformation must be performed, perhaps numerically.

The real problem comes, of course, when $\lambda(t)$ cannot be assumed to be smooth over suitably short intervals and no simple parametric form for $\lambda(t)$ is available.

2) If the analysis is based on the conditional distribution of the t_i's, given $N(t_0) = n$, then the transformation (5.31) is seen from (5.2) to be just proportional to the probability integral transform for the density $\lambda(t)/\Lambda(t)$ with estimated parameters.

3) There is also a possibility of looking at (5.31) as a filter in the spectral domain.

Formal methods for analyzing the τ_i are difficult and will be discussed elsewhere. This is simplest for $\lambda(t;\underline{\theta})$ when sufficient statistics for $\underline{\theta}$ are available. The probability structure of the τ_i process, given the sufficient statistics, is independent of $\underline{\theta}$. Of course, in the simplest case of $\lambda(t) = \lambda$, with $\lambda = n/t_0$, it is well known that the χ_i's are the gap statistics for independent uniform random variables and are asymptotically exponentially distributed and independent (Pyke, 1972). For small samples, they will generally be correlated random variables.

Informal, graphical methods are useful and are discussed in the context of the arrival data given in Section 1. The first section of this data was transformed, using (5.31) and an exponential linear trend, and discussed in Lewis (1970). No test for Poisson processes performed in the SASE IV program gave significant indication of departure. The spectrum of the transformed data is shown in Lewis (1970).

Clearly the (exponential)linear trend plus day cycle is not adequate for the complete set of data, as we saw in Section 1. We examine its adequacy further here. Figure 9 shows a plot of the average of successive groups of 20 X_i's after detrending with the (exponential) linear trend plus day cycle. The model for $\lambda(t)$ is clearly inadequate.

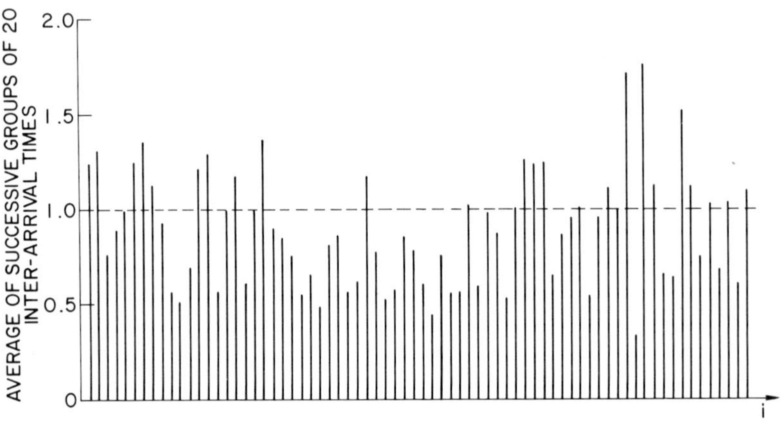

Figure 9. *Arrival of Patients at an Intensive Care Unit.* Averages of successive groups of 20 inter-arrival times after detrending with $\tau = \int_0^t \exp\{\hat{\alpha}+\hat{\beta}u+\hat{k}\sin(2\pi u+\hat{\theta})\}du$, where $\hat{\alpha}$, $\hat{b}$, $\hat{k}$, $\hat{\theta}$ are maximum likelihood estimates. n = 1458 arrivals in 1829 days from 4 February 1963 to 8 February 1968. Overall average after detrending: 1.0088. Homogeneity of variance statistic = 309.07 (χ^2_{71}: mean 71, σ = 11.92).

Figure 10 shows the spectrum of counts after the linear transformation. There is no spike corresponding to the one day period and no indication of departure from a Poisson process. Similarly, the tests discussed in Section 3 for Poisson processes do not give any great indication of departures. However, the estimated asymptotic slope of the variance-time curve is approximately 8, and since this quantity is

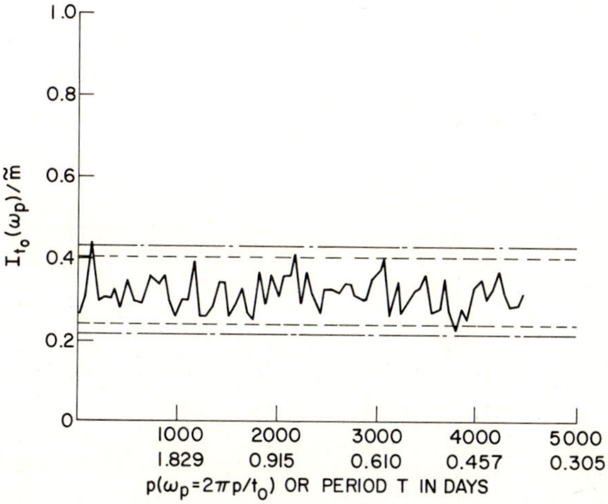

Figure 10. *Arrival of Patients at an Intensive Care Unit.* Spectrum of counts for complete record after detrending with $\tau = \int_0^t \exp\{\hat{\alpha}+\hat{\beta}u+\hat{k}\sin(2\pi u+\hat{\theta})\}du$, where $\hat{\alpha}$, $\hat{\beta}$, $\hat{k}$, $\hat{\theta}$ are maximum likelihood estimates. n = 1458 arrivals in 1829 days from 4 February 1963 to 8 February 1968. Bands are approximately 0.95 and 0.99 confidence regions for individual values.

related to the rate of change of the low frequency end of the spectrum, there is again an indication of inadequacy of the form of $\lambda(t)$ or the hypothesis of a nonhomogeneous Poisson process.

A quadratic term was added to the rate $\lambda(t)$, giving the transformation

$$\tau_i = \int_0^{t_i} \exp\{\hat{\alpha}+\hat{\beta}u+\hat{\gamma}u^2+\hat{k}\sin(2\pi u+\hat{\theta})\}du, \tag{5.33}$$

where the values of the estimated parameters are given in Section 5.4. The plot corresponding to Fig. 9 is shown for the (exponential) quadratic detrending in Fig. 11. This and other results give a much clearer picture of a possible cycle or quasi-cycle in the data with the period of about a year and a half. The example is discussed in more detail in Lewis (1971), where local smoothing is used to estimate $\lambda(t)$ for this quasi-cycle and test the nonhomogeneous Poisson hypothesis.

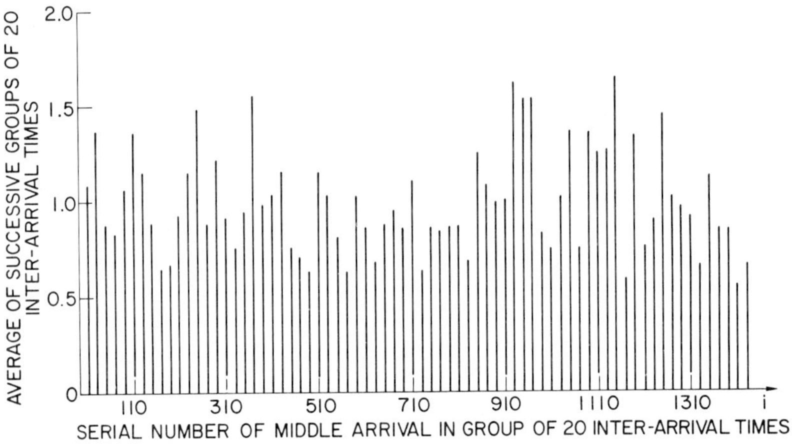

Figure 11. *Arrival of Patients at an Intensive Care Unit.* Averages of successive groups of 20 inter-arrival times after detrending with $\tau = \int_0^t \exp\{\hat{\alpha}+\hat{\beta}u+\hat{\gamma}u^2+\hat{k}\sin(2\pi u+\hat{\theta})\}\,du$, where $\hat{\alpha}$, $\hat{\beta}$, $\hat{\gamma}$, $\hat{k}$, are maximum likelihood estimates. $n = 1458$ arrivals in $t_0 = 1829$ days from 4 February 1963 to 8 February 1968. Overall average after detrending: 1.0000. Homogeneity of variance statistic: 287.06 (χ^2_{71}: mean 71, $\sigma = 11.92$).

6. CONCLUSIONS

Much further work is needed in the statistical analysis of series of events. Particular areas that come to mind are the analogues of the procedures in Section 5 for modulated renewal processes. Likelihood functions can be written down, but results are fairly difficult to obtain from the expressions (see Cox, 1972, for details). Both for these processes and the nonhomogeneous Poisson process there is a great deal of work to be done to justify estimation and testing procedures based on random sample sizes. Some of these justifications have been given by Brown (1972).

Finally, it is hoped that new methods will be developed which will allow some of the non-renewal point processes to be analyzed in a more systematic way.

ACKNOWLEDGMENT

I wish to thank Mr. A. S. Goodman of IBM for continuing help in obtaining the simulation results reported in Section 2. I also wish to thank Mr. A. Barr for making the intensive care unit data available.

REFERENCES

Bar-David, I. (1969). Communication under the Poisson regime. IEEE Trans. Info. Th. 15, 31-37.

Bartlett, M. S. (1954). Problèmes de l'analyse spectrale des séries temporelles stationnaires. Publ. Inst. Statist. Univ. de Paris 3, 119.

Bartlett, M. S. (1963). The spectral analysis of point processes. J. R. Statist. Soc. B 25, 264-96.

Bartlett, M. S. (1966). *An Introduction to Stochastic Processes,* 2nd ed.. Cambridge University Press.

Bartlett, M. S. (1967). Some remarks on the analysis of time-series. Biometrika 54, 25-38.

Bedard, G. (1966). Photon counting statistics of Gaussian light. Phys. Rev. 151, 1038-39.

Bickel, P. J. (1969). Tests for monotone failure rate II. Ann. Math. Statist. 40, 1216-35.

Bickel, P. J. and Doksum, K. S. (1969). Tests for monotone failure rate based on normalized spacings. Ann. Math. Statist. 40, 1216-35.

Blumenthal, S., Greenwood, J. A., and Herbach, L. (1971). Superimposed non-stationary renewal processes. J. Appl. Prob. 8, 184-92.

Boswell, M. T. (1966). Estimating and testing trend in a stochastic process of Poisson type. Ann. Math. Statist. 37, 1564-73.

Boswell, M. T. and Brunk, H. D. (1969). Distribution of likelihood ratio in testing against trend. Ann. Math. Statist. 40, 371-80.

Brown, M. (1972). Statistical analysis of non-homogeneous Poisson processes. This volume.

Brillinger, D. R. (1972). The spectral analysis of stationary interval functions. *Proc. 6th Berkeley Symposium,* to appear.

Clark, J. R. and Hoversten, E. (1970). Poisson process as a statistical model for photodetectors excited by Gaussian light. Mass. Institute Technology, Research Lab. Electronics, Quart. Prog. Rpt. 98, 95-101. MIT, Cambridge, Mass.

Cooley, J. W., Lewis, P. A. W. and Welch, P. D. (1970). The application of the fast Fourier transform algorithm to the estimation of spectra and cross-spectra. J. Sound Vib. 12, 339-52.

Cooley, J. W. and Tukey, J. W. (1965). An algorithm for the machine calculation of complex Fourier series. Maths. Comput. 19, 297-301.

Cox, D. R. (1972). The statistical analysis of dependencies in point processes. This volume.

Cox, D. R. (1970). *The Analysis of Binary Data*. Methuen: London.

Cox, D. R. (1966). The null distribution of the first serial correlation coefficient. Biometrika 53, 623-6.

Cox, D. R. (1965). On the estimation of the intensity function of a stationary point process. J. R. Statist. Soc. B 27, 332 -7.

Cox, D. R. (1955). Some statistical methods connected with series of events. J. R. Statist. Soc. B 17, 129-64.

Cox, D. R. and Lewis, P. A. W. (1966). *The Statistical Analysis of Series of Events*. Methuen: London, and Barnes and Noble: New York.

Cox, D. R. and Lewis, P. A. W. (1972). Multivariate point processes. *Proc. 6th Berkeley Symposium*, to appear.

Durbin, J. (1967). The probability that the sample distribution function lies between two parallel straight lines. Ann. Math. Statist. 39, 398-411.

Durbin, J. (1969). Tests of serial independence based on the cumulated periodogram. Bull. I. S. I. 42, 1039-1048.

Fisher, R. A. (1953). Dispersion on a sphere. Proc. Roy. Soc. London, A 217, 295-305.

Forrest, J. S. (1950). Variations in thunderstorm severity in Great Britain. Quart. J. Roy. Meteo. Soc. 76, 277-86.

Gaver, D. P. (1972). Point process problems in reliability. This volume.

Gaver, D. P. (1970). Multivariate gamma distributions generated by mixture. Sankhyā A 32, 123-26.

Gaver, D. P. (1963). Random hazard in reliability problems. Technometrics 5, 211-26.

Gilles, D. C. and Lewis, P. A. W. (1967). The spectrum of intervals of a geometric branching Poisson process. J. Appl. Prob. 4, 201-5.

Goodman, A. S. and Lewis, P. A. W. (1972). The null distributions of test statistics based on sample serial correlations and the periodogram for renewal processes with exponential and 1/2 Weibull intervals. IBM Research Report, to appear.

Govier, L. J. and Lewis, T. (1967). Estimation of the dispersion parameter of an (A, B) process in queueing theory. *Recent Developments and Applications*. R. Gruan, ed. English Universities Press.

Grandell, J. (1971). On stochastic processes generated by a stochastic intensity function. Skand. Aktuar. Tidskr, to appear.

Grandell, J. (1972). Statistical inference for doubly stochastic Poisson processes. This volume.

Greenwood, J. A. and Durand, D. (1955). The distribution of length and components of the sum of n random unit vectors. Ann. Math. Statist. 26, 233-46.

Griffiths, R. C. (1969). The canonical correlation coefficients of bivariate gamma distributions. Ann. Math. Statist. 40, 1401-8.

Gumbel, E. J., Greenwood, J. A. and Durand, D. (1953). The circular normal distribution: Theory and tables. J. Am. Statist. Ass. 48, 131-52.

Hannan, E. J. (1970). *Multiple Time Series*. New York: Wiley.

Hellstrom, C. W. (1964). The distribution of photoelectric counts from partially polarized Gaussian light. Proc. Phys. Soc. 83, 777-82.

Jackson, O. A. Y. (1967). An analysis of departures from the exponential distribution. J. R. Statist. Soc. B 29, 540-49.

Jakeman, E., Pyke, E. R. and Swain, S. (1971). Statistical accuracy in the digital autocorrelation of photon count fluctuations. J. Phys. A. 4, 517-34.

Karp, S. and Clark, J. R. (1970). Photon counting: A problem in classical noise theory. IEEE Trans. Info. Th. IT-16, 672-80.

Koppel, D. E. (1971). Analysis of Gaussian light by clipped photocount autocorrelation: The effect of finite sampling times and incomplete spatial coherence. J. Applied Physics 42, 3216-25.

Lampard, D. G. (1968). A stochastic process whose successive intervals between events from a first order Markov chain. I. J. Applied Prob. 5, 648-68.

Lawrance. A. J. (1972). Some models for stationary series of univariate events. This volume.

Lewis, P. A. W. (1971). A statistical analysis of a series of arrivals of patients at an intensive care unit. IBM Research Report, to appear.

Lewis, P. A. W. (1970). Remarks on the theory, computation and application of the spectral analysis of series of events. J. Sound Vib. 12, 353-75.

Lewis, P. A. W. (1969). Asymptotic properties and equilibrium conditions for branching Poisson processes. J. Appl. Prob. 6, 355-71.

Lewis, P. A. W. (1966). A computer program for the statistical analysis of series of events. IBM Syst. J. 5, 202-25.

Lewis, P. A. W. (1965). Some results on tests for Poisson processes. Biometrika 52, 67-77.

Lewis, P. A. W. (1964). A branching Poisson process model for the analysis of computer failure patterns. J. R. Statist. Soc. B 26, 398-456.

Lewis, P. A. W. and Goodman, A. S. (1970). The null distribution of the first three product-moment statistics for exponential, half-gamma, and normal scores. In *Selected Tables in Mathematical Statistics,* H. L. Harter and D. B. Owen, eds., Markham Publishing Company: Chicago.

Lewis, P. A. W. and Goodman, A. S. (1969). Sampling results on the distributions of score product-moment statistics under a null permutation hypothesis. IBM Research Report, RC 2544. To appear in Selected Tables in Mathematical Statistics, Vol. II, H. L. Harter and D. B. Owen, eds.

Lewis, P. A. W., Katcher, A. M. and Weiss, A. H. (1969). SASE IV -- An improved program for the statistical analysis of series of events. IBM Research Report, RC 2365.

McNeill, I. B. (1971). Censored overlapping Poisson counts. To appear.

Moore, G. P., Segundo, J. P., Perkel, D. H. and Levitan, H. (1970). Statistical signs of synaptic interaction in neurons. Biophysical J. 10, 867-900.

Moran, P. A. P. (1967a). Testing for correlation between non-negative variates. Biometrika 54, 385-94.

Moran, P. A. P. (1967b). Testing for serial correlation with exponentially distributed variates. Biometrika 54, 395-401.

Moran, P. A. P. (1970). A note on serial correlation coefficients. Biometrika 57, 670-73.

Proschan, F. and Pyke, R. (1967). Tests for monotone failure rate. *Proc. Fifth Berkeley Symp.* 3, 293-312.

Perkel, D. H., Gerstein, G. L. and Moore, G. P. (1967a). Neuronal spike trains and stochastic point processes I. The single spike train. Biophysical J. 7, 391-418.

Perkel, D. H., Gerstein, G. L. and Moore, G. P. (1967b). Neuronal spike trains and stochastic point processes II. Simultaneous spike trains. Biphysical J. 7, 419-40.

Pusey, P. N. (1971). Calculation of the signal-to-noise ratio for a photon correlation experiment. IBM Research Report, RW-135.

Pyke, R. (1965). Spacings. J. R. Statist. Soc. B 27, 395-449.

Pyke, R. (1972). Spacings revisted. *Proc. 6th Berkeley Symposium,* to appear.

Quenouille, M. H. (1956). Notes on bias in estimation. Biometrika 43, 353-60.

Qureishi, A. S. (1964). The discrimination between two Weibull processes. Technometrics 6, 57-76.

Reiffen, B. and Sherman, H. (1963). An optimum demodulator for Poisson processes: Photon source detectors. Proc. IEEE 51, 1316-20

Ross, S. M. (1970). Identifiability in Gl/G/k queues. J. Appl. Prob. 7, 776-80.

Shlien, S. and Toksöz, M. N. (1970). A clustering model for earthquake occurrences. Bull. Seismol. Soc. Amer. 60, 1765-87.

Snyder, D. L. (1971). Statistical analysis of dynamic tracer data. Proceedings of Workshop on Processing of Dynamic Tracer Data from an Anger Scintillation Camera, Biomedical Computer Laboratory, Washington University, St. Louis, Mo.

Snyder, D. L. (1972). Filtering and detection for doubly stochastic Poisson processes. IEEE Trans. Inf. Theory IT-18, to appear.

Stephens, M. A. (1969a). Tests for the von Mises distribution. Biometrika 56, 149-160.

Stephens, M. A. (1969b). Tests for randomness of directions against two circular alternatives. J. Am. Statist. Ass. 64, 280-89.

Vere-Jones, D. (1970). Stochastic models for earthquake occurrence. J. R. Statist. Soc. B 32, 1-62.

Vere-Jones, D. (1967). The infinite divisibility of bivariate gamma distribution. Sankhyā A 29, 421-22.

Vere-Jones, D. and Davies, R. D. (1966). A statistical survey of earthquakes in the main seismic region of New Zealand, Part II. Time series analysis. N. Z. J. Geol. Geophys. 9, 251-84.

Vere-Jones, D., Turnovsky, S. and Eiby, G. A. (1964). A statistical survey of earthquakes in the main seismic region of New Zealand Part I. Time trends in the pattern of recorded activity. N. Z. J. Geol. Geophys. 7, 722-44.

Wold, H. (1948). On stationary point processes and Markov chains. Skand. Aktuar. Tidskr. 31, 229-40.

Walloe, L., Jansen, J. K. S. and Nygaard, K. (1969). A computer simulated model of a second order sensory neuron. Kybernetik 6, 130-40.

Watson, G. S. and Williams, E. J. (1956). On the construction of significance tests on the circle and the sphere. Biometrika 43, 344-52.

Wilk, M. B. and Gnanadesikan, R. (1968). Probability plotting methods for the analysis of data. Biometrika 55, 1-17.

Yang, G. L. (1970). Asymptotic tests for serial correlation with exponentially distributed random variables. Ann. Math. Statist. 41, 1390.

THE STATISTICAL ANALYSIS OF DEPENDENCIES IN POINT PROCESSES

D. R. Cox
Imperial College, University of London

SUMMARY

A class of point processes is considered in which the simple hazard function of a renewal process or of a Poisson process is perturbed by a factor depending on quantities thought to influence the probability of occurrence, each quantity being combined with an unknown regression coefficient. Inference about the unknown regression coefficients is achieved via a conditional likelihood, leading in particular to asymptotic significance tests for the presence of dependence.

1. INTRODUCTION

Methods for the statistical analysis of point processes are of a number of types. Some are adaptations of standard second-order techniques applied either to the counting process or to the intervals between successive events. Others, particularly connected with renewal processes, are essentially one-sample tests of distributional form. In a relatively few cases, such as a parametric semi-Markov process or a parametric Markov process of intervals (Cox and Lewis, 1971), a likelihood can be obtained in useful form and standard techniques such as maximum likelihood applied.

In the present paper a fairly general method is outlined for assessing dependencies in point processes, especially in point processes closely related to renewal processes. The methods lead to significance tests and confidence limits that will be asymptotically efficient under rather artificial assumptions. Unfortunately the procedures, while computationally quite feasible, are a bit clumsy, except for testing global null hypotheses. The main restriction to the methods is their close tie to renewal processes.

We deal for simplicity with processes in continuous time so that the problem of ties can be ignored. In practice, a small proportion of ties will cause no serious difficulty, but if the proportion were to become appreciable an extension of the arguments to discrete time would be required and is possible.

2. MODULATED RENEWAL AND POISSON PROCESSES

Consider first a renewal process with hazard function $\lambda_0(x)$, i.e., if X is a typical interval between successive events

$$\lambda_0(x) = \ell im \frac{pr(x \leq X < x + \Delta x | x \leq X)}{\Delta x} . \tag{1}$$

Thus if at time t the backward recurrence time to the preceding event is u(t) the probability of an event in $(t, t + \Delta t)$ given the whole previous history of the process is $\lambda_0\{u(t)\}\Delta t + o(\Delta t)$.

Now suppose that at time t, there is a column vector $\underline{z}(t) = \{z_1(t),\ldots,z_p(t)\}^T$ such that the hazard is

$$\exp\{\beta_1 z_1(t) +\ldots+ \beta_p z_p(t)\}\lambda_0\{u(t)\} = \exp\{\underline{\beta}\underline{z}(t)\}\lambda_0\{u(t)\}. \tag{2}$$

Here $\underline{\beta} = (\beta_1,\ldots,\beta_p)$ is a row vector of unknown parameters. If the z's are values of stochastic processes we argue conditionally on their observed values. We can call a process defined by (2) a renewal process modulated by the vector $\underline{z}(t)$. If $\lambda_0(\cdot)$ is constant we can call the process a modulated Poisson process.

A wide range of processes can be represented in this way and there follow just a few examples.

Example 1. Trend. The presence of trend can be examined in various ways. The simplest is probably to include a single term $z_1(t) = t$; more generally a set of polynomial terms could be taken. Alternatively $z_1(t)$ may be the serial number of the event preceding t. When trend on time itself is under consideration, it may as an approximation be convenient to replace t by the instant of occurrence of the preceding event so that the regressor variable remains constant between each pair of events.

Example 2. Sinusoidal variation. By taking $z_1(t) = \cos(\omega_0 t)$ and $z_2(t) = \sin(\omega_0 t)$, we can represent sinusoidal fluctuations of known wavelength.

Example 3. Bivariate point process. In a bivariate point process we may be interested in the dependence of one type of event, say type A, on the other events, called type B. We may then argue conditionally on the observed type B events. Let $u_b(t)$ be the backward recurrence time from t to the preceding type B event and $v_b(t)$ be the corresponding forward recurrence time. We might then take $z_1(t) = u_b(t)$ and $z_2(t) = v_b(t)$, although there are many other possibilities. Note that in this approach to the study of dependence in bivariate point processes a viewpoint analogous to regression theory is taken; we do not consider the stochastic properties of the process of type B events.

Example 4. Point process with quantitative variables. Similarly if we have a real variable attached to each event we can modify the hazard depending, for example, on the preceding quantity.

Example 5. Serially correlated process. A rather more complicated application arises if we have one type of event and take $z_1(t)$ to be the preceding interval between successive events. For most of the discussion however, we suppose that the functions are 'fixed' and do not depend on the point process which is the 'dependent' variable of the analysis.

Often it will be useful to combine some of these models. For example in the problem of Example 3, it may be wise to allow for possible time trends by including terms of the type mentioned in Example 1. This is roughly parallel to using trend eliminating filters before studying the relation between two or more real-valued series.

To summarise, the model (2) provides a simple flexible and general way of introducing dependencies into point processes; it is, however, not a substitute for a special stochastic model developed on physical grounds for a particular problem.

3. TYPES OF ANALYSIS

We now concentrate on analysis of the unknown regression coefficients in the model (2). There are several approaches that may be relevant. In some ways the simplest is to assume that $\lambda_0(t)$ is constant, i.e., that we are dealing with a modulated Poisson process. A second possibility is to take some rather more general parametric form for $\lambda_0(t)$; one possibility is to take the Weibull distribution, i.e., to have a power law for $\lambda_0(t)$. Next there is the approach of putting some broad rather qualitative restriction on $\lambda_0(t)$, such as that it is nondecreasing. While such distributions have been quite extensively studied, they probably do not lend themselves so easily to the analysis of the associated regression coefficients. Finally we may leave $\lambda_0(t)$ entirely arbitrary. This, while not entirely sensible, leads to fairly simple analysis for the regression coefficients.

4. ANALYSIS FOR MODULATED POISSON PROCESS

We first deal with the modulated Poisson process in which the function $\lambda_0(t) = \rho$, an unknown constant. Provided that $\underline{z}(t) = \{z_1(t),\ldots, z_p(t)\}^T$ can be treated as non random, the analysis is in principle straightforward. If events in $(0,t_0)$ occur at times $t_1,\ldots,t_n$, the full likelihood is

$$\rho^n \exp(\underline{\beta}\underline{s}) \exp[-\rho\int_0^{t_0} \exp\{\underline{\beta}\underline{z}(t)\}dt], \tag{3}$$

where

$$s_\xi = \sum_{i=1}^{n} z_\xi(t_i) \qquad (\xi = 1,\ldots,p) \tag{4}$$

and can be considered as a column vector. We are regarding ρ as a nuisance parameter and for a given vector $\underline{\beta}$ the total number of occurrences n is sufficient for ρ. The likelihood conditionally on n is obtained by dividing by the Poisson probability of n and except for a constant the log likelihood is

$$L_P(\underline{\beta}) = \underline{\beta}\,\underline{s} - n \log [\int_0^{t_0} \exp\{\underline{\beta}\,\underline{z}(t)\}dt]. \tag{5}$$

In particular we can calculate the efficient score vector

$$U_{P\xi}(\underline{\beta}) = \frac{\partial L_P(\underline{\beta})}{\partial \beta_\xi} = s_\xi - n A_P\{z_\xi(t); \underline{\beta}\}, \tag{6}$$

where

$$A_P\{z_\xi(t); \underline{\beta}\} = \frac{\int_0^{t_0} z_\xi(t)\exp\{\underline{\beta}\,\underline{z}(t)\}dt}{\int_0^{t_0} \exp\{\underline{\beta}\,\underline{z}(t)\}dt}. \tag{7}$$

The matrix of observed and expected second derivatives is, independently of $t_1, \ldots, t_n$,

$$I_{P,\xi,\eta}(\underline{\beta}) = -\frac{\partial^2 L_P(\underline{\beta})}{\partial \beta_\xi \partial \beta_\eta} = E\{-\frac{\partial^2 L_P(\underline{\beta})}{\partial \beta_\xi \partial \beta_\eta}\}$$

$$= n[A_P\{z_\xi(t) z_\eta(t); \underline{\beta}\} - A_P\{z_\xi(t); \underline{\beta}\}A_P\{z_\eta(t); \underline{\beta}\}], \tag{8}$$

a formal covariance.

There are several types of question about the parameters $\underline{\beta}$ that can now be considered; in particular we can obtain

(i) a test of the global null hypothesis $\underline{\beta} = \underline{0}$;

(ii) a confidence region for $\underline{\beta}$;

(iii) a test of the vanishing of one or more components of the vector $\underline{\beta}$;

(iv) a confidence region for one or more components of $\underline{\beta}$.

In both (iii) and (iv) we regard the other components as nuisance parameters.

Asymptotic answers to all these follow by standard methods based on maximum likelihood theory, but it is really only the first that gives an answer that is reasonably explicit. Some analytical conditions will be required for asymptotic theory to hold. From a physical point of view an essential requirement is that the observed process should not contain too small a number of events; from a more theoretical standpoint the asymptotic theory can hardly apply to a process which comes to an end after a finite number of events have occurred.

The results (6) - (8) express in fairly general terms methods developed for the special case of simple trend (Example 1) by Cox (1955) and by Cox and Lewis (1966, pp. 45-51) and for the sinusoidal case (Example 2) by Lewis (1970).

In some very special cases 'exact' results, always possible in principal, can be obtained in a useful form. That this is difficult even for fairly simple instances of problems of type (iii) is illustrated by the case $z_1(t) = t$ and $z_2(t) = t^2$, where it might be required to test whether $\beta_2 = 0$ as a test of the adequacy of an assumption of simple exponential trend. An appropriate conditional distribution is available, but it does not seem easy to get results in a simple usable form.

In the special case $\underline{\beta} = \underline{0}$, the operator A_p reduces to a simple averaging over $(0,t_0)$ and an asymptotic test is obtained by treating

$$\underline{U}_p^T(0)\ \{\underline{I}_p(\underline{0})\}^{-1}\underline{U}_p(0)$$

as a chi-squared variable with p degrees of freedom.

As a special case consider Example 3 of Section 2 in which, in particular, the suspected dependence is on the backward recurrence time $u_b(t)$ to the preceding event of type B. Then p = 1, and dropping the suffix on z, we have that $z(t) = u_b(t)$. If we make the simplification for purposes of exposition that the interval of observation $(0,t_0)$ starts and ends with a type B event, we describe the configuration of type B events by the sequence of intervals in the type B process, say $x_1^{(b)},\ldots,x_m^{(b)}$. We then have that

$$A_p\{z(t);\beta\} = \frac{\sum_{j=1}^{m}\int_0^{x_j^{(b)}} xe^{\beta x}dx}{\sum_{j=1}^{m}\int_0^{x_j^{(b)}} e^{\beta x}dx}, \qquad (9)$$

$$A_p\{z(t);\beta\} = \frac{\Sigma\{\beta x_j^{(b)} e^{\beta x_j^{(b)}} - e^{\beta x_j^{(b)}} + 1\}/\beta^2}{\Sigma(e^{\beta x_j^{(b)}} - 1)/\beta}, \tag{10}$$

$$A_p\{z^2(t);\beta\} = \frac{\Sigma\{\beta^2(x_j^{(b)})^2 e^{\beta x_j^{(b)}} - 2\beta x_j^{(b)} e^{\beta x_j^{(b)}} + 2e^{\beta x_j^{(b)}} - 2\}/\beta^3}{\Sigma(e^{\beta x_j^{(b)}} - 1)/\beta}. \tag{11}$$

and by (8), $I_p(\beta)/n$ is the second of these minus the square of the first.

Thus asymptotic $1 - \alpha$ confidence limits for β are obtained from

$$\{\beta; \left|\sum_{i=1}^{n} u_b(t_i) - nA_p\{z(t);t\}\right| < k_{\frac{1}{2}\alpha}\sqrt{I_p(\beta)}\},$$

where $\Phi(-k_\alpha) = \alpha$, denoting the standard normal integral by $\Phi(\cdot)$.

This is not so formidable computationally as it might seem, since usually it will be enough to compute a critical ratio for a few values of β and to interpolate on a normal integral scale. In principle more refined calculations are possible using the techniques of Bartlett (1953).

The significance test of the null hypothesis $\beta = 0$ of independence was previously obtained by a less formal argument (Cox and Lewis, 1972). In fact when $\beta = 0$ the operation A_p is a simple unweighted averaging over $(0,t_0)$, so that

$$A_p\{z(t);0\} = \frac{1}{t_0}\{\sum_{j=1}^{m}\int_0^{x_j^{(b)}} x\,dx\} = \frac{1}{2t_0}\Sigma\{x_j^{(b)}\}^2 = \frac{1}{2}m_2^{(b)}, \tag{12}$$

$$A_p\{z^2(t);0\} = \frac{1}{3t_0}\Sigma\{x_j^{(b)}\}^3 = \frac{1}{3}m_3^{(b)}, \tag{13}$$

say. Thus the critical ratio for testing whether $\beta = 0$ is

$$\frac{\sum_{i=1}^{n} u_b(t_i) - \frac{1}{2} n\, m_2^{(b)}}{[\frac{1}{3} m_3^{(b)} - \frac{1}{4} \{m_2^{(b)}\}^2]^{\frac{1}{2}} \sqrt{n}} . \tag{14}$$

Note especially that the sum over i is for the n events in the dependent process under analysis, whereas the sum over j concerns the intervals between successive events in the process of type B events, which for the present purpose is regarded as fixed.

As a further example of a test of significance for a null hypothesis of independence, suppose that the dependence to be tested is on one or both of the backward recurrence-time, $u_b(t)$, to the previous event of type B and the forward recurrence-time, $v_b(t)$, to the next event of that type. That is, $p = 2$, $z_1(t) = u_b(t)$, $z_2(t) = v_b(t)$.

Then

$$\underline{U}_p(\underline{0}) = \begin{pmatrix} \Sigma u_b(t_i) \\ \Sigma v_b(t_i) \end{pmatrix} - \frac{1}{2} n m_2^{(b)} \begin{pmatrix} 1 \\ 1 \end{pmatrix}, \tag{15}$$

$$\underline{I}_p(\underline{0}) = n \begin{pmatrix} \frac{1}{3} m_3^{(b)} - \frac{1}{4} \{m_2^{(b)}\}^2 & \frac{1}{6} m_3^{(b)} - \frac{1}{4} \{m_2^{(b)}\} \\ \frac{1}{6} m_3^{(b)} - \frac{1}{4} \{m_2^{(b)}\}^2 & \frac{1}{3} m_3^{(b)} - \frac{1}{4} \{m_2^{(b)}\}^2 \end{pmatrix} \tag{16}$$

and the test statistic for independence is

$$\underline{U}_p^T(\underline{0}) \{\underline{I}_p(\underline{0})\}^{-1} \underline{U}_p(\underline{0}) ; \tag{17}$$

under the null hypothesis its asymptotic distribution is chi-squared with 2 degrees of freedom.

A central assumption in the discussion of this section is that the vector $\underline{z}(t)$ can be treated as fixed, and in particular does not involve the point process the dependencies in which are under investigation. For example if the vector $\underline{z}$ included the backward recurrence time in the process itself, as in a renewal process, integrals of z(t) over $(0,t_0)$ are no longer constant, but are statistics. While a direct maximum likelihood analysis can be made from (3), the subsequent reduction and simplification by conditioning does not occur.

5. ANALYSIS FOR MODULATED RENEWAL PROCESS

In the previous section it was assumed that $\lambda_0(t)$ is constant. We now develop methods for handling the corresponding problems when $\lambda_0(t)$ is arbitrary. Of course the assumption of an arbitrary function here is to be regarded as one way of proceding with cautious conservatism; some reasonable smoothness in $\lambda_0(t)$ is in fact to be expected.

It will be assumed for simplicity that the intervals between successive events in the process under analysis $x_1,\ldots,x_n$ are all different. Also it is supposed that the process $\underline{z}(t)$ can be taken as fixed.

We can then write down the likelihood of the observed process. No information about $\underline{\beta}$ can be provided except from those values of the interval between successive events that are observed, i.e., the values $x_1,\ldots,x_n$. For it is conceivable that at all other values $\lambda_0(t)$ is identically zero. The nuisance function $\lambda_0(t)$ is thus eliminated by arguing conditionally on the set of observed intervals. The argument follows closely that used for the generalized life-table analysis of a set of independent individuals (Cox, 1972).

It is convenient to suppose that the intervals $x_1,\ldots,x_n$ are arranged in order of increasing size, i.e., are order statistics. A fuller notation would write $x_{(1)},\ldots,x_{(n)}$, but this is unnecessary at this stage. For $i \geq j$, define $\underline{z}_{ij}$ to be the vector of explanatory variables associated with the interval x_i, but calculated a time x_j after the start of the interval. Thus $\underline{z}_{ii}$ is the vector calculated at the end of the interval x_i.

We can build up a likelihood for the data conditionally on the magnitudes of the intervals, by considering contributions in order starting with the smallest interval. Thus the contribution from the first interval is

$$\frac{\exp(\underline{\beta}\underline{z}_{11})}{\sum_{\ell=1}^{n} \exp(\underline{\beta}\underline{z}_{\ell 1})} . \tag{18}$$

Conditionally on the first interval, the contribution from the second interval is

$$\frac{\exp(\underline{\beta}\underline{z}_{22})}{\sum_{\ell=2}^{n} \exp(\underline{\beta}\underline{z}_{\ell 2})} ,$$

and so on. Thus the log likelihood is

$$L_R(\underline{\beta}) = \underline{\beta}\, \Sigma \underline{z}_{ii} - \sum_{i=1}^{n} \log \{ \sum_{\ell=i}^{n} \exp(\underline{\beta}\underline{z}_{\ell i}) \}. \tag{19}$$

This is to be compared with (5) for the modulated Poisson process. We can proceed as before to calculate the efficient score vector

$$U_{R\xi}(\underline{\beta}) = \frac{\partial L_R(\underline{\beta})}{\partial \beta_\xi} = s_\xi - n\, A_R(z_\xi ; \underline{\beta}) , \tag{20}$$

where $\underline{s}$ is the same as in the Poisson case and where

$$A_R(z_\xi ; \underline{\beta}) = \frac{1}{n} \sum_{i=1}^{n} \left[\frac{\sum_{\ell=i}^{n} z_{\ell i \xi} \exp(\underline{\beta}\underline{z}_{\ell i})}{\sum_{\ell=i}^{n} \exp(\underline{\beta}\underline{z}_{\ell i})} \right]. \tag{21}$$

Further

$$I_{R\xi\eta}(\underline{\beta}) = - \frac{\partial^2 L_R(\underline{\beta})}{\partial \beta_\xi \partial \beta_\eta} = E\left[- \frac{\partial^2 L_R(\underline{\beta})}{\partial \beta_\xi \partial \beta_\eta} \right]$$

$$= n \left[A_R(z_\xi z_\eta ; \underline{\beta}) - \frac{1}{n} \sum_{i=1}^{n} \left(\frac{\sum_{\ell=1}^{n} z_{\ell i \xi} \exp(\underline{\beta}\underline{z}_{\ell i}) \sum_{\ell=i}^{n} z_{\ell i \eta} \exp(\underline{\beta}\underline{z}_{\ell i})}{\{ \sum_{\ell=i}^{n} \exp(\underline{\beta}\underline{z}_{\ell i}) \}^2} \right) \right]. \tag{22}$$

Again the only case where fairly direct explicit answers are possible is when $\underline{\beta} = \underline{0}$; then

$$A_R(\underline{z};\underline{0}) = \frac{1}{n} \sum_{i=1}^{n} \frac{\sum_{\ell=i}^{n} z_{\ell i}}{n-i+1} . \tag{23}$$

Thus $\underline{U}_R(\underline{0})$ compares the sum of the z's at the events with the sum to be expected with the $\underline{z}$ chosen at random from the intervals 'at risk' at that point.

It seems possible that under suitable restrictions on the z's the operators A_R and A_P are asymptotically equivalent and hence the tests of this and the preceding section become asymptotically equivalent. This aspect will not be investigated here.

A very special case arises when the explanatory variable z is exactly or approximately constant in each interval x_i. The problem then becomes essentially that discussed by Cox (1972). Equation (23) simplifies to

$$A_R(z_\xi;\underline{0}) = \frac{1}{n} \sum_{i=1}^{n} \left(\frac{\sum_{\ell=i}^{n} z_{\ell\xi}}{n-i+1} \right) = \frac{1}{n} \sum_{i=1}^{n} z_{i\xi} e_{ni}, \tag{24}$$

where $e_{ni} = 1/n + \ldots + 1/(n-i+1)$ is the expected order statistic in samples of size n from the unit exponential distribution. Standard types of rank tests, efficient for exponential distributions, are recovered.

Finally consider the possibility explored in Section 5 that the hazard depends on the backward recurrence-time in a process of events of type B, regarded as fixed. The test statistic, although similar to (14), is more complicated in that we have to examine in addition to the backward recurrence-times at the observed events the backward recurrence times at many other points, especially in the longer intervals x_i.

6. DISCUSSION

There follow brief comments on a number of further points.

(i) It would be useful to know the asymptotic efficiency of the procedures based on the renewal process when in fact the Poisson model is appropriate.

(ii) The small sample properties of the procedures would be of interest, and in particular the possibility of 'exact' tests in both renewal and Poisson cases.

(iii) More detailed methods for examining the form of the function $\lambda_0(\cdot)$ would be of value.

(iv) The above procedures are formal tests and confidence intervals. It would be useful also to have more exploratory techniques, including graphical ones.

(v) If the explanatory variables change relatively slowly compared with the intervals between successive events, the very flexible, even if somewhat inefficient, methods based on the least squares analysis of the log intervals between sets of events are available (Cox and Lewis, 1966, pp. 39-43).

(vi) This paper has explicitly concentrated on the analysis of a single homogeneous set of data. However the methods in fact cover the comparison of sets of data, the search for invariant parameters, etc. In particular the comparison of two samples in its simplest form is covered by introducing a binary variable z taking the values 0 and 1 in the two samples.

(vii) It would be of interest to study the probabilistic properties of modulated renewal processes under various assumptions on the z's.

REFERENCES

Bartlett, M. S. (1953). Approximate confidence intervals. Biometrika 40, 12-19.

Cox, D. R. (1955). Some statistical methods connected with series of events. J. R. Statist. Soc. B 17, 129-64.

Cox, D. R. (1972). Regression models and life tables. J. R. Statist. Soc. B 34, to appear.

Cox, D. R. & Lewis, P. A. W. (1966). *The Statistical Analysis of Series of Events*. London: Methuen.

Cox, D. R. & Lewis, P. A. W. (1972). Multivariate point processes. *Proc. 6th Barkeley Symposium*. To appear.

Lewis, P. A. W. (1970). Remarks on the theory, computation and application of the spectral analysis of series of events. J. Sound Vib. 12, 353-75.

STATISTICAL ANALYSIS OF NON-HOMOGENEOUS POISSON PROCESSES

M. Brown
Cornell University, Ithaca, New York

SUMMARY

Events occur in time according to a non-homogeneous Poisson process with unknown mean measure μ_θ assumed to belong to a parametric family of measures $\{\mu_\theta, \theta \in \Phi\}$. The general problem of interest is to develop statistical inference procedures for θ based on the data in $[0,t]$ and asymptotic results as $t \to \infty$. We obtain results for the simple vs. simple case, for one-sided hypothesis tests under a monotonicity structure and for exponential families.

1. INTRODUCTION

Suppose that events occur in time according to a Poisson process with unknown mean measure μ_θ assumed to belong to a parametric family of measures $\{\mu_\theta, \theta \in \Phi\}$. Based on the data in $[0,t]$ we wish to estimate or test hypotheses about θ and to study the asymptotic $(t \to \infty)$ properties of our procedures.

The motivation for studying this problem is that in recent years several results have appeared relating to convergence of point processes to a Poisson limit. These results suggest that Poisson processes may often be good approximations to point processes of interest. Some of these results may be found in Breiman (1963), Goldman (1968), Kennedy (1969), Stone (1968) and Thedeen (1964). In addition Shorrock (1971) has demonstrated a close connection between non-homogeneous Poisson processes and record values from identically and independently distributed (i.i.d.) observations and it follows from his results that the statistical analysis of record data is equivalent to that of non-homogeneous Poisson processes.

Our starting point is the simple vs simple hypothesis. Since the Radon-Nikodym derivative of one Poisson process with respect to another is easily computable, Neyman-Pearson tests are simple to derive. If we let $t \to \infty$ the Neyman-Pearson tests will be consistent if and only if the Poisson processes on $[0,\infty)$ are singular. Necessary and sufficient conditions for singularity were derived by the author (Brown, 1971) and are reviewed here. The exact distribution of the Neyman-Pearson statistic although known in terms of its characteristic function is difficult to work with. We show (Theorem 2) that under weak conditions this statistic is asymptotically normal.

In the infinite parameter case we consider a few classes of problems in which there is no convenient sufficient statistic but nevertheless "reasonable" procedures are available. One class is characterized by the property that $(d\mu_\theta/d\nu)(x) = f_\theta(x)$ is monotone increasing in θ for each x. This structure yields, for all θ_1, θ_2, a stochastic ordering in θ of the distribution of the statistic

$$\sum_{1}^{N_t} \log(f_{\theta_2} / f_{\theta_1})$$

under P_{μ_θ} (P_{μ_θ} is the Poisson process with mean measure μ_θ). This allows one to derive one-sided level α tests with monotone power function consistent over a certain range of alternatives. Another class we consider satisfies a condition under which θ can be consistently estimated through sample intensities.

We also consider a specific problem involving the estimation of a weak trend and in spite of the absence of a convenient sufficient statistic derive an asymptotically efficient estimator.

In Section 7 we study an exponential structure for $f_\theta = d\mu_\theta/d\nu$, similar to the familiar exponential family widely used in the i.i.d. case. In the presence of this structure there is a simple complete sufficient statistic and a monotone likelihood ratio property. From these properties composite vs composite hypothesis testing results which parallel the i.i.d. case are easily derived. Some of the ideas of this section are implicit in Cox and Lewis (1966, Chapter 3); see also Cox (1972) and Lewis (1972).

Although we work with Poisson processes on the positive real line our results are applicable to Poisson processes on arbitrary spaces. This is discussed in Section 10.

2. DEFINITIONS

Suppose that events are randomly occurring in time. Let N(I) be the number of events which occur in an interval I. A Poisson process with mean measure μ is characterized by the property that for every finite collection of non-overlapping intervals $I_1,\ldots,I_m$, $N(I_1),\ldots, N(I_m)$ are independent Poisson distributed random variables with parameters $\mu(I_1),\ldots, \mu(I_m)$. By convention a Poisson distribution with parameter $0(\infty)$ is the distribution which assigns mass 1 to $0(\infty)$. One can prove the existence of a Poisson process with mean measure μ for every σ-finite measure μ on $([0,\infty),\mathcal{B})$ where $\mathcal{B}$ is the collection of Borel sets (Brown, 1970, p. 1939).

Frequently it is assumed that μ is absolutely continuous with respect to Lebesgue measure and one works with the density μ' which is called the intensity function. We won't assume the absolute continuity of μ with respect to Lebesgue measure. We usually will assume that $\{\mu_\theta, \theta \in \Phi\}$ is a dominated family and that $\mu_\theta(t) < \infty$ for all θ, t. We will work with $f_\theta = d\mu_\theta/d\nu$ where ν is a dominating measure for the family. The reader who feels more at home with intensity functions can think of f_θ as a generalized intensity function which coincides with the usual intensity function when ν = Lebesgue measure and $\mu_\theta << \nu$ for all $\theta \in \Phi$.

In working with Radon-Nikoydm derivatives of Poisson processes under a varying t we have to interpret a Poisson process as a probability measure on an appropriate measure space, $(\Omega,\mathcal{A})$, this measure space being independent of t. A convenient selection is:

Ω = {all countable collections of non-negative real numbers including those with multiple occurrences of individual reals}.

A = smallest σ-algebra containing all subsets of Ω of the form $N(I,\omega) = k$ as I ranges through all intervals and k through all non-negative integers.

$N(I,\omega)$ is defined as the number of points in $I \cap \omega$ with the interpretation that if a given real number belongs to I and is included k times in ω then it is counted k times in $N(I,\omega)$.

We use the notation $P_\mu^{(t)}$ for a Poisson process on $[0,t]$ with mean measure μ, and P_μ for a Poisson process on $[0,\infty)$ with mean measure μ. The notation $dP_\mu^{(t)}/dP_\nu^{(t)}$ will be used for the Radon-Nikodym derivative of the Poisson process $P_\mu^{(t)}$ with respect to the dominating Poisson process $P_\nu^{(t)}$; $P_\mu \perp P_\nu$ means that P_μ and P_ν are mutually singular as measures over (Ω, A); $P_\mu << P_\nu$ means that P_μ is dominated by P_ν, i.e., that $P_\nu(A) = 0 \rightarrow P_\mu(A) = 0$ for $A \in A$.

3. STRUCTURE OF NEYMAN-PEARSON TESTS

Suppose that we observe a Poisson process on $[0,t]$, the mean measure μ of the process assumed to be either μ_1 or μ_2 where $\mu_i[0,t] < \infty$, $i=1,2$. We wish to test the null hypothesis $\mu = \mu_1$ vs the alternative $\mu = \mu_2$. This is a simple vs simple hypothesis testing problem and the Neyman-Pearson lemma applies. In order to derive the form of Neyman-Pearson tests we have to be able to compute densities. Towards this end let ν be any measure which dominates both μ_1 and μ_2 with $\nu[0,t] < \infty$, and let $f_1 = d\mu_1/d\nu$, $f_2 = d\mu_2/d\nu$. Let $P_{\mu_1}^{(t)}$, $P_{\mu 2}^{(t)}$, $P_\nu^{(t)}$ represent Poisson processes on $[0,t]$ with mean measures μ_1, μ_2 and ν, respectively. It then follows from Brown (1971, p. 774) that $P_{\mu_i}^{(t)} << P_\nu^{(t)}$, $i = 1,2$, and

$$g_i(t) = \frac{dP_{\mu_i}^{(t)}}{dP_\nu^{(t)}} = \{\prod_{i=1}^{N_t} f_i(x_i)\} \exp[-\int_0^t \{f_i(x)-1\}\,\nu(dx)] \qquad (i = 1, 2),$$

where N_t is the number of events in $[0,t]$ and $x_1, x_2, \ldots$ are the epochs at which events occur. The Neyman-Pearson lemma tells us to reject if $[\prod_{i=1}^{N_t} \{f_2(x_i)/f_1(x_i)\}] \exp[-\int_0^t \{f_2(x) - f_1(x)\}\,\nu(dx)]$ is large, the cutoff point being determined by the desired α level of the test. This is equivalent to rejecting if $W_t = \Sigma_1^{N_t} \log\{f_2(x_i)/f_1(x_i)\}$ is large. Under P_{μ_1}, $W_t = -\infty$ with probability $1 - \exp[-\mu_1(B_t)]$ where $B_t = \{x \in [0,t]: f_2(x) = 0\}$; conditional on $W_t > -\infty$, W_t has a compound Poisson distribution with

$$\log E_{P_{\mu_1}}\{\exp(iuW_t)\,|W_t > -\infty\}$$

$$= \int_{\tilde{B}_t} [\exp[(iu \log\{f_2(x)/f_1(x)\}] - 1]\; f_1(x)\; \nu(dx)$$

where $\tilde{B}_t = \{x \in [0,t]$ such that $f_2(x) > 0\}$. In principle this gives sufficient information to determine cutoff points for α level tests. We shall see in Section 5 that if P_{μ_1} and P_{μ_2} are singular as Poisson processes on $[0,\infty)$, then under weak conditions W_t is asymptotically normal as $t \to \infty$. The normal approximation provides a more convenient method for computing cutoff points than does the above compound Poisson distribution.

Under P_{μ_2}, $W_t = \infty$ with probability $1 - \exp\{\mu_2(A_t)\}$ where $A_t = \{x \in [0,t]$ such that $f_1(x) = 0\}$; the conditional distribution of W_t given $W_t < \infty$ is compound Poisson with

$$\log E_{P_{\mu_2}}\{\exp(iuW_t)\,|W_t < \infty\}$$

$$= \int_{\tilde{A}_t} [\exp[iu \log\{f_2(x)/f_1(x)\}] - 1]\; f_2(x)\; \nu(dx),$$

where $\tilde{A}_t = \{x \in [0,t]$ such that $f_1(x) > 0\}$. Using this fact, the power of Neyman-Pearson tests may be computed, but as before it will be simpler to use a normal approximation. A normal approximation can be justified for large t in many cases (see Section 5).

4. CONSISTENCY OF NEYMAN-PEARSON TESTS

Let P_{μ_1} and P_{μ_2} be two Poisson processes on $[0,\infty)$ with $\mu_i[0,t] < \infty$ for $i = 1, 2$, $t \to \infty$. In Section 3 we studied the Neyman-Pearson test of $\mu = \mu_1$ vs $\mu = \mu_2$ for a fixed value of t. Now we let $t \to \infty$ and study the asymptotic properties of this test.

Suppose that

$$P_{\mu_{\theta_1}} \perp P_{\mu_{\theta_2}} .$$

It then follows from Neveu (1965, p. 144) that

$$\sum_{1}^{N_t} \log f_{\theta_2}/f_{\theta_1} - \int_0^t (f_{\theta_2} - f_{\theta_1})\, d\nu$$

diverges a.s. to $+\infty$ under $P_{\mu_{\theta_2}}$ and diverges a.s. to $-\infty$ under $P_{\mu_{\theta_1}}$. Thus every level $\alpha > 0$ Neyman-Pearson test of $\theta = \theta_1$ vs $\theta = \theta_2$ or of $\theta = \theta_2$ vs $\theta = \theta_1$ has power approaching 1 as $t \to \infty$ and is thus consistent. On the other hand, if every level $\alpha > 0$ Neyman-Pearson test of $\theta = \theta_1$ vs $\theta = \theta_2$ is consistent then $P_{\mu_{\theta_1}} \perp P_{\mu_{\theta_2}}$ and thus every level $\alpha > 0$ test of $\theta = \theta_2$ vs $\theta = \theta_1$ is consistent. This can be shown by using the consistent tests and the a.s. limiting behavior of the liklihood ratio to construct two disjoint sets each containing the support of $P_{\mu_{\theta_1}}$ or $P_{\mu_{\theta_2}}$.

Necessary and sufficient conditions for singularity were derived by the author (Brown, 1971, p. 775).

Theorem 1. $P_{\mu_1} \perp P_{\mu_2}$ iff at least one of the following two conditions hold:

(i) $$\int_{\{|f_1/f_2-1| > c\}} |f_1 - f_2| \, d\nu = \infty \qquad \text{for some } c > 0$$

(ii) $$\int_{\{|f_1/f_2-1| < c\}} (f_1/f_2 - 1)^2 f_2 \, d\nu = \infty \qquad \text{for all } c > 0.$$

We remark that condition (ii) is the non-trivial part of Theorem 1, condition (i) being a rather obvious sufficient condition for singularity. Theorem 1 is a slight rephrasing of Theorem 2 of Brown (1971) where a condition implying condition (i) was included.

The least obvious cases of singularity involve examples in which condition (i) does not hold. For example let ν = Lebesgue measure, $f_1(x)=x$, $f_2(x) = x + 1$; condition (i) does hold not hold but (ii) does and therefore $P_{\mu_1} \perp P_{\mu_2}$. On the other hand, choose $\delta > 0$ and let $f_2(x) = x^{1+\delta}+1$, $f_1(x) = x^{1+\delta}$. In this case $P_{\mu_1} \equiv P_{\mu_2}$.

Note that if P_{μ_1} and P_{μ_2} are not singular then the other possibilities are:

(a) $P_{\mu_1} \equiv P_{\mu_2}$;

(b) $P_{\mu_1} << P_{\mu_2}$ but P_{μ_1} does not dominate P_{μ_2};

(c) $P_{\mu_2} << P_{\mu_1}$ but P_{μ_2} does not dominate P_{μ_1};

(d) Neither dominates the other.

It turns out that (a) holds iff $\mu_1 \equiv \mu_2$ and both conditions (i) and (ii) do not hold: (b) holds iff $\mu_1 << \mu_2$, $0 < \mu_2\{f_1 = 0\} < \infty$ and both conditions (i) and (ii) do not hold; similarly for (c); (d) holds iff $0 < \mu_1\{f_2 = 0\} < \infty$, $0 < \mu_2\{f_1 = 0\} < \infty$ and both (i) and (ii) do not hold.

Note that the Neyman-Pearson tests are based on the log likelihood ratio $W_t^* = \Sigma_1^{N_t} \log\{f_2(x_i)/f_1(x_i)\} - \int_0^t (f_2 - f_1)\, d\nu$. If P_{μ_1} and P_{μ_2} are not singular then under P_{μ_1}, W_t^* converges a.s. as $t \to \infty$ to a (possibly improper) random variable W with $\text{prob}(W = -\infty) = 1 - \exp(-\mu_1\{f_2 = 0\})$ and

$$\log E_{P_{\mu_1}} \{\exp(iuW | W > -\infty)\}$$

$$= \int_{f_2>0} [\exp\{iu \log(f_2/f_1)\} - iu\{(f_2/f_1) - 1\} - 1]\, f_1\, d\nu;$$

under P_{μ_2}, W_t^* converges a.s. to a (possibly improper) random variable W^* with $P(W^* = \infty) = 1 - \exp(-\mu_2\{f_1 = 0\})$ and

$$\log E_{P_{\mu_2}} (\exp iuW^* | W^* < \infty)$$

$$= \int_{f_1>0} [\exp\{iu \log(f_2/f_1)\} - iu\{(f_2/f_1) - 1\} - 1]\, f_2\, d\nu.$$

We also note that if $P_{\mu_1} << P_{\mu_2}$ then $W = \log(dP_{\mu_1}/dP_{\mu_2})$. Moreover the Neyman-Pearson test of $\mu = \mu_1$ vs $\mu = \mu_2$ based on the data from the Poisson process on all of $[0,\infty)$ rejects $\mu = \mu_1$ if W is large.

5. ASYMPTOTIC NORMALITY OF W_t

Consider the simple vs simple test $\mu = \mu_1$ vs $\mu = \mu_2$ where $\mu_i(t) < \infty$, $i = 1,2$, $t < \infty$. The Neyman-Pearson test based on the data in $[0,t]$ depends directly on the statistic $W_t = \Sigma_1^{N_t} (\log f_2/f_1)(x_i)$. In this section we derive conditions under which W_t is asymptotically normal.

Theorem 2. Assume that either (i), (ii) and (iii) or (i), (ii) and (iv) below hold. Then under P_{μ_1}, $(W_t - a_{t,M}^{(1)})/b_{t,M}^{(1)} \xrightarrow{L} N(0,1)$, where

$$A_{t,M}^{(1)} = \int_{S_{t,M}} (\log f_2/f_1)\, f_1\, d\nu,$$

$$(b_{t,M}^{(1)})^2 = \int_{S_{t,M}} (\log f_2/f_1)^2\, f_1\, d\nu,$$

$$S_{t,M} = \{x \in [0,t] \text{ such that } |\log f_2/f_1(x)| \le M\}.$$

(i) $\mu_1 << \mu_2$;

(ii) there exists M such that $\mu_1\{x: |\log f_2/f_1(x)| > M\} < \infty$;

(iii) $\int_{|f_2/f_1 - 1| > c} (f_2/f_1 - 1)^2\, f_1\, d\nu = \infty$ for all $c > 0$;

(iv) there exists $c > 0$ such that $\mu_1\{x: |\log f_2/f_1| > c\} = \infty$.

Proof. Let $\bar{W}_t = \sum_{S_{t,M}} \log f_2/f_1$. It follows from conditions (i) and (ii) that under P_{μ_1}, $W_t - \bar{W}_t = 0_p(1)$ as $t \to \infty$. Assume that (iii) holds; then since $\log f_2/f_1$ and $f_2/f_1 - 1$ are of the same order of magnitude near $f_2/f_1 = 1$ it follows that

$$\int_{|f_2/f_1 - 1| > c} (\log f_2/f_1)^2\, f_1\, d\nu = \infty \qquad \text{for all } c > 0$$

and thus that

$$(b_{t,M}^{(1)})^2 = \int_{S_{t,M}} (\log f_2/f_1)^2\, f_1\, d\nu \to \infty \text{ as } t \to \infty.$$

Therefore $(W_t - \bar{W}_t)/b_{t,M}^{(1)} = o_p(1)$ and the conclusion will follow if we can show that $(\bar{W}_t - A_{t,M}^{(1)})/b_{t,M}^{(1)} \xrightarrow{L} N(0,1)$. Using Feller's representation for infinitely divisible distributions and his convergence results, (Feller, 1971, p. 464), this is equivalent to showing that:

(1) $\mu_1\{x \in [0,t]$ such that $|\log f_2/f_1| > b_{t,M}^{(1)}\} \to 0$ as $a \to \infty$ uniformly in t.

(2) $$\int_{\tilde{S}_{t,ab^{(1)}_{t,M}} \cap S_{t,M}} (\log f_2/f_1)^2 f_1 \, d\nu/(b^{(1)}_{t,M})^2 \to 0 \text{ for all } a > 0.$$

Condition (1) is verified by Chebyshev's inequality and condition (ii) by recalling that $b^{(1)}_{t,M} \to \infty$.

Now suppose (i), (ii) and (iv) hold. It follows that there exists a $d \in (0,M]$ such that $\mu_1\{d \leq |\log f_2/f_1| \leq M\} = \infty$ and thus that $(b^{(1)}_{t,M})^2 = \int_{S_{t,M}} (\log f_2/f_1)^2 f_1 \, d\nu \to \infty$. The argument now proceeds as in the previous case.

Remarks.

(1) Condition (ii) of Theorem 2 is trivially satisfied if $|\sup f_2/f_1| < \infty$.

(2) The following example indicates that under conditions (i), (iii) and (iv) W_t can have a non-normal limiting distribution.

Let ν = Lebesgue measure, $f_1 \equiv 1$, $\log f_2(x) = e^x$. Then $(w_t - e^t)\sqrt{2/e^t}$ converges in distribution to a random variable with log characteristic function given by $\int_{y=1}^{\sqrt{2}} \{\exp(iuy) - iuy - 1\} y^{-1} dy$.

Note that in this example $\mathrm{var}(W_t) < \infty$ for all t. Therefore the existence of the second moment of W_t does not imply asymptotic normality. Also, W_t may be asymptotically normal even though $E(W_t)$ may not exist. This follows from condition (iii) of Theorem 2.

(3) If the conditions of Theorem 2 hold and $\int_0^t \{\log^2(f_2/f_1)\} f_i \, d\nu < \infty$ for all t then $(W_t - a^{(i)}_{t,\infty})/b^{(i)}_{t,\infty} \xrightarrow{L} N(0,1)$ under P_{μ_i}.

(4) If P_{μ_1} and P_{μ_2} are not singular then W_t will not be asymptotically normal. In the presence of condition (i) (with a straight-forward modification if condition (i) does not hold) $W_t - \int_0^t (f_2 - f_1) \, d\nu$ converges a.s. to a proper random variable with log characteristic function given by $\int_0^\infty [\exp\{iu \log(f_2/f_1)\} - iu\{(f_2/f_1) - 1\} - 1] \, d\nu$.

(5) In the following example $\mu_1 << \mu_2$, $P_{\mu_1} \perp P_{\mu_2}$ and condition (ii) holds, but W_t is not asymptotically normal: ν = Lebesgue measure + counting measure on $I = \{0, 1, 2, \ldots\}$, μ_1 = Lebesgue measure, $\mu_2 = \nu$, $f_2 \equiv 1$, $f_1 = 0$ on I, $f_1 = 1$ elsewhere. Then, under P_{μ_1}, $W_t \equiv 0$. Note that in this case (iii) and (iv) do not hold.

By a similar technique, again under $\mu_1 << \mu_2$, $P_{\mu_1} \perp P_{\mu_2}$ and (ii), we can construct examples in which W_t converges a.s. to a non-degenerate non-normal limit.

6. MONOTONE DENSITIES

Suppose that the family of measures, $\{\mu_\theta, \theta \in I\}$ where I is a (possibly infinite) interval, is dominated by a σ-finite measure ν and that $f_\theta(x) = (d\mu_\theta/d\nu)(x)$ is monotone non-decreasing in θ for each x. Also assume that $\mu_\theta(t) < \infty$ for all θ, t. We shall show that the level α Neyman-Pearson test of $\theta = \theta_1$ vs $\theta = \theta_2$ (where $\theta_1 < \theta_2$) is also a level α test of $\theta \le \theta_1$ vs $\theta > \theta_1$ and has a monotone non-decreasing power function. Thus if

$$P_{\mu_{\theta_1}} \perp P_{\mu_{\theta_2}},$$

then the power of a fixed level α Neyman-Pearson test approaches 1 as $t \to \infty$ uniformly for $\theta \ge \theta_2$.

The above result will follow easily from Lemma 2:

Lemma 2. Let $P_{\mu_1}^{(t)}$ and $P_{\mu_2}^{(t)}$ be Poisson processes with $\mu_i(t) < \infty$, i=1,2 and $f_1(x) = (d\mu_1/d\nu)(x) \le (d\mu_2/d\nu)(x) = f_2(x)$ for all $x \in [0,t]$; let k be a non-negative function on [0,t]. Then for all a, $P_{\mu_1}\{\Sigma_1^{N_t} k(x_i) > a\} \le P_{\mu_2}\{\Sigma_1^{N_t} k(x_i) > a\}$.

Proof. Let $A_{j,n} = \{x \in [0,t]$ such that $k(x) \in [j/2^n, j+1/2^n)\}$, $j = 0,\ldots, 4^n-1$, and $c_n = \Sigma_{j=0}^{4^n-1} (j/2^n)\, N(A_{j,n})$. Now, under P_{μ_i}, $N(A_{j,n})$ is Poisson distributed with parameter $\mu_i(A_{j,n})$, i = 1,2. Note that

$\mu_1(A_{j,n}) = \int_{A_{j,n}} f_1\, d\nu \leq \int_{A_{j,n}} f_2\, d\nu = \mu_2(A_{j,n})$. Construct independent random variables $\{M_{j,n},\ j = 0,\ldots,\ 4^n-1\}$ so that $M_{j,n}$ is Poisson distributed with parameter $\mu_2(A_{j,n}) - \mu_1(A_{j,n})$, and the collection $\{M_{j,n},\ j = 0,\ldots,\ 4^n-1\}$ is independent of $\{N(A_{j,n}),\ j = 0,\ldots,\ 4^n-1\}$. Then $c_n + \Sigma_{j=0}^{4^n-1} j/2^n M_{j,n}$ has the same distribution under P_{μ_1} as does c_n under P_{μ_2}. Since $\Sigma_{j=0}^{4^n-1} (j/2^n) M_{j,n} \geq 0$ it follows that $P_{\mu_1}(c_n > a) \leq P_{\mu_2}(c_n > a)$ for all a; also, $c_n \uparrow \Sigma_1^{N_t} k(x_i)$ a.s. under P_{μ_1} and P_{μ_2}. Therefore

$$P_{\mu_1}\{\Sigma_1^{N_t} k(x_i) > a\} = \lim_{n\to\infty} P_{\mu_1}(c_n > a) \leq \lim_{n\to\infty} P_{\mu_2}(c_n > a)$$
$$= P_{\mu_2}\{\Sigma_1^{N_t} k(x_i) > a\}.$$

Now, consider a Neyman-Pearson test of $\theta = \theta_1$ vs $\theta = \theta_2$ where $\theta_1 < \theta_2$. This test rejects if $\Sigma_1^{N_t} \log(f_{\theta_2}/f_{\theta_1}) > c$. Since f_θ is monotone non-decreasing in θ, $\log f_{\theta_2}/f_{\theta_1} \geq 0$. Applying Lemma 2 we see that $P_{\mu_\theta}\{\Sigma_1^{N_t} \log(f_{\theta_2}/f_{\theta_1}) > c\}$ is monotone non-decreasing in θ. Therefore for $\theta \leq \theta_1$, $P_{\mu_\theta}\{\Sigma_1^{N_t} \log(f_{\theta_2}/f_{\theta_1}) > c\} \leq \alpha$ so that the above test is a level α test of $\theta \leq \theta_1$ vs $\theta > \theta_1$. Also for $\theta > \theta_2$, $P_{\mu_\theta}\{\Sigma_1^{N_t} \log(f_{\theta_2}/f_{\theta_1}) > c\} \geq P_{\mu_\theta}\{\Sigma_1^{N_t} \log(f_{\theta_2}/f_{\theta_1}) > c\}$ and therefore the power at $\theta > \theta_2$ is at least as great as the power at θ_2. This implies that if $P_{\mu_{\theta_1}} \perp P_{\mu_{\theta_2}}$ (see Theorem 1 for conditions) then the power of the above test approaches 1 uniformly for $\theta \geq \theta_2$ as $t \to \infty$.

It would be tempting to conclude that if $\theta_1 < \bar{\theta} < \theta_2$ and $P_{\mu_{\theta_1}} \perp P_{\mu_{\bar{\theta}}} \perp P_{\mu_{\theta_2}}$ then the Neyman-Pearson test of $\theta = \theta_1$ vs $\theta = \theta_2$ has power approaching 1 at $\bar{\theta}$. Unfortunately this need not be the case. A counter-example is provided by: $f_{\theta_1}(x) \equiv 1$, $f_{\bar{\theta}}(x) = 1 + m^{-1/2}$ for $x \in [m-1,m)$, $m = 1, 2, \ldots$; $f_{\theta_2}(x) = e^m$ for $x \in [m-1,m)$, $m = 1, 2, \ldots$. Then $P_{\mu_{\theta_1}} \perp P_{\mu_{\bar{\theta}}} \perp P_{\mu_{\theta_2}}$ (Theorem 1) but the test statistic

$$Z_k = \sum_{1}^{N_{k+1}} \log f_{\theta_2}/f_{\theta_1} = \sum_{1}^{k} mN[m,m+1) \; ,$$

has the property that $(Z_m - m)/m^{3/2}$ converges in distribution to a non-degenerate limiting law under both θ_1 and $\bar{\theta}$.

If $f_\theta(x)$ is differentiable in θ for each x then under mild conditions the test which rejects if

$$Z_t = \sum_{1}^{N_t} \frac{\partial}{\partial\theta} \log f_\theta(x_i)\Big|_{\theta=\theta_0} > c,$$

where $P_{\mu_{\theta_0}}(Z_t > c) = \alpha$, is a level α test of $\theta \leq \theta_0$ vs $\theta > \theta_0$ with monotone power function and is consistent over all of H_1.

This suggests the concept of level α tests which are locally most powerful against certain specified families of alternatives. We will pursue this in a future paper.

7. EXPONENTIAL STRUCTURE

We will say that $\{\mu_\theta, \theta \in I\}$ has exponential structure if $f_\theta(x) = (d\mu_\theta/d\nu)(x) = \exp[p(\theta)\, k(x) + S(x)]$ where $p(\cdot)$, $k(\cdot)$ and $S(\cdot)$ are real-valued functions. Without loss of generality we set $S = 0$ as this is equivalent to taking Radon-Nikodym derivatives with respect to $d\nu' = e^S d\nu$. Hereafter when we refer to exponential structure it will be assumed that $S = 0$. Lemma 3 below provides us with sufficient equipment to apply many classical (i.i.d.) results and techniques to Poisson processes with exponential structure.

Theorem 3. Let $\{\mu_\theta, \theta \in I\}$ have exponential structure. Then if condition 1 (below) holds, then $\Sigma_1^{N_t} k(x_i)$ is a minimal sufficient statistic for the family $\{P_{\mu_\theta}^{(t)}, \theta \in I\}$. If conditions 1 and 2 (below) hold then $\Sigma_1^{N_t} k(x_i)$ is complete and the family $\{P_{\mu_\theta}^{(t)}, \theta \in I\}$ has monotone likelihood ratio in $\Sigma_1^{N_t} k(x_i)$. If conditions 1, 2', 3 and 4 hold then $P_{\mu_{\theta_1}} \perp P_{\mu_{\theta_2}}$ for all $\theta_1 \neq \theta_2$. If 1, 2', 3, 4 and 5 hold, then under P_{μ_θ}

$$\{\Sigma_1^{N_t} k(x_i) - a_{t,M,\theta}\}/b_{t,M,\theta} \xrightarrow{L} N(0,1),$$

where

$$a_{t,M,\theta} = \int_{S_{t,M}} k(x)\ f_\theta(x)\ \nu(dx)$$

$$b^2_{t,M,\theta} = \int_{S_{t,M}} k^2(x)\ f_\theta(x)\ \nu(dx)\ ,$$

$$S_{t,M} = \{x \in [0,t] \text{ such that } \{k(x)\} \leq M\}.$$

Moreover, if $b_{t,\infty,\theta} < \infty$ for all t then $\{\Sigma_1^{N_t} k(x_i) - a_{t,\infty,\theta}\}/b_{t,\infty,\theta} \xrightarrow{L} N(0,1)$ under P_{μ_θ}.

(1) $\mu_\theta(t) < \infty$ for all θ, t.

(2) $p(\theta)$ is continuous and monotone increasing.

(2') $p(\theta)$ is one to one.

(3) k is not a.e. $[\nu]$ equal to 0.

(4) For each θ either there exists a c_θ such that $\mu_\theta\{|k| > c_\theta\} = \infty$ or else for all $a > 0$, $\int_{S_{\infty,a}} k^2 f_\theta\, d\nu = \infty$.

(5) There exists an M such that $\mu_\theta\{|k| > M\} < \infty$.

Proof.

(a) Sufficiency.

The exponential structure implies that $\mu_{\theta_1} \equiv \mu_{\theta_2}$ for all θ_1, θ_2. If condition 1 holds then $g_\theta = d\mu_\theta/d\mu_{\theta_0} = \exp[\{p(\theta) - p(\theta_0)\}\, k]$. Thus we have exponential structure with respect to a measure μ_{θ_0} satisfying $\mu_{\theta_0}(t) < \infty$ for all t. Therefore without loss of generality we can assume that $\nu[0,t] < \infty$ for all t. It follows from 1, $\nu[0,t] < \infty$ and Brown (1971, p. 774) that $P^{(t)}_{\mu_\theta} << P^{(t)}_\nu$ with

$$\log dP^{(t)}_{\mu_\theta}/dP^{(t)}_\nu = p(\theta) \sum_1^{N_t} k(x_i) - \int_0^t (f_\theta - 1)\, d\nu, \tag{7.1}$$

this expression holding even if 3 does not hold or if $p(\theta) \not\equiv 0$. From (7.1) it follows that $\Sigma_1^{N_t} k(x_i)$ is a minimal sufficient statistic.

(b) Monotone likelihood ratio.

We have seen that 1 implies (7.1). Obviously (7.1) and 2 imply that $dP^{(t)}_{\mu_\theta}/dP^{(t)}_\nu$ is a non-decreasing function of $\Sigma_1^{N_t} k(x_i)$. This is of course true even if 3 does not hold or if $p \equiv 0$.

(c) Completeness.

Assume that $E_{P_{\mu_\theta}}[h\{\Sigma_1^{N_t} k(x_i)\}] = 0$. Now condition 1 implies that

$$E_{P_{\mu_\theta}}[h\{\sum_1^{N_t} k(x_i)\}] = \sum_{m=0}^{\infty} \{\ell_m\, \mu_\theta^m(t)\, e^{-\mu_\theta(t)}\}/m!$$

where

$$\ell_m = E_{P_{\mu_\theta}}[h\{\sum_1^{N_t} k(x_i)\} | N_t = m] .$$

Now conditions 1 and 2 and Lehman (1959, p. 52) imply that the range of $\{\mu_\theta(t),\ \theta \in I\}$ contains an interval. Since the Poisson distribution belongs to the exponential family it follows from Lehman (1959, p. 132) that $\ell_m = 0$ for all m. Now conditional on $N_t = m$ for $m \geq 1$, $x_1, \ldots, x_m$ are distributed as the order statistics of an i.i.d. sample of size m with exponential density $f_\theta/\mu_\theta^{(t)}$. Condition 2 implies that the range of $\{p(\theta), \theta \in I\}$ contains an interval and it thus follows from Lehman (1959, p. 132) that h = 0 a.e. with respect to $\{\lambda_{m,\theta}, \theta \in I\}$ where $\lambda_{m,\theta}$ is the conditional distribution of $\sum_1^{N_t} k(x_i)$ given $N_t = m$. Also $h(0) = \ell_0 = 0$. Therefore h = 0 a.e. with respect to $\{\lambda_\theta, \theta \in I\}$ where λ_θ is the unconditional distribution of $\sum_1^{N_t} k(x_i)$ under P_{μ_θ}, and thus $\sum_1^{N_t} k(x_i)$ is complete.

(d) $P_{\mu_{\theta_1}} \perp P_{\mu_{\theta_2}}$.

If 2' and 3 hold then the measures $\{\mu_\theta, \theta \in I\}$ are distinct. If 1 holds then $\mu_{\theta_1} \equiv \mu_{\theta_2}$ for all θ_1, θ_2. It thus follows from Theorem 1 that if 4 holds then $P_{\mu_{\theta_1}} \perp P_{\mu_{\theta_2}}$ for all $\theta_1 \neq \theta_2$.

(e) Asymptotic normality

Follows from our previous remarks and Theorem 2.

Remarks.

1) When $\sum_1^{N_t} k(x_i)$ is a complete sufficient statistic the Rao-Blackwell-Lehman-Scheffe theorem (Rao, 1965, p. 261) holds. This is useful for obtaining uniformly best unbiased estimates under convex loss.

2) Under the conditions of Theorem 3 which imply monotone likelihood ratio it follows from Lehman (1959, p. 68, p. 126), that UMP tests of one-sided hypotheses exist as well as UMP unbiased tests for two-sided hypotheses. The one-sided tests reject if $\sum_1^{N_t} k(x_i) > c$ (or less than c depending on the direction of monotonicity). Under the additional

presence of condition 4 these tests (for $\alpha > 0$) are consistent as $t \to \infty$.

3) We could also introduce m dimensional exponential families involving m parameters $\theta^{(1)}, \ldots, \theta^{(m)}$, and derive UMP unbiased tests for various hypotheses concerning one parameter, the others treated as nuisance parameters.

For example if $f^{(x)}_{\theta,\phi} = (d\mu_{\theta,\phi}/d\nu)(x) = h(\phi) \exp\{p(\theta)\, k(x)\}$ where h and p are continuous and monotone increasing, and $\mu^{(t)}_{\theta,\phi} < \infty$ for all θ, ϕ, t, then the UMP unbiased test of $\theta \leq \theta_0$ vs $\theta > \theta_0$ rejects if

$$\sum_1^{N_t} k(x_i) > c_{N_t} \qquad \text{where} \qquad P_{\theta_0}\{\sum_1^{N_t} k(x_i) > c_{N_t} | N_t\} = \alpha$$

(the conditional distribution of $\Sigma_1^{N_t} k(x_i)$ given N_t is independent of ϕ). The UMP unbiased test of $\phi \leq \phi_0$ vs $\phi > \phi_0$ rejects if

$$N_t > c\{\Sigma_1^{N_t} k(x_i)\} \quad \text{where} \qquad P_{\phi_0}[N_t > c\{\Sigma_1^{N_t} k(x_i)\} | \Sigma_1^{N_t} k(x_i)] = \alpha$$

(of course randomization at the boundary may be necessary). See Cox and Lewis (1966, Section 3.3) for a discussion of the case $f^{(x)}_{\theta,\phi} = \exp[\phi + \theta x]$, and Lewis (1972) for higher order exponential polynomials.

More generally if

$$f^{(x)}_{\theta^{(1)},\ldots,\theta^{(m)}} = \exp[\sum_{i=1}^{m} \theta^{(i)} k^{(i)}(x)] \tag{7.2}$$

with $\mu^{(t)}_{\theta^{(1)},\ldots,\theta^{(k)}} < \infty$ for all θ, t, then the conditional distribution of $\Sigma_{j=1}^{N_t} k_i(x_j)$ given $\{\Sigma_{j=1}^{N_t} k_\ell(x_j), \ell \neq i\}$ is independent of $\{\theta^{(\ell)}, \ell \neq i\}$ and the UMP unbiased test of $\theta^{(i)} \leq \theta_0^{(i)}$ vs $\theta^{(i)} > \theta_0^{(i)}$ rejects if

$$\sum_{1}^{N_t} k_i(x_j) > c\ \{\sum_{j=1}^{N_t} k_\ell(x_j),\ \ell \neq i\}$$

where

$$P_{\theta_0^{(i)}}[\sum_{j=1}^{N_t} k_i(x_j) > c\ \{\sum_{j=1}^{N_t} k_\ell(x_j),\ \ell \neq i\} \big| \sum_{j=1}^{N_t} k_\ell(x_j),\ \ell \neq i] = \alpha.$$

Cox (1972) has discussed the form (7.2).

4) We could have required p to be decreasing rather than increasing with trivial modification.

8. ESTIMATING θ THROUGH THE SAMPLE INTENSITY

From the homogeneous Poisson process the sample intensity $N(t)/t$ is from many points of view a good estimate of λ, the intensity of the process. For non-homogeneous Poisson processes the rate of occurrence of events is often rather uninformative.

In this section we give conditions on a family of measures under which one can estimate θ consistently by use of sample intensities.

Lemma 3. Suppose that there exists an increasing family of Borel sets $\{A_t,\ t \geq 0\}$ and a measure λ such that:

(i) $A_t \subset [0,t]$ for all t;

(ii) $\lambda[0,t] < \infty$ for all t;

(iii) $\lim_{t\to\infty} \lambda(A_t) = \infty$;

(iv) $\lim_{t\to\infty} \mu_\theta(A_t)/\lambda(A_t) = c(\theta)$ for all θ where c is a strictly monotone function.

Then $P_{\mu_\theta}[c^{-1}\{N(A_t)/\lambda(A_t)\} \to \theta] = 1$ for all θ.

Proof. $N(A_t)/\mu_\theta(A_t)$ can be represented as $N^*\{\mu_\theta(A_t)\}/\mu_\theta(A_t)$ where N^* is a homogeneous Poisson process of intensity one. Since $\mu_\theta(A_t) \to \infty$ (except perhaps at one value of θ) it follows from the a.s. convergence of $N^*(t)/t$ to 1 that $N(A_t)/\mu_\theta(t) \xrightarrow{a.s.} 1$ under P_{μ_θ}. Therefore $N(A_t)/\lambda(t) \xrightarrow{a.s.} c(\theta)$ (if $\lim_{t\to\infty} \mu_\theta(A_t) < \infty$ then $N(A_t)/\lambda(t) \xrightarrow{a.s.} 0$). Since c is strictly monotone, c^{-1} exists and is continuous. Therefore $c^{-1}\{N(A_t)/\lambda(t)\} \xrightarrow{a.s.} \theta$.

Lemma 4. Suppose that $[\mu_\theta(A(t)) - a(t)]/b(t) \to c(\theta)$ as $t \to \infty$ for all θ, where c is strictly monotone. Suppose also that $a(t) \to \infty$, $\mu_\theta(A_t)/a(t) \to 1$, $\mu_\theta(A_t)/b_t^2 \to 0$, $A_t \subset [0,t]$ for all t. Then $c^{-1}[(N(A_t) - a(t))/b(t)] \xrightarrow{prob} \theta$ under P_θ, for all θ.

Proof. $E_\theta[(N(A_t) - a(t)/b(t)] \to c(t)$, $\text{var}_\theta[(N(A_t) - a(t))/b(t)] \to 0$. Therefore $\{N(A_t) - a(t)\}/b(t) \xrightarrow{prob} c(t)$. Since c is strictly monotone, c^{-1} is 1-1 and continuous and thus $c^{-1}[\{N(A_t) - a(t)\}/b(t)] \xrightarrow{prob} \theta$.

Example. ν = Lebesgue measure, $f_1 \equiv 1$, $f_2 = 1 + x^{-1/4}$ on $(1,\infty)$, $f_2 = 1$ on $[0,1]$. Then $\mu_\theta(t) = t + (4/3)\,\theta t^{3/4}$, $a(t) = t$, $b(t) = (4/3)t^{3/4}$, $c(\theta) = \theta$.

9. ESTIMATION OF A WEAK TREND

Let ν = Lebesgue measure and

$$f_\theta(x) = \begin{cases} 0 & 0 \le x \le 1 \\ 1 + \theta x^{-1/2} & x > 1 \end{cases} \qquad \text{for } \theta \in [0,\infty).$$

We refer to this example as a weak trend since $x^{-1/2}$ is small so that when $\theta > 0$ departures from homogeneity ($\theta = 0$) are difficult to detect.

It turns out that condition (ii) of Theorem 1 holds so that $P_{\mu_{\theta_1}} \perp P_{\mu_{\theta_2}}$ for all $\theta_1 \neq \theta_2$. Condition (i) of Theorem 1 does not hold.

In this example there is no simple sufficient statistic as

$$\log dP_{\mu_{\theta}}^{(t)}/dP_{\mu_{\theta}}^{(t)} = \sum_{1}^{N_t} \log\{(1 + \theta_2 x_i^{-1/2})/(1 + \theta_1 x_i^{-1/2})\}.$$

However since $1 + \theta x^{-1/2} \approx \exp(\theta x^{-1/2})$ it is plausible that $\sum_1^{N_t} x_i^{-1/2}$ is in some sense asymptotically sufficient.

An intuitive estimator is obtained by noting that $\mu_\theta(t) = E_\theta\{N(t)\} = t + 2\theta(t^{1/2} - 1)$. Therefore set $N(t) = t + 2\theta(t^{1/2} - 1)$ and solve for θ obtaining $\hat{\theta}_t = \{N(t) - t\}/2(t^{1/2} - 1)$. It turns out that $\mathrm{var}_\theta\, \hat{\theta}_t \to 1/4$ as $t \to \infty$, and that $\hat{\theta}_t$ does not converge in probability to θ, and is thus not a consistent estimator.

The estimator $\{N(t) - t\}/2(t^{1/2} - 1)$ is an unbiased estimator. A general form for unbiased estimators is

$$\{\sum_1^{N_t} h(x_i^{-1/2}) - \int_1^t h(x^{-1/2})\, dx\}/\{\int_1^t h(x^{-1/2})\, x^{-1/2}\, dx\}.$$

The above estimator $\hat{\theta}_t$ had $h(x) \equiv 1$. Another choice $h(x) = x^2$ yields the estimator $\theta_t^* = (\sum_1^{N_t} x_i^{-1} - \log t)/2(1 - t^{-1/2})$, with $\mathrm{var}_\theta\, \theta_t^* \to 1/4$ as $t \to \infty$; θ_t^* is also inconsistent. The estimator corresponding to $h(x) = x$ is $\bar{\theta}_t = \{\sum_1^{N_t} x_i^{-1/2} - 2(t^{1/2} - 1)\}/\log t$. The variance of $\bar{\theta}_t$ under θ is equal to $(\log t)^{-1} + 2\theta(1 - t^{-1/2})(\log t)^{-2}$. It turns out that $\bar{\theta}_t$ is strongly consistent $\{P_{\mu_\theta}(\bar{\theta}_t \to \theta) = 1\}$ and it follows from Theorem 2 that $(\log t)^{1/2}\, (\bar{\theta}_t - \theta) \to N(0,1)$. The Cramér-Rao lower bound for the variance of an unbiased estimator is $\{\int_1^t (x^{-1}/1 + \theta x^{-1/2})\, dx\}^{-1}$. The ratio of var $\bar{\theta}_t$ to the lower bound approaches 1 as $t \to \infty$. Therefore, in this sense $\bar{\theta}_t$ is a BAN estimator. In fact, a lot more can be said about the efficiency of $\bar{\theta}_t$. It follows from Weiss (1971), that for any estimator $\tilde{\theta}_t$ having the property that $(\log t)^{1/2}\, (\tilde{\theta}_t - \theta)$ converges in distribution uniformly in θ as $t \to \infty$, $\lim_{t\to\infty} P_\theta\{(\log t)^{1/2}\, (\hat{\theta}_t - \theta) \le c\} \ge \lim_{t\to\infty} P_\theta\{(\log t)^{1/2}\, (\tilde{\theta}_t - \theta) \le c\}$ for all c. This property and the asymptotic normality of $\bar{\theta}_t$ implies that $\bar{\theta}_t$ is BAN. It also follows from Weiss that if θ_t is any estimator satisfying $(\log t)^{1/2}(\theta_t - \theta) = O_p(1)$ then $\theta_t + \{\Sigma_1^{N_t}(x_i^{-1/2}/1 + \theta_t x_i^{-1/2}) - \int_1^t k(x)\,dx\}/\log t$ is asymptotically efficient in the same (above) sense as $\bar{\theta}_t$.

The fact that this problem worked out nicely in the absence of a convenient sufficient statistic suggests the following questions.

(i) In some sense $\Sigma_1^{N_t} x_i^{-1/2}$ is asymptotically sufficient. How can one in general recognize an asymptotically sufficient statistic?

(ii) The family $f_\theta(x) = 1 + \theta x^{-1/2}$ is "close" to the family $f_\theta^*(x) = \exp(\theta x^{-1/2})$ in the sense that $1 + \theta x^{-1/2} \approx \exp(x^{-1/2})$ and also in that

$$P_{\mu_{\theta_1}} \perp P_{\mu_{\theta_2}}, \qquad P^*_{\mu_{\theta_1}} \perp P^*_{\mu_{\theta_2}} \qquad \text{for } \theta_1 \neq \theta_2,$$

and $P_{\mu_\theta} \equiv P^*_{\mu_\theta}$ for all θ. It follows that $P_{\mu_\theta}(Z_t \to \theta) = 1 \iff P^*_{\mu_\theta}[Z_t \to \theta] = 1$, so that consistent estimators are interchangeable. This suggests developing a notion of asymptotically equivalent parametric families of Poisson processes.

The author believes that recent work of Weiss (1971) which is closely related to a series of papers of Weiss and Wolfowitz is potentially significant in estimation problems arising in Poisson processes.

10. EXTENSIONS

Instead of observing a Poisson process on [0,t] and letting $t \to \infty$ we may be observing a spatial Poisson process and looking at points falling in a sphere of radius r and letting $r \to \infty$. In general we can take an arbitrary measure space (X,C) and have a Poisson process with unknown mean measure μ assumed to lie in a family of σ-finite measures on (X,C), $\{\mu_\theta, \theta \in \Phi\}$ (see Brown, 1971, for the definition of a Poisson process over a general measure space). We then take an increasing family X_t of measureable subsets of X with $\lim_{t\to\infty} X_t = X$. The tth experiment consists of looking at those "events" of the Poisson process which fall in X_t.

Everything we have done for the real line easily generalizes to this case.

ACKNOWLEDGMENTS

This paper was written with support from the National Science Foundation under contract GK-5758 with Cornell University.

REFERENCES

Breiman, L. (1963). The Poisson tendency in traffic distribution. Ann. Math. Statist. 34, 308-11.

Brown, M. (1970). A property of Poisson processes and its application to macroscopic equilibrium of particle systems. Ann. Math. Statist. 41, 1935-41.

Brown, M. (1971). Discrimination of Poisson processes. Ann. Math. Statist. 42, 773-6.

Cox, D. R. (1972). The statistical analysis of dependencies in point processes. This volume.

Cox, D. R. and Lewis, P.A.W. (1966). *The Statistical Analysis of Series of Events*. London: Methuen and New York: Barnes & Noble.

Feller, W. (1971). *An Introduction to Probability Theory and its Applications*. Vol. 2, 2nd ed. New York: Wiley.

Gnedenko, B. V. and Kolmorogov, A. N. (1954). *Limit Distributions for Sums of Independent Random Variables*. (English translation). New York: Addison Wesley.

Goldman, J. R. (1968). Stochastic point processes: Limit theorems. Ann. Math. Statist. 39, 771-9.

Kennedy, D. P. (1969). A functional limit theorem for k-dimensional renewal theory. Technical report No. 9, Stanford University.

Lehman, E.L. (1959). *Testing Statistical Hypotheses*. New York: Wiley.

Lewis, P.A.W. (1972). Recent results in the statistical analysis of univariate point processes. This volume.

Neveu, J. (1965). *Mathematical Foundations of the Calculus of Probability*. (English translation). New York: Addison-Wesley.

Rao, C.R. (1965). *Linear Statistical Inference and its Applications*. New York: Wiley.

Shorrock, R. (1971). Caravans of traffic flow. To appear in J. Appl. Prob.

Stone, C. (1968). On a theorem of Dobrushnin. Ann. Math. Statist. 39, 1391-402.

Thedeen, T. (1964). A note on the Poisson tendency in traffic distribution. Ann. Math. Statist. 35, 1823-4.

Weiss, L. (1971). Asymptotic properties of maximum likelihood estimators in some nonstandard cases. J. Amer. Statist. Assoc. 66, 345-50.

STATISTICAL INFERENCE FOR DOUBLY STOCHASTIC POISSON PROCESSES

Jan Grandell
University of Stockholm

SUMMARY

A realization of a doubly stochastic Poisson process is assumed to be observed. Different inference situations in terms of the underlying intensity process are studied. In Section 2 examples of such situations are given. The problem treated in this paper is how to use the observed realization in order to make inference about the intensity process.

1. INTRODUCTION

Let $\{\lambda(t);\ t \in [0,T]\}$ be a real-valued and nonnegative stochastic process defined on (Ω, A, Π). The process $\lambda(t)$ will be assumed to be almost surely Riemann integrable. A doubly stochastic Poisson process $\{N(t);\ t \in [0,T]\}$ defined on (Ω_N, A_N, Π_N) is a generalization of an ordinary Poisson process obtained by replacing the intensity with the process $\lambda(t)$. If $\Pi_N(A_N|\omega)$ is A-measurable for every $A_N \in A_N$ and for almost every $\omega \in \Omega$ is the probability measure on A_N corresponding to a non-homogeneous Poisson process with intensity function $\lambda(t,\omega)$, then

$$\Pi_N(A_N) = \int_\Omega \Pi_N(A_N|\omega)\ \Pi(d\omega) .$$

For details we refer to Grandell (1971).

In the discrete parameter case a doubly stochastic Poisson sequence is defined in a similar way. Let $\{\ell_k;\ k = 1, 2, \ldots, n\}$ be a real-valued and nonnegative stochastic sequence defined on (Ω, A, Π). A doubly stochastic Poisson sequence $\{N_k;\ k = 1, 2, \ldots, n\}$ is then defined on (Ω_N, A_N, Π_N) where

$$\Pi_N[\bigcap_{j=1}^{m} \{N_{k_j} = n_{k_j}\}] = \int_\Omega \prod_{j=1}^{m} \frac{\{\ell_{k_j}(\omega)\}^{n_{k_j}}}{n_{k_j}!} \exp\{-\ell_{k_j}(\omega)\} \, \Pi(d\omega)$$

for any integer m ($1 \leq m \leq n$), any integers $1 \leq k_1 < \cdots < k_m \leq n$ and for any nonnegative integers $n_{k_1}, \ldots, n_{k_m}$.

Observe that in the continuous parameter case N(t) is a model for the number of events that occurs in (0,t] but in the discrete parameter case N_k is a model for the number of events occurring at "time" k.

2. SOME GENERAL REMARKS ON THE INFERENCE PROBLEM

Suppose that $\lambda(t)$ has one of the measures Π^α where α is a parameter taking values in a parameter space A. From the definition of N(t) it follows that N(t) then has one of the measures Π_N^α. The general problem of how to form a functional of an observed realization of N(t) in order to get a reasonable estimate of α was first studied by Grenander (1950).

In many cases it is not natural, or at least not practical, to use such a general formulation in order to derive estimates of α. It is therefore of interest to study special inference situations. It is also of interest to restrict the estimates to certain classes which do not require complete specification of the measures Π^α or Π_N^α.

We now give some examples of such situations. The examples also serve as a description of the contents of this paper.

2.1. Linear estimation

The class of linear estimates plays a certain role since only second order properties have to be specified. Grenander (1950, 1951a) studies in the case when $\lambda(t) = m + \varepsilon(t)$ where m is unknown and $\varepsilon(t)$ is a stochastic process with $E\{\varepsilon(t)\} = 0$ and $E\{\varepsilon(s)\ \varepsilon(t)\}$ known, unbiased linear estimates of m based on $\lambda(t)$. An important fact is that $\mathrm{cov}\{\lambda(t),\lambda(s)\}$ does not depend on m. $\mathrm{Cov}\{N(s),N(t)\}$ depends on m, however, and the situation changes when linear estimates of m based on N(t) are studied. It will be seen that it may be suitable to consider a somewhat reformulated situation.

2.2. A regression model

An extension of the model mentioned in Section 2.1 is the regression model. This model will be studied only in the discrete parameter case. In this model

$$\ell_m = \sum_{i=1}^{k} \gamma_i y_{i,m} + \varepsilon_m$$

where $y_{i,m}$ ($i = 1,\ldots, k$ and $m = 1,\ldots, n$) are assumed to be known and where ε_m is a stochastic process with $E(\varepsilon_m) = 0$ and $E(\varepsilon_j\varepsilon_k)$ known. The problem is to find good linear estimates of $\gamma_1,\ldots, \gamma_k$.

2.3. Estimates of the covariance sequence

Assume that $E(\ell_k) \equiv 1$ and that $\mathrm{cov}(\ell_j,\ell_k) = r_{j-k}$. In the analysis of time series estimates r_k^* of r_k based on $\ell_1,\ldots, \ell_n$ are investigated. Grandell (1971) has compared these estimates with estimates $\hat{r}_k$ based on $N_1,\ldots, N_n$. In Section 7 of this paper these comparisons are supplemented by some numerical illustrations.

Some comments will also be given on the estimation of the coefficients of an autoregressive process.

3. LINEAR ESTIMATION OF THE MEAN VALUE

Consider for simplicity the discrete parameter case. Put $\ell_k = m + \varepsilon_k$ where $\{\varepsilon_k;\ k = 1,\ldots, n\}$ is a stochastic process with $E(\varepsilon_k) = 0$ and $E(\varepsilon_k\varepsilon_j) = r_{k,j}^{(\varepsilon)}$ known and m is to be estimated. Let

$$m^* = \alpha + \sum_{k=1}^{n} \beta_k N_k$$

be a linear estimate of m. In this case we have

$$E(N_k) = E\{E(N_k|A)\} = E(m + \varepsilon_k) = m ,$$

$$\begin{aligned}\mathrm{var}(N_k) &= E\{\mathrm{var}(N_k|A)\} + \mathrm{var}\{E(N_k|A)\} = E(m + \varepsilon_k) + \mathrm{var}(m+\varepsilon_k)\\ &= m + r_{k,k}^{(\varepsilon)}\end{aligned}$$

and, for $k \neq j$,

$$\mathrm{cov}(N_k, N_j) = r_{k,j}^{(\varepsilon)}.$$

From this it follows that

$$E(m^*) = \alpha + m \sum_{k=1}^{n} \beta_k$$

and thus m* is an unbiased estimate of m if and only if $\alpha = 0$ and $\sum_{k=1}^{n} \beta_k = 1$. In order to find the best linear unbiased estimate m* of m we have to minimize

$$\mathrm{var}(m^*) = \mathrm{var}\left(\sum_{k=1}^{n} \beta_k N_k\right) = m \sum_{k=1}^{n} \beta_k^2 + \sum_{k=1}^{n} \sum_{j=1}^{n} \beta_k \beta_j \, r_{k,j}^{(\varepsilon)}$$

under the side condition $\sum_{k=1}^{n} \beta_k = 1$.

Put

$$\underline{\beta}^T = (\beta_1, \ldots, \beta_n), \qquad \underline{N}^T = (N_1, \ldots, N_n), \qquad \underline{1}^T = (1, \ldots, 1)$$

and

$$\underline{R}^{(\varepsilon)} = \{r_{i,j}^{(\varepsilon)}\} \qquad (1 \leq i, j \leq n).$$

With these notations

$$m^* = \underline{\beta}^T \underline{N}$$

and

$$\mathrm{var}(m^*) = m\underline{\beta}^T \underline{\beta} + \underline{\beta}^T \underline{R}^{(\varepsilon)} \underline{\beta} = \underline{\beta}^T (m\underline{I} + \underline{R}^{(\varepsilon)}) \, \underline{\beta} \, ,$$

where I is the (n×n) identity matrix. By ordinary matrix calculation it follows that var m* is minimized under the side condition $\underline{\beta}^T \underline{1} = 1$ by

$$\underline{\beta} = \frac{1}{\underline{1}^T(m\underline{I} + \underline{R}^{(\varepsilon)})^{-1}\,\underline{1}}\,(m\underline{I} + \underline{R}^{(\varepsilon)})^{-1}\,\underline{1}$$

and for this choice of β

$$\text{var}(m^*) = \frac{1}{\underline{1}^T(m\underline{I} + \underline{R}^{(\varepsilon)})^{-1}\,\underline{1}} \quad .$$

It is thus seen that the best linear unbiased estimate depends on m. It is possible to use some kind of mini-max-principle and choose the estimate as if m were very large. Then $\text{var}(m^*) \approx m\underline{\beta}^T\underline{\beta}$ and the estimate will be $\frac{1}{n}\sum_{k=1}^{n} N_k$. In practice this principle may be too pessimistic since because of some prior information it may be reasonable to believe that the true value of m lies in the neighborhood of some m_0. It is then reasonable to choose the estimate m* which minimizes $m_0\underline{\beta}^T\underline{\beta} + \underline{\beta}^T\underline{R}^{(\varepsilon)}\underline{\beta}$. If it is possible to formulate this prior information in such a way that m is regarded as a random variable uncorrelated with $\{\varepsilon_k\}_1^n$ and with

$$E(m) = m_0$$

and

$$\text{var}(m) = \sigma^2$$

we are led to study another class of linear estimates.

4. LINEAR BAYES ESTIMATION OF THE MEAN VALUE

Consider the continuous parameter case. Assume that

$$E\{\lambda^2(t)\} < \infty \qquad \text{for all } t \in [0,T]$$

and that

$$\int_0^T E\{\lambda^2(t)\}\,dt < \infty \quad .$$

Put $\lambda(t) = m + \varepsilon(t)$ where m is a random variable with $E(m) = m_0$ and $\text{var } m = \sigma^2$ uncorrelated with the stochastic process $\{\varepsilon(t);\ t \in [0,T]\}$ with $E\{\varepsilon(t)\} \equiv 0$ and $E\{\varepsilon(t)\ \varepsilon(s)\} = r^{(\varepsilon)}(t,s)$. Then

$$E\{\lambda(t)\} = E(m) + E\{\varepsilon(t)\} = m_0 + 0 = m_0$$

and

$$r(s,t) = \text{cov}\{\lambda(s),\lambda(t)\} = \sigma^2 + r^{(\varepsilon)}(s,t) ,$$

since $E\{m\cdot\varepsilon(t)\} = 0$.

Assume further that $\lambda(t)$ is integrable in quadratic mean.

Definition 1. The best linear Bayes estimate m_B^* of m is the linear estimate m^* based on $\{N(t);\ t \in [0,T]\}$ which minimizes $E(m^*-m)^2$.

First we give two lemmata from which most of the results in this section will follow.

Lemma 1. (The projection theorem). Let ξ and a family $\{\xi(s);\ s \in S\}$, defined on the same probability space, be random variables with mean values 0 and finite variances. In the closed (with respect to convergence in mean) set of all linear combinations of $\{\xi(s);\ s \in S\}$ there exists an unique element ξ^* such that $E(\xi^*-\xi)^2$ is minimized with respect to ξ^*. This element ξ^* is further the unique solution of $E\{(\xi^*-\xi)\xi(s)\} = 0$ for all $s \in S$.

Proof. Let H be the Hilbert space spanned by ξ and $\{\xi(s);\ s \in S\}$ and M be the Hilbert space spanned by $\{\xi(s);\ s \in S\}$. M is then a Hilbert subspace of H. In both H and M the product moment is used as inner-product. For details about Hilbert spaces in connection with stochastic processes we refer to Parzen (1959). A linear estimate of ξ based on $\{\xi(s);\ s \in S\}$ is given by $a + \eta$ where a is a constant and $\eta \in M$.

Since $E\{(\xi-a-\eta)^2\} = a^2 + E\{(\xi-\eta)^2\}$ for all $\eta \in M$ it is enough to study linear estimates $\eta \in M$ of ξ. Since M is a Hilbert subspace of

H it follows from the projection theorem (cf. Parzen, 1959) that there exists a unique element $\xi^* \in M$ such that $E\{(\xi^*-\xi)^2\} = \min_{\eta \in M} E\{(\eta-\xi)^2\}$ and that ξ^* is the unique solution of $E\{(\xi^*-\xi)\ \xi(s)\} = 0$ for all $s \in S$.

Consider the process $N(t)$. In Grandell (1971) it is shown that

$$E\{N(t)\} = m_0 t$$

and

$$\operatorname{cov}\{N(s),N(t)\} = m_0 \cdot \min(s,t) + \int_0^s \int_0^t r(\sigma,\tau)\, d\sigma\, d\tau .$$

Lemma 2. Let M be the Hilbert space spanned by $\{N(t) - m_0 t;\ t \in [0,T]\}$. $\eta \in M$ if and only if

$$\eta = \int_0^T f(s)\ d\{N(s) - m_0 s\}$$

for some square integrable function $f(s)$, $s \in [0,T]$.

Proof. For every

$$\eta = \int_0^T f(s)\ d\{N(s) - m_0 s\}$$

we have that $\eta \in M$ since

$$E(\eta^2) = m_0 \int_0^T f^2(s)\ ds + \int_0^T \int_0^T f(s)\ r(s,t)\ f(t)\ ds\ dt$$
$$\leq \int_0^T f^2(s)\ ds\ \{m_0 + \int_0^T r(t,t)\ dt\},$$

which follows by applying Cauchy-Schwarz inequality twice.

For every $\eta \in M$ there exists a sequence $\{\eta_k\}$ where

$$\eta_k = \sum_{j=1}^{n_k} f_j^{(k)}\ \{N(t_j^{(k)}) - m_0 \cdot t_j^{(k)}\} = \int_0^T f_k(s)\ d\{N(s) - m_0 s\}$$

where $f_k(s) = \sum_{\{j; t_j^{(k)} \geq s\}} f_j^{(k)}$ and such that $\lim_{k\to\infty} E\{(\eta-\eta_k)^2\} = 0$. Thus

$$\lim_{k,j\to\infty} E\{(\eta_k-\eta_j)^2\} = 0 .$$

Since

$$\lim_{k,j\to\infty} E\{(\eta_k-\eta_j)^2\} = \lim_{k,j\to\infty} E[\int_0^T \{f_k(s) - f_j(s)\} d\{N(s) - m_0 s\}]^2$$

$$= \lim_{k,j\to\infty} [m_0 \int_0^T \{f_k(s) - f_j(s)\}^2 ds$$

$$+ \int_0^T \int_0^T \{f_k(s) - f_j(s)\} r(s,t) \{f_k(t) - f_j(t)\} ds\, dt],$$

it follows that

$$\lim_{k,j\to\infty} \int_0^T \{f_k(s) - f_j(s)\}^2 ds = 0$$

and thus there exists a function f(s) such that

$$\lim_{k\to\infty} \int_0^T \{f_k(s) - f(s)\}^2 ds = 0$$

since the set of square integrable functions forms a Hilbert space. Put

$$\xi = \int_0^T f(s) d\{N(s) - m_0 s\} .$$

Since

$$E\{(\eta-\xi)^2\} = \lim_{k\to\infty} E\{(\eta_k-\xi)^2\} = \lim_{k\to\infty} E[\{\int_0^T (f_k(s)-f(s))\, d(N(s)-m_0 s)\}^2]$$

$$\leq \{m_0 + \int_0^T r(t,t)\, dt\} \lim_{k\to\infty} \int_0^T \{f_k(s)-f(s)\}^2 ds = 0$$

the lemma is proved.

Theorem 1. The best linear Bayes estimate m_B^* of m is given by

$$m_B^* = m_0 + \int_0^T \beta(s)\, d\{N(s) - m_0 s\}$$

where $\beta(s)$ is the unique solution of

$$\beta(s)\cdot m_0 + \int_0^T \beta(\sigma)\, r(\sigma,s)\, d\sigma = \sigma^2 .$$

Proof. From Lemma 1 and Lemma 2 it follows that the best linear Bayes estimate m_B^* of m is given by

$$m_B^* = m_0 + \int_0^T \beta(s)\, d\{N(s) - m_0 s\}$$

where $\beta(s)$ is the unique solution of

$$E\{(m_B^* - m)\,(N(t) - m_0 t)\} = 0 \qquad \text{for all } t \in [0,T] .$$

Since

$$E\{(m_B^* - m)\,(N(t) - m_0 t)\}$$

$$= E\left[\,[\int_0^T \beta(s)\, d\{N(s) - m_0 s\} - (m - m_0)]\,[\int_0^T d\{N(s) - m_0 s\}]\right]$$

$$= m_0 \int_0^t \beta(s)\, ds + \int_0^t \int_0^T \beta(\sigma)\, r(\sigma,s)\, d\sigma\, ds - \int_0^t \sigma^2\, ds ,$$

$\beta(s)$ is the solution of

$$\beta(s)\, m_0 + \int_0^T \beta(\sigma)\, r(\sigma,s)\, d\sigma = \sigma^2$$

and the proof is ended.

For the estimate m_B^* we have

$$E\{(m_B^* - m)^2\} = E[\int_0^T \beta(s)\, d\{N(s) - m_0 s\} - (m - m_0)]^2$$

$$= m_0 \int_0^T \beta^2(s)\, ds + \int_0^T \int_0^T \beta(s)\, r(s,t)\, \beta(t)\, ds\, dt + \sigma^2$$

$$- 2\sigma^2 \int_0^T \beta(s)\, ds$$

$$= m_0 \int_0^T \beta^2(s) + \int_0^T \beta(s)\, \{\sigma^2 - m_0 \beta(s)\}\, ds$$

$$+ \sigma^2 - 2\sigma^2 \int_0^T \beta(s)\, ds$$

$$= \sigma^2\, \{1 - \int_0^T \beta(s)\, ds\} .$$

The estimate m_B^* is not unbiased in the sense that $E[\{E(m_B^*|m) - m\}^2] = 0$ since

$$E(m_B^*|m) = m_0 + \int_0^T \beta(s)\, (m - m_0)\, ds$$

and thus

$$0 = E[\{(m - m_0) \int_0^T \beta(s)\, ds - (m - m_0)\}^2] = \sigma^2\, \{\int_0^T \beta(s)\, ds - 1\}^2$$

would imply that

$$E\{(m_B^* - m)^2\} = 0$$

for every T which is of course impossible.

The problem of finding linear Bayes estimates $\lambda^*(t)$ of $\lambda(t)$ is treated by Grandell (1971). There m_0 is taken as 1 but it follows from Lemma 1 and Lemma 2 that $E[\{\lambda^*(t) - \lambda(t)\}^2]$ is minimized if

$$\lambda^*(t) = m_0 + \int_0^T \beta_t(s)\, d\{N(s) - m_0 s\}$$

where $\beta_t(s)$ is the unique solution of

$$m_0 \beta_t(s) + \int_0^T \beta_t(\sigma)\, r(\sigma,s)\, d\sigma = r(t,s) .$$

The function $\beta(t)$ can be given in terms of $\beta_t(s)$. Put

$$\beta(t) = \frac{\sigma^2}{m_0}\{1 - \int_0^T \beta_t(s)\, ds\} .$$

Then we have

$$m_0 \frac{\sigma^2}{m_0}\{1 - \int_0^T \beta_t(s)\, ds\} + \frac{\sigma^2}{m_0}\int_0^T \{1 - \int_0^T \beta_s(\sigma)\, d\sigma\}\cdot r(t,s)\, ds$$

$$= \frac{\sigma^2}{m_0}[m_0 - m_0 \int_0^T \beta_t(s)\, ds + \int_0^T r(t,s)\, ds - \int_0^T \{r(t,\sigma) - m_0\beta_t(\sigma)\}d\sigma]$$

$$= \sigma^2$$

since $\beta_t(s) = \beta_s(t)$ (cf. Grandell, 1971).

From this it follows that

$$m_B^* = m_0 + \int_0^T \beta(t)\, d\{N(t) - m_0 t\}$$

$$= m_0 + \frac{\sigma^2}{m_0}\int_0^T \{1 - \int_0^T \beta_t(s)\, ds\}\, d\{N(t) - m_0 t\}$$

$$= m_0 + \frac{\sigma^2}{m_0} N(T) - \frac{\sigma^2}{m_0}\int_0^T \lambda^*(s)\, ds .$$

At the first glance this form of m_B^* may seem strange, but the following theorem gives a natural interpretation.

Theorem 2. Let $\hat{m}_B$ be the best linear Bayes estimate of m based on $\{\lambda(t);\ t \in [0,T]\}$. Then m_B^* is the best linear Bayes estimate of $\hat{m}_B$ based on $\{N(t);\ t \in [0,T]\}$.

Proof. As it follows from Lemma 1 we have to show that

$$E[(m_B^* - \hat{m}_B)\{N(t) - m_0 t\}] = 0 \qquad \text{for all } t \in [0,T].$$

Since

$$E[(m_B^* - m)\{N(t) - m_0 t\}] = 0$$

and

$$E[(\hat{m}_B - m)\{\lambda(t) - m_0\}] = 0$$

for all $t \in [0,T]$, it follows that

$$E[(m_B^* - \hat{m}_B)\{N(t) - m_0 t\}]$$

$$= E[(m_B^* - m)\{N(t) - m_0 t\}] - E[(\hat{m}_B - m)\{N(t) - m_0(t)\}]$$

$$= 0 - E(E[(\hat{m}_B - m)\{N(t) - m_0 t\} | m, \{\lambda(t);\ t \in [0,T]\}])$$

$$= -E((\hat{m}_B - m)\ [\int_0^t \{\lambda(\tau) - m_0\}\ d\tau]) = 0\ ,$$

since

$$\int_0^t \{\lambda(\tau) - m_0\}\ d\tau$$

is for every $t \in [0,T]$ an element in the Hilbert space spanned by $\{\lambda(t) - m_0;\ t \in [0,T]\}$.

Example. Consider the case where

$$r^{(\varepsilon)}(s,t) = \exp(-\alpha|s-t|).$$

It is easy to verify that

$$\hat{m}_B = \frac{(2m_0/\sigma^2) + \lambda(0) + \lambda(T) + \alpha \int_0^T \lambda(\tau)\ d\tau}{2 + 2/\sigma^2 + \alpha T}\ .$$

From Theorem 2 it thus follows that

$$m_B^* = \frac{(2m_0/\sigma^2) + \lambda^*(0) + \lambda^*(T) + \alpha \int_0^T \lambda^*(\tau)\ d\tau}{2 + 2/\sigma^2 + \alpha T}\ .$$

Consider now the discrete parameter case. Put $\ell_k = m + \varepsilon_k$ where $E(\ell_k) = E(m) + E(\varepsilon_k) = m_0 + 0 = m_0$ and $r_{k,j} = \text{cov}(\ell_k, \ell_j) = \sigma^2 + r_{k,j}^{(\varepsilon)}$. Like in the continuous parameter case we assume that $\text{cov}(m, \varepsilon_k) = 0$ for all k.

From Lemma 1 it follows that $m_B^* = m_0 + \sum_{k=1}^{n} \beta_k (N_k - m_0)$ is the best linear Bayes estimate of m if and only if $E\{(m_B^* - m)(N_j - m_0)\} = 0$ for all j.

Put

$$\underline{R} = \{r_{k,j}\} \qquad (1 \le k, j \le n)$$

and define $\underline{1}$ and $\underline{\beta}$ as in Section 3.

Since $E\{(m_B^* - m)(N_j - m_0)\} = m_0\beta_j + \sum_{k=1}^{n} \beta_k r_{k,j} - \sigma^2$, we get $(m_0\underline{I} + \underline{R})\underline{\beta} = \sigma^2\underline{1}$ or

$$\underline{\beta} = \sigma^2 (m_0\underline{I} + \underline{R})^{-1} \underline{1} \, .$$

We further have

$$E\{(m_B^* - m)^2\} = \sigma^2\{1 - \sum_{j=1}^{n} \beta_j\} = \sigma^2\{1 - \sigma^2 \underline{1}^T (m_0\underline{I} + \underline{R})^{-1} \underline{1}\} \, .$$

5. A COMPARISON BETWEEN BAYES AND NON-BAYES ESTIMATION OF THE MEAN VALUE

Consider the discrete parameter case.

In section 3 the best unbiased linear estimate m* of m was shown to be given by

$$m^* = \sum_{k=1}^{n} b_k N_k \, ,$$

where

$$\underline{b} = \frac{1}{\underline{1}^T (m_0\underline{I} + \underline{R}^{(\varepsilon)})^{-1} \underline{1}} (m_0\underline{I} + \underline{R}^{(\varepsilon)})^{-1} \underline{1} \, .$$

(In order to avoid confusion we use in this section $\underline{b}$ instead of $\underline{\beta}$ in connection with non-Bayes estimation.)

For this estimate

$$\mathrm{var}(m^*) = \frac{1}{\underline{1}^T(m_0\underline{I} + \underline{R}^{(\varepsilon)})^{-1}\,\underline{1}} .$$

In section 4 is was shown that the best linear Bayes estimate m_B^* of m is given by

$$m_B^* = m_0 + \sum_{k=1}^{n} \beta_k(N_k - m_0) ,$$

where

$$\underline{\beta} = \sigma^2(m_0\underline{I} + \sigma^2\underline{1}\underline{1}^T + \underline{R}^{(\varepsilon)})^{-1}\,\underline{1} .$$

For this estimate

$$E\{(m_B^* - m)^2\} = \sigma^2\{1 - \sigma^2\underline{1}^T(m_0\underline{I} + \sigma^2\underline{1}\underline{1}^T + \underline{R}^{(\varepsilon)})^{-1}\,\underline{1}\} .$$

Theorem 3. The relation between $E\{(m_B^* - m)^2\}$ and $\mathrm{var}(m^*)$ is given by

$$E\{(m_B^* - m)^2\} = \frac{\sigma^2\ \mathrm{var}(m^*)}{\sigma^2 + \mathrm{var}(m^*)}$$

Proof. We will give a probabilistic proof.

Consider a stochastic process $\{X_k;\ k = 1,\ldots, n\}$ with unknown mean value μ and covariance matrix $m_0\underline{I} + \underline{R}^{(\varepsilon)}$ and a stochastic variable Y with $E(Y) = 0$ and $\mathrm{var}(Y) = \sigma^2$. Assume that $E(X_kY) = 0$ for all k and that $Z_k = X_k + Y$ is observed.

For unbiased linear estimates $\mu^* = \sum_{k=1}^{n} a_kZ_k$ of μ we have

$$\mathrm{var}(\mu^*) = \sigma^2 + \mathrm{var}\sum_{k=1}^{n} a_kX_k .$$

Especially we get the best unbiased estimate if $\mathrm{var}(\sum_{k=1}^{n} a_kZ_k)$ is minimized Since we equally well may minimize $\mathrm{var}(\sum_{k=1}^{n} a_kX_k)$ it follows from Section 3 that

$$\frac{1}{\underline{1}^T(m_0\underline{I} + \sigma^2\underline{1}\underline{1}^T + \underline{R}^{(\varepsilon)})^{-1}\underline{1}} = \sigma^2 + \frac{1}{\underline{1}^T(m_0\underline{I} + \underline{R}^{(\varepsilon)})^{-1}\underline{1}} .$$

Thus we have

$$E\{(m_B^* - m)^2\} = \sigma^2[1 - \sigma^2/\{\sigma^2 + \text{var}(m^*)\}]$$

and the theorem is proved.

Corollary. The relation between m_B^* and m^* is given by

$$m_B^* = \frac{m_0 \text{ var}(m^*) + m^*\sigma^2}{\text{var}(m^*) + \sigma^2} .$$

Proof. Put $\underline{N}^T = (N_1, \ldots, N_n)$ as in Section 3. We have

$$m_B^* = m_0 + \sigma^2\underline{1}^T(m_0\underline{I} + \sigma^2\underline{1}\underline{1}^T + \underline{R}^{(\varepsilon)})^{-1} (\underline{N} - m_0\underline{1})$$

$$= m_0(1 - \sigma^2\underline{1}^T(m_0\underline{I} + \sigma^2\underline{1}\underline{1}^T + \underline{R}^{(\varepsilon)})^{-1}\underline{1}) + \sigma^2\underline{1}^T(m_0\underline{I} + \sigma^2\underline{1}\underline{1}^T + \underline{R}^{(\varepsilon)})^{-1}\underline{N}$$

and

$$m^* = \frac{\underline{1}^T(m_0\underline{I} + \underline{R}^{(\varepsilon)})^{-1}\underline{N}}{\underline{1}^T(m_0\underline{I} + \underline{R}^{(\varepsilon)})^{-1}\underline{1}} .$$

From the arguments in the proof of Theorem 3 it follows that

$$\frac{\underline{1}^T(m_0\underline{I} + \underline{R}^{(\varepsilon)})^{-1}}{\underline{1}^T(m_0\underline{I} + \underline{R}^{(\varepsilon)})^{-1}\underline{1}} = \frac{\underline{1}^T(m_0\underline{I} + \sigma^2\underline{1}\underline{1}^T + \underline{R}^{(\varepsilon)})^{-1}}{\underline{1}^T(m_0\underline{I} + \sigma^2\underline{1}\underline{1}^T + \underline{R}^{(\varepsilon)})^{-1}\underline{1}} .$$

Thus

$$m_B^* = m_0\{1 - \sigma^2 1^T(m_0\underline{I} + \sigma^2\underline{1}\underline{1}^T + \underline{R}^{(\varepsilon)})^{-1}\underline{1}\}$$

$$+ \sigma^2\underline{1}^T(m_0\underline{I} + \sigma^2\underline{1}\underline{1}^T + \underline{R}^{(\varepsilon)})^{-1}\underline{1}\cdot m^*.$$

Since

$$E\{(m_B^* - m)^2\} = \sigma^2\{1 - \sigma^2\underline{1}^T(m_0\underline{I} + \sigma^2\underline{1}\underline{1}^T + R^{(\varepsilon)})^{-1}\ \underline{1}\}$$

we get from Theorem 3 that

$$\sigma^2\underline{1}^T(m_0\underline{I} + \sigma^2\underline{1}\underline{1}^T + \underline{R}^{(\varepsilon)})^{-1}\ \underline{1} = \frac{\sigma^2}{\sigma^2 + \text{var}(m^*)} .$$

Thus we have

$$m_B^* = \frac{m_0\ \text{var}(m^*)}{\sigma^2+\text{var}(m^*)} + \frac{\sigma^2 m^*}{\sigma^2 + \text{var}(m^*)}$$

and the proof is ended.

Consider the case where $\{\varepsilon_k\}$ is stationary. Then $r_{k,j}^{(\varepsilon)} = r_{k-j}^{(\varepsilon)}$ and

$$r_k^{(\varepsilon)} = \int_{-\pi}^{\pi} e^{ikx}\ dF(s)$$

where F(x) is a nondecreasing bounded function.

The elements of the matrix $m_0\underline{I} + \underline{R}^{(\varepsilon)}$ have then the representation

$$\int_{-\pi}^{\pi} \exp\{i(k-j)x\}\ dG(x)$$

where $G(x) = (m_0 x)/(2\pi)+F(x)$.

From e.g., Grenander and Rosenblatt (1956, Section 2.5), it follows that

$$\lim_{n\to\infty} \text{var}(m^*) = F(0) - F(0-) .$$

Comparing this with the estimate $\bar{N} = \frac{1}{n}\sum_{k=1}^{n} N_k$ we see that

$$\lim_{n\to\infty} \text{var}(m^*) = \lim_{n\to\infty} \text{var}(\bar{N}) .$$

If $F(x)$ is continuous at $x = 0$ we have $\lim_{n\to\infty} \operatorname{var}(m^*) = 0$ and thus

$$\lim_{n\to\infty} \frac{E\{(m_B^* - m)^2\}}{\operatorname{var}(m^*)} = 1 .$$

If further $F(x)$ is assumed to be absolutely continuous in a neighborhood of $x = 0$ and if $f(x) = F'(x)$ is continuous in $x = 0$ and elsewhere piecewise continuous it follows from Grenander and Rosenblatt (1956, Section 7.3) and Grenander (1951b, Section 5.3) that

$$\lim_{n\to\infty} \frac{\operatorname{var}(m^*)}{\operatorname{var}(\overline{N})} = 1 .$$

Under these conditions we further have

$$\lim_{n\to\infty} n \operatorname{var}(m^*) = m_0 + 2\pi f(0) .$$

Under rather general conditions it is thus seen that $\overline{N}$ may be used instead of the more complicated estimates m^* and m_B^* if n is large.

Finally we have calculated $\operatorname{var}_1(m^*)$ if in fact m_1 is the true value of m and $E\{(m_B - m)^2 | m = m_1\}$. After some calculations we get

$$\operatorname{var}_1(m^*) = \frac{1}{\underline{1}^T(m_0\underline{I} + \underline{R}^{(\varepsilon)})^{-1}\,\underline{1}} + (m_1 - m_0)\frac{\underline{1}^T(m_0\underline{I} + \underline{R}^{(\varepsilon)})^{-2}\,\underline{1}}{\{\underline{1}^T(m_0\underline{I} + \underline{R}^{(\varepsilon)})^{-1}\underline{1}\}^2}$$

and

$$E\{(m_B^* - m)^2 | m = m_1\} = \frac{\operatorname{var}_1(m^*)}{\left\{1 + \frac{\operatorname{var}_0(m^*)}{\sigma^2}\right\}^2} + \left\{\frac{(m_1 - m_0)\operatorname{var}_0(m^*)}{\sigma^2 + \operatorname{var}_0(m^*)}\right\}^2 ,$$

where $\operatorname{var}_0(m^*)$ is the variance of m^* if m_0 is the true value of m.

6. ESTIMATION OF PARAMETERS IN A REGRESSION MODEL

Consider a linear regression model

$$\ell_m = \sum_{j=1}^{k} \gamma_j y_{m,j} + \varepsilon_m \qquad (m = 1,\ldots,n) ,$$

where $\underline{\gamma}^T = (\gamma_1, \ldots, \gamma_k)$ are unknown regression parameters and where $y_{m,j}$ $(1 \leq j \leq k,\ 1 \leq m \leq n)$ are known constants. In the case where there are no disturbance factors ε_m present this model of the intensities has been applied to accident data by Erlander, Gustavsson and Lárusson (1968).

Let $\underline{Y}$ be the $(n\times k)$ matrix of the known constants $y_{m,j}$ and $\underline{\varepsilon}^T = (\varepsilon_1, \ldots, \varepsilon_n)$ the disturbance factors. Assume the $E(\underline{\varepsilon}) = 0\cdot\underline{1}$ and that $E(\underline{\varepsilon}\underline{\varepsilon}^T) = \underline{R}^{(\varepsilon)}$. Put further $\underline{\ell}^T = (\ell_1, \ldots, \ell_n)$ and $\underline{N}^T = (N_1, \ldots, N_n)$. With these notations we have

$$\underline{\ell} = \underline{Y}\underline{\gamma} + \underline{\varepsilon} .$$

In e.g., Grenander and Rosenblatt (1956) least square estimates and minimum variance linear unbiased estimates of $\underline{\gamma}$ based on ℓ are derived.

Assume now that $\underline{N}$ is observed. Under the assumptions made we have $E(\underline{N}) = \underline{Y}\underline{\gamma}$ and $E\{(\underline{N} - \underline{Y}\underline{\gamma})(\underline{N} - \underline{Y}\underline{\gamma})^T\} = \underline{D} + \underline{R}^{(\varepsilon)}$ where $\underline{D}$ is a diagonal $(n\times n)$ matrix given by

$$\underline{D} = \{\delta_{i,m} \sum_{j=1}^{k} \gamma_j y_{m,j}\} \qquad (1 \leq i,\ m \leq n) ,$$

$$\text{where } \delta_{i,m} = \begin{cases} 1 & \text{if } i = m , \\ 0 & \text{if } i \neq m . \end{cases}$$

The least square estimate $\underline{\gamma}_L^*$ is then given by minimizing $(\underline{N} - \underline{Y}\underline{C})^T(\underline{N}-\underline{Y}\underline{C})$ with respect to the vector $\underline{C}$. If $\underline{Y}^T\underline{Y}$ is assumed to be nonsingular it is well known that $\underline{\gamma}_L^* = (\underline{Y}^T\underline{Y})^{-1}\ \underline{Y}^T\underline{N}$ and that $E\{(\underline{\gamma}_L^* - \underline{\gamma})(\underline{\gamma}_L^* - \underline{\gamma})^T\} = (\underline{Y}^T\underline{Y})^{-1}\ \underline{Y}^T(\underline{D} + \underline{R}^{(\varepsilon)})\ \underline{Y}(\underline{Y}^T\underline{Y})^{-1}$.

Remark. If $k = 1$ and if $y_{m,1} = 1$ for $m = 1,\ldots,n$ then $\gamma_L^* = \frac{1}{n}\sum_{k=1}^{n} N_k$. In this case γ_1 corresponds to m in the problem treated in Section 3. Thus $\bar{N}$ is the least square estimate of m.

In order to find estimates of $\underline{\gamma}$ corresponding to the minimum variance linear unbiased estimates mentioned above it is natural to consider Bayes estimates in a similar way as done in Section 4. Therefore we assume that $\underline{\gamma}$ may be regarded as a random vector with $E(\underline{\gamma}) = \underline{\gamma}_0$ and

$E\{(\underline{\gamma} - \underline{\gamma}_0)(\underline{\gamma} - \underline{\gamma}_0)^T\} = \underline{\Gamma}$. Further $\underline{\gamma}$ and $\underline{\varepsilon}$ are assumed to be uncorrelated.

Definition 2. If in a class of matrices $\underline{M}$ there exists a matrix $\underline{M}_0$ such that $\underline{M} - \underline{M}_0$ is nonnegative definite for all $\underline{M} \in \underline{M}$ then $\underline{M}_0$ is said to minimize $\underline{M}$.

Consider $\underline{\gamma}_B^* = \underline{\alpha} + \underline{C}(\underline{N} - \underline{Y}\underline{\gamma}_0)$ where $\underline{\alpha}^T = (\alpha_1, \ldots, \alpha_k)$ and $\underline{C}$ is a $(k \times n)$ matrix with elements $c_{i,j}$.

Definition 3. If $\underline{\alpha}$ and $\underline{C}$ is chosen such that $E\{(\underline{\gamma}_B^* - \underline{\gamma})(\underline{\gamma}_B^* - \underline{\gamma})^T\}$ is minimized, then $\underline{\gamma}_B^*$ is called the best linear Bayes estimate of $\underline{\gamma}$.

In this model we have $E(\underline{N}) = \underline{Y}\underline{\gamma}_0$ and $E\{(\underline{N} - \underline{Y}\underline{\gamma}_0)(\underline{N} - \underline{Y}\underline{\gamma}_0)^T\} = \underline{D}_0 + \underline{Y}\underline{\Gamma}\underline{Y}^T + \underline{R}^{(\varepsilon)}$, where

$$\underline{D}_0 = \{\delta_{i,m} \sum_{j=1}^{n} \gamma_{0,j} y_{m,j}\} \qquad (1 \leq i,\ m \leq n)$$

and where $\gamma_{0,j} = E(\gamma_j)$.

Since $\underline{D}_0 \underline{1} = \underline{Y}\underline{\gamma}_0 = E(\underline{N})$ it follows that $\underline{D}_0$ is nonnegative definite. The matrix $\underline{D}_0 + \underline{Y}\underline{\Gamma}\underline{Y}^T + \underline{R}^{(\varepsilon)}$ is assumed to be positive definite.

Theorem 4. The best linear Bayes estimate $\underline{\gamma}_B^*$ of $\underline{\gamma}$ is given by

$$\underline{\gamma}_B^* = \underline{\gamma}_0 + \underline{\Gamma}\underline{Y}^T(\underline{D}_0 + \underline{Y}\underline{\Gamma}\underline{Y}^T + \underline{R}^{(\varepsilon)})^{-1}(\underline{N} - \underline{Y}\underline{\gamma}_0) .$$

For this estimate

$$E\{(\underline{\gamma}_B^* - \underline{\gamma})(\underline{\gamma}_B^* - \underline{\gamma})^T\} = \underline{\Gamma} - \underline{\Gamma}\underline{Y}^T(\underline{D}_0 + \underline{Y}\underline{\Gamma}\underline{Y}^T + \underline{R}^{(\varepsilon)})^{-1}\underline{Y}\underline{\Gamma} .$$

The proof is simple and is therefore omitted.

7. A NUMERICAL ILLUSTRATION OF ESTIMATES OF THE COVARIANCE SEQUENCE

Let $\{\ell_k;\ k = 1,\ldots, n\}$ be a stationary sequence with mean value 1. The two unbiased estimates

$$r_k^* = \frac{1}{n-k} \sum_{j=1}^{n-k} (\ell_j - 1)(\ell_{j+k} - 1)$$

and

$$\hat{r}_k = \frac{1}{n-k} \sum_{j=1}^{n-k} (N_j - 1)(N_{j+k} - 1) - \delta_{k,0}$$

of $r_k = E\{(\ell_j - 1)(\ell_{j+k} - 1)\}$ are compared by Grandell (1971).

In order to calculate $\mathrm{var}(r_k^*)$ and $\mathrm{var}(\hat{r}_k)$ moments of $\{\ell_k\}$ up to the fourth order must be known.

Consider the following special case. Let ℓ_k be exponentially distributed with parameter 1, i.e., $\mathrm{prob}\{\ell_k \le x\} = 1 - e^{-x}$ for $x \ge 0$. Let further ℓ_k and ℓ_{k+1} be equal with probability ρ and independent with probability $1 - \rho$. For such a process we have, for $0 \le k \le j \le m$,

$$E(\ell_i) = 1,$$

$$E\{(\ell_i - 1)(\ell_{i+k} - 1)\} = \rho^k,$$

$$E\{(\ell_i - 1)(\ell_{i+k} - 1)(\ell_{i+j} - 1)\} = 2\rho^j$$

and

$$E\{(\ell_i - 1)(\ell_{i+k} - 1)(\ell_{i+j} - 1)(\ell_{i+m} - 1)\} = 8\rho^m + \rho^{m-j+k} .$$

Suppose that n is so large that terms equal to O(1) may be neglected in comparison with terms equal to O(n). As an example we then get

$$\mathrm{var}\{\frac{1}{n} \sum_{k=1}^{n} \ell_k\} = \frac{1}{n^2} \sum_{k=1}^{n} \sum_{j=1}^{n} \rho^{|k-j|} = \frac{1}{n^2} \sum_{s=-n+1}^{n-1} (n - s)\, \rho^{|s|}$$

$$\approx \frac{1}{n} \sum_{s=-n+1}^{n-1} \rho^{|s|} \approx \frac{1}{n} \sum_{s=-\infty}^{\infty} \rho^{|s|} = \frac{1}{n} \frac{1 + \rho}{1 - \rho} .$$

With this approximation we find after some calculations that

$$\operatorname{var}(r_k^*) \approx \frac{1}{n}\{8\rho^k \frac{1+\rho}{1-\rho} + \frac{(1+\rho^2)(1-\rho^{2k})}{1-\rho^2} - 2k\,\rho^{2k}\}.$$

From the results by Grandell (1971) we get

$$E\{(\hat{r}_k - r_k^*)^2\} \approx \begin{cases} \frac{1}{n}\{\frac{1+\rho}{1-\rho} + 21\} & (k = 0) \\ \frac{1}{n}\{3 + 5\rho^k + 6\rho^{2k}\} & (k \geq 0) \end{cases}$$

From this and Grandell (1971) it follows that

$$E\{(\hat{r}_k - r_k)^2\} \approx \begin{cases} \frac{1}{n}\{13\,\frac{1+\rho}{1-\rho} + 21\} & (k = 0) \\ \frac{1}{n}\{8\rho^k \frac{1+\rho}{1-\rho} + \frac{(1-\rho^2)(1-\rho^{2k})}{1-\rho^2} + 3 + 5\rho^k \\ \quad + (6-2k)\,\rho^{2k}\} & (k > 0)\,. \end{cases}$$

We have made three random generations for $n = 500$ and for $\rho = 0.0$ and $\rho = 0.75$ respectively. In each of these six generations we have calculated r_k^* and $\hat{r}_k$ for $k = 0,\ldots, 10$. We have further calculated $\bar{\ell} = \frac{1}{n}\sum_{k=1}^{n} \ell_k$, $\bar{N} = \frac{1}{n}\sum_{k=1}^{n} N_k$ and $\frac{1}{n}\sum_{k=1}^{n} (N_k - \ell_k)^2$. For these quantities we have

$$E(\bar{\ell}) = 1, \qquad \operatorname{var}(\bar{\ell}) \approx \frac{1}{n}\frac{1+\rho}{1-\rho}, \qquad E(\bar{N}) = 1, \qquad \operatorname{var}(\bar{N}) \approx \frac{1}{n}\frac{2}{1-\rho};$$

$$E\{\frac{1}{n}\sum_{k=1}^{n}(N_k - \ell_k)^2\} = 1 \text{ and } \operatorname{var}\{\frac{1}{n}\sum_{k=1}^{n}(N_k - \ell_k)^2\} \approx \frac{1}{n}(5 + \frac{1+\rho}{1-\rho})\,.$$

In Tables 1 and Tables 2 we have calculated some theoretical values which may be of interest in the study of the random generations. These values have been calculated for $n = 500$.

Table 1. *Standard deviations of estimates of the covariances*

k	$\rho = 0.0$			$\rho = 0.75$		
	r_k	$\sqrt{E\{(r_k^*-r_k)^2\}}$	$\sqrt{E\{(r_k-r_k)^2\}}$	r_k	$\sqrt{E\{(r_k^*-r_k)^2\}}$	$\sqrt{E\{(r_k-r_k)^2\}}$
0	1	0.126	0.210	1.000	0.335	0.473
1	0	0.045	0.089	0.750	0.291	0.324
2	0	0.045	0.089	0.563	0.256	0.284
3	0	0.045	0.089	0.422	0.226	0.252
4	0	0.045	0.089	0.316	0.201	0.225
5	0	0.045	0.089	0.237	0.179	0.203
6	0	0.045	0.089	0.178	0.162	0.185
7	0	0.045	0.089	0.133	0.147	0.170
8	0	0.045	0.089	0.100	0.134	0.158
9	0	0.045	0.089	0.075	0.124	0.149
10	0	0.045	0.089	0.056	0.115	0.141

Table 2. *Standard deviations of some characteristics of the generations*

	$\rho = 0.0$	$\rho = 0.75$
$\sqrt{\mathrm{var}(\bar{\ell})}$	0.045	0.118
$\sqrt{\mathrm{var}(\bar{N})}$	0.063	0.126
$\sqrt{\mathrm{var}\{\frac{1}{n}\sum_1^n (N_k-\ell_n)^2\}}$	0.110	0.160

Call the three generations made for $\rho = 0.0$ for G1, G2 and G3 and the three made for $\rho = 0.75$ for G4, G5 and G6. As a compensation for publishing the generations completely the values of $\bar{\ell}$, $\bar{N}$ and $\frac{1}{n}\sum_{k=1}^{n} (N_k - \ell_k)^2$ are given in Table 3.

Table 3. *Some characteristics of the generations*

Generation		$\bar{\ell}$	$\bar{N}$	$\frac{1}{n}\sum_{1}^{n}(N_k - \ell_k)^2$
	G1	0.993	0.982	0.951
$\rho = 0.0$	G2	1.025	1.052	1.074
	G3	1.018	1.046	0.856
	G4	0.929	0.884	0.717
$\rho = 0.75$	G5	0.878	0.788	0.809
	G6	0.933	0.976	0.978

From Table 1 it is seen that we can not expect to get estimates r_k^* and $\hat{r}_k$ which are close to the true values r_k. Furthermore it is known that estimates of r_k are correlated for different values of k when $\rho > 0$. From the calculations which will be presented in Section 8 it follows that $\text{corr}(r_0^*, r_1^*) = 0.985$ and $\text{corr}(\hat{r}_0, \hat{r}_1) = 0.841$ for $\rho = 0.75$ and $n = 500$. It is thus difficult to get a good idea of how the estimates will behave. This difficulty was in fact the main reason for making these random generations.

In Table 4 and Table 5 we give the values of the estimates r_k^* and $\hat{r}_k$.

Table 4. *Estimates of covariances*

$\rho = 0.0$	G1		G2		G3	
k	r_k^*	$\hat{r}_k$	r_k^*	$\hat{r}_k$	r_1^*	$\hat{r}_k$
0	0.905	0.762	1.032	1.512	1.158	1.030
1	0.035	0.130	0.046	0.032	-0.038	-0.016
2	0.067	0.131	0.010	-0.016	-0.020	0.078
3	-0.006	0.010	-0.039	-0.163	-0.049	-0.016
4	0.043	-0.083	-0.044	0.002	0.001	0.052
5	-0.048	-0.040	0.011	0.071	-0.032	0.036
6	-0.054	-0.115	-0.029	0.071	-0.001	-0.077
7	0.035	-0.028	-0.028	-0.067	0.043	0.061
8	0.097	0.189	-0.020	-0.098	0.094	0.033
9	0.005	0.022	-0.011	-0.065	0.054	0.027
10	0.039	-0.012	-0.003	0.004	-0.025	-0.202

Table 5. *Estimates of covariances*

$\rho = 0.75$	G4		G5		G6	
k	r_k^*	$\hat{r}_k$	r_k^*	$\hat{r}_k$	r_k^*	$\hat{r}_k$
0	0.847	0.592	0.709	0.312	0.793	0.896
1	0.560	0.491	0.419	0.463	0.565	0.679
2	0.392	0.317	0.340	0.335	0.383	0.440
3	0.240	0.217	0.270	0.205	0.230	0.244
4	0.128	0.179	0.203	0.073	0.149	0.179
5	0.054	0.069	0.192	0.176	0.081	0.042
6	0.005	-0.034	0.155	0.134	0.019	0.087
7	-0.051	-0.128	0.114	0.093	-0.015	0.002
8	-0.050	-0.083	0.113	0.071	-0.015	-0.035
9	-0.046	-0.020	0.090	0.134	-0.047	-0.075
10	-0.058	-0.039	0.083	0.078	-0.059	-0.029

A comparison of Table 1 with Table 4 and Table 5 shows that although 500 observations were generated the true and the estimate values differ a lot. As an example it is seen that in order to get $\sqrt{E\{(\hat{r}_0 - r_0)^2\}} < 0.1$ for $\rho = 0.75$ more than 11,200 observations are required.

8. A NUMERICAL ILLUSTRATION OF ESTIMATES OF THE COEFFICIENTS OF AN AUTOREGRESSIVE PROCESS

Let $\{\ell_k;\ k = 0, \pm 1, \ldots\}$ be a stationary sequence with mean value 1 satisfying the difference equations

$$\sum_{k=0}^{p} \alpha_k (\ell_{m-k} - 1) = \varepsilon_m ,$$

where $\alpha_0 = 1$ and where $E(\varepsilon_m) = 0$ and $E(\varepsilon_j \varepsilon_m) = \delta_{j,m} \sigma^2$. In order to get the parameters identifiable we assume that the roots of the equation

$$\sum_{k=0}^{p} \alpha_k z^k = 0$$

have modulus larger than one.

In e.g., Hannan (1960) it is shown that the least square estimates of $\alpha_1, \ldots, \alpha_p$ and σ^2 based on $\ell_1, \ldots, \ell_n$ are given by (after a slight modification)

$$\sum_{j=0}^{p} \alpha_j^* \sum_{s=p+1}^{n} (\ell_{s-j} - 1)(\ell_{s-k} - 1) = 0 \qquad (k = 1, \ldots, p)$$

and

$$(\sigma^2)^* = \frac{1}{n-p} \sum_{s=1}^{n-p} \{\sum_{j=0}^{p} \alpha_j^*(\ell_{s-j} - 1)\}^2 .$$

(Of course $\alpha_0^* = 1$.)

We will only consider a first order autoregressive scheme. Put $\alpha_1 = -\rho$. Then we get

$$\sum_{s=2}^{n} (\ell_s - 1)(\ell_{s-1} - 1) = \rho^* \sum_{s=2}^{n} (\ell_{s-1} - 1)^2$$

and

$$(\sigma^2)^* = \frac{1}{n-1} \sum_{s=1}^{n-1} \{(\ell_s - 1) - \rho^*(\ell_{s-1} - 1)\}^2 .$$

Since n is assumed to be large (which is assumed in the modification mentioned above) we have after another slight modification

$$\rho^* = \frac{r_1^*}{r_0^*}$$

and

$$(\sigma^2)^* = r_0^* - \frac{(r_1^*)^2}{r_0^*} ,$$

where r_0^* and r_1^* are defined in Section 7.

Consider now the case when $N_1, \ldots, N_n$ are observed. It is then natural to study the estimates

$$\hat{\rho} = \frac{\hat{r}_1}{\hat{r}_0}$$

and

$$(\hat{\sigma}^2) = \hat{r}_0 - \frac{(\hat{r}_1)^2}{\hat{r}_0},$$

where $\hat{r}_0$ and $\hat{r}_1$ are defined in Section 7.

Consider now the special process which was treated in Section 7. Then we have

$$(\ell_m - 1) - \rho(\ell_{m-1} - 1) = \varepsilon_m .$$

Thus

$$E(\varepsilon_m) = 0$$

and

$$\operatorname{cov}(\varepsilon_m, \varepsilon_{m+k}) = \begin{cases} 1 - \rho^2 & (k = 0) \\ 0 & (k \neq 0) . \end{cases}$$

Note that the ε_m are uncorrelated variables, but they are not independent since e.g., $E(\varepsilon_m^2 \varepsilon_{m+k}^2) \neq (1 - \rho^2)^2$. This explains why the variance formulae in Section 7 differ from the variance formulae for linear processes given by e.g., Hannan (1960).

By a Taylor expansion we get

$$\rho^* = \rho + (r_1^* - \rho) - \rho(r_0^* - 1) + O\{(r_1^* - \rho)^2 + (r_0^* - 1)^2\}$$

and

$$(\sigma^2)^* = 1 - \rho^2 - 2\rho(r_1^* - \rho) + (1 + \rho^2)(r_0^* - 1) + O\{(r_1^* - \rho)^2 + (r_0^* - 1)^2\} .$$

Thus we have

$$E(\rho^*) = \rho + O(\frac{1}{n})$$

and

$$E\{(\sigma^2)^*\} = 1 - \rho^2 + O(\frac{1}{n}) .$$

Under certain assumptions (which are fulfilled in this case) on the eight moments (cf. e.g., Hannan, 1960) we have

$$\mathrm{var}(\rho^*) = \mathrm{var}(r_1^*) + \rho^2\,\mathrm{var}(r_0^*) - 2\rho\,\mathrm{cov}\,r_0^* r_1^* + O(n^{-2})$$

and

$$\mathrm{var}\{(\sigma^2)^*\} = 4\rho^2\,\mathrm{var}\{r_1^*\} + (1+\rho^2)^2\,\mathrm{var}\{r_0^*\}$$
$$- 4\rho(1+\rho^2)\,\mathrm{cov}\{r_0^*, r_1^*\} + O(n^{-2}).$$

Similar expressions hold for $\hat{\rho}$ and $(\hat{\sigma}^2)$.

After some calculations similar to the calculations in Section 7 we get

$$\mathrm{cov}(r_0^*, r_1^*) \approx \frac{1}{n}\,\frac{16\rho}{1-\rho}$$

and

$$\mathrm{cov}(\hat{r}_0, \hat{r}_1) \approx \frac{\rho}{n}\,(14 + \frac{18}{1-\rho}) \; .$$

Thus we have

$$\mathrm{var}(\rho^*) \approx \frac{1}{n}\,(1-\rho)(1+9\rho),$$

$$\mathrm{var}(\hat{\rho}) \approx \frac{1}{n}\,(4 + 7\rho - 15\rho^2 + \frac{6\rho}{1-\rho}) \; ,$$

$$\mathrm{var}\{(\sigma^2)^*\} \approx \frac{1}{n}\,(8 + 16\rho - 28\rho^2 + 16\rho^3 - 12\rho^4)$$

and

$$\mathrm{var}\{(\hat{\sigma}^2)\} \approx \frac{1}{n}\,(34 + 2\rho - 42\rho^2 + 34\rho^3 - 28\rho^4 + \frac{24\rho}{1-\rho}) \; .$$

In Table 6 we have for $\rho = 0.0$ and $\rho = 0.75$ and for $n = 500$ calculated the square root of the variances given above.

Table 6. *Standard deviations of estimates of* ρ *and* σ^2

	$\rho = 0.0$	$\rho = 0.75$
$\sqrt{\text{var}(\rho^*)}$	0.04	0.06
$\sqrt{\text{var}(\hat{\rho})}$	0.09	0.19
$\sqrt{\text{var}\{(\sigma^2)^*\}}$	0.13	0.13
$\sqrt{\text{var}\{(\hat{\sigma}^2)\}}$	0.26	0.41

In Table 7 the values of the different estimates of ρ and σ^2 in the generations G1,..., G6 described in Section 7, are given.

Table 7. *Estimates of* ρ *and* σ^2

	ρ	σ^2	ρ^*	$\hat{\rho}$	$(\sigma^2)^*$	$(\hat{\sigma}^2)$
G1	0.0	1.0	0.04	0.17	0.90	0.74
G2	0.0	1.0	0.04	0.02	1.03	1.51
G3	0.0	1.0	-0.03	-0.02	1.16	1.03
G4	0.75	0.44	0.66	0.83	0.48	0.18
G5	0.75	0.44	0.59	1.48	0.46	-0.38
G6	0.75	0.44	0.71	0.76	0.39	0.38

9. ESTIMATION OF THE AVERAGE INTENSITY

In Sections 3-5 linear estimation of the mean value was studied. We now consider estimates of the average intensity $\bar{\ell} = \frac{1}{n} \sum_{k=1}^{n} \ell_k$. From Lemma 1 in Section 4 and Grandell (1971) it follows that the best linear Bayes estimate $\bar{\ell}^*$ of $\bar{\ell}$ is given by

$$\bar{\ell}^* = m_0 + \sum_{k=1}^{n} \beta_k (N_k - m_0) ,$$

where

$$\underline{\beta} = \frac{1}{n} \underline{R}(m_0 \underline{I} + \underline{R})^{-1} \underline{1}$$

($\bar{\ell}^* = \frac{1}{n}\sum_{k=1}^{n} \ell_k^*$ where ℓ_k^* is the best linear Bayes estimate of ℓ_k.)

Both m_0 and $\underline{R}$ are assumed to be known. Further we have

$$E\{(\bar{\ell}^* - \bar{\ell})^2\} = \frac{m_0}{n^2} \underline{1}^T \underline{R}(m_0 I + R)^{-1} \underline{1} = \frac{m_0}{n^2} \underline{1}^T (\underline{I} - m_0(m_0 + \underline{R})^{-1} \underline{1}$$

$$= \frac{m_0}{n} - \frac{m_0^2}{n} \cdot \frac{1}{n} \underline{1}^T (m_0 \underline{I} + \underline{R})^{-1} \underline{1} .$$

Suppose now that the intensity process ℓ_k is stationary, i.e.,

$$r_k = \int_{-\pi}^{\pi} e^{ikx} \, dF(x) .$$

If $F(0) - F(0-) > 0$ it follows from Section 5 that

$$\lim_{n\to\infty} \underline{1}^T (m_0 \underline{I} + \underline{R})^{-1} \underline{1} = \frac{1}{F(0) - F(0-)}$$

and thus

$$\lim_{n\to\infty} nE\{(\bar{\ell}^* - \bar{\ell})^2\} = m_0 .$$

If $F(x)$ is absolutely continuous in a neighborhood of $x = 0$ and if $f(x)$ is continuous in $x = 0$ and elsewhere piecewise continuous then

$$\lim_{n\to\infty} \frac{1}{n} \underline{1}^T (m_0 \underline{I} + \underline{R})^{-1} \underline{1} = \frac{1}{m_0 + 2\pi\, f(0)}$$

and thus

$$\lim_{n\to\infty} nE\{(\bar{\ell}^* - \bar{\ell})^2\} = m_0 - \frac{m_0^2}{m_0 + 2\pi\, f(0)} = \frac{2\pi\, f(0)\, m_0}{m_0 + 2\pi\, f(0)} .$$

Since

$$E\{(\bar{N} - \bar{\ell})^2\} = \frac{m_0}{n} ,$$

we have

$$\lim_{n\to\infty} \frac{E\{(\bar{\ell}^* - \bar{\ell})^2\}}{E\{(\bar{N} - \bar{\ell})^2\}} = 1$$

if $F(0) - F(0-) > 0$.

If on the other hand $f(x)$ fulfills the regularity conditions given we may consider

$$\frac{m_0^2 + 2\pi\ f(0)\ \bar{N}}{m_0 + 2\pi\ f(0)} .$$

For this expression we obtain after some calculations

$$\lim_{n\to\infty} nE\left\{\left(\frac{m_0^2 + 2\pi f(0)\bar{N}}{m_0 + 2\pi\ f(0)} - \bar{\ell}\right)^2\right\} = \frac{2\pi\ m_0\ f(0)}{m_0 + 2\pi\ f(0)} .$$

Thus

$$\lim_{n\to\infty} \frac{E\{(\bar{\ell}^* - \bar{\ell})^2\}}{E\left\{\left(\frac{m_0^2 + 2\pi\ f(0)\ \bar{N}}{m_0 + 2\pi\ f(0)} - \bar{\ell}\right)^2\right\}} = 1 ,$$

which implies that

$$\frac{m_0^2 + 2\pi\ f(0)\ \bar{N}}{m_0 + 2\pi\ f(0)}$$

serves as an approximation of $\bar{\ell}^*$ when n is large.

We will now use the generations G1,..., G6 as a numerical illustration of this estimate. In this case $m_0 = 1$ and $2\pi\ f(0) = \frac{1 + \rho}{1 - \rho}$. Thus we have

$$\lim_{n\to\infty} nE\{(\bar{\ell}^* - \bar{\ell})^2\} = \frac{1 + \rho}{2}$$

and as an approximation of $\bar{\ell}^*$ we have

$$\frac{1 - \rho + (1 + \rho)\,\bar{N}}{2} .$$

For n = 500 we thus have

$$\sqrt{E\{(\bar{\ell}^* - \bar{\ell})^2\}} \approx \begin{cases} 0.032 & (\rho = 0.0) , \\ 0.042 & (\rho = 0.75) . \end{cases}$$

In Table 8 we give the values of the approximation which is to be compared with $\bar{\ell}$ and $\bar{N}$ given in Table 3 in Section 7.

Table 8. *Value of the approximation to* $\bar{\ell}^*$

		Estimate
$\rho = 0.0$	G1	0.991
	G2	1.026
	G3	1.023
$\rho = 0.75$	G4	0.899
	G5	0.815
	G6	0.979

ACKNOWLEDGMENTS

The author wishes to thank Bengt von Bahr, Jan Gustavsson, Lars Peiram, Rolf Sundberg and Olof Thorin for many valuable comments.

REFERENCES

Erlander, S., Gustavsson, J. & Lárusson, E. (1968). Some investigations on the relationship between road accidents and estimated traffic. Accid. Anal. & Prev. 1, 17-64.

Grandell, J. (1971). On stochastic processes generated by a stochastic intensity function. Skand. Aktuar. Tidskr. To appear.

Grenander, U. (1950). Stochastic processes and statistical inference. Arkiv. för matematik 1, 195-277.

Grenander, U. (1951a). On empirical spectral analysis of stochastic processes. Arkiv för matematik 1, 508-31.

Grenander, U. (1951b). On Toeplitz forms and stationary processes. Arkiv för matematik 1, 555-71.

Grenander, U. & Rosenblatt, M. (1956). *Statistical analysis of stationary time series*. Stockholm: Almqvist & Wiksell.

Hannan, E.J. (1960). *Time series analysis*. London: Methuen.

Parzen, E. (1959). Statistical inference on time series by Hilbert space methods, I. Published in Parzen, E. (1967). *Time series analysis papers*. San Francisco: Holden-Day.

THE USE OF DISTANCE METHODS FOR THE ANALYSIS OF SPATIAL DISTRIBUTION OF POINTS

P. Holgate
Birkbeck College, University of London

SUMMARY

Distance methods are those in which the estimation and testing of parameters of a spatial point process are based on the measurement of distances from sample points to points in the realization of the process. The development of these methods is surveyed, and attention drawn to their advantages and disadvantages as estimators, tests of randomness, and measures of pattern.

I. INTRODUCTION

Botanists are frequently interested in both the density and pattern of occurence of a plant species in a habitat. Since the 1920's it has been usual to take the Poisson process as a basis and describe processes as 'overdispersed' or 'underdispersed' according to whether the variance (i.e., dispersion) of the numbers of plants in a quadrat of specified size was greater than or less than the mean.

This crude classification has several drawbacks.

(i) The ratio of variance to mean depends on the quadrat size.

(ii) If attention is concentrated on a fixed size of quadrat, there are many processes which can produce the same distribution of numbers per quadrat.

The above points are illustrated by considering the Poisson process with rate λ, and the process in which clusters of points occur at the vertices of a square lattice of unit spacing, the number of points in each cluster having a zero modified Poisson distribution

$$p_0 = a,\ p_j = (1-a)\lambda^j e^{-\lambda}/\{j!(1-e^{-\lambda})\} \qquad (j = 1,2,\ldots),$$

and sampling with a circular 'quadrat' of radius ρ determined by

$$\rho^2 = (1-e^{-\lambda})/\{\pi(1-a)\}.$$

In this latter case the variance to mean ratio for general ρ is dependent on quadrat size.

(iii) The classification into overdispersed and underdispersed is not sufficiently fine.

(iv) The method gives no indication as to whether overdispersion arises from heterogeneity in the density (e.g., as in a doubly stochastic Poisson process) or from the occurrence of clusters of points.

These difficulties led botanists to take an interest in distance methods, based on choosing instead of a sample of quadrats, a sample of n points, and measuring the distance from each to the nearest, second nearest,...,j^{th} nearest plant in a randomly orientated sector of θ radians. The case of $\theta = 2\pi$ is common, and the n points may be chosen randomly or systematically, or taken to be the locations of n randomly chosen plants. The distance measurements are used to estimate density, test for randomness, and characterize the pattern of the plant distribution.

2. ESTIMATION BASED ON DISTANCE METHODS FOR THE POISSON FOREST

In forestry as distinct from ecology, distance methods were developed primarily to facilitate estimation of density, on the grounds that they required considerably less effort.

Suppose that the trees arise from a Poisson point process with density λ. The mean distance from a fixed point to the nearest plant was derived by Hertz (1909) in the three dimensional case, in a physical context. Nothing further seems to have appeared on the subject until the independently written papers of Skellam (1952), Moore (1954) and Thompson (1956) who derived the probability density function. For a fixed sample point and an angular sector of θ radians the joint density of distances $x_1,\ldots,x_k$, to the nearest,...,k^{th} nearest tree is

$$(\lambda\theta)^k \prod_{i=1}^{k} x_i e^{-\frac{1}{2}\lambda\theta x_k^2}. \qquad (1)$$

If a new set of variables $y_i = \pi x_1^2$ is introduced, their joint probability density is

$$(\tfrac{1}{2}\lambda\theta/\pi)^k e^{-\frac{1}{2}\lambda\theta y_k/\pi}. \qquad (2)$$

This formula takes its simplest form when $\theta = 2\pi$, namely

$$\lambda^k e^{-\lambda y_k}. \qquad (3)$$

If $y_1,\ldots,y_{k-1}$ are integrated out from (3), the marginal density of y_k is found to be of gamma type

$$\lambda^k e^{-\lambda y_k}/\Gamma(k). \qquad (4)$$

If n sample points are taken, and the distance from the ith of them to its jth nearest tree is denoted by $x_{ij}(y_{ij} = \pi x_{1j}^2)$, the likelihood of the sample is

$$(\tfrac{1}{2}\lambda\theta/\pi)^{kn} \exp\{-\tfrac{1}{2}\lambda\theta \sum_{i=1}^{n} y_{ik}/\pi\}.$$

From this it can be seen that the maximum likelihood estimator of λ is

$$\hat{\lambda}_k = 2\pi kn/\{\theta \sum_{i=1}^{n} y_{ik}\}$$

and that it is sufficient. It is slightly biased with

$$E(\hat{\lambda}) = kn\lambda/(kn-1)$$

and the variance of the unbiased version is

$$\mathrm{var}\,\{(kn-1)\hat{\lambda}/(kn)\} = \lambda^2/(kn-2).$$

Since factors other than the usual ideas of statistical efficiency are relevant, a comparison of variances for equal sample sizes as against the quadrat count method is not necessarily of value. Moore (1954) gives a bivariate table of the ratios of variances of estimators based on n sample points, and on quadrats having a total area of m units. In cases where searching the ground for the individuals whose distribution is being studied is the main expenditure of effort it is reasonable to choose n so that the expectation of $\Sigma\pi x_{ik}^2 = n$, and in this case, as in sequential estimation in more familiar situations, the asymptotic relative efficiency of the methods is unity (Holgate, 1964).

Some of the early forestry workers proposed basing the estimation of λ not on the sufficient statistic Σy_{ik}, but on Σx_{ik} (for k=1). It is easy to show that the asymptotically unbiased estimator based on Σx_{ik} has efficiency 0.9149 (Kendall and Moran, 1962, p.38). An unusual method suggested in the literature has been to choose a point at random, determine the nearest tree, and measure the distance from it to its nearest neighbor. If this is denoted by ξ_1, then $E(\xi_1) = 1.191\lambda^{-\frac{1}{2}}$, and hence $(1.191\ n/\Sigma\xi_{i1})^2$ provides an asymptotically unbiased estimator of λ, although again an inefficient one.

While estimation of density based on quadrat counts is highly robust, that based on distance methods is not, and the question arises of whether it can be improved if slightly inefficient estimators are used.

Persson (1964, 1965, 1971) has studied estimators based on the sufficient statistic

$$\hat{\lambda} = cn^2/\{\Sigma x_{ik}^2\} \tag{6}$$

for k = 1,2,3 when measurements are made from a random point, a random tree, and the same statistic (no longer sufficient) when they are from the tree nearest a random point to its nearest neighbor; the estimator based on the mean distance, that is

$$\hat{\lambda} = cn^2/\{\Sigma x_{ik}\}^2, \tag{7}$$

for k = 1,2,3 for measurements from a random point; and also estimators

$$\hat{\lambda} = cn^{-1}\Sigma x_{i3}^{-2}, \tag{8}$$

$$\hat{\lambda} = cn^{-1}\sum_{i=1}^{n}\{4/\sum_{j=1}^{4} x_{i3j}^2\}, \tag{9}$$

where in (9) x_{i3j} is the distance from the ith sample point to the third nearest plant in the jth quadrant formed by placing orthogonal axes at random. These estimators, with slightly different notation, are listed in Persson (1971, Table 1) and the values of the constant c required to produce asymptotic unbiasedness are also listed. Their biases are then investigated for cases where instead of arising from a Poisson process, the trees form (i) a square lattice, (ii) a triangular lattice, and when they form a realization of a Poisson process of clusters, the distributions of cluster size being (iii) degenerate, clusters all containing k plants, for k = 3,4,5 and (iv) geometric with $p_i = (\frac{1}{2})^i$, i = 1,2,... Table 3 of the same paper gives the factors by which the estimator of the density will be biased, although only for situations where measurements are made from a random point. In general it appears that where the population has regular structure, quite serious bias can occur. The inefficient estimator (7) does seem better than (6), while (8) differs by underestimating the density while (6) and (7) overestimate it. For particular values of k, the bias in (7) can be made quite small against certain alternatives. With k = 2, it overestimates the density by only 6% when the plants form a triangular lattice, while for k = 3 the figure is only 7% when they form a square lattice.

When plants occur in clusters, robustness is almost impossible to achieve, since distance methods inevitably estimate the density of clusters.

3. ESTIMATION FOR REGULAR PATTERNS

The early papers by foresters did not assume that the trees formed a Poisson field. Although not written mathematically these papers repay careful study to elucidate the ideas of the practical workers. If the trees are packed closely in an equilateral triangular lattice of side a, their density is $2/(\sqrt{3}a^2)$. Bauersachs (1942), Cottam (1947) and Cottam and Curtis (1949) all argue that if the trees are placed in some pattern approximating to a triangular lattice, then their density may be estimated by

$$\hat{\lambda} = cn^2/\{\Sigma x_{i1}^2\}$$

for an appropriate value of c, not too different from $2/\sqrt{3} = 1.155$. The actual value to be used was determined by these and later writers including Essed (1954) on the basis of empirical studies in the field and by simulation experiments.

Consider a model in which trees are initially placed at the points of an equiangular triangular lattice, and then an independent random movement is applied to each tree independently. An easily tractable form for this movement is to assume that each tree takes a step of length ε in the direction of one of the six trees in the surrounding hexagon, with equal probabilities. The probability distribution of the distance from a chosen tree to its nearest neighbor can then be determined by exhaustive enumeration. From this, the expectation of x_1^2 is found to be

$$E(x_1^2) = (1-2.67\varepsilon)a^2$$

if terms in ε^2 can be ignored, which would lead to a positive bias in the estimator of λ. If however measurements are restricted to an angular sector it is possible to eliminate the effect of observing the smallest of six dependent, and for some pairs, negatively correlated measurements. The expected squared distance between a fixed pair of neighbors after perturbation is a^2, if terms of order ε^2 are ignored. If however an observation angle θ is chosen so that $\pi/3 < \theta < 2\pi/3$, it will include one neighbor from the surrounding hexagon, and possibly two, with respective probabilities

$$p_1 = 2 - 3\theta/\pi \ , \quad p_2 = 3\theta/\pi - 1.$$

Conditional on the sector including two trees, the distribution of nearest neighbor distance can again be evaluated by exhaustive enumeration, and putting all results together it is found that for θ in this range,

$$E(x_1^2) = 1-0.7427\theta\varepsilon + 0.7778\varepsilon.$$

Thus by appropriate choice of θ the bias can be reduced to a negligible quantity.

On the other hand if $\pi/6 < \theta < \pi/3$ there is a possibility that the sector will include no trees of the surrounding hexagon, though it will always include one at the next nearest distance and in fact the distribution of x_1 in this case is

$$\text{prob}(x_1=a) = 3\theta/\pi, \quad \text{prob}(x_1= \sqrt{3}a) = 1 - 3\theta/\pi$$

and hence

$$E(x_1^2) = 3a^2(1-2\theta/\pi),$$

and this will cause a drastic negative bias.

Similar computations are possible for a square lattice. If the angle of observation is taken as 2π, then

$$E(x_1^2) = (1-2.22\varepsilon)a^2,$$

while if it is restricted to the range $(\pi/2,\pi)$ the expectation becomes

$$E(x_1^2) = (1-0.72\theta\varepsilon + 1.13\varepsilon)a^2$$

which can be made close to a^2. Empirical discussion of the effects of choice of observation angle will be found in the papers of Cottam, Curtis and Hale (1953), Shanks (1954), Rice and Penfound (1955) and Cottam and Curtis (1955).

Another effect of a suitable choice of observation angle is to produce an estimator which is robust as between a Poisson process and a regular lattice. The family of estimators

$$\hat{\alpha}_k = \theta\Sigma x_{ik}^2/(2nk) \qquad (k = 1,2,\ldots), \tag{10}$$

is unbiased for $\alpha = \lambda^{-1}$ when the trees form a Poisson forest.

If the trees form an exact square lattice, a single measurement from one tree to its nearest neighbor will provide the value of a^2. If there is a slight perturbation, the value may be estimated by

$$\hat{\alpha} = \Sigma x_{i1}^2/n, \tag{11}$$

with a bias of order ε. But if $k = 1$ the choice $\theta = 2(=114^\circ\ 35')$ makes (10) agree with (11), consequently providing an estimator unbiased both for the Poisson process and the square lattice.

Similar reasoning applied to the equilateral triangular lattice leads to a choice of $\theta = \sqrt{3}/2(=99^\circ\ 14')$.

Another way of obtaining unbiased estimators is to take linear combinations of Σx_{i1}^2 and Σx_{i2}^2. Results in this direction have recently been obtained by Miss Susan Thompson as part of her M.Sc. project at Imperial College, London.

5. TESTS OF RANDOMNESS BASED ON DISTANCE METHODS

Tests of the hypothesis that the trees constitute a realization of a Poisson process, given n essentially independent observations on

$x_1,\ldots,x_k$ can be classified into three groups, based on the following ideas:

(i) if the field is Poisson, both the quadrat estimator and the nearest neighbor estimator (5) provide asymptotically unbiased estimates of λ,

(ii) in a Poisson field, measurements from a random point have the same distribution as measurements from a random plant, and

(iii) from (3), the distribution of $y_1,\ldots,y_{k-1}$ conditional on y_k is that of the ordered values of k-1 points thrown independently and uniformly on the real segment $(0,y_k)$.

In class (i) Pielou (1959) suggested considering the ratio of $\hat{\lambda}_1 = z/A$, where z is the number of plants found in a set of quadrats of total area A, and $\hat{\lambda}_2$ defined by (5) for k=1. It was shown by Mountford (1961) that

$$E(\hat{\lambda}_1/\hat{\lambda}_2) = 1,\ \mathrm{var}(\hat{\lambda}_1/\hat{\lambda}_2) = 1/n + 1/(\lambda A) + 1/(nA\lambda).$$

As $A \to \infty, \hat{\lambda}_1$ converges to λ, so that for fixed $n, \hat{\lambda}_1/\hat{\lambda}_2$ is proportional to a chi squared variable with 2n degrees of freedom. As $n \to \infty$ the standardized version

$$n^{\frac{1}{2}}\ (\hat{\lambda}_1/\hat{\lambda}_2 - 1)$$

tends to have a normal distribution with mean 0 and variance 1.

The phenomenon mentioned in (ii) was exploited by Hopkins (1954) who considered the ratio $\hat{\lambda}_2/\hat{\lambda}_3$ where λ_2 is based on measurements from n random points, and $\hat{\lambda}_3$ is a statistic of the same type, but based on measurements from m random plants. If the points do arise from a Poisson process, this statistic will have an F distribution with (2m, 2n) degrees of freedom.

I have investigated the powers of these two tests against two families of alternatives (Holgate, 1965a), taking the special version of Pielou's statistic which arises as $A \to \infty$, when it coincides with a statistic discussed by Skellam (1952) and Moore (1954). The first alternative was that plants occur in tight clusters, the clusters themselves being a realization of a Poisson process. The number of plants per cluster has a distribution which is "one plus a Poisson variable". The power curves of both tests have the usual ogive shape. Hopkins' test appears to be more powerful than that of Skellam and Moore, which agrees with intuition, since the effect of clustering on Hopkins' statistic is to increase the sum of squared distances based on point to plant measurements, and reduce that based on plant to plant measurements, with the quadrat estimator which appears in Pielou's statistic retaining the same expectation.

The other class of alternatives considered was a generalization of the square lattice distribution, in which each lattice point contained a cluster having random size "one plus a Poisson variable". Against this alternative both tests gave rise to U-shaped power curves, with certain parameter values leading to values of the test statistic which would make rejection of the Poisson null hypothesis almost impossible, even though it is untrue. This is related to the phenomenon mentioned in Section 1, and will be discussed further in the next section.

In a subsequent paper (Holgate, 1965b), I introduced two further tests which depended on measurements from sample points to the ith and j^{th} nearest plants $(i < j)$. In the Poisson field, the correlation between x_i^2 and x_j^2 is $(i/j)^{\frac{1}{2}}$, while the expected value of the ratio x_i^2/x_j^2 is i/j. The sample values of each of these quantities provide statistics by which randomness may be tested. Clustering tends to increase both the correlation coefficient and the ratio, while a tendency to regular spacing reduces them for $i = 1$, $j = 2$. The paper referred to contains discussion of the power of these tests against clustered and regular alternatives. The conclusion was that the correlation test was likely to be most powerful against clustering, and the ratio test against regular spacing.

The correlation statistic is of little practical value, since the sample would have to be extremely large to secure approximate normality in its distribution, which is not known for moderate fixed values of n. The ratio test can be generalized by noting that as a consequence of (iii), any test for the randomness of a number of points on a segment provides a test of the Poisson property of a spatial point process. The essential difference is that here there are n replicates of the experiment.

6. ALTERNATIVE DISTRIBUTIONS

Future research on nearest neighbor methods needs to be concerned with the situation for point processes other than the Poisson or completely regular. Warren (1962, 1971) has obtained interesting results for what he calls the "centre-satellite" process. In this, 'centres' occur as the points of a Poisson process, and the plants are then distributed on the circumference of a circle centered on the 'centre'. In work which I am currently carrying out with Miss Susannah Brown of the London School of Economics, a model of a 'thinned plantation' is being studied. In this, trees occur first at the points of a lattice, and are then allowed to survive or die with probabilities p, 1 - p. A study of the powers of tests will shortly be submitted for publication. A model of this type was studied using simulation techniques by Persson (1964).

Models involving a Poisson field of tight clusters are relatively easy to deal with since distributions of the distance to the jth tree are obtained as mixtures of simple families of gamma distributions. More precisely if p_{js} is the probability that the jth nearest plant lies in the sth nearest cluster, a quantity deducible from renewal theoretic type arguments, then y_i has probability density function

$$e^{-\lambda y} \sum_s p_{js} \lambda^s y^{s-1}/\Gamma(s).$$

If $p_{ij,st}$ is the probability that the ith and j^{th} nearest plants belong respectively to the sth and t^{th} nearest clusters, the ratio y_i/y_j has probability density function

$$\sum_{s,t} P_{ij,st} z^{s-1}(1-z)^{t-s-1}/B(s,t).$$

Unfortunately the calculation of $P_{ij,st}$ although simple in concept is cumbersome.

Some other relevant references on distance methods are given in the next section along with the cited works.

REFERENCES

Bauersachs, E. (1942). Bestandesmassenaufnahme nach dem Mittelstammverfahren des zweitkleinsten Stammabstandes. Forstwiss. Cbl. 64, 182-6.

Blackith, R. E. (1958). Nearest neighbour distance measurements for the estimation of animal populations. Ecology 39, 147-50.

Catana, A. J. (1963). The wandering quarter method of estimating population density. Ecology 44, 349-60.

Clark, P. J. & Evans, F. C. (1954). Distance to nearest neighbour as a measure of spatial relationships in populations. Ecology 35, 445-53.

Clark, P. J. & Evans, F. C. (1955). On some aspects of spatial patterns in biological populations. Science 121, 397-8.

Craig, C. E. (1953). On a method of estimating biological populations in the field. Biometrika 40,216-8.

Cottam, G. (1947). A point method for making rapid surveys of woodlands. Bull. Ecol. Soc. America 28, 60.

Cottam, G. & Curtis, J. T. (1949). A method for making rapid surveys of woodlands by means of pairs of randomly selected trees. Ecology 30, 101-4.

Cottam, G., Curtis, J. T., & Hale, B. W. (1953). Some sampling characteristics of a population of randomly dispersed individuals. Ecology 34, 741-57.

Cottam, G. & Curtis, J. T. (1955). Correction for various exclusion angles in the random pairs method. Ecology 36, 767.

Cottam, G. & Curtis, J. T. (1956). The use of distance measures in phytosociological sampling. Ecology 37, 451-60

Dacey M. F. (1960). A note on the derivation of nearest neighbour distances. Journ. Reg. Sci. 2, 81-7.

Dacey, M. F. (1962). Analysis of central place and point patterns by a nearest neighbour method. Lund Studies in Geog. B 24, 55-75.

Dacey, M. F. (1963). Order neighbour statistics for a class of random patterns in multidimensional space. Ann. Assn. Am. Geog. 53, 505-15.

Dacey, M. F. (1965). Order distance in an inhomogeneous random point pattern. Canad. Geographer 9, 144-53.

Deltheil, R. (1926). *Probabilitiés Géometriques*. Paris: Gauthier-Villars.

Dice, L. R. (1952). Measure of the spacing between individuals within a population. Contr. Lab. Vert. Biol. No. 55, Univ. of Michigan.

Eberhardt, L. L. (1967). Some developments in 'Distance Sampling'. Biometrics 23, 207-16.

Essed, F. E. (1954). A quick, simple and at the same time accurate method for estimating the total volume and the increment percent of evenaged stands. Indian For. 82, 260-3.

Essed, F. E. (1957). Estimation of standing timber. Meded. Landbouwhogesch. Wageningen 57 (5), 1-60.

Ghosh, B. (1951). Random distances within a rectangle, and between two rectangles. Bull. Calcutta Math. Soc. 43, 17-24.

Goodall, D. W. (1952). Quantitative aspects of plant distribution. Biol. Rev. 27, 194-245.

Greig-Smith, P. (1964). *Quantitative Plant Ecology (2nd Ed.)* Butterworth. London.

Hammersley, J. M. (1950). The distribution of distance in a hypersphere. Ann. Math. Statist. 21, 447-52.

Hammersley, J. M. (1951). A theorem on multiple integrals. Proc. Camb. Phil. Soc. 47, 274-8.

Hammersley, J. M. (1951). On a certain type of integral associated with circular cylinders. Proc. Roy. Soc. A 210, 98-110.

Hammersley, J. M. (1952). Lagrangian integration coefficients for distance functions taken over right circular cylinders. J. Math. Phys. 31, 139-50.

Hausburg, H. (1962). Stammabstandsverfahren. Allg. Forst. und Jagd-Zeitung 133 19-27.

Hertz, P. (1909). Über die gegenseitigen Durchshnittlichen Abstand von Punkten, die mit bekannter mittlerer Dichte im Raum angeordnet sind. Math. Ann. 67, 387-98.

Holgate, P. (1964). The efficiency of nearest neighbour estimates. Biometrics 20, 647-9.

Holgate, P. (1965). Tests of randomness based on distance methods. Biometrika 52, 345-53.

Holgate. P. (1965). Some new tests of randomness. J. Ecol. 53, 261-6.

Holgate, P. (1965). The distance from a random point to the nearest point of a closely packed lattice. Biometrika 52, 261-3.

Hopkins, B. (1954). A new method of determining the type of distribution of plant individuals. Ann. Bot. 18. 213-26.

Kendall, M. G. & Moran, P. A. P. (1963). *Geometrical probability.* Griffin: London.

Keuls, M., Over, H. J. & DeWit, C. T. (1963). The distance method of estimating densities. Statist. Neerlandica 17, 71-91.

Kurth, H. & Asmus, F. (1967). Kurzdarstellung der Abstandmethoden in der Forstwirschaft und Ergebnisse von Modelluntersuchungen an einem Kiefernbestand. Arch. Forstu. 16, 873-877.

Leibherr, H. G. (1952). A comparison of rapid botanical sampling methods on an artificial plant population showing various degrees of contagion. Unpublished M.A. thesis, Univ. of Wisconsin.

Lord, R. D. (1954). The distribution of distance in a hypersphere. Ann. Math. Statist. 25, 794-8.

Matérn, B. (1959). Några tillämpringar av teorin for geometriska sannolikheter. Svenska skogsvårdsföreningens tidskr. 3, 453-8.

Matérn, B. (1971). Doubly stochastic Poisson processes in the plane. In *Statistical Ecology,* ed. G. P. Patil and others, vol. 1, 195-213.

Mawson, J. C. (1968). A Monte Carlo study of distance measures in sampling for spatial distribution in forest stands. For. Sci. 14, 127-39.

Moore, P. G. (1954). Spacing in plant populations. Ecology 35, 222-7.

Morisita, M. (1954). Estimation of density by spacing method. Mem. Fac. Sci. Kyushu Univ. El, 187-97.

Morista, M. (1957). A new method for the estimation of density by the spacing method applicable to non randomly distributed populations. (in Japanese). Physiology and Ecology 7, 134-44 USDA Forest Service Translation No. 11116.

Mountford, M. D. (1961). On E. C. Pielou's Index of Non Randomness. J. Ecol. 49, 271-5.

Persson, O. (1964). *Distance methods.* Studia Forestalia Suecica, nr. 15 68 pp.

Persson, O. (1965). Distance methods II. Forest Research Inst. of Sweden, Dept. of Forest Biometry, Research Report 6, 23 pp.

Persson, O. (1971). The robustness of estimating density by distance measurements. In *Statistical Ecology*, ed. G. P. Patil and others, vol. 2, 175-90.

Pielou, E. C. (1959). The use of point to plant distances in the study of the pattern of plant populations. J. Ecol. 47, 607-13.

Pielou, E. C. (1960). A single mechanism to account for regular random and aggregated populations. J. Ecol. 48, 575-84.

Pielou, E. C. (1961). Segregation and symmetry in two species populations as studied by nearest neighbour relationships. J. Ecol. 49, 255-69.

Pielou, E. C. (1962). The use of plant-to-neighbour distances for the detection of competition. J. Ecol. 50, 357-67.

Rice, F. L. & Penfound, W. T. (1955). An evaluation of the variable radius and paired-tree methods in the blackjack-post oak forest. Ecology 36, 315-20.

Shanks, R. E. (1954). Plotless sampling trials in appalachian forest types. Ecology 35, 237-44.

Skellam, J. G. (1952). Studies in statistical ecology. I. Spatial pattern. Biometrika 39, 346-362.

Stoffels, A. (1955). Die Genauigkeit der Bestimmung der Stammzahl pro Hektar durch Messung von Stammabstanden. Forstwiss. Cbl. 74, 211-8.

Strand, L. (1953). Mål for fϕrdelingen av individer over et område. Medd fra det Norske skogsfϕrsϕks - vesen 12, 195-207.

Thompson, H. R. (1956). Distribution of distance to nth neighbour in a population of randomly distributed individuals. Ecology 37, 391-4.

Viktorov, S. V. (1947). A study of the distribution and dispersion of plants by aerial photographs. Bull. Soc. Nat Moscou (sect. biol) N.S. 52, 71-8.

Warren, W. G. (1962). Contributions to the study of spatial point processes. Unpublished Ph.D. thesis, Univ. of N. Carolina.

Warren, W. G. (1971). The center-satellite concept as a basis for ecological sampling. In *Statistical Ecology*, ed. G. P. Patil and others, vol. 2, 87-118.

SOME APPLICATIONS OF MULTIVARIATE POINT PROCESSES

M. S. Bartlett
Department of Biomathematics, University of Oxford

SUMMARY

The statistical analysis of multivariate point processes is illustrated by two examples. The first is similar to a previous spectral analysis of vehicle times at a fixed point, with associated vehicle speeds, but refers to three-lane one-way traffic. The spectral analysis is broadly consistent, in contrast with clumping effects observed under other road conditions, with spacing effects between vehicles. Further details of the results and possible models are discussed. The second type of application occurs in the area of plant ecology and is illustrated by the nested type of analysis of variance and covariance of counts of two or more species.

1. INTRODUCTORY REMARKS

A general theoretical survey of point processes was given by me many years ago (Bartlett, 1954), but subsequent interest by various authors (see, for example, Cox and Lewis, 1966) in the statistical analysis of data of point process type has included multi-dimensional and multivariate applications which merit specific discussion. My own work (1963) on the spectral analysis of point processes was extended in my (1964) Biometrika paper to *two-dimensional* point processes, and, more incidentally, in my discussion of the analysis of traffic data in a review of time-series analysis (1967b), to *multivariate* point processes.

A detailed theoretical discussion of multivariate point processes has recently been given by Cox and Lewis (1972), and they mention several possible applications including physiological, geophysical, industrial, and to traffic studies. My own purpose in this paper will therefore be to discuss one or two specific applications; and I shall start with an analysis of traffic data which was made in 1967 but not yet recorded in print.

2. SPECTRAL ANALYSIS OF TRAFFIC DATA

The joint spectral analysis of the point process dN(t) defined by the times vehicles pass a fixed point on a road, and the associated process $U_t dN(t)$, where U_t is the velocity of the vehicle passing at time t, is in effect a bivariate analysis of these two processes (cf. Cox and Lewis, 1972); but the data in question were classifiable as a multivariate point process in a more immediate sense, as they consisted of times vehicles were passing a fixed point on a road with three lanes. Indeed, as velocities were also available, the data were classified by the six processes:

$$dN_i(t), \quad U(t)dN_i(t), \qquad (i=1,2,3).$$

It has been pointed out in my previous papers that the interpretation of the spectral analysis of traffic data may be carried out in relation to a model of more than one type. For example, the joint $\{dN(t), U_t dN(t)\}$ analysis has previously been considered in relation to (i) a specific queueing-type model (1967c) (ii) a general speed-density law (1967b). The same possibilities exist for multi-lane data, and the linear models given in the Appendix may be postulated as rather simpler exploratory relations than those used in my (1967b) Biometrika paper. However, it became clear that the multi-lane data available for analysis, referring to traffic on a three-lane one-way highway out of New Haven, Connecticut, would be more realistically related to spacing or 'headway' models than to macroscopic 'densities', so that the regression models in the Appendix (I) were not in fact made use of, and are merely noted for reference.

Any 'analysis of variance'-type study of a *multivariate* stationary process $\underline{X}(t)$ can, as I have previously pointed out, be conveniently discussed in terms of the analysis of a *univariate* process $\alpha'X(t)$. For example,

$$[\alpha_1 X_1(t)+\alpha_2 X_2(t)][\alpha_1 X_1(t+\tau)+\alpha_2 X_2(t+\tau)]^*$$

$$= \alpha_1\alpha_1^* X_1(t)X_1^*(t+\tau)+\alpha_1\alpha_2^* X_1(t)X_2^*(t+\tau)$$

$$+ \alpha_2\alpha_1^* X_1^*(t+\tau)X_2(t)+\alpha_2\alpha_2^* X_2(t)X_2^*(t+\tau)$$

Of course, the number of separated terms rapidly becomes large, and for six variables is 36. Some selection is almost inevitable, and the following reduction, at least in this preliminary investigation, was made. Firstly, for the three Lanes 1, 2 and 3, the number of vehicles on the 'slowest' Lane 1 was 448 in a period of about 50 minutes (a mean time interval between vehicles of about 6.7 secs, variance 31.2) and a (time) mean speed of 48.8 m.p.h. In the next lane the number of vehicles was about the same (434, with a mean time interval of 6.8, variance 36.1) and with a mean speed of 57.7 m.p.h. On the fastest lane there were only 130 vehicles with a mean speed of 62.2 m.p.h., so that provided the effect of this Lane was not forgotten, it seemed reasonable to study the first two Lanes 1 and 2,reducing the number of terms to 16.

Next, for the properties of Lanes 1 and 2 separately it was recalled that in the previous spectral analysis of speed-density traffic relations (Bartlett, 1967b) a significant cross-spectral asymmetric component (Q) had been obtained between $dN(t)$ and $U_t dN(t)$, as well as a significant symmetric component (P), so that it seemed advisable to calculate both components. These together with spectral estimates (I) for $dN(t)$ and $\Delta U_t dN_t$ are shown in Table 1 (for Lane 1) and Table 2 (for Lane 2). (For computing convenience, the estimates for $dN(t)$ are linearly multiplied by the mean speed $\bar{U}$, or quadratically multiplied by $\bar{U}^2$.) These two Tables thus record 8 terms.

For the interrelation of Lanes 1 and 2, provisional consideration was given to the *symmetric* cross-spectral terms, yielding 4 terms, or 12 in all out of the total 36. These further P terms are given in Table 3 (in this case no factor $\bar{U}$ has been introduced).

Table 1. *Spectral and Cross-Spectral Estimates* (sums of 32 terms). $N_1 = 448$, $\bar{U} = 48.83036$ m.p.h.. Expectation of $I(\bar{U})$ (for high frequency) is $64\bar{U}^2 = 152.600$. Frequency steps ($\omega_p = 2\pi p/N_1$) range from 0 to 2π.

Lane 1

p	$I(\bar{U})$	$I(\Delta U)$	$P(\bar{U},\Delta U)$	$Q(\bar{U},\Delta U)$
1-	108640	5197.8	-3101.4	905.5
33-	135106	4867.4	-4636.6	3387.2
65-	148308	3604.9	-5210.7	2373.9
97-	145428	4515.6	-3354.3	2528.3
129-	117007	3978.9	-4012.5	-998.4
161-	142304	3098.9	-5424.9	1167.3
193-	118288	2692.2	-3018.1	-2601.5
225-	142518	2433.5	1269.2	3043.1
257-	117237	2043.8	58.7	1532.1
289-	129791	1939.9	1423.0	626.7
321-	72326	2284.3	-3130.9	326.3
353-	125623	1659.8	1875.9	1521.3
385-	125676	2656.8	5338.4	3864.0
417-	136920	3066.1	-1148.4	397.8

When the spectral estimates $I(\bar{U})$ of Tables 1 and 2 are compared with their expectation on the null hypothesis of a Poisson stream, the systematic negative discrepancies suggest a spacing effect, in contrast with a positive discrepancy near the origin in examples analyzed previously, associated with heterogeneity or clumping. Both heterogeneity and spacing effects may of course be needed to account for the observed results.

It has been remarked more than once in relation to previous spectral analysis of this kind that it should not become an obsession. It has proved to be a useful technique in situations where clumping or clustering models are relevant, sometimes to an extent where an alternative

Table 2. *Spectral and Cross-Spectral Estimates* (sums of 32 terms). $N_2 = 434$, $\bar{U} = 57.66751$ m.p.h. Expectation of $I(\bar{U})$ (for high frequency) is $64\bar{U}^{-2}\frac{434}{448} = 206{,}197$. Frequency steps and range as in Table 1.

Lane 2

p	$I(\bar{U})$	$I(\Delta U)$	$P(\bar{U},\Delta U)$	$Q(\bar{U},\Delta U)$
1-	125460	2131.0	-1181.3	-433.1
33-	178043	1575.3	-2583.5	-327.9
65-	194683	1667.3	-4447.7	2857.6
97-	104745	1501.6	1751.7	-2106.9
129-	148856	1682.2	408.8	1958.3
161-	202053	1573.9	-896.4	-1684.0
193-	153015	1556.0	-868.7	669.8
225-	148816	1155.8	-122.1	-1583.5
257-	154920	1407.8	1890.8	2582.9
289-	144266	1578.5	-2046.3	149.5
321-	202567	1275.1	-2413.4	-2145.7
353-	237713	1738.6	2045.9	120.1
385-	136321	1184.5	-69.4	-444.2
417-	200111	1296.7	-637.2	821.3

analysis of comparable power and precision is difficult to find. In the present problem it has also seemed useful in appraising the type of data under analysis, but once that it had emerged that spacing or headway models were likely to be relevant it seemed more logical to look at these directly, and this has been undertaken by G. Wright. Such more direct analyses might be expected to suggest the finer details of any quantitative models fitted to the data. Nevertheless, any such models should of course correspond to theoretical spectra consistent with the results of the spectral analysis, so that a start has also been made with deriving the spectra of simple spacing models.

Thus the P and Q components between $dN(t)$ and $U_t dN(t)$, given in Tables 1 and 2 for Lanes 1 and 2 respectively, suggest a negative dip

Table 3. *Cross-Spectral Estimates* (Lane 1 and Lane 2).

$N_1 = 448$, $N_2 = 434$. Time scale chosen so that estimated mean intensity $\lambda_1 = 1$ (and hence estimated mean intensity $\lambda_2 = 434/448$). $\bar{U}_1 = 48.83$ m.p.h. (variance 42.04). $\bar{U}_2 = 57.67$ m.p.h. (variance 21.55).

Lanes 1 and 2

p	2P(1,2)	$2P(\Delta U_1, \Delta U_2)$	$2P(1, \Delta U_2)$	$2P(2.\Delta U_1)$
1-	17.055	-1228.9	-360.44	103.71
33-	27.108	-1386.5	-337.81	243.25
65-	34.194	-2161.6	-440.18	208.28
97-	5.646	-502.1	-101.20	22.87
129-	27.980	-2559.5	-381.46	197.06
161-	27.462	-2134.2	-380.29	168.84
193-	8.854	-1828.3	-307.48	-0.30
225-	-7.622	514.2	-3.14	-81.55
257-	14.629	-2121.2	-167.47	249.80
289-	19.705	-1974.1	-188.12	257.63
321-	17.042	-3025.6	-299.73	195.21
353-	-5.290	245.4	30.02	-54.83
385-	33.287	-2982.0	-277.89	331.54
417-	12.130	-1545.1	-40.95	229.30

in the P components, more clearly for Lane 1, near the origin. In Appendix II a simple relation between U_t and the 'time headway' I(t) is shown to lead to such a result, taking a renewal process as a simple (too simple!) model for the point process dN(t).

The same model with the incorporation of an effect from the second lane gives a similar effect for the cross-spectral component between $U_t dN_1(t)$ for one lane, and $dN_2(t)$ for the second; but the theory is very provisional, as it is based on the (dubious) neglect of interdependence between the two processes $dN_1(t)$ and $dN_2(t)$.

Looking at the actual results in Table 3, we infer that β_{12} in our model (as defined in AII) is positive, but β_{21} appears if anything negative. However, the assumption of independence of $N_1(t)$ and $N_2(t)$ seems contradicted by P(1,2), which is predominantly positive; this is in contrast with $P(\Delta U_1, \Delta U_2)$, which appears negative.

3. PLANT COUNTS WITH MORE THAN ONE SPECIES

The second type of application I would like to refer to, at least briefly, refers to the multivariate extension of multi-dimensional point processes, and in particular to the possible use of techniques discussed in my (1964) Biometrika paper for ecological (or other) applications involving more than one type of plant (or individual) distributed over a two-dimensional area.

In view of the complexity which we have seen arises in the case of several variables, let me discuss this for simplicity in the case of two types, so that our point process is $\{dN_1(\underline{r}), dN_2(\underline{r}\}$. Then, as already recalled, any quadratic-type analysis of such a process is equivalent to the quadratic analysis of

$$\alpha' dN(\underline{r}) = \alpha_1 dN_1(\underline{r}) + \alpha_2 dN_2(\underline{r}),$$

and any complete analysis of $dN_1(\underline{r})$, $dN_2(r)$ would automatically include the covariance of $dN_1(\underline{r})$ and $dN_2(\underline{r})$.

Such analyses include not only spectral analysis, but any 'analysis-of-variance'-type analysis, including the nested type used by Greig-Smith. This latter type of analysis has been criticized, in my opinion rather unfairly, by Pielou (1969) in her book on Mathematical Ecology, on the grounds that any 'squared' analysis of 0,1 variables cannot discriminate between presence and absence. She gives two (fictitious) diagrams, one showing large uninhabited areas for a single species, the other where the species occupies just those gaps, and she (correctly) points out that both situations would yield a similar analysis. Such criticism would, however, apply to any analysis of variance of any type of variable, and seems to be easily eliminated in any real example. At least two possibilities might arise. In one the areas uninhabited by one species might be *completely* uninhabited, and a physical explanation would be needed. If the areas were rock or swamp, then a *multivariate* analysis including a variable to describe the terrain would be useful. In the second possibility these areas might be inhabited as in the second diagram, but with other (for simplicity say one other) species. Thus the complete picture would be given by

$$N(t) = N_1(t) + N_2(t)$$

where N_1 and N_2 relate to two different species. Suppose, again for simplicity, that N(t) is random, so that in the quadratic analysis

$$N^2(t) = N_1^2(t) + 2N_1(t)N_2(t) + N_2^2(t)$$

no differentiation of note appears in the separation of terms on the left-hand side for both species together. Two possibilities now arise on the right-hand-side. If $N_1(t)$ and $N_2(t)$ are independent, no term $N_1(t)N_2(t)$ arises on average, so that the analysis of $N_2(t)$ would (contrary to what was claimed above) have to lead to negative terms where $N_1(t)$ gave positive terms. In fact for the situation where $N_1(t)$ and $N_2(t)$ are envisaged as the points corresponding to Pielou's two separate diagrams, we shall have the same quadratic analysis for $N_1(t)$ and $N_2(t)$, and hence this must be compensated by the perfect negative covariance term arising from $2N_1(t)N_2(t)$, which would be at once revealed by the full bivariate analysis.

We may illustrate these remarks by analyzing the data for two species represented by two Figures (23a and 23b) given by Pielou in her book (Pielou, 1969). The data are given as accurately as possible as counts in a grid 6 x 8, where x denotes the first species and y the second (Tables 4 and 5).

The first of these examples represents the situation suggested above where a large negative correlation exists between x and y, the second has no such effect but a broad common clumping effect. The analysis of variance and covariance for x and y, with convenient aggregation of the counts of Tables 4 and 5 into square sets of four, and then into rows and columns, gave the results shown in Tables 6 and 7 respectively. It will be seen that there is consistent negative covariance in Table 6,

Table 4. *Counts in a 6 x 8 grid for two species (x and y)* (from Pielou's Fig. 23a).

x,y	x,y	x,y	x,y	x,y	x,y	x,y	x,y
3,0	1,1	0,1	0,2	1,0	3,0	5,0	1,0
3,0	2,2	1,1	0,4	5,0	3,0	0,0	1,6
3,0	0,0	2,0	2,0	1,0	4,1	0,2	0,1
2,0	0,1	2,2	3,0	1,0	2,1	0,3	0,2
1,0	3,2	0,4	3,0	3,0	1,0	1,0	1,0
1,0	3,0	0,0	5,0	3,0	3,0	4,0	2,0

Table 5. *Second example of counts in a 6 x 8 grid* (from Pielou's Fig. 23b).

x,y	x,y	x,y	x,y	x,y	x,y	x,y	x,y
-	-	3,1	4,2	3,2	4,1	-	-
-	-	2,1	3,2	2,2	1,0	-	-
-	-	-	2,0	1,1	-	-	-
-	1,0	-	-	-	2,0	2,2	3,2
1,1	6,3	1,0	-	-	3,1	2,2	3,1
0,1	1,1	-	-	1,0	2,2	3,0	2,2

Table 6. *Analysis of variance and covariance of data in Table 4.*

	Sx^2	Sxy	Sy^2	$S(x+y)^2$	D.F.
Within 4's	70.7	+0.5	30.5	102.2	36
Rows	4.6	-2.35	3.9	3.8	2
Columns	10.6	-9.25	9.0	1.1	3
Remainder	18.6	-24.65	37.6	6.9	6
Total	104.5	-35.75	81.0	114.0	47

Mean x: 1.77 Mean y: 0.75 Mean (x+y): 2.52

Table 7. *Analysis of variance and covariance of data in Table 5.*

	Sx^2	Sxy	Sy^2	$S(x+y)^2$	D.F.
Within 4's	49.0	+16.75	17.0	99.5	36
Rows	6.8	+4.2	2.6	17.8	2
Columns	4.2	+1.3	0.7	7.5	3
Remainder	39.9	+23.55	14.9	101.9	6
Total	99.9	+45.8	35.2	226.7	47

Mean x 1.21 Mean y 0.625 Mean (x+y): 1.83

apart from the item 'within 4's'. In Table 7 the covariance is positive and a highly significant clumping effect is apparent from the size of the 'remainder' term for x + y.

In my previous paper discussing the spectral analysis of two-dimensional point processes, it was noted that analyses of the above type were convenient partial analyses of the detailed spatial record when available; for example, the extension of a univariate spectral analysis would be a bivariate (or multivariate) spectral analysis in two (or more) dimensions.

ACKNOWLEDGMENTS

I am greatly indebted to D. Walley for programming the spectral analyses, which were carried out while I was still at University College, London; I wish to thank also G. Wright for some preliminary results I have made use of from his more direct analyses of the traffic data. The latter were collected by M. J. Huber, to whom and to H. L. Seal, who arranged their availability to me, I am consequently also indebted. (For reasons associated with my move to Oxford in 1967, a more exhaustive analysis and interpretation of the traffic data was interrupted; but I still hope that G. Wright will report his own results for these data at some stage.)

REFERENCES

Bartlett, M. S. (1954). Processus stochastiques ponctuels. Ann. Inst. H. Poincaré, 14, Fasc. 1, 35-60.

Bartlett, M. S. (1963). The spectral analysis of point processes. J. R. Statist. Soc. B 25, 264-96.

Bartlett, M. S. (1964). The spectral analysis of two-dimensional point processes. Biometrika 51, 299-311.

Bartlett, M. S. (1967a). Fine structure of traffic flow indicated by spectral analysis (unpublished paper given at Symposium on Aspects of Road Traffic, organized by Institute of Mathematics and its Applications, January 3, University College, London).

Bartlett, M. S. (1967b). Some remarks on the analysis of time-series. Biometrika 54, 25-38.

Bartlett, M. S. (1967c). The spectral analysis of line processes. *Proc. 5th Berkeley Symposium on Mathematical Statistics and Probability*, 3, 133-54.

Cox, D. R. and Lewis, P. A. W. (1966). *The Statistical Analysis of Series of Events*. Methuen: London and Barnes & Noble: New York.

Cox, D. R. and Lewis, P. A. W. (1972). Multivariate point processes. To appear in *Proc. 6th Berkeley Symposium on Mathematical Statistics and Probability*.

Greig-Smith, P. (1957). *Quantitative Plant Ecology*. Butterworth: London.

Pielou, E. C. (1969). *An Introduction to Mathematical Ecology*. Wiley: New York.

APPENDIX

AI. Simplified Regression Models Between Velocity and Density

a) *Single lane*. As noted in Section II of the paper, when the relations assumed are linear and exploratory, it seems justifiable to make them as simple as possible. The linear relation

$$E[\Delta U_t | \{\Lambda(t)\}] = \alpha + \beta\Lambda(t),$$

where ΔU_t denotes the difference of U_t from its (time) average, assumed in an earlier paper (1967b), leads to a quadratic relation between $\Delta U_t dN(t)$ and $dN(t)$, and might be adequately replaced by the linear relation

$$E[\Delta U_t dN(t) \,|\, \{\Lambda(t)\}] = \lambda\{\eta_0+\eta_1\Lambda(t)\}dt,$$

where

$$E\{\Delta U_t dN(t)\} = \lambda(\eta_0+\eta_1\lambda) = 0 \quad \text{if} \quad \eta_0 = -\lambda\eta_1,$$

and

$$E\{\Delta U_t dN(t)\cdot dN(t+\tau)\} = E[\lambda\{\eta_0+\eta_1\Lambda(t)\}\Lambda(t+\tau)]dtd\tau$$

$$= \lambda\{\eta_0\lambda + \eta_1(\lambda^2+\mu_{11})\}dtd\tau$$

$$= \lambda\eta_1\mu_{11}dtd\tau,$$

$$E\{\Delta U_t dN(t)\cdot\Delta U_{t+\tau}dN(t+\tau)\} = \lambda^2\eta_1^2\mu_{11}dtd\tau.$$

b) *Two lanes with speed interaction with both densities.* Similarly for two lanes it might be assumed that

$$E[\Delta_1 U_t dN_1(t) \,|\, \{\Lambda_1(t),\Lambda_2(t)\}]$$

$$= \lambda_1\{\eta_0+\eta_{11}\Lambda_1(t) + \eta_{12}\Lambda_2(t)\}dt,$$

where $\Delta_1 U_t$ refers to $dN_1(t)$. Now

$$E\{\Delta_1 U_t dN_1(t)\} = 0 \quad \text{if} \quad \eta_0 + \eta_{11}\lambda_1 + \eta_{12}\lambda_2 = 0.$$

Then

$$E\{\Delta_1 U_t dN_1(t)\cdot dN_1(t+\tau)\} = \lambda_1\{\lambda_1\eta_0+\eta_{11}(\lambda_1^2+\mu_{11}) + \eta_{12}(\lambda_1\lambda_2+\mu_{12})\}dtd$$

$$= \lambda_1(\eta_{11}\mu_{11}+\eta_{12}\mu_{12})\ dtd\tau,$$

$$E\{\Delta_1 U_t dN_1(t)\cdot dN_2(t+\tau)\} = \lambda_1(\eta_{11}\mu_{12}+\eta_{12}\mu_{22})dtd\tau,$$

$$E\{\Delta_1 U_t dN_1(t)\cdot\Delta_1 U_{t+\tau}dN_1(t+\tau)\} = \lambda_1^2\{\eta_{11}^2\mu_{11}+\eta_{11}\eta_{12}(\mu_{12}+\mu_{21})+\eta_{12}^2\mu_{22})dtd\tau,$$

$$E\{\Delta_1 U_t dN_1(t)\cdot\Delta_2 U_{t+\tau}dN_2(t+\tau)\} = \lambda_1\lambda_2(\eta_{11}\eta_{21}\mu_{11}+\eta_{12}\eta_{21}\mu_{21}$$

$$+ \eta_{12}\eta_{22}\mu_{22}+\eta_{11}\eta_{22}\mu_{12})dtd\tau,$$

etc.

AII. Cross-Spectrum Between $\Delta U_t dN_t$ and dN_t in Relation to 'Headway'

a) *Single lane.* Suppose we assume an approximate linear relation

$$E_U\{\Delta U_t\} = \beta\{I(t)-m\},$$

where I(t) is the 'time headway' in appropriate units (if its mean m in seconds = $1/\lambda$, then choose $\lambda \approx 1$ in new units, and the coefficient β must be multiplied by λ to convert back to m.p.h.-sec. units). As noted in the main text, whether the above relation holds may be investigated directly, but here we consider the effect of such a relation on the $\Delta U_t dN_t, dN_t$ cross-spectrum. This has to be considered with some further model for dN_t, and it is derived below in the case of dN_t a renewal process.

We have, for t reckoned negatively so that $I_r = t_r - t_{r-1}$,

$$J = (2/N)^{\frac{1}{2}}\sum_r e^{it_r\omega}, \quad K=(2/N)^{\frac{1}{2}}\sum_r e^{it_r\omega}\Delta U_r,$$

$$E_U\{JK^*\} = (2/N)\sum_r \sum_s e^{i(t_r-t_s)\omega}\beta[t_s-t_{s-1})-m]$$

$$\approx 2\beta E_I[\sum_r e^{-iI_r\omega}(I_r-m)\{1+e^{-iI_{r-1}\omega}+e^{-i(I_{r-1}+I_{r-2})\omega}+\ldots\}]$$

$$\approx 2\beta\left[\frac{1}{1-L(i\omega)}\left(\frac{\partial}{\partial(-i\omega)}-m\right)L(i\omega)\right], \quad (\lambda-1),$$

where $E_I(e^{-I\psi}) = L(\psi)$. For low frequency ($\omega$ small), this becomes

$$2\beta\left[\frac{1}{mi\omega-\frac{1}{2}\mu_2'\omega^2\ldots}\left((m-\mu_2'(i\omega)-m+m^2(i\omega)\ldots\right)\right]$$

$$\to -2\beta\sigma_1^2/m \text{ as } \omega \to 0.$$

(As this is real, it will appear in the P, and not the Q, component.)

For Lane 1, it is estimated, for a value of P_0 of about -4000 on the scale of Table 1, that $\beta \approx 0.14$ m.p.h.-sec.

For Lane 2, with $P_0 \approx -1500$, $\beta \approx 0.038$ m.p.h.-sec.

These values may be compared with *direct* regression estimates (calculated by G. Wright) of 0.11 and 0.014 respectively.

b) *Two lanes with speed interaction.* If we extend the above linear relation to two lanes, it would be more realistic to do this for 'space headway', but for simplicity the relation is still assumed for 'time headways' in the form

$$E_U(\Delta U_1) = \beta_{11}\{I_1(t)-m_1\} + \beta_{12}\{I_{12}(t)-m_{12}\},$$

$$E_U(\Delta U_2) = \beta_{21}\{I_{21}(t)-m_{21}\} + \beta_{22}\{I_2(t)-m_2\},$$

where U_2 and I_2 (or I') relate to the second lane, and, if $I_{12} = t_r - t_r'$, $t_r' < t_r$ when t is reckoned negatively. Notice that if $dN_1(t)$, $dN_2(t)$ are *independent* renewal processes, the interval I_{12} is the interval for the renewal process $dN_1(t)$ from an *arbitrary* time-point, so that the renewal distribution f(u), say, is replaced by $\{1-F(u)\}/m$, with Laplace transform $[1-L(\psi)]/(m\psi)$. Hence for the quantity

$$E(J_2K_1^*) = (\frac{2}{N})E[\sum_r \sum_s e^{i(t_r'-t_s)\omega}\{\beta_{11}(t_s-t_{s-1}-m_1)+\beta_{12}(t_s'-t_s-m_{12})\}],$$

we write

$$t_s'-t_s = -[(t_s-t_s')+(t_s'-t_{s-1}')+\dots(t_{r-1}'-t_r')] \qquad (r<s),$$

and obtain

$$2\beta_{12}E_I\{\sum_r e^{-iI_{12}\omega}(I_{12}-m_{12})(1+e^{-iI_{r-1}'\omega}+e^{-i(I_{r-1}'+I_{r-2}')\omega}+\dots)\}$$

$$\approx 2\beta_{12}[\frac{1}{1-L'(i\omega)}\{\frac{\partial}{\partial(-i\omega)}-m_{12}\}\{\frac{L(-i\omega)-1}{i\omega m_{12}}\}],$$

where $m_{12} = \frac{1}{2}(m_1+\sigma_1^2/m_1)$. As $\omega \to 0$, this reduces to $-2\beta_{12}\sigma_{12}/m_2$, where

$$\sigma_{12}^2 = \frac{\mu_3}{3m_1} + \frac{\sigma_1^2}{2}\left(1-\frac{\sigma_1^2}{2m_1^2}\right) + \frac{m_1^2}{12}.$$

For Lane 1, with a value 256.9 for μ_3, the cross-coefficient factor σ_{12}^2/m_1m_2 is estimated to be 0.59, compared with factors $\sigma_1^2/m_1^2 \approx 0.70$ for Lane 1 by itself, and $\sigma_2^2/m_2^2 \approx 0.78$.

ESTIMATION OF TRAFFIC DENSITIES AT THE LINCOLN TUNNEL FROM TIME SERIES OF FLOW AND SPEED DATA

D. C. Gazis
IBM Thomas J. Watson Research Center, Yorktown Heights, New York

M. W. Szeto
Massachussets Institute of Technology, Cambridge, Massachusetts

SUMMARY

An algorithm given in a previous paper for a single lane is adapted to the case of a multi-lane highway. The algorithm is based on obtaining a rough estimate of the density, and then filtering out the error in this estimate by using a relationship which must be satisfied between successive count estimates and flow measurements at the input and output points of a roadway section.

I. INTRODUCTION

The purpose of this paper is to discuss a basic problem in traffic control: How to derive a dependable estimate of a variable which is not directly measurable, given a set of measurements of other variables. The problem arose in the course of an application of computer control methodologies for optimizing the traffic flow through the Lincoln Tunnel which connects Manhattan with New Jersey (Gazis & Foote, 1969). The control algorithm used required a fairly accurate estimate of the number of cars which were at a given instant between detection stations located about one-half mile apart. The only information obtainable at those detection stations was the speed, length, and the time of arrival of each vehicle. The earliest attempts to estimate count of vehicles, or traffic densities, were based on phenomenological flow versus concentration relationships, and they were not very successful since they could be in error by as much as a factor of two either way. In the control experiments mentioned above, the estimation of densities was done by means of a fairly elaborate scheme of matching the patterns of lengths of vehicles over consecutive detection points. The scheme was very successful, but it had two disadvantages: It required a fair amount of computing; and it depended heavily on the fact that lane changes were illegal in the tunnel and consequently limited in number, (although by no means unusual). As a result, the pattern recognition procedure would very likely be inapplicable in the case of a freeway.

Gazis and Knapp (1971) proposed an algorithm, subsequently called the G-K algorithm, for estimating counts of cars between observation points using the time-series of speed and flow data. The algorithm was tested using as a benchmark the available tunnel data for a single lane of traffic, obtained with the pattern-recognition scheme. It was found to give an error less than 10% at least 99% of the time and an error less than 15% all the time. In this paper we extend those results to a two-lane highway, and test it against available tunnel data. We also indicate possible modifications of the density estimation algorithm with an eye toward economy in computer power and storage requirements.

2. THE G-K ALGORITHM

Figure 1 shows the important parameters of the problem. It illustrates typical trajectories of cars moving along one lane, in a space-time domain. (The broken trajectories indicate cars changing lanes). It is assumed that we have observation stations, hereafter referred to as "traps", at points x_1 and x_2. At those traps we can determine the speed and time of arrival of each vehicle. The G-K algorithm is based on the following two self-evident propositions:

1) At the time of exit of a car from the section, the count of cars in the section must be consistent with the travel time of the existing car over the section and the inventory of car arrivals at the entrance of the section.

2) Successive count estimates must be compatible with flow data over the traps. That is, the increase, over a given time interval, of

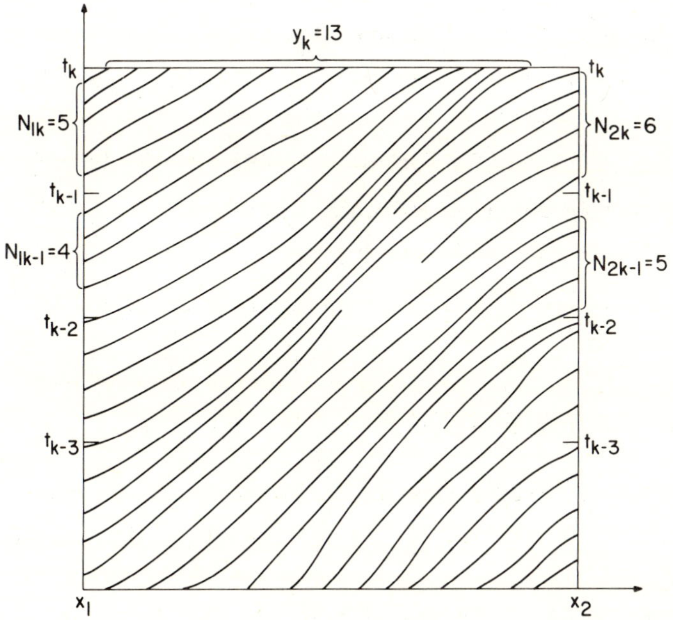

Fig. 1. Typical trajectories in the space-time plane.

the number of vehicles on the roadway should equal the number of vehicles entering minus the number leaving during the interval.

The first proposition is used to derive a rough count of cars. The second one is used to "filter out" the noise contained in the rough count. The complete estimation procedure comprises the following:

1) An estimate of vehicle travel times is obtained, at regular time intervals, from the time series of speeds at the entrance and exit points of the roadway section.

2) A rough count of vehicles is obtained from the inventory of car arrivals at the entrance of the section and the travel times of the existing cars.

3) The rough counts are adjusted by means of a "sequential estimator", to insure the compatibility of successive counts with flow data.

3. ESTIMATION OF TRAVEL TIME AND ROUGH COUNT

The sequence of speeds of successive vehicles over each trap is smoothed exponentially to eliminate high frequency noise in the speed measurement. Then, every T seconds, at times T_1, T_2,... T_k, the two average speeds are stored as well as the number of vehicles counted entering and leaving the section. Referring to Fig. 1, let

$T(x,t_k)$ = Time required to reach x_2 at time t_k starting at point $x \le x_2$.

Assuming that the speed is a function of space and time, $V(x,t)$, we can now write

$$T(x,t_k) = \int_x^{x_2} \frac{ds}{V[s,t_k - T(s,t_k)]} \ . \tag{1}$$

Equation (1) states that a vehicle which reaches x_2 at time t_k must have been at point $s < x_2$ at time $t_k - T(s,t_k)$ traveling with speed $V[s,t_k-T(s,t_k)]$.

We are interested in the travel time $T(x_1,t_k)$ which is hereafter designated T_{12}. It may be seen that (1) is a transcendental, integral equation which cannot be easily solved even if $V(x,t)$ is a known function of s and t. In Gazis and Knapp (1971) it is shown that if we assume that $\frac{1}{V(x,t)}$ varies linearly with x and t, we can reduce the integral equation (1) to an algebraic transcendental equation; namely

$$T_{12} = \left(\frac{b}{2} + [\max\{0, (\frac{b}{2})^2 - c\}]^{\frac{1}{2}}\right)(x_2 - x_1), \tag{2}$$

where

$$b = \frac{1}{12}\left[\frac{5}{V(x_1,t_k-T_{12})} + \frac{5}{V(x_2,t_k)} + \frac{1}{V(x_1,t_k)} + \frac{1}{V(x_2,t_k-T_{12})}\right],$$

$$c = \frac{1}{12}\left[\frac{1}{V^2(x_1,t_k-T_{12})} + \frac{1}{V^2(x_2,t_k)} + \frac{1}{V(x_1,t_k-T_{12})V(x_1,t_k)} - \frac{1}{V(x_2,t_k)V(x_2,t_k-T_{12})}\right]. \tag{3}$$

Equation (2) can be solved by iteration, assuming some initial value of T_{12}. In practice, a reasonably good approximation to T_{12} is obtained by computing b and c from (3), on the basis of a fixed value of T_{12}, or by using the last computed value of T_{12}. Once T_{12} is computed, the rough count is estimated by the relationship

$$z_k = \sum_{j=0}^{m-1} N_{1,k-j} + \frac{\tau}{T} N_{1,k-m} \tag{4}$$

where N_{1r} is the number of cars entering the section during the rth observation period, and

$$m = T_{12} (\text{modulo } T), \qquad \tau = T_{12} - mT. \tag{5}$$

Equation (4) is based on the assumption of a uniform arrival rate of vehicles during any observation period, T.

4. SEQUENTIAL ESTIMATOR (FILTERING)

Let y_k denote the number of cars on $[x_1,x_2]$ at time t_k, and N_{1k} and N_{2k} be the number of cars crossing x_1 and x_2 respectively during the interval t_{k-1} to t_k. Assuming that N_{1k} and N_{2k} are measured with a certain error, we have the relationship between successive values of y_k,

$$y_k = y_{k-1} + \Delta N_k + w_k, \tag{6}$$

where

$$N_k = N_{1k} - N_{2k} \tag{7}$$

and w_k is the (random) error in measuring N_k. If we had a good initial count and enough information about the character of w_k, we could use (6) to obtain successive values of y_k. In order to obtain a self-starting algorithm we use the rough count in conjunction with (6) and seek the best estimate of y_k, say $\hat{y}_k$, which is compatible with the rough count and equation (6). In effect we use a Kalman-type filtering technique (Kalman, 1960 and Sage, 1968), which yields an "optimum" estimate according to the formula

$$\hat{y}_k = B_k z_k + (1-B_k)(\hat{y}_{k-1} + \Delta N_k) \tag{8a}$$

or

$$\hat{y}_k = \hat{y}_{k-1} + \Delta N_k + B_k(z_k - \hat{y}_{k-1} - \Delta N_k). \tag{8b}$$

Equations (8a) and (8b) yield an optimum estimate in the sense that

$$C_k + E[\hat{y}_k - y_k)^2] \tag{9}$$

is minimized provided that

$$E(w_k) = E(u_k) = 0, \tag{10a}$$

$$E(w_k w_j) = \begin{matrix} Q, & k=j \\ 0, & k \neq j \end{matrix} = Q\delta_{jk}, \tag{10b}$$

$$E(u_k u_j) = R\delta_{jk}, \tag{10c}$$

$$E(w_k u_j) = 0, \tag{10d}$$

where E signifies expectation, and

$$u_k = z_k - y_k, \tag{11}$$

the difference between rough and true counts. Furthermore, the requirements on B_k, and the conditions on y_o and C_o are

$$B_k = \frac{Q+C_{k-1}}{R+Q+C_{k-1}} = 1 - \frac{R}{R+Q+C_{k-1}}, \tag{12}$$

$$C_k = \frac{R(Q+C_{k-1})}{R+Q+C_{k-1}} = B_k R, \tag{13}$$

$$y_o = E(y_o),\ C_o = E\ \{\hat{y}_o - E(y_o)\}^2. \tag{14}$$

According to (12), B_k lies between 0 and 1 depending on the variances of u and w, and the error of the previous estimate. Referring to (8), it may be pointed out that $B_k = 0$ would be the appropriate value for B_k if one had a very good estimate for y_{k-1} and the variance of w_k were known to be small. If, on the other hand the variance of u_k is small, then $B_k = 1$ is the appropriate value for B_k.

As explained in Gazis and Knapp (1971), the above estimating procedure was slightly revised in order to account for the fact that (10c) is generally not satisfied; that is, the errors in the successive rough estimates of the count are not uncorrelated. (A very likely explanation is that the causes of errors in the rough counts, such as irregularities of the speed profile between traps, do not vanish instantly between successive observations.) In fact, the correlation of successive counts may be approximated by the function

$$\frac{E\{u(t)u(t-\tau)\}}{R} = e^{-\alpha|\tau|}, \qquad (0.01 \le \alpha \le 0.2). \tag{15}$$

Accordingly, the procedure was revised as follows:

Using the approximation

$$E(u_k u_j) \approx Rd^{k-j} \qquad (k \ge j), \tag{16}$$

where

$$d = e^{-\alpha T}, \tag{17}$$

we can ascertain that

$$E\{(u_k - du_{k-1})(u_j - du_{j-1})\} = R(1-d^2)\,\delta_{jk}. \tag{18}$$

Consequently, if a modified rough count, z_k^*, is defined according to

$$z_k^* = \frac{z_k - dz_{k-1} - d\Delta N_k}{1-d}, \tag{19}$$

the variances are given by

$$E(u_k^* u_j^*) = R\{(\frac{1+d}{1-d}) + Q\frac{d^2}{(1-d)^2}\}\,\delta_{jk} = R^*\delta_{jk}, \tag{20a}$$

$$E(u_k^* w_j^*) = \frac{d}{1-d}\,Q\delta_{jk}, \tag{20b}$$

where

$$u_k^* = z_k^* - y_k = \frac{u_K - du_{k-1}}{1-d} + \frac{d}{1-d}\,w_k. \tag{21}$$

Equations (20a) and (20b) replace (10c) and 10d), meaning that we have satisfied the condition for uncorrelated errors in the modified rough count at the expense of introducing some cross correlation between the modified rough count errors and the input measurement errors. The latter, however, is generally acceptably small.

The final form of the estimation algorithm consists of computing y_k according to

$$\hat{y}_k = \hat{y}_{k-1} + \Delta N_k + B_k^* \{z_k - dz_{k-1} - (1-d)\hat{y}_{k-1} - \Delta N_k\}, \tag{22}$$

where

$$B_k^* = \frac{Q + C_{k-1}}{R(1+d) + (C_{k-1}+Q)(1-d) + Qd^2/(1-d)},$$

$$C_K = B_k^* k^* (1-d) = B_k^* \{R(1+d) + \frac{Qd^2}{1-d}\} \tag{23}$$

and the starting conditions are given by (14). The constant d is to be obtained by appropriate "calibration" of the system. As was pointed out in Gazis and Knapp (1971) such a calibration is necessary in order to insure fast convergence of the algorithm and minimize bias errors. It can be accomplished if one has a way of estimating the exact count; for example, through photographs.

5. RESULTS FOR A SINGLE LANE

The G-K algorithm was tested using the measurements of an early phase of the Lincoln tunnel experiments during which only one tunnel lane was monitored and controlled. The results are summarized in Table 1. A graphical comparison of the exact and estimated count is given in Figs. 2 to 4. There were four traps in the tunnel located about one-half mile apart, at the entrance, front of the downgrade, front of the upgrade, and exit. They divided the tunnel into three sections, referred to as the Downgrade, Level and Upgrade sections in the Figs. 2 to 4, and in Table 1.

Table 1. *Summary of Results for a Single Lane of Lincoln Tunnel Traffic.*

Average and Standard Deviation of Estimation Errors in No. of Cars

Date	11/15/67		11/28/67	
Error	Mean	Std. Dev.	Mean	Std. Dev.
Downgrade Section	0.73	0.42	-0.32	0.61
Level Section	-2.70	2.50	0.39	0.91
Upgrade Section	0.04	1.25	0.02	1.40

Inspection of Table 1 and Figs. 2 to 4 shows excellent agreement between the exact and estimated count. Another measure of the goodness of the estimate is shown in Table 2 which shows the percent of time during which the error was within 10%, and within 15% of the exact values for the 11/28/67 data. To our knowledge such accuracy has not been previously obtained using flow and speed data.

Table 2. *Percent of Time When Error is Below 10% and 15% of Exact Values for 11/28/67 Data*

Downgrade Section	Level Section	Upgrade Section	Error
99.5	99.2	99.2	$\leq$ 10%
100.0	99.9	100.0	$\leq$ 15%

One interesting observation has been that the speed measurements for individual vehicles need not be very accurate. It is sufficient, for example, to estimate the speed of individual vehicles by assuming a standard length, using the signal from a single detector only. The standard length is properly chosen during calibration to minimize the bias error of the rough count. Experience showed that the results thus obtained were generally as accurate as those obtained using the signals from two detectors per trap to obtain a more accurate speed estimate.

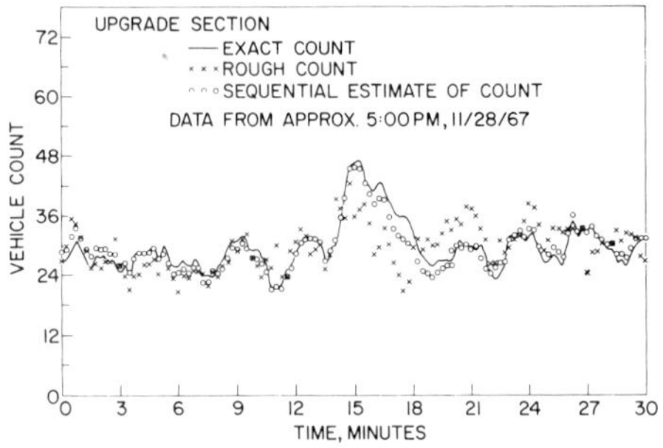

Fig. 2. Vehicle Count in One Lane of the Upgrade Section Versus Time.

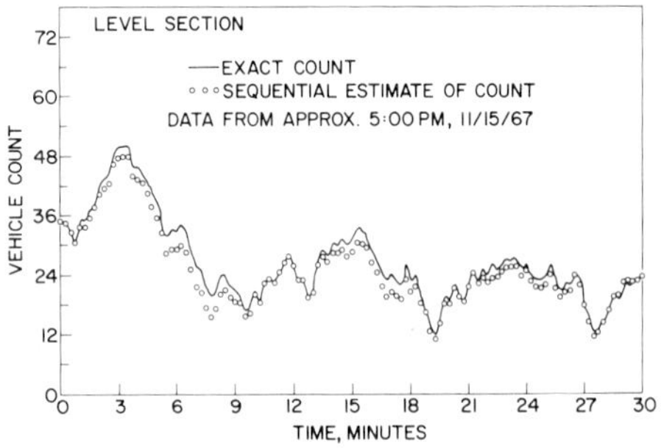

Fig. 3. Vehicle Count in One Lane of the Level Section Versus Time.

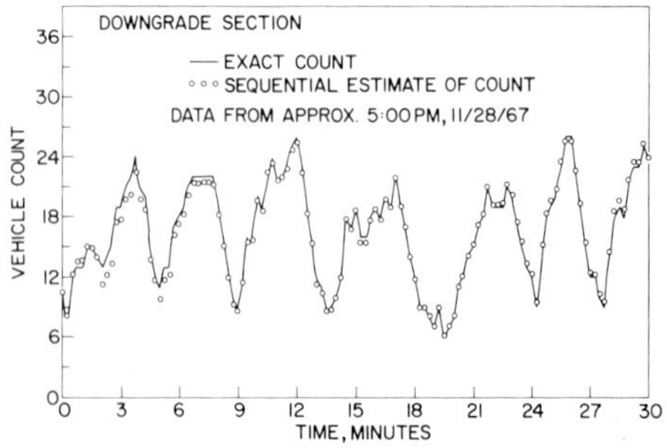

Fig. 4. Vehicle Count in One Lane of the Downgrade Section Versus Time.

This is not so surprising, considering the extensive averaging of the speed data before it is used in the algorithm.

6. EXTENSION TO THE CASE OF A MULTI-LANE HIGHWAY

For the multi-lane highway one can, in principle, apply the G-K algorithm for each lane individually. This involves an expense in detection and data processing requirements, which increases with the number of lanes. If the count of cars in individual lanes is required for control purposes, then this cost is unavoidable. However, there are certain situations when only the total count of cars in all lanes is needed for control purposes. In those cases, it is appropriate to economize by combining the estimation of the count of all lanes into one.

In this section we assume that the detection of cars is done separately for the two lanes, and only the estimation of the count of cars between traps is combined. This is probably the most useful case, since individual detectors for each lane are generally needed for reliable measurement of flows. In any case, it is desirable to combine the data of the two lanes for two reasons: Reduction of storage requirements and reduction of computation requirements.

If we can generate some satisfactory aggregate statistics of the traffic over all lanes, the combination of lanes can be viewed as a single lane with "virtual sensors" measuring the aggregate count and average velocity. Then the G-K algorithm can be applied to estimate the count between virtual sensors.

The aggregate count is, of course, the sum of counts for the individual lanes, for each observation period. For the average velocity, we have used two aggregation schemes. The first one is in the form

$$V = \left(\sum_{i=1}^{n} N_i V_i^2\right) / \left(\sum_{i=1}^{n} N_i V_i\right), \tag{24}$$

where n is the number of lanes, N_i is the flow over the detector of the ith lane, and V_i the corresponding exponentially smoothed speed. Equation (24) is derived from the condition of minimizing a normalized, weighted measure of speed error, ε, given by

$$\varepsilon = \sum_{i=1}^{n} N_i (1-V_i/\bar{V})^2 . \tag{25}$$

The second scheme is of the form

$$\bar{V} = (\sum_{i=1}^{n} N_i)/(\sum_{i=1}^{n} \frac{N_i}{V_i}) \tag{26}$$

and is based on the condition of minimizing the weighted squared error of the rough count computation, for constant speeds and constant arrival rates. It is seen that (26) corresponds to a weighted harmonic average of the speeds V_i.

Both schemes were tested using data obtained during the second phase of the Lincoln tunnel experiment (1968-1969) when both lanes of the eastbound tunnel tube were monitored and controlled. The data were used to estimate the counts in each lane individually, and in aggregation. Both schemes performed satisfactorily. Typically the mean error of the estimator can be reduced to 4 or 5 cars, and the standard deviation of the error to something between 1 and 1.5 cars. While the scheme of equation (24) consistently gave a smaller standard deviation, the scheme of equation (26) gave a smaller mean error. It may be pointed out that since we are considering two lanes which typically have a total of at least 50 cars between traps, (compared to about 25 for one lane), the algorithm appears to give even better results for the case of two lanes than it did for one lane. It was observed that the performance of the estimator did not deteriorate even when the speed in one lane was twice as high as that in the other lane.

Table 3 summarizes the results obtained for the level section of the tunnel using the data of about twenty minutes of experimentation. For the purpose of comparison, we give both the estimation of the aggregate count and the estimation of the count of the individual lanes. A graphical display of the same data is shown in Fig. 5, where the exact

Table 3. *Typical Results for the Estimation of the Two-Lane Count of Cars in the Level Section of the Lincoln Tunnel.*

	Mean Speed (f.p.s.)	Initial Values for the Estimator				Errors	
		Y_o	C_o	Q	R	Mean	Standard Deviation
Lane 1	40.82	20	30	0	50	3.62	0.51
Lane 2	39.65	15	30	0	50	0.63	0.57
Lane 1&2	40.30	35	80	0	100	3.78	0.95

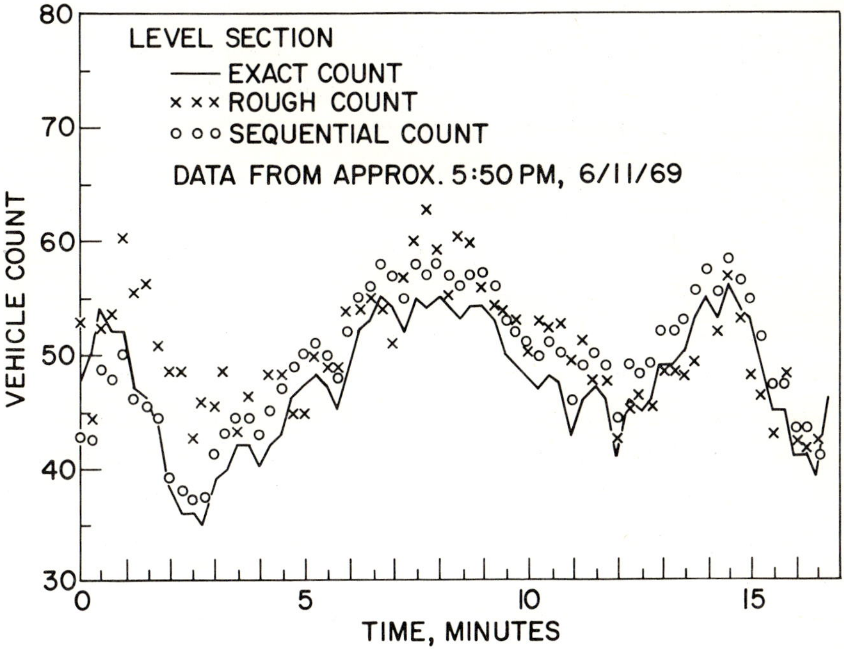

Fig. 5. Vehicle Count in the Combined Two Lanes of the Level Section Versus Time.

count is plotted together with an indication of the rough count and the sequential estimate using the G-K algorithm and the speed averaging scheme of (24). The accuracy of the algorithm appears to be more than satisfactory for control purposes, and far better than obtained heretofore using point measurements in conjunction with phenomenological flow versus concentration relationships.

5. CONCLUDING REMARKS

The algorithm given here has been tested so far using the Lincoln tunnel data. These data are limited in two ways: First, lane changing is relatively rare in the tunnel, since it is presumably illegal. And second, the distance between traps is fixed at about one-half mile. It would be interesting to test the algorithm in situations such as those of multi-lane highways, where lane-changing is frequent, and investigate its effectiveness as a function of trap distance. The latter question has a bearing on the trade-off between accuracy of density estimation and cost of instrumentation and computation. It is hoped that the above tests will be carried out, using data obtained by aerial photographs to obtain both benchmark density estimates and speed and flow measurements for a variety of trap locations.

Another current investigation re-examines the method of computation of rough counts. In view of the fact that the sequential estimator, or filtering, corrects for a substantial error of the rough count, we decided to test whether or not a rough count based on a phenomenological flow versus concentration relationship might not be satisfactory. The advantage of such a scheme would be that the rough count would be an analytic rather than transcendental function of the speed. Thus, questions of optimization of trap distance could be answered by an analytical rather than numerical investigation. Preliminary results of such an investigation, using the same tunnel data as above, appear to be very promising. In fact, they appear to indicate that it may be possible to obtain at the same time the parameters of the phenomenological data and the vehicle count, by treating all of them together as a set of state variables and finding an optimum set which minimizes an aggregate squared error.

REFERENCES

Gazis, D. C. & Foote, R. S. (1969). Surveillance and Control of Tunnel Traffic by an On-Line Digital Computer. Transportation Science, 3, 255-75.

Gazis, D. C. & Knapp, C. H. (1971). On-Line Estimation of Traffic Densities from Time-Series of Flow and Speed Data. Transportation Science, 5, (in press).

Kalman, R. E. (1960). A New Approach to Linear Filtering and Prediction Problems. Trans. ASME, Series D, J. Basic Eng., 82, 34-45.

Sage, A. P. (1968). *Optimum Systems Control*. Englewood Cliffs, N. J.: Prentice-Hall.

STATISTICAL PROPERTIES OF TRAFFIC COUNTS

G. F. Newell
Division of Transportation Engineering
University of California, Berkeley

G. A. Sparks
Division of Transportation Engineering
University of California, Berkeley

SUMMARY

This paper describes some of the problems associated with the interpretation and analysis of observations on point processes generated by vehicles (particularly cars or airplanes) passing some location. We are particularly concerned with counting arrivals over periods of time (hours) in which there is a significant variation in demand. In order to obtain data on stochastic properties, one must observe the process over many days. If the process were stationary one could have observed it only over a sufficiently long time on a single day.

1. INTRODUCTION

We will be concerned here only with point processes in time generated by vehicles (particularly cars or aircraft) passing some location. In analyzing such processes, it is natural to postulate a mathematical structure which is as simple as can be reasonably justified, for example, a Poisson process, renewal process, stationary process, etc.. What can be reasonably justified, however, depends upon what one wishes to evaluate from the mathematical model. If one is concerned with merging or crossing traffic streams as, for example, in evaluating delays to cars at a stop sign trying to find suitable gaps in the cross traffic, or cars on an expressway ramp trying to merge into the main stream, or aircraft trying to take-off on a runway between landings, then one wishes to accurately represent the distribution of headways (time between points) and perhaps the short range probability structure, i.e., statistical dependences between consecutive pairs or triplets of headways. On the other hand, if one is concerned with congestion or queueing at a bottleneck, one usually wishes to accurately represent the properties of the point process on a coarser time scale so as to evaluate correctly the fluctuations in gross counts which cause the formation of large queues. It is not important that one also have an accurate representation of the headway distributions or other features involving only a few points.

Volumes of data have been collected for headway distributions of cars on a highway (May, 1965, and Buckley, 1967). People have even made statistical tests to show that consecutive headways are nearly independent (Breiman, et al, 1968). For light traffic, the probability distribution of headways is nearly exponential, except for the obvious lack of short headways which would correspond to cars in collision. It is, therefore, common practice in the analysis of such things as queueing at a highway intersection to postulate that the arrival process is a Poisson process or a renewal process.

Similar properties have been observed for the process of aircraft arrivals at an airport and similar postulates have been made in calculating landing queues at airports. Delay formulas based upon a stationary Poisson arrival process appear in standard handbooks for airport design. (Airborne Instruments Lab., 1969).

One should not conclude from this, however, that the postulates of a Poisson or a renewal process are typically reasonable for the analysis of traffic counts involving many vehicles (five or more minute counts of cars or half hour counts of aircraft arrivals, for example). There are many situations in which this is grossly incorrect. Obviously the number of aircraft arrivals in 24 hours is not expected to be a Poisson distributed random variable (as it would be if arrivals really formed a Poisson process); one should know quite accurately from the published airline schedule how many aircraft will arrive in 24 hours.

If, on the other hand, one observes consecutive one minute flows of cars on an expressway during a time when the traffic demand is moderately heavy but appears to be stationary, and one examines the distribution of these counts, one will usually find that the sample variance is larger than the mean by a factor typically about 1.5 to 2.0 (Miller, 1970). This is counter to the widely held misconception that congested traffic should be more "regularly spaced" than light traffic. Qualitatively, this large variance is due to clustering and is not well represented by a renewal hypothesis.

For any point process, a convenient way to represent experimental observations is to draw a graph of the cumulative number of points (arrivals) to time t vs t as in Fig. 1, or to draw a graph of counts during successive suitably chosen time intervals as in Fig. 2. We expect that such graphs will show various irregularities due to rush hours, stochastic variations, irregular schedules, etc.. Unless one can a priori assume that the process in question is stationary, ergodic or whatever, one can infer nothing about the probability structure from a single such realization. To measure experimentally a probability distribution for anything, one must make repeated observations "under identical conditions," whatever that means, or make some observations which by some inference are equivalent to doing this.

Unfortunately most real processes which we encounter in transportation studies are influenced by all sorts of things, some predictable (i.e., deterministic) and some accidental (i.e., stochastic). To separate these one must look at many realizations to see which irregularities in the graphs are reproducible and which are not. Presumably, for such things as traffic counts, one should repeat the observations on similar days, but this presents problems also. We recognize that traffic may show weekly, seasonal and yearly trends and daily variations associated with weather, baseball games, etc..

2. AN AIRLINE EXAMPLE

Figure 1 shows the cumulative number of aircraft arrivals at gate positions operated by United Airlines at Chicago O'Hare Airport.

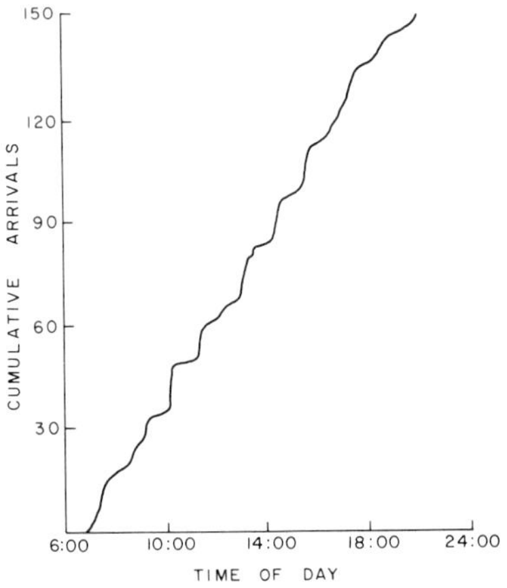

Figure 1. Cumulative number of aircraft arrivals at gate positions of Chicago O'Hare Airport over a twelve hour period.

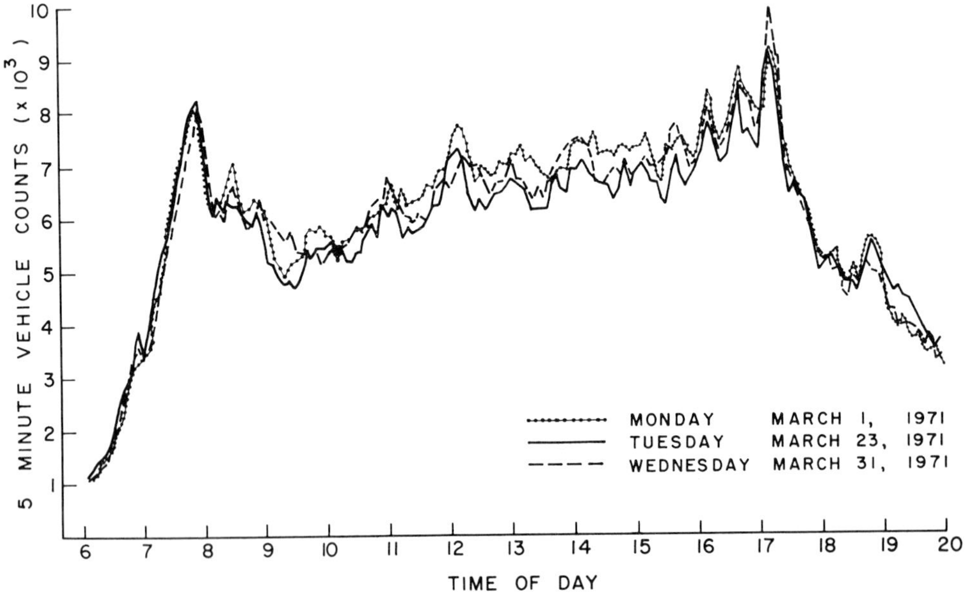

Figure 2. Combined five minute counts from all vehicle detectors in the San Jose computer controlled signal system. Data for three different days.

Steuart (1969) used data like this as part of a study to investigate how the number of gate positions needed to accommodate the aircraft depended upon the strategy of use, the schedule, etc.. In essence, one wanted to know how more flights could be served without building more gates. At that time, United Airlines liked to bring in 15 or 20 airplanes in a bunch, transfer passengers and baggage, and send them out in a bunch. That, of course, explains the very irregular form of this graph. A graph of the scheduled arrival times would show even sharper surges. Lateness of arrivals has a tendency to smooth the curve.

The schedule is dominant in determining the shape of Fig. 1. If one wished to make a stochastic model, it is natural to assume the aircraft have some random lateness relative to a known schedule. The simplest postulate is that the lateness of aircraft are independent, identically distributed random variables, although aircraft are not identical and the latenesses are usually casually related. The final test of such a model is whether or not it correctly predicts the day to day variations in curves such as Fig. 1. The actual sample variances evaluated over many days were larger by a factor of two or more than those computed from the observed lateness distribution along with the hypothesis of independent lateness of different aircraft. This we did not interpret as a failure of the model. It is rather a warning that one should not expect too much and be satisfied if predictions are of the correct order of magnitude. Traffic engineers are usually afraid to interpret experiments done on different days even though the value of any stochastic model is questionable if it can not be used to predict something about the future. Some model, even a poor one, is better than none.

3. A ROAD TRAFFIC EXAMPLE

Figure 2 gives five minute counts of highway traffic for the total traffic crossing all detectors in the entire computerized signal system of the city of San Jose, California (about 400 detectors). The graph shows data for three different days with five minute counts in the range of 10^3 to 10^4. These graphs show well-defined peaks spaced about a half-hour apart in the morning and evening, which obviously arise from the fact that most people start work on the hour or half hour. There are also peaks at about noon and 1:00 P.M., occurring at nearly the same time every day.

If we had drawn corresponding graphs for counts at only a few detectors, so that the scale was only 100 or so, it would have been rather difficult to see whether these irregularities were due to stochastic or deterministic causes. It would be necessary to average data over many days to see the underlying time dependence of the mean.

These data are quite unusual. There are not many places in the world where one can observe such high rates of counts (1/2 million per hour). Also the traffic in San Jose is relatively uncongested compared with most large cities, otherwise these peaks would have been flattened out by the capacity constraint of the highway system.

Figure 2 also shows, at least qualitatively, that the variation in flow from day to day is smaller during morning and evening peaks than during the middle of the day. One would expect the morning and evening traffic to be mostly commuters who make the same trip every day, but the midday traffic will be influenced by store sales, meetings, weather, etc.,

The variations from day to day in Fig. 2 are quantitatively much larger than one would expect from the usual types of stochastic models (Poisson traffic, for example). For counts of order 10^4 one would normally expect a standard deviation of order 10^2, which would be barely visible on a graph of this scale. One possible explanation for this large variance is that any car which enters the signal system is likely to be detected many times (quite likely as many as 10 times). Thus the scale of counts is artificially inflated by a sizeable factor. Some of the variance may also be due to fluctuations in the number of faulty detectors.

In addition to the above counts on the entire system, we have also looked at some counts at individual intersections and been able to identify certain locations where traffic counts are highly reproducible from day to day.

Imagine a hypothetical place where all traffic is commuter traffic composed of people who pass this location at nearly the same time every day. There is an underlying schedule, and this pattern differs from that of an airline only in that the schedule is not published. If one were to make traffic counts over a time which is large compared with the uncertainty in an individual's arrival time, the variance in counts from one day to the next should be quite small.

In an attempt to observe this phenomena, we analyzed some traffic counts at several individual traffic intersections in San Jose. Although these are signalized intersections, the counts are for an integer number of cycles, the flows are undersaturated, and the cycle time is fixed. The signal itself should have no effect on the observations.

The traffic counts can be represented in a matrix, where n_{ij} is the traffic count for the jth cycle of the ith day, the jth cycle being at the same time each day. From this we can also form corresponding matrices of counts using 2, 3, --- (disjoint or overlapping) cycle times as the basic time unit.

The traffic counts over the rows showed several peaks but the counts down the columns were quite stationary except for holidays and some erratic behavior over most of December and January. As a measure of the variation from day to day, we evaluated the sample variance divided by the sample mean. For a Poisson process this should have a value close to 1 but, for highway traffic, one seldom sees a value less than 1.

This variance to mean ratio was evaluated for each column of n_{ij} and then averaged over j. A graph was then drawn of variance/mean as a function of the duration of the counting interval (1, 2, --- cycles). We hoped to obtain a value close to 1 for short counting intervals and see the variance/mean decrease as the counting interval increased to values comparable with a likely value of the uncertainty in a person's time of arrival.

We did not succeed in finding this until we located an intersection which bordered a working-class residential area. We observed this intersection for about nine months. Except for the erratic behavior during December and January, these data were quite consistent and produced the results shown in Fig. 3. The variance/mean does decrease

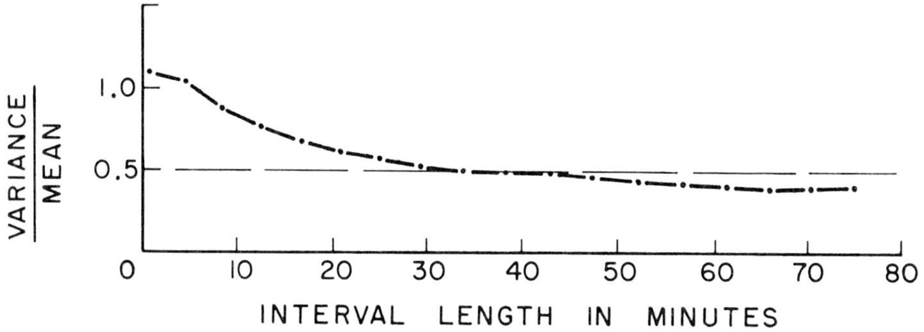

Figure 3. The variance over many days of vehicle counts divided by the mean count is shown as a function of the duration of the counting interval. These are average values over the morning rush hour at a location where the traffic is primarily commuter traffic.

from about 1 for short time counts to a value of about 0.4 for counts over about an hour. We were disappointed that the curve did not drop more rapidly. The shape is probably influenced by shifting demands between half-hour peaks due possibly to seasonal variation in working hours. If this shape were due to random lateness from a schedule, we would have to conclude that people have an uncertainty in arrival of the order 1/2 hour, which is not plausible.

We had originally hoped to obtain traffic counts at many different types of locations (city streets, residential streets, expressways, etc.) and to see how the variances depend upon certain qualitative characteristics of the traffic. We have not, however, been able yet to find suitable sources of reliable data from other people and are not prepared to develop the equipment necessary to collect our own data.

ACKNOWLEDGMENT

The authors are grateful to Mr. Eugene E. Mahoney, Assistant Traffic Engineer, San Jose, California, who furnished the data in Fig. 2, assisted in the collection of the data from which Fig. 3 was constructed, and gave many helpful suggestions based upon his extensive practical knowledge of the traffic behavior in San Jose. One of the authors [G.A.S.] gratefully acknowledges financial support under a Canadian Transport Commission Fellowship. Some miscellaneous expenses for equipment were also supported by a grant from General Motors Corporation.

REFERENCES

Airborne Instruments Lab. (1969). *Airport Capacity Handbook, 2nd Ed.*, Dearport, Long Island, Report #1167-H-1.

Breiman, L., Gafarian, A. J., Lawrence, R. & Murthy, V. K. (1968). An experimental analysis of single-lane headways in freely flowing traffic. *4th International Symposium on the Theory of Traffic Flow*, Karlsruhe.

Buckley, D. J. (1967). Road traffic counting distributions. Transportation Research 1 (2), 105-16.

May, A. D. (1965). Gap availability studies. Highway Research Record #72, 101-36.

Miller, A. J. (1970). An empirical model for multi-lane road traffic. Transportation Science 4, 164-86.

Steuart, G. N. (1969). Utilization of gate position at metropolitan airports. Dissertation Series, The Institute of Transportation and Traffic Engineering, University of California, Berkeley.

TIME INFORMATION IN NUCLEAR PHYSICS EXPERIMENTS

M. W. Sachs
Yale University, New Haven, Connecticut, U. S. A.

SUMMARY

This paper describes the nature of the time information in data acquired in nuclear physics experiments and how it is used in the analysis of the data. Some specific problems in the analysis of coincidence counting are discussed. The examples are drawn primarily from the author's work at the Wright Nuclear Structure Laboratory at Yale University.

1. INTRODUCTION

The purpose of this paper is to give an introduction to the collection and analysis of nuclear physics data, emphasizing the nature and use of the time information in the data. It concentrates primarily on measurement of accelerator-induced reactions and entirely on phenomena in the "low energy" (several MeV to several tens of MeV) range, where phenomena involving the structure of the normal atomic nucleus predominate. The examples are drawn primarily from the author's work at the Wright Nuclear Structure Laboratory at Yale University, using a 10 MV "Emperor" Tandem Van de Graaff Accelerator.

The basic problem of data analysis is the extraction of information on the yield of a particular reaction product from the total, and very large, body of data on many reactions which occur simultaneously during bombardment of a typical target by a typical beam. The yield is usually expressed as the cross section for the production of the reaction in question, which is simply a measure of the probability that one incident beam particle would suffer the specified interaction while passing through a target foil containing one target nucleus per square centimeter

(Evans, 1955, Appendix A). Typical cross section values are in the range 10^{-30} to 10^{-20} cm^2, corresponding to event rates in the raw data ranging from a few per hour to thousands or even tens of thousands per second.

Information required from an experiment includes energies, angular momenta, and symmetries of the nuclear states under study, quantum mechanical functions for these states, gross distribution of matter in the nucleus, and the interrelations between these nuclear states. A common method of analysis consists of hypothesizing a mathematical model for the process under study, expressing the model in terms of physically meaningful adjustable parameters, and using the model to calculate (i.e., predict) the features of the observed data, adjusting the parameters for the best fit to the actual observations. As this model is refined as a consequence of repeated confrontation with actual observations of the real life natural phenomenon, unless it is eventually shown to have some major flaw and thus discarded, it approximates ever more closely the actual physical situation under study.

The uncertainty principle in quantum mechanics tells us that we can never have exact knowledge. Thus statistical interpretation of experiments may reflect either a fundamental property of nature or an expression of our current incomplete state of knowledge. Many nuclear phenomena are expressible as statistical quantities. These include energy widths of nuclear states, mean lifetimes of these states, and distributions of the energy levels of the nucleus.

One of the oldest and most familiar examples of a statistical process in nuclear physics is that of radioactive decay. If N is the number of disintegrations occurring per second in a given sample of material at time t, and N_0 is the number of disintegrations per second at an earlier time $t = 0$, then

$$N = N_0 e^{-at}, \tag{1}$$

where a is the probability that any particular atom will disintegrate in unit time. The basic physical facts in this law are:

a) The probability of decay, a, is the same for all atoms of a given species (though it may be different for different species);

b) The probability of decay is independent of the age of the particular particle and of its surroundings.

These facts are exactly the contitions which define a Poisson process, and indeed it has been verified experimentally that the fluctuations in the measured decay rates are exactly those predicted by the Poisson distribution.

In an ideal accelerator bombardment of an amorphous target, the occurrences of collisions between beam particles and target particles is expected to be a stationary point process in time. This is because both the beam and target particles are randomly distributed in space. It is also worth noting that the nuclear diameter is some 100,000 times smaller than the diameter of its nearest electronic shell, so that in truth the atom is mostly empty space. Only approximately one beam particle per million interacts with a target nucleus (the rest passing through the target foil with no interaction), with the impact parameters (distances of closest approach) also being randomly distributed. These physical facts are the basis of the definition of reaction cross section mentioned previously.

Evans (1955) provides an excellent survey of the statistical methods commonly used in nuclear physics data reduction and analysis with some reference citations to primary sources. See especially Chapters 26-28 and Appendix G of Evans (1955).

The elemental measurement in a nuclear physics experiment is the detection, and usually measurement of the energy, of either energetic nuclear particles (protons, neutrons, alpha particles, heavier nuclei, electrons, etc.) or electromagnetic radiation (X rays and gamma rays).

Most detection methods involve production, by the radiation under study, of electrical charge in matter (the detector) by such processes as ionization, photoelectric effect, electron/hole production in crystals, etc., and subsequent measurement of the charge produced. These methods can be subdivided into three general categories:

1. Counter techniques: The total quantity of charge is measured by converting it to a light flash, as in a scintillation counter, or an

electrical impulse, as in an ionization chamber or semiconductor detector. This type of measurement yields some information on each of many events.

2. Bubble, spark, and cloud chamber: The ionization cause bubbles, sparks, or droplets (respectively) which make visible the actual path of the primary radiation. This method, usually used in elementary particle physics, yields very detailed information on each of a very few events.

3. Nuclear emulsions: The particles are stopped in a photographic emulsion. The ionization produces a track in the emulsion which is made visible by photographic development. Although in principle this is the same type of measurement which is made with a bubble chamber, usually in nuclear physics experiments the tracks are too short for detailed measurements of their structure, and what is of interest is the number of tracks in each region of the emulsion.

The examples in this paper are drawn from counter experiments, although similar principles apply to the other detection methods.

To the extent that the bombarding beam intensity is constant (this point will be discussed later) and the number of atoms in the target also remains constant, the output of each detector is a stream of electrical impulses forming a stationary point process in time. In some cases we are interested in the distributions of amplitudes of the pulses, which provides information on the energy of the radiation. In other cases the quantity of interest is the number of events per incident quantity of beam particles as a function of an auxiliary quantity such as a magnetic field through which the detected particles passed. Depending on the nature of the experiment, the time information may or may not be significant. In this paper we first consider examples of different types of experiments classified by the nature of the time information in the data. In the examples we assume that time intervals involved in the measuring process itself are negligible compared to the shortest expected intervals between events. This is in fact not always true, and in a later section we consider some aspects of this distortion problem.

2. DECAY OF A RADIOACTIVE NUCLIDE

In the study of the radiation emitted by a fixed quantity of a single radioactive nuclide, we are dealing with a pure Poisson process. As each atom emits its radiation (decays) it is changed into, in the simplest case, a stable nuclide, therefore no longer contributing to the radiation. The radiation intensity thus decreases with time according to the exponential law given in equation (1). The basic problem in this type of measurement is to measure the disintegration probability, $\underline{a}$ (equation 1), or in more conventional terminology, the half-life,

$$T = \frac{\ln 2}{a} . \tag{2}$$

This in turn can be used to help identify the radiation emitter or may be of intrinsic interest in the case of a newly discovered nuclide.

3. REACTIONS INDUCED BY A STEADY ACCELERATOR BEAM

An ideally steady accelerator beam, i.e., one in which the number of particles per second passing through the target is constant in time, produces interactions with the target material which occur randomly in time with a constant average rate, and no interaction is influenced by a previous one. The events in a detector therefore form a steady time series. As long as the characteristic times involved in the detection and measurement process are small compared to the shortest expected interval between events, the times of occurence of the events are of no interest. Rather, the quantity of interest is the additional information, such as the energies of the detected particles, associated with each event as well as the average rate of events relative to the intensity of the accelerator beam (from which the reaction cross sections are obtained). The time information in this type of experiment becomes of interest only when conditions deviate from ideal, in which case instrumental distortions can occur. This will be discussed in section 7.

4. REACTIONS INDUCED BY A PULSED ACCELERATOR BEAM

Many types of accelerators produce pulsed beams. In these the beam is turned on and off at a regular rate. In different accelerators the rate may vary from a few per minute to several thousand per second and the burst durations from nanoseconds to seconds. In some cases the pulsing is an undesireable effect which is inherent in the acceleration process. In other cases it is introduced deliberately in order to make possible certain types of experiments.

Ideally the beam intensity is constant during the duration of the beam burst, and zero outside it as is shown in the upper half of Fig. 1.

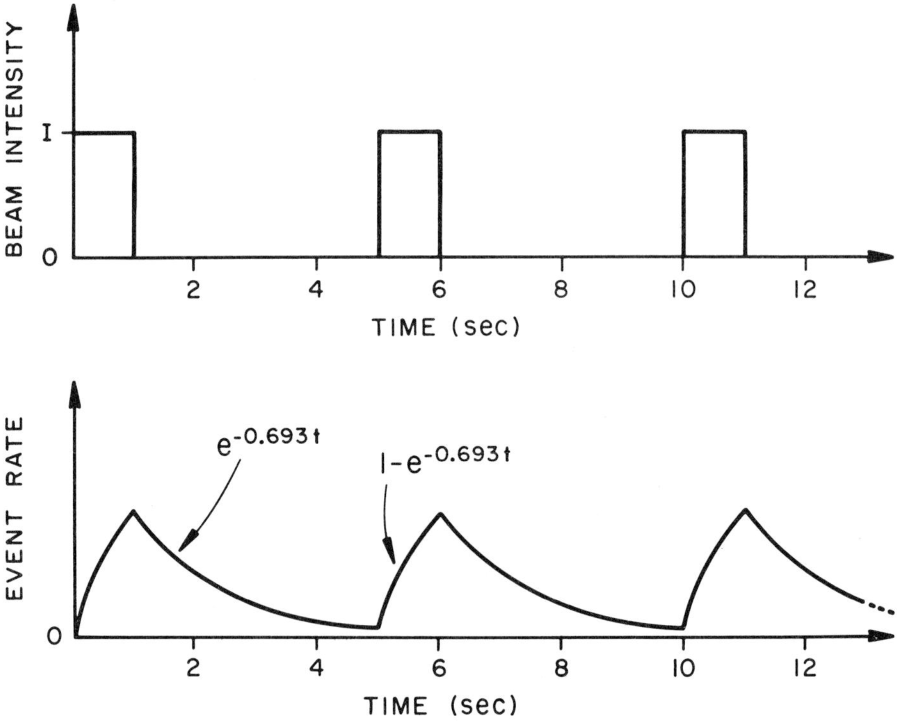

Figure 1. Upper half: Intensity (in arbitrary units)of a pulsed accelerator beam plotted as a function of time for several cycles.

Lower half: Event rate (in arbitrary units) resulting from buildup of an activity with a 1 second half-life in a target bombarded by the beam shown in the upper half.

Depending on the particular beam and target, the events in a radiation detector may or may not form a steady time series during the beam pulse (with no events between beam pulses). If, for example, the reactions induced lead to the production of radioactive nuclides with half-lives comparable to the beam pulse duration, then the event rate in the detector (for constant beam intensity during the burst) increases as the burst continues, reflecting buildup of the radioactive product, and then, following the end of the burst, dies out with the same time dependence as the decay of this beam-produced radioactive species. This cycle repeats with each beam burst, as illustrated in the lower half of Fig. 1, for production of a radioactive species with a half-life of 1 sec. by a beam of 1 sec. bursts every 5 seconds.

An example of an experiment designed for pulsed beams is the measurement of neutron velocities. Since neutrons are electrically neutral, they do not produce ionization in matter in the same way as charged particles. Although counters can be used to detect their presence, they cannot be used to make accurate energy measurements. The use of a pulsed beam enables direct measurement of neutron velocities from which the energies can be calculated. The detector is placed a large distance from the target (typically many meters), and the beam burst is short compared to the transit times of the neutrons from the target to the detector. The occurrence of the beam burst provides a starting time and the times of occurrence of the individual output pulses from the detectors are recorded. Since the detector is at a known distance from the target, the velocity distribution of the neutrons (and hence the energy distribution) may be inferred directly from the distribution of times of occurrence of the detector events. Modern measurements of this type are made over an energy range which may be typically 0.5 MeV to 25 MeV, corresponding to velocities extending up to 30% of the speed of light. With a flight path of 30 meters the range of times to be measured is 2×10^{-7} to 2×10^{-6} seconds, the measurements being performed with an accuracy of about 3×10^{-9} sec. Such measurements are described in more detail by Firk (1966).

5. COINCIDENCES BETWEEN EVENTS IN TWO DETECTORS

In a large fraction of nuclear reaction or scattering experiments, it is not adequate to simply measure the radiation falling on a single detector. In a typical bombardment by an accelerator beam, many different types of reactions occur and the process of interest may be obscured by other processes. For example, the following is a partial list of processes which occur when a carbon-12 target is bombarded by a beam of 115 MeV boron-11 nuclei:

$C^{12} + B^{11} \rightarrow C^{12} + B^{11}$ (elastic scattering);

$C^{12} + B^{11} \rightarrow C^{12} + B^{11}$ (inelastic excitation of one or both);

$C^{12} + B^{11} \rightarrow C^{13} + B^{10}$ (direct reactions);

$C^{12} + B^{11} \rightarrow Be^{9} + N^{14}$;

$C^{12} + B^{11} \rightarrow Be^{10} + N^{13}$;

$C^{12} + B^{11} \rightarrow Na^{23} \rightarrow Na^{22} + n$ (compound nucleus reactions);

$C^{12} + B^{11} \rightarrow Na^{23} \rightarrow Ne^{22} + p$;

$C^{12} + B^{11} \rightarrow Na^{23} \rightarrow F^{19} + He^{4}$.

All these occur simultaneously during the bombardment, and therefore the radiation from the target is a mixture of the products of all the above reactions plus others not listed. See Sachs (1964, 1965, 1966) for a description of the measurements and the apparatus.

It is therefore necessary to apply a constraint which causes the measuring apparatus to be sensitive to only the desired events. Many processes result in the simultaneous occurrence of two (or more) radiations. In these cases it is possible to use a coincidence technique to select the events of interest. Two detectors are used, arranged such that whenever the process of interest occurs, events occur simultaneously in both detectors. The presence of this coincidence is detected and, when it occurs, the energy information associated with the events in the two detectors is processed further. All other events are rejected.

Assuming the ideal instrumentation and accelerator conditions noted previously, the stream of events in each detector is separately a steady time series. In this case even if there were no processes producing radiation detected simultaneously by the two detectors, there would still be occasional coincidences detected because of the random occurrence of pulses nearly simultaneously in the two radiation detectors. The number of coincidences measured can be derived as follows:

Let t_1, the resolving time of channel 1, be the time following receipt of an event in detector 1 during which the coincidence detector is sensitive to the occurrence of an event in detector 2. Similarly let t_2 be the resolving time of channel 2. If M_1 events occur in detector 1 during measuring time T which contains many t_1 intervals, then the total time available for responding to events in detector 2 is just $M_1 t_1$, and the fraction of the measuring time is $M_1 t_1/T$. Then if M_2 events occur in detector 2 during the same measuring period, the Poisson distribution tells us that the number M_{21} of those events occurring during the intervals t_1 following receipt of events in detector 1, is estimated as

$$M_{21} = M_2(M_1 t_1/T). \tag{3}$$

Similarly the apparatus is sensitive to the detection of a coincidence resulting from the occurrence of a pulse from detector 1 during each interval t_2 following the receipt of an event from detector 2. Using the same arguments, the number of coincidences of this type is

$$M_{12} = M_1(M_2 t_2/T) \tag{4}$$

and the total number of coincidences detected is just

$$M_r = M_{12} + M_{21} = M_1 M_2(t_1 + t_2)/T \tag{5}$$

or, dividing by T to obtain the average rates N_r, N_1, N_2:

$$N_r = N_1 N_2(t_1 + t_2). \tag{6}$$

In the foregoing, the subscript r is used to denote the fact that these coincidences are totally random or accidental.

The above derivation has ignored the fact that any time an event occurs in detector i within t_j following an event in detector j, it cannot itself be the start of a t_i interval since a t_j interval is in progress. A rigorous derivation can be performed; however in most cases of interest, the additional terms which would result in equation 6 are negligible compared to $N_1N_2(t_1 + t_2)$ since the fractions t_1/T and t_2/T are very small.

If each time interval during which the coincidence detector is sensitive is subdivided into smaller intervals and the distribution of actual times recorded is plotted, it will on the average be a straight line, with fluctuations given by the Poisson process. In other words all intervals are equally probable.

If, in addition to the above-mentioned random process, there is a reaction being produced which results in radiation simultaneously detected by both detectors, then one time interval (actually a range of intervals because of instrumental resolution problems) will be more probable than the rest, resulting in a peak in the time interval distribution. Such a distribution is shown in Fig. 2, from Sie (1971). Events in the peak region are a mixture of random and "real" coincidences as will be shown later. The detailed shape of the peak is determined by instrumental factors.

The output of the coincidence apparatus is thus a mixture of the true coincidences and a Poisson process.

As an example of the use of this technique we consider a study of the elastic scattering of an O^{16} beam with energy approximately 50 MeV by an O^{16} target similar to the measurements of Maher *et al* (1969). By elastic scattering we mean collisions in which kinetic energy is conserved as well as momentum, i.e., no excitations are produced in the colliding nuclei.

Since oxygen is a gas at all reasonable temperatures, there are experimental complications in making a target of it. The method used in these experiments consisted of making a self-supporting foil of silicon monoxide (SiO). Within the accuracy required in most nuclear physics experiments, the fact that the oxygen is chemically bound to the silicon does not affect the results since the chemical binding energies are negligible compared to the nuclear energies involved. The silicon

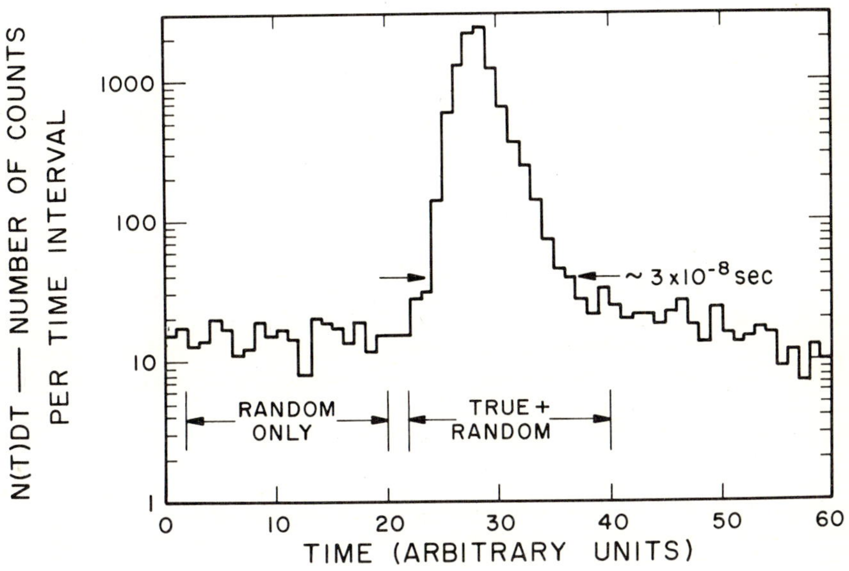

Figure 2. Distribution of relative times between near-coincident events in two detectors, with some of the events (the peak) being true coincidences. Regions used for determination of true coincidence events and random background are indicated.

therefore serves as simply a convenient carries matrix. Of course, some of the beam particles will interact with silicon nuclei, but these radiations can be distinguished from those produced by the interaction with the target oxygen.

A schematic view of the situation is shown in Fig. 3. (This figure differs slightly from the actual experiment in order to make the essential points. Were the experiment to be repeated, it would probably be done as shown here.) The beam is shown entering from the left and striking the target. A small fraction of the beam particles collide with target oxygen nuclei and are scattered into different angles; all angles between 0 and 90 degrees from the beam direction are possible. The

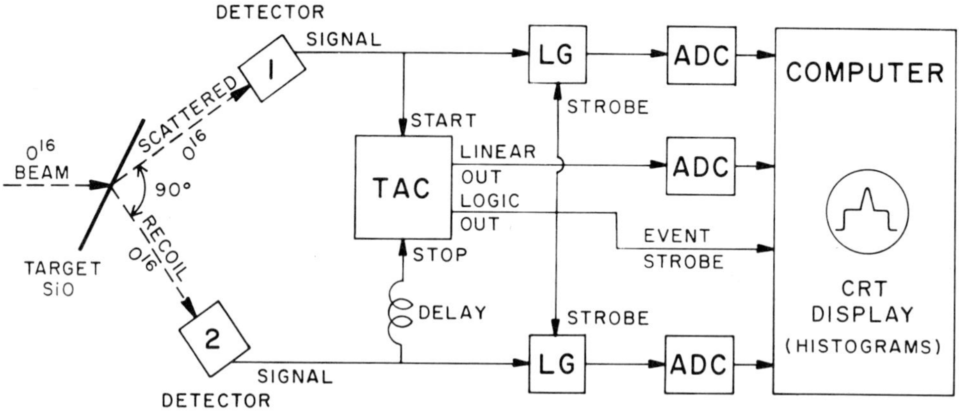

Figure 3. Experimental arrangement for identical particles (in this case oxygen-16 beam and oxygen-16 target) elastic scattering experiment. Abbreviations: TAC = time-amplitude converter; LG = linear gate; ADC = analog-to-digital converter.

struck particle recoils out of the target material and is constrained, by conservation of energy and momentum, to recoil in the plane determined by the entering and scattered beam particles, and because the colliding nuclei are identical, to always recoil at exactly 90 degrees from the direction of the scattered particle.

The object of the experiment is to measure the probability of scattering into each of the permissible angles (within an angular range of $\pm$ 0.5 degrees), i.e., to measure the angular distribution of the scattering, and to measure this at a number of different beam energies. From these data can be deduced gross properties of the nuclei, such as their radii and the distribution of matter within them, and of the interaction between them. These results in turn have bearing on the nature of the fundamental nucleon-nucleon interaction and on one of the fundamental goals of nuclear physics -- the derivation of the interactions between complex nuclei from the nucleon-nucleon interaction.

Because other processes (reactions, inelastic scattering, interactions with the silicon and contaminants in the target) can also occur, it is not adequate simply to sample the output of a single particle detector. However since the only interaction resulting in the scattered and recoil particles being 90° apart which can occur in this bombardment is elastic scattering of O^{16} by O^{16}, requiring a coincidence between two detectors 90 degrees apart serves to eliminate most of the non-elastic events and interactions between O^{16} and silicon. Since the detectors have finite apertures (in this case 1°), some other events will also be detected. Because the elastic scattered particles have a known energy which is constant at each angle, though it varies with angle (determined by conservation of energy and momentum), measurement of the energies of the particles causing the coincidence events serves to eliminate the remaining non-elastic events.

Shown in Fig. 3 are two particle detectors, 90 degrees apart. The signals from the detectors are fed to a time-amplitude converter (TAC). The output of this instrument is a pulse whose amplitude is proportional to the difference between the arrival times of the two input signals. The pulse from detector 2 is delayed in order that the time corresponding to a true coincidence be in the middle of the measuring range of the TAC. In addition the energy signals from the two detectors are sent to linear gates. These instruments prevent passage of the signals unless their gates have been opened by a logic signal from the TAC indicating that a coincidence within the time range of the TAC has occurred. When such a signal occurs, the energy information from the two detectors is digitized by a pair of analog-to-digital converters and presented to the computer along with the digitized time information. The logic signal from the TAC informs the computer that information of interest is present, and the energy and time information are read into the computer memory.

At the start of the measurement, the computer prepares a histogram showing the distribution of time intervals, similar to Fig. 2, and displays it on a cathode ray tube. With a light pen, the experimenter indicates to the computer the limits of the region containing the peak and an equal-width region containing only random coincidences as shown in Fig. 2. The number of random coincidences within the peak region is equal to the number in the random region, within the statistical uncertainty, and therefore the latter information is used to determine how to subtract the background due to the random coincidences.

Once these regions are indicated to the computer, the actual measurement begins. There are several ways in which the computer can treat the data, the goal in all cases being the energy spectra (number of counts in each energy interval in the range of interest) of the outputs of the two detectors, prepared as histograms for inspection on the cathode ray display. In one method, separate histograms are prepared for the peak region and the random region in the time information. The computer first inspects the time information in each event. If the time falls in the region indicated as containing the peak, a count is added to one energy histogram. If it falls in the random region, a count is added to a second histogram. Otherwise the event is rejected. At the end of the measurement, two energy spectra are presented, one containing the real coincidences plus some randoms, the other containing randoms only. The point-by-point difference between these spectra is the real coincidence spectrum. Clearly the same measurement can be made by having the computer prepare a single energy spectrum to which it adds a count if the time information is in the peak region and subtracts a count if it is in the random region.

6. A TRIPLE COINCIDENCE EXPERIMENT

Many interesting reactions result in the production of more than two radiations and can be studied only be using several detectors, looking for events occurring simultaneously in all detectors. Bromley *et al.* (1971) have recently studied a particularly interesting example of such a reaction. If a carbon-12 target is bombarded by an oxygen-16 beam, one of the reactions which can occur produces magnesium-24 and an alpha particle (helium-4 nucleus). In some cases the Mg^{24} can decay almost instantenously (in a time too short to measure) into two or more particles.

A study of the various possible modes of breakup yields information on the structure of the Mg^{24} nucleus. It is known that most nuclei are not simply homogeneous distributions of nucleons, but may be described as consisting of well-defined complex clusters. The specific cluster configuration may be different in different energy levels. One way to determine the cluster configuration is to excite the nucleus to the energy level of interest and then measure the relative probabilities of different modes of breakup.

One possibility is the breakup into two C^{12} nuclei. This is of interest since there is evidence, first discovered by Bromley et al. (1961) that the interaction between two C^{12} nuclei to form Mg^{24} in certain energy levels has the same appearance as the interaction between members of a diatomic molecule except that the force involved is the nuclear force rather than the electromagnetic force. These states are called quasi-molecular states or "nuclear molecules".

Such a two-step reaction is written as follows:

$$C^{12} + O^{16} \rightarrow Mg^{24} + He^{4} ; \tag{7a}$$

$$Mg^{24} \rightarrow C^{12} + C^{12}. \tag{7b}$$

It is of interest to search for this reaction by detection of the two C^{12} nuclei in coincidence with an alpha particle. Analysis of the data will then yield information of the nature of the energy levels in Mg^{24} which decay into two C^{12} nuclei. Since the two steps occur too rapidly to be separated in time by the measuring apparatus, the reaction may be confused with the single step reaction

$$C^{12} + O^{16} \rightarrow C^{12} + C^{12} + He^{4} \tag{8}$$

which does not involve Mg^{24} at all. A detailed study of the data is expected to resolve the question.

In this experiment the beam strikes the target, emitting various particles in all directions. Directly beyond the target is a gas cell which stops all particles emitted in the forward direction except the energetic alpha particles (helium nuclei) which penetrate through the gas cell and strike a particle detector directly forward of the target. On either side, covering the angular range of approximately 15° to 70° are two particle detectors of a variety known as "position-sensitive". These detectors produce two separate signals for each particle.

One is proportional to the particle energy and the other a measure of the distance from one end of the detector to the point at which the particle struck, and hence of the scattering angle of the particle.

The coincidence detector consists of a pair of time-amplitude converters which measure the time intervals between events in the forward detector and in each of the side detectors. When events occur approximately simultaneously in all three detectors (triple coincidence), the three energy, two position, and two time signals are digitized and sent to the computer.

One of the major problems in this type of experiment is the proper correction for random coincidences. In the days before computers and high resolution time-amplitude converters were available in nuclear physics laboratories, this correction had to be done on an *ad-hoc* basis using the physicist's intuition to decide which of several categories of random coincidences is the most important. In the present experiment, however, it has been possible to do a detailed event-by-event computer analysis and make a rigorous (within counting statistics) correction.

Before discussing the random coincidences, it must be emphasized that in the case of a true coincidence, all detectors are struck simultaneously and this time is independent of angle, within the accuracy of the measuring equipment. This is because the transit times from target to detectors are so short (10^{-11} to 10^{-10} sec.) that differences among them (due to differing particle velocities) are negligible on the time scales used in the measurement (10^{-7} to 10^{-6} sec.). Any spreads in these times are due to instrumental effects.

In a triple coincidence experiment, there are four kinds of random coincidences. In this experiment all four are present and account for most of the measured event rate. In fact the data of interest were not even visible above the background until the events were reanalyzed with the proper random coincidence correction applied. The four kinds of random coincidences are as follows:

a) Triple random coincidences, i.e., coincidence indication in which the events in the three detectors are totally unrelated to each other.

b) Events consisting of a true coincidence between detectors 1 and 2 with an unrelated event in detector 3.

c) Events consisting of a true coincidence between detectors 1 and 3 with an unrelated even in detector 2.

d) Events consisting of a true coincidence between detectors 2 and 3 with an unrelated event in detector 1.

During the actual measurement, all the information about all triple-coincident events was recorded on magnetic tape. Subsequent analysis steps were carried out on the same set of events by replaying the magnetic tape.

The first step in the measurement was to produce a two dimensional histogram of the numbers of events as a function of the two times, displaying it on the computer's cathode ray tube for further processing. Such a histogram is shown in Fig. 4. The vertical coordinate is t_{12}, the time between detectors 1 and 2. The horizontal coordinate is t_{13}, the time between detectors 1 and 3. Detector 1 is the alpha particle detector at 0°, the others are the symmetrically placed side detectors. Each dot represents a number of events having the time coordinates given by projecting the dot down to each of the axes. The intensity of the dot represents the number of events. A number of distinct features stand out quite clearly in this figure. Events in the horizontal band in the lower part of the figure are random coincidences consisting of a true coincidence between detectors 1 and 2 in random coincidence with an event in detector 3. They may have any t_{13} but have a unique t_{12} (true coincidence). Similarly the vertical band near the left contains events with a true coincidence between detectors 1 and 3 (unique t_{13}) in random coincidence with an event in detector 2. The 45° band running from lower left to upper right cointains those events having a true coincidence between detectors 2 and 3 in random coincidence with an event in detector 1. Since it is detector 1 which provides the start pulse to both TACs, t_{13} and t_{12} must be equal for all such events (detectors 2 and 3 struck simultaneously) but since the detector 1 events are random, the absolute values of t_{13} and t_{12} may have any value. Hence these events lie along the 45° diagonal. The final type of coincidence, in which all three events are unrelated to each other, may lie anywhere in the plane and is represented by the uniform background scatter of dots throughout the figure.

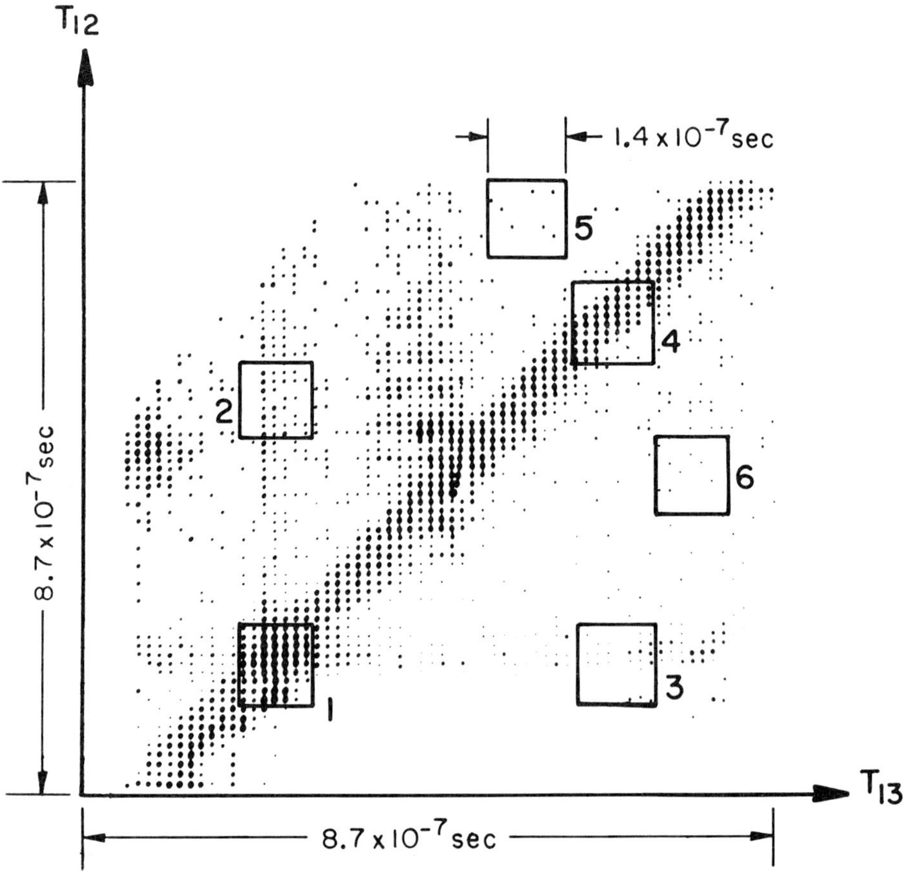

Figure 4. The 2-dimensional distribution of times in the triple coincidence measurement in the reaction $O^{16} + C^{12} \rightarrow Mg^{24} + He^4$; $Mg^{24} \rightarrow C^{12} + C^{12}$. The vertical coordinate is time between detector 1 and detector 2 events; the horizontal coordinate is the time between detector 1 and detector 3 events. Dot intensity represents the number of triple coincidence events having each pair of time coordinates. True and random coincidence selection regions are superimposed. See the text for explanations of the labels.

The region at the intersection of the three above-mentioned loci contains the true triple coincidences mixed with some of each type of random.

The feature near the top center (the additional vertical locus extending upward from the 45° locus) and the blob at the far left were caused by instrumental aberrations. They may be ignored for purposes of this discussion.

To extract the events of interest the computer must select only those in the region indicated as containing the true coincidences, subtracting the background due to the random coincidences. As an example, one part of the analysis consisted of producing an energy distribution histogram for the detector 1 (alpha particle)events in the true triple coincidences. This was done as follows:

With the light pen, the experimenter indicated to the computer the locations, on the (t_{12}, t_{13}) plot, of equal-area regions containing the true coincidences and samples of the background of random coincidences. This was done by first drawing the desired boundary around the true coincidence region and then commanding the computer to place a region of the same shape and area at points indicated by pointing the light pen.

Six regions of equal area were so indicated and are shown in Fig. 4. The six regions are:

1. The true coincidence region (with its background of all types of random coincidences).

2. A region containing only type *b* and *a* (triple) randoms.

3. A region containing only type *c* and *a* randoms.

4. A region containing only type *d* and *a* randoms.

5 and 6. Two regions containing only type *a* randoms.

The total number of true coincidences is (within statistical error) equal to the number of counts in region 1 minus the sum of the number of counts in regions 2, 3, and 4, plus the sum of the number of counts in regions 5 and 6. The reason that regions 5 and 6 must be included and *added* is that regions 2, 3, and 4 each contain type *a* randoms as well as (respectively) types *b*, *c*, and *d*. The total number of type *a* randoms in regions 2, 3, and 4 is three times the number in region 1. Thus only a third of this number of type *a* randoms should be subtracted from region 1, and this correction is accomplished by adding back two thirds (i.e., regions 5 and 6).

To produce the desired histogram of detector 1, the event tape was then replayed. The computer first analyzed the time information in each event. If it was not within one of the six regions, the event was rejected. If it was, the detector 1 energy information was analyzed to determine the correct bin in its histogram. Then if the time information was in regions 1, 5, or 6 a count was added to the correct bin in the histogram; if the time information was in regions 2, 3, or 4, a count was subtracted. When all the events had been replayed a histogram of the detector 1 energy distribution, corrected for random coincidences, was displayed.

7. INSTRUMENTAL DISTORTIONS

In the foregoing discussions of time information, instrumental distortions have been ignored to simplify the discussion of basic principles. In fact, distortions are almost always present and must be accounted for in data analysis. In this section we consider some of the chief kinds of distortion in these experiments. In most cases the effects of the distortions are treated by means of recording of auxiliary data from which the distortions may be corrected rather than by applying a theoretical treatment.

One source of distortion is counting losses caused by instrumental deadtime. Most measuring instruments of the kind used in nuclear physics experiments are insensitive for a period following the response to a given event. This period is known as resolving time or dead time. Evans (1955, Chapter 28) has given a rather detailed survey of the basic statistical principles in the analysis of such deadtime.

In analyzing deadtime it is proper to consider that the measuring apparatus is actually counting intervals between events. An output from the instrument signifies that it has recorded the presence of an interval ending with the instrument's output signal.

The effect of deadtime is to eliminate the shortest time intervals from the measurement. Although the phenomenon being measured may obey the Poisson distribution, the output of the measuring instrument does not because of the insensitivity to the shortest time intervals. This must be taken into account in calculating properties of the actual phenomenon from the output of the instruments.

Although deadtime can be calculated theoretically in an ideal case, there are complicating factors which make theoretical correction of the data for deadtime difficult if not impossible. Among these are:

(i) The deadtime may not be constant for each event;

(ii) The average event rate may vary during a measurement.

The usual method of coping with deadtime is to design apparatus with as low a deadtime as is possible, and to design it to provide a means of monitoring the instantaneous deadtime. Measurement of the deadtime (or directly the fractional counting efficiency) is then accomplished by means of a second experiment having negligible deadtime carried out simultaneously with the main measurement.

A second type of distortion which is essentially a time problem is known as "pileup" or distortion of energy information. Because the data pulses have a significant duration (typically of the order of 10^{-6} sec.), there is always a finite probability that a second pulse will occur before the first has completely died out. In that case the residual amplitude of the first pulse is added to the amplitude of the second (hence the term "pileup"), thus introducing an error in the energy information. Although the various energies are distributed randomly in time, distortion of the measured energy spectrum results, the nature of the distortion depending on the details of the shape of the spectrum and the electronic shapes of the pulses.

Where pileup is serious, special electronic equipment known as pileup rejectors can be brought into play. A pileup rejector detects the presence of a pileup condition and prevents passage of the distorted pulse onward in the measuring system. Effectively, then, it makes the apparatus insensitive to the shortest time intervals and keeps it dead until the start of an interval with no pulse present in the system. It thus replaces spectral distortion by increased deadtime for which the data can easily be corrected.

A third type of distortion concerns the coincidence timing information. When a coincidence measurement is made, the variations in pulse arrival times due to physical processes are quite small. Nonetheless when the time distribution is plotted, as in Figs. 2 and 4, the peak representing true coincidences is significantly wider than would be expected on physical grounds. The reason is that there is additional uncertainty in the time-measuring process, due to variation in pulse shapes and related parameters.

The effect of this width in the time distribution is to set a lower limit on the effective resolving time which may be used without loss of true events (must be at least as wide as the peak). This effectively sets lower bounds on the sizes of the "gating" regions shown in Figs. 2 and 4. The wider the peak region, the more random coincidences are included in it and hence the larger the statistical error in the correction for them.

8. CONCLUSIONS

In some nuclear physics experiments, the times of occurrence of events in the detectors are the primary data. In others, the streams of events in the detectors are stationary point processes in time. In the latter category it is the information associated with each event (such as the energy of the radiation), along with the total number of events, which is of interest. Even in the latter category, however, the time information is important in identification of events of interest (coincidences between events in two or more detectors) and in assessing the distortions introduced by the measuring instruments.

During the past 40 years improvements in the technology of measuring short times and detection of coincidences between events in different instruments have been among the most important technological advances making possible the evolution of nuclear physics from the very crude "atom-smashing" experiments of 40 years ago to the delicate measurements, such as those described herein, characterizing modern experimentation. There has been a real symbiosis between the physics and the technology, with each in turn feeding upon the advances in the other and then inspiring further advances in it. The modern high speed digital computer, in fact, owes its existence in part to this symbiosis since one of its genealogical lines may be traced back to the early electronic logic circuits used by nuclear physicists to perform the first coincidence experiments in the nineteen-thirties.

9. ACKNOWLEDGMENTS

The author wishes to express sincere thanks to Professor D. A. Bromley of the Yale Nuclear Structure Laboratory and Professor P. A. W. Lewis of the IBM Research Center, Yorktown Heights, New York, (now at the Naval Postgraduate School, Monterey, California) for their invaluable assistance in selection of material for this paper. This work was supported by the United States Atomic Energy Commission under contract AT(30-1)3223.

REFERENCES

Bromley, D. A., Kuehner, J. A. & Almqvist, E. (1961). Elastic Scattering of Identical Spin-Zero Nuclei. Phys. Rev. 123, 878-93.

Bromley, D. A., Chua, L., Gobbi, A., Maurenzig, P. A., Parker, P.D., Sachs, M. W., Shapira, D., Stokstad, R. & Wieland, R. (1971). In *Proceedings of a Symposium on Heavy Ion Reactions and Many-Particle Excitations,* Saclay, France, Sept. 8-14, 1971, to be published.

Evans, R. D. (1955). *The Atomic Nucleus.* New York: McGraw-Hill.

Firk, F. W. (1966). Fast Neutron Spectroscopy Using the Yale Electron Linac and a New Nanosecond Time of Flight System. Nuclear Instruments and Methods 43, 312-16.

Maher, J. V., Sachs, M. W., Siemssen, R. H., Weidinger, A. & Bromley, D. A. (1969). Nuclear Interaction of Oxygen with Oxygen. Phys. Rev. 188, 1665-1682.

Sachs, M. W. (1964). *Nuclear Spectroscopy with High Energy Boron Ions.* Doctoral Dissertation: Yale University.

Sachs, M. W., Chasman, C. & Bromley, D. A. (1965). Nuclear Spectroscopic Studies with High Energy Boron Ions. Phys. Rev. 139, B92-B109.

Sachs, M. W., Chasman, C. & Bromley, D. A. (1966). Heavy Ion Reaction Product Identification by Measurement of dE/dx and E. Nuclear Instruments and Methods. 41, 213-225.

Sie, S. (1971). *Magnetic Moments of Excited States in* Os^{188} *and* Os^{192}. Doctoral Dissertation: Yale University.

SOME MODELS FOR STATIONARY SERIES OF UNIVARIATE EVENTS

A. J. Lawrance
University of Birmingham,England and IBM Research Center, New York

SUMMARY

In this paper results for three types of stationary stochastic point processes are surveyed. These are (1) non-recurrent point processes with conditionally independent interevent intervals, (2) doubly stochastic Poisson processes, and (3) point processes with cluster structure; several processes of each type are discussed and the treatments are self-contained with the minimum of formality. Processes of these types have found applications in the modeling of series of events: for instance, the failure times of computers, the emission times of neurons, the occurrence times of earthquakes, etc. The general approach for each process is to give (a) the arbitrary time and arbitrary event initial conditions, (b) the stationary counting distribution and its second order properties, and (c) the marginal and joint distributions of interevent intervals, the intensity function and serial correlations. Some results complement earlier work. Areas requiring further investigation are outlined; other results and generalizations are discussed.

CONTENTS

1. *INTRODUCTION AND GENERAL APPROACH TO STATIONARY POINT PROCESSES*

2. *NON-RECURRENT POINT PROCESSES WITH CONDITIONALLY INDEPENDENT INTEREVENT INTERVALS*

2.1 *Two-state and two-interval Markov renewal point processes*
2.2 *Moran's pair-wise interval dependent point process*
2.3 *Processes with first order Markov dependent interevent intervals*
2.4 *Wold's and Cox's particular results for Markov dependent processes*
2.5 *Lampard's results for a particular Markov dependent process*

3. *DOUBLY STOCHASTIC POISSON PROCESSES*

3.1 *Introduction and general theory for continuous rate processes*
3.2 *The shot-noise doubly stochastic Poisson process*
3.3 *The Ornstein-Uhlenbeck doubly stochastic Poisson process*
3.4 *The random hazard doubly stochastic Poisson process*
3.5 *The doubly stochastic Poisson process driven by a point process*
3.6 *Some related processes*

4. *STATIONARY POINT PROCESSES WITH CLUSTER STRUCTURE*

4.1 *Structure of the clustering*
4.2 *Description of specific cluster processes*
4.3 *The Bartlett-Lewis process*
4.4 *The Neyman-Scott process*
4.5 *The process of displaced regular events*
4.6 *Comments on generalizations of the cluster processes*

1. INTRODUCTION AND GENERAL APPROACH TO STATIONARY POINT PROCESSES

In this paper we survey results on three types of stationary point processes. These are

(1) non-recurrent point processes with conditionally independent interevent intervals,

(2) doubly stochastic Poisson processes, and

(3) point processes with cluster structure.

This choice is partly personal, but to some degree reflects particular processes which have found applications and where explicit results have been obtained. The treatment of each process is self-contained and does not employ very general results or dwell on very technical aspects. Mainly we aim at results which describe salient features of the models and which would be useful in statistical analyses. Our general approach is to give

(a) the arbitrary time and arbitrary event initial conditions,

(b) the stationary counting distribution and its second order properties, and

(c) the marginal and joint distributions of interevent intervals, the intensity function and serial correlations.

Unfortunately not all of these results can explicitly be obtained for most point processes and we will indicate where further investigations seem to be required.

Our treatment of each particular process will make use of several basic general ideas about stationary point processes, and so here we give a brief and simplified account of these; we also introduce terminology as this has not yet become stable in the area. For fuller details on the basic ideas and general theory we refer to McFadden (1962), Cox and Lewis (1966, Chap. 4) and Crámer and Leadbetter (1967, Section 3.8).

We always take a point process with fixed origin and consider events occurring one after the other along a one-dimensional axis, customarily regarded as time. We assume that the events are univariate (all of the same type). The aspects of interest are therefore mainly related to their times of occurrence. A further assumption is that a stationary point process commences at time $t = 0$ according to a set I_0 of initial conditions, these having a specified probability distribution; this feature is not usually included in the general theory, but is useful in applications to particular processes. The point processes we consider are stationary in the structural sense; we mean by this that they do not have features defining their mode of continuation which progressively alter with the passing of time or events. It will generally also be assumed that the processes are orderly, meaning that the simultaneous occurrence of two or more events is ruled out.

Basic to any point process are both the counting and interevent properties, and these very much depend on how the initial point of observation is chosen. This usually means choosing an *arbitrary time* or choosing an *arbitrary event*, and then observing the process for a specified time or number of events. Such procedures will be referred to as *asynchronous* and *synchronous* sampling, respectively. For our purposes, an arbitrary time is defined as the time t' as t' tends to infinity. Initial conditions which are appropriate to an arbitrary time are obtained by considering the behavior of the process subsequent to a time t', and deriving new initial conditions $I_{t'}$; then, by letting t' tend to infinity in $\text{prob}\{I_{t'}\}$ the probabilistic *arbitrary time initial conditions* I_S are defined. It will then generally be assumed that a process started at $t=0$ with initial conditions I_S will be immediately *stationary*, in the sense that the marginal initial conditions at time t have the same probability as I_S for any time t chosen without prior knowledge of the positions of events - any such time is then an arbitrary time. This would imply technical stationarity which is usually taken to mean that the distribution of $N(t, t + \tau)$, the number of events in $(t, t + \tau]$, is independent of t (≥ 0), and similarly for the higher order distributions. The intervals between events following an arbitrary time will be denoted by $\chi_1, \chi_2, \ldots$, and termed the *asynchronous sequence*; in particular, χ_1 is the forward recurrence time.

In *synchronous sampling* we observe the process after choosing an arbitrary event; this can be interpreted mathematically in several ways, and we shall take it here to mean conditioning the future behavior of the process on the event $N(0,t') \geq 1$ and then taking the limit as t' tends to zero. *Arbitrary event initial conditions* are then determined in terms of the arbitrary time initial conditions by the limit of $\text{prob}\{I_{t'} | N(0,t') \geq 1\}$ as t' tends to zero; a process can be started at an arbitrary event by using these conditions at t = 0. The intervals following an arbitrary event will be denoted by $X_1, X_2, \ldots$, and are termed the *synchronous sequence*. It will generally be assumed that these are a stationary sequence of random variables, and moreover, that the arbitrary event initial conditions repeat themselves at each subsequent event. Analogously to the asynchronous case, an arbitrary event is then any event in the synchronous sequence chosen without regard to the intervals between the events. There are many delicate features of asynchronous and synchronous sampling which have not been mentioned here.

The notions of arbitrary times and arbitrary events may be traced back to Sections 5c and 5d of Wold (1948). These ideas were then used very perceptively by McFadden (1962) to give intuitive proofs of many general stationary point results. The described limiting procedures in the field of point processes appear to have been first demonstrated by Cox (1962) for arbitrary times, and by Khintchine (1960) for arbitrary events. Arbitrary event initial conditions give practical constructions of "Palm distributions" as first discussed in Khintchine (1960). The arbitrary event limiting procedure does not appear to have been used previously to produce initial conditions for particular point processes.

This is probably because, for many processes, both the arbitrary time and arbitrary event initial conditions can be obtained by inspection or guess work. However, when the intuition is weak, as it will be for complex or unusual processes, the limiting procedures are very constructive. Finally, we note that in the paper we will generally proceed from the asynchronous to the synchronous aspects; often this is easiest, but for some processes it may be more natural to reverse that order.

2. NON-RECURRENT POINT PROCESSES WITH CONDITIONALLY INDEPENDENT INTER-EVENT INTERVALS

2.1 Two-State and Two-Interval Markov Renewal Point Processes

The first point process we consider with conditionally independent intervals is formed from a two-state and two-interval Markov renewal process (semi-Markov process) by associating an identical event with each transition and treating all intervals in its realizations as being distributionally alike. In this process, which was first considered by Cox (1963) and is also discussed in Cox and Lewis (1966, Chap. 7), the successive intervals are sampled independently from either of the two distributions with probability density functions (p.d.f.'s) $f_1(x)$ and $f_2(x)$; the sequence of types of intervals is determined by the transition matrix

$$\begin{array}{ccc} \text{state} & 1 & 2 \\ 1 & \multicolumn{2}{c}{\multirow{2}{*}{}} \end{array}$$

$$\begin{matrix} \text{state} \\ 1 \\ 2 \end{matrix} \quad \begin{matrix} \;1 \qquad\quad 2 \\ \begin{pmatrix} p_1 & 1-p_1 \\ 1-p_2 & p_2 \end{pmatrix} \end{matrix} \qquad (0 \le p_1,\ p_2 \le 1). \tag{2.1.1}$$

This is a particularly simple form of Markov renewal process because the interval distribution depends only on the present state and not on the state to which it next enters. We let the survivor functions of the two interval distributions be $R_1(x)$ and $R_2(x)$ and the associated means and variances be ν_1, ν_2 and σ_1^2, σ_2^2, respectively. We derive all the main stationary point properties using our general approach; the second order properties are in agreement with the derivations of Cox (1963).

Arbitrary time initial conditions for the process involve specifying jointly the current type of interval and the forward recurrence time to the next transition. At time t, let these two quantities be denoted by S_t and L_t. Let us suppose for simplicity that the process starts with an interval of type 1. Then noting that the process alternates recurrently between runs of type 1 and type 2 intervals, and letting their p.d.f.'s be $r_1(t)$ and $r_2(t)$, we have

$$\text{prob}(S_t = 1,\ L_t \geq x)$$
$$= \sum_{m=0}^{\infty} \sum_{n=0}^{\infty} \{r_1(t)*r_2(t)\}^{(m)}*\{p_1 f_1(t)\}^{(n)}*R_1(t+x) \tag{2.1.2}$$

making use of the usual notations for convolution. From Markov-chain theory the number of intervals in runs of each type are geometrically distributed, and hence we have

$$r_i(t) = (1-p_i) \sum_{m=1}^{\infty} p_i^{m-1}\{f_i(t)\}^{(m)} \qquad (i = 1,2)\ . \tag{2.1.3}$$

Letting t tend to infinity in (2.1.2) we may apply renewal theory (Cox, 1962, Chap. 2), to obtain for the arbitrary time initial conditions

$$\text{prob}\{S = i,\ L \geq x\} = \frac{(1-p_{3-i})\nu_i}{(1-p_2)\nu_1 + (1-p_1)\nu_2}\ \nu_i^{-1} \int_{u=x}^{\infty} R_i(u)\,du \tag{2.1.4}$$
$$(i = 1,2)\ ,$$

absence of the suffix on S and L denoting an initial arbitrary time. This expression may be interpreted in two parts: the first gives the probability that an arbitrary time in the process falls in a type i interval, and the second gives the probability that the time to the next event is greater than x; this latter probability is just the survivor function of the time forward to the next event from an arbitrary time in a renewal process with type i intervals.

For the arbitrary event initial conditions we apply our general mathematical approach; hence, denoting an arbitrary event by the suffix A, we define

$$\text{prob}(S_A = i,\ L_A \geq x) = \lim_{t \downarrow 0} \text{prob}\{S_t = i,\ L_t \geq x \mid N(0,t) \geq 1\}. \quad (2.1.5)$$

Writing this in terms of unconditional probabilities we obtain

$$\text{prob}(S_A = i,\ L_A \geq x)$$

$$= \lim_{t \downarrow 0} \left[\frac{\text{prob}\{S=i,\ L \geq x\} - \text{prob}\{N(0,t)=0,\ S_t=i,\ L_t \geq x\}}{\text{prob}\{N(0,t) \geq 1\}} \right], \quad (2.1.6)$$

where we have omitted the dependency on t in the first expression in view of the fact that we start with a process which is stationary. Dividing both numerator and denominator of (2.1.6) by t and taking the limit as $t \downarrow 0$, we find

$$\text{prob}(S_A=i,\ L_A \geq x) = \frac{[D_t\ \text{prob}\{N(0,t)=0,\ S_t=i,\ L_t \geq x\}]\big|_{(t=0)}}{[D_t\ \text{prob}\{N(0,t) = 0\}]\big|_{(t=0)}}$$

$$(i = 1,2) \ . \quad (2.1.7)$$

Here D_t denotes differentiation with respect to t. Mathematically, it is the right-hand derivative which always occurs in the arbitrary event limiting procedure, but for most processes, and always in this paper, it does not need to be distinguished from the ordinary derivative. Now from the arbitrary time initial conditions (2.1.4) we see

$$\text{prob}\{N(0,t) = 0\}$$

$$= [(1-p_2)\nu_1 + (1-p_1)\nu_2]^{-1}\{(1-p_2) \int_{u=t}^{\infty} R_1(u)\,du + (1-p_1) \int_{u=t}^{\infty} R_2(u)\,du\}$$

(2.1.8)

and

$$\text{prob}\{N(0,t) = 0,\ S_t=i,\ L_t \geq x\}$$

$$= [(1-p_2)\nu_1 + (1-p_1)\nu_2]^{-1}\ (1-p_{3-i}) \int_{u=t+x}^{\infty} R_i(u)\,du \ , \quad (2.1.9)$$

for i = 1,2. These last three equations then lead easily to

$$\text{prob}(S_A=i,\ L_A \geq x) = \pi_i R_i(x)\ , \qquad (i = 1,2) \qquad (2.1.10)$$

where

$$\pi_i = (2 - p_1 - p_2)^{-1}\ (1 - p_{3-i}).$$

This result could, of course, be directly noted, as in Cox (1963), in terms of the equilibrium distribution associated with the transition matrix (2.1.1). The present derivation mainly serves to illustrate a typical use of the arbitrary event procedure.

From (2.1.10) the survivor function of an arbitrary interevent interval X is given by

$$R_X(x) = \pi_1 R_1(x) + \pi_2 R_2(x)\ . \qquad (2.1.11)$$

We can similarly obtain the joint distribution of X_1 and X_{k+1} $(k \geq 1)$, and hence find the serial correlations, first given by Cox (1963), as

$$\rho_k = \text{corr}(X_1, X_{k+1}) = \frac{(\nu_1-\nu_2)^2 \pi_1\pi_2}{\pi_1\sigma_1^2+\pi_2\sigma_2^2+\pi_1\pi_2(\nu_1-\nu_2)^2}\ (p_1 + p_2 - 1)^k. \qquad (2.1.12)$$

This leads to the *interval spectrum* of the process, defined by

$$\pi^{-1}\{1 + 2 \sum_{k=1}^{\infty} \rho_k \cos(k\omega)\} \qquad (0 \leq \omega \leq \pi)$$

in Cox and Lewis (1966, Section 4.4), and useful for statistical analysis. A further synchronous feature which is often useful in statistical analysis is the *intensity function* defined as the derivative of the mean of the synchronous counting distribution (Cox & Lewis, 1966, Section 4.5). For the present process, Cox (1963) first obtained this as

$$m^*(s) = \frac{(1-p_2)\ f_1^*(s)+(1-p_1)\ f_2^*(s) + (1-p_1-p_2)(2-p_1-p_2)\ f_1^*(s)\ f_2^*(s)}{(2-p_1-p_2)\{1 - p_1 f_1^*(s)-p_2 f_2^*(s)-(1-p_1-p_2)\ f_1^*(s)\ f_2^*(s)\}}.$$

(2.1.13)

Finally in this section we give our result for the asynchronous (stationary) counting distribution for the number of events in (0,t]. This is built up using the arbitrary time initial conditions (2.1.4) and counting distributions initially synchronized with events of specified type; these can be obtained from two simple transition based equations. Let $\phi(z;t)$ denote the probability generating function (p.g.f.) of the asynchronous counting distribution. Then for its Laplace transform we have

$$\begin{aligned}\phi^*(z;s) = s^{-1} &+ (1-z)[(1-p_2)\nu_1 + (1-p_1)\nu_2]^{-1} s^{-2}[-(2 - p_1 - p_2) \\ &+ p_1\{(1-p_1)z + (1-p_2)\} f_1^*(s) + p_2\{(1-p_2)z + (1-p_1)\} f_2^*(s) \\ &- (p_1 + p_2 -1)(2 - p_1 - p_2)zf_1^*(s)f_2^*(s)]\{1-p_1zf_1^*(s) \\ &- p_2zf_2^*(s) + (p_1 + p_2 -1)z^2f_1^*(s)f_2^*(s)\}^{-1} . \qquad (2.1.14)\end{aligned}$$

Most stationary point results may be obtained from (2.1.14); in particular, expressions for the mean and variance of N(t) are

$$E\{N(t)\} = (2 - p_1 - p_2)\ \{(1-p_2)\nu_1 + (1-p_1)\nu_2\}^{-1}\ t \qquad (2.1.15)$$

and

$$\begin{aligned}\mathrm{var}\{N(t)\} = \frac{2 - p_1 - p_2}{\alpha_1}&\{(2 - p_1 - p_2)\frac{\alpha_2}{\alpha_1^2} + 2(p_1 + p_2 - 1)\frac{\beta_1}{\alpha_1} - 1\}t \\ &+ (2 - p_1 - p_2)\ \{(2 - p_1 - p_2)\frac{\alpha_2^2}{2\alpha_1^4} - \frac{1}{3}(2-p_1-p_2)\frac{\alpha_3}{\alpha_1^3} \\ &+ (p_1 + p_2 - 1)\frac{\alpha_2\beta_1}{\alpha_1^3} - (p_1 + p_2 - 1)\frac{\beta_2}{\alpha_1^2}\} + o(1), \end{aligned}$$

(2.1.16)

where α_1, α_2 and β_1, β_2 are the first two moments about the origin from p.d.f.'s $\alpha(t)$ and $\beta(t)$ defined as

$$\alpha(t) = p_1 f_1(t) + p_2 f_2(t) + (1 - p_1 - p_2)\ f_1(t) * f_2(t)\ ,$$

$$\beta(t) = (2 - p_1 - p_2)^{-1}[(1-p_1)\ f_1(t) + (1-p_2)\ f_2(t)]\ . \qquad (2.1.17)$$

Further asynchronous features which are often useful in statistical analysis are the *covariance density function* of the stationary counting process, and the associated spectrum--the so-called *Bartlett spectrum*. Cox and Lewis (1966, Section 4.5) show that both of these are most conveniently calculated in terms of the intensity function, the relevant formulae being

$$\gamma_+(t) = \mu\{m(t) - \mu\} \qquad (t > 0),$$

and

$$g_+(\omega) = \mu\pi^{-1}\{1 + m^*(i\omega) + m^*(-i\omega)\} \qquad (\omega \geq 0), \qquad (2.1.18)$$

respectively, where μ denotes the rate of the process. Easy explicit results are obtainable in the present case with exponential interval distributions.

This model has been used by Cox and Lewis (1966), Lewis (1966) to examine the traffic data of Bartlett (1963). Ekholm and Hyvärinen (1970) attempted to fit this model to some neuronal spike data. This was not always successful and they developed a "pseudo-Markov" model in which runs of type 1 and type 2 intervals alternate with non-geometric distributions of run lengths; the analysis is very similar to that given here, since it is the alternation of runs which is the main factor and not the distributions of run lengths.

2.2 Moran's Pair-Wise Interval Dependent Point Process

In this section we deal with a process which is constructed from independent pairs of jointly and identically distributed intervals. We denote the pair-wise interval p.d.f. by $f(x,y)$ and its correlation by ρ; further we let the common marginal p.d.f. be $f(x)$ with mean μ and variance σ^2. The joint p.d.f. of a sequence of intervals, each identified with their serial number, will be, say, $f(x_1,x_2), f(x_3,x_4), \ldots$. The process becomes a univariate point process when there is no information about the pairs, and the intervals are treated alike. A particular process of this type was constructed by Moran (1967) as an example for which the intervals between events were exponentially distributed, a counting distribution of events was Poisson, and yet the overall process was not Poisson; he used the terminology, "a non-Markovian quasi-Poisson process". We derive here the main stationary point properties for the more general version of Moran's process.

Arbitrary time initial conditions specify that the process is in the first or second interval of a pair, denoted by $S = 1$ or 2, and the time or times forward to the completion of a pair, denoted by L_1, L_2. The arbitrary time distribution of these quantities, derived from the limiting process "time t' as t' tends to infinity," is found to be

$$\text{prob}(S=1,\ x < L_1 < x + dx,\ y < L_2 < y + dy)/dx\,dy$$
$$= (2\mu)^{-1} \int_{u=x}^{\infty} f(u,y)\,du, \tag{2.2.1}$$

$$\text{prob}(S=2,\ y < L_2 < y + dy)/dy = (2\mu)^{-1} \int_{v=y}^{\infty} f(v)\,dv\ . \tag{2.2.2}$$

The arbitrary event initial conditions are of course evident, but for verification, may be calculated as

$$\text{prob}(S_A=1,\ x < L_{A,1} < x + dx,\ y < L_{A,2} < y + dy)/dx\,dy$$
$$= \frac{1}{2}\, f(x,y), \tag{2.2.3}$$

$$\text{prob}(S_A=2,\ y < L_{A,2} < y + dy)/dy = \frac{1}{2}\, f(y)\ . \tag{2.2.4}$$

It follows from (2.2.3) and (2.2.4) that the arbitrary interval p.d.f of X in the point process is f(x), and that the joint interval p.d.f. of X_1 and X_2 is given by

$$f_{X_1,X_2}(x_1,x_2) = \frac{1}{2}\{f(x_1,x_2) + f(x_1)\ f(x_2)\}. \tag{2.2.5}$$

Thus we have $\rho_1 = \text{corr}(X_1,X_2) = \frac{1}{2}\rho$, and consequently that $-\frac{1}{2} \leq \rho_1 \leq \frac{1}{2}$ is an implicit restriction of this model. For the higher order distributions we find

$$f_{X_1,\ldots,X_{2r}}(x_1,\ldots, x_{2r}) = \frac{1}{2}\{\prod_{i=1}^{r} f(x_{2i-1},x_{2i}) + f(x_1)\prod_{i=1}^{r-1} f(x_{2i},x_{2i+1})\ f(x_{2r})\}, \tag{2.2.6}$$

$$f_{X_1,\ldots,X_{2r+1}}(x_1,\ldots, x_{2r+1}) = \frac{1}{2}\{\prod_{i=1}^{r} f(x_{2i-1},x_{2i})\ f(x_{2r+1}) + f(x_1)\prod_{i=1}^{r} f(x_{2i},x_{2i+1})\}, \tag{2.2.7}$$

and thus it is possible to write down the likelihood for this model. This is quite unusual for non-recurrent point processes. From (2.2.6) and (2.2.7) it may be seen that the intervals X_1,X_{i+1} for $i \geq 2$ are independently distributed and that the interval spectrum is just $\pi^{-1}(1+\rho\cos\omega)$. We may also use (2.2.6) and (2.2.7) to derive the intensity function and related quantities for the process.

The stationary asynchronous counting distribution may be obtained from the initial conditions (2.2.1) and (2.2.2). The result is most easily written in terms of Laplace transforms. Let therefore $R^*(s)$ denote the transform of the survivor function corresponding to $f(x)$, $f_+^*(s)$ denote the transform corresponding to the convolution $f_+(x) = \int_{y=0}^{x} f(y,x-y)\ dy$, and let

$$G^*(s) = \int_{u=0}^{\infty} du \int_{v=0}^{\infty} dv\ \exp\{-s(u+v)\}\{\int_{a=u}^{\infty} f(a,v)\ da\}$$

$$= s^{-1}\{f^*(s) - f_+^*(s)\}. \tag{2.2.8}$$

Then considering the various possible configurations of events and letting $P_j^*(s)$ denote the Laplace transform of $\text{prob}\{N(0,t) = j\}$, we find

$$P_j^*(s) = \begin{cases} (\mu s)^{-1}\{\mu - R^*(s)\} & (j=0) \\ (2\mu s)^{-1}\{R^*(s)-G^*(s)\}+ (2\mu)^{-1}\{R^*(s)\}^2 & (j=1) \\ \mu^{-1}G^*(s)\ R^*(s)\{f_+^*(s)\}^{n-1} & (j=2n,\ n \geq 1) \\ (2\mu)^{-1}\ \{G^*(s)\}^2\{f_+^*(s)\}^{n-1}+(2\mu)^{-1}\{R^*(s)\}^2\{f_+^*(s)\}^n & (j=2n+1,\ n \geq 1). \end{cases} \quad (2.2.9)$$

We note that the stationary counting distribution depends on $f(x,y)$ through its convolution density $f_+(x)$. For the mean of the stationary counting distribution (2.2.9) yields $\mu^{-1}t$ and for the transformed variance-time function it gives

$$\begin{aligned} V^*(s) = {} & (\mu s^2)^{-1}+\mu^{-1}\{1-f_+^*(s)\}^{-3}[2G^*(s)R^*(s)+\{G^*(s)\}^2+f_+^*(s)\{R^*(s)\}^2] \\ & -\mu^{-1}\{1-f_+^*(s)\}^{-2}[6G^*(s)R^*(s)+\{G^*(s)\}^2 \\ & + f_+^*(s)\{R^*(s)\}^2] - 2(\mu^2 s^3)^{-1}\ . \end{aligned} \quad (2.2.10)$$

Asymptotically, $(t \to \infty)$, the variance time curve is linear, and its slope and intercept may be obtained from renewal expansions of (2.2.10). The asymptotic slope is simply $(1 + \rho)\sigma^2/\mu^3$, which is the factor $(1 + \rho)$ different from a renewal process.

Moran (1967) constructed a dependent form of $f(x,y)$ such that the marginal densities were e^{-x}, and the density of their sum was xe^{-x}, the "independent" convolution of two of these. Thus, the resulting process had exponential interevent intervals, a Poisson counting distribution, and yet was not a Poisson process; other constructions may also be used to produce such effects. The joint distribution chosen by Moran has the form

$$f(x,y) = \exp\{-(x + y)\} + f_1(x,y)\ , \quad (2.2.11)$$

where $f_1(x,y)$ is defined to equal ε $(0 < \varepsilon < e^{-6})$ in the squares (0,2), (1,3), (2,1) and (3,0), and $-\varepsilon$ on the squares (0,3), (1,2), (2,0) and

(3,1), where (m,n) denotes the square ($m \leq x \leq m+1$, $n \leq y \leq n+1$). This density is of course rather artificial and not surprisingly has a zero correlation coefficient. It would be interesting to know if there can be any realistic joint densities with the same properties for which the correlation is not zero; our impression is that predetermining both the marginal and sum distributions makes this unlikely.

In its more general formulation, the model may be useful as a simple alternative to a renewal process in which ρ_1, with $\frac{1}{2} \leq \rho_1 \leq \frac{1}{2}$, is the only non-zero serial correlation. Bivariate analogues of exponential and gamma renewal distributions are discussed in Kibble (1941), Vere-Jones (1967), Downton (1970) and Gaver (1970).

2.3 Processes With First Order Markov Dependent Interevent Intervals

Stationary point processes with Markov-dependent intervals are most appropriately defined through the synchronous intervals $X_1, X_2, \ldots$; a general form of the Markov property is the condition, stated in terms of conditional p.d.f.'s, that for any fixed positive r,

$$f_{X_j, X_{j+1}, \ldots | X_{j-1}, X_{j-2}, \ldots,}(x_1, x_2, \ldots | y_1, y_2, \ldots)$$

$$= f(x_1, x_2, \ldots | y_1, \ldots, y_r) \ . \qquad (2.3.1)$$

This equation assumes that $X_1, X_2, \ldots$ is a stationary sequence of random variables and states that the intervals $X_j, X_{j+1}, \ldots$ are conditionally independent of $X_{j-r-1}, X_{j-r-2}, \ldots$ given the values of $X_{j-1}, \ldots, X_{j-r}$. This property may be termed the rth order Markov property; we only discuss processes with the first order Markov property. In this section we summarize the general results of Wold (1948), pointing out the limitations of their tractability, and in Sections 2.4 and 2.5 we specialize them with results obtained by Wold (1948), Cox (1955) and Lampard (1965, 1968) for individual processes. All this work is theoretical.

We first note that the (unconditional) p.d.f. of a synchronous interval X, denoted by f(x), must satisfy the equation

$$f(x) = \int_{y=0}^{\infty} f(x|y)\, f(y)\, dy \tag{2.3.2}$$

if it exists and refers to a stationary synchronous sequence. In principle, $f(x)$ and $f(x|y)$ are all we need to know about the process since the joint p.d.f. of any subsequence from $X_1, X_2, \ldots$ can then be written down. However, in practice, the equation (2.3.2), which is of the Fredholm type, is not very tractable. The basic counting process is one giving the number of events $\tilde{M}(t)$ in $(0,t]$ conditional on (a) an event at zero and (b) the time y back to the previous event. Writing $F(x|y)$ for the conditional distribution function corresponding to $f(x|y)$, we then clearly have by considering the time forward to the next event

$$\text{prob}\{\tilde{M}(t) = 0|y\} = 1 - F(t|y) \ , \tag{2.3.3}$$

$$\text{prob}\{\tilde{M}(t) = n|y\} = \int_{x=0}^{t} \text{prob}\{\tilde{M}(t-x) = n-1|x\}\, dF(x|y) \quad (n \geq 1). \tag{2.3.4}$$

The usual synchronous counting distribution follows from (2.3.3) and (2.3.4), once their solution has been obtained, as

$$\text{prob}\{M(0,t) = n\} = \int_{y=0}^{\infty} \text{prob}\{\tilde{M}(t) = n|y\}\, f(y)\, dy \qquad (n \geq 0) \ . \tag{2.3.5}$$

For the (stationary) asynchronous counting distribution we argue that at an arbitrary time the joint p.d.f. of the backward and forward recurrence time is $f(x + y)/\mu$ where μ is the mean corresponding to $f(x)$. It then follows, using $F(x)$ for the distribution function associated with $f(x)$, that

$$\begin{aligned}\text{prob}\{N(0,t) = 0\} &= \int_{x=0}^{\infty} dx \int_{y=t}^{\infty} \mu^{-1} f(x + y)\, dy \\ &= \int_{y=t}^{\infty} \mu^{-1}\{1 - F(y)\}\, dy \ ,\end{aligned} \tag{2.3.6}$$

and

$$\text{prob}\{N(0,t) = n\}$$

$$= \int_{x=0}^{\infty} \int_{y=0}^{t} \mu^{-1} f(x+y)\ \text{prob}\{\tilde{M}(t-y)=n-1 \mid x+y\}\ dx\ dy \quad (n \geq 1). \quad (2.3.7)$$

It is interesting to note that Wold's derivation of the general and important stationary result (2.3.6) appears to be the first (Wold, 1948, Remark 4d). This appears to be the extent to which general results for this model can be developed.

2.4 Wold's and Cox's Particular Results for Markov Dependent Processes

Wold (1948) and Cox (1955) considered the first order Markov dependent case in which

$$F(x|y) = 1 - e^{-k(y)x} , \quad (2.4.1)$$

with $k(y)$ taking specific forms. Thus the interevent intervals are exponential, given the length of the previous interval. The integral equation (2.3.2) becomes in this case

$$f(x) = \int_{y=0}^{\infty} k(y)\ e^{-k(y)x}\ f(y)\ dy . \quad (2.4.2)$$

An explicit solution is still difficult, but letting μ_r denote the rth moment of $f(x)$, and taking moments of each side of (2.4.2) we obtain

$$\mu_r = \Gamma(1 + r)\ E[\{k(x)\}^{-r}] \qquad (r \geq 1) . \quad (2.4.3)$$

Wold's particularly convenient expression for $k(y)$ was $y^{-1/2}$, and then (2.4.3) gives

$$\mu_r = \Gamma(1 + r)\mu_{r/2} = \prod_{n=0}^{\infty} \Gamma(1 + r/2^n) . \quad (2.4.4)$$

This produces $\mu_1 = 0.707\ldots$, $\mathrm{var}(X) = 0.914\ldots$, and a coefficient of variation squared of about 1.8. Wold used (2.4.4) and a Mellin inversion formula to obtain

$$f(x) = \sum_{n=0}^{\infty} x^n \sum_{m=0}^{n} A_{n,m}(\log x)^m . \qquad (2.4.5)$$

No general expressions appear to be available for the $A_{n,m}$, but Wold gave the first few numerically and remarked that they appear to converge quickly. A numerical approximation for the first serial correlation should be possible using (2.4.1) and (2.4.5). There is clearly much further work which needs to be done in relation to the counting properties of the process.

Cox (1955) took $k(y) = \lambda_0(1 + \lambda_1 y)$, where $\lambda_0 > 0$, and $\lambda_1 > 0$ or small enough for negative values of $k(y)$ to be neglected; his approximate solution of (2.4.2) is the p.d.f.

$$f(x) = \lambda_0 \exp(-\lambda_0 x) \{1 + \lambda_1(\lambda_0^{-1} - x)\} + o(\lambda_1^2) . \qquad (2.4.6)$$

The use of this result in conjunction with (2.3.3)-(2.3.7) is not particularly tractable, although it does not seem hopeless for numerical work. Cox obtained an approximate test of $\lambda_1 = 0$ based on the likelihood of a series of observations for his form of $k(y)$.

2.5 Lampard's Results for a Particular Markov Dependent Process

Lampard (1965, 1968) and Lee (1966) have discussed a stochastic point process in which the intervals have the first order Markov property, and which is rather more tractable than Wold's or Cox's examples; they remarked that it was being used in neurophysiological experiments. Two Poisson processes A and B are considered; process A generates pulses at rate α and process B generates pulses at rate β. At time zero a counter for process B is set to register a point process event at the (N_0+r+1)th pulse in process B; the N_0 may be regarded as an initial condition and r is a parameter of the process. During this time another counter records the number of pulses N_1 which have occurred in process A. The counter B is then automatically reset to register the next event at the (N_1+r+1)th pulse in process B following the previous event (the $\{N_0+N_1+2(r+1)\}$ pulse in process B). A series of events is generated in this manner, so that the jth synchronous interevent interval is the sum of $N_{j-1}+r+1$ interpulse intervals in process B. The first order Markov property is evident from this description.

To calculate $f(x|y)$ of (2.3.2) let us consider further this density function conditional on the number n of pulses from process A in the interval of length y, and denote the new conditional p.d.f by $f(x|y,n)$. Then

$$f(x|y) = \sum_{n=0}^{\infty} f(x|y,n)\, e^{-\alpha y}(\alpha y)^n/n! \,, \tag{2.5.1}$$

and it is clear from the preceding description that

$$f(x|y,n) = e^{-\beta x}\, \beta(\beta x)^{n+r}/(n+r)! \,, \tag{2.5.2}$$

thus not involving y.

Hence,

$$\begin{aligned} f(x|y) &= \sum_{n=0}^{\infty} \{e^{-\beta x}\, \beta(\beta x)^{n+r}/(n+r)!\}\{e^{-\alpha y}(\alpha y)^n/n!\} \,, \\ &= \beta e^{-\beta x-\alpha y} \sum_{n=0}^{\infty} \frac{(\alpha y)^n\, (\beta x)^{n+r}}{n!\, (n+r)!} \,, \\ &= \beta e^{-\beta x-\alpha y} \left[\frac{\beta x}{\alpha y}\right]^{r/2} I_r\{2(\alpha\beta xy)^{1/2}\} \,, \end{aligned} \tag{2.5.3}$$

where $I_r(x)$ is the modified Bessel function of the first kind of order r and argument x. Lampard then asserts that the unique solution to (2.3.2) in this case is given by

$$f(x) = (\beta-\alpha)\, \frac{\{(\beta-\alpha)x\}^r}{r!}\, e^{-(\beta-\alpha)x} \,. \tag{2.5.4}$$

This is of course a gamma distribution of index r+1 and parameter $\beta-\alpha$; since the parameter must be positive it is evident that the process will not attain equilibrium unless $\alpha/\beta < 1$.

The joint p.d.f. of adjacent synchronous intervals, f(x,y), may be obtained from (2.5.3) and (2.5.4) and becomes

$$f(x,y) = \beta(\beta-\alpha)\frac{\{(\beta-\alpha)/\alpha\}^r}{r!} e^{-\beta(x+y)}(\alpha\beta xy)^{r/2} I_r\{2(\alpha\beta xy)^{1/2}\}. \tag{2.5.5}$$

This is the same bivariate gamma distribution as discussed by Kibble (1941), Vere-Jones (1967) and Gaver (1970). It has double-Laplace transform

$$f^*(s_1,s_2) = \left[\frac{\beta(\beta-\alpha)}{\beta(\beta-\alpha)+\beta s_1+\beta s_2+s_1 s_2}\right]^{r+1}, \tag{2.5.6}$$

and correlation coefficient

$$\rho_1 = \alpha/\beta . \tag{2.5.7}$$

This is an important result. It shows that the interval correlation does not depend on the shape parameter (r) of the gamma interevent distribution. Further it will allow us to estimate α, β and r using maximum likelihood or moments for the gamma distribution. Also by setting $s_1=s_2$ in (2.5.6) we may obtain the distribution of the sum of two adjacent intervals; we note that this is not also a gamma distribution. The distribution for the sum of more than two intervals is extremely complicated, and has been found after voluminous calculations by Lee (1966, section 8) who discusses this process in great detail. There does not seem to be any hope of using it to obtain the intensity function or variance-time function for the process. In fact, all the asynchronous properties of the process seem impossible to obtain, either directly or indirectly.

3. DOUBLY STOCHASTIC POISSON PROCESSES

3.1 Introduction and General Theory for Continuous Rate Processes

Cox (1955) considered a Poisson process whose rate of events varied according to a stochastic process itself; such point processes, considered unconditionally on the rate process, have been called *doubly stochastic Poisson processes* (d.s.P. processes) by Bartlett (1963). Elementary results for a process of this type were first given by Cox (1955) and by Bartlett in the discussion to Cox's paper. Further results and references are to be found in Bartlett (1963), Cox and Lewis (1966), Westcott (1970), Grandell (1971) and Serfozo (1972). Some particular uses of d.s.P. processes have been in textile technology: Cox (1955); the theory of noise; Darling and Siegert (1957), Siegert (1957); photon counting of Gaussian light: Bédard (1966), Karp and Clark (1970), Macchi (1971); the reliability of equipment: Gaver (1963); and biological variation: Barndorff-Nielson and Yeo (1969).

The class of d.s.P. processes is a wide one, covering both stationary and non-stationary point processes, and processes which are generated by continuous rate processes, or non-continuous rate processes (such as when the rate changes its value only at particular instants; see Gaver, 1963). When there is no distributional knowledge of the underlying rate process, the point process becomes stochastic in two ways. Firstly, the random variation of the Poisson rate, and secondly, the usual Poisson process variability; it should further be noted that conditional on the rate process, a d.s.P. process is a time-dependent Poisson process. We shall restrict ourselves here to stationary d.s.P. processes, as appears to be the case with much of the earlier work. The general theory is basically restricted to the continuous rate case and will be given in this section; the rate will then be explicitly specified by (continuous: non-continuous) processes, and individual results given in Sections 3.2 to 3.5.

We denote the rate of events at time τ by λ_τ and assume that λ_τ is given by a continuous, non-negative, stationary stochastic process beginning at $\tau = 0$. In particular we let $E(\lambda_\tau) = \lambda$ $(0 < \lambda < \infty)$,

$\mathrm{var}(\lambda_\tau) = \sigma^2$ $(0 < \sigma^2 < \infty)$ and $\mathrm{corr}(\lambda_\tau, \lambda_{\tau+t}) = \rho(t)$ with $-1<\rho(t) < 1$. It will be clear that the stationarity of the d.s.P. process depends directly on the stationarity of the rate process.

Arbitrary time initial conditions for d.s.P. processes just specify the stationarity of the rate process for $\tau = 0$ onwards. To obtain the asynchronous counting distribution for the number of events $N(0,t)$ in $(0,t]$ we follow the conditional arguments of Cox (1955), that given λ_τ for $0 < \tau < t$, $N(0,t)$ has a Poisson distribution with mean

$$m(t) = \int_{\tau=0}^{t} \lambda_\tau \, d\tau . \tag{3.1.1}$$

Thus, explicitly,

$$\begin{aligned} &\mathrm{prob}\{N(0,t) = k | \lambda_\tau \text{ for } 0 < \tau < t\} \\ &\quad = e^{-m(t)}\{m(t)\}^k/k! \qquad (k \geq 0) . \end{aligned} \tag{3.1.2}$$

Here $m(t)$ is the integral of λ_τ considered as an ordinary continuous function. Specification of a stochastic process for λ_τ implies that $m(t)$, now regarded as a stochastic integral, is a stochastic process, with for example, a distribution for each t $(0 \leq t < \infty)$. The determination of this distribution is the main objective in analyzing d.s.P. processes. Now assuming the distribution of $m(t)$, we may write (3.1.2) unconditionally as

$$\mathrm{prob}\{N(0,t) = k\} = E[e^{-m(t)}\{m(t)\}^k/k!] \qquad (k \geq 0) . \tag{3.1.3}$$

For the probability generating function (p.g.f) of $N(0,t)$ we then have the useful result

$$\phi(z;t) = E[\exp\{(z-1)\, m(t)\}] . \tag{3.1.4}$$

Thus, in words, the p.g.f. of $N(0,t)$ evaluated at z is equal to the m.g.f. of $m(t)$ evaluated at $z-1$; generalizing to the intervals $(0,t_1]$,

$(t_1, t_1+t_2]$, a similar result may be written down. Important point process results follow easily from (3.1.4); in particular

$$E\{N(0,t)\} = E\{m(t)\} = \lambda t , \tag{3.1.5}$$

$$\operatorname{var}\{N(0,t)\} = \lambda t + \operatorname{var}\{m(t)\} . \tag{3.1.6}$$

The result (3.1.1) suitably redefined, and (3.1.2)-(3.1.6) are also valid for non-stationary and non-continuous rate processes. In stationary continuous cases, var{m(t)} can easily be written in terms of the autocorrelation function $\rho(u)$ of the λ_τ process, and is

$$\operatorname{var}\{N(0,t)\} = \lambda t + 2\sigma^2 \int_{u=0}^{t} (t-u)\, \rho(u)\, du . \tag{3.1.7}$$

We see from (3.1.6) or (3.1.7), as noted by Cox (1955), that a general property of a d.s.P. is that it is over dispersed as compared with the Poisson process. General properties of the interevent intervals, both asynchronously and synchronously, do not appear to be available. However, a possible general approach to the asynchronous intervals is to take expectations over the corresponding Poisson process results, just as we do to the asynchronous counting distribution; this has been adopted in an application by Macchi (1971). For the synchronous properties, the arbitrary event initial condition approach does not yield much in general, mainly because the appropriate initial behavior of the rate process is difficult to specify. However, with particular processes, and a basic synchronous, asynchronous result following from the arbitrary event definition, the synchronous interevent distribution is at least sometimes available; we give two instances in Sections 3.2 and 3.3 of the present paper.

We continue this survey of general results for stationary d.s.P. processes by mentioning the connections between the N(t) and λ_t processes. The most fundamental relation is in terms of characteristic functionals and was noted by Bartlett in the discussion of Cox (1955); amplification is given in Bartlett (1963) and in Westcott (1972) the equivalent probability generating functional is discussed. Bartlett (1963) noted in particular that the covariance density function of N(t) and the autocorrelation function of λ_t are identical; he also gave a sampling theory for the estimated spectrum of a d.s.P. process. Grandell (1970) has studied the connections between N(t) and m(t); he showed that m(t) and N(t) - m(t) are uncorrelated, and related the asymptotic behavior of N(t) to the variance of m(t). Specifically, he shows that when t^{-1} var{m(t)} $\to 0$ as $t \to \infty$, the standardized N(t) tends to normality; however, if t^{-1} var{m(t)} $\to k$ (> 0) and the distribution function of the standardized m(t) $\to G(x)$, then the standardized N(t) tends to $G\{x\surd(1+k^{-1})\} * \Phi\{x \surd(1+k)\}$, where $\Phi(x)$ is the standardized normal distribution function and an asterisk denotes convolution. Serfozo (1972) has related in more mathematical detail the behavior of m(t) and N(t). Westcott (1972) has obtained mixing results for d.s.P. processes, indicating the asymptotic independence of the number of events in widely separated intervals when the λ_t process also satisfies suitable mixing conditions. The questions of when particular classes of stationary point process can be represented as d.s.P. processes are mostly unanswered at present, although Kingman, in the discussion of Bartlett (1963) and in Kingman (1964), has provided the answer for renewal processes; we touch on his result in Section 3.4. Special cases of d.s.P. processes are equivalent to special cases of point processes with cluster structure, as will be mentioned in Sections 3.2 and 4.4. We have not attempted to be exhaustive here with our references and further work may, for instance, be found and traced through these proceedings.

3.2 The Shot-Noise Doubly Stochastic Poisson Process

In this section we consider the d.s.P. process when λ_t is given by a stationary shot-noise process. The shot-noise process in its simplest form consists of a Poisson process, each event of which occurring prior to time τ contributes a "shot" to λ_τ; the contributions are additive and for an event occurring at time $\tau - x$, the shot magnitude is $g(x)$ at time τ. The function $g(x)$ needs to be non-negative and tend to zero sufficiently rapidly as $x \to \infty$ for several infinite integrals of powers of $g(x)$ to be finite. We do not know of any applications of this d.s.P. process.

Letting $N_\alpha(u)$ be a Poisson counting process of rate α which starts at $\tau = 0$, we have

$$\lambda_\tau = \int_{u=0}^{\tau} g(\tau-u)\, dN_\alpha(u) . \qquad (3.2.1)$$

The random fluctuations in λ_τ are referred to as the "shot effect". The stationary form of this process is obtained as $\tau \to \infty$, and in this case it is well known (Cox and Miller, 1965, Chap. 9, for instance) that the m.g.f. of λ_τ is given by

$$E[\exp\{(z-1)\}\lambda_\tau] = \exp[\,\alpha \int_{u=0}^{\infty} [\exp\{(z-1)g(u)\}-1]\, du] , \qquad (3.2.2)$$

and thus that the mean and variance are

$$\lambda = \alpha \int_{u=0}^{\infty} g(u)\, du, \qquad \sigma^2 = \alpha \int_{u=0}^{\infty} \{g(u)\}^2\, du . \qquad (3.2.3)$$

The corresponding joint m.g.f. of λ_τ and $\lambda_{\tau+t}$ is

$$E[\exp\{(z_1-1)\lambda_\tau + (z_2-1)\lambda_{\tau+t}\}]$$

$$= \exp\left[\alpha \int_{u=0}^{\infty} [\exp\{(z_1-1)g(u)+(z_2-1)g(t+u)\}-1]\, du + \alpha \int_{u=0}^{t} [\exp\{(z_2-1)g(u)\}-1]\, du\right] , \qquad (3.2.4)$$

and this gives the autocovariance function of the stationary process as

$$c(t) = \alpha \int_{u=0}^{\infty} g(u)\, g(u+t)\, du . \qquad (3.2.5)$$

Our main concern now in the d.s.P. process application is the distribution of $m(t)$, the integrated process, with an arbitrary time beginning. Thus using the shot-noise process beginning at $t = 0$ we consider the distribution of

$$m(t;T) = m(T + t) - m(t) \tag{3.2.6}$$

as $T \to \infty$. From (3.2.1) we have

$$m(t;T) = \int_{u=T}^{T+t} du \int_{x=0}^{u} g(u-x)\, dN_{\alpha}(x) . \tag{3.2.7}$$

Now interchanging the order of integrations in (3.2.7) gives

$$m(t;T) = \int_{x=0}^{T+t} K(t,x;T)\, dN_{\alpha}(x) , \tag{3.2.8}$$

where

$$K(t,x;T) = \begin{cases} G(T+t-x) - G(T-x) & (0 < x < T) , \\ G(T+t-x) & (T < x < T+t) , \end{cases}$$

and

$$G(u) = \int_{v=0}^{u} g(v)\, du .$$

To evaluate the m.g.f. of $m(t;T)$ we use (3.2.8) and the conditional property of the Poisson process that the events in $(0,T+t]$ are uniform and independently distributed over this interval, given their number. Writing x_i for the times of the events in $(0,T+t]$, $i=1,2,\dots,N_{\alpha}(T+t)$, we have

$$m(t;T) = \sum_{i=0}^{N_{\alpha}(T+t)} K(t,x_i;T) \tag{3.2.9}$$

and thus

$$E[\exp\{(z-1)\, m(t;T)\}]$$

$$= \sum_{n=0}^{\infty} e^{-\alpha(T+t)} \{\alpha(T+t)\}^{n} (n!)^{-1} [\, (T+t)^{-1} \int_{x=0}^{T+t} \exp\{(z-1)K(t,x;T)\}dx]^{n}$$

$$= \exp\left[\alpha \int_{x=0}^{T+t} [\exp\{(z-1)\, K(t,x;T)\} - 1]\, dx\right]$$

$$= \exp\left[\alpha \int_{u=0}^{T} \{\exp[(z-1)\{G(t+u)-G(u)\}]-1\}\, du\right.$$

$$\left. + \alpha \int_{u=0}^{t} [\exp\{(z-1)G(u)\}-1]du\right] . \tag{3.2.10}$$

Letting $T \to \infty$ in (3.2.10) the m.g.f. of $m(t)$ with an arbitrary time beginning is obtained; by (3.1.4) this is equivalently $\phi(z;t)$, the p.g.f. of the stationary shot-noise d.s.P. process. Explicitly for the mean and variance of $N(0,t)$ we have from (3.2.10) or by the general results in Section 3 and (3.2.5),

$$E\{N(0,t)\} = \alpha G(\infty) t , \tag{3.2.11}$$

and

$$\text{var}\{N(0,t)\} = \alpha G(\infty) t + 2\alpha \int_{u=0}^{t} (t-u)\{ \int_{v=0}^{\infty} g(v)\, g(v+u)\, dv\}\, du . \tag{3.2.12}$$

For the synchronous properties we use a result due to Khintchine (1960, Section 10) which relates the synchronous and asynchronous counting distributions for $(0,t]$. Letting $\psi(z;t)$ denote the p.g.f. of the former, the result, which depends on the finiteness of λ and orderliness, is

$$\psi(z;t) = [\lambda(z-1)]^{-1}\, D_t \phi(z;t) . \tag{3.2.13}$$

Applying this to (3.2.10) we obtain

$$\psi(z;t) = \{G(\infty)\}^{-1} [\int_{u=0}^{\infty} g(t+u)\, e^{(z-1)\{G(t+u)-G(u)\}} du$$

$$+ (z-1)^{-1} \{e^{(z-1)G(t)} - 1\}]\, \phi(z;t) \tag{3.2.14}$$

The survivor function of a synchronous interval is obtained by setting $z = 0$ in (3.2.14); we regret that we cannot report a numerical assessment of this result. The mean of (3.2.14) is

$$\alpha G(\infty)t + \frac{1}{2}G(\infty) - \{G(\infty)\}^{-1}\int_{u=0}^{\infty} g(t+u)G(u)\,du \qquad (3.2.15)$$

and will lead immediately to the intensity function and Bartlett spectrum. Nothing appears to be known about the serial correlations of the synchronous intervals.

At this point, we remark that the shot-noise d.s.P. process is also a particular Neyman-Scott cluster process. This type of process is also considered in Section 4.4, but we can observe here that each event in the shot-noise Poisson process may be regarded as initiating an independent, finite, time-dependent Poisson process whose instantaneous rate after a time x is g(x); superposition of these processes then gives a particular Neyman-Scott process. This observation was first made by Bartlett (1964) who completely identified, by means of characteristic functionals, the shot-noise d.s.P. process with the Neyman-Scott process, in which the cluster size distribution is Poisson with mean $G(\infty)$ and the associated density function is $g(x)/G(\infty)$. The explanation of this is that with a cluster defined as a Poisson number of events with mean $G(\infty)$, each located at distances independently and identically distributed with p.d.f. $g(x)/G(\infty)$, the resulting superposition of events is a time-dependent Poisson process of rate g(x) at distance x from the cluster center. Thus, for instance, our result (3.2.10) may be compared with (4.4.8).

3.3 The Ornstein-Uhlenbeck Doubly Stochastic Poisson Process

In this section we consider the d.s.P. process for which the rate process is the square of a stationary Ornstein-Uhlenbeck process (O.U. process). Cox and Lewis (1966, Section 7.2) briefly mention the unsquared O.U. process for the rate, although note that it is not strictly positive. In considering the square of an O.U. process for the rate we can make use of earlier results on the distribution of the integral of the square of an O.U. process, our m(t) process. This distribution has been discussed by Kac and Siegert (1947), Darling and Siegert (1957) and

Siegert (1957) in relation to noise phenomena in radio receivers and by Bédard (1966), Karp and Clark (1970), amongst others, in relation to photon emissions by sensitive surfaces subject to light (in this application it should be noted that the intensity function is the sum of squares of two independent O.U. processes, this resulting from the complex envelope of the light). A general treatment in relation to biological variation is given by Barndorff-Nielson and Yeo (1970). It is fortunate that the work of Siegert (1957) applies since this gives an explicit result for the stationary counting distribution of the d.s.P. process; the other work gives less explicit results under more general assumptions.

We let U(x) be a stationary O.U. process, as defined in Cox and Miller (1965, Chap. 5), for instance, in which

$$E\{ U(x)\} = 0, \qquad \mathrm{var}\{U(x)\} = \sigma^2 ,$$
$$\mathrm{corr}\{U(x),\ U(x+y)\} = e^{-\beta|y|} , \tag{3.3.1}$$

and we assume the process begins at time zero; this process is both Gaussian and Markovian.

We are interested in the distribution of

$$m(t) = \alpha \int_{x=0}^{t} \{U(x)\}^2\, dx . \tag{3.3.2}$$

For its mean and variance, and hence for the mean and variance of N(t), we have by direct expectation methods from (3.3.1) and (3.3.2) that

$$E\{m(t)\} = \alpha\sigma^2 t , \tag{3.3.3}$$

$$\mathrm{var}\{m(t)\} = (\alpha^2\sigma^4/2\beta^2)(2\beta t - 1 + e^{-2\beta t}) . \tag{3.3.4}$$

Unfortunately, the distribution of m(t) is not so readily available; there have been two approaches in the mentioned works. The first is due to Darling and Siegert (1957) and has antecedents in Kac and Siegert

(1947, section VII.2). This requires that U(x) be expanded in a Karhunen-Loève series and leads to an expression for the m.g.f. of m(t) as the reciprocal square root of a Fredholm determinant of an integral operator (this is an infinite product of terms each involving a different eigenvalue of a Fredholm equation. These are determined by Kac and Siegert (1947). The Karhunen-Loève expansion is also used by Karp and Clark (1970) to obtain the corresponding result for the sum of squares of two O.U. processes, in agreement with Bédard (1966). The second method is due to Siegert (1957) and also has antecedents in Kac and Siegert (1947); a differential equation is derived for the m.g.f. of m(t) and the explicit solution obtained is given, in our notation, by

$$E\{e^{zm(t)}\}$$
$$= e^{\beta t/2}\{\cosh(\beta t\eta)+\eta^{-1}(1+2\alpha\sigma^2 z/\beta)\ \sinh(\beta t\eta)\}^{-1/2}\ , \qquad (3.3.5)$$

where $\eta = (1 + 4\alpha\sigma^2 z/\beta)^{1/2}$.

It is tempting to think that a direct derivation of the simple looking result (3.3.5) is possible. Replacing z by z-1 in (3.3.5) we obtain $\phi(z;t)$ and then by (3.2.13) we have the synchronous counting distribution, and in particular the synchronous interevent distribution. Nothing seems to be known about this, and thus some numerical calculations of the density function and moments would be informative. The same is true for the joint distributions and their serial correlations and spectrum.

With general Gaussian processes, instead of the O.U. process, the first method applies, as indicated in Darling and Siegert (1957), Karp and Clark (1970) and Macchi (1971); the practical difficulty is generally that the eigenvalues cannot be obtained, and so no explicit results are possible; two exceptions are considered in Karp and Clark (1970). Barndorff-Nielson and Yeo (1969) have considered a "gamma" rate process, defined as the sum of squares of several independent and identical O.U. processes, and obtained (3.3.5) raised to the appropriate power; they call this kind of process a "negative binomial" process, although the counting process is only negative binomial in a special case. These authors also briefly consider negative binomial processes over the plane, and obtain the first two moments of the counting distribution. Finally, we note that in the communications context, it is of interest to add a deterministic process to U(x) which is then the "noise" process; these aspects are mentioned in Kac and Siegert (1947) and Karp and Clark (1970), and would lead to time-dependent d.s.P. processes.

3.4 The Random Hazard Doubly Stochastic Poisson Process

In considering the failure behavior of equipment subject to a fluctuating environment, Gaver (1963) supposed that the rate of a Poisson process alternated between two levels λ_1 and λ_2. The two levels endure for times forming an alternating renewal process with interval p.d.f.'s $f_1(\cdot)$ and $f_2(\cdot)$, respectively. Since it is supposed that the change over instants of the levels are not observable, the resulting failure process becomes an interesting point process, and it may identified as being of the doubly stochastic Poisson type. It differs from the shot-noise and O.U. —d.s.P. process in the significant way that its rate process is not continuous; thus the general theory of Section 3.1 either does not immediately apply or does not give the simplest approach. Fairly complete results for this type of d.s.P. process can be derived, and will be given elsewhere; here we shall outline the marginal stationary point properties of counts and intervals. The tractability of the process may be attributed to the simple switching of the rate levels, in contrast to the difficulties of a continuously changing rate. Our results are obtained from direct renewal techniques based on the instants of level change which are regeneration points in the process.

Arbitrary time initial conditions for the random hazard d.s.P. process are just the arbitrary time initial conditions for an alternating renewal process; this is a specific case of a Markov renewal process, and its initial conditions have been given in Section 2.2. Letting ν_1 and ν_2 be the means corresponding to $f_1(\cdot)$ and $f_2(\cdot)$, respectively, and denoting the type and length of initial interval by S and L respectively, the result (2.1.4) with $p_1 = p_2 = 0$ gives

$$\text{prob}\{S=i,\ L \geq x\} = \frac{\nu_i}{\nu_1 + \nu_2} \int_{u=x}^{\infty} \frac{R_i(u)}{\nu_i}\, du \qquad (i = 1,2)\ . \qquad (3.4.1)$$

For the stationary counting distribution of the interval (0,t] we have that the conditional Poisson mean m(t) is given by

$$m(t) = \lambda_1 A(t) + \lambda_2 [t - A(t)]\ , \qquad (3.4.2)$$

where A(t) denotes the total length of type 1 renewal intervals in (0,t]; this was noted in Gaver (1963). We now use (3.4.2) to obtain the m.g.f. of m(t) conditional on the type of the first renewal interval as given by the initial conditions (3.4.1). Thus, given a type 1 interval first, which has probability $\nu_1(\nu_1 + \nu_2)^{-1}$, it is evident from considering the possible configurations of the renewal intervals and the conditional Poisson m.g.f. of m(t), that the unconditional m.g.f. of m(t) evaluated at z-1 may be written

$$\int_{x=0}^{t} \exp[(z-1)\{\lambda_1 x+\lambda_2(t-x)\}]\ \nu_1^{-1} R_1(x)\ R_2(t-x)\ dx$$
$$+ \sum_{r=1}^{\infty} \int_{x=0}^{t} \exp[\,(z-1)\{\lambda_1 x+\lambda_2(t-x)\}]$$
$$\times\{\nu_1^{-1} R_1(x) * f_1^{(r)}(x)\ f_2^{(r)}(t-x) * R_2(t-x)$$
$$+ \nu_1^{-1}\ R_1(x) * f_1^{(r-1)}(x) * R_1(x)\, f_2^{(r)}(t-x)\}]\ dx\ . \qquad (3.4.3)$$

A similar result holds when the initial conditions specify a type 2 interval first. Combining the results for each case, the Laplace transform of $\phi(z;t)$, the p.g.f. of N(t), becomes, using (3.1.4),

$$\phi^*(z;s) = \frac{1}{\nu_1+\nu_2}\left(\frac{\nu_1}{s+\lambda_1(1-z)} + \frac{\nu_2}{s+\lambda_2(1-z)}\right) - \frac{(\lambda_1-\lambda_2)^2}{\nu_1+\nu_2}$$

$$\times \frac{(1-z)^2}{\{s+\lambda_1(1-z)\}\{s+\lambda_2(1-z)\}} \; \frac{R_1^*\{s+\lambda_1(1-z)\}\,R_2^*\{s+\lambda_2(1-z)\}}{1-f_1^*\{s+\lambda_1(1-z)\}f_2^*\{s+\lambda_2(1-z)\}} . \tag{3.4.4}$$

The mean and variance of the stationary counting process may then be calculated as

$$E\{N(0,t)\} = (\nu_1+\nu_2)^{-1}(\lambda_1\nu_1 + \lambda_2\nu_2)t , \tag{3.4.5}$$

and

$$LT[\mathrm{var}\{N(0,t)\}]$$

$$= (\nu_1+\nu_2)^{-1}(\lambda_1\nu_1+\lambda_2\nu_2)s^{-2} + \{2(\nu_1+\nu_2)^{-2}(\lambda_1-\lambda_2)^2\nu_1\nu_2$$

$$\times s^{-1}[s^{-1}\{1-f_1^*(s)f_2^*(s)\} - (\nu_1\nu_2)^{-1}(\nu_1+\nu_2)\; R_1^*(s)\; R_2^*(s)]$$

$$\times [s\{1-f_1^*(s)f_2^*(s)\}]^{-1}\} . \tag{3.4.6}$$

We may use (3.4.6) to obtain a linear asymptotic expansion for the variance time curve. Some numerical values of its slope and intercept are given in Tables 1 and 2; the three entries at each location correspond to type 1 and type 2 intervals which are both exponential, gamma 5 and periodic. We see quantitatively that the slope is largest in the exponential case and least in the periodic case, and vice versa for the intercept.

Arbitrary event initial conditions may be obtained by our general method, and specify the type S_A and duration L_A of the first renewal interval; their joint distribution is given by

$$\mathrm{prob}(S_A=i,\; L_A \geq x) = \frac{\lambda_i\nu_i}{\lambda_1\nu_1+\lambda_2\nu_2}\;\frac{R_i(x)}{\nu_i} \qquad (i = 1,2) . \tag{3.4.7}$$

Table 1. *The Asymptotic Slope of the Variance-Time Curve for Random Hazard d.s.P. Processes. Cases Tabulated are* $f_1(x)$, $f_2(x)$ *= exponential, gamma 5 and periodic with* $\nu_1 = 1$, $\nu_2 = 3$.

$\lambda_1 =$	1	3	5	8	10	15
$\lambda_2 = 1$	1.0000	2.6250	6.5000	16.5312	25.0312	59.6250
	1.0000	1.7250	2.0000	5.5062	7.8062	15.5250
	1.0000	1.5000	2.0000	2.7500	3.2500	4.5000
$\lambda_2 = 3$	3.6250	3.0000	4.6250	11.2812	18.5312	46.5000
	2.7250	3.0000	3.7250	5.6562	7.5062	14.1000
	2.5000	3.0000	3.5000	4.2500	4.7500	6.0000
$\lambda_2 = 5$	8.5000	5.6250	5.0000	8.2812	13.2812	35.6250
	4.9000	4.7230	5.0000	6.2562	7.6562	13.1250
	4.0000	4.5000	5.0000	5.7500	6.2500	7.5000

Table 2. *The Asymptotic Intercept of the Variance-time Curve for Random Hazard d.s.P. Processes. Cases Tabulated are* $f_1(x)$, $f_2(x)$ *= exponential, gamma 5 and periodic with* $\nu_1 = 1$, $\nu_2 = 3$.

$\lambda_1 =$	1	3	5	8	10	15
$\lambda_2 = 1$	0.0000	-0.8437	-3.3750	-10.3359	-17.0859	-41.3437
	0.0000	0.3262	1.3050	3.9966	6.6066	15.9862
	0.0000	0.3750	1.5000	4.5937	7.5937	18.3750
$\lambda_2 = 3$	-0.8437	0.0000	-0.8437	- 5.2734	-10.3359	-30.3750
	0.3262	0.0000	0.3262	2.0391	3.9966	11.7450
	0.3750	0.0000	0.3750	2.3437	4.5937	13.5000
$\lambda_3 = 5$	-3.3750	-0.8437	0.0000	- 1.8984	- 5.2734	-21.0937
	1.3050	0.3262	0.0000	0.7341	2.0391	8.1562
	1.5000	0.3750	0.0000	0.8437	2.3437	9.3750

The marginal distribution of a synchronous interval X then follows, after considerable calculation, to have the transformed p.d.f.

$$f_X^*(s) = \frac{1}{\lambda_1\nu_1+\lambda_2\nu_2}\left(\frac{\lambda_1^2\nu_1}{s+\lambda_1} + \frac{\lambda_2^2\nu_2}{s+\lambda_2}\right) - \frac{(\lambda_1-\lambda_2)^2}{\lambda_1\nu_1+\lambda_2\nu_2}$$

$$\times\left(\frac{s}{(s+\lambda_1)(s+\lambda_2)}\right)^2 \frac{\{1 - f_1^*(s+\lambda_1)\}\{1 - f_2^*(s+\lambda_2)\}}{1 - f_1^*(s+\lambda_1)\ f_2^*(s+\lambda_2)} . \quad (3.4.8)$$

Some values of the coefficient of variation squared of this distribution are given in Table 3, and indicate its robustness to this type of environment change.

Table 3. *The Coefficient of Variation Squared of the Times-between-events in random hazard d.s.P. processes. Cases tabulated are $f_1(x)$, $f_2(x)$ = exponential, gamma 5 and periodic with $\nu_1 = 1$, $\nu_2 = 3$.*

$\lambda_1 =$	1	3	5	8	10	15
$\lambda_2 = 1$	1.0000	1.3000	1.7826	2.5750	3.1192	4.5000
	1.0000	1.2247	1.6375	2.5518	2.8512	4.1285
	1.0000	1.1983	1.5958	2.2969	2.7868	4.0376
$\lambda_2 = 3$	1.2368	1.0000	1.0763	1.3160	1.5057	2.0189
	1.1689	1.0000	1.0700	1.2997	1.4845	1.9881
	1.1488	1.0000	1.0691	1.2984	1.4832	1.9867
$\lambda_2 = 5$	1.5806	1.0714	1.0000	1.0708	1.1607	1.4412
	1.4348	1.0638	1.0000	1.0684	1.1564	1.4355
	1.3909	1.0620	1.0000	1.0682	1.1563	1.4333

Gaver (1963) calculated an interval distribution which was conditional on the beginning of a particular type of renewal interval. Further results on the serial correlations and joint distribution of synchronous intervals will be given elsewhere.

There is some related work by Serfozo (1972, Section 4) which concerns Poisson processes with Markovian environments; this treats a many level rate process where the transitions occur according to a Markov chain and the durations at each level are exponential variates. The analysis is in terms of an imbedded two-dimensional Markov process. The asymptotic behavior of the rate and counting processes are related in a number of precisely stated theorems; in particular, several normality and convergence results are obtained.

An interesting result related to the random hazard d.s.P. process is given by Kingman (1964); he showed that a renewal process with interevent distribution F(x) is a d.s.P. process if and only if there exists a strictly positive λ and a non-decreasing function K such that for all $\theta > 0$,

$$\int_0^\infty e^{-\theta x}\, dF(x) = \lambda\{\lambda + \theta + \int_0^\infty (1 - e^{-\theta x})\, dK(x)\}^{-1} . \qquad (3.4.9)$$

The corresponding rate process can take only the values λ and 0 and the 0 to λ transitions are regenerative events. Further, when $k = K(\infty)-K(0) < \infty$, Kingman shows that the d.s.P. process is of the random hazard type with $f_1(x) = \nu_1 \exp(-x/\nu_1)$ and $\lambda_2 = 0$. In this case we find from (3.4.7) that

$$f_X^*(s) = \lambda_1\nu_1/\{\nu_1(s+\lambda_1) + 1 - f_2^*(s)\}, \qquad (3.4.10)$$

which agrees with (3.4.9).

It can also be shown that when both types of renewal distributions are exponential, the random hazard d.s.P. process is also equivalent to a two-state Markov renewal point process with two exponential interval distributions.

3.5 The Doubly Stochastic Poisson Process Driven by a Point Process

We have so far given the theory for d.s.P. processes when the rate process is a general continuous process, and in particular for the shot-noise and O.U. processes, and also for the non-continuous rate process which alternates between two levels. A more radical form of non-continuous rate process which suggests itself is when each event in a point process contributes an instantaneous shot of magnitude α; the profile of this process is thus a sequence of spikes. Let us denote the rate point process by R and assume that it is stationary. We thus have in terms of (3.1.1) that $m(t) = \alpha N_R(t)$, where α is a positive constant and $N_R(t)$ denotes the number of events from the rate process in $(0,t]$. Further from (3.1.2), and in a clear p.g.f. notation,

$$\phi\{z;t|N_R(t) = k\} = e^{\alpha k(z-1)} \qquad (k \geq 0) \, , \tag{3.5.1}$$

and hence

$$\phi(z;t) = \phi_R\{e^{\alpha(z-1)};t\} \, , \tag{3.5.2}$$

where $\phi_R(z;t)$ is the p.g.f. of $N_R(t)$. It is evident from this result that the d.s.P. process here is a compound form of the rate process, where the multiplicity of events at each R-event is independently Poisson distributed; a multiplicity of 0 thus represents a deletion. The synchronous interval distribution may be written

$$F_X(x) = \sum_{r=1}^{\infty} (1 - e^{-\alpha})(e^{-\alpha})^r \, F_R^{(r)}(x) \, , \tag{3.5.3}$$

where $F_R^{(r)}(x)$ is the distribution function of r successive synchronous intervals in the rate process. This is perhaps a somewhat degenerate type of d.s.P. process, and one which is always non-orderly.

3.6 Some Related Processes

The mixed Poisson process, introduced by Bartlett (1955), is sometimes regarded as a type of d.s.P. process. Here the rate is a random variable independent of time, and not a time-dependent stochastic process. This process has been studied as a stationary point process by McFadden (1965) and Grandell (1971). The main practical drawback of the process is its lack of ergodicity. More general types of d.s.P. process have so far received little attention. Whittle, in the discussion of Bartlett (1963), noted the characteristic function of a d.s.P. clustering process (in the sense of Section 4 of the present paper), and Serfozo (1971) generalized independent increment processes in a doubly stochastic way.

4. STATIONARY POINT PROCESSES WITH CLUSTER STRUCTURE

4.1 Structure of the Clustering

In the last part of the paper we shall discuss processes of a particular class which can provide specific interpretations to clustering or bunching effects in a stationary point process; this class of processes give suitable models for quite a variety of physical phenomena, such as the bunching in traffic (Bartlett, 1963), the breakdowns of a computer (Lewis, 1964) and the occurrences of earthquakes (Vere-Jones, 1970). Each process consists basically of two separate components. These are (1) a stationary point process of main events (cluster centers) and (2) a stationary or non-stationary point process of following subsidiary events (of finite mean number), forming a cluster. Each main event sets off a subsidiary process, and the different subsidiary processes are independent. The clustering point process can be defined as the superposition of events from both the main and subsidiary processes, or as the superposition of events just from the subsidiary processes. It is clear that in most cases the structure will be complicated, and in particular, nearly always non-recurrent as far as the intervals between events are concerned. Perhaps the main physical features are that of subsidiary processes continually beginning, overlapping and ending in a stochastic fashion. Here we shall review and coordinate the available work on particular processes of the cluster type, mainly emphasizing where useful explicit results can be obtained; these, it will be seen often involve a Poisson main process. A few additional results will also be given.

4.2 Descriptions of Specific Cluster Processes

a) *The Bartlett-Lewis Process.* In this process the main events follow a Poisson process of rate α, and are included in the superpositions. The subsidiary structure is that each main event generates S (≥ 0) following subsidiary events with distribution $\{p_S(r), r \geq 0\}$, and that the intervals between these S events, $Y_1, Y_2, \ldots, Y_S$ are positive and independent and identically distributed according to the distribution function $F_Y(x) = 1 - R_Y(x)$. In particular, we note that S may be zero with probability $p_S(0)$, and then only the main event is counted. We shall for convenience assume that α, E(Y) and E(S) are finite, and that $F_Y(0+) = 0$ to avoid complications with multiple events. This process was introduced by Bartlett (1963) as a model for the spectral analysis of some traffic data, and by Lewis (in the discussion to that paper and more fully in Lewis, 1964) as a "branching Poisson process model" for the analysis of computer failure sequences. In the computer application it was thought that the basic failure process should be Poisson —arguing from the superposition of the failure processes of the large number of individual components in a computer, and well known Poisson limit theorems for superposition, e.g., Khintchine (1960, Chap. 5). The observed departures from a Poisson process were thought due to faulty repair and reoccurrences of the same failure at later times. These failures then constitute subsidiary processes. Differences between the basic failure rates and the repair characteristics of different machines were studied by Lewis(196 using the branching Poisson process model.

b) *The Neyman-Scott Process.* Here again the main events follow a Poisson process of rate α, and each main event generates S (≥ 0) follow ing subsidiary events with distribution $\{p_S(r), r \geq 0\}$. However, now the subsidiary events are at positive distances $D_1, D_2, \ldots, D_S$ from their cluster center, and these distances are independent and identically distributed according to the subsidiary distribution function $F_D(x)=1-R_D(x)$. We assume that α, E(S) and E(D) are finite and that $F_D(0+) = 0$. A rather general process of this form was suggested by Neyman and Scott (1958) in relation to statistical cosmology, and Bartlett (1964) used the process over the plane as an ecological model. However, its use as a one-dimensional tractable point process (in which main events were not included) is primarily due to Vere-Jones (1970) in connection with the occurrences of New Zealand earthquakes. Vere-Jones also considered the cluster process with main events, and with subsidiary events allowed to precede their cluster center. The model in which main events are included and the subsidiary distribution is one-sided can be used for the earthquake situation in which a main shock is followed by a number of after shocks. This is the model which will be developed here. Vere-Jones was able to draw quantitative conclusions about the different clustering effects of shallow, intermediate and deep shocks.

c) *The Process of Displaced Regular Events.* Here the main process is a regular sequence of interval a, and each main event is independently displaced from its origin by a distance B which has the distribution function $F_B(x)$, $-\infty < x < \infty$. It is assumed that E(B) is finite. Only the displaced (subsidiary) events are counted. T. Lewis (1961) introduced this process in connection with the scheduled equispaced arrivals of oil tankers at a terminal; early or later arrivals are then assumed to deviate from the planned times, according to the distribution $F_B(x)$. The model has been used by Govier and Lewis (1963) to assess the storage capacity required at a tanker terminal. The model has been further investigated in some particular cases by Nelson and Williams (1970). Ten Hoopen and Reuver (1967) have described applications of the model for neuron firing when the natural order of subsidiary events is preserved.

These are the three main types of cluster process which will be discussed; in addition, we will review more general versions of the processes.

4.3 The Bartlett-Lewis Process

We continue our general approach of giving first the arbitrary time initial conditions. These have been derived indirectly in Lewis (1969), and so here we shall use "$t \to \infty$" arguments. It is clear that the future behavior of the process after time t may be specified in terms of the joint distribution of

(a) the time X_t to the next main event after time t,

(b) the number Z_t of subsidiary processes running at time t,

(c) the number of events $Q_t(i)$ remaining at time t in the ith of the (unordered) Z_t subsidiary process, $1 \leq i \leq Z_t$,

(d) the time to the first event $Y_t(i)$ in the ith of the Z_t subsidiary processes after time t, $1 \leq i \leq Z_t$.

In (d) we are assuming the finite renewal structure for subsequent events. We also temporarily assume as a matter of convenience that the process begins at t = 0 without a main event or any subsidiary processes.

We next give two expressions for subsidiary processes in terms of which the joint distributions of (a) to (d) can be written. Let $F_{SY}(t)$ denote the probability that the length of a subsidiary process (which may be zero with probability $p_S(0)$) is less than t. Further, let $F_{SY}(t;r:y)$ denote the probability that a subsidiary process terminates after a time t, together with the distribution of the number r (≥ 1) of remaining events after t, and the survivor function of the time to the first event after t, evaluated at y. These two quantities are then explicitly

$$F_{SY}(t) = p_S(0) + \sum_{j=1}^{\infty} p_S(j)\, F_Y^{(j)}(t) , \tag{4.3.1}$$

$$F_{SY}(t;r:y) = p_S(r)\, R_Y(t+y) + \sum_{j=1}^{\infty} p_S(r+j) \int_{u=0}^{t} R_Y(t-u+y)\, dF^{(j)}(u) ,$$

where $F_Y^{(j)}(t)$ denotes the j-fold convolution of $F_Y(t)$ with itself.

The joint distribution of (a) to (d) may then be written as

$$\text{prob}\{X_t > x,\, Z_t = k,\, Q_t(i) = r_i,\, Y_t(i) > y_i,\, 1 \leq i \leq k\}$$

$$= e^{-\alpha x} \sum_{r=k}^{\infty} \binom{r}{k} \{t^{-1} \int_{v=0}^{t} F_{SY}(v)\, dv\}^{r-k}$$

$$\times \left[\prod_{i=1}^{k} \{t^{-1} \int_{v=0}^{t} F_{SY}(v;r_i:y_i)\, dv\}\right] e^{-\alpha t} (\alpha t)^r / r! . \tag{4.3.2}$$

This follows by assuming the conditional property of the Poisson process for the uniform distribution of the start of the r subsidiary processes over (0,t], and that r-k of these must have terminated before time t. After summation (4.3.2) leads to

$$(\alpha^k/k!)\ \exp[-\alpha x-\alpha \int_{u=0}^{t} \{1-F_{SY}(u)\}du] \prod_{i=1}^{k} [\int_{v=0}^{t} F_{SY}(v;r_i:y_i)\ dv] \ . \tag{4.3.3}$$

The arbitrary time initial conditions may then be evaluated as

$$\lim_{t\to\infty} \text{prob}\{X_t > x,\ Z_t = k,\ Q_t(i) = r_i,\ Y_t(i) \geq y_i,\ 1 \leq i \leq k\}$$

$$= e^{-\alpha x-\mu}\ \frac{\mu^k}{k!} \prod_{i=1}^{k} \left[\frac{\sum_{j=r_i}^{\infty} p_S(j)}{E(S)} \int_{u=y_i}^{\infty} \left\{ \frac{R_Y(u)}{E(Y)} \right\} du \right] \tag{4.3.4}$$

where $\mu = \alpha E(S)\ E(Y)$ is assumed finite. They may be thought of in terms of a Poisson number of "modified" (finite renewal) subsidiary processes. In each modified subsidiary process the time to the first event L has survivor function

$$R_L(y) = \int_{u=y}^{\infty} \{R_Y(u)/E(Y)\}\ du = 1 - F_L(y) \ , \tag{4.3.5}$$

and the times between subsequent events all have $F_Y(y)$ as their distribution function. Further, the total number of events Q ($\geq$ 1) has the distribution

$$p_Q(r) = \sum_{j=r}^{\infty} p_S(j)/E(S) \ . \tag{4.3.6}$$

The survivor functions of S and Q will be denoted by

$$R_S(r) = \sum_{j=r+1}^{\infty} p_S(j) \qquad \text{and} \qquad R_Q(r) = \sum_{j=r+1}^{\infty} p_Q(j) .$$

The (stationary) asynchronous counting distribution for the Bartlett-Lewis process was obtained by Lewis (1964) for S geometrically distributed, and conjectured generally. Franken and Richter (1965) indirectly derived the correct general result, although there is a slip in their Laplace transformed expression, which has been corrected in Lewis (1969). Here we give a simple derivation of the untransformed p.g.f. which makes use of the arbitrary time initial conditions. The required p.g.f. $\phi(z;t)$ is the product of two independent parts: the first, $\phi_0(z;t)$, gives the count of events in $(0,t]$ which come from the initial modified subsidiary processes, and the second, $\phi_1(z;t)$, gives the count of main and subsidiary events in $(0,t]$ associated with main events in $(0,t]$.

For $\phi_1(z;t)$ we first need the p.g.f. $\phi_{SY}(z;t)$ for the number of events in $(0,t]$ from a finite renewal subsidiary process with origin at $t = 0$; it may be shown, as in Lewis (1964), that is is explicitly

$$\phi_{SY}(z;t) = 1 + (z-1) \sum_{i=0}^{\infty} z^i R_S(i)\, F_Y^{(i+1)}(t) . \tag{4.3.7}$$

Then using again the conditional uniform property of the main Poisson process, we have

$$\begin{aligned} \phi_1(z;t) &= \sum_{n=0}^{\infty} z^n \{t^{-1} \int_{v=0}^{t} \phi_{SY}(z;v)\, dv\}^n e^{-\alpha t} \frac{(\alpha t)^n}{n!} \\ &= \exp[\alpha t(z-1) + \alpha z \int_{v=0}^{t} \{\phi_{SY}(z;v) - 1\}\, dv] . \end{aligned} \tag{4.3.8}$$

For $\phi_0(z;t)$ we first give the count p.g.f. $\phi_{QL}(z;t)$ for the number of events in $(0,t]$ from a modified finite renewal subsidiary process with origin at $t = 0$; the result is

$$\phi_{QL}(z;t) = 1 + (z-1) \sum_{i=0}^{\infty} z^i R_Q(i)\{E(S)\}^{-1} \int_{u=0}^{t} \{F_Y^{(i)}(u) - F_Y^{(i-1)}(u)\}du \tag{4.3.9}$$

The initial conditions (4.3.4) then give

$$\phi_0(z;t) = \sum_{k=0}^{\infty} e^{-\mu}(\mu^k/k!)\,\{\phi_{QL}(z;t)\}^k ,$$

$$= \exp[\mu\{\phi_{QL}(z;t) - 1\}] . \qquad (4.3.10)$$

Now for the p.g.f. of the total number of events we take the product of (4.3.8) and (4.3.10) to obtain after a little simplification

$$\phi(z;t)$$

$$= \exp[\alpha\{1+E(S)\}t(z-1)+\alpha(z-1)^2 \sum_{k=1}^{\infty}\{\sum_{i=k}^{\infty} R_S(i-1)\}z^{k-1}\int_{u=0}^{t} F_Y^{(k)}(u)\,du\,] . \qquad (4.3.11)$$

This is the elegant form of the result due to Franken and Richter (1965); setting $z = 0$ we would obtain the survivor function of the forward recurrence time. Our assumption that $E(Y) < \infty$ is not necessary for this result, although the derivation without it is not straightforward, and relates to the existence of initial conditions. The mean and variance of the stationary counting variable $N(0,t)$ follow from (4.3.11) as

$$E\{N(0,t)\} = \alpha\{1 + E(S)\}t \qquad (4.3.12)$$

and

$$\mathrm{var}\{N(0,t)\} = \alpha E\{(S+1)^2\}t - 2\alpha \sum_{k=1}^{\infty}\{\sum_{i=k}^{\infty} R_S(i-1)\}\int_{u=0}^{t} R_Y^{(k)}(u)\,du . \qquad (4.3.13)$$

It follows from (4.3.13) that the asymptotic intercept of the variance-time curve is $-\frac{1}{3}\alpha E\{S(S+1)(S+2)\}E(Y)$, in agreement with Lewis (1964). Results on the higher cumulants are given by Smith in the discussion to Lewis (1964) and more fully in Lewis (1969); in particular, they are shown to be asymptotically linear in t. Lewis (1969) also establishes the asymptotic normality of $N(0,t)$ as $t \to \infty$ under reasonable conditions; use is made of the device of shifting each subsidiary event to its cluster center, a procedure which does not alter many asymptotic properties but simplifies calculations.

The synchronous stationary point properties for the Bartlett-Lewis process are rather complicated, although much can be obtained explicitly. The arbitrary event initial conditions have been derived in Lawrance (1972) and are quoted here in our usual type of notation as

$$\text{prob}\{X_A > x,\ Z_A = k,\ Q_A(i) = r_i,\ Y_A(i) > y_i,\ 1 \le i \le k\}$$

$$= \{1 + E(S)\}^{-1} \exp(-\alpha x-\mu)\{\mu^k/k!\} \prod_{i=1}^{k} p_Q(r_i)\ R_L(y_i)$$

$$+ E(S)\{1 + E(S)\}^{-1} \exp(-\alpha x-\mu)\{\mu^{k-1}/(k-1)!\}$$

$$\times k^{-1} \sum_{i=1}^{k} p_Q(r_i)\ R_Y(y_i) \prod_{\substack{j=1\\ j\neq i}}^{k} p_Q(r_j)\ R_L(y_j). \tag{4.3.14}$$

These may be interpreted in terms of the distributions of a number of initial processes. To obtain the marginal survivor function of the synchronous intervals from (4.3.14) we sum the $r_i \ge 1$, set $x = t$, $y_i = t$ and sum $k \ge 0$. This gives the result

$$R_X(t) = \{1+E(S)\}^{-1}\{1+E(S)R_Y(t)\} \exp\{-\alpha t-\alpha E(S) \int_{u=0}^{t} R_Y(u)\,du\}, \tag{4.3.15}$$

as derived indirectly in Lewis (1964); Lewis also shows that the coefficient of variation of the distribution is greater than one, and that it has an exponential tail. By using the initial conditions (4.3.14) or the general result (3.2.13) we may obtain the synchronous counting distribution as

$$\psi(z;t) = [1+\{1+E(S)\}^{-1}(z-1) \sum_{k=1}^{\infty} \{ \sum_{i=k}^{\infty} R_S(i-1)\}z^{k-1}F_Y^{(k)}(t)]$$

$$\times \exp[\alpha\{1+E(S)\}t(z-1)+\alpha(z-1)^2 \sum_{k=1}^{\infty} \{ \sum_{i=k}^{\infty} R_S(i-1)\}z^{k-1} \int_{u=0}^{t} F_Y^{(k)}(u)\,du \tag{4.3.16}$$

This leads to the intensity function, covariance density function and Bartlett spectrum of the process, as previously derived by Bartlett (1963) and Lewis (1964).

The joint distributions of synchronous intervals have not so far been obtained, although Lewis (1964), using results of McFadden (1962), has reached the first serial correlation coefficient, and Gilles and Lewis (1967) have derived a neat summation formula for the interval spectrum when S is geometrically distributed. The joint distribution of adjacent synchronous intervals can probably be derived, although is bound to be messy. Further comments on generalizations will be given in Section 4.6 of the present paper.

4.4 The Neyman-Scott Process

The Neyman-Scott process differs from the Bartlett-Lewis process only in part of its subsidiary structure; like the latter it has S (≥ 0) subsidiary events following each main event and S has a distribution $\{p_S(r), r \geq 0\}$, independently for each main event. However, the subsidiary events are now located from their main events (cluster centers) at positive distances D which are independent and identically distributed with subsidiary distribution function $F_D(x)$, $F_D(0+) = 0$; a subsidiary process will not be stationary in the sense of a finite renewal process. In fact, as remarked in Section 3.2, when S follows a Poisson distribution the subsidiary structure is that of a time-dependent Poisson process. Nevertheless, we shall see that the methods used for the Bartlett-Lewis process also apply with appropriate modifications to the Neyman-Scott process.

The arbitrary-time initial conditions parallel those for the Bartlett-Lewis process, except that for an initial modified subsidiary process we must specify jointly the time to each of its events, and not just the time to its first event. Thus, at time t in the process, let X_t, Z_t and $Q_t(i)$ be as defined in Section 4.3; further let $D_t(i,1),\ldots,$ $D_t(i,r_i)$ denote the (unordered) times to the remaining subsidiary events after time t in the ith of the Z_t processes.

We next define $F_{SD}(t)$, analogously to $F_{SY}(t)$ of the Bartlett-Lewis process, as the distribution function for the length of a subsidiary process, and let $F_{SD}(t;r:y_1,\ldots,y_r)$ denote the probability that a subsidiary process terminates after time t together with the distribution of the number r (≥ 1) of remaining events at time t and the survivor functions of the (unordered) times to the events after time t, evaluated at $y_1,\ldots,y_r$. In view of the subsidiary structure it is evident that

$$F_{SD}(t) = \sum_{j=0}^{\infty} p_S(j) \{F_D(t)\}^j ,$$

$$F_{SD}(t;r:y_1,\ldots,y_r) = \sum_{j=r}^{\infty} p_S(j) \binom{j}{r} \{F_D(t)\}^{j-r} \prod_{k=1}^{r} R_D(t+y_k). \qquad (4.4.1)$$

Now, as for the Bartlett-Lewis process at (4.3.3), we have by considering the future behavior of the process at time t,

$$\text{prob}\{X_t > x,\ Z_t=k,\ Q_t(i)=r_i,\ D_t(i,1) > y_{i,1},\ldots,D_t(i,r_i) > y_{i,r_i},\ 1\leq i\leq k\}$$

$$= (\alpha^k/k!) \exp[-\alpha x-\alpha \int_{u=0}^{t} \{1-F_{SD}(u)\} du]$$

$$\times \prod_{i=1}^{k} \{ \int_{v=0}^{t} F_{SD}(v;r_i:y_{i,1},\ldots,y_{i,r_i}) dv\} . \qquad (4.4.2)$$

Letting $t \to \infty$ in (4.4.2) gives the arbitrary-time initial conditions. The result is simplified by writing $\mu = \alpha\kappa$, which will be the mean of the Poisson number of initial modified subsidiary processes, where

$$\kappa = \int_{u=0}^{\infty} [1 - \phi_S\{F_D(u)\}] du , \qquad (4.4.3)$$

and $\phi_S(z)$ is the p.g.f. of S; the assumption is made that $\kappa < \infty$. We also write

$$F_{BG}(r:y_1,\ldots,y_r)$$

$$= \kappa^{-1} \sum_{j=r}^{\infty} p_S(j) \binom{j}{r} \int_{u=0}^{\infty} [\{F_D(u)\}^{j-r} \prod_{k=1}^{r} R_D(u+y_k)] \, du \tag{4.4.4}$$

for the joint distribution of the number of events, B, evaluated at r, in a modified initial subsidiary process, together with the survivor functions of their locations, G, evaluated at $y_1,\ldots,y_r$. Thus from (4.4.2) the arbitrary time initial conditions become

$$e^{-\alpha x-\mu}(\mu^k/k!) \prod_{i=1}^{k} F_{BG}(r_i:y_{i,1},\ldots,y_{i,r_i}) \ . \tag{4.4.5}$$

These bear comparison with those given at (4.3.4) for the Bartlett-Lewis process.

The derivation of the (stationary) asynchronous counting distribution also parallels the derivation in the Bartlett-Lewis case. The p.g.f.'s giving the number of events in (0,t] for a subsidiary and an initial modified subsidiary processes with origin at t = 0 will be denoted by $\phi_{SD}(z;t)$ and $\phi_{BG}(z;t)$ respectively, and their explicit forms are

$$\phi_{SD}(z;t) = \phi_S\{1 + (z-1)\, F_D(t)\}, \tag{4.4.6}$$

$$\phi_{BG}(z;t) = \kappa^{-1} \int_{u=0}^{\infty} [\phi_S\{1+(z-1)(F_D(t+u)-F_D(u))\}-\phi_S\{F_D(u)\}]du. \tag{4.4.7}$$

Equations very similar to (4.3.8) and (4.3.10) hold in this case also, and hence taking their product, we have for the p.g.f. $\phi(z;t)$ of the asynchronous counting distribution

$$\phi(z;t) = \exp\Bigg[\alpha t(z-1) + \alpha z \int_{u=0}^{t} [\phi_S\{1 + (z-1)\, F_D(u)\} - 1] \, du$$

$$+ \alpha \int_{u=0}^{\infty} [\phi_S[1+(z-1)\{F_D(t+u) - F_D(u)\}]-1]du \Bigg] \ . \tag{4.4.8}$$

Setting z = 0 we obtain the survivor function of the forward recurrence time. The mean and variance of N(0,t) follow from (4.4.8) as

$$E\{N(0,t)\} = \alpha\{1 + E(S)\}t \tag{4.4.9}$$

and

$$\begin{aligned} \operatorname{var}\{N(0,t)\} \\ &= \alpha E\{(S+1)^2\}t - 2\alpha E(S^2)\int_{u=0}^{t} R_D(u)\,du - 2\alpha E\{S(S-1)\}\int_{u=0}^{\infty} R_D(u)R_D(t+u)\,du \\ &\quad + 2\alpha E\{S(S-1)\}\int_{u=0}^{\infty}\{R_D(u)\}^2\,du. \end{aligned} \tag{4.4.10}$$

The last three terms of (4.4.10) yield the intercept of the asymptotically linear variance-time curve as

$$-2\alpha E(S^2)\,E(D) + 2\alpha E\{S(S-1)\}\int_{u=0}^{\infty}\{R_D(u)\}^2\,du \quad . \tag{4.4.11}$$

Most of these counting results have been obtained by Vere-Jones (1970) using a probability generating functional (p.g.fl.) for the process. The present derivations are more elementary. As remarked in Section 3.2, (4.4.8) with a Poisson distribution for S is equivalent to the shot-noise d.s.P. process result (3.2.10) in which main events are not counted.

Finally, we turn briefly to the synchronous properties. The arbitrary event initial conditions can be obtained quite easily by our general approach, although the result is long and can not be given here. Instead we just note that they follow after setting t = 0 in the derivative with respect to t of

$$\begin{aligned} &\operatorname{prob}\{N(0,t)=0, Z_t=k, Q_t(i)=r_i, D_t(i,1) > y_{i,1}, \ldots, D_t(i,r_i) > y_{i,r_i}, 1\le i\le k\} \\ &= \exp\{-\alpha(t+x)-\mu\}(\mu^k/k!)\prod_{i=1}^{k}\{\int_{v=0}^{\infty}\kappa^{-1}F_{SY}(v;r_i:y_{i,1}+t,\ldots,y_{i,r_i}+t)\,dv\} \; . \end{aligned}$$

(4.4.12)

The arbitrary event initial conditions lead in the usual way to the synchronous interval distribution and synchronous counting distribution. For the survivor function of a synchronous interval we find

$$R_X(t) = \{1 + E(S)\}^{-1}[1 + \int_{u=0}^{\infty} f_D(t+u)\ \phi_S'\{1 - F_D(t+u) + F_D(u)\}du]$$

$$\times \exp\{-\alpha t-\alpha \int_{u=0}^{\infty} [1 - \phi_S\{1-F_D(t+u)+F_D(u)\}]du \cdot \qquad (4.4.13)$$

Analogously to the Bartlett-Lewis process (Lewis, 1964) we find that the log-survivor function for large t is

$$\log R_X(t) \simeq -\log\{1 + E(S)\} - \mu - \alpha t\ , \qquad (4.4.14)$$

although $\mu = \alpha\kappa$, defined at (4.4.3) is rather more complicated. Here we have made the assumption which will sometimes be difficult to verify, that the first integral in (4.4.13) converges to zero as $t \to \infty$. Following Lewis (1964) we may set $F_D(u) = 1$ in the second moment to show that the coefficient of variation is necessarily greater-equal one. The analogous result for this process when main events are not counted is much more complicated; Vere-Jones (1970) has derived a condition for this from the hazard function and showed that it is satisfied when D is exponentially distributed. Physically, the complication arises because cluster centers are lost, and, technically, the middle term in (4.4.8) does not vanish when z = 0.

There is perhaps a need for further study of all the synchronous results, both counting and not counting main events. In particular, nothing is known about the serial correlations. Further comments concerned with generalization of the process are given in Section 4.6.

4.5 The Process of Displaced Regular Events

The process of displaced regular events is a specific type of cluster process where the successive main events, which are not counted, occur at equispaced distances a; the subsidiary processes are independent and consist of just one event located at a distance B on either side of its main point. We let B have the distribution function $F_B(x)$, $-\infty < x < \infty$, with $E(B) < \infty$.

This process was one of the first stationary point processes to be investigated (T. Lewis, 1961), but the available results are still ad hoc, approximate and incomplete. This is because of removal of the assumption that the main process of a cluster point process is a Poisson process. In this section we merely summarize the work in T. Lewis (1961), Cox and Lewis (1966, Section 7.6) and Nelson and Williams (1968, 1970). It has not been found possible to bring the different results together under a common approach. Arbitrary time and arbitrary event initial conditions are not available to enable us to use our previous approach.

We consider first the asynchronous counting distribution discussed by T. Lewis (1961). Let the regular events be located at ia $(-\infty \le i \le \infty)$, and consider the count of displaced events in (u, u+t]. The probability that the displaced event associated with the ith main event occurs in (u, u+t] is

$$F_B(t+u-ia) - F_B(u-ia) . \qquad (4.5.1)$$

Using the independence of the displacements for the entire sequence of regular events, the p.g.f. evaluated at z for the total number of displaced events in (u, u+t] is thus

$$\prod_{i=-\infty}^{\infty} [1 - (1-z) \{F_B(t+u-ia) - F_B(u-ia)\}] . \qquad (4.5.2)$$

In our interpretation of asynchronous counting we should now let $u \to \infty$ in (4.5.2); however, it is not known how to carry out the evaluation. T. Lewis (1961) considers the log of (4.5.2),

$$\sum_{i=-\infty}^{\infty} \log[1 - (1-z) \{F_B(t+u-ia) - F_B(u-ia)\}] . \qquad (4.5.3)$$

Then making the approximate assumption that when $a^{-1}\,\mathrm{var}(B)$ is large enough the dependence on u can be neglected, the summation in (4.5.3) can be approximated by the integral

$$a^{-1} \int_{u=-\infty}^{\infty} \log[1 - (1-z)\{F_B(t+u) - F_B(u)\}]\, du \,. \tag{4.5.4}$$

This is still not a particularly tractable result, and not a good approximation when a^{-1} var(B) is small.

The synchronous counting distribution has been obtained by Cox and Lewis (1966, Section 7.6) using a conditional form of the above argument. Suppose that an initial arbitrary event (in a non-technical sense) is chosen; let its corresponding regular event define i = 0 of the regular events and be at a distance u. It is assumed that u is a realization of the $F_B(u)$ distribution. Conditionally on u, the required p.g.f. is (4.5.2), and thus unconditionally the synchronous p.g.f. is

$$\psi(z;t) = \int_{u=-\infty}^{\infty} \prod_{i=-\infty}^{\infty} [1 - (1-z)\{F_B(t+u-ia)-F_B(u-ia)\}]\, dF_B(u) . \tag{4.5.5}$$

Cox and Lewis (1966) have managed to obtain the mean from this distribution and also the intensity function, both exactly. They have also given good approximations for the covariance density and variance of the asynchronous counting process using (4.5.5) and general stationary point results. A property of this process is that the variance time function tends, as $t \to \infty$, to a constant value. Processes with this property have been called "controlled variability" processes by T. Lewis and Govier (1964). By setting $z = 0$ in (4.5.5) and differentiating with respect to t an expression for the synchronous interval can be obtained.

A different approach to the synchronous intervals is given by Nelson and Williams (1970), and they obtain exact results in one case. They consider 2n+2 main events situated at -n,...,0,...,n,n+1. The zeroth displaced event is then defined as that displaced event such that there are n displaced events before it and n+1 after. The first displaced event is similarly that for which there are n+1 before and n after. A synchronous interval is the distance between the zeroth and first displaced events and we let $n \to \infty$ for the complete process. The probability that there are n displaced events before time u and n after time τ (given that u and τ are the times of the displaced events associated with the ith and jth regular events) is observed to be the coefficient of θ^{n+1} in

$$\prod_{\substack{k=-n \\ k\neq i,j}}^{n+1} [F_B(u-ka) + \theta\{1 - F_B(\tau - ka)\}] \quad . \tag{4.5.6}$$

The joint p.d.f. of the times to the zeroth and first displaced events is thus

$$\begin{aligned} f(u,\tau) &= \sum_{i=-\infty}^{\infty} \sum_{j=-\infty}^{\infty} f_B(u-ia)\, f_B(\tau-ja) \\ &\quad \times \lim_{n\to\infty} \operatorname{coeff}_{\theta^{n+1}} \prod_{\substack{k=-n \\ k\neq i,j}}^{n+1} [F_B(u-ka) + \theta\{1 - F_B(\tau-ka)\}] \\ &= -\frac{\partial^2}{\partial\tau\partial u} \lim_{n\to\infty} \operatorname{coeff}_{\theta^{n+1}} \prod_{k=-n}^{n+1} [F_B(u-ka)+\theta\{1-F_B(\tau-ka)\}] \ . \end{aligned} \tag{4.5.7}$$

Using this approach Nelson and Williams obtained explicit results for the interval mean, variance and first serial correlation when B is exponentially distributed. The interval coefficient of variation is less or equal to one, and the first serial correlation lies between -1/2 and 0. Further results lead to severe computational difficulties.

4.6 Comments on Generalizations of the Cluster Processes

In this section we briefly review results for generalizations of the three cluster processes. Our observations may be separated into four parts.

(1) For the Bartlett-Lewis process, Lewis (1967) has generalized the main process to a time-dependent Poisson process; this extension is possible because the time-dependent Poisson process also possesses a simple conditional property; most results on the counting distribution can still be obtained, although of course, the interevent intervals

are not a stationary sequence. A similar generalization for the Neyman-Scott process should also be tractable.

(2) The main Poisson process and independent subsidiary structure are retained, but the subsidiary processes are allowed to be quite general provided they are well defined and the expected number of events increases monotonically with time from zero to $E(S) < \infty$. Such processes have been called "generalized branching Poisson processes" by Lewis (1969), and "Poisson cluster processes" by Vere-Jones (1970). These two papers give results on the moments of the counting distributions, the forward recurrence distribution, the synchronous interval distribution, and the interval hazard function. No results on the serial correlations have been obtained. Our results on the two types of initial conditions, and in particular the Poisson number of subsidiary processes running at an arbitrary time, may also be generalized to this situation. Milne and Westcott (1972) have shown that the Gauss-Poisson process of Newman (1970) is a Poisson cluster process. Milne and Westcott discuss several representations of its subsidiary structure, and note in particular that a subsidiary process consists of at most two events.

(3) The main process is allowed to be any non-regular renewal process and the subsidiary structure is quite general, as in (2). Most of the results in this case are derived or summarized in Lewis (1970); special forms of some of the results are given for Bartlett-Lewis processes. The time origin in this work is taken at the beginning of an interval in the main process when there are no subsidiary processes running. Results are obtained for the mean number of events, the intensity function, and the asymptotic counting distribution. These require delicate applications of renewal theorems. There are no results for most of the asynchronous or synchronous stationary point properties, such as the forward recurrence time, the interevent intervals, or serial correlations.

(4) The broadest generalization which has been considered deals with a general stationary point main process, and the previous independent general subsidiary structure. Vere-Jones (1970) refers to this as a "general cluster process", and gives formulae for the mean and

covariance function of the stationary counting process. Daley (1972) establishes asymptotic properties of the stationary counting distribution. Some results apply also to two-sided subsidiary processes.

(5) A very powerful tool for use with cluster point process, and point processes in general, is the probability generating functional (p.g.fl.). The relevant definitions and theory may be found in Vere-Jones (1968) and Westcott (1972). The p.g.fl. has been used for cluster processes by Vere-Jones (1970), Westcott (1970, 1971), Newman (1970) and Milne and Westcott (1972). It is particularly good for identifying similarities between apparently different processes. Multivariate p.g.fl.'s offer possibilities of dealing with events of several types, and aspects additional to the times of events.

ACKNOWLEDGMENTS

I would like to thank Dr. P. A. W. Lewis for various helpful discussions during the preparation of this paper. I am very grateful to several conference participants for constructive comments on an earlier version. The paper was prepared during a leave from the University of Leicester, England, at the Mathematical Sciences Department of the IBM Thomas J. Watson Research Center, to whom I am grateful for support.

REFERENCES

Barndorff-Nielsen, O. & Yeo, G. F. (1969). Negative binomial processes. J. Appl. Prob. 6, 633-47.

Bartlett, M. S. (1955). *An Introduction to Stochastic Processes.* Cambridge University Press.

Bartlett, M. S. (1963). The spectral analysis of point processes. J. R. Statist. Soc. B 25, 264-96.

Bartlett, M. S. (1964). The spectral analysis of two-dimensional point processes. Biometrika 51, 299-311.

Bédard, G. (1966). Photon counting statistics of Gaussian light. Phys. Rev. 151, 1038-39.

Cox, D. R. (1955). Some statistical methods connected with series of events. J.R. Statist. Soc. B 17, 129-164.

Cox, D. R. (1962). *Renewal Theory*. London: Methuen.

Cox, D. R. (1963). Some models for series of events. Int. Statist. Inst. 40, 737-746.

Cox, D. R. & Miller, H. D. (1965). *The Theory of Stochastic Processes*. London: Methuen.

Cox, D. R. & Lewis, P. A. W. (1966). *The Statistical Analysis of Series of Events*. London: Methuen and New York: Barnes and Noble.

Cramér, H. & Leadbetter, M. R. (1967). *Stationary and Related Stochastic Processes*. New York: Wiley.

Daley, D. J. (1972). Asymptotic properties of stationary point processes with generalized clusters. Z. Wahrscheinlichkeitstheorie verw. 18, to appear.

Darling, D. A. & Siegert, A. J. F. (1957). A systematic approach to a class of problems in the theory of noise and other random phenomena -Part I. IRE Trans. Information Theory IT-3, 32-36.

Downton, F. (1970). Bivariate exponential distributions of reliability theory. J. R. Statistic. Soc. B 32, 408-417.

Ekholm, A. & Hyvärinen, J. (1970). A pseudo-Markov model for series of neuronal spike events. Biophysic. J. 10, 773-796.

Franken, P. & Richter, G. (1965). Über eine Klasse von züfalligen Punktfolgen. Wiss. Z. Friedrich-Schiller-Univ., Jena, Math. Naturwiss. Reihe 14, 247-249.

Gaver, D. P. (1963). Random hazard in reliability problems. Technometrics 5, 211-226.

Gaver, D. P. (1970). Multivariate gamma distributions generated by mixture. Sankhyā A 32, 123-126.

Gilles, D. C. & Lewis, P. A. W. (1967). The spectrum of intervals of a geometric branching Poisson process. J. Appl. Prob. 4, 201-205.

Govier, L. J. & Lewis, T. (1963). Stock levels generated by a controlled variability arrival process. Operat. Res. 11, 693-701.

Grandell, J. (1971). On stochastic processes generated by a stochastic intensity function. Skand. Aktuartidskr. 54, to appear.

Kac, M. & Siegert, A. J. F. (1947). On the theory of noise in radio receivers with square-law detectors. J. Appl. Phys. 18, 383-397.

Karp, S. & Clark, J. R. (1970). Photon counting: a problem in classical noise theory. IEEE Trans. Information Th. IT-16, 672-680.

Khintchine, A. Y. (1960). *Mathematical Methods in the Theory of Queuing*. London: Griffins.

Kibble, W. F. (1941). A two-variate gamma type distribution. Sankhyā 5, 137-150.

Kingman, J. F. C. (1964). On doubly stochastic Poisson processes. Proc. Camb. Phil. Soc. 60, 923-932.

Lampard, D. G. (1965, 1968). A stochastic process whose successive intervals between events form a first order Markov chain-I. Report No. MEE 65-1, Electrical Engineering Dept., Monash University, Australia, and J. Appl. Prob. 5, 648-668.

Lawrance, A. J. (1972). Arbitrary event initial conditions for branching Poisson processes. J. R. Statistic. Soc. B, 34 to appear.

Lee, P. A. (1966). A stochastic process whose successive intervals between events form a first order Markov chain-II. Report No. MEE 66-2, Electrical Engineering Dept., Monash University, Australia.

Lewis, P. A. W. (1964). A branching Poisson process model for the analysis of computer failure patterns. J. R. Statist. Soc. B 26, 398-456.

Lewis, P. A. W. (1966). A computer program for the statistical analysis of series of events. IBM Systems J. 5, 202-225.

Lewis, P. A. W. (1967). Non-homogeneous branching Poisson processes. J. R. Statist. Soc. B. 29, 343-354.

Lewis, P. A. W. (1969). Asymptotic properties and equilibrium conditions for branching Poisson processes. J. Appl. Prob. 6, 355-371.

Lewis, P. A. W. (1970). Asymptotic properties of branching renewal processes. IBM Research Report, RC 2878, Yorktown Heights, N.Y.

Lewis, T. (1961). The intervals between regular events displaced in time by independent random deviations of large dispersion. J. R. Statist. Soc. B 23, 476-483.

Lewis, T. & Govier, L.J. (1964). Some properties of counts of events for certain types of point process. J. R. Statist. Soc. B. 26, 325-337.

Macchi, O. (1971). Distribution statisque des instants d'emission des photoélectrons d'une lumière thermique. C. R. Acad. Sci. Paris A 272, 437-440.

McFadden, J. A. (1962). On the lengths of intervals in a stationary point process. J. R. Statist. Soc. B 24, 364-382.

McFadden, J. A. (1965). The mixed Poisson process. Sankhyā A 27, 83-92.

Moran, P. A. P. (1967). A non-Markovian quasi-Poisson process. Studia Sci. Math. Hungar. 2, 425-429.

Milne, R. K. & Westcott, M. (1972). Further results for Gauss-Poisson processes. Adv. Appl. Prob. 4, to appear.

Nelson, R. B. & Williams, T. (1968). Randomly delayed appointment streams. Nature 219, 573-574.

Nelson, R.B. & Williams, T. (1970). Random displacements of regularly spaced events. J. Appl. Prob. 7, 183-195.

Newman, D. S. (1970). A new family of point processes which are characterized by their second order moment properties. J. Appl. Prob. 7, 338-358.

Neyman J. & Scott, E. L. (1958). Statistical approach to the problems of cosmology. J. R. Statist. Soc. B. 20, 1-43.

Serfozo, R. F. (1971). Processes with conditional stationary independent increments. J. Appl. Prob. 8, to appear.

Serfozo, R. F. (1972). Conditional Poisson processes. J. Appl. Prob. 9, to appear.

Siegert, A. J. F. (1957). A systematic approach to a class of problems in the theory of noise and other random phenomena: Pt. II-Examples. IRE Trans. Information Theory IT-3, 38-43.

Ten-Hoopen, M. & Reuver, H. A. (1967). Analysis of sequences of events with random displacements applied to biological systems. Math. Biosci. 1, 599-617.

Vere-Jones, D. (1967). The infinite divisibility of a bivariate gamma distribution. Sankhyā A, 29, 421-422.

Vere-Jones, D. (1968). Some applications of probability generating functionals to the study of input-output streams. J. R. Statist. Soc. B 30, 321-333.

Vere-Jones, D. (1970). Stochastic models for earthquake occurrence. J. R. Statist. Soc. B 32, 1-62.

Westcott, M. (1970). Identifiability in linear processes. Z. Wahrscheinlichkeitstheorie verw. Geb. 16, 39-46.

Westcott, M. (1971). On existence and mixing results for cluster point processes. J. R. Statist. Soc. B 33, 290-300.

Westcott, M. (1972). The probability generating functional. J. Aust. Math Soc. 13, to appear.

Wold, H. (1948). Sur les processus stationnaires ponctuels. Colloques Int. Cent. Natn. Rech. Scient. 13, 75-86.

SUMMARY OF "FURTHER RESULTS FOR GAUSS-POISSON PROCESSES"

R. K. Milne
Australian National University, Canberra, Australia

M. Westcott
Imperial College, University of London

SUMMARY

In a recent paper, Newman (1970) introduced an interesting new class of stochastic point processes which he christened *Gauss-Poisson* (G-P). These processes are defined through their probability generating functionals (p.g.fls) as a natural generalization of the Poisson process. In Milne & Westcott (1972) we show that the class of G-P processes on the real line is a 2-parameter family of point processes, the parameters being the first two factorial cumulant measures, and investigate further aspects of such processes while answering some of the questions which Newman left open.

Newman studied G-P processes in general spaces, whereas we are content with a treatment for the real line $R^1 = R$. However we employ the theory of point processes and p.g.fls expounded recently by Vere-Jones (1968, 1970) and Westcott (1972), which enables us to specialize our results easily to stationary point processes on the whole of R.

We now outline some of the results obtained, referring readers to Milne & Westcott (1972) for further details.

In common with Newman, we first ask: can one determine the class of possibly signed Borel measures $\lambda(\cdot)$ and $H(\cdot)$ for which the functional $G[\xi]$ defined by

$$G[\xi] = \exp[\int\{\xi(t) - 1\}\lambda(dt) + \frac{1}{2}\iint\{\xi(t) - 1\}\{\xi(u) - 1\}H(dt \times du)] \quad (1)$$

is the p.g.fl of a point process in R? Using a known characterization theorem for p.g.fls, we prove that it is necessary and sufficient that

(i) $\lambda(\cdot)$ be non-negative,

(ii) $H_s(\cdot)$ be non-negative, and

(iii) $H_s(A\times B) \leq \min\{\lambda(A), \lambda(B)\}$ for all bounded Borel sets A, B,

where $H_s(\cdot)$, the symmetrized version of $H(\cdot)$, is defined by $H_s(A\times B) = \frac{1}{2}\{H(A\times B) + H(B\times A)\}$.

A direct consequence of (iii), which leads to other important properties, is that for all bounded Borel sets A, $H_s(A\times R) < \infty$.

When λ and H are absolutely continuous, it is shown that their Radon-Nikodym derivatives are non-negative almost everywhere. This is essentially the case considered by Newman (1970).

We next prove that for G-P processes all modes of stationarity (i.e., simple, weak, p^{th} order and strict) are equivalent, and show that in the stationary case the p.g.fl (1) can be rewritten as

$$(2) \quad G[\xi] = \exp[\lambda\int\{\xi(t) - 1\}dt + \frac{1}{2}\iint\{\xi(t) - 1\}\{\xi(t+u) - 1\}K(du)dt] ,$$

where $0 < \lambda < \infty$ and $K(\cdot)$ is non-negative and totally finite. When K has a density ρ this is the covariance density of the process, and when K has no atom at zero the process is orderly.

It is shown that the stationary G-P process is mixing, and hence ergodic, using definitions first stated by Matthes (1963) (see also Westcott, 1972), and that the factorial moment measures are all finite and 'ergodically consistent'; this is a term due to Newman (1970). Newman originally established a similar result for the associated densities, when they exist; however we show that the additional assumption $\lim_{u\to\infty} \rho(u) = 0$ is required in this case.

Further, we prove that G-P processes are infinitely divisible, and explicitly determine the associated canonical or KLM measure (the existence of such a measure is proved in Kerstan & Matthes, 1964, and Lee, 1967). From this we can show that such processes are in fact regular infinitely divisible (see Kerstan & Matthes for the terminology), and hence that they are expressible as Poisson clustering processes. We indicate several such formulations, all being variants of the basic

Bartlett-Lewis model (Bartlett, 1963; Lewis, 1964). This correspondence enables the statistical theory developed for such processes to be applied directly to G-P processes.

We prove that the class of G-P processes is closed under the operations of superposition, translation and deletion. Some other situations in which G-P processes arise are then given, one such being as the superposition of the marginals of certain bivariate point processes. In particular, the superposed input and output of an M/G/∞ system is a stationary G-P process. However, we conjecture that a G-P process cannot be a renewal process unless it is in fact Poisson, and a heuristic proof of this is presented. Also, we mention the curious fact that, at least formally, some G-P processes can be obtained as doubly stochastic Poisson processes with a Gaussian mean process.

It is clear from (1) that G-P processes are the natural generalization of the Poisson process to processes with only the first <u>two</u> factorial cumulant measures not identically zero (for a Poisson process of course $H \equiv 0$). The obvious question is whether one can make further extensions, to processes with p.g.fls of the form

$$(3)\quad G[\xi] = \exp[\sum_{i=1}^{k} \int_{\underline{R}^i} \prod_{j=1}^{i} \{\xi(t_j) - 1\} H_i(dt_1 \times \ldots \times dt_i)],$$

where $H_i(\cdot)$ is a possibly signed Borel measure in R^i. We prove that (3) can indeed be a p.g.fl in some circumstances, using a class of multivariate point processes developed by Milne (1971), and point out some of the difficulties involved in determining necessary and sufficient conditions for this in general.

Finally, we look at some results in estimation and hypothesis testing for G-P processes. The likelihood of the process is given, and shown to be highly intractable, and applications of several other methods for estimating covariance densities (e.g. Cox & Lewis, 1966, p.121; Vere-Jones, 1970) are mentioned in relation to estimation of ρ. The uses of the mixing properties and cluster point process representations of G-P processes are discussed. However, as with much of the statistical theory for point processes, rather ad hoc procedures are generally all that are available.

REFERENCES

Bartlett, M.S. (1963). The spectral analysis of point processes. J. R. Statist. Soc. B 25, 264-96.

Cox, D. R. & Lewis, P. A. W. (1966). *The Statistical Analysis of Series of Events*. Methuen: London, and Barnes & Noble: New York.

Kerstan, J. & Matthes, K. (1964). Stationäre zufällige Punktfolgen II. J-ber. Deutsch. Math. Verein. 66, 106-18.

Lee, P. M. (1967). Infinitely divisible stochastic processes. Z. Wahrscheinlichkeitstheorie verw. Geb. 7, 147-60.

Lewis, P. A. W. (1964). A branching Poisson process model for the analysis of computer failure patterns. J. R. Statist. Soc. B 26, 398-456.

Matthes, K. (1963). Unbeschränkt teilbare Verteilungsgesetze stationärer zufälliger Punktfolgen. Wiss. Z.Hochsch. Electro. Ilmenau 9, 235-8.

Milne, R. K. (1971). Stochastic Analysis of Multivariate Point Processes. Ph.D. Thesis, The Australian National University: Canberra.

Milne, R. K. & Westcott, M. (1972). Further results for Gauss-Poisson processes. Adv. Appl. Prob. 4 (to appear).

Newman, D. S. (1970). A new family of point processes which are characterized by their second moment properties. J. Appl. Prob. 7, 338-58.

Vere-Jones, D. (1968). Some applications of probability generating functionals to the study of input-output streams. J. R. Statist. Soc. B 30, 321-33.

Vere-Jones, D. (1970). Stochastic models for earthquake occurrence. J. R. Statist. Soc. B 32, 1-62.

Westcott, M. (1972). The probability generating functional. J. Aust. Math. Soc. (To appear)

SPECTRA OF SOME MUTUALLY EXCITING POINT PROCESSES WITH ASSOCIATED VARIABLES

A. G. Hawkes
University of Durham, England

SUMMARY

A model is postulated in which point processes have variables or marks associated with each event, and in which the intensities of the point processes are linearly dependent on the previous history of the process itself and of a second driving process. The Bartlett spectral matrices of the process are derived.

1. INTRODUCTION

Hawkes (1971a, 1971b) introduced a class of stationary point processes in which the current intensity is linearly dependent on the past, and obtained the point spectrum (or Bartlett spectrum). The simplest case is the self-exciting process N(t) for which

$$\text{prob}\{dN(t) = 1 \mid N(s),\ s \leq t\} = \Lambda(t)dt + o(dt),$$

where

$$\Lambda(t) = \nu + \int_{-\infty}^{t} \gamma(t-a)dN(a).$$

For many problems it is of interest to consider a variable associated with each event. Matthes (1963) calls this a mark and provides a measure-theoretic foundation for the study of marked point processes. Vere-Jones (1970) considers the energies associated with earthquakes and suggests a suitable form of spectral analysis. Bartlett (1967) proposed a theoretical model to study the velocities associated with vehicles passing a point on a road and fitted this to some data, making use of a different form of spectral analysis.

In this paper we introduce a model for a marked process in which the values taken by the marks influence the future intensity $\Lambda(t)$ in a linear fashion, both marks and point process being multivariate. A simple expression for the Bartlett spectral matrices is obtained.

2. THE MODEL

We consider k point processes $N_i(t)$, i = 1 to k, each with an associated vector valued additive process $\underline{Z}_i(t)$ such that $d\underline{Z}_i(t) = \underline{0}$ if $dN_i(t) = 0$ and if $dN_i(t) = 1$ then $d\underline{Z}_i(t)$ is equal to the vector mark associated with the event occurring at time t.

Let

$$\underline{X}_0^T(t) = \{\underline{Z}_1^T(t), \underline{Z}_2^T(t), \ldots, \underline{Z}_k^T(t)\}$$

and

$$\underline{X}_1^T(t) = \{N_1(t), N_2(t), \ldots, N_k(t)\},$$

and let $\underline{X}_2(t)$ be a general vector valued point process which may include components of both the above types. We assume that these processes are jointly weakly stationary with mean densities

$$E\{d\underline{X}_i(t)\} = \underline{m}_i \, dt.$$

In particular $\underline{m}_1 = \underline{\lambda}$ represents the vector of intensities of the point processes $N_i(t)$.

We shall be interested in the cross-covariance density matrices

$$\underline{\mu}_{ij}(\tau) = E\{d\underline{X}_i(t+\tau) \, d\underline{X}_j^T(t)\}/(dt)^2 - \underline{m}_i\underline{m}_j^T.$$

In general there may be delta functions at the origin and it is convenient to write the complete covariance density matrices as

$$\underline{\mu}_{ij}^{(c)}(\tau) = \underline{D}_{ij}\delta(\tau) + \underline{\mu}_{ij}(\tau), \tag{1}$$

where $\underline{\mu}_{ij}(\tau)$ is continuous at the origin and $\underline{D}_{ij}$ is a matrix of suitable constants.

Throughout we use the convention that for any matrix of functions denoted by a Greek letter the matrix of Fourier transforms is denoted by the corresponding Roman capital, for example

$$\underline{M}_{ij}(\omega) = \int_{-\infty}^{\infty} e^{-i\omega\tau} \underline{\mu}_{ij}(\tau)d\tau.$$

Then the spectral density matrices are

$$\underline{f}_{ij}(\omega) = \frac{1}{2\pi}\int_{-\infty}^{\infty} e^{-i\omega\tau}\underline{\mu}_{ij}^{(c)}(\tau)d\tau$$

$$= \tfrac{1}{2}\underline{M}_{ij}^{(c)}(\omega)/\pi = \tfrac{1}{2}\{\underline{D}_{ij} + \underline{M}_{ij}(\omega)\}/\pi. \qquad (2)$$

It should be noted that these definitions imply

$$\begin{aligned} \underline{\mu}_{ij}(-\tau) &= \underline{\mu}_{ji}^{T}(\tau) \qquad (i \neq j), \\ \underline{\mu}_{ii}(-\tau) &= \underline{\mu}_{ii}^{T}(\tau), \end{aligned} \qquad (3)$$

which implies

$$\begin{aligned} \underline{M}_{ij}(-\omega) &= \underline{M}_{ji}^{T}(\omega) \qquad (i \neq j), \\ \underline{M}_{ii}(-\omega) &= \underline{M}_{ii}^{T}(\omega). \end{aligned} \qquad (4)$$

Our model involves conditioning on the sample paths of $\underline{X}_0$, $\underline{X}_1$ prior to time t and the whole of the sample path of $\underline{X}_2$. If we write

$$C(t)=[\{\underline{X}_0(s), \underline{X}_1(s)\}, s \leq t; \underline{X}_2(s), -\infty < s < \infty],$$

then our assumptions are

A.1 $E\{d\underline{X}_1(t)|C(t)\} = \underline{\Lambda}(t)dt + o(dt)$;

where

$$\underline{\Lambda}(t) = \underline{\nu} + \sum_{i=0}^{2}\int_{-\infty}^{t} \underline{\gamma}_i(t-a)d\underline{X}_i(a).$$

A.2 The elements of $\underline{\Lambda}(t)$ are almost certainly > 0;

A.3 $E\{d\underline{Z}_i(t)|C(t), dN_i(t) = 1\} = \underline{0}$;

A.4 $E\{d\underline{Z}_i(t)d\underline{Z}_i^T(t)\,|C(t),\ dN_i(t) = 1\} = \underline{V}_i$;

A.5 The processes are weakly stationary.

Thus the intensity $\underline{\Lambda}(t)$ of the point process is linearly dependent on the history of the three processes $\underline{X}_i(t)$. Here $\underline{X}_2(t)$ may be considered as a driving process whose spectral density matrix $\underline{f}_{22}(\omega)$ is assumed known.

Assumptions A3, A4 hold if the mark associated with any event is distributed independently of the history of the process prior to that event, and independently of the whole driving process $\underline{X}_2(t)$, and has mean zero and dispersion $\underline{V}_i$. Assumption A3 involves no loss of generality since if we have a process $\underline{Z}_i^1(t)$ with mean mark $\underline{b}$ we can construct a new process $\underline{Z}_i(t) = \underline{Z}_i^1(t) - \underline{b}N_i(t)$ which satisfies A3 while the form of $\underline{\Lambda}(t)$ in A1 will still be correct if $\underline{\gamma}_1(t-a)$ is suitably modified. One may then if necessary use such expressions as

$$E\{d\underline{Z}_i^1(t+\tau)dN_i(t)\} = E\{d\underline{Z}_i(t+\tau)\ dN_i(t)\} + \underline{b}E\{dN_i(t+\tau)dN_i(t)\}$$

to find the quantities of interest. A simple example of this procedure is included in Section 4. Assumption A2 may be satisfied if $\underline{\gamma}_i(\nu) \geq \underline{0}$ and the marks are non-negative before being transformed to satisfy A3 in the manner described above, and if $d\underline{X}_2(t) \geq 0$ almost certainly. Assumptions A1, A3, A5 lead immediately to

$$\underline{m}_1 = \underline{\lambda} = E\{\underline{\Lambda}(t)\} = \underline{\nu} + \underline{G}_1(0)\underline{m}_1 + \underline{G}_2(0)\underline{m}_2,$$

so that

$$\underline{m}_1 = \{\underline{I} - \underline{G}_1(0)\}^{-1}\{\underline{\nu} + \underline{G}_2(0)\underline{m}_2\}, \tag{5}$$

and we must have the elements of this finite and positive.

3. DERIVATION OF THE SPECTRAL MATRICES

The fundamental equations are derived by taking conditional expectations. In general

$$E\{d\underline{X}_i(t+\tau)d\underline{X}_j^T(t)\} = E[E\{d\underline{X}_i(t+\tau)d\underline{X}_j^T(t)\,|\,C(t+\tau)\}] \tag{6}$$

and in particular

$$E\{d\underline{X}_1(t+\tau)d\underline{X}_j^T(t)\} = E\{dt\underline{\Lambda}(t+\tau)d\underline{X}_j^T(t)\} \qquad (\tau > 0).$$

The form of $\underline{\Lambda}(t)$ assumed in A1 then leads to

$$\underline{\mu}_{11}(\tau) = E\{dX_1(t+\tau)d\underline{X}_1^T(t)\}/(dt)^2 - \underline{m}_1\underline{m}_1^T$$

$$= \sum_{i=0}^{2} \int_{-\infty}^{\tau} \underline{\gamma}_i(\tau-\nu)\, \underline{\mu}_{i1}^{(c)}(\nu)d\nu \qquad (\tau > 0). \tag{7}$$

Similarly we obtain

$$\underline{\mu}_{10}(\tau) = \sum_{i=0}^{2} \int_{-\infty}^{\tau} \underline{\gamma}_i(\tau-\nu)\underline{\mu}_{i0}^{(c)}(\nu)d\nu \qquad (\tau > 0), \tag{8}$$

while the form in which $\underline{X}_2(s)$ enters the condition $C(t+\tau)$ implies that the similar equation

$$\underline{\mu}_{12}(\tau) = \sum_{i=\nu}^{2} \int_{-\infty}^{\tau} \underline{\gamma}_i(\tau-\nu)\underline{\mu}_{i2}^{(c)}(\nu)d\nu \tag{9}$$

holds for all values of τ. The use of A3 with (6) implies that

$$\underline{\mu}_{00}(\tau) = \underline{\mu}_{01}(\tau) = 0 \qquad (\tau > 0),$$

$$\underline{\mu}_{02}(\tau) = 0 \qquad (-\infty < \tau < \infty).$$

Then (3) implies that we can write this as

$$\begin{aligned} &\underline{\mu}_{01}(\tau) = 0 && (\tau > 0), \\ &\underline{\mu}_{00}(\tau) = \underline{\mu}_{02}(\tau) = 0 && (-\infty < \tau < \infty), \end{aligned} \tag{10}$$

while $\underline{\mu}_{22}(\tau)$ is assumed known.

To complete the covariance matrices as in (1) we note that

$$\underline{D}_{11} = \mathrm{diag}(\lambda_1, \lambda_2, \ldots, \lambda_k),$$

$$\underline{D}_{00} = \mathrm{diag}(\lambda_1 V_1,\ \lambda_2 V_2, \ldots, \lambda_k V_k), \tag{11}$$

$$\underline{D}_{ij} = 0 \qquad\qquad (i \neq j).$$

We now take the Fourier transform of (9), noting that (10) implies $\underline{M}_{02}(\omega) = \underline{0}$, obtaining

$$\underline{M}_{12}(\omega) = \underline{G}_1(\omega)\underline{M}_{12}(\omega) + \underline{G}_2(\omega)\underline{M}_{22}^{(c)}(\omega),$$

so that

$$\underline{M}_{12}(\omega) = \{\underline{I}-\underline{G}_1(\omega)\}^{-1}\underline{G}_2(\omega)\underline{M}_{22}^{(c)}(\omega). \tag{12}$$

Using (1), (3) and (10) we observe that (8) may be written

$$\underline{\mu}_{10}(\tau) = \underline{\gamma}_0(\tau)\underline{D}_{00} + \int_{-\infty}^{\tau} \underline{\gamma}_1(\tau-\nu)\underline{\mu}_{10}(\nu)d\nu$$

for all τ, being identically zero for $\tau < 0$, and so we may take its Fourier transform and obtain

$$\underline{M}_{10}(\omega) = \{\underline{I} - \underline{G}_1(\omega)\}^{-1}\underline{G}_0(\omega)\underline{D}_{00}. \tag{13}$$

We now define an auxiliary matrix

$$\underline{\beta}(\tau) = \sum_{i=0}^{2} \int_{-\infty}^{\tau} \underline{\gamma}_i(\tau-\nu)\underline{\mu}_{i1}^{(c)}(\nu)d\nu - \underline{\mu}_{11}(\tau)$$

and observe that (7) implies $\underline{\beta}(\tau) = 0$ for all $\tau > 0$. Then

$$\underline{\beta}(\omega) = \underline{G}_0(\omega)\underline{M}_{10}^{T}(-\omega) + G_1(\omega)\{\underline{D}_{11} + \underline{M}_{11}(\omega)\} + \underline{G}_2(\omega)\underline{M}_{12}^{T}(-\omega) - \underline{M}_{11}(\omega),$$

which may be rearranged to give

$$\underline{M}_{11}(\omega) = \{\underline{I}-\underline{G}_1(\omega)\}^{-1}\{\underline{G}_1(\omega)\underline{D}_{11} - \underline{B}(\omega) + \underline{G}_0(\omega)\underline{M}_{10}^{T}(-\omega) + \underline{G}_2(\omega)\underline{M}_{12}^{T}(-\omega)\}. \tag{14}$$

From (4) we see that $\underline{M}_{11}(\omega) = \underline{M}_{11}^{T}(-\omega)$, so that substituting (12), (13) into (14) and using this relation we find

$$\underline{G}_1(\omega)\underline{D}_{11} + \{\underline{I} - \underline{G}_1(\omega)\}\underline{B}^T(-\omega) = \underline{D}_{11}\underline{G}_1^T(-\omega) + \underline{B}(\omega)\{\underline{I} - \underline{G}_1^T(-\omega)\}$$

Analytic arguments given in Hawkes (1971b) imply that the elements of the LHS of this equation are regular in the lower half ω plane and the elements of the RHS are regular in the upper half plane and we are able to infer that both are identically zero for values of ω for which they exist, in particular for ω real. Thus

$$\underline{B}(\omega) = -\underline{D}_{11}\underline{G}_1^T(-\omega)\{\underline{I} - \underline{G}_1^T(-\omega)\}^{-1},$$

so that substituting this and (12), (13) into (14) we have an expression for $\underline{M}_{11}(\omega)$ which with (2) gives the spectral matrix

$$\underline{f}_{11}(\omega) = \{\underline{I} - \underline{G}_1(\omega)\}^{-1}[\tfrac{1}{2}\{\underline{D}_{11} + \underline{G}_0(\omega)\underline{D}_{00}\underline{G}_0^T(-\omega)\}/\pi + \underline{G}_2(\omega)\underline{f}_{22}(\omega)\underline{G}_2^T(-\omega)] \cdot \{\underline{I} - \underline{G}_1^T(-\omega)\}^{-1}. \quad (15)$$

Equations (11), (12) and (13) imply

$$\underline{f}_{10}(\omega) = \tfrac{1}{2}\{\underline{I} - \underline{G}_1(\omega)\}^{-1}\underline{G}_0(\omega)\underline{D}_{00}/\pi, \quad (16)$$

and

$$\underline{f}_{12}(\omega) = \{\underline{I} - \underline{G}_1(\omega)\}^{-1}\underline{G}_2(\omega)\underline{f}_{22}(\omega). \quad (17)$$

Equations (10), (11) imply

$$\underline{f}_{02}(\omega) = \underline{0} \quad (18)$$

$$\underline{f}_{00}(\omega) = \tfrac{1}{2}\underline{D}_{00}/\pi. \quad (19)$$

These give us all the spectral matrices, using (4), in terms of $\underline{f}_{22}(\omega)$ which is assumed known. Notice particularly that for any $i \neq j \neq k$ the cross spectral matrix $\underline{f}_{ij}(\omega)$ is independent of the parameters of the third process $d\underline{X}_k(t)$.

It should be noted that some of the results are still valid under slightly more general assumptions. If $\underline{\psi}(t)$ is a positive valued stationary process and we have the condition

$$C'(t) = (\{\underline{X}_0(s), \underline{X}_1(s)\}, s \leq t; \quad \underline{\psi}(s), -\infty < s < \infty).$$

Then we consider the assumption

A.1' $\text{prob}\{d\underline{N}_1(t) = 1 | C'(t)\} = \underline{\Lambda}(t)dt + o(dt),$

where

$$\underline{\Lambda}(t) = \underline{\nu} + \sum_{i=0}^{1} \int_{-\infty}^{t} \underline{\gamma}_i(t-u)d\underline{X}_i(u) + \underline{\psi}(t)$$

with the other assumptions remaining unchanged except replacing $C(t)$ by $C'(t)$. Then the results (15), (16) and (19) remain valid if we write

$$\underline{G}_2(\omega)\ \underline{f}_{22}(\omega)\underline{G}_2^T(-\omega) = \underline{f}_{\psi\psi}(\omega),$$

the spectral matrix of the process $\underline{\psi}(t)$.

4. THE UNIVARIATE CASE

In this section we illustrate the general results by reference to the simple case of a single point process $N(t)$ with an associated univariate mark process $X(t)$, for example earthquakes and their associated energies. The main assumptions are that the mark is standardized to mean zero and variance σ^2, marks are independent of the past while

$$\text{prob}\{dN(t) = 1 | C'(t)\} = \Lambda(t)dt + o(dt),$$

where

$$\Lambda(t) = \nu + \int_{-\infty}^{t} \gamma_0(t-u)dX(u) + \int_{-\infty}^{t} \gamma_1(t-u)dN(u) + \psi(t).$$

Then we have

$$f_{dXdX}(\omega) = \tfrac{1}{2}\lambda\sigma^2/\pi,$$

$$f_{dNdX}(\omega) = \lambda\sigma^2 G_0(\omega)/[2\pi\{1-G_1(\omega)\}]$$

$$f_{dNdN}(\omega) = \frac{\tfrac{1}{2}\lambda(1 + \sigma^2|G_0(\omega)|^2)/\pi + f_{\psi\psi}(\omega)}{|1 - G_1(\omega)|^2}.$$

As special cases we consider

(a) a process which is unmarked, or where the mark does not affect the occurrence rate, given by $\gamma_0(u) = 0$, for which we get

$$f_{dNdN}(\omega) = \frac{\frac{1}{2}\lambda/\pi + f_{\psi\psi}(\omega)}{|1 - G_1(\omega)|^2},$$

which is a simple generalization of the well known result for a doubly stochastic process which we obtain by letting $\gamma_1(u) = G_1(\omega) = 0$.

(b) A further special case of (a) without a driving term gives the result for the simple self-exciting process for which

$$\Lambda(t) = \nu + \int_{-\infty}^{t} \gamma_1(t-u)dN(u),$$

having spectral density

$$f_{dNdN}(\omega) = \frac{\lambda}{2\pi|1 - G_1(\omega)|^2}.$$

This bears a close resemblance to an autoregression, perhaps it might be called an autoregressive point process.

(c) As a simple example of a situation in which the average mark is non-zero, suppose that $X'(t)$ is a process of positive marks with mean m and variance σ^2 and that

$$\Lambda(t) = \nu + \int_{-\infty}^{t} \gamma_0'(t-u)dX'(u) + \int_{-\infty}^{t} \gamma_1'(t-u)dN(u) + \psi(t).$$

Then $X(t) = X'(t) - mN(t)$ will have marks of zero mean and

$$\Lambda(t) = \nu + \int_{-\infty}^{t} \gamma_0(t-u)dX(u) + \int_{-\infty}^{t} \gamma_1(t-u)dN(u) + \psi(t),$$

where

$$\gamma_0(v) = \gamma_0'(v), \quad \gamma_1(v) = \gamma_1'(v) + m\gamma_0'(v).$$

Then the general results of this section apply to the transformed problem. If necessary one may then use

$$f_{dNdX'}(\omega) = f_{dNdX}(\omega) + mf_{dNdN}(\omega),$$

$$f_{dX'dX'}(\omega) = f_{dXdX}(\omega) + m\{f_{dXdX}(\omega) + f_{dNdX}(\omega)\} + m^2 f_{dNdN}(\omega).$$

5. UNMARKED PROCESSES

If the marks do not exist or do not affect the intensity we set $\underline{\gamma}_0(u) = \underline{G}_0(\omega) = \underline{0}$ and the results obtained in section three reduce to those derived in Hawkes (1971b). We note here some simple special cases.

(a) If in addition $\underline{\gamma}_0(u) = \underline{0}$ we have a k dimensional doubly stochastic process with spectrum

$$f_{11}(\omega) = \tfrac{1}{2}\underline{D}_{11}/\pi + \underline{G}_2(\omega)\underline{f}_{22}(\omega)\ \underline{G}_2^T(-\omega)$$

$$= \tfrac{1}{2}\underline{D}_{11}/\pi + \underline{f}_{\Lambda\Lambda}(\omega),$$

which is a simple generalization of the one dimensional case.

(b) If $\underline{\gamma}_2(u) = \underline{0}$ the driving process is inoperative and $\underline{X}_1(t)$ forms what might be called a set of mutually exciting point processes with spectrum

$$\underline{f}_{11}(\omega) = \tfrac{1}{2}\{\underline{I} - \underline{G}_1(\omega)\}^{-1}\underline{D}_{11}\{\underline{I} - \underline{G}_1^T(-\omega)\}^{-1}/\pi.$$

(c) If $\underline{\gamma}_1(u) = \underline{\gamma}_2(u) = \underline{0}$ we have k independent Poisson processes.

6. APPLICATIONS

The model does not allow for any inhibiting effect, which restricts its application somewhat, but it would still seem to be a plausible model for quite a wide range of problems. Unmarked processes could be usefully used in epidemiology to study the occurrence of cases in a number of regions, schools, etc. A marked process model is being fitted to earthquake data by L. Adamopoulos at Durham.

REFERENCES

Bartlett, M. S. (1963). The spectral analysis of point processes. J. R. Statist. Soc. B 25, 264-96.

Bartlett, M. S. (1967). Some remarks on the analysis of time series. Biometrika 54, 25-38.

Cox, D. R. & Lewis, P. A. W. (1966). *The statistical analysis of series of events*. London: Methuen and New York: Barnes & Noble.

Hawkes, A. G. (1971a). Spectra of some self-exciting and mutually exciting point processes. Biometrika 58, 83-90.

Hawkes, A. G. (1971b). Point spectra of some mutually exciting point processes. J. R. Statist. Soc. B 33, 438-443.

Mathes, K. (1963). Stationare zufallige Punktfolgen, I. Jber. Deutsch. Math-Verein 66, 66-78.

Vere-Jones, D. (1970). Stochastic models for earthquake occurrence. J. R. Statist. Soc. B 32, 1-62.

A GENERALIZATION OF THE TWO-STATE TWO-INTERVAL SEMI-MARKOV MODEL

A. Ekholm
The State Alcohol Monopoly, Helsinki, Finland

SUMMARY

A model which can be used for the statistical analysis of series of events is proposed. The physical system producing the events is supposed to alternate between two states. The numbers of consecutive events produced in each state are assumed to be independent integer-valued random variables. Given the state, the intervals between the events are independent real-valued random variables. The widely used two-state two-interval semi-Markov model is obtained as a special case when the distributions of the integer-valued random variables are geometric. The first and second-order properties of both the stationary sequence of intervals between events and the stationary stochastic counting process can be derived in a form suitable for numerical work. A set of traffic data is analyzed as an example of the numerical methods.

1. INTRODUCTION

This paper is a summary of results published recently (Ekholm, 1971) in a journal rather difficult to obtain. Copies of that paper can be obtained by request from the author. The impetus to this work was an attempt to fit the two-state two-interval semi-Markov model to some spike trains recorded from nerve cells. The background and some statistical work with the present model is presented in Ekholm & Hyvärinen (1970).

Hyvärinen (1966) had collected a fairly large material of spike trains — about 300, of which almost 100 did not have the statistical characteristics of a renewal process. The two-state two-interval semi-Markov model is then a natural choice as a model, for the two states can be thought of as a bursting and a resting state of the neuron. The two-state two-interval semi-Markov model implies that the serial correlations decrease exponentially with increasing lags (Cox and Lewis, 1966, p. 196). However, Hyvärinen's material contained very few spike trains showing this kind of correlogram.

2. DESCRIPTION OF THE PROCESS

In the papers just referred to the model to be described has been called a pseudo-Markov (PM) model. The name is not very good but we shall use it for consistency. A realization of the two-state PM process could be constructed in the following way. We agree to let the origin of the time-axis coincide with an event and to start in state 1. We begin by choosing a positive integer in accordance with a discrete distribution; we shall refer to this discrete distribution as $\{p_1\}$. Let the integer chosen be k. We then sample k independent observations from a distribution — denoted F_1 — on the positive real axis. Call the observations $x_1,\ldots,x_k$. The time from the origin to the first event after the origin is x_1, and the time between the first and second x_2. At the time point $t = x_1+\ldots+x_k$ we consult another discrete distribution — denoted $\{p_2\}$ — on how many observations to choose next. Let the result be the positive integer j. We then choose j independent observations from a second distribution — denoted F_2 — on the positive real axis. Call the new observations $x_{k+1},\ldots,x_{k+j}$. At the time point $t = x_1+\ldots+x_{k+j}$ we again turn to $\{p_1\}$ and so on alternatingly. Note that all samplings from the distributions $\{p_1\}$, $\{p_2\}$, F_1 and F_2 are independent of each other. The part of a realization just simulated is illustrated in Fig. 1 with k = 3 and j = 2.

The requisite of the PM model is thus a pair of discrete distributions $\{p_1\}$ and $\{p_2\}$ on the positive integers and a pair of (continuous) distributions F_1 and F_2 on the positive real axis. The discrete distributions regulate the number of events of like kind in a run and the continuous distributions the actual times between events. If the run distributions

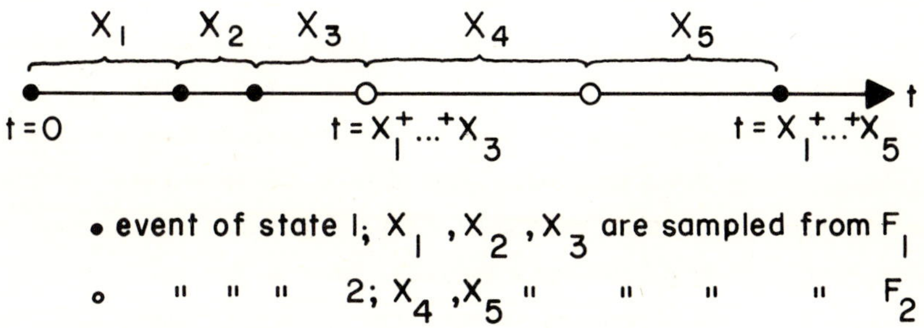

Figure 1. The realization described in Section 2.

$\{p_v\}$, v = 1,2 are chosen to be geometric, then the PM model reduces to the two-state two-interval semi-Markov model. The use of the semi-Markov model in the statistical analysis of series of events is thoroughly described by Cox & Lewis (1966, p. 194). In the construction of a realization we would then choose the type of each event separately at the preceding event. From the point of view of applications the most important feature of the PM model as a generalization of the semi-Markov model is to give the process a "memory". This will fundamentally modify the structure of the serial correlations though the intuitive set up is very much like that for the semi-Markov model.

3. MAIN RESULTS

The empirical series of events which we have had in mind when studying the properties of the PM model are assumed to have two fundamental features. First, they are assumed to show no evolutionary tendencies. The problem of how to start the process to make the relevant aspects of it formally stationary is not self-evident but will not be commented upon here (cf. Ekholm, 1971). Secondly, all events are originally indistinguishable. While we propose a model where the process alternates between two different states — e.g an excited and an inhibited state of the neuron — this does not mean that we are generally able to observe when the switching from one state to the other occurs. If this were the case, the statistical analysis could be based on the two separate series, and would be much easier. Under the present circumstances the two states represent an implicit hypothesis, and we will have to draw inferences about their existence and their properties from the combined single series.

Our aim has accordingly been to derive expressions for the second-order properties of the stationary sequence of times between events and the second-order properties of the stationary process of counts. For the definitions of these concepts we refer to Cox and Lewis (1966, Chap. 4). To report the main results we need the following symbols. Let λ_v be the mean of the run distributions $\{p_v\}$, $v = 1,2$ and μ_v and σ_v^2 be the mean and variance respectively of the interval distribution F_v, $v = 1,2$. Let further $\pi_v = \lambda_v/(\lambda_1 + \lambda_2)$, $v = 1,2$. Let F stand for the marginal distribution of the times between events in the stationary sequence. It is intuitively clear that

$$F(x) = \pi_1 F_1(x) + \pi_2 F_2(x) \tag{1}$$

and the marginal mean (μ) and variance (σ^2) are then

$$\mu = \pi_1\mu_1 + \pi_2\mu_2, \tag{2}$$

$$\sigma^2 = \pi_1\sigma_1^2 + \pi_2\sigma_2^2 + \pi_1\pi_2(\mu_1 - \mu_2)^2. \tag{3}$$

We need the following notation:

$$\Lambda = (\lambda_1 + \lambda_2)/(\lambda_1\lambda_2), \tag{4}$$

$$C = \pi_1\pi_2(\mu_1 - \mu_2)^2/\sigma^2. \tag{5}$$

The second-order properties of the stationary sequence of times between events are in the case of the PM model most easily given by the serial correlation coefficient. Let $\rho(k)$ be the ordinary serial correlation coefficient (Cox & Lewis, p. 70) of lag k. We have the following

Proposition 1. Let the convolution of the run distributions be $p(k) = p_1 * p_2(k)$, and let the tails of these three distributions, for $k = 0,1,\ldots$, be $q_v(k) = \sum_{j=k+1}^{\infty} p_v(j)$, $v = 1,2$, and $q(k) = \sum_{j=k+1}^{\infty} p(j)$, and define $\xi(k) = \rho(k+1)$ for $k = 0,1,\ldots$. Then the serial correlation coefficient $\xi(k)$ can be found from the following recursive formula:

$$\xi(k) = C\{q(k) - \Lambda q_1 * q_2(k)\} + p*\xi(k). \tag{6}$$

This formula is really recursive since the $\{p_v\}$'s are restricted to the positive integers. It provides a handy way of computing the serial correlations once the $\{p_v\}$'s are known in numerical form. The following facts are easy consequences of this formula:

a) if the convolution $\{p\}$ of the run distributions is not periodic, then $\lim_{k\to\infty} \rho(k) = 0$;

b) if further the probability generating functions of the $\{p_v\}$'s are regular at $z = 1$, then for $r = 1,2,\ldots$ the series $\sum_k k^r|\rho(k)|$ is convergent;

c) if the $\{p_v\}$'s are finite, that is their p.g.f.'s are polynomials, then the correlogram $\rho(k)$, $k = 1,2,\ldots$ is a linear combination of oscillations.

The second-order properties of the stationary counting process for the PM model can best be obtained from a derivation of the Laplace-transform of the intensity function (cf. Cox & Lewis, 1966, Equ. 4.5.12).

Proposition 2. Suppose the interval distributions F_v, $v = 1,2$, have continuous density functions and let $\hat{f}_v(s)$ stand for the Laplace-transform of these p.d.f's, and let further $P_v(\cdot)$ and $Q_v(\cdot)$ respectively be the g.f.'s of the run distributions $\{p_v\}$ and their tails $\{q_v\}$. The Laplace-transform of the intensity function $\hat{h}(s)$ is then

$$\hat{h}(s) = \frac{\pi_1\hat{f}_1(s)}{1-\hat{f}_1(s)} + \frac{\pi_2\hat{f}_2(s)}{1-\hat{f}_2(s)}$$

$$- \frac{1}{\lambda_1+\lambda_2} \cdot \frac{Q_1\{\hat{f}_1(s)\}Q_2\{\hat{f}_2(s)\}}{1-P_1\{\hat{f}_1(s)\}P_2\{\hat{f}_2(s)\}} \cdot \frac{\{\hat{f}_1(s)-\hat{f}_2(s)\}^2}{\{1-\hat{f}_1(s)\}\{1-\hat{f}_2(s)\}}.$$

The first two terms are readily recognized as the Laplace-transforms of the renewal densities of two renewal processes with p.d.f.'s respectively $f_1(t)$ and $f_2(t)$. A comparison with Proposition 1 shows that the second-order properties of counts are more complex than the second-order properties of the intervals. The latter involve information only about the means and variances of the distributions F_v, $v = 1,2$. But in the expression for $\hat{h}(s)$ all four distributions F_v, $\{p_v\}$ are involved in full. With regard to the serial correlations the PM model behaves as the basic renewal model already when $\mu_1 = \mu_2$, while $\hat{h}(s)$ is reduced to the corresponding renewal result only when $\hat{f}_1(s) = \hat{f}_2(s)$.

The expression for $\hat{h}(s)$ is clearly very difficult to invert. It is, however, useful in another way; consider a bivariate power series expansion of $\hat{h}(s)$ in terms of the $\hat{f}_v(s)$'s. From the expression of $\hat{h}(s)$ follows that the coefficients of this expansion can be computed recursively from the $\{p_v\}$'s in a way closely related to the recursive relation we found for the serial correlation coefficient. This result makes it feasible to find the spectral density of the counts in numerical form. The spectral density $g(\omega)$ is thus, up to constants and normalizations, given by a double series

$$\sum_k \sum_h a(k,h) r_1^k(\omega) r_2^h(\omega) \cos\{k\phi_1(\omega) + h\phi_2(\omega)\},$$

where the $a(k,h)$'s are computed recursively from the run distributions and the $r_v(\omega)$'s and $\phi_v(\omega)$'s are defined by

$$\int_0^\infty e^{i\omega t} f_v(t)\,dt = r_v(\omega)\exp\{i\phi_v(\omega)\}.$$

Once the spectral density is found in numerical form it is a matter of standard Fourier integral programs to find the intensity function in numerical form.

4. A NUMERICAL EXAMPLE

It seems to me that the crucial point in applications of the PM model is whether one can find a natural and physically meaningful hypothesis for the run distributions. An ad hoc hypothesis for the interval distributions might be acceptable but the run distributions are the more important part of the hidden mechanism assumed and should have a clear-cut interpretation if any proper insight is bound to arise from fitting the PM model to real data. In the neurophysiological applications we have not so far been able to find any such hypotheses (Ekholm & Hyvärinen, 1970) for the run distributions but in the less sophisticated connection of a traffic stream a relevant hypothesis has been advanced (Ekholm, 1971). Consider the traffic data which Cox & Lewis (1966, p. 197) have reproduced as their Example 7.2. These data were first presented by Bartlett (1963) and they consist of the intervals between the moments when vehicles pass a point on a road. Bartlett fitted a branching Poisson model to these data and Cox & Lewis fitted a two-state two-interval semi-Markov model to the data. These two fitted models are very much alike with regard both to their spectrum of intervals and their spectrum of counts. It will accordingly be enough to compare the fit of the PM model with the fit of the semi-Markov model.

To motivate the choice of run distributions consider first the runs of long intervals. Suppose a car — call it number one — passes the fixed point at a speed high enough to leave the road free. That is, the interval following this car is sampled from the distribution of long intervals. Finally the car number two emerges. Whether number two is followed by a long or a short interval should be independent of the number of long intervals preceding number two as long as we know that there was at least one long interval preceding number two. This argument amounts to the statement that the distribution of runs of long intervals should have no "memory". Therefore we took $\{p_2\}$ to be geometric, that is for $k = 1,2,\ldots$ and $0 < \theta_2 < 1$

$$p_2(k) = (1-\theta_2)\theta_2^{k-1}. \tag{7}$$

Next, consider the distribution $\{p_1\}$. Note that if the number of short intervals in a run is k, then the size of the corresponding cluster of cars is $k + 1$. We shall advance the hypothesis that $\{p_1\}$ is the equilibrium distribution of a simple time-discrete random walk on the positive integer. Let n be a fixed integer, $n \geq 2$, and α and β real numbers subject to the conditions $0 < \beta < \alpha < 1$, $\alpha + \beta \leq 1$. The random walk is such that in the stages above n there is probability β for a step to the right, probability α for a step to the left, and probability $1-\alpha - \beta$ for a stay. For the states below n — except state 1 — the roles of α and β are interchanged. In state 1 there is probability α for a step to the right and $1 - \alpha$ for a stay. Finally, in state n there is probability β for steps both to the right and to the left and $1 - 2\beta$ for a stay. Briefly, the state 1 is a reflecting barrier and the state n is a partly reflecting, partly permeable barrier. There is a drift towards the state n from the states on both sides of state n.

The physical considerations behind this random walk hypothesis are, of course, that when there is a cluster on the road the cluster is led by a slow car. The queue behind the slow car will grow quickly but soon there will also be break-outs from the cluster. And when the queue gets longer than some tolerable limit the tendency for break-outs may become larger than the tendency for new joinings.

One finds easily that the equilibrium distribution of this random walk is a bilateral geometric distribution. That is, writing $\beta/\alpha = \theta_1$, for k = 1,2,...

$$p_1(k) = A\theta_1^{|n-k|},$$

where A is a normalization factor.

This derivation of the run distribution $\{p_1\}$ suggests a simple and appealing theory of the increase and decrease of traffic clusters. The theory is, of course, unrealistic in that formally no cluster can disappear. On the other hand in reality new clusters emerge and both these features are summarized by the peculiar behaviour in the state 1. Another and more seriously unrealistic feature of the theory is that it disregards the possibility of one cluster joining up behind another, that is the possibility of transitions to other than neighbouring states. Under some road conditions such joinings may be rare enough to be disregarded, under others the random walk would have to be modified into a full Markov chain. From the traffic engineer's point of view the most important feature of the present theory is perhaps that clusters smaller than the mode value have the same probability to increase in a unit of time as clusters bigger than the mode value to decrease. This feature could be modified and we would then end up with an asymmetric bilateral geometric distribution. Anyhow, these arguments should show that there are ample possibilities to derive other hypotheses for the distribution $\{p_l\}$ in this frame.

In fitting the theoretical correlogram to the data we had four parameters to estimate: n, θ_1, θ_2 and the amplitude factor C (see equations (5), (6), (7) and (8)). The "estimation" was done by very rough methods. The final run distribution in numerical form are visualized in Fig. 2. In Fig. 3 the empirical correlogram is shown together with the theoretical correlograms of the PM model and the semi-Markov model fitted by Cox & Lewis (1966, pp. 197-204). The general shape of the PM correlogram is amazingly similar to that of the empirical one. There are occasional large deviations — the most striking one for the lag 18 — but it should be kept in mind that the number of observations is small,

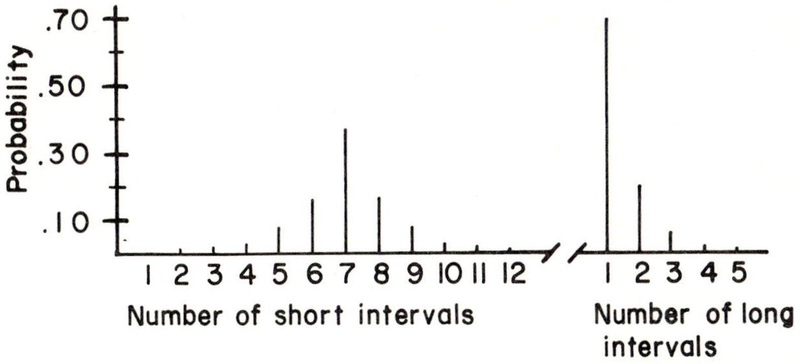

Figure 2. The run distributions of the PM model fitted to the traffic data.

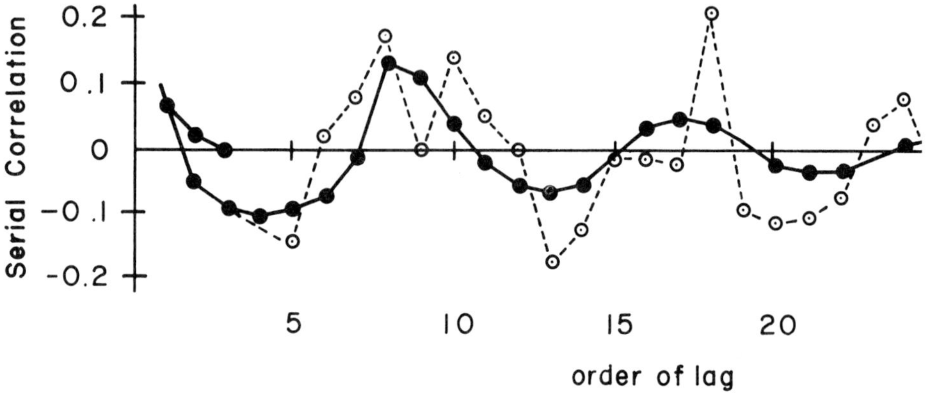

Figure 3. *The correlograms.* The circles connected by broken lines indicate the empirical estimates and the dots connected by solid lines the correlogram of the fitted PM model. The thin solid line which differs from zero only for lags one and two is the correlogram of the semi-Markov model.

viz. 128, and the standard error of the estimates $\tilde{\rho}(k)$ of the serial correlations are of the order 0.1. The fit of the semi-Markov correlogram is poor.

The theoretical and empirical spectra of intervals are shown in Fig. 4. The good fit of the PM spectrum could be expected on the basis of the correlogram but the dominance, in the theoretical spectrum,of the peak at $\omega = 0.23\pi$ is unexpectedly heavy.

To check the fit of the PM model in regard to the second-order properties of counts we had to choose an analytically specified hypothesis concerning the distributions of intervals. Without any substantial reasonings we took the distribution of short intervals (F_1) to be exponential and the distribution of long intervals (F_2) to be a gamma distribution of index 2. The values of the parameters were solved by equating moments of the marginal distribution.

The theoretical and empirical spectra of counts are visualized in Fig. 5. The main conclusion to be drawn from Fig. 5 is that both the PM and the semi-Markov model well reproduce the high initial value of the empirical spectrum. There is reason to believe that the high

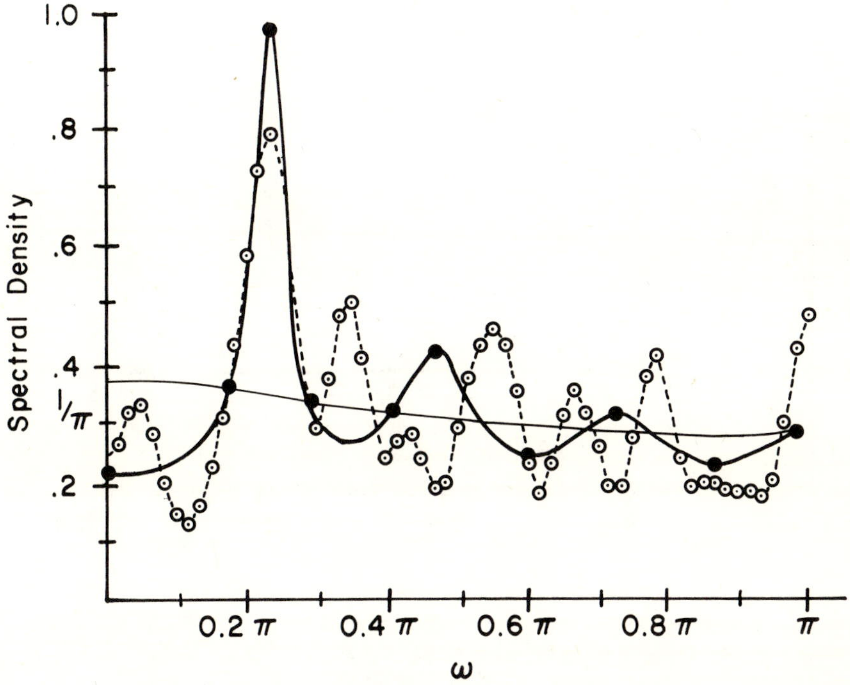

Figure 4. *The spectra of intervals.* The circles connected by broken lines indicate the empirical estimate. The solid line with the high peak is the theoretical spectrum of the PM model, the other solid line is the spectrum of the semi-Markov model.

initial value is the only important feature of the empirical spectrum (Cox & Lewis, 1966, p. 133). Still, the two theoretical curves have important dissimilarities but the number of observations is too small to obtain any resolution between them. In this example no more resolution can be obtained by the use of the intensity functions and we leave them out.

As an overall summary of the fit of the two competing models, it is perhaps fair to say that they are equal in explaining the observed second-order properties of counts while the PM model provides a much better explanation of the observed second-order properties of intervals. It would then be premature to give much importance to the relative merits of the distributions chosen for F_1 and F_2, but some attention should be given to the implications of the run distributions. The main point of

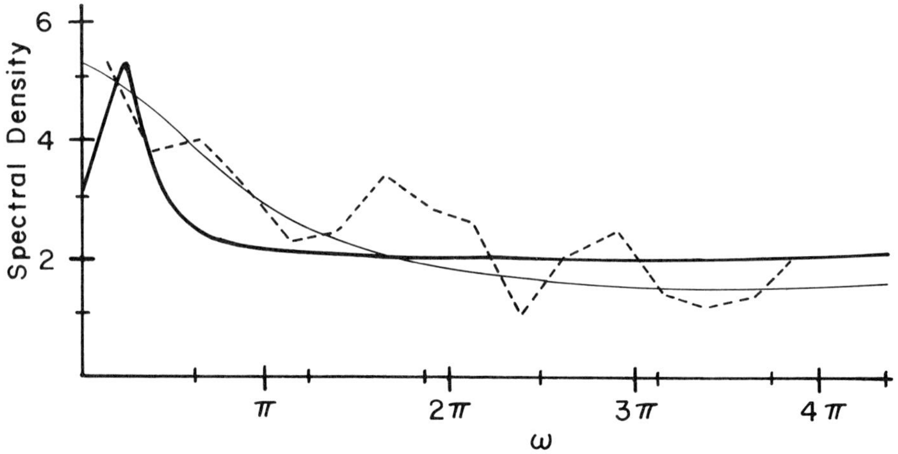

Figure 5. *The spectra of counts.* The dashed broken lines indicate the empirical estimate of the spectral density. The solid line with the lower initial value is the spectral density of the PM model and the other solid line the spectral density of the semi-Markov model.

our application of the PM model to these traffic data has been to show that the form of the run distributions can be given an interpretation in terms of the behaviour of traffic clusters. By use of the PM model such hypotheses can be tested on simple interval data of the kind analyzed here. In this way some qualitative information about the dynamic features of traffic clusters may be reached from cross-section data. Note, finally, that this approach does not call for a specification in distance or time of the limits of a cluster.

REFERENCES

Bartlett, M. S. (1963). The spectral analysis of point processes. J. R. Statist. Soc. B 25, 264-96 (with discussion).

Cox, D. R. & Lewis, P. A. W. (1966). *The Statistical Analysis of Series of events*. London: Methuen & New York: Barnes and Noble.

Ekholm, A. (1971). A pseudo-Markov model for stationary series of events. Commentationes Physico-Mathematicae 41, 73-120.

Ekholm, A. and Hyvärinen, J. (1970). A pseudo-Markov model for series of neuronal spike events. Bioph. J. 10, 773-96.

Hyvärinen, J. (1966). Analysis of spontaneous spike potential activity in developing rabbit diencephalon. Acta Physiologica Scandinavica 68, Supplementum 278.

CERTAIN CHANCE MECHANISMS RELATED TO BIRTH AND DEATH PROCESSES WHICH HAVE CLASSICAL DISCRETE DISTRIBUTIONS AS SPECIAL CASES

M. T. Boswell
Department of Statistics, The Pennsylvania State University

G. P. Patil
Department of Statistics, The Pennsylvania State University

SUMMARY

This paper is a study of birth and death processes. Little known results are brought together with well known results and some new results are found. A formula for the solution of a general birth and death process is found and the case where each birth rate factors into a function of the sizes of the population components and a function of time is solved. For the r-dimensional birth and death process where the birth and death rates factor as above and where all rates are linear, the solution is found to have either the negative multinomial, independent Poisson, or the multinomial distribution. A few related processes are studied.

1. INTRODUCTION

This paper started out with the idea of finding chance mechanisms which yield the negative binomial, Poisson, and the binomial distributions or their r-dimensional analogues. In this sense it was to be a continuation of our previous papers (Boswell and Patil, 1970, 1971). What has resulted is more a study of birth and death processes. Recently Cannings (1971) found the solution to a one-dimensional birth process whose birth rate $\lambda(x,t)$ factors into a product $\lambda(x) \cdot h(t)$ of a function of x and a function of time. As mentioned by Lundberg (1939, p. 64) this process can be transformed into a homogeneous birth process by the change of time scale $s(t) = \lambda(0)H(t)$, where $H(t) = \int_0^t h(u)du$. The re-resulting process has a birth rate $\lambda(x)/\lambda(0)$ and was solved by Lundberg (1939, p. 65). Section 2 generalizes Lundbergs work to r-dimensions and presents some special cases resulting in classical discrete distributions. Section 3 formulates a pure death process as a pure birth process. Section 4 extends a few known results on birth and death processes. This work relies heavily on the work of Darwin (1953) and Mode (1962). Section 5 studies birth and death processes at equilibrium. Finally Section 6 studies some models for group size, where the major feature is the transformation of these models to birth or birth and death processes.

2. BIRTH PROCESSES

Let $\{\underline{X}(t): t\varepsilon R^+\}$ be an integer-valued vector random variable, $\underline{X}(t) = \{X_1(t),\ldots,X_r(t)\}$. Then $\{(\underline{X}(t)\}$ is a birth process with birth rate $\underline{\lambda}(\underline{x},t) = \{\lambda_1(\underline{x},t),\ldots,\lambda_r(\underline{x},t)\}$ if $\text{prob}\{\underline{X}(t+h) = \underline{x} + i_j|\underline{X}(t) = \underline{x}\}$ $= \lambda_j(\underline{x},t)\cdot h + o(h), j = 1,\ldots,r$, and $\text{prob}\{\underline{X}(t+h) = \underline{x}|\underline{X}(t) = \underline{x}\}$ $= 1 - \sum_{j=1}^{r} \lambda_j(\underline{x},t)\cdot h + o(h)$, where i_j is a vector with a 1 in the jth position and 0's elsewhere. The differential equations for the processes are

$$\frac{\partial}{\partial t} p(\underline{x},t) = \sum_{j=1}^{r} \lambda_j(\underline{x}-i_j,t)p(\underline{x}-i_j,t) - \lambda(\underline{x},t)p(\underline{x},t),$$

where

$$\lambda(\underline{x},t) = \underline{\lambda}(\underline{x},t) \cdot \underline{1} = \sum_{j=1}^{r} \lambda_j(\underline{x},t),$$

or equivalently

$$\frac{\partial}{\partial t}\{p(\underline{x},t)e^{\Lambda(\underline{x},t)}\} = \sum_{j=1}^{r} \lambda_j(\underline{x}-i_j,t)p(\underline{x}-i_j,t)e^{\Lambda(\underline{x},t)}, \tag{2.1}$$

where $\Lambda(\underline{x},t) = \int_0^t \lambda(\underline{x},s)ds$. Birth processes can be solved iteratively. This leads to Theorem 2.1. First we make some definitions. We say $\underline{P}$

is a path to $\underline{x}$ if $\underline{P} = (P_0, \ldots P_{\underline{x}\cdot\underline{1}})$ where $P_0 = \underline{0}$, $P_{\underline{x}\cdot\underline{1}} = \underline{x}$, and $P_k - P_{k-1} = i_j$ for some j (depending on k). Let $P_{\underline{x}}$ be the collection of all paths to $\underline{x}$.

Theorem 2.1. For the process described above, with $\underline{X}(0) = \underline{0}$,

$$p(\underline{0},t) = e^{-\Lambda(\underline{0},t)}$$

and

$$p(\underline{x},t) = e^{-\Lambda(\underline{x},t)} \sum_{\underline{P}\varepsilon P_{\underline{x}}} \int_0^t \int_0^{s_{\underline{x}\cdot\underline{1}}} \cdots \int_0^{s_2} \prod_{k=1}^{\underline{x}\cdot\underline{1}} u(k,\underline{P},s_k)\,ds_1,\ldots,ds_{\underline{x}\cdot\underline{1}},$$

where

$$u(k,\underline{P},s_k) = (P_k - P_{k-1})\cdot\lambda(P_{k-1},s_k)e^{\Lambda(P_k,s_k)-\Lambda(P_{k-1},s_k)}.$$

Proof. Plugging into (2.1) yields

$$\frac{\partial}{\partial t}\{p(\underline{x},t)e^{\Lambda(\underline{x},t)}\}$$

$$= \sum_{\underline{P}\varepsilon P_{\underline{x}}} \frac{\partial}{\partial t}\int_0^t [u(\underline{x}\cdot\underline{1},\underline{P},s_k)\{\int_0^{s_{\underline{x}\cdot\underline{1}}} \cdots \int_0^{s_2} \prod_{k=1}^{\underline{x}\cdot\underline{1}-1} u(k,\underline{P},s_k)ds_1\ldots ds_{\underline{x}\cdot 1-1}\}]ds_{\underline{x}\cdot\underline{1}}$$

$$= \sum_{j-1}^{r} \sum_{\underline{P}\varepsilon P_{\underline{x}-i_j}} \{\underline{x}-(\underline{x}-i_j)\}\cdot \underline{\lambda}(\underline{x}-i_j,t)\int_0^t \int^{s_{\underline{x}\cdot\underline{1}-1}} \prod_{k=1}^{\underline{x}\cdot\underline{1}-1} u(k,\underline{P},s_k)ds_1\ldots ds_{\underline{x}\cdot\underline{1}-1}$$

$$= \sum_{j=1}^{r} \lambda_j(\underline{x}-i_j,t)p(\underline{x}-i_j,t),$$

which is the desired result. Since $p(\underline{0},t)$ is given by a special form, to finish the proof one needs check $p(\underline{0},t)$ and $p(i_j,t)$ in (2.1), which is straightforward.

Theorem 2.2. For a birth process with birth rates given by $\lambda_k(\underline{x},t) = \lambda_k(\underline{x})\cdot h(t)$ and $\underline{x}(0) = \underline{0}$, where $\lambda(\underline{y}) \equiv \sum\lambda_k(\underline{y})$ are distinct for all $\underline{y}$ for which $\underline{y}\cdot 1 \leq \underline{x}\cdot\underline{1}$, we have $p(\underline{0},t) = e^{-\lambda(\underline{0})H(t)}$, and

$$p(\underline{x},t) = \sum_{\underline{P}\varepsilon P_{\underline{x}}} \{ \prod_{k=1}^{\underline{x}\cdot\underline{1}} (P_k - P_{k-1}) \cdot \underline{\lambda}(P_{k-1}) \} \sum_{k=0}^{\underline{x}\cdot\underline{1}} e^{-\lambda(P_k)H(t)}$$

$$/ \prod_{\substack{j=0 \\ (j\neq k)}}^{\underline{x}\cdot\underline{1}} \{\lambda(P_j) - \lambda(P_k)\},$$

where $H(t) = \int_0^t h(s)ds$ and $\lambda(P_k) = \underline{\lambda}(P_k)\cdot\underline{1}$.

Proof. Either (2.1) can be checked or Theorem 2.1 can be applied.

Theorem 2.3. For a birth process with $\lambda_k(\underline{x},t) = \lambda_k(\underline{x})\cdot h(t)$ and $\underline{X}(0) = \underline{0}$:

(1) If $\lambda(\underline{x},t) \equiv \underline{\lambda}(\underline{x},t)\cdot\underline{1} = \{a\underline{x}\cdot\underline{1}+b\}h(t)$, then

$$p(\underline{x},t) = \frac{1}{(\underline{x}\cdot\underline{1})!} e^{-bH(t)} \{(1-e^{-aH(t)})/a\}^{\underline{x}\cdot\underline{1}}$$

$$\times \sum_{\underline{P}\varepsilon P_{\underline{x}}} \prod_{k=1}^{\underline{x}\cdot\underline{1}} (P_k - P_{k-1}) \cdot \underline{\lambda}(P_{k-1}).$$

(2) If $\lambda_j(\underline{x},t) = a_j(\delta\underline{x}\cdot\underline{1}+\gamma)h(t)$ where δ is either -1, 0, or 1, and

$\sum a_j = 1.$

Then $\underline{X}(t)$ has the

(A) negative multinomial distribution

$$p(\underline{x},t) = \binom{\gamma+\underline{x}\cdot\underline{1}}{\underline{x}} \{e^{H(t)}\}^{\gamma} \prod_{j=1}^{r} \{a_j(1-e^{-H(t)})\}^{x_j},$$

if $\delta = +1$,

(B) multiple Poisson distribution with parameters $\{a_j\gamma H(t)\}$, if $\delta = 0$, and

(C) the multinomial distribution

$$p(\underline{x},t) = \binom{\gamma}{\underline{x}} [e^{H(t)}]^{\gamma-\underline{x}\cdot\underline{1}} \prod_{j=1}^{r} \{-a_j(1-e^{H(t)})\}^{x_j},$$

for $\underline{x}$ satisfying $\underline{x}\cdot\underline{1} \leq \gamma$ (see Note 2.1 below), and $\delta = -1$.

Proof. These follow from Theorem 2.2 with proper substitutions.

Note 2.1. If γ is not an integer, formula (C) is still correct. However, it does not give all the probabilities since $\underline{x}$ may assume points on the hyperplane $\underline{x} \cdot \underline{1} = [\gamma] + 1$. Let $d = [\gamma]$; then $p(\underline{x},t)$ where $\underline{x} \cdot \underline{1} = d + 1$ are positive and can be found iteratively from the differential equations. For these points we have

$$\frac{\partial}{\partial t} p(\underline{x},t) = \sum_{j=1}^{r} \lambda_j(\underline{x}-\underline{i}_j)h(t)p(\underline{x}-\underline{i}_j,t)$$

$$= \sum_{j=1}^{r} a_j(\gamma-\underline{x}\cdot\underline{1}+1)h(t)\binom{\gamma}{\underline{x}-\underline{i}_j}\{e^{H(t)}\}^{\gamma-\underline{x}\cdot\underline{1}+1}$$

$$\times\{\prod_{i=1}^{r}(-a_i)^{x_i}\}\left(\frac{-1}{a_j}\right)\{1-e^{H(t)}\}^{\underline{x}\cdot\underline{1}-1}$$

$$= \sum_{j=1}^{r}(-1)x_j\binom{\gamma}{\underline{x}}\{\prod_{i=1}^{r}(-a_i)^{x_i}\}\{e^{H(t)}\}^{\gamma-\underline{x}\cdot\underline{1}+1}$$

$$\times\{1-e^{H(t)}\}^{\underline{x}\cdot\underline{1}-1}h(t)$$

$$= (-1)(\underline{x}\cdot\underline{1})\binom{\gamma}{\underline{x}}\{\prod_{i=1}^{r}(-a_i)^{x_i}\}\cdot\{e^{H(t)}\}^{\gamma-\underline{x}\cdot\underline{1}+1}$$

$$\times\{1-e^{H(t)}\}^{\underline{x}\cdot\underline{1}-1}h(t).$$

Integrating from 0 to t would give a formula for $p(\underline{x},t)$ on the hyperplane $\underline{x} \cdot \underline{1} = d + 1$. Only the last three terms depend on t, and they do not depend on which point of the hyperplane is used. The ratio of probabilities of points would therefore not depend on t. Thus

$$p(\underline{x},t) = \binom{d+1}{\underline{x}}\{\prod_{i=1}^{r}(a_i/a_1)^{x_i}\}p\{(d+1)\underline{i}_1,t\}.$$

Summing over the hyperplane gives

$$\text{prob}\{X(t) = d+1\} = (-a_1)^{-(d+1)}p\{(d+1)\underline{i}_1,t\},$$

where $X(t) \equiv \sum_{i=1}^{r} X_i(t)$ is the total population size and has a birth rate $\lambda(x,t) = (ax+b)h(t)$, $a < 0$. Therefore this probability is given by Theorem 2.2, and

$$p\{(d+1)i_1,t\} = (-1/a_1)^{-(d+1)}[1-\sum_{j=0}^{d}\binom{-b}{j}\{1-e^{H(t)}\}^j\{e^{H(t)}\}^{-(b+j)}].$$

Therefore on the hyperplane

$$p(\underline{x},t) = \binom{d+1}{\underline{x}}[1-\sum_{j=0}^{d}\binom{-b}{j}\{1-e^{H(t)}\}^j\{e^{H(t)}\}^{-(b+j)}]\prod_{i=1}^{r}(-a_i)^{x_i}.$$

The model resulting in (2) of Theorem 2.3 we like to think of as a model for handedness. At least in two dimensions $X_1(t)$ could be the number of left-handed individuals and $X_2(t)$ the number of right-handed. Any individual would produce either a left-handed or right-handed individual independent of his being right- or left-handed.

Another interesting model results if an individual reproduces itself with a certain probability or produces an individual of one of the other types with equally likely probabilities. That is $\lambda_j(\underline{x},t) = (ax_j + b\sum_{i\neq j} x_i + c)h(t)$. Then $\lambda(\underline{x},t) = \{a+b(r-1)\}\sum x_j + rc$, which is a special case of the model of (1) of Theorem 2.3.

Note 2.2. For a birth process with boundary condition $\underline{X}(0) = \underline{x}_0$, a completely different distribution results. These processes can, however, be transformed by $Y(t) = X(t) - x_0$ into a birth process with boundary condition $\underline{Y}(0) = \underline{0}$. The birth rates for $\underline{Y}(t)$ are $\{\lambda_k(\underline{y}+\underline{x}_0,t)\}$, $\{\lambda_t(\underline{x},t)\}$ being the birth rates for $\underline{X}(t)$.

3. PURE DEATH PROCESS

The pure death process starts with n individuals at time 0 and they die as time progresses. This process can be formulated as a birth process and we believe it often is formulated this way without realizing that it is a death process. If one counts the number of living individuals, then deaths occur; if on the other hand one counts the number of deaths (and calls them births), then one has a pure birth process. Let $X(t)$ be the number alive and $Y(t)$ be the number dead at time t.

Then $X(t) + Y(t) = n$. If the death rate is $\mu(x,t)$, then the birth rate is $\lambda(y,t) = \mu(n-y,t)$.

4. BIRTH AND DEATH PROCESSES

Let $\{\underline{X}(t): t\varepsilon R^+\}$ be an integer-valued vector stochastic process satisfying

$$\text{prob}\{\underline{X}(t+h) = \underline{x} + i_j | \underline{X}(t) = \underline{x}\} = \lambda_j(\underline{x},t)\cdot h+o(h) \qquad (j=1,\ldots,r),$$

$$\text{prob}\{\underline{X}(t+h) = \underline{x} - i_j | \underline{X}(t) = \underline{x}\} = \mu_j(\underline{x},t)\cdot h+o(h) \qquad (j=1,\ldots,r)$$

and

$$\text{prob}\{\underline{X}(t+h) = \underline{x} | \underline{X}(t) = \underline{x}\} = 1 - \{\lambda(\underline{x},t) + \mu(\underline{x},t)\}\cdot h+o(h),$$

where $\lambda(\underline{x},t) = \sum_{j=1}^{r} \lambda_j(\underline{x},t)$ and $\mu(\underline{x},t) = \sum_{j=1}^{r} \mu_j(\underline{x},t)$ and death rates $\mu_j(\underline{x},t)$, $j = 1,\ldots,r$. The differential equations for this process are

$$\frac{\partial}{\partial t} p(\underline{x},t) = \sum_{j=1}^{r} \lambda_j(\underline{x}-i_j,t)p(\underline{x}-i_j,t) - \{\lambda(\underline{x},t) + \mu(\underline{x},t)\} p(\underline{x},t)$$

$$+ \sum_{j=1}^{r} \mu_j(\underline{x}+i_j,t)p(\underline{x}+i_j,t).$$

Theorem 4.1. Let $\{\underline{X}(t): t\varepsilon R^+\}$ be an r-dimensional birth and death process with birth rates $\lambda_k(\underline{x},t) = a_k(\gamma+ \delta\underline{x}\cdot\underline{1})h(t)$ and death rates $\mu_k(\underline{x},t) = x_k g(t)$, $k = 1,\ldots,r$, where $\sum_{k=1}^{r} a_k=1$, and $\delta = -1$, 0, or 1.

Then

$$E\{\underline{X}(t) \cdot \underline{1}\} = \gamma e^{H(t)-G(t)} \int_0^t h(s)e^{-\{H(s)-G(s)\}}ds.$$

Further $\{\underline{X}(t)\}$ has the negative multinomial distribution

$$p(\underline{x},t) = \binom{\underline{x}\cdot\underline{1}+\gamma-1}{\underline{x}} p_0^\gamma(t) \prod_{k=1}^{r} p_k^{x_k}(t)$$

with $p_k(t) = a_k\{1-p_0(t)\}$ and $p_0(t) = 1/[1+E\{\underline{X}(t)\cdot\underline{1}\}/\gamma]$, has a multiple Poisson distribution with parameters $E(X_k) = a_k E\{\underline{X}(t)\cdot\underline{1}\}$, or has the multinomial distribution (γ an integer)

$$p(\underline{x},t) = \binom{\gamma}{\underline{x}} p_0^{\gamma - \underline{x}\cdot\underline{1}}(t) \prod_{k=1}^{r} p_k^{x_k}(t)$$

with $p_k(t) = a_k\{1-p_0(t)\}$ and $p_0(t) = 1 - E\{\underline{X}(t)\cdot\underline{1}\}/\gamma$ according as $\delta = 1, 0,$ or -1.

Proof. The Poisson case is obvious. For either the multinomial or the negative multinomial case we observe that $\prod_{k=1}^{r} p_k^{x_k}(t) = \{1-p_0(t)\}^{\underline{x}\cdot\underline{1}} \prod_{k=1}^{r} a_k^{x_k}$. The conclusion follows by checking the differential equations and calculating the expected value.

Note 4.1. For birth and death processes with boundary condition $\underline{X}(0) = \underline{x}_0$, completely different distributions result. If the one dimensional process has a birth rate of $xh(t)$ and the death rate of $xg(t)$, then the result is somewhat complicated but is known (see Bailey, 1964, p. 94 and p. 112). For $x_0 = 1$, a zero-modified geometric distribution results instead of a geometric negative(binomial distribution with $k = 1$), and for $x_0 > 1$ the process is the sum of n independent such zero-modified geometric processes. Now if the birth rate is $(\gamma+x)h(t)$ one can think of this process as immigration plus reproduction where each individual reproduces with rate $h(t)$ and new individuals arrive at a rate $\gamma\cdot h(t)$. Thus the process is the sum $X(t) + Y(t)$ of independent processes with birth rates $\lambda(x,t) = (\gamma+x)h(t)$ and $\lambda(y,t) = yh(t)$, death rates $\mu(x,t) = x\cdot g(t)$, $\mu(y,t) = y\cdot g(t)$, and boundary conditions $X(0) = 0$, $Y(0) = n$. The solution for the case with birth rate $(\gamma-x)h(t)$ is not available.

5. LINEAR BIRTH AND DEATH PROCESS AT EQUILIBRIUM.

We have the following result at equilibrium.

Theorem 5.1. Consider a process with birth rates $\lambda_k(\underline{x},t)=a_k(\gamma+\delta\underline{x}\cdot\underline{1})h(t)$ and death rates $\mu_k(\underline{x},t) = x_k g(t)$, $\sum a_k = 1$. If $\{G(t) - \delta H(t)\}\to\infty$, $h(t)\to\lambda$, and $g(t)\to\mu$ as $t\to\infty$, then equilibrium is reached and the distribution is

(1) the negative multinomial distribution

$$p(\underline{x}) = \binom{\underline{x}\cdot\underline{1}+\gamma-1}{\underline{x}} p_0^{\gamma} \prod_{k=1}^{r} p_k^{x_k},$$

where

$$p_k = a_k(1-p_0)$$

and

$$p_0 = 1/\{1+m(\infty)/\gamma\},$$

if $\delta = +1$;

(2) that of multiple Poisson random variables with parameters $a_k m(\infty)$ if $\delta = 0$, and

(3) the multinomial distribution (γ an integer)

$$p(\underline{x}) = \binom{\gamma}{\underline{x}} p_0^{\gamma-\underline{x}\cdot\underline{1}} \prod_{k=1}^{r} p_k^{x_k},$$

where

$$p_k = a_k(1-p_0) \text{ and } p_0 = 1-m(\infty)/\gamma,$$

if $\delta = -1$,

where $m(\infty) = \lim E\{\underline{X}(t)\cdot\underline{1}\} = \gamma\lambda/(\mu-\delta\lambda)$ as t goes to infinity.

Proof. The proof follows by taking limits in Theorem 3.1.

For one-dimensional birth and death process at equilibrium it is well known that

$$p(x) = \frac{\lambda(0)\ldots\lambda(x-1)}{\mu(1)\ldots\mu(x)}\, p(0) \quad \text{or} \quad \frac{p(x)}{p(x-1)} = \frac{\lambda(x-1)}{\mu(x)},$$

where $\lambda(x) = \lim \lambda(x,t)$ and $\mu(x) = \lim \mu(x,t)$ as t goes to infinity are assumed to be finite. So if one is looking for a birth and death process at equilibrium which has a particular distribution or vice versa one need equate the ratio of probabilities to the ratio of birth rate to death rate. For example, consider the case where $\lambda(x,t) = (\gamma-x)\cdot h(t)$ and $\mu(x,t) = x\cdot g(t)$, where γ is not necessarily an integer. We assume $\lim h(t) = \lambda$ and $\lim g(t) = \mu$. Then $\lambda(x-1)/\mu(x) = (\lambda/\mu)(\gamma-x+1)/x$. The ratio of probabilities for the binomial distribution is $p(x)/p(x-1) = \{p/(1-p)\}(\gamma-x+1)/x$. Therefore we have the following result.

Theorem 5.2. If $\lambda(x,t) = (\gamma-x)h(t) \to (\gamma-x)\lambda$ and $\mu(x,t) = xg(t) \to x\mu$ as $t \to \infty$, then the distribution at equilibrium is given by

$$p(x) = \frac{\binom{\gamma}{x}p^x(1-p)^{\gamma-x}}{\sum_{j=0}^{\delta+1}\binom{\gamma}{j}p^j(1-p)^{\gamma-j}} \qquad (x = 0, 1, \ldots, \delta + 1),$$

where $\delta = [\gamma]$ and $p = (\lambda/\mu)\{1+\lambda/\mu\}$. We observe that this binomial distribution for non-integer parameter $n = \gamma$ is from the one given in Theorem 2.3, (2), (C).

As another example consider the hypergeometric distribution $p(x) = \binom{r}{x}\binom{w}{n-x}/\binom{r+w}{n}$. The ratio of probabilities is $p(x)/p(x-1) = (r-x+1)(n-x+1)/\{(w-n+x)x\}$. Let $\lambda(x) = (r-x)(n-x)g(x+1)$ and $\mu(x) = x(w-n+x)g(x)$ for an arbitrary positive function $g(x)$. We observe that $\lambda(x) = 0$ if $x = \min\{r,n\}$ and $\mu(x) = 0$ if $x = \max\{0,n-w\}$. Therefore $\max\{0,n-w\} \leq X < \min\{r,n\}$, the usual range for the hypergeometric distribution. The above development needs to be modified if $n > w$ in which case $\lambda(0),\ldots,\lambda(n-w-1)$ and $\mu(0),\ldots,\mu(n-2)$ are zero. The result needed is

$$p(x) = \frac{\lambda(n-w)\ldots\lambda(x-1)}{\mu(n-w+1)\ldots\mu(x)}\, p(n-w).$$

The ratio remains the same for the restricted range. Thus we have the following theorem.

Theorem 5.3. If $\lambda(x,t) \to (\gamma-x)(n-x)g(x+1)$ and $\mu(x,t) \to x(w-n+x)g(x)$ for an arbitrary positive function $g(x)$, then at equilibrium

$$p(x) = \binom{\gamma}{x}\binom{w}{n-x}/\binom{\gamma+w}{n}.$$

From Theorem 2.4.1 and (2.4.1) for any birth and death process no change in the equilibrium distribution will result if the birth rate $\lambda(x,t)$ is multiplied by $g(x+1)$ and the death rate $\mu(x,t)$ is multiplied by $g(x)$. Thus we have the following theorem.

Theorem 5.4. Let $\lambda(x,t) = (\gamma+\delta x)\cdot g(x+1)\cdot\lambda(t)$ and $\mu(x,t) = x\cdot g(x)\cdot\lambda(t)$, $\delta = 1, 0,$ or -1. Then under the assumptions of Theorem 5.1 the conclusions remain true.

For example if $g(x) = 1/x$ in the above theorem, the birth rate is $\frac{\gamma+ax}{x+1}\lambda(t)$ and the death rate is $\mu(t)$. This process can be interpreted as a single server queue with exponential service time (departures are according to the non-homogenous Poisson process with parameters $M(t)$, and individuals arrive at the locality of the queue according to a Poisson process with parameter $\Lambda(t)$. Further individuals in the locality of the queue join the queue with probability $\beta(x) = (\gamma+ax)/(x+1)$ or leave the locality. Note $\lambda(t)$ and $\beta(x)$ are always less than one.

6. MODELS FOR GROUP SIZE.

Cohen (1971) has introduced essentially a deterministic model for the number of individuals in a randomly chosen sleeping group of monkeys. He assumes that changes in size are proportional to the number of individuals and groups involved and that they come about by a single individual joining or leaving a group. Interpreting the resulting size at equilibrium as the expected size he finds the truncated negative binomial distribution. This model is modified in Boswell and Patil (1970) by making similar assumptions on the expected group size, and the truncated binomial, Poisson, or negative binomial distributions result depending on the signs of the constants of proportionality. This model appears to be a Markov process similar to a birth and death process. Consider the following formulation.

Let $X_k(t)$ be the number of groups of size k at time t where groups change in size by acquiring or losing a single individual. That is assume

$$\text{prob}\{\underline{X}(t+h) = \underline{x}+\underline{i}_k-\underline{i}_1 | \underline{X}(t)=\underline{x}\} = \lambda_k(\underline{x},t)\cdot h+o(h) \qquad (k = 2,\ldots,r),$$

$$\text{prob}\{\underline{X}(t+h) = \underline{x}+\underline{i}_1 | \underline{X}(t) = \underline{x}\} = \lambda_1(\underline{x},t)\cdot h+o(h),$$

$$\text{prob}\{\underline{X}(t+h) = \underline{x}+\underline{i}_k+\underline{i}_1 | \underline{X}(t) = \underline{x}\} = \mu_k(\underline{x},t)\cdot h+o(h) \qquad (k = 2,\ldots,r)$$

and

$$\text{prob}\{\underline{X}(t+h) = \underline{x}-\underline{i}_1 | \underline{X}(t) = \underline{x}\} = \mu_1(\underline{x},t)\cdot h+o(h).$$

Further the probability of any other changes is o(h). For Cohen's model we assume that the total population size $N(t) = \sum jX_j(t)$ is constant. Then $X_1(t) = N(0) - \sum jX_j(t)$ and can be dropped from the formulation, and the process $\{Y_k(t)\}$ where $Y_k(t) = X_{k+1}(t)$, $k = 1,\ldots,r-1$, is an ordinary birth and death process. The birth and death rates for Cohen's model would be the linear rates of Theorem 4.1, which explains why the marginals have the binomial, Poisson, or negative binomial distributions.

The more general case where the population size can change (single individuals arrive or leave the population) can be transformed into a birth and death process by the transformation $Y_1(t) = N(t)$, $Y_k(t) = x_k(t)$, $k = 2,\ldots,r$. The resulting birth and death rates are $\lambda_k(\underline{x},t)$ and $\mu_k(\underline{x},t)$ where $\underline{x}$ is a function of $\underline{y}$ defined by the above transformation. The distribution of $\underline{X}(t)$ can easily be found from the distribution of $\underline{Y}(t)$ which hopefully can be found from the equations of a birth and death process.

Another model of group size was worked on by Gani (1965) and later modified by Gani and Srivastava (1968). They considered antibodies attaching themselves to bacteria and in the modified model detaching themselves from bacteria. We consider arbitrary "birth and death" rates. Let $X_k(t)$ be the number of bacteria with k attached antibodies (since the number of bacteria is fixed we ignore $X_0(t)$), $k = 1,\ldots,r$. Assume

$$\text{prob}\{\underline{X}(t+h) = \underline{x}+\underline{i}_k-\underline{i}_{k-1} | \underline{X}(t) = \underline{x}\} = \lambda_k(\underline{x},t)\cdot h+o(h),$$

and

$$\text{prob}\{\underline{X}(t+h) = \underline{x}-\underline{i}_k+\underline{i}_{k-1} | \underline{X}(t) = \underline{x}\} = \mu_k(\underline{x},t)\cdot h+o(h),$$

for $k = 1,\ldots,r$, where $\underline{i}_0$ is $\underline{0}$. This process can be transformed into a birth ($\mu_k(\underline{x},t) \equiv 0$) or a birth and death process by the transformation $Y_k(t) = \sum_{j=k}^{r} X_j(t)$. The resulting birth and death rates are $\lambda_k(\underline{x}, t)$ and $\mu_k(\underline{x},t)$ where $\underline{x}$ is a function of $\underline{y}$ defined by the above transformation. Again the distribution of $\underline{X}(t)$ can be found from the distribution of

$\underline{y}(t)$ which hopefully can be found from the birth or birth and death process. For the attachment model Gani assumes the death rates are zero and the birth rates are $\lambda_k(\underline{x},t) = a_k x_{k-1}(\gamma-\sum jx_j)$ where γ is the number of antibodies in the system. Thus the birth rates for $\underline{Y}(t)$ are $\lambda_k(\underline{y},t) = a_k(y_{k-1}-y_k)(\gamma-\underline{y}\cdot 1)$.

ACKNOWLEDGMENTS

Research for this paper was sponsored in part by the United States Army under contract No.: DA-31-124-ARO-D-462 at the Mathematics Research Center while on leave from the Pennsylvania State University.

REFERENCES AND BIBLIOGRAPHY OF RELATED PAPERS

Adke, S. R. (1964). A multi-dimensional birth and death process. Biometrics 20, 213-6.

Arley, N. (1967). On the general birth-and-death-with immigration stochastic process. Skand. Aktuarietidskr. 50, 175-82.

Bailey, N. T. J. (1964). *The Elements of Stochastic Processes*. New York: Wiley.

Boswell, M. T. & Patil, G. P. (1970). Chance mechanisms generating the negative binomial distribution. In *Random Counts in Scientific Work, Vol. 1. Random Counts in Models and Structures*. Pennsylvania: The Pennsylvania State University Press, 3-22.

Boswell, M. T. & Patil, G. P. (1971). Chance mechanisms generating the logarithmic series distribution used in the analysis of number of species and individuals. *Statistical Ecology, Vol. 1. Spatial Patterns and Statistical Distributions*. Pennsylvania: The Pennsylvania State University Press, 97-127.

Cannings, C. (1971). Binomial and negative binomial distributions arising in pure birth process. Note to appear.

Cohen, J. E. (1971). *Casual Groups of Monkeys and Men: Stochastic Models of Elementary Social Systems*. Cambridge: Harvard Univ. Press.

Darwin, J. H. (1953). Population differences between species growing according to simple birth and death processes. Biometrika 40, 370-82.

Gani, J. (1965). Stochastic phage attachments to bacteria. Biometrics 21, 134-9.

Gani, J. and Srivastava, R. C. (1968). A stochastic model for the attachment and detachment of antibodies to virus. Math. Bios. 3, 307-21.

Goodman, L. A. (1953). Population growth of the sexes. Biometrics 9, 212-25.

Holgate, P. (1966). The size of elephant herds. Math.Gazette LI. 302-4.

Karlin, S. & McGregor, J. (1958). Linear growth, birth and death processes. J. of Math. and Mechanics 7, 643-61.

Kendall, D. G. (1948). On the generalized "birth-and-death" process. Ann. Math. Stat. 19, 1-15.

Kendall, D. G. (1949). Stochastic processes and population growth. J. R. Statist. Soc. B 11, 230-82.

Lundberg, Ove (1939, reprinted 1964). *On random processes and their application to sickness and accident statistics*. Uppsala.

Mode, C. J. (1962). Some multi-dimensional birth and death processes and their applications in population genetics. Biometrics 18, 543-67.

Patil, G. P. & Boswell, M. T. (1972). On a class of multivariate birth and death processes. Sankhyā A 34, to appear.

A SUMMARY OF THE THEORY OF POINT PROCESSES

D. J. Daley
Statistics Department (IAS), Australian National University, Canberra

D. Vere-Jones
Mathematics Department, Victoria University of Wellington, New Zealand

SUMMARY: TABLE OF CONTENTS

1. *INTRODUCTION*

2. *BASIC PROPERTIES*

2.1 *Counting properties*
2.2 *Interval properties*
2.3 *Finite processes*
2.4 *Generalizations involving the state space*
2.5 *Generalizations including random measures*

3. *MOMENT STRUCTURE AND GENERATING FUNCTIONALS*

3.1 *Moment measures*
3.2 *Probability generating functional*
3.3 *Generalizations concerning moments*
3.4 *Generalizations concerning generating functionals*

4. *SPECTRAL THEORY OF SECOND ORDER STATIONARY POINT PROCESSES*

4.1 *Representation of the covariance measure*
4.2 *Representation of the process*

5. *OPERATIONS ON POINT PROCESSES*

5.1 *Compounding*
5.2 *Clustering*
5.3 *Random deletions and random shifts*
5.4 *Randomization*

6. *ORDERLINESS AND THE INTENSITY MEASURE OF A POINT PROCESS*

6.1 *Definitions concerning orderliness*
6.2 *Korolyuk type results*
6.3 *Extensions and generalizations*

7. *PALM-KHINCHIN THEORY*

7.1 *General comments*
7.2 *Existence of Palm functions and measures*
7.3 *Palm-Khinchin equations*

7.4 *One-one correspondence of stationary measures*
7.5 *Some complementary results*

8. *LIMIT THEOREMS AND INFINITELY DIVISIBLE POINT PROCESSES*

8.1 *Ergodicity and mixing conditions*
8.2 *Infinitely divisible point processes*
8.3 *Some convergence theorems*

1. INTRODUCTION

The title of this paper amply describes its general aim. Other papers in this volume deal with various applications of point processes, so we shall not dwell on the practical motivation for the mathematical details to be described. Nor shall we give any detailed analysis of particular examples of point processes, although we have felt it worthwhile for the sake of reference to include brief descriptions of some important models known to us. We have not attempted to provide a historical survey, nor are the references intended to indicate priority.

The existing literature on point processes is mostly still scattered. So far as books or monographs are concerned, there is none dealing exclusively with the subject. Part I of Khinchin (1955) remains a valuable reference. A moderately comprehensive account of certain aspects, especially the Poisson process, is in Chapters 1 and 2 of Gnedenko and Kovalenko's (1965) eminently readable *Introduction to Queueing Theory*. Cox and Lewis (1966) are concerned chiefly with questions of statistical analysis, and from this point of view develop some aspects of the theory and describe several models. Cramer and Leadbetter (1967) discuss point processes in the context of axis crossings, and in the appendix to the Russian edition of this book, Belyaev (1969) has partly complemented their work by placing in a general setting some basic theorems in point process theory (see also Leadbetter, 1971). Harris (1963) provides a general framework for point processes in the context of population models while Jirina (1966) gives a more general treatment applicable also to random measures. Matern (1960) treats some aspects of spatial point processes, with reference to forestry problems. An account of product moment densities (cf. Section 3) with many references is in Srinivasan (1969).

A stochastic point process is the mathematical abstraction which arises from considering such phenomena as a *randomly located population* or a *random sequence of events in time*. Typically, there is envisaged a *state space*, X, and a set of points $\{x_n\}$ from X representing the locations of the different members of the population, or the times at which the "events" occur. If X is the plane, the x_n may be the points in a garden plot where weeds grow; if it is the real line they may be the instants at which a piece of equipment fails and is replaced, or at which customers arrive at a queueing facility; if it is the surface of a sphere they may represent the epicentres of major earthquakes. Because a realization (or sample path) of any of these phenomena is just a set of points in time or space, a family of such realizations has come to be called a *point process*. (In the Russian literature a point process is usually described as a *stream* or *flow* of (homogeneous) events.)

Four features may be noted at the outset:

(a) In almost all the applications we shall consider, X will be a finite-dimensional Euclidean space, or built up in a simple way from such spaces. We should therefore obtain adequate generality by supposing X to be a complete, separable, σ-compact metric space, and there is little difficulty in presenting most of the formal theory in this context. Nevertheless, to avoid overburdening an introductory talk with mathematical technicalities, we have preferred in the sequel to develop the theory for the special case where X is the real line, and to indicate in brief subsequent remarks the extent to which the results stated can be generalized, along with any difficulties which lie in the way of such generalizations. By this compromise, we hope to make the contents more accessible to readers whose main interest is in the applications, without unduly disappointing those more mathematically inclined.

(b) We rarely have occasion to deal with situations where more than a finite number of the points $\{x_n\}$ are contained within any bounded region of the state space, and a corresponding assumption will be built into the theory throughout. In view of our assumption that the state space is to be σ-compact, this implies that the populations we shall consider will be at most countably infinite.

(c) While a realization $\{x_n\}$ may be regarded as a sequence, *the order in which the points are presented is immaterial in most applications*. Thus we shall think of the realizations as *unordered sets*. Of course, if X is the real line, it is customary and convenient to list the points according to their natural order, and this is also likely to be the order in which they are generated. However, it is worth stressing from the outset that mostly the mathematical theory is a theory of unordered point sets; notable exceptions occur in Section 2.2 and much of Section 7.

(d) We do not have to assume (though we may wish to do so) that all the elements in $\{x_n\}$ are distinct. The restriction that $x_m \neq x_n$ whenever $m \neq n$ is the theme of Section 6 on orderliness.

The properties of a point process fall roughly into two categories. On the one hand, there are those properties which relate to the numbers of points falling within specified subsets of the state space; these may be called *counting properties*. On the other hand, there are properties relating to the relative spacings between points; in view of our emphasis on the real line case, we shall refer to these as *interval properties*.

As far as the counting properties are concerned, the point nature of the process plays a subsidiary role. We are essentially concerned with *random measures*, the measure of a set being in this case the number of the x_n's lying in the set. Most of the theorems about counting properties turn out to be specializations of more general results which hold true for random measures or (usually with some restrictions) for random signed measures. This is particularly true of the existence theorems and the theory relating to the moments of the counting process. To a lesser extent it is also true of the theory relating to the probability transforms of a point process (characteristic and generating functionals) and their applications to ergodic theorems, infinite divisibility, and limit theorems. The relationship between point processes and random measures is here similar to the relationship between discrete and continuous random variables: the discrete case can in principle be subsumed under the more general theory, but retains its own special character and its own special techniques.

When X is the real line, both point processes and random measures are subsumed under the more general concept of a (perhaps not countably additive) random interval function, and this in turn can be subsumed under the theory of random distributions (in the sense of Schwartz; see Gel'fand and Vilenkin, 1964, *Generalized Functions, Vol. IV*). Following a similar philosophy to that expressed in (a) above, our approach here will be to state the results for point processes, and then to indicate as best we are able, the relevant generalizations.

The peculiar features of point process theory, which give it some claim to be considered a mathematical theory in its own right, lie in the interval properties, rather than the counting properties, since the former cannot be translated readily into the context of general random measures. An important problem of point process theory is to discover means of expressing interval properties in terms of counting properties and vice versa. This is the essential theme of Section 7 on Palm functions and their extensions. The relationship is not a simple one, however; we cite as a typical instance the fact that the moments of a distribution relating to an interval property will require knowledge of all the distributions of related counting properties. This difficulty has its repercussions for the applications, where the properties of interest are often counting properties, while for many models the natural specification (which also has the practical advantage of being constructive) is in terms of interval properties.

2. BASIC DEFINITIONS

2.1 Counting properties

The formal definition of a stochastic point process is most conveniently given in terms of its counting properties. To do this, let N be the space of all non-negative integer-valued measures $N(\cdot)$ defined on the σ-algebra $B(X)$ of all Borel sets of the state space X. We assume here that X is the real line R, writing $B = B(R)$, and impose on N the following restriction.

Assumption 2.1. $N(A) < \infty$ for all bounded $A \in \mathcal{B}$.

Let $\mathcal{S}$ be the σ-algebra generated by all subsets of the form

$$\{N: N(A) \leq k\} \qquad (k \in Z_+ \equiv \{0,1,\ldots\},\ A \in \mathcal{B}) \ . \tag{2.1.1}$$

Definition 2.2. A stochastic point process is a measurable mapping from a probability space $(W,\mathcal{A},\mathbb{P})$ into $(\mathcal{N},\mathcal{S})$.

In particular therefore, *any probability measure on* $(\mathcal{N},\mathcal{S})$ *yields a (stochastic) point process.* From here onwards, every point process is to be understood to be stochastic.

Supposing that $w \in W$ is mapped into $N \in \mathcal{N}$, with value $N(A,w)$ on $A \in \mathcal{B}$, note the following properties.

(1°) For every fixed bounded $A \in \mathcal{B}$, $N(A,\cdot)$ is a non-negative integer-valued (i.e., Z_+-valued) random variable.

(2°) For every pair of disjoint sets $A, B \in \mathcal{B}$,

$$\mathbb{P}\{w: N(A \cup B,w) = N(A,w) + N(B,w)\} = 1 \ . \tag{2.1.2}$$

(3°) If $\{A_n\}$, $n \in Z_+$, is a decreasing sequence of bounded $A_n \in \mathcal{B}$ with $A_n \downarrow \emptyset$, then

$$\mathbb{P}\{w: N(A_n,w) \downarrow 0\} = 1 \ . \tag{2.1.3}$$

The properties at (2.1.2) and (2.1.3) are of course corollaries of the stronger property, inherent in the definition, that

$$\mathbb{P}\{w: N(\cdot,w) \text{ is a measure on } \mathcal{B}\} = 1 \ . \tag{2.1.4}$$

From these definitions and remarks it follows that the question of the existence of a point process reduces to the problem of constructing a measure on $\mathcal{S}$. Just as for ordinary stochastic processes, the problem is solved by introducing the finite-dimensional distributions of the process (abbreviated to fidi), and invoking an extension theorem, leading to the following result (Moyal, 1962; Harris, 1963, p. 55; Nawrotzki, 1962).

Theorem 2.3. If for every positive integer r, all $k_i \in Z_+$, and bounded $A, A_i \in \mathcal{B}$, $i = 1,\dots,r$, there exists a non-negative function $p(.;.)$ such that

(i) $p(A_1,\dots,A_r;k_1,\dots,k_r) = p(A_{i_1},\dots,A_{i_r};k_{i_1},\dots,k_{i_r})$ for every permutation $(i_1,\dots,i_r)$ of $(1,\dots,r)$:

(ii) $\sum_{k=0}^{\infty} p(A_1,\dots,A_r,A;k_1,\dots,k_r,k) = p(A_1,\dots,A_r;k_1,\dots,k_r);$

(iii) $\sum_{k=0}^{\infty} p(A;k) = 1;$

(iv) whenever $A_1,\dots,A_r$ are mutually disjoint,

$$p(\bigcup_{i=1}^{r} A_i;k) = \sum_{k_1+\dots+k_r=k} p(A_1,\dots,A_r;k_1,\dots,k_r) \text{ and}$$

$$p(\bigcup_{i=1}^{r} A_i,A_1,\dots,A_r;k,k_1,\dots,k_r) = 0, \text{ unless } \sum_{i=1}^{r} k_i = k,$$

in which case it equals $p(A_1,\dots,A_r;k_1,\dots,k_r);$

(v) $p(A_n;0) \uparrow 1$ whenever $A_n \downarrow \emptyset;$

then there exists a uniquely determined probability measure $\mathbb{P}$ on $(\mathcal{N},\mathcal{S})$ such that for all r, k_i, A_i as above,

$$\mathbb{P}\{N:N(A_i) = k_i,\ i=1,\dots,r\} = p(A_1,\dots,A_r;k_1,\dots,k_r).$$

The first three conditions are the familiar Daniell-Kolmogorov consistency conditions which here are necessary and sufficient to ensure that $p(.;.)$ should define fidi distributions of a family of Z_+-valued random variables indexed by the Borel sets. The remaining conditions then ensure that these random variables satisfy (2.1.2) and (2.1.3). However this is not enough to prove the theorem as we require the stronger assertion (2.1.4), implying in particular that (cf. (2.1.2) and (2.1.3))

$$\mathbb{P}\{N:N(A\cup B)=N(A)+N(B) \text{ for all disjoint bounded } A, B \in \mathcal{B}\} = 1, \quad (2.1.5)$$

$$P\{N:N(A_n) \downarrow 0 \text{ for all sequences of bounded Borel sets } A_n \downarrow \emptyset\} = 1. \quad (2.1.6)$$

To obtain these stronger results we use the compactness properties of the real line, the non-negativity of N(.), and the fact that the Borel sets are countably generated. We cannot go into detail here; see the cited references and Jirina (1966).

In questions of convergence and characterization of set functions like p(.;.) the following technical concept is needed (∂A denotes the boundary of the Borel set A, i.e., $\partial A = A^{cl} \setminus A^{0}$ where A^{cl} and A^{0} denote the closure and open hull respectively of A).

Definition 2.4. The Borel set A is a p-continuity set for the set function p(B;k), $B \in \mathcal{B}$, $k \in Z_+$, if

$$\sum_{k=1}^{\infty} p(\partial A;k) = 0\,; \tag{2.1.7}$$

the point x is a fixed atom for p(.;.) if

$$\sum_{k=1}^{\infty} p(\{x\};k) > 0\ . \tag{2.1.8}$$

For the process $(N,S,\mathbb{P})$, $A \in \mathcal{B}$ is a $\mathbb{P}$-continuity set when

$$\mathbb{P}\{N(\partial A) > 0\} = 0\ , \tag{2.1.9}$$

while x is a fixed atom of the process when

$$\mathbb{P}\{N(\{x\}) > 0\} > 0\ . \tag{2.1.10}$$

The following characterization result depends for its validity on the σ-compactness of R and on the rational intervals forming a base for the Borel sets $\mathcal{B}$ (cf. Billingsley's, 1968, p. 15, determining class); note the absence of (v) of Theorem 2.3.

Theorem 2.5. Let the function p(.;.) satisfy conditions (i)-(iv) of Theorem 2.3 when the sets A, A_i therein are in the ring $\mathcal{R}$ generated by finite (open, closed, or half-open) rational intervals. Then there is a uniquely determined point process $(N,S,\mathbb{P})$ such that on $\mathbb{P}$-continuity sets $A_i \in R$,

$$\mathbb{P}\{N(A_i) = k_i,\ i = 1,\dots,r\} = p(A_1,\dots,A_r;k_1,\dots,k_r) .$$

Thus, a point process can be defined simply by specifying its fidi distributions on rational intervals.

Example 2.1.A. The Poisson process with parameter measure $\mu(.)$ is defined for any Borel measure $\mu(.)$ on the real line by the two conditions

(i) For every bounded $A \in B$, $N(A)$ has a Poisson distribution with mean $\mu(A)$;

(ii) For any finite family of mutually disjoint bounded sets $A_i \in B$ the random variables $N(A_i)$ are mutually independent.

The fidi distributions of such a family of random variables can be written down immediately, and that the conditions of Theorem 2.3 are then satisfied is easily verified.

In the Russian literature especially, a point process satisfying condition (ii) is said to be without after-effects; Kingman (1967) calls (ii) the completely random property.

It is easily shown that $A \in B$ is a $\mathbb{P}$-continuity set for a Poisson process with parameter measure $\mu(.)$ if and only if $\mu(\partial A) = 0$. In particular, Khinchin (1956a) showed that $\mu(\{x\}) = 0$ (all $x \in R$), which implies that $\mathbb{P}\{N(\{x\}) = 0\} = 1$ (all $x \in R$), characterizes what is called the *simple Poisson process*.

Example 2.1.B. We obtain the *stationary Poisson process* by setting the parameter measure above equal to a multiple of Lebesgue measure, $\mu(A) = \lambda|A|$ say, for some $\lambda < \infty$. (We use $|.|$ to denote Lebesgue measure.) λ is called the *parameter* of the process, and is a measure of the mean rate of occurrence of points (see Section 6); sometimes then it is called the rate parameter or just simply the rate. A stationary Poisson process is necessarily simple.

Example 2.1.C. The *mixed Poisson process*, logically a special case of the doubly stochastic Poisson process to be described in Section 5 with other processes stemming from the Poisson process, is defined in terms of a family of Poisson processes indexed by the non-negative random variable Y with distribution function G(.) and with parameter measures $Y\mu(.)$. If $p(.;.|y)$ are the fidi distributions as in Theorem 2.3 for a Poisson process with parameter measure $y\mu(.)$, the fidi distributions of the mixed Poisson process are

$$\mathbb{P}\{N(A_i) = k_i,\ i = 1,\ldots,r\} = \int_0^\infty p(A_1,\ldots,A_r;k_1,\ldots,k_r|y)\ dG(y).$$

Then the multivariate *probability generating function* (p.g.f.) of the random vector $\{N(A_1),\ldots,N(A_r)\}$ for disjoint $A_1,\ldots,A_r$ is

$$E\left\{z_1^{N(A_1)}\cdots z_r^{N(A_r)}\right\} = \int_0^\infty \exp\{-y\sum_{i=1}^{r}(1-z_i)\ \mu(A_i)\}\ dG(y). \qquad (2.1.10)$$

2.2 Interval properties

Consider now the family Ω of sequences $\omega = \{x_n\}$, $n\varepsilon Z \equiv \{0,\pm 1,\ldots\}$, defined on N as follows. If $N(R) = 0$, then $\omega = \emptyset$. Otherwise, either $N[0,\infty) > 0$ or $N(-\infty,0) > 0$ or both, and for such $n \varepsilon Z$ as the definitions are meaningful set

$$x_n = \begin{cases} y \geq 0 : N[0,y) \leq n < N[0,y] & (n=0,1,\ldots), \\ y < 0 : N(y,0) < -n \leq N[y,0) & (n=-1,-2,\ldots). \end{cases} \qquad (2.2.1)$$

(When the argument of a set function is an interval, we usually write (for example) N(a,b] instead of N((a,b]), etc.) Note that Ω is simply the family of all finite or countably infinite subsets of R, multiple points being allowed, indexed according to the convention

$$\cdots \leq x_{-1} < 0 \leq x_0 \leq x_1 \leq \cdots \leq x_n \leq x_{n+1} \leq \cdots$$

and subject to the restriction of having no finite limit points.

Now the function

$$N(A,\omega) \equiv \text{card}\{n: x_n \in \omega \cap A\} \qquad (\omega \in \Omega,\ A \in \mathcal{B}) \tag{2.2.2}$$

is readily seen to be the inverse of (2.2.1), with $N(.,\omega)$ a measure on $\mathcal{B}$, and if we introduce the σ-algebra $\mathcal{F}$ in Ω generated by the sets

$$\{\omega: N(A,\omega) \leq k\} \qquad (A \in \mathcal{B},\ k \in Z_+)\ ,$$

the mapping defined by (2.2.1) is seen to be one-one and measurable both ways. Thus, *any probability measure on* $(\Omega,\mathcal{F})$ *defines a point process, and conversely*. The important point here is that irrespective of whether we start from $(\mathcal{N},\mathcal{S})$ or $(\Omega,\mathcal{F})$, the quantities x_n (defined by (2.2.1) if the initial space is $\mathcal{N}$ and not Ω), as well as $N(A) \equiv N(A,\omega)$ (defined by (2.2.2) when the initial space is Ω), are well-defined random variables.

To construct a (finite portion of a) sample realization of a point process it is required to specify (some of) the points $\{x_n\}$, and Theorem 2.3 does not really help in doing this. An equivalent construction is the specification of $\{x_0, -x_{-1}, x_n - x_{n-1}, n \in Z\}$ (or such of these quantities as are defined by ω) with the appealing interpretation that $\{\eta_n\} \equiv \{x_n - x_{n-1}\}$ are the times between successive events. Denote by H the space of all such $\eta = \{\xi_0, \xi_{-1}, \eta_n, n \in Z\} \equiv \{x_0, -x_{-1}, \eta_n, n \in Z\}$, so that $\eta \in H$ has its elements, where defined, satisfying

$$\xi_0 \geq 0 > -\xi_{-1}, \qquad \eta_0 = \xi_0 + \xi_{-1}, \qquad \eta_n \geq 0\ . \tag{2.2.3}$$

Then the mapping of Ω to H that has been described is one-one, and accordingly the subsets $\mathcal{Y}$ of H defined as the images of $\mathcal{F}$ constitute a σ-algebra, and by definition the mapping is measurable. It can be verified that $\mathcal{Y}$ is the σ-algebra generated by the sets $\{\eta: \xi_i \leq z_i\}$, $i = 0,-1;\ z_i \in [0,\infty) \equiv R_+$, and $\{\eta: \eta_i \leq y_i\}$, $i \in Z$, $y_i \in R_+$; also, the set $\{\omega: N([0,x],\omega) \leq r\} = \{\omega: x_r > x\}$ for example is mapped onto

$\{\eta:\xi_0 + \eta_1 + \cdots + \eta_{r-1} > x\}$, so specifying a consistent family of fidi distributions in H does define a point process in the sense of Definition 2.2.

The simplest example of a point process defined in this way is the renewal process, which because of its simple probabilistic structure and practical relevance where components are replaced, has acquired an extensive literature of its own (cf. Smith, 1958; Cox, 1962; Cox and Lewis, 1966).

Example 2.2.A. The *renewal process* is usually defined on R_+ by taking in H, $x_0 = 0$ and $\{\eta_i\}$, $i = 1,2,\ldots,$ independent random variables with $\mathbb{P}(\eta_i \le y) = F(y)$ (all i) for some distribution function $F(.)$ on R_+. The renewal process can be defined on R simply by extending these definitions to all $i \in Z$. A *delayed renewal process* on R_+ is obtained by allowing x_0 to be a non-degenerate random variable. A delayed renewal process is defined on the whole of R by specifying the joint distribution $H(u,v) = \mathbb{P}\{\xi_0 \le u, \xi_{-1} \le v\}$ and setting $\mathbb{P}\{\eta_i \le y\} = F(y)$ (all $i \ne 0$), these η_i being mutually independent and independent of the ξ_i. Consequently, if $H(u,v) = F(v)$, the renewal process on R is recovered. Alternatively, if $F(y) = 1-e^{-\lambda y}$ and $H(u,v) = F(u) \times F(v)$, then the stationary Poisson process with parameter λ is recovered.

Definition 2.6. A point process $(\Omega, F, \mathbb{P})$ is *stationary* when its fidi distributions

$$\mathbb{P}\{N(A_i + y) = k_i,\ i = 1,\ldots,r\} \qquad (A_i \in B,\ k_i \in Z_+)$$

are independent of $y \in R$. (Here and elsewhere, $A + y = (x : x - y \in A)$.)

Definition 2.7. A point process $(H, Y, \mathbb{P})$ has *stationary intervals* when its fidi distributions

$$\mathbb{P}(\eta_{i+n} \le y_i,\ i = 0, \pm 1, \ldots, \pm r) \qquad (y_i \in R_+)$$

are independent of $n \in Z$.

For example, a renewal process on R has stationary intervals; furthermore, the methods of Section 7 show that a delayed renewal process on R is stationary if and only if the mean $\mu = \int_0^\infty x\, dF(x)$ is finite and

$$H(u,v) = \mu^{-1} \int_0^u \{F(w+v) - F(w)\}\, dw . \qquad (2.2.4)$$

A pertinent question is to ask whether any point process can be stationary and also have stationary intervals. The answer is "almost no": seeking for any such processes in the narrower class of stationary delayed renewal processes, it is required that

$$F(y) = \mathbb{P}(\eta_0 \le y) = \mathbb{P}(\xi_0 + \xi_{-1} \le y) = \mu^{-1} \int_0^y \{F(y) - F(u)\}\, du$$

$$= \{\int_0^\infty x\, dF(x)\}^{-1} \int_0^y x\, dF(x) . \qquad (2.2.5)$$

The only distribution function solution of this integral equation is $\mathbb{P}(\eta_0 = \mu) = 1$, implying that the process has all consecutive pairs of points equidistant. Such a delayed renewal process we propose calling a *deterministic point process*. More generally, results cited in Section 7 lead to equation (2.2.5) for any stationary point process, with F(.) the distribution function for the time between an "arbitrary" event and its successor. Hence the theorem below.

Theorem 2.8. The only stationary point process on R which also has stationary intervals is the deterministic point process.

Of course to be stationary, a deterministic point process must have ξ_0 uniformly distributed on $[0,\mu]$ where μ is the interval length. The theorem can be justified intuitively by observing that if the intervals are not of fixed length, the origin (or any other fixed point in R) is more likely to be in an interval longer than the average (cf. Cox, 1962, pp. 65-66; Cox & Lewis, 1966, pp. 60-61).

In spite of Theorem 2.8, there is a natural relation between point processes with $\xi_0 = 0$ and stationary intervals, and what are called finite stationary point processes (see Section 7).

2.3 Finite Processes

A completely different method of specifying a point process is possible when with probability one the total number of points is finite, i.e., $\mathbb{P}\{N(R) < \infty\} = 1$. Since this is a typical situation in some population studies, such processes are sometimes called *population processes*. To describe the probability structure of such a process, two basic components need to be specified:

(i) the distribution of the total population size;

(ii) the joint distribution of the positions of the population members conditional on a given population size.

We shall denote these by $\{p_n\}$, $n \in Z_+$, and $P_n(t_1,..., t_n)$ respectively, noting that we assume (conditional on n) neither that the points are ordered (as in $\omega = \{x_n\}$) nor that $P_n(.)$ is symmetric in its arguments. The fidi distributions needed in Theorem 2.3 can in principle be written down directly in terms of these two components, though in practice combinatorial difficulties analogous to those encountered in computing the distributions of order statistics arise. Conversely, these fidi distributions suffice only to determine a symmetrized version of $P_n(.)$ unless additional assumptions be made. All these points are discussed in detail in Moyal (1962).

Two particular examples of such processes will be especially useful later in our discussion of cluster processes.

Example 2.3.A. In a *Neyman-Scott cluster* the individual population members are supposed independently distributed about the origin with common probability distribution G(.) say. If $\Pi(z) = \Sigma_{n=0}^{\infty} p_n z^n$ denotes the p.g.f. of the total population size and $\Pi_A(z)$ the p.g.f. for the number of the population falling in the Borel set A, then $\Pi_A(z) = \Pi\{1-(1-z) \int_A dG(x)\}$. Similar expressions can be obtained for the multivariate p.g.f.s of the higher order distributions, but these are better treated in terms of the probability generating functional (see Section 3). Note that when N(R) has a Poisson or mixed Poisson distribution, the process reduces to the Poisson or mixed Poisson process respectively with parameter measure proportional to dG(x); see Neyman and Scott (1958) for further details.

Example 2.3.B. As an alternative, the *Bartlett-Lewis cluster* was suggested in Bartlett (1963) and used also by Lewis (1964a,b) in the analysis of computer failure times. This finite process is assumed to be a finite segment of a renewal process with its initial point at the origin and terminating after n steps with probability p_n. See Lewis (1964a) for some explicit results. A natural generalization, particularly for processes in 2- or 3-dimensional space which no longer has the simple interpretation of time, would be to take a random walk process in place of the renewal process. However, even in R the combinatorial problems are formidable.

2.4. Generalizations Involving the State Space

Doob (1953) succinctly describes any stochastic process as an indexed family of random variables. From this viewpoint a point process is a family of Z_+-valued random variables indexed by the Borel sets in R (and indexed so that a realization is also a measure on $\mathcal{B}$). We propose indicating two generalizations: in the next section the random variables are no longer restricted to be Z_+-valued, while in this section we suppose R replaced by any complete separable σ-compact metric space. With such a state space X our Definition 2.2 and Theorem 2.3 concerning existence continue to make sense and remain valid, and may even do so under weaker hypotheses (cf. Jirina, 1966; Le Cam,1961).

Such an extension clearly embraces *multidimensional point processes* where the real line is replaced by 2- or 3-dimensional Euclidean space: examples arise in ecology, particle physics, cosmology, and elsewhere (cf. the paper in this volume by Fisher, 1972.). The concept of a Poisson process, and hence also of the family of doubly stochastic processes, can be extended immediately to these more general spaces.

Not so obviously, the generalization also embraces *multivariate point processes*. By this we envisage a jointly distributed family $\{N^{(\gamma)}(.),\ \gamma \in \Gamma\}$ of point processes defined on a common state space X, here taken to be R. Let Z_+^m denote the space of m-vectors with coordinates non-negative integers. When $\Gamma = (1,\ldots,m)$ such a process can be described as a Z_+^m-valued point process in R, or equally well as Z_+-valued point processes on the Borel subsets of the Cartesian product $R \times (1,\ldots,m)$ (or equivalently, m copies of R). Such processes arise as models for times of occurrence of m different kinds of events: arrivals or departures (or both) of different types of customers for example, or occurrences of earthquakes in different energy ranges. For fixed γ we call $N^{(\gamma)}(.)$ a *marginal* process of the given multivariate point process, and a common (statistical) question concerns the independence of the marginal processes of a multivariate point process.

Definition 2.9. A finite family of point processes $\{N^{(\gamma)}(.),\ \gamma = 1,\ldots,m\}$ defined on a common probability triple is *independent* when the families of random variables $\{N^{(\gamma)}(A),\ A \in \mathcal{B}\}$, $\gamma = 1,\ldots,m$, are mutually independent. An arbitrary family of point processes is independent when all finite subfamilies are independent.

A necessary and sufficient condition for independence is the obvious factorization of the joint fidi distributions.

Bearing in mind the possible dependence between the marginal processes of a multivariate point process, we choose to define a multivariate Poisson process not as a process on state space $R \times (1,\ldots,m)$ with the complete randomness property (ii) of Example 2.1.A, but as follows.

Example 2.4.A. A *multivariate Poisson process* is a multivariate point process for which every marginal process is a Poisson process. For example, the process consisting of the input and output of an infinite server queueing system with Poisson arrivals (i.e., $M/G/\infty$) is a bivariate Poisson process. Since there is no known simple canonical form even for bivariate distributions with Poisson marginals, it is hardly surprising to find that there is no simple characterization of multivariate Poisson processes as we have defined them. However Milne (1971) has given a characterization of infinitely divisible multivariate Poisson processes; for example, an infinitely divisible bivariate Poisson process is characterized by three measures.

Example 2.4.B. The *compound Poisson process* can be regarded as being derived from a multivariate Poisson process on $R \times (1,2,..)$ with independent marginal processes having non-atomic parameter measures $\{\mu_i(.)\}$, i = 1,2,... , and sample realization of the compound Poisson process given by $N(A) = \Sigma_{i=1}^{\infty} iN^{(i)}(A)$. Equivalently--and better--the process may be defined as in Example 2.1.A excpet that the Poisson distributions of (i) there are replaced by compound Poisson distributions with p.g.f.s

$$P_A(z) = \exp\{- \sum_{i=1}^{\infty} \mu_i(A)\ (1 - z^i)\}\ . \tag{2.4.1}$$

Khinchin (1956a,b) and Prekopa (1957a,b) have shown that the most general point processes satisfying merely the complete randomness condition (ii) of Example 2.1.A is this compound Poisson process.

More extensive accounts of theoretical and statistical work on multivariate point processes can be found in Milne (1971), Cox and Lewis (1972), and Bartlett (1972).

The multivariate point process is subsumed under the more general concept of a *marked point process* introduced in König and Matthes (1963) and Matthes et al. (1963, 1964, 1965) (markierte zufällige Punktfolgen, also translated as labeled point process or sometimes labeled random point sequence) and also in Bartlett (1967) where it is called a point process with ancillary variable. It is envisaged that with each point x_i in the basic point process there is associated an auxiliary variable y_i lying in a *mark space* Y endowed (if need be) with a measure structure $\mathcal{Y}$. In general these (y_i) need not be independent, either amongst themselves or of the (x_i). As examples, y_i may represent the speed of a vehicle passing an observer at a time x_i, or the energy of an earthquake occurring at a particular time and location x_i, or the history of the point process up to time x_i.

Definition 2.10. A *marked point process* is a point process defined on the product space X × Y.

In order to apply the previous theory, we need the more specific assumption that Y, like X, is a complete separable σ-compact metric space, and that $\mathcal{Y}$ is the class of Borel sets in Y.

Bartlett (1966, p. 215; 1967) has used such processes to describe a process of randomly distributed lines, illustrating the more general property that any collection of randomly distributed geometrical objects may be treated as a point process in a suitably complicated coordinate space (cf. also Miles, 1969, 1970). Whether there be any advantage in so doing depends of course on the particular context: in many examples the process is homogeneous with respect to the X-space but not the Y-space, in which case the process may be more usefully analyzed as a random measure on the X-space.

2.5. Generalizations Including Random Measures

Another possible generalization is to consider more general random measures than point processes (which are necessarily Z_+-valued). For example, certain multivariate processes above can be regarded as Z_+^m-valued random measures; the amount of water flowing into a dam over a period of time may be viewed as a measure; the yield of grain from an area of land may be thought of as a (random) measure defined on the Borel subsets of the given region. For simplicity we shall suppose that the underlying space X is always the real line R.

In this context there is little difficulty generalizing Definition 2.2 and Theorem 2.3 into the class M_+ of Borel measures on R satisfying the finiteness Assumption 2.1, endowed with the σ-algebra G_+ generated by sets of the form

$$\{\Lambda:\Lambda(A) \leq y\} \qquad (A \in B,\ \Lambda \in M_+) \ . \tag{2.5.1}$$

(Any measure is always to be understood to be non-negative, unless it be said specifically that it is a signed measure.) Thus we define a random measure as follows.

Definition 2.11. A *random measure* is a measurable mapping from a probability space $(W, A, \mathbb{P})$ into (M_+, G_+).

A theorem precisely analogous to Theorem 2.3 can then be formulated and proved, with only the trivial modifications necessitated by the requirement that in general the fidi distributions are no longer concentrated on the lattice of integers.

The situation for random signed (real or complex) measures is not as simple. Formally, they may be introduced in a similar way, replacing (M_+, G_+) by (M, G) where M is the space of all random signed measures satisfying Assumption 2.1 and G is the σ-algebra generated by the analogues of sets as at (2.5.1). Then the probability measure $\mathbb{P}$ of such a process satisfies conditions analogous to (i)-(v) of Theorem 2.3, but Λ being no longer necessarily non-negative, these conditions on the fidi distributions are not of themselves sufficient to ensure that sample realizations of the process will be of bounded variation on bounded intervals (cf. Example 2.5.A below).

One way out of this difficulty is to renounce the rather strict requirement (2.1.4), and even the continuity property (2.1.3), and consider simply families of random variables, indexed by the Borel sets, and satisfying the finite additivity property (2.1.2) (for given sets A, B, not for all such sets simultaneously, as in (2.1.5)). In the case of the real line, it is more convenient to replace Borel sets by intervals in the weakened form of the definition, and we are therefore led to the following concept.

Definition 2.12. A *random interval function* is a family of random variables X(I) indexed by the family I of all finite intervals I, and satisfying for such intervals the finite additivity property (2.1.5).

The properties of such functions are developed in Chapters 5 and 6 of Bochner (1955). In particular, conditions analogous to (i)-(iv) of Theorem 2.3 are necessary and sufficient to characterize the fidi distributions of such a process. The most important examples are the Gaussian random interval functions, where all the fidi distributions are multivariate normal.

Example 2.5.A. A *white noise process* is defined to be a random interval function with the random variables X(I) having Gaussian distributions with zero mean, variance $|I|$, and independent for disjoint intervals. The fidi distributions satisfy conditions analogous to (i)-(v) of Theorem 2.3, but if I_{kn} denotes the interval $((k-1)/n, k/n]$ it is also verified easily that $S_n = \Sigma_{k=1}^{n} |X(I_{kn})|$ diverges to infinity in probability, so the realizations cannot be of locally bounded variation with probability one (see also Doob, 1953, section 11.11).

Finally, we mention that random measures, and hence also point processes, can be considered as random generalized functions (or random distributions, in the sense of Schwartz) on suitable spaces of test functions. In the point process case, this amounts to treating the process as a sequence of Dirac delta functions. From the formal point of view, two approaches to this theory are again possible, since we may define a random distribution either by analogy with Definition 2.11, as a mapping into the space of distributions endowed with a suitable measure structure, or by analogy with Definition 2.12 as a family of random variables indexed by the test function and satisfying certain appropriate additivity conditions. It is the latter approach which has proved more popular, since it embraces both random interval functions and other derivatives of the Gaussian process (Ito, 1953, and Gel'fand and Vilenkin, 1964). These authors have been mainly concerned with developing a second order theory, so that the contact with point process theory comes through the spectral properties (see Section 4).

3. MOMENT STRUCTURE AND GENERATING FUNCTIONALS

3.1. Moment Measures

The moments of a point process to be discussed here relate to our Definition 2.2 in Section 2.1; moments of the interval properties of Section 2.2 are outlined briefly in Section 7.

Definition 3.1. The *first moment measure* or *expectation measure*

$$M(A) \equiv E\{N(A)\} \qquad (A \in \mathcal{B}) \tag{3.1.1}$$

is said to exist when it is finite for all bounded Borel sets A.

Observe that if M(A) does exist then it is indeed a measure because N(.) is non-negative and σ-additive.

In general, higher order moments can be defined by

$$M_r(A_1,\ldots,A_r) \equiv E\{N(A_1) \cdots N(A_r)\} \qquad (A_i \in \mathcal{B}) \;. \tag{3.1.2}$$

If these moments are finite for bounded A_i--and this will be assumed whenever $M_r(.)$ appears unless the contrary be stated--then the usual extension procedure for measures on product sets starting from

$$M_r(A_1 \times \cdots \times A_r) = M_r(A_1,\ldots,A_r)$$

leads to the conclusion that $M_r(.)$ is a measure in R^r. Alternatively, $M_r(.)$ can be regarded as the first moment measure of the multidimensional point process in R^r consisting, for each $\omega \in \Omega$, of the set of all points $(x_{n_1},\ldots,x_{n_r}) \in R^r$ where each x_{n_i} ranges over all $x_n \in \omega$. These measures $M_r(.)$ are the analogues of the usual cross-moments of an ordinary stochastic process. Except for r = 1 and 2, $M_r(.)$ cannot be recovered from the elementary moment functions (measures only for r = 1)

$$M_r^*(A) = E\{N^r(A)\} = M_r(A,\ldots,A) \qquad (A \in \mathcal{B}) \,. \tag{3.1.3}$$

Cumulant measures can be defined in terms of the random variables $N(A_1)$, ... , $N(A_r)$ in the usual manner. For example

$$C_2(A_1 \times A_2) = M_2(A_1 \times A_2) - M(A_1)\, M(A_2)$$
$$= \operatorname{cov}\{N(A_1),\, N(A_2)\} \,. \tag{3.1.4}$$

In general $C_r(.)$ exists if and only if $M_r(.)$ does, but unlike the latter, $C_r(.)$ may be a signed measure.

As with ordinary random variables the second moment measures $M_2(.)$ and $C_2(.)$ can indicate dependence between $N(A_1) \equiv N(A_1,\omega)$ and $N(A_2) \equiv N(A_2,\omega)$, $\omega \in \Omega$. Because of the relation amongst the random variables X, Y, Z

$$2\ \mathrm{cov}(X,Z) = \mathrm{var}(X + Y + Z) - \mathrm{var}(X + Y) - \mathrm{var}(Y + Z) + \mathrm{var}(Y)\ ,$$

and because the Borel sets in R are generated by the rational intervals, the cumulant measure $C_2(.)$ of a point process in R is determined from a knowledge of the variance function

$$V(I) = \mathrm{var}\{N(I)\} \qquad (I \in \mathcal{I}) \tag{3.1.5}$$

on the family $\mathcal{I}$ of all finite rational intervals.

$M_2(.)$ and $C_2(.)$ enjoy the property of being *positive definite*.

Definition 3.2. A Borel measure $K(.)$ on R^2 is *positive definite* if it is symmetric and for all bounded Borel functions ϕ of compact support in R,

$$\int \phi(x)\ \phi(y)\ K(dx \times dy) \geq 0\ . \tag{3.1.6}$$

To show that $M_2(.)$ and $C_2(.)$ enjoy the property (3.1.6), suppose first that $\phi(.)$ is an indicator function $\chi_A(.)$ for some $A \in \mathcal{B}$. Then the random integral $\int_R \chi_A(x)\ N(dx) = N(A)$, and hence

$$E\{\int_R \chi_A(x)\ N(dx) \int_R \chi_B(y)\ N(dy)\}$$

$$= M_2(A \times B) = \int_{R^2} \chi_A(x)\ \chi_B(y)\ M_2(dx \times dy)\ .$$

These equations extend by linearity to all simple functions, and hence to all bounded Borel functions of compact support. In particular,

$$\int \phi(x)\ \phi(y)\ M_2(dx \times dy) = E[\{\int_R \phi(x)\ N(dx)\}^2] \geq 0\ ,$$

and similarly

$$\int \phi(x)\ \phi(y)\ C_2(dx \times dy) = \mathrm{var}\{\int \phi(x)\ N(dx)\} \geq 0\ .$$

Factorial moments are used in discussing Z_+-valued random variables because of their connection with the p.g.f. So it is with point processes: for example

$$M^*_{[r]}(A) \equiv E[N(A)\{N(A)-1\}\cdots\{N(A)-(r-1)\}] ,$$

so that $M^*_{[1]}(A) = M^*(A) = M(A)$. Analogues $M_{[r]}(.)$ of $M_r(.)$ exist and are measures, for like $M_r(.)$ they can be regarded as the first moment measure of a certain multidimensional point process in R^r, namely that consisting, for each $\omega \in \Omega$, of the set of all points $(x_{n_1},\ldots, x_{n_r}) \in R^r$ where each x_{n_i} ranges over all $x_n \in \omega$ subject to the constraint $n_i \neq n_j$, $i, j = 1,\ldots, r$; $i \neq j$. The main asset of factorial moments in point process theory is in discussing *orderly* point processes; they arise naturally and usefully in level crossing problems (see Leadbetter, 1972).

Definition 3.3. A point process $(\Omega, F, \mathbb{P})$ is *orderly* if

$$\mathbb{P}\{\omega : x_n \neq x_m \text{ for all } x_n, x_m \in \omega, (n \neq m)\} = 1 . \qquad (3.1.7)$$

(In view of (3.1.7), we may sometimes for emphasis describe this property as *almost sure orderliness*, or a.s. orderliness for short.)

The factorial moment measures of orderly point processes, unlike the ordinary moment measures, have no concentration along the diagonal subspaces $(x,\ldots,x) \in R^r$, $(x \in R)$. These concentrations arise when it is possible to have $x_{n_i} = x_{n_j}$ for some $i \neq j$ in $(x_{n_1},\ldots, x_{n_r})$ with positive probability. In many important applications the processes are orderly and have absolutely continuous factorial moment measures. The corresponding densities $m_{[r]}(.)$, also known as *product densities*, are then linked to the moments of $N(.)$ by expressions of which a general type is

$$E\{N^r(A)\} = \sum_{i=1}^{r} S_i^{(r)} \int_A \cdots \int_A m_{[r]}(x_1,\ldots, x_r)\, dx_1 \cdots dx_r$$

where the $S_i^{(r)}$ are Stirling numbers (Ramakrishnan, 1950; for an account of further developments see Srinivasan, 1969).

Definition 3.4. The rth *moment measure* $M_r(.)$ is *stationary* when $M_r\{(A_1 + x) \times \cdots \times (A_r + x)\}$ is independent of $x \in R$.

For example, a point process $(\Omega, F, \mathbb{P})$ with $\mathbb{P}$ stationary (as in Definition 2.5) has $M_r(.)$ stationary whenever $M_r(.)$ exists. The structure of stationary moment measures simplifies a little on account of the following lemma (simple proofs are given in Brillinger, 1972, and Vere-Jones, 1971, for example).

Lemma 3.5. Let K(.) be a Borel signed measure in R^r with the property that $K(T_y A) = K(A)$, $A \in B(R^r)$, where the operator T_y maps $x = (x_1, \ldots, x_r)$ into $(x_1 + y, \ldots, x_r + y)$. Regard R^r as the direct sum of the diagonal ray $\{x: x_1 = \cdots = x_r\}$ and its orthogonal subspace. Then K is expressible as a product measure $K = L \times K_{r-1}$ where L is Lebesgue measure down the diagonal ray and K_{r-1} is a Borel measure on the orthogonal subspace.

Applied to the moment measures of a point process we obtain in particular (using differential notation)

$$E\{N(dt)\} = M(dt) = m\,dt , \tag{3.1.8}$$

$$\operatorname{cov}\{N(t + dt), N(t + u + du)\} = C(du)\,dt .$$

We call m the *mean density* (or just *density*) and C(.) the *reduced covariance measure* as distinct from the *complete covariance measure* $C_2(.)$ defined at (3.1.4). The measure C(.) always has an atom at the origin, but may be absolutely continuous elsewhere with *covariance density* c(.) given by

$$c(u)\,du = C(du) . \tag{3.1.9}$$

In the orderly case the reduced factorial covariance measure, if absolutely continuous, has c(.) as its density function.

Writing for $y \geq 0$, $V(y) = V\{(0,y]\} = \mathrm{var}\{N(0,y]\}$, it follows from the above that

$$V(y) = \int_{-y}^{y} (y - |u|)\ C(du) = \alpha y + 2 \int_{0+}^{y} (y-u)\ C(du)\ , \tag{3.1.10}$$

where $\alpha = C(\{0\}) = V'(0+)$ and we have used the symmetry of $C(.)$. Formal differentiation leads to

$$V'(y) = \alpha + 2 \int_{0+}^{y} C(du) = C\{(-y,y)\} \tag{3.1.11}$$

for all continuity points y of the right hand side. Thus, $V(y)$ is everywhere continuous with left- and right-hand derivatives existing everywhere, being equal except perhaps at the points of a countable set. Daley (1971) used the convexity of $M_2^*(t) \equiv E\{N^2(0,t]\}$ to establish these properties of $V(.)$, and examples there show that the exceptional countable set may be everywhere dense.

Example 3.1.A (cont. from 2.1.A). The first moment measure of a Poisson process equals the parameter measure $\mu(.)$. The rth factorial moment measure is $\mu^r(.)$, while all factorial cumulant measures beyond the first vanish. In particular it follows that $\mathrm{cov}\{N(A), N(B)\} = \mu(A \cap B)$, so the complete covariance measure vanishes apart from its atom at the origin; this behavior is of course characteristic of any process with the completely random property (ii) of Example 2.1.A.

Example 3.1.B (cont. from 2.2.A). The first moment measure is determined from the *renewal function* $H(t) = E\{N(0,t]\}$. This function also determines higher moments: for example, $E\{N^2(0,t]\} = H(t) + 2H^{2*}(t) = H(t) + 2 \int_0^t H(t-u)\ H(du)$, and more generally, $M^*_{[r]}\{(0,t]\} = r!\ H^{r*}(t)$. For a stationary renewal process with $F(0+) = 0$, it is easily shown that its complete covariance measure is

$$C[-y,y]) = m + 2m\ H(y) - 2m^2 y\ .$$

3.2. Probability Generating Functional

The probability generating functional (p.g.fl.) defined below is a natural generalization of the probability generating function (p.g.f.) from a Z_+-valued random variable to an infinite family $N(.)$ of Z_+-valued random variables. We mostly follow the recent exposition concerning the p.g.fl. by Westcott (1972); for earlier work see Moyal (1962) and Vere-Jones (1968, 1970). Analogous extensions of the Laplace-Stieltjes transform (or the moment generating functional) and the characteristic functional are appropriate for discussing random measures and random signed measures respectively.

Definition 3.6. The *probability generating functional* (p.g.fl.) of a point process $(\Omega, F, \mathbb{P})$ is defined by

$$G[\xi] = E[\exp\{\int_R \log \xi(t)\ N(dt)\}] = \int_\Omega \{\prod_{x_n \varepsilon \omega} \xi(x_n)\}\ \mathbb{P}(d\omega) \ , \qquad (3.2.1)$$

where $\xi(.)$ belongs to a suitable class of functions and, for consistency, $\exp\{\int_R \log \xi(t)\ N(dt)\}$ is set equal to zero if $\xi(x_n) = 0$ for some $x_n \varepsilon \omega$, and to one if $N[\{t:\xi(t) \neq 1\},\omega] = 0$.

In this definition the class of functions is deliberately left unstated as various classes have been used. Certainly though, whatever class is used it should include all step functions of the form

$$\xi(t) = 1 - \sum_{i=1}^{r} \chi_{A_i}(t)(1-z_i) \qquad (0 \le z_i \le 1) \ , \qquad (3.2.2)$$

where r is a positive integer and $\{A_i,\ i = 1,\dots,r\}$ are pairwise disjoint bounded Borel sets, for with such ξ, $G[\xi]$ reduces to the joint p.g.f. of the random variables $N(A_i)$. These ξ determine the fidi distributions of the point process, and hence the process by Theorem 2.3. The extent to which the class may be extended is perhaps of academic interest: we define some useful classes.

Definition 3.7. Let the function $\xi(t) \ \varepsilon \ U$ when it is measurable and satisfies $0 \le \xi(t) \le 1$, (all $t \ \varepsilon$ R). Then

(a) $\xi \in V$ if $1-\xi$ has bounded support;

(b) $\xi \in V_S$ if it is of the form at (3.2.2);

(c) $\xi \in L_1(\nu)$ for a given Borel measure ν if $\int_R \{1-\xi(t)\}\, \nu(dt) < \infty$;

(d) $\xi \in L(\nu)$ for a given Borel measure ν if $\int_R |\log \xi(t)|\, \nu(dt) < \infty$.

Obviously, $V_S \subset V$ and $L(\nu) \subset L_1(\nu)$. There then follows the basic result.

Theorem 3.8. For any point process the p.g.fl. is well-defined for $\xi \in V$, and also for $\xi \in L(M)$ if the expectation measure $M(.)$ exists. Conversely, the point process is uniquely specified in terms of its fidi distributions by a knowledge of $G[\xi]$, $\xi \in V_S$, and a fortiori when $\xi \in V$ or $\xi \in L(M)$ when M exists.

There now arises naturally the question of characterizing those functionals $G[\xi]$, $\xi \in V$, which are p.g.fls. (of a point process). To do so it is first requisite that continuity properties of p.g.fls. be investigated. Given a sequence $\xi_n \in U$, we write $\xi_n \to \xi$ when for every $t \in R$, $\lim_{n\to\infty} \xi_n(t)$ exists (say, $\xi(t)$), in which case $\xi \in U$ also.

Lemma 3.9. If $G[.]$ is the p.g.fl. of a point process and $\xi_n \to \xi$, and $G[\xi_n] \to G[\xi]$, $n \to \infty$, whenever either

(i) $\xi_n \in V$ and $1-\xi_n$ vanish outside some finite interval I for all n; or

(ii) $M(.)$ exists and $\xi_n \to \xi$ in $L_1(M)$, i.e., $\xi_n \in L_1(M)$ and $\int_R |\xi_n(t) - \xi(t)|\, M(dt) \to 0$ as $n \to \infty$.

This continuity property of p.g.fls. reflects the continuity condition (v) of Theorem 2.3, which then admits the following characterization of p.g.fls.

Theorem 3.10. A functional $G[\xi]$, $\xi \in V$, is a p.g.fl. if and only if

(i) for $\xi \in V_S$, $G[\xi]$ is the p.g.f. of a Z_+^r-valued random variable; and

(ii) for $\xi_n \to \xi$ and $1-\xi_n$ vanishing outside some common finite interval, $G[\xi_n] \to G[\xi]$.

The proof consists in verifying that the fidi p.g.f.s defined by $G[\xi]$ for $\xi \in V_S$ correspond to a family of fidi distributions satisfying (i)-(v) of Theorem 2.3. There is a corresponding version of Theorem

2.5; it asserts that from any functional defined on the subset V_I of V_S for which the sets A_i are rational intervals, and reducing there to a multivariate p.g.f., there can be defined a unique functional on V which is a p.g.fl. agreeing with the original functional except possibly for certain $\xi \in V_I$ where $\mathbb{P}$-continuity set complications arise.

The p.g.fl. can be linked, by an expansion generalizing the moment expansion of an ordinary p.g.f., to the factorial moment measures when they exist. The asymptotics require careful handling; Westcott (1972) gives the following result.

Theorem 3.11. Let $(\Omega, F, \mathbb{P})$ be a point process for which $M_{[r]}(.)$ exists. Then for $\xi \in V$ and $0 < \rho < 1$,

$$G[1-\rho\xi] = 1 + \sum_{i=1}^{r} \frac{(-\rho)^i}{i!} \int_{R^i} \xi(t_1)\ldots\xi(t_i)\ M_{[i]}(dt_1\times\cdots\times dt_i) + o(\rho^r)$$

$$(\rho \to 0) \ . \qquad (3.2.3)$$

Further, if $M_{[r+1]}(.)$ exists, then the term $o(\rho^r)$ is bounded by

$$\frac{\rho^{r+1}}{(r+1)!} \int_{R^{r+1}} \xi(t_1)\ldots\xi(t_{r+1})\ M_{[r+1]}(dt_1\times\cdots\times dt_{r+1}) \ . \qquad (3.2.4)$$

The proof can be accomplished by using analogues of Bonferroni's inequalities for the coefficients in expansions of products $\Pi_{i=1}^{r}\{1 - \rho\xi(t_i)\}$. Westcott (1972) gives further details. Results analogous to (3.2.3) and (3.2.4) can be stated for the expansion of $\log(G[1 - \rho\xi])$ in terms of the factorial cumulants, although it does not seem possible to cast the error bound in as convenient a form as (3.2.4). Some applications of these results are discussed by Vere-Jones (1968) and Westcott (1972). An important limitation on the usefulness of these results in proving limit theorems is that the error term (3.2.4) is not uniform in ξ. It should also be noted here that when all moments exist with $M_{[r]}(A) = o(r!)$, $(r \to \infty)$, for bounded $A \in B$, an expansion like (3.2.3) with $\rho = 1$ and $r = \infty$ is valid. In fact it follows from preceding results that the moments must then determine the point process.

Suppose the point processes $N^{(1)}(.)$ and $N^{(2)}(.)$ are independent, and that we form their superposition $N(.) \equiv N^{(1)}(.) + N^{(2)}(.)$. Then it is easily shown that the respective p.g.fls. are related by

$$G[\xi] = G_1[\xi]\, G_2[\xi] \qquad (\xi \in V) . \tag{3.2.5}$$

To describe the convergence properties of p.g.fls. it is first necessary to define convergence of a sequence of point processes. We phrase this in terms of measures (not necessarily probability measures) on the space $(\Omega,\mathcal{F})$ as this more general definition is needed in discussing convergence of infinitely divisible point processes.

Definition 3.12. The measure triples $(\Omega,\mathcal{F},Q_n)$, $n = 1,2,\ldots$, satisfying

$$Q_n(\emptyset) = 0 \tag{3.2.6}$$

and

$$Q_n\{N(A) > 0\} < \infty \qquad (\text{all bounded } A \in \mathcal{B}) , \tag{3.2.7}$$

converge in distribution to the measure triple $(\Omega,\mathcal{F},Q)$ satisfying (3.2.6) and (3.2.7) if and only if for every positive integer r, all $k_i \in Z_+$ such that

$$k_1 + \cdots + k_r \geq 1 , \tag{3.2.8}$$

and all bounded Q-continuity sets $A_i \in \mathcal{B}$, $i = 1,\ldots, r$,

$$Q_n\{N(A_i) \geq k_i,\ i = 1,\ldots,r\} \to Q\{N(A_i) \geq k_i,\ i = 1,\ldots,r\}. \tag{3.2.9}$$

We write $Q_n \xrightarrow{d} Q$ for such convergence which is indeed the natural definition of convergence in distribution for point processes when Q and all Q_n are probability measures. In this case the restriction (3.2.8) is not required, and the convergence in (3.2.9) can be relaxed to the extent of requiring it only for intervals. The following result is proved by using methods as in Billingsley (1968).

Lemma 3.13. For the sequence $(\Omega, F, \mathbb{P}_n)$ to converge in distribution to $(\Omega, F, \mathbb{P})$, it is necessary and sufficient that the convergence at (3.2.9) hold merely when the $\mathbb{P}$-continuity sets A_i are disjoint finite intervals.

In terms of p.g.fls. Definition 3.12 is the statement that

$$G_n[\xi] \to G[\xi] \qquad (\text{all } \xi \in V_S), \tag{3.2.10}$$

where $G_n[.]$ corresponds to $\mathbb{P}_n$. The extension of this property to the larger class V requires a continuity-set proviso: the *fixed atoms* of $\mathbb{P}$ are the points of the set

$$\{x : \mathbb{P}[N(\{x\}) > 0] > 0\} . \tag{3.2.11}$$

Theorem 3.14. In order that $\mathbb{P}_n \xrightarrow{d} \mathbb{P}$ for given $(\Omega, F, \mathbb{P}_n)$, $(\Omega, F, \mathbb{P})$, it is necessary and sufficient that $G_n[\xi] \to G[\xi]$ for all $\xi \in V$ whose discontinuities avoid the fixed atoms of $\mathbb{P}$.

Westcott (1972) has weakened the hypotheses of these results by showing that if $\lim_{n\to\infty} \mathbb{P}_n\{N(A_i) \geq k_i,\ i = 1, \ldots, r\}$ exist for all r, A_i, and k_i as in Definition 3.12, and if these limits, $p\{.\}$ say, constitute a probability distribution, then necessarily (Ω, F, p) is a point process. Thus the existence of $\lim_{n\to\infty} G_n[\xi]$, $\xi \in V$, already implies that the limit functional is a p.g.fl. provided only that it is continuous in the sense of Theorem 3.10 for $\xi_m(t) \uparrow 1$, $t \in R$, for $m \to \infty$.

Example 3.2.A (cont. from 2.1.A and 3.1.A). From the form of the multivariate p.g.f. of $N(A_i)$ it follows that the p.g.fl. is

$$G[\xi] = \exp\left[\int_R \{\xi(t) - 1\}\, \mu(dt)\right] , \tag{3.2.12}$$

while

$$G[1 - \xi] = 1 + \sum_{i=1}^{\infty} \left\{-\int_R \xi(t)\, \mu(dt)\right\}^i / i! \, . \tag{3.2.13}$$

Example 3.2.B (cont. from 2.4.B). Similarly, following on from (2.4.1), we get for the compound Poisson process

$$G[\xi] = \exp[-\sum_{i=1}^{\infty} \int_R \{\xi^i(t) - 1\} \mu_i(dt)] . \tag{3.2.14}$$

Example 3.2.C. Newman (1970) introduced a class of point processes which he calls Gauss-Poisson precisely because of the form of the p.g.fl. Milne and Westcott (1972a) have continued his work and characterized the class completely as those point processes with p.g.fl.

$$G[\xi] = \exp[\int_R \{\xi(t) - 1\} M(dt) + \frac{1}{2} \int_{R^2} \{\xi(s) - 1\}\{\xi(t) - 1\} C(ds \times dt)], \tag{3.2.15}$$

where M(.) is any Borel measure and the measure C(.) can without loss of generality be taken to be symmetric and must satisfy $C(A \times B) \leq \min\{M(A), M(B)\}$. Included amongst Milne and Westcott's observations is the interpretation of a Gauss-Poisson process as a Poisson cluster process in which a cluster has at most two elements. For a summary see Milne and Westcott (1972b) in this volume.

3.3. Generalizations Concerning Moments

The extension of the definition of moment measures of point processes on state spaces more general than R presents no formal difficulties, and accordingly we confine discussion to more general types of processes with real line as state space.

The first extension of this kind is to random measures and random signed measures. The non-negativity of random measures ensures the σ-additivity of the moments of such objects when defined analogously to (3.1.1) and (3.1.2). In the case of a random signed measure however, there is no longer any direct implication (either way) between σ-additivity of the realizations and σ-additivity of their moments (when they exist). Consequently, we say that the first moment measure of a random signed measure $\Lambda(.)$ exists if and only if the expectations $E\{\Lambda(A)\}$

exist for all bounded $A \in \mathcal{B}$ and the set function so defined is σ-additive. Such σ-additivity obtains if and only if the set function is of bounded variation on finite intervals, for which it suffices that $E\{\Lambda(A)\}$ be bounded for all subsets $A \in \mathcal{B}$ of any given finite interval. For random interval functions (Definition 2.13), when the realizations themselves need no longer be σ-additive, it makes sense to speak of moment *measures* when the set function $E\{X(I)\}$, $I \in \mathcal{I}$, exists and can be extended to a σ-additive function on $\mathcal{B}$. For example, the white noise process of Example 2.5.A has a first moment measure without itself being σ-additive.

Example 3.3.A. Any random interval function whose fidi distributions are jointly normal is a *Gaussian interval function*. Bochner (1955) showed that given any Borel measure $M(.)$ on R and any positive definite measure $C_2(.)$ on $R \times R$, there exists a Gaussian interval function having $M(.)$ as first moment measure and $C_2(.)$ as its complete covariance measure.

In view of this result it is natural to enquire as to what conditions on $M(.)$ and $C_2(.)$ ensure that the realizations may be taken to be σ-additive with probability one. We are not aware of any such conditions: presumably they entail appropriate smoothness in $C_2(.)$ just as degrees of smoothness in the correlation function of a Gaussian stochastic process entails its differentiability etc. (see e.g. Cramer and Leadbetter, 1967).

Extensions to the moments of random integrals and, more generally, of random generalized functions, can be given with suitable provisos concerning finiteness (existence) and continuity (σ-additivity); for random generalized functions see Gel'fand and Vilenkin (1964).

3.4. Generalizations Concerning Generating Functionals

The appropriate extension of the p.g.fl. for dealing with a random measure $\Lambda(.)$ is the Laplace functional defined for any non-negative bounded Borel function ξ by

$$L[\xi] = E[\exp\{-\int_R \xi(t)\ \Lambda(dt)\}] . \tag{3.4.1}$$

For step functions of the form

$$\xi(t) = \sum_{i=1}^{r} s_i \chi_{A_i}(t) \qquad (0 < s_i < \infty) , \qquad (3.4.2)$$

with bounded sets $A_i \in \mathcal{B}$, the functional reduces to the multivariate Laplace-Stieltjes transform of the random variables $\Lambda(A_i)$, whence its domain of definition extends by continuity to any non-negative bounded Borel function. Effectively, the functional was introduced in early work by Kendall (1949) and Bartlett and Kendall (1951); Jirina (1964, 1966) gives a detailed discussion of its properties. Shiraev (1960) discusses another version of the Laplace functional in which $\Lambda(.)$ is required to be absolutely continuous with density $\lambda(.)$ and (the integral of) ξ is replaced by any totally finite Borel measure $\nu(.)$, thus replacing the integral in (3.4.1) by $\int_R \lambda(t)\,\nu(dt)$.

For the most general contexts, the appropriate functional is the characteristic functional, defined for random signed measures by

$$\Phi[\xi] = E[\exp\{i \int_R \xi(t)\, X(dt)\}], \qquad (3.4.3)$$

which may be written in the general form

$$\Phi[\xi] = E[\exp\{i(X,\xi)\}] , \qquad (3.4.4)$$

where X is a random element from a linear space and ξ is an element in the dual space. Regarded in this way the idea goes back to Kolmogorov (1935); Prohorov (1960) gives a more recent survey. It is clear that as such the characteristic functional is the direct extension of the idea of a characteristic function from a finite to an infinite dimensional space, and hence is a basic tool in the discussion of probability measures on abstract spaces (see for example Parthasarathy, 1967). In the more particular applications to stochastic processes the concept was introduced by Le Cam (1947) and for random interval functions by Bochner (1947); in the latter context Bochner (1955) contains a comprehensive account. For applications to random distributions see Gel'fand and

Vilenkin (1964). We give no further details here except to note the obvious relation between the three functionals (when defined)

$$\Phi[\xi] = L[-i\xi] = G[e^{i\xi}] \ .$$

4. SPECTRAL THEORY OF SECOND ORDER STATIONARY POINT PROCESSES

Our aim in this section is to outline some extensions of the familiar spectral theory of second order stationary stochastic processes appropriate to the point process context. Practical applications of the extended theory, including some discussions of inference and estimation problems, have been given by Bartlett (1963), Cox and Lewis (1966), and Lewis (1970), and hardly need emphasizing here. On the theoretical side, Doob's (1953) adaptation of Kolmogorov (1940) to processes with stationary increments, and the work of Ito (1953), Gel'fand (1955) and Gel'fand and Vilenkin (1964) on stationary random distributions, provide a mathematical framework of sufficient generality to embrace the applications to point processes, stationary random measures, and stationary random interval functions: for this last topic, see also Bochner (1955) and the recent survey by Brillinger (1972).

It follows that in principle no new ideas are needed to furnish a spectral theory for point processes. Nevertheless, as in many such situations, passing from the general theory to particular contexts produces its own special problems. For example, the existence of a second moment measure for point processes implies a bounded variation property stronger than the mean square continuity condition required for the Kolmogorov-Doob theory, and hence there is a corresponding restriction on the growth of the spectral measure at infinity. There is an attempt in Daley (1971) to work out the implications of this extra condition, but it remains an open problem to determine necessary and sufficient conditions on a measure that will ensure that it is the spectral measure of a point process.

In the first section below we give a Bochner-type representation for the reduced covariance measure C(.) (see (3.1.8)) of a stationary point process, while in the second we give the corresponding representation of the process itself.

4.1. Representation of the Covariance Measure

In the last chapter we introduced the notion of positive definiteness as applied to symmetric measures on the plane. In this section we are concerned throughout with processes that are second order stationary in the following sense.

Definition 4.1. A random measure $\Lambda(.)$ is *second order stationary* when for all $t \in R$ and bounded $A \in \mathcal{B}$, the moments below exist with

$$M(A) = E\{\Lambda(A)\} = M(A + t)$$

and (4.1.1)

$$M_2(A) = E\{\Lambda^2(A)\} = M_2(A + t) \ .$$

It is obvious how the definition is amended for point processes or random signed measures or random interval functions.

It is convenient to begin by reframing the definition of a positive definite measure so that it applies directly to the one-dimensional reduced covariance measure C(.) defined at (3.1.8).

Definition 4.2. A (complex-valued) Borel measure $\gamma(A)$, $A \in \mathcal{B}$, is *positive definite* if for all bounded Borel functions ϕ of compact support in R,

$$\int_R \gamma(dx) \int_R \overline{\phi(x + u)}\ \phi(u)\ du \geq 0. \qquad (4.1.2)$$

It is clear that C(.) enjoys this property.

Let P_+ denote the class of positive measures on $\mathcal{B}$ (i.e., take non-negative values and are not identically zero) which are also positive definite: we call them p.p.d. measures for short. Because

$$C(du)dt + m^2\ du\ dt = E(\Lambda(t + dt)\ \Lambda(t + u + du)) \ .$$

the following is also true.

Lemma 4.3. The reduced covariance measure of a second order stationary point process or (non-negative) random measure differs from a p.p.d. measure by at most a multiple of Lebesgue measure. Explicitly, $C(.) + m^2|.| \in P_+$.

It is an open question whether the converse to the lemma is true: we state this question because such properties as we know for covariance measures are in fact consequences of their being related to p.p.d. measures. These properties are given in the following theorem, which states inter alia that P_+ is closed under the Fourier transform mapping indicated by (4.1.3). We follow Gel'fand and Vilenkin (1964) in denoting by S the class of all functions ϕ infinitely differentiable with $x^n \phi^{(m)}(x) \to 0$ ($|x| \to \infty$) for all m, $n \geq 0$.

Theorem 4.4. Given $\gamma \in P_+$, there exists one and only one measure μ such that for all $\phi \in S$,

$$\int_R \phi(x)\ \gamma(dx) = \int_R \tilde{\phi}(\theta)\ \mu(d\theta)\ , \tag{4.1.3}$$

where $\tilde{\phi}(\theta) = \int_R e^{i\theta u}\ \phi(u)\ du$. This measure $\mu \in \mathbb{P}_+$. In general $\gamma \in \mathbb{P}_+$ need not be totally finite, but $\gamma(x,\ x + 1]$ is uniformly bounded on $x \in R$, and the limit

$$\lim_{x\to\infty} \gamma(-x,x]/2x = \mu(\{0\}) \tag{4.1.4}$$

exists with μ as defined in (4.1.3).

The Parseval relation (4.1.3) is in fact true for a wider class of functions than $\phi \in S$, including in particular all continuous functions of bounded support, integrable with respect to μ. These results may be found in embryonic form in Daley (1971) and Vere Jones (1971); a treatment more in line with the present discussion will be given in Vere-Jones (1972). The argument starts from the observation that for all functions ϕ such that

$$\phi^{(2)}(x) \equiv \int_R \overline{\phi(u + x)}\ \phi(u)\ du \tag{4.1.5}$$

is continuous and integrable with respect to a given p.p.d. measure γ, the function

$$\gamma_\phi(y) \equiv \int_R \phi^{(2)}(x+y)\,\gamma(dx) \tag{4.1.6}$$

is continuous and positive definite. Thus on the one hand the familiar Bochner theorem applies, while taking ϕ first to be the indicator function of the unit interval yields the uniform boundedness of $\gamma(x, x+1]$, and then to be a gamma density leads ultimately to the Parseval relation (4.1.3). Finally an inversion formula yields (4.1.4).

It is worthwhile stating these results in terms of the reduced covariance measure and variance function. Note that since the random measure $\Lambda(.)$ of Definition 4.1 is real, $C(.)$ is symmetric, implying that the measure corresponding to it via (4.1.3) is also symmetric so it could be taken to be concentrated on $[0,\infty)$, as in Daley (1971). Equation (4.1.7) below follows from the Parseval relation by setting $\phi(u)=(y-|u|)^+$ in (4.1.3).

Theorem 4.5. For a second order stationary random measure $\Lambda(.)$ there exists a unique symmetric measure $\nu(A)$, $A \in \mathcal{B}$, such that for all $y \geq 0$,

$$\begin{aligned} V(y) = \operatorname{var}\{\Lambda(0,y)\} &= \int_R \left(\frac{\sin(\theta y/2)}{\theta/2}\right)^2 \nu(d\theta) \\ &= y^2\,\nu(\{0\}) + 4\int_{(0,\infty)} \frac{1-\cos\theta y}{\theta^2}\,\nu(d\theta)\,. \end{aligned} \tag{4.1.7}$$

The measure ν differs from a p.p.d. measure by at most an atom at the origin, and

$$\lim_{\theta\to\infty} \frac{2\nu(0,\theta)}{\theta} = \frac{V'(0+)}{\pi} = \frac{C(\{0\})}{\pi}, \tag{4.1.8}$$

$$\nu(\{0\}) = \lim_{y\to\infty} \frac{V(y)}{y^2} = \lim_{y\to\infty} \frac{C(-y,y)}{2y}\,. \tag{4.1.9}$$

From (4.1.9) and the identity $E\{|x^{-1} N(0,x] - m|^2\} = V(x)/x^2$, it follows that the weak law of large numbers with mean square convergence applies to a point process N(.) if and only if $\nu(\{0\}) = 0$, so the asymptotic behavior of the variance function is intimately related to the ergodic nature of the process. In Section 8 we indicate how stronger conditions on C(.) lead to stronger ergodic results.

The situation for random signed measures and other random interval functions is not so easily handled as the above, even when the existence of a second moment measure is postulated. For one thing, decomposing a positive definite measure into its positive and negative parts does not in general yield two positive definite measures. Consequently the uniform boundedness property will not be true in general, but at worst the mean square continuity (which is implied by the existence of a covariance measure) ensures via the Kolmogorov-Doob theory the existence of a spectral measure integrating $(1 + \theta^2)^{-1}$. The representation (4.1.8) will still hold but results concerning the existence of $\lim_{\theta\to\infty} \mu(-\theta,\theta]/2\theta$ need not be true.

4.2. Representation of the Process

For a second order stationary random measure $\Lambda(.)$ consider the stochastic integral

$$X_\phi = \int_R \phi(u)\ \Lambda(du) \tag{4.2.1}$$

for a certain class (as yet unspecified) of functions $\phi(.)$. Then

$$\begin{aligned} E(|X_\phi|^2) &= E\{\iint \phi(u)\ \overline{\phi(v)}\ \Lambda(du)\ \Lambda(dv)\} \\ &= \iint \phi(u)\ \overline{\phi(v)}\ M_2(du,\ dv) \\ &= \iint \phi(u)\ \overline{\phi(u + v)}\ \{C(dv) + m^2 dv\}\ du, \end{aligned}$$

so (4.1.5) and the Parseval relation (4.1.3) yield

$$E(|X_\phi|^2) = \int \phi^{(2)}(v) \{C(dv) + m^2 dv\}$$
$$= \int |\tilde{\phi}(\theta)|^2 \{m^2 \delta_0(d\theta) + \nu(d\theta)\} , \qquad (4.2.2)$$

where $\delta_0(A) = 1$ or 0 as $0 \in A$ or $0 \notin A$ $(A \in \mathcal{B})$. Bochner's result on Gaussian interval functions (Example 3.3.A) shows that there exists a random interval function $Z(I)$, $I \in \mathcal{I}$ with mean $m\delta_0(I)$ and second moment measure

$$E\{Z(I_1) \overline{Z(I_2)}\} = (m^2\delta_0 + \nu)(I_1 \cap I_2)$$

vanishing off the diagonal, i.e., $Z(I_1)$ and $Z(I_2)$ are uncorrelated for disjoint I_1 and I_2. Thus (4.2.2) can be written as

$$E\{|\int_R \phi(u) \ \Lambda(du)|^2\} = E\{|\int \tilde{\phi}(\theta) dZ(\theta)|^2\} . \qquad (4.2.3)$$

Now (4.2.3) makes plausible the following result, which is essentially given in Doob (1953, Section 11.11); it can also be simply proved from the Cramer representation of a smoothed version of the point process (see Vere-Jones, 1972).

Theorem 4.6. Let $N(.)$ be any second order stationary point process (or random measure). Then there exists a random interval function $Z(.)$ with uncorrelated increments such that

$$E\{|Z(I)|^2\} = m^2 \delta_0(I) + \nu(I) \qquad (4.2.4)$$

where m is the density of $N(.)$ and $\nu(.)$ its spectral measure, and such that with probability one

$$N(x,y] = \int_R \{(e^{ix\theta} - e^{iy\theta})/i\theta\} \, dZ(\theta) \qquad (-\infty < x < y < \infty) . \qquad (4.2.5)$$

Further, for $-\infty < \theta < \theta' < \infty$,

$$Z(\theta,\theta') + \frac{1}{2}[Z\{\theta\} + Z\{\theta'\}]$$

$$= \underset{T\to\infty}{\text{l.i.m.}} \int_{-T}^{T} \{(e^{-ix\theta} - e^{-ix\theta'})/2\pi ix\}\, N(dx) \quad , \tag{4.2.6}$$

and for functions $\phi \in S$

$$\int_R \phi(x)\, N(dx) = \int_R \tilde{\phi}(\theta)\, dZ(\theta) \, . \tag{4.2.7}$$

Note that the integrals at (4.2.5) and (4.2.7) are to be understood as mean square integrals. A similar extension as for the Parseval formula (4.1.3) can be given here also.

It will be recognized that (4.2.4) is a special case of (4.2.7). The theorem, with some modification to the Parseval relation (4.2.6), remains true with N(.) replaced by second order stationary random signed measures provided they are mean square continuous (Doob, 1953). Similar representations exist for stationary random distributions (Gel'fand and Vilenkin, 1964). Bochner (1955) has extended the theory to non-stationary interval functions by imposing an extra condition that he calls $L_{2,2}$-boundedness; in spite of the elegance of the theory it does not apply directly to point processes because the $L_{2,2}$-boundedness can be shown to imply absolute continuity of the covariance measure.

The application of results on linear least squares prediction theory as extended to stationary random distributions and processes with stationary increments (e.g., Yaglom, 1955; Rozanov, 1959, Balagangadharan, 1960) is summarized in the paper in this volume by Jowett & Vere-Jones (1972).

One final question may be briefly mentioned. Regarding N(.) at (4.2.7) as a generalized function, the left hand side equals (ϕ,N), and then by the Parseval relation for generalized functions it equals $(\tilde{\phi},\tilde{N})/2\pi$ where $\tilde{N}$ is the Fourier transform of N in the generalized function sense. This does not mean however that Z can be identified with $\tilde{N}/2\pi$: rather, it is a version of Z, and the problem is to determine what regularity properties this particular version possesses. The question is far from easy; for discussion of this and related problems we refer to Dudley (1969) and further references contained therein.

5. OPERATIONS ON POINT PROCESSES

In this Section we discuss several operations for transforming point processes. A common feature of these operations is that they are readily described in terms of the machinery set up in the preceding chapters. On the other hand they form only a restricted selection of the types of operations we should like to be able to handle. Essentially we are restricted to processes and operations in which, in some appropriate sense, the individual points are treated independently (e.g., the Poisson process and its compounds). Even a mild degree of interdependence, such as that between one point and its successor in a renewal problem, or between the operations carried out on one point and its successor, is sufficient to place the process or operation out of convenient reach of the tools presently available.

5.1. Compounding

The operation of compounding consists in replacing each point in the original process N(.) by an associated random variable. In general this transforms a point process into a marked point process, though when the associated random variables are Z_+-valued, the result may also be regarded as another point process (non-orderly in general). The operation is succinctly summarized by the equation

$$X(dt) = Y(t)\ N(dt) \ , \qquad (5.1.1)$$

where Y(.) is to be thought of as a continuous parameter stochastic process. (In detail, (5.1.1) abbreviates the equation between sample realizations,

$$X(A) = \sum_{x_i \varepsilon A} Y(x_i) \qquad (\text{bounded } A \ \varepsilon \ \mathcal{B}) \ , \qquad (5.1.2)$$

where N(.) defines the points $\{x_n\}$ as in Section 2.2). The process X(.) will exist as a (purely atomic) random measure if for example Y(.) is finite-valued and N(.) satisfies the boundedness Assumption 2.1. Here we shall always suppose that the two processes Y(.) and N(.) are independent (i.e., their joint fidi distributions factorize appropriately). This then enables us to write the characteristic functional of X(.) formally as

$$\begin{aligned}\Phi_X[\xi] &\equiv E[\exp\{i\textstyle\int\xi(t)\ X(dt)\}] \\ &= E[\exp\{i\textstyle\int\xi(t)\ Y(t)\ N(dt)\}] \\ &= E\{E(\exp[\textstyle\int \log\{e^{i\xi(t)\ Y(t)}\}\ N(dt)]\,|\,Y(.))\} \\ &= E(G[e^{i\xi Y}])\ ,\end{aligned} \tag{5.1.3}$$

where G[.] is the p.g.fl. of N(.) and $(\xi Y)(t) = \xi(t)\ Y(t)$.

There are at least two cases of interest, the first where both Y(.) and N(.) are stationary and our concern is with the spectral properties of X(.), and the second where Y(.) is a process of independent variables.

If the first and second moments of both Y(.) and N(.) exist and Y(.) is mean square continuous then the first and second moment measures of the compound process X(.) exist with

$$E\{X(dt)\} = m_Y(t)\ M(dt)\ , \tag{5.1.4}$$

$$\begin{aligned}&\operatorname{cov}\{X(dt),\ X(du)\} \\ &\quad = c(t,u)\ \{C_2(dt\times du) + M(dt)\ M(du)\} + m_Y(t)\ m_Y(u)\ C_2(dt\times du)\ ,\end{aligned} \tag{5.1.5}$$

where $m_Y(t) = E\{Y(t)\}$ and $c(t,u) = \operatorname{cov}\{Y(t),\ Y(u)\}$. The appearance of the first-order moments in (5.1.5) is linked to the non-linearity of the compounding operation. Although it may be possible to assume $m_Y(t) = 0$ for all $t \in R$, the corresponding assumption for M(.) would imply N(A) = 0 a.s.

When both processes are stationary, (5.1.4) and (5.1.5) simplify and the representation theory of Section 4 is applicable. For example, with $m_Y = E\{Y(t)\}$, $c(t, t+u) = \int_R \cos \theta u \, F_Y(d\theta)$, and $\mathrm{var}\{X(0,u]\} = \int_R (\sin \frac{1}{2} \theta u / \frac{1}{2}\theta)^2 \, \nu_X(d\theta)$,

$$\nu_X(A) = \int_R \nu(A-\theta) \, F_Y(d\theta) + m_Y^2 \, \nu(A) + m^2 \, F_Y(A) \qquad (A \in \mathcal{B}). \quad (5.1.6)$$

The convergence of the integral in (5.1.6) is assured by the uniform boundedness property of the p.p.d. measure $\nu(.)$. Brillinger (1972) discusses (5.1.6).

If $Y(.)$ is a process of independent variables with $\phi_t(\theta) = E\{e^{i\theta Y(t)}\}$, then by modifying the argument leading to (5.1.3) we get

$$\begin{aligned}\Phi_X[\xi] &= E(E[\exp\{i\int \xi(t) \, Y(t) \, N(dt)\} | N(.)]) \\ &= E\{\exp(\int \log[\phi_t\{\xi(t)\}] \, N(dt))\} \\ &= G[\phi \circ \xi] \qquad (5.1.7)\end{aligned}$$

where $(\phi \circ \xi)(t) \equiv \phi_t\{\xi(t)\}$. In particular, when $N(.)$ is Poisson with parameter measure $\mu(.)$, and the support of $F_t(y) \equiv \mathrm{prob}\{Y(t) \leq y\}$ is contained in Z_+, then we obtain for $\Phi_X[\xi]$ equation (3.2.13) of the compound Poisson process.

When $N(.)$ in (5.1.7) is stationary with finite first and second moments and the $Y(t)$ are identically distributed with mean m_Y and variance σ^2 (such a process $Y(.)$ with independent variables is not mean square continuous unless $\sigma^2 = 0$), the analogues of (5.1.5) (cf. also (3.1.8)) and (5.1.6) are

$$\begin{aligned}&\mathrm{cov}\{X(t, t+dt), X(t+u, t+u+du)\} \equiv C_X(du) \, dt \\ &\quad = \{m_Y^2 \, C(du) + m\sigma^2 \, \delta(u) \, du\} \, dt, \qquad (5.1.8)\end{aligned}$$

where $\delta(.)$ is the Dirac delta function, and

$$\nu_X(d\theta) = m_Y^2\, \nu(d\theta) + m\sigma^2 d\theta/2\pi \qquad (\theta \in R) . \tag{5.1.9}$$

5.2. Clustering

A more general operation than compounding is that of clustering in which, loosely speaking, each point in the original process N(.) is replaced by a string or cluster of points (or by a general (signed) measure). Analogous to (5.1.1) it is summarized by

$$X(dt) = \int_R \Lambda_u(dt-u)\; N(du) , \tag{5.2.1}$$

where if $N(\{u\}) = 1$, such a parent point is replaced by the measure $\Lambda_u(.)$ with cluster center u. Little progress in the analysis of (5.1.1) has been made except where the clusters $\Lambda_u(.)$, $u \in R$, form an independent family independent of N(.). In such a case X(.) has p.g.fl.

$$G_X[\xi] = G[G_\Lambda[\xi|.]] \tag{5.2.2}$$

when $\Lambda_{\cdot}(.)$ is a point process, and (cf. (3.2.1))

$$G_\Lambda[\xi|t] \equiv E[\exp\{\int_R \log \xi(u)\; \Lambda_t(du-t)\}] \qquad (t \in R) . \tag{5.2.3}$$

There is a corresponding relationship when $\Lambda_t(.-t)$ is a family of (signed) random measures with characteristic functional $\Phi_\Lambda[\xi|t]$, namely

$$\Phi_X[\xi] = G[\Phi_\Lambda[\xi|.]] . \tag{5.2.4}$$

It is no longer an easy matter to determine necessary and sufficient conditions on N(.) and $\Lambda_{\cdot}(.)$ to ensure that the resultant process X(.) should satisfy the boundedness Assumption 2.1 (or its analogue when $\Lambda_{\cdot}(.)$ is a measure).

Various sufficient conditions for particular cases have been given by (among others) Neyman (1963), Franken and Richter (1965), Liemant (1969), and Westcott (1971). Westcott's paper contains a survey and an attempt at a more systematic approach.

Relations between the moment measures of $N(.)$, $\Lambda(.)$ and $X(.)$ may be evaluated directly or from the p.g.fl. equation. In the general case,

$$E\{X(dt)\} = \int M_\Lambda(dt-u|u)\ M(du) \ , \tag{5.2.5}$$

$$\operatorname{cov}\{X(dx),\ X(dt)\}$$

$$= \int_{u\varepsilon R} c_2(ds \times dt|u)\ M(du) + \int\int_{u,v\varepsilon R} M_\Lambda(ds|u)\ M_\Lambda(dt|v)\ C_2(du \times dv), \tag{5.2.6}$$

where $M_\Lambda(.)$ and $c_2(.)$ refer to $\Lambda(.)$, $M(.)$ and $C_2(.)$ to $N(.)$. The equations simplify in the stationary case, and yield the simple identity between the spectral measures

$$\nu_X(d\theta) = |\beta(\theta)|^2\ \nu(d\theta) + m\{\gamma(\theta) - |\beta(\theta)|^2\}d\theta/2\pi \qquad (\theta\ \varepsilon\ R)\ , \tag{5.2.7}$$

where $\beta(\theta) = E\{\int_R e^{i\theta u}\ \Lambda_0(du)\}$, $\gamma(\theta) = E\{|\int_R e^{i\theta u}\ \Lambda_0(du)|^2\}$. Equation (5.1.9) is a special case of (5.2.7). Notice that the spectrum of $X(.)$ is the sum of two terms, one being a filtered version of the spectrum of the parent process $N(.)$, the other arising as an average of an instantaneous spectrum of a subsidiary process.

Despite the limitation implied by the independence assumption, cluster processes have been applied in a wide variety of fields. The development above is similar to Moyal (1962a,b) though he concentrated on the case $N(R)$ a.s. finite; extensions to the general case are in Vere-Jones (1968, 1970). Detailed explicit results are available essentially only for a Poisson parent process with either Bartlett-Lewis clusters (Example 2.3.B) in Bartlett (1963) and Lewis (1964), or Neyman-Scott clusters (Example 2.3.A) in Neyman and Scott (1958), Bartlett (1964), and Vere-Jones and Davies (1966). Extensions to the cases where the parent process is a renewal process or a general stationary orderly point process are given in Lewis (1972) and Daley (1972a).

5.3. Random Deletions and Random Shifts

Two special cases of the preceding operations warrant brief separate mention. By a randomly deleted point process X(.) we shall understand that each of the points $\{x_n\}$ of a realization of N(.) is retained with probability p and otherwise omitted, independently of all other points. It is thus a special case of compounding with independent identically distributed variables Y(.). Thus $\text{prob}\{Y(t) = 1\} = p = 1 - \text{prob}\{Y(t) = 0\} = 1 - q$ for $t \in R$.

By a *randomly shifted* or *randomly translated* point process X(.), derived from an initial point process N, we shall understand the process with sample realizations $\{x_n + Y_n\}$ where $\{Y_n\}$ is a family of independent identically distributed random variables independent of the realization $\{x_n\}$ of N(.). If $B(y) = \text{prob}(Y_n \leq y)$, this reduces to a special case of clustering, with $G_\Lambda[\xi|t] = \int \xi(t + x)\, dB(x)$.

Relations between the moment and spectral measures for these two operations may be derived by appropriate specialization, and we omit the details.

A randomly deleted process is appropriate to contexts where points may or may not be observed. The randomly translated process occurs in the G/G/∞ system of queueing theory, or as a system of particles moving with random velocities (Dobrushin, 1956, Thedeen, 1964), or as the "jitter process" in engineering contexts (Beutler and Leneman, 1966, 1968). We defer to Section 8 our brief discussion of invariance and limit properties under these operations.

5.4. Randomization

Any process or distribution depending on a parameter or family of parameters can be generalized by treating such parameters as random variables. This is the general operation we have in mind when using the term "randomization"; another term is "mixture" (hence the mixed Poisson process or distribution, etc.), while Bochner (1955) and Le Cam (1961) speak of a "subordination" operation in a similar sense. (Gurland, 1957, advocates the term "compound", but we prefer the present distinction between compounding, as in Section 5.1, and mixing.)

In the point process context a most important example of this kind of operation occurs when the quantity to be randomized is the parameter measure of a Poisson process, leading to what are now known as *doubly stochastic Poisson processes*. Supposing that the random measure $\Lambda(.)$ is the parameter measure of a Poisson process and has Laplace functional $L[.]$, the p.g.fl. of the doubly stochastic Poisson process is (cf. (3.2.11) and (3.4.1))

$$G[\xi] = E(\exp[-\int_R \{1 - \xi(t)\}\ \Lambda(dt)]) = L[1 - \xi] \ . \tag{5.4.1}$$

In particular then the p.g.f. $P_I(.)$ of the number of points in a finite interval I is that of a mixed Poisson distribution (cf. Example 2.1.C)

$$P_I(z) = \int_{[0,\infty)} e^{-\lambda(1-z)}\ F_I(d\lambda) \ , \tag{5.4.2}$$

where $F_I(A) = \mathbb{P}\{\Lambda(I) \in A\}$, $A \in \mathcal{B}$, is a probability measure. Consequently the distribution of $N(I)$ is always over-dispersed relative to the Poisson distribution, whence the importance of such processes as models for contagion and the impossibility of incorporating inhibitory effects within the framework of (5.4.1) (cf. Hawkes, 1971 and Lawrence, 1972, for details).

The moment measures are easily and simply related, for by (5.4.1) the *rth factorial moment measure of a doubly stochastic Poisson process equals the rth ordinary moment measure for the parameter process*. In particular, if $\Lambda(.)$ is stationary with spectral measure $\nu_\Lambda(.)$, then

$$\nu(d\theta) = \nu_\Lambda(d\theta) + m d\theta/2\pi, \tag{5.4.3}$$

where $m dt = E\{\Lambda(dt)\}$. When $\Lambda(.)$ is absolutely continuous, $\nu_\Lambda(.)$ gives the ordinary spectral measure for the density process.

6. ORDERLINESS AND THE INTENSITY MEASURE OF A POINT PROCESS

6.1. Definitions Concerning Orderliness

In simple models of a point process there is interest in knowing conditions for the approximations

$$M(x, x + h] = E\{N(x, x + h]\} \simeq \mathbb{P}\{N(x, x + h] = 1\}$$
$$\simeq \mathbb{P}\{N(x, x + h] \geq 1\} \tag{6.1.1}$$

to be valid. Observe that for the first approximation to be meaningful it is necessary for the first moment measure M(.) to be finite, while the second approximation requires that

$$\mathbb{P}\{N(x, x + h] \geq 2\} = o(\mathbb{P}\{N(x, x + h] = 1\}) \qquad (h \downarrow 0), \tag{6.1.2}$$

suggesting that the points $\{x_n\}$ of a generic sample realization ω of $(\Omega, F, \mathbb{P})$ should be distinct. There exist in the literature various definitions and results stemming from this general idea, so we start by reviewing terminology. Daley (1972b) contains further details.

The property that the points $\{x_n\}$ of ω should be distinct, that is, for $x_i, x_j \in \omega$, $x_i \neq x_j$ whenever $i \neq j$, or equivalently,

$$N(\{x\}) = 0 \text{ or } 1 \qquad (\text{all } x \in R), \tag{6.1.3}$$

is usually described by saying that ω has no multiple points. Since our interest is confined to events of probability one, a *point process without multiple points* is one for which

$$\mathbb{P}\{N(\{x\}) = 0 \text{ or } 1; (\text{all } x \in R)\} = 1 ; \tag{6.1.4}$$

Leadbetter (1972a) and others use this terminology. Other terminology that has been used includes *calling time process* (Belyaev, 1969), and processes for which the distances between the points of ω are all positive (with probability one) (Vasil'ev, 1965), since (6.1.4) implies that

$$\mathbb{P}(|x_i - x_j| > 0 \text{ for } i \neq j) = 1 . \tag{6.1.5}$$

By means of the mappings in Section 2.2 relating (N,S) and (Ω,F), for a process satisfying (6.1.4) and Assumption 2.1 there is a labeling of the points of $\{x_n\}$ such that the orderings of the labels and of the points correspond. Presumably it is for this reason that the term *orderliness* (or, a.s. orderliness) as in Definition 3.3 has come into being: clearly it is equivalent to (6.1.4) or (6.1.5).

The definition of a.s. orderliness makes no mention of infinitesimals as implied in (6.1.1). What is most commonly referred to as orderliness or ordinariness is the property, possessed by certain stationary point processes, that

$$\mathbb{P}\{N(x, x+h] \geq 2\} = \mathbb{P}\{N(0,h] \geq 2\} = o(h) \qquad (h \downarrow 0). \tag{6.1.6}$$

(The Russian word ордина́рный (= ordinary, single) has been translated into English as both ordinary and orderly. In Daley (1972b) orderliness is used as a general term with appropriate qualifying adjectives, and this usage is followed here.)

Korolyuk's theorem (Theorem 6.5 below) shows that (6.1.6) implies (6.1.4), and the converse, known as Dobrushin's lemma (Volkonskii, 1960), holds when the first moment measure exists. Observe that (6.1.6) refers to stationary point processes: there have been various attempts to generalize the property so as to include non-stationary processes (and processes with state space X more general than R), but none is entirely satisfactory.

Definition 6.1. A point process is *Khinchin orderly* when for every $\varepsilon > 0$ and for each x contained in R there exists $\delta \equiv \delta(\varepsilon,x)$ such that

$$\mathbb{P}\{N(x - h_1, x + h_2] \geq 2\} < \varepsilon\mathbb{P}\{N(x - h_1, x + h_2] \geq 1\} \tag{6.1.7}$$

for all h_1, h_2 in $[0,\delta)$. If for every $\varepsilon > 0$ there exists $\delta \equiv \delta(\varepsilon)$ such that

$$\mathbb{P}\{N(x, x+h] \geq 2\} < \varepsilon \mathbb{P}\{N(x, x+h] \geq 1\}$$

$$(\text{all } 0 < h < \delta, \text{ all } x \text{ in } R) \;, \qquad (6.1.8)$$

then the process is *uniformly Khinchin orderly*.

Definition 6.2. A point process is *analytically orderly* when

$$\lim_{h_1, h_2 \downarrow 0} \frac{\mathbb{P}\{N(x - h_1, x + h_2] \geq 2\}}{h_1 + h_2} = 0 \qquad (\text{every } x \in R) \;. \qquad (6.1.9)$$

The process is *uniformly analytically orderly* (*unif. anal. orderly*) when

$$\lim_{h \downarrow 0} \sup_{x \in R} \frac{P\{N(x, x+h] \geq 2\}}{h} = 0 \;. \qquad (6.1.10)$$

Khinchin (1956a) and Fieger (1964) used the term ordinary to describe a property close to unif. Khinchin orderliness. Belyaev (1969) and Leadbetter (1972a) using the terms ordinary and regular respectively, have given a definition applicable to point processes on more general state spaces X which, with X = R, reduces essentially to unif. Khinchin orderliness.

The point process consisting of the birth epochs of a linear pure birth process can be shown to be Khinchin and anal. orderly but not uniformly so. An anal. orderly process is always a.s. orderly but even with finite first moment measure an a.s. orderly point process need not be anal. orderly. A Khinchin orderly point process need not be a.s. orderly, though it will be when the first moment measure is finite; the converse need not be true. Nor need a unif. Khinchin orderly process be anal. orderly nor a unif. anal. orderly process be Khinchin orderly, even with finite first moment measure. However, we can assert the following.

Theorem 6.3. For stationary processes with finite first moment measure, orderliness in one sense implies orderliness in all other senses.

In this case then the process can be called orderly without fear of ambiguity.

All these assertions are justified by the requisite proof or counterexample (or reference thereto) in Daley (1972b). Analytic orderliness is always a sufficient condition for a.s. orderliness, and in this it is better than Khinchin orderliness as a generalization of (6.1.6).

Daley (1972b) defines yet another kind of orderliness which, for want of any distinctive name, is there called *ordinariness:* it is the property that

$$\inf_{\{I_j\}\varepsilon\mathcal{D}(I)} \sum_j \mathbb{P}\{N(I_j) \geq 2\} = 0 , \tag{6.1.11}$$

where $\mathcal{D}(I)$ is as in Definition 6.4 below. Its place amongst the other definitions of orderliness is intermediate between anal. and a.s. orderliness, for anal. orderliness implies ordinariness which implies a.s. orderliness; in the converse direction neither of these implications need hold except when the first moment measure is finite in which case ordinariness and a.s. orderliness are equivalent.

6.2. Korolyuk Type Results

The most important theorem in this area is the generalized Korolyuk theorem, notable both for its own sake and for yielding a necessary and sufficient condition for a.s. orderliness of point processes with finite first moment measure. To describe it we need to define *intensity measure* (called parametric measure in Belyaev, 1969 and Leadbetter, 1972a).

Definition 6.4. The *intensity measure* $\lambda(A)$, $\mathbf{A}\ \varepsilon\ \mathcal{B}$, of a point process is the possibly infinite-valued measure induced by the set function defined on intervals I by

$$\lambda(I) = \sup_{\{I_j\}\varepsilon\mathcal{D}(I)} \sum_j \mathbb{P}\{N(I_j) > 0\} , \tag{6.2.1}$$

where $\mathcal{D}(I)$ denotes the family of all finite partitions of I into mutually disjoint sub-intervals I_j.

Theorem 6.5. (Generalized Korolyuk Theorem). For an a.s. orderly point process

$$\lambda(A) = M(A) \qquad (A \in \mathcal{B}) . \tag{6.2.2}$$

Conversely, when M(.) is a σ-finite measure and (6.2.2) holds, the process is a.s. orderly.

Belyaev (1969) proves the theorem in the case where λ(.) and M(.) are absolutely continuous with respect to Lebesgue measure. A general proof is given in Leadbetter (1972a) and is based on techniques in Leadbetter (1968) utilizing indicator random variables for any partition $\{I_j\} \in \mathcal{D}(I)$, monotone convergence, and Fatou's lemma.

In its original form in Section 11 of Khinchin (1955) Korolyuk's theorem asserts that for anal. orderly stationary point processes

$$\lambda = m \quad , \tag{6.2.3}$$

where the intensity $\lambda = \lim_{h\downarrow 0} \mathbb{P}\{N(0,h] > 0\}/h$ and $m = E\{N(0,1]\} = E\{N(0,h]\}/h$ as at (3.1.8). Here the existence of λ is assured by Khinchin's existence theorem (Khinchin, 1955, Sect. 7). Zitek (1957) showed that in this original form the converse holds (i.e., $\lambda = m$ implies anal. orderliness) when $m < \infty$, while if $m = \infty$ then the process may have $\lambda = m = \infty$ without being anal. orderly.

Equation (6.2.2) is not the only generalization in the literature of the original Korolyuk equation (6.2.3). Retaining stationarity rather than orderliness, it can be shown (e.g., by sub-additivity methods) that when $\lambda < \infty$, $\mathbb{P}\{N(0,h] \geq k\}/h$ tends to a limit λ_k as $h \downarrow 0$, and that

$$m = \sum_{k=1}^{\infty} \lambda_k \ . \tag{6.2.4}$$

The finiteness assumption on $\lambda = \lambda_1$ is fairly important, for it is shown (e.g., in Slivnyak, 1962, 1966) that without it, there need not be convergence of $\mathbb{P}\{N(0,h] \geq k\}/h$ as $h \downarrow 0$ for $k = 2,3, \ldots$.

Fieger (1965) relaxed both assumptions of orderliness and stationarity, showing that when the intensity measure $\lambda(.)$ is σ-finite there exist functions

$$p_k(x) = \lim_{h \downarrow 0} \mathbb{P}\{N(x - h, x] = k \mid N(x - h, x] \geq 1\} ,$$

defined almost everywhere with respect to $\lambda(.)$, and that when $M(.)$ is σ-finite,

$$M(A) = \int_A \sum_{k=1}^{\infty} k p_k(x) \; \lambda(dx) \; . \tag{6.2.5}$$

To us it seems simpler to introduce the restricted intensity measures based (as is $\lambda(.)$) on extending the interval functions

$$\lambda_k(I) = \sup_{\{I_j\} \varepsilon \mathcal{D}(I)} \sum_j \mathbb{P}\{0 < N(I_j) \leq k\} \; . \tag{6.2.6}$$

It is not difficult to establish (Daley, 1972b) that what may be termed the *generalized Korolyuk equation* then holds:

$$\begin{aligned} M(A) &= \lambda(A) + \sum_{k=1}^{\infty} \{\lambda(A) - \lambda_k(A)\} \\ &= \sum_{k=1}^{\infty} k\{\lambda_k(A) - \lambda_{k-1}(A)\} \qquad (A \,\varepsilon\, \mathcal{B}), \end{aligned} \tag{6.2.7}$$

where $\lambda_0(A) \equiv 0$ for $A \,\varepsilon\, \mathcal{B}$. It follows from the definitions that $\lambda_k(.)$ is absolutely continuous with respect to $\lambda(.)$, so (6.2.5) can be recovered from (6.2.7) via the Radon-Nikodym theorem, except that the identification of $p_k(.)$ as a limit of conditional probabilities is lacking.

6.3. Extensions and Generalizations

The main point that emerges in extending the preceding ideas to point processes on more general state spaces X is the basic importance of the topological nature of X (Belyaev, 1969). There is of course no difficulty in altering R to X in our Definition 3.3 of a.s. orderliness.

To generalize the other definitions we adapt Belyaev (1969) and Leadbetter's (1972a) work in assuming that in place of *I*, X is furnished with a sub-family *A* countably generating the σ-algebra *B*(X) on which N(.) is defined, *A* being such that for any A ε *B*(X) there exist families $\{A_i^n\} \subset A$, $i = 1,\ldots, k_n$; $n = 1,2,\ldots$, for which

(a) $A \supset \bigcup_i A_i^n \to A \qquad (n \to \infty)$;

(b) $A_i^n \cap A_j^n = \emptyset \qquad (1 \le i < j \le k_n)$;

(c) given x, y ε A, x ≠ y, there exists n such that for distinct i and j, $x \in A_i^n$, $y \in A_j^n$;

(d) for distinct i and j, $A_i^{n-1} \subset A_{i_1}^n \cup \cdots \cup A_{i_r}^n$ and $A_j^{n-1} \subset A_{j_1}^n \cup \cdots \cup A_{j_s}^n$, these two unions being mutually disjoint.

We call such $\{A_i^n\}$ a *nested dissecting system* for A.

The intensity measure of a point process on X with such dissecting systems can be defined as in Definition 6.4 with A ε *A* replacing I and D(A) the family of all dissecting systems contained in A as at (a) above. Provided A is such that $\mathbb{P}\{N(A) < \infty\} = 1$, the generalized Korolyuk Theorem 6.5 can then be proved as in Leadbetter (1972a). Similarly, restricted intensity measures and ordinariness as at (6.2.6) and (6.1.11) can be defined more generally, with the generalized Korolyuk equation (6.2.7) remaining true, and ordinariness a sufficient condition for a.s. orderliness, necessary as well when λ(.) is σ-finite.

To generalize Khinchin-and analytic orderliness is not as simple. What amounts to *uniform* Khinchin orderliness is defined in Belyaev (1969) and Leadbetter (1971a), who show that with λ(.) σ-finite it is a sufficient condition for a.s. orderliness. Analytic orderliness incorporates two notions not present in the idea of dissecting systems, namely a measure on (elementary) sets in X, and the choice of an interior point in such a set.

Unlike much of the material in previous chapters, orderliness is a concept peculiar to point processes and not related in general to random measures. However the idea of generalized intensity measures carries over to (non-negative) random measures on R, being defined from

$$\lambda(I,y) = \sup_{\{I_j\}\varepsilon\mathcal{D}(I)} \sum_j \mathbb{P}\{0 < \Lambda(I_j) \le y\} . \tag{6.3.1}$$

Then (cf. (6.2.7))

$$M(A) = E\{\Lambda(A)\} = \int_0^\infty y \, d_y\lambda(A,y) . \tag{6.3.2}$$

7. PALM-KHINCHIN THEORY

7.1. General Comments

In the introductory section we spoke of interval properties as being peculiar to point processes as distinct from random measures, and this section is devoted to problems in this area. The most important problem concerns stationarity, as indicated around Theorem 2.8.

Intuitively it seems reasonable that the stationarity property that

$$\mathbb{P}\{N(A_i + y) = k_i, \ i = 1,\ldots,r\} \text{ is invariant over } y \ \varepsilon \ R, \tag{7.1.1}$$

should imply some equivalent stationarity property for distances

$$\{Y_n\} \equiv \{x_{j+n+1} - x_{j+n}\} \qquad (n \ \varepsilon \ Z) \tag{7.1.2}$$

between consecutive points starting from a "typical" point x_j, assuming of course, that *some typical point can be chosen appropriately*. Mathematically this proviso is crucial, and there have been various attempts to meet it. Equivalently, we ask whether there exists (and if so, we seek to specify) a one-one relation between $(\Omega, F, \mathbb{P})$ when $\mathbb{P}$ is stationary as at (7.1.1) and (H_0, Y_0, Q) when Q is a stationary probability measure on (H_0, Y_0) defined as follows: H_0 is the family of doubly infinite sequences $\{Y_n\}$ of non-negative finite real numbers with $\Sigma_{n=0}^{\infty} Y_n = \Sigma_{n=-\infty}^{0} Y_n = \infty$, and Y_0 is the minimal σ-algebra with respect to which the subsets $\{Y_k \leq y_i, i = 0, \pm 1, \ldots, \pm r\}$ are measurable for all $y_i \geq 0$; (H_0, Y_0, κ) is stationary when

$$\kappa\{Y_{i_1+n} \leq y_1, \ldots, Y_{i_r+n} \leq y_r\} \tag{7.1.3}$$

is invariant over $n \in Z$.

The answer to this existence question is in the affirmative provided that either we restrict attention to point processes with finite first moment (i.e., $m < \infty$; we call this the *finite case* below), or else we allow possibly σ-finite measures K on Y_0 with the invariance property (7.1.3). In this generality (Theorem 7.7 below) the solution is due to Slivnyak (1962, 1966).

Kaplan (1955) appears to be the first to have recognized the problem and in his Theorems 2.1 and 2.2 he claims to have solved it in the finite case, though so far we have not been able to follow his arguments satisfactorily. Rigorous arguments have been given by Ryll-Nardzewski (1961) for the finite case with the added assumption of a.s. orderliness; by Matthes (1963a) in the finite case, by Neveu (1968) assuming a.s. orderliness; and by Papangelou (1972) in the finite case.

7.2. Existence of Palm Functions and Measures

In all work subsequent to Kaplan's, reference is made to Khinchin's (1955) monograph where earlier work of Palm (1943) is extended in showing the existence of what are now known as *Palm functions* (throughout this section, except where stated otherwise, $\mathbb{P}$ is stationary)

$$q_k(y) = \lim_{x \downarrow 0} \mathbb{P}\{N(0,y] = k \mid N(-x,0] > 0\} \qquad (y > 0,\ k \in Z_+) \tag{7.2.1}$$

when the *intensity*

$$\lambda = \lim_{x\downarrow 0} \mathbb{P}\{N(0,x] > 0\}/x \tag{7.2.2}$$

is finite. Both these limits are shown to exist by observing that

$$\phi(x) \equiv \mathbb{P}\{N(0,x] > 0\} \quad \text{and} \quad \psi_k(y,x) \equiv \mathbb{P}\{N(0,y] \le k,\ N(-x,0] > 0\}$$

are sub-additive in x and applying Lemma 7.1.

Lemma 7.1. (Sub-additive function lemma). If $0 = f(0) \le f(x)$ for $x > 0$, $f(x + y) \le f(x) + f(y)$ and $f(x) \to 0,\ x \downarrow 0$, then $\lim_{x\downarrow 0} f(x)/x$ exists and equals $\sup_{x>0}\{f(x)/x\}$, finite or infinite.

In fact, the functions

$$\phi_j(x) \equiv \mathbb{P}\{0 < N(0,x] \le j\}$$

and

$$\psi_{k|j}(y,x) \equiv \mathbb{P}\{N(0,y] \le k,\ 0 < N(-x,0] \le j\}$$

are sub-additive in x, so for $\lambda < \infty$ it follows by taking differences that we can set

$$\pi_j = \lim_{x\downarrow 0} \frac{\mathbb{P}\{N(0,x] = j\}}{\lambda x} = \lim_{x\downarrow 0} \mathbb{P}\{N(0,x] = j \mid N(0,x] > 0\}\ , \tag{7.2.3}$$

and use the supremum expression of Lemma 5.1 in

$$\lambda = \sup_{x>0}\sup_{j>0} \frac{\phi_j(x)}{x} = \sup_{j>0}\sup_{x>0} \frac{\phi_j(x)}{x} = \sup_{j>0} \lambda(\pi_1 + \cdots + \pi_j), \tag{7.2.4}$$

to deduce that $\{\pi_j\}$, $j = 1,2,\ldots$, is a probability distribution. The process is orderly provided $\pi_1 = 1$, while more generally,

$$m = \lambda \sum_{j=1}^{\infty} j\pi_j\ , \tag{7.2.5}$$

finite or infinite. Further, still from Lemma 7.1, when $\lambda < \infty$,

$$Q_{k|j}(y) = \lim_{x\downarrow 0} \mathbb{P}\{N(0,y] \leq k | N(-x,0] \equiv j\} \qquad (7.2.6)$$

exists for j such that $\pi_j > 0$, while defining $\pi_j Q_{k|j}(.) = 0$ if $\pi_j = 0$, the supremum expression yields

$$Q_k(y) \equiv \lim_{x\downarrow 0} \mathbb{P}\{N(0,y] \leq k | N(-x,0] > 0\} = \sum_{j=1}^{\infty} \pi_j Q_{k|j}(y) . \qquad (7.2.7)$$

Fieger (1964) deduced these results using different methods. More generally, the arguments just sketched prove the following result.

Theorem 7.2. In a stationary point process with finite intensity, for any event B of the form

$$\{N(0,y_i] \leq k_i,\ i = 1,\ldots, r\} \qquad (7.2.8)$$

for finite positive integers r, $y_i \geq 0$, $k_i \in Z_+$, the limits below exist;

$$Q(B) = \lim_{x\downarrow 0} \mathbb{P}\{B | N(-x,0] \geq 0\} , \qquad (7.2.9)$$

$$Q(B|j) = \lim_{x\downarrow 0} \mathbb{P}\{B | N(-x,0] = j\} \qquad (j \text{ such that } \pi_j > 0) . \qquad (7.2.10)$$

Obviously the theorem remains true on replacing B by any element in the ring generated by the family of events as at (7.2.8). However these simple arguments do not show us that Q(.) is a measure on this ring, which assertion can be shown in the orderly case much as in Leadbetter (1972a) using a construction of Matthes (1963). Moreover, the interpretation of Q(B) as a version of the conditional probability

$$\mathbb{P}\{B | N(\{0\}) > 0\} \qquad (7.2.11)$$

is thereby justified, so we give it below. Note that a conditional probability measure like (7.2.11) is called a *Palm measure*.

Associate with each event $B \in F$ a new point process whose sample realizations $\omega_B^* \equiv \omega_B^*(\omega)$ are determined from the realizations ω of a given finite stationary orderly point process by taking each point of ω as origin and retaining it as a point of ω_B^* only when the translated set of points satisfies the conditions defining the event B. In other words, if $\omega^x = \{y:y - x \in \omega\}$, then

$$\omega_B^*(\omega) = \{x:x \in \omega;\ \omega^{-x} \in B\} . \tag{7.2.12}$$

Observe that

$$\omega_{\Omega_0}^*(\omega) = \omega \text{ for } \Omega_0 \equiv \{\omega:N(\{0\}) > 0\} . \tag{7.2.13}$$

Also, irrespective of the given process being orderly or not, the new process is necessarily orderly.

Define

$$\lambda_B = E\{N_B^*(0,1]\} , \tag{7.2.14}$$

where for any $A \in \mathcal{B}$, $N_B^*(A) = \text{card}\{\omega_B^* \cap A\}$. Then by Korolyuk's theorem, $\lambda = \lambda_\Omega$, so λ_B is finite, and

(1°) $N_B^*(A) \le N^*(A)$ $(A \in \mathcal{B})$;

(2°) for disjoint $B_i \in \mathcal{F}$, with union B, $\Sigma_i N_{B_i}^*(A) = N_B^*(A)$ $(A \in \mathcal{B})$;

(3°) if $B_n \downarrow \emptyset$, then $\omega_{B_n}^* \downarrow \emptyset$, $N_{B_n}^*(A) \downarrow 0$ for bounded $A \in \mathcal{B}$, and $\lambda_{B_n} \downarrow 0$.

Theorem 7.3. The set function $\lambda_\cdot$ is a measure on $\mathcal{F}$, and

$$\mathbb{P}_0(B) \equiv \lambda_B/\lambda \tag{7.2.15}$$

defines a probability measure on $(\Omega,\mathcal{F})$.

By (7.2.13) $\mathbb{P}_0$ has the property that

$$\mathbb{P}_0\{\Omega_0\} = 1, \tag{7.2.16}$$

so conditional probabilities like (7.2.11) are meaningful for $\mathbb{P}_0$. However, as an example in Leadbetter (1972a) shows, it is only for certain $B \varepsilon F$ that

$$\lim_{x \downarrow 0} \mathbb{P}\{B | N(-x,0] > 0\} = \mathbb{P}_0(B) \ . \tag{7.2.17}$$

Specifically, call the event B right-continuous when for each $\omega \varepsilon \Omega$,
$\chi_B(\omega^{-x}) \to \chi_B(\omega^{-y}) \qquad (x \downarrow y).$

Theorem 7.4. For a stationary orderly point process with finite parameter (7.2.17) is true for right-continuous events B.

The events at (7.2.8) and the ring generated therefrom are right-continuous, and Palm functions exist for events in this ring. Then, by Theorems 7.3 and 7.4, $Q(.)$ is a measure on the ring with $Q = \mathbb{P}_0$, and therefore it extends to a measure on the generated σ-ring with $Q = \mathbb{P}_0$.

In this way we can obtain probabilistic interpretations for $Q_k(.)$ such as "$1 - Q_k(x)$ is the distribution function for the time between an arbitrary point and the kth subsequent point". Leadbetter (1966) has given another justification of this interpretation based on analytical arguments and the Palm-Khinchin equations to which we now turn.

7.3. Palm-Khinchin Equations

Khinchin (1955) proves the following result.

Theorem 7.5. For a stationary, orderly point process with finite intensity λ

$$P_k(x) \equiv \mathbb{P}\{N(0,x] \leq k\} = 1 - \lambda \int_0^x q_k(u)\, du$$

$$= \lambda \int_x^\infty q_k(u)\, du \qquad (x > 0,\ k \varepsilon Z_+) \ , \tag{7.3.1}$$

or in derivative form,

$$D_+ P_k(x) = -\lambda q_k(x) \ , \tag{7.3.2}$$

where D_+ denotes the right-hand derivative operator.

Equation (7.3.1) shows *inter alia* that as a function of x, $1 - P_k(x)$ is absolutely continuous with Radon-Nikodym derivative $q_k(x)$. A neat argument showing this absolute continuity without identifying the derivative rests on an observation of Beutler and Leneman (1966) that

$$S_n(x) = \sum_{k=1}^{n} \mathbb{P}\{N(0,x] \geq k\} = E[\min\{n,N(0,x]\}] \tag{7.3.3}$$

is a concave function of x in $(0,\infty)$. To us a more appealing derivation of the Palm-Khinchin equations like (7.3.1) and (7.3.2) remains that of Khinchin, utilizing stationarity, orderliness, finiteness of λ, and the existence of the Palm functions (7.2.1). Generalizations of (7.3.1) lead ultimately to the sought-for (see Section 7.1) one-one relation between $(\Omega,F,\mathbb{P})$ and (H_0,Y_0,κ). In another direction Fieger (1965) has generalized (7.3.1) to the case of a not necessarily orderly, non-stationary process with finite intensity measure; the simpler non-stationary Khinchin orderly case is given in Fieger (1964) where it is shown for example that

$$\mathbb{P}\{N(x,y] \leq k\} = 1 - \int_x^y q_k(u,y]\ \lambda(du) , \tag{7.3.4}$$

where $q_k(u,y] = \lim_{v\downarrow 0} \mathbb{P}\{N(u,y] = k|N(u - v, u] > 0\}$ and $\lambda(.)$ is the intensity measure of Section 6. It is clear that (7.3.4) reduces to (7.3.1) in the stationary case.

7.4. One-one Correspondences of Stationary Measures

In the case of a stationary, orderly point process with finite intensity, the Palm-Khinchin equations provide a convenient algorithm for computing the $\mathbb{P}_0$-measure of certain basic sets. For example, writing $B_k(y) = \{\omega:N(0,y] \leq k\}$ and $E_k(y) = \{\omega:0 \in \omega,\ N(0,y] = k\}$, (7.3.2) implies that $\mathbb{P}_0\{E_k(y)\} = -\lambda^{-1}D_+\mathbb{P}\{B_k(y)\}$. The question then arises as to whether this relationship can be extended to provide a one-one mapping between measures $\mathbb{P}_0$ on Ω_0 and $\mathbb{P}$ on Ω. Since there is a natural correspondence between Ω_0 and H_0 got by putting $x_0 = 0$ and $Y_n = x_n - x_{n-1}$, this leads to the question concerning stationary measures on Ω and H_0 foreshadowed in Section 7.1.

There are at least four approaches to these questions discernible in the literature. They can be described very briefly as follows.

1°. Slivnyak (1962, 1966) shows that with B as at (7.2.8) and

$$S_x B = \{N(0,x] = 0,\ N(0,y_i + x] \le k_i,\ i = 1,\dots, r\}\ , \qquad (7.4.1)$$

$\mathbb{P}(S_x B)$ is not only sub-additive but also convex in $x > 0$, and hence defines a set-function ρ by

$$\rho(B) = -D_+ \mathbb{P}(S_x B)\Big|_{x=0} \ . \qquad (7.4.2)$$

Extension arguments are then invoked to show that ρ is in fact a measure on the σ-algebra generated by such B.

2°. Ryll-Nardzewski (1961) and Papangelou (1970, 1972) use the Radon-Nikodym theorem to demonstrate the existence of Palm measures (for this they require a finite first moment measure of the point process but not stationarity), and then the Fubini theorem, exploited to the full by Papangelou. A similar approach was used by Mecke (1967) to describe Palm measures for processes on a locally compact Abelian group.

3°. Marked point processes have been used by Matthes and co-workers (e.g., Matthes, 1963a, Kerstan and Matthes, 1964, Section 5); cf. also Theorem 7.3 above.

4°. Neveu (1968) sketches a method based on measure isomorphisms of quotient spaces.

In 2° and 3° the finiteness of the first moment measure is basic to the argument. Almost sure orderliness is assumed in 2° and 4°. It is essential to Ryll-Nardzewski's later development, while Papangelou believes that with certain additional arguments his conclusions can be justified without the assumption. It is irrelevant to Neveu's method.

To state Slivnyak's result we need to define a measurable mapping $L: H_0 \to \Omega_\infty$ where $\Omega_\infty = \{\omega \in \Omega: N(-\infty, 0] = N(0,\infty) = \infty\}$ with measurable inverse $L^{-1}: \Omega_\infty \to H_0^+$ where $H_0^+ = H \cap \{Y_0 > 0\}$. With $Y = \{Y_n\}$ these are given by

$$LY = \{ \sum_{i=0}^{n} Y_i, - \sum_{i=-n}^{-1} Y_i \} \qquad (n \in Z), \tag{7.4.3}$$

$$\begin{aligned} Y_0 + \cdots + Y_n &= \sup\{y > 0 : N(0,y) \leq n\} \quad (n=0,1,\ldots), \\ -(Y_{-1} + \cdots + Y_{-n}) &= \inf\{y \leq 0 : N(y,0] < -n\} \quad (n=-1,-2,\ldots). \end{aligned} \tag{7.4.4}$$

Let us emphasize that

$$(L^{-1}\omega)_0 \equiv Y_0(\omega) = \inf\{x > 0 : x \in \omega\},$$

and that in the definition of H_0, unlike H_0^+ and H (see (2.2.3)), no accommodation is made for relating the intervals to the interval amongst $\{x_n\} = \omega$ containing the origin.

Theorem 7.6. Between the families of stationary (7.1.1) probability measures $(\Omega, F, \mathbb{P})$ and of stationary (7.1.3) measures (H_0, Y_0, κ) satisfying

$$\int_0^\infty Y_0 \, \kappa(dY_0) = 1 \tag{7.4.5}$$

there exists the one-one relation $\kappa = D\mathbb{P}$ as follows: for events B' in the ring R generated by events $S_x B$ at (7.4.1),

$$\kappa(L^{-1}B') = -D_+ \mathbb{P}(S_x B')\Big|_{x=0}, \tag{7.4.6}$$

while for $A \in Y_0$,

$$\mathbb{P}(LA) = \int_{H_0} \kappa(dY) \int_0^{Y_0} \chi_{LA}\{(LY)^{-u}\} \, du. \tag{7.4.7}$$

Further, if $\kappa = D\mathbb{P}$ then

$$\mathbb{P}\{ \lim_{x \to \infty} x^{-1} N(0,x] = 0\} = 0 \tag{7.4.8}$$

if and only if

$$\kappa\{\lim_{n\to\infty} n^{-1}(Y_1 + \cdots + Y_n) = 0\} = 0 . \tag{7.4.9}$$

For a stationary orderly point process with finite intensity λ, (7.4.6) is essentially a restatement of Theorem 7.4 with $\kappa = \lambda\mathbb{P}_0$. Note (cf. (7.2.2)) that $\lambda = \kappa(L^{-1}\Omega_\infty)$, so $Q(.) = \kappa(.)/\lambda$ is a probability measure for orderly processes with $\lambda < \infty$. (In general, $\lambda = \kappa\{Y_0 > 0\}$ while $m = \kappa(H_0)$.) In the case $\lambda < \infty$ Slivnyak also shows that the expression (7.4.7) may be replaced by the intuitively more appealing form

$$\mathbb{P}(B) = \lim_{T\to\infty} \frac{1}{T}\int_0^T Q(B^{-u})\, du . \tag{7.4.10}$$

This equation has the interpretation that $Q(.)$ is the probability measure of an interval stationary point process with a point at the origin. It may be regarded as describing the transient behavior of a stationary point process with probability measure $\mathbb{P}(.)$. As (7.4.10) shows, $\mathbb{P}(.)$ is the Cesaro mean got by translating the origin through an arbitrarily large distance. $\mathbb{P}(B)$ is of course equal to $\lim_{T\to\infty} Q(B^{-T})$ whenever this limit happens to exist, in which case we can approximate (part of) a sample realization of a stationary point process with probability measure $\mathbb{P}(.)$ by using $Q(.)$ to construct the intervals between successive points, and then locate the first of these at a new origin distributed uniformly on $(-T,0)$ for some sufficiently large T.

Slivnyak also claims that a metrically transitive stationary point process (see Section 8.1) necessarily has finite intensity, but by considering a cluster process with Poisson parent process and Neyman-Scott clusters of infinite mean size and contained in a bounded interval, Matthes (private communication) has shown this to be incorrect.

The argument based on the Radon-Nikodym theorem is sufficiently important to warrant a brief summary, although it leads to results which are essentially a restatement of Theorem 7.6. (Much of Ryll-Nardzewski's (1961) argument is devoted to showing how the simple argument, sketched in Breiman (1968, Section 6.10), for a discrete time stationary point process carries over to the continuous time process.)

The argument starts with the observation that with

$$M(A) = \int_\Omega N(A,\omega)\ \mathbb{P}\{d\omega\} \qquad \text{(bounded } A \in \mathcal{B}) , \tag{7.4.11}$$

the set function with argument A ($B \in \mathcal{F}$ being fixed)

$$\int_\Omega \chi_B(\omega)\ N(A,\omega)\ \mathbb{P}\{d\omega\} \tag{7.4.12}$$

is absolutely continuous with respect to the measure M(.), assumed finite on bounded A. Thus it equals

$$\int_A P_u(B)\ M(du)$$

for some functions (Radon-Nikodym derivatives) $P_{\cdot}(B)$. The stationarity of $\mathbb{P}(.)$ implies that $M(du) = m\ du$ for some finite $m > 0$, and that for all $B \in \mathcal{F}$ and bounded $A \in \mathcal{B}$

$$\int_A P_u(B)\ M(du) = m \int_A P_u(B)\ du = m \int_A P_{u+t}(B^t)\ du , \tag{7.4.13}$$

where $B^t = \{\omega : \omega^{-t} \in B\}$. An argument based on the Fubini theorem then shows that we can set

$$P_u(B) = P_0(B^{-u}) \qquad \text{(all } u \in R,\ B \in \mathcal{F}),$$

and take $P_0(.)$ to be a regular conditional probability measure identifiable with the Palm measure $\mathbb{P}_0$ of Theorem 7.3.

7.5. Some Complementary Results

If a stationary point process is to be interval stationary then Q(.) must exist as a probability measure and hence $\kappa(.)$ must be totally finite. Since $\kappa(H_0) = E\{N(0,1]\} = m$, finite or infinite, we must have both $\lambda < \infty$ and $m < \infty$. Thus if $(\Omega,\mathcal{F},\mathbb{P})$ is both stationary and interval stationary, then (cf. (2.2.4))

$$H(u,v) \equiv \mathbb{P}(\xi_0 \le u,\ \xi_{-1} \le v) = \mathbb{P}\{N(0,u] > 0,\ N(-v,0] > 0\}$$

$$= \mathbb{P}\{N(0,u] > 0\} - \mathbb{P}\{N(0,v] = 0\} + \mathbb{P}\{N(0,\ u+v] = 0\}$$

$$= \lambda \int_0^u \{q_0(y) - q_0(y+v)\}$$

and

$$\mathbb{P}(\eta_0 \le x) = \mathbb{P}(\xi_0 + \xi_{-1} \le x) = \int_{u=0}^{x} d_u H(u,v)\Big|_{v=x-u}$$

$$= \lambda \int_0^x \{q_0(u) - q_0(x)\}\, du ,$$

which gives equation (2.2.5) as asserted in Section 2.

In Section 3.1 we deferred discussing moments of intervals. Their relation with the moments of count properties now follow from Palm-Khinchin equations, but are mostly not expressible in terms of the moment measures $M_r(.)$ (see e.g., Lewis, 1961 and McFadden, 1962 for early work, and Cox and Lewis, 1966, Chapter 4). An exception is that (for a stationary point process)

$$E(Y_n) = E(Y_0) = \int_0^\infty q_0(x)\, dx = \lambda^{-1} .$$

The analogue of the renewal function for interval stationary processes is the function

$$H(x) \equiv E\{N(0,x] \mid N(\{0\}) > 0\},$$

equal in the orderly case to $\Sigma_{k=1}^{\infty} kq_k(x) = \Sigma_{k=0}^{\infty}\{1 - Q_k(x)\}$, and related to the reduced covariance measure $C(.)$ of the corresponding stationary process $(\Omega,F,\mathbb{P})$ by $C(-y,y) = \lambda\{1 + 2H(y) - 2\lambda y\}$ for all continuity points y. Other properties of $H(.)$ are reviewed in Daley (1971,1972a), while moment analogues of Theorem 7.4 as well as some inequalities between $\mathbb{P}(.)$ a $Q(.)$ are given in Cramer, Leadbetter, and Serfling (1971).

8. LIMIT THEOREMS AND INFINITELY DIVISIBLE POINT PROCESSES

A survey of any field of stochastic processes would be incomplete without a discussion of asymptotic and limit properties. A far more comprehensive review in this area is given (see Matthes, 1972, this volume, for a summary) by Matthes whose own work is a most important contribution to the theory, so for this reason, amongst others, our account here is of necessity brief. Nevertheless we hope to indicate the main directions of such theory and give some unity of purpose to several aspects developed earlier in the paper.

8.1. Ergodicity and Mixing Conditions

Let N(.) be a stationary point process with finite first moment. Then by applying the usual discrete parameter version of the ergodic theorem (e.g., Breiman, 1968, p. 113) to the sequence of random variables $X_n = N(n-1,n]$, and noting that the difference $|N(0,T]/T - N(0,[T])/[T]|$ $\to 0$ a.s. $(T \to \infty)$ because $E\{N(0,1]\} < \infty$, we obtain the following result.

Theorem 8.1. For a stationary point process N(.) with finite mean density $m = E\{N(0,1]\}$ the limit

$$U = \lim_{T\to\infty} N(0,T]/T \tag{8.1.1}$$

exists a.s., defining a random variable with $E(U) = m$ and characterized as the conditional expectation of N(0,1] with respect to the σ-field of invariant events.

As usual the invariant events here are those $B \in \mathcal{S}$ for which $B = T_y B$ (all $y \in R$) where the shift operator T_y operates on each $N \in B$ according to $T_y N(A) = N(A + y)$ for $A \in \mathcal{B}$.

Definition 8.2. A point process is *ergodic* (or, *metrically transitive*) when the only invariant events have probability 0 or 1.

Theorem 8.3. For a metrically transitive process satisfying Theorem 8.1

$$\mathbb{P}\{N(0,T]/T \to m(T \to \infty)\} = 1 . \tag{8.1.2}$$

In practice, other sufficient conditions for ergodicity are often easier to verify, in particular, various mixing conditions which can be viewed as expressing asymptotic independence.

Definition 8.4. A process is *mixing* if

$$\mathbb{P}(A \cap T_y B) \to \mathbb{P}(A)\ \mathbb{P}(B) \qquad (y \to \infty) \qquad (\text{all } A, B \in \mathcal{S}), \tag{8.1.3}$$

and *weakly mixing* if

$$T^{-1} \int_0^T \left| \mathbb{P}(A \cap T_y B) - \mathbb{P}(A)\ \mathbb{P}(B) \right| dy \to 0 \qquad (T \to \infty) \qquad (\text{all } A, B \in \mathcal{S}). \tag{8.1.4}$$

A third condition like (8.1.3) and (8.1.4) is that

$$T^{-1} \int_0^T \mathbb{P}(A \cap T_y B)\ dy \to \mathbb{P}(A)\ \mathbb{P}(B) \qquad (T \to \infty) \qquad (\text{all } A, B \in \mathcal{S}), \tag{8.1.5}$$

but from Rosenblatt (1962, p. 110) it follows fairly easily that (8.1.5) holds if and only if the point process is ergodic. Of course (8.1.3) implies (8.1.4) implies (8.1.5). Westcott (1972) characterizes (8.1.5) in terms of p.g.fls as follows, where for $\xi \in \mathcal{U}$ we write $(S_u \xi)(t) = \xi(t + u)$.

Theorem 8.5. A point process with p.g.fl. G[.] is ergodic if and only if

$$T^{-1} \int_0^T G[\xi_1 S_u \xi_2]\ du \to G[\xi_1]\ G[\xi_2] \qquad (T \to \infty) \qquad (\text{all } \xi_1, \xi_2 \in V). \tag{8.1.6}$$

Similar characterizations for mixing and weak mixing can be given. These results can be used to show for example that a doubly stochastic Poisson process with stationary random parameter measure $\Lambda(.)$ is mixing if and only if $\Lambda(.)$ is mixing (Westcott, 1972), and that a stationary cluster process, assumed to exist, is mixing if and only if the parent process is mixing (Westcott, 1971). In particular then all stationary cluster processes with stationary renewal parent processes (which includes the Poisson case) are mixing.

With stronger assumptions stronger results, like a central limit theorem, can be proved. For example, let F_t^- be the sub-σ-algebra of F generated by the sets

$$\{N(A) = k\} \qquad (k \in Z_+,\ A \in B \cap (-\infty,t]) ,$$

and define F_t^+ analogously for $A \in B \cap (t,\infty)$. Then (cf. Billingsley, 1968, Chap. 4) define $(\Omega,F,\mathbb{P})$ to be ψ-mixing for a given function $\psi(.)$ if

$$|\mathbb{P}(A \cap B) - \mathbb{P}(A)\ \mathbb{P}(B)| \le \psi(u)\ \mathbb{P}(A) \qquad (8.1.7)$$

whenever $A \in F_t^-$, $B \in F_{t+u}^+$, $t \in R$, $u > 0$. Note that for A such that $\mathbb{P}(A) > 0$, (8.1.7) is equivalent to $|\mathbb{P}(B|A) - \mathbb{P}(B)| \le \psi(u)$. The following theorem can be proved in a manner analogous to that of Theorem 20.1 of Billingsley (1968) where further details may be found.

Theorem 8.6. Let the stationary point process N(.) be ψ-mixing where $\psi(t)$, $t \in R_+$, is monotonic, non-negative, and $\int_0^\infty \{\psi(t)\}^{1/2}\,dt < \infty$, and let N(.) have finite first and second moment measures with totally finite reduced covariance measure C(.) having $\sigma^2 = \int_R C(du) > 0$. Then with $m = E\{N(0,1]\}$ and

$$Y_n(t) \equiv \{N(0,nt] - mnt\}/(\sigma n^{1/2}) \qquad (0 \le t \le 1) , \qquad (8.1.8)$$

Y_n converges weakly in the Skorokhod topology to the Wiener process on (0,1).

Note that in discussing the asymptotic behavior of cluster processes we are interested in expressing the conditions for a given type of limit behavior in terms of the component processes (see e.g., Lewis, 1969, Daley, 1972a). Such formulations as have been given are rather different from Theorem 8.6, concluding simply convergence of $Y_n(1)$ as $n \to \infty$.

There are two other problems concerning asymptotic behavior. One is to give conditions sufficient for the factorial cumulants of the process to be asymptotically linear, analogous to the discussion in Chap. 4 of Cox and Lewis (1966). The other is to give conditions that a process N(.) evolving in time (e.g., a renewal process starting at 0) should have some limit behavior in the sense of convergence in distribution of the processes N(.+t), ($t \to \infty$). (The discussion around (7.4.10) is a special case.)

8.2. Infinitely Divisible Point Processes

Definition 8.7. A point process (or random measure or random signed measure) is *infinitely divisible* if for every positive integer n it can be represented as the superposition of independent identically distributed components.

In terms of transforms, and using the p.g.fl. as the most convenient example, this means that if the given point process has p.g.fl. G[.], there exists for every n, a p.g.fl. $G_n[.]$ such that

$$G[\xi] = (G_n[\xi])^n . \tag{8.2.1}$$

The most important problem concerning such processes is to find the analogue of the Lévy-Khinchin representation of the characteristic function of real-valued infinitely divisible random variables. This can be done in various ways: by adapting the general representation theory of infinitely divisible distributions on a locally compact second countable Abelian group as in Parthasarathy (1967), or by using the theory of infinitely divisible stochastic processes as in Lee (1967, 1968), or by extension from the p.g.f. of infinitely divisible Z_+^m-valued random variables (e.g., in Feller, 1957; Milne, 1971, gives this approach), or by using marked point processes as in Kerstan and Matthes (1964). In any case we arrive at the following result, first given (for the stationary case) by Lee (1967) and Kerstan and Matthes (1964); the setting here, which seems the natural one, is in terms of the p.g.fl.

Theorem 8.8. The p.g.fl. of an infinitely divisible point process can be uniquely represented in the form

$$G[\xi] = \exp[\int_{N} [\exp\{\log \int_{R} \xi(t)\, N(dt) - 1\}]\, \tilde{\mathbb{P}}(dN)] , \tag{8.2.2}$$

where the not necessarily totally finite measure $\tilde{\mathbb{P}}$ on (N,S) satisfies

(i) for every finite interval I, $\tilde{\mathbb{P}}\{N:N(I) > 0\} < \infty$, (8.2.3)

(ii) $\tilde{\mathbb{P}}\{N(R) = 0\} = 0.$ (8.2.4)

Conversely, for every $\tilde{\mathbb{P}}$ satisfying (8.2.3) and (8.2.4), (8.2.2) is the p.g.fl. of an infinitely divisible point process.

It is useful to name $\tilde{\mathbb{P}}$: we follow Westcott (1972) in calling it the KLM (= Kerstan-Lee-Matthes) measure.

Definition 8.9. An infinitely divisible point process is *regular* if $\tilde{\mathbb{P}}\{N(R) = \infty\} = 0$, and it is *singular* if $\tilde{\mathbb{P}}\{N(R) < \infty\} = 0$.

Various properties of infinitely divisible processes $(\Omega,F,\mathbb{P})$ are expressed simply when the KLM measure is used:

1°. $\mathbb{P}$ is stationary if and only if $\tilde{\mathbb{P}}$ is stationary;

2°. $\mathbb{P}$ is mixing if and only if for every pair of finite intervals I_1, I_2,

$$\tilde{\mathbb{P}}\{N:N(I_1) > 0,\ N(I_2 + t) > 0\} \to 0 \qquad (t \to \infty) ,$$

and hence all regular infinitely divisible processes are mixing;

3°. The superposition of $(\Omega,F,\mathbb{P}_1)$ and $(\Omega,F,\mathbb{P}_2)$ has KLM measure $\tilde{\mathbb{P}}_1 + \tilde{\mathbb{P}}_2$;

4°. $\tilde{\mathbb{P}}$ is expressible in the form $\tilde{\mathbb{P}} = \tilde{\mathbb{P}}_1 + \tilde{\mathbb{P}}_2$ where $\tilde{\mathbb{P}}_1$ and $\tilde{\mathbb{P}}_2$ are the KLM measures of mutually independent singular and regular processes respectively;

5°. The class of regular $(\Omega, F, \mathbb{P})$ coincides with the class of cluster processes with Poisson parent process and a.s. finite clusters.

For these and further details we refer to the original papers by Matthes (and with co-authors) and Lee, and to Westcott (1972), Matthes (1972) and the forthcoming monograph by Kerstan, Matthes and Mecke (1972).

8.3. Some Convergence Theorems

It is convenient in this section to abbreviate "the point process $(\Omega, F, \mathbb{P})$" to $\mathbb{P}$. Our subject is then convergence in distribution $\mathbb{P}_n \xrightarrow{d} \mathbb{P}$ (see around Definition 3.12).

The family of infinitely divisible $\mathbb{P}$ is a natural family of limit point processes, and we start with the fairly trivial result that $\mathbb{P}_n \xrightarrow{d} \mathbb{P}$ and $\mathbb{P}_n$ infinitely divisible implies the infinite divisibility of $\mathbb{P}$.

Matthes and co-workers, in particular in Kerstan and Matthes (1964, 1965), have given results relating the KLM measures $\tilde{\mathbb{P}}_n$ of a sequence of infinitely divisible $\mathbb{P}_n$. Thus, supposing each $\mathbb{P}_n$ is stationary (an assumption that can be dispensed with), $\mathbb{P}_n \xrightarrow{d} \mathbb{P}$ if and only if $\tilde{\mathbb{P}}_n \xrightarrow{d} \tilde{\mathbb{P}}$. If each $\mathbb{P}_n$ is stationary and orderly with finite intensities $\lambda_n \to \lambda < \infty$, then $\mathbb{P}_n \xrightarrow{d} \mathbb{P}$ is equivalent to $(\tilde{\mathbb{P}}_n)_0 \xrightarrow{d} (\tilde{\mathbb{P}})_0$ where for any stationary measure triple (Ω, F, Q), the Palm measure Q_0 is defined as in Section 7.3.

Motivation for discussing infinitely divisible distributions in classical probability theory comes from studying sums of a large number of elements all small in the appropriate sense (cf. Feller, 1966).

Definition 8.10. A triangular array $\{\mathbb{P}_{ni}\}$, $i = 1, \ldots, n$; $n = 1, 2, \ldots$, of point processes is *uniformly asymptotically negligible*, abbreviated to *uan*, when

$$\lim_{n \to \infty} \sup_{1 \le i \le n} \mathbb{P}_{ni}\{N(A) > 0\} = 0 \qquad \text{(all bounded } A \in B). \quad (8.3.1)$$

Letting

$$\mathbb{P}_n = \mathbb{P}_{n1} * \cdots * \mathbb{P}_{nn} , \qquad (8.3.2)$$

or in terms of sample realizations N_{ni} of $\mathbb{P}_{ni}$, $N_n = N_{n1} + \cdots + N_{nn}$, the general theory of infinitely divisible stochastic processes gives the following theorem.

Theorem 8.11. If $\mathbb{P}_n \xrightarrow{d} \mathbb{P}$ and $\{\mathbb{P}_{ni}\}$ is a uan family, then $\mathbb{P}$ is infinitely divisible.

Historically, Palm (1943) studied such superpositions and concluded correctly (from incomplete arguments) that the Poisson process can arise from such an operation. Khinchin (1955, Chap. 5) and Ososkov (1956) completed Palm's work in the case of stationary orderly point processes, and Grigelionis (1963) (see e.g., Gnedenko and Kovalenko, 1968, Section 2.7) developed the following definitive result.

Theorem 8.12. For a uan triangular array $\{\mathbb{P}_{ni}\}$ and $\mathbb{P}_n$ as at (8.3.2), $\mathbb{P}_n \xrightarrow{d} \mathbb{P}$ where $\mathbb{P}$ is a simple Poisson process with parameter measure $\mu(.)$ if and only if for all bounded $A \in \mathcal{B}$,

$$\sum_{i=1}^{n} \mathbb{P}_{ni}\{N(A) \geq 2\} \to 0 \qquad (n \to \infty) \qquad (8.3.3)$$

and for all bounded $A \in \mathcal{B}$ with $\mu(\partial A) = 0$,

$$\sum_{i=1}^{n} \mathbb{P}_{ni}\{N(A) \geq 1\} \to \mu(A) \qquad (n \to \infty) . \qquad (8.3.4)$$

Khinchin and Ososkov discussed the case where each $\mathbb{P}_{ni}$ is stationary and orderly with finite intensity λ_{ni}. Replacing the uan condition with a similar condition on the intensities that

$$\lim_{n \to \infty} \sup_{1 \leq i \leq n} \lambda_{ni} = 0 , \qquad (8.3.5)$$

Ososkov proved the following, where $\mathbb{P}_\lambda$ denotes the stationary Poisson process with intensity $\lambda < \infty$.

Theorem 8.13. If the triangular array $\{\mathbb{P}_{ni}\}$ of stationary orderly point processes has finite intensities λ_{ni} uniformly small as at (8.3.5), then $\mathbb{P}_n \xrightarrow{d} \mathbb{P}_\lambda$ if and only if for every fixed $y \in R_+$,

$$\lim_{n\to\infty} \sum_{i=1}^{n} \lambda_{ni} = \lim_{n\to\infty} \sum_{i=1}^{n} \lambda_{ni}\, q_{ni}(y) = \lambda , \tag{8.3.6}$$

where

$$q_{ni}(y) = \lim_{x\downarrow 0} \mathbb{P}_{ni}\{N(0,y] = 0 \mid N(-x,0] > 0\} . \tag{8.3.7}$$

It is easily shown that (8.3.5) implies (8.3.1), i.e., that under the conditions of Theorem 8.13, $\{P_{ni}\}$ is uan. Matthes (private communication) has shown that a uan array of stationary orderly cluster point processes converging to $\mathbb{P}_\lambda$ may have finite intensities λ_{ni} with

$$\lim_{n\to\infty} \sup_{1\le i\le n} = \infty .$$

It is worth stressing that just as the central limit theorem explains the frequent occurrence of the normal distribution, so Grigelionis theorem shows why the Poisson process should often occur as an approximation--and a good one--in various contexts. Khinchin (1955) discusses this further in his Chapter 5. See also the paper by Çinlar (1972) in this volume for further details on superpositions.

The Poisson process also arises as the limit of the random translation operation of Section 5.3. It was first shown (incompletely, as Goldman, 1967 and Stone, 1968, point out) by Dobrushin (1956). As given below the result is adapted from Stone's paper where multi-dimensional point processes are treated. Debes et al. (1970) have generalized Dobrushin's work further, including questions of cluster invariance outlined in Liemant (1969).

To give the theorem, let $\mathbb{P}$ have sample realization $\{x_n\}$, and with families $\{Y_n(\tau)\}$, $0 < \tau < \infty$, of mutually independent identically distributed random variables independent of $\{x_n\}$ with $F_\tau(y) = \text{prob}\{Y_n(\tau) \le y\}$, let $\mathbb{P}^\tau$ denote the point process with realization $\{x_n + Y_n(\tau)\}$.

Theorem 8.14. In order that for all point processes $\mathbb{P}$ satisfying

$$\lim_{y \to \infty} E|y^{-1} N(x, x + y] - \lambda| = 0 \quad \text{uniformly for} \quad x \in R \tag{8.3.8}$$

it should follow that $\mathbb{P}^\tau \xrightarrow{d} \mathbb{P}_\lambda$ $(\tau \to \infty)$, it is necessary and sufficient that for every $h > 0$

$$\lim_{\tau \to \infty} \sum_{n=-\infty}^{\infty} |\text{prob}\{nh < Y(\tau) \le (n+1)h\} - \text{prob}\{(n-1)h < Y(\tau) \le nh\}|$$

$$= \lim_{\tau \to \infty} \sum_{n=-\infty}^{\infty} |F_\tau\{(n+1)h\} - 2F_\tau(nh) + F_\tau\{(n-1)h\}| = 0 . \tag{8.3.9}$$

Theorem 8.15. If instead of (8.3.8) we have

$$\mathbb{P}\{y^{-1} N(x, x + y] \to \Lambda \quad \text{uniformly in } x\} = 1 \tag{8.3.10}$$

for some random variable Λ with distribution $G(.)$ and (8.3.9) holds, then $\mathbb{P}^\tau$ converges in distribution to a mixed Poisson process with mixing distribution $G(.)$ (i.e., $\mathbb{P}^\tau \xrightarrow{d} \int_0^\infty \mathbb{P}_\lambda\, G(d\lambda)$).

Another important question concerning random translations is that of invariance of $\mathbb{P}$ with respect to the operation. The simplest result is in Dobrushin (1956) (stemming in part from Doob, 1953, p. 404); see also Goldman (1967), Thedeen (1967), and Harris (1971), and for non-independent translation variables $\{Y_n(.)\}$ see e.g., Brown (1970) and Spitzer (1970).

Theorem 8.16. If $\mathbb{P}$ is a stationary mixed Poisson process then $\mathbb{P}^\tau = \mathbb{P}$ for every family of independent random translations with distribution F_τ. Conversely, if $\mathbb{P}^\tau = \mathbb{P}$ for all such translations then $\mathbb{P}$ is a stationary mixed Poisson process.

Another operation leading to a Poisson process limit is that of independent random deletion of points (Section 5.4). Starting from a process satisfying (8.3.8) and with sample realization $\{x_n\}$, points are independently omitted with probability $1 - p$ and retained with probability p to form a new point process $\{x'_n\} = \{px_n\}$ for such points as are retained. As $p \downarrow 0$ this rescaled process converges in distribution to $\mathbb{P}_\lambda$ (Renyi, 1956; Nawrotzki, 1962; Belyaev, 1963; Goldman, 1967; see also Mecke, 1968).

Definition 3.12 is not a satisfactory definition for the convergence in distribution of random measures or random signed measures (Section 2.5) for which the following seems more appropriate.

Definition 8.17. The random signed measures $(M,G,\mathbb{P}_n)$ converge in distribution to $(M,G,\mathbb{P})$ when for every positive integer r, all bounded $A_i \in B$ which are $\mathbb{P}$-continuity sets, and all $y_i \in R$ that are continuity points of $\mathbb{P}\{\Lambda(A_i) \leq y_i\}$,

$$\mathbb{P}_n\{\Lambda(A_i) \leq y_i,\ i=1,\ldots,r\} \to \mathbb{P}\{\Lambda(A_i) \leq y_i,\ i=1,\ldots,r\} \qquad (n\to\infty).$$

An appropriate setting for the stronger concept of weak convergence of point processes or random signed measures would appear to lie in the theory of weak convergence on the function space $D[0,\infty)$ as discussed in Whitt (1971) (cf. Stone, 1963, for an earlier briefer treatment). From this point of view each random signed measure $\Lambda(.)$ on the Borel subsets of R_+ determines uniquely a random function $\Lambda[0,t]$ in the space $D[0,\infty)$ which with the appropriate definition is a complete separable metric space. For further detail we refer to Whitt's work; Jirina's (1966) work on random totally finite measures may also be consulted. Applications to point processes of weak convergence theory have been made in Kennedy (1970) and Jagers (1972). Whitt has pointed out to us that for point processes the concepts of weak convergence and convergence in distribution coincide (see e.g., Straf, 1972). Whitt (1972) gives further work and more references on the general setting of weak convergence for point processes.

ACKNOWLEDGMENT

We thank innumerable people for access to as yet unpublished material and Professor M. R. Leadbetter, Professor K. Matthes, and Dr. R. K. Milne for comments on parts of the paper. We also acknowledge the stimulus of the Conference organizers: our original draft was double the present size. In a paper of such length oversights and inaccuracies are inevitable: we should welcome notes on them.

REFERENCES

Balagangadharan, K. (1960). The prediction theory of stationary random distributions. Mem. Coll. Sci. Univ. Kyoto Ser. A 33, 244-56.

Bartlett, M. S. (1963). The spectral analysis of point processes. J. R. Statist. Soc. B 25, 264-96.

Bartlett, M. S. (1964). The spectral analysis of two-dimensional point processes. Biometrika 51, 299-311.

Bartlett, M. S. (1966). *An Introduction to Stochastic Processes*. 2nd ed. Cambridge: Cambridge Univ. Press.

Bartlett, M. S. (1967). The spectral analysis of line processes. Proc. 5th Berkeley Symp. 3, 135-52.

Bartlett, M. S. (1972). Some applications of multivariate point processes. This volume.

Bartlett, M. S. & Kendall, D.G. (1951). On the use of the characteristic functional in the analysis of some stochastic processes occurring in physics and biology. Proc. Camb. Phil. Soc. 47, 65-76.

Belyaev, Yu. K. (1963). Limit theorems for dissipative flows. Teor. Veroyatnost. i Primenen. 8, 175-84 (English translation in Theory Prob. Appl. 8, 165-73).

Belyaev, Yu. K. (1969). Elements of the general theory of random streams. Appendix 2 to Russian ed. (MIR, Moscow) of Cramer & Leadbetter (1967). (Translation: Dept. of Statistics, Univ. North Carolina, Chapel Hill, Mimeo Ser. No. 703.)

Beutler, F. J. & Lenenman, O. A. Z. (1966). The theory of stationary point processes. Acta Math. 116, 159-97.

Beutler, F. J. & Leneman, O. A. Z. (1968). The spectral analysis of impulse processes. Information and Control 12, 236-58.

Billingsley, P. (1968). *Convergence of Probability Measures.* New York: Wiley.

Bochner, S. (1947). Stochastic processes. Ann. Math. 48, 1014-61.

Bochner, S. (1955). *Harmonic Analysis and the Theory of Probability.* Berkeley: Univ. California Press.

Breiman, L. (1968). *Probability.* Reading, Mass.: Addison-Wesley.

Brillinger, D. R. (1972). The spectral analysis of stationary interval functions. Proc. 6th Berkeley Symp. (to appear).

Brown, M. (1970). A property of Poisson processes and its application to macroscopic equilibrium of particle systems. Ann. Math. Statist. 41, 1935-41. Corrigendum, Ann. Math. Statist. 42, 1777.

Cinlar, E. (1972). Superposition of point processes. This volume.

Cox, D. R. (1962). *Renewal Theory.* London: Methuen.

Cox, D. R. & Lewis, P. A. W. (1966). *The Statistical Analysis of Series of Events.* London: Methuen & Barnes and Noble: New York.

Cox, D. R. & Lewis, P.A.W. (1972). Multivariate point processes. Proc. 6th Berkeley Symp. (to appear).

Cramer, H. & Leadbetter, M. R. (1967). *Stationary and Related Stochastic Processes.* New York: Wiley.

Cramer, H., Leadbetter, M. R. & Serfling, R. J. (1971). On distribution function-moment relationships in a stationary point process. Z. Wahrscheinlichkeitstheorie verw. Geb. 18, 1-8.

Daley, D. J. (1971). Weakly stationary point processes and random measures. J. R. Statist. Soc. B (to appear).

Daley, D. J. (1972a). Asymptotic properties of stationary point processes with generalized clusters. Z. Wahrscheinlichkeitstheorie verw. Geb. 21, 65-76.

Daley, D. J. (1972b). Various concepts of orderliness for point processes. Stochastic Geometry and Analysis (Rollo Davidson Memorial Volume). New York: Wiley (to appear).

Debes, H., Kerstan, J., Liemant, A. & Matthes, K. (1970). Verallgemeinerungen eines Satzes von Dobruschin, I. Math. Nachr. 47, 183-244.

Dobrushin, R. L. (1956). On Poisson laws for distributions of particles in space. Ukrain. Mat. Z. 8, 127-34

Doob, J.L. (1953). *Stochastic Processes.* New York: Wiley.

Dudley, R. M. (1969). Fourier transforms of stationary processes are not functions. J. Math. Mech. 19, 309-16.

Feller, W. (1957). *An Introduction to Probability Theory and its Applications*. Vol. 1, 2nd ed. New York: Wiley.

Feller, W. (1966). *An Introduction to Probability Theory and its Applications*. Vol. 2. New York: Wiley.

Fieger, W. (1964). Zwei Verallgemeinerungen der Palmschen Formeln. *Trans. 3rd Prague Conf. on Information Theory*, 107-22.

Fieger, W. (1965). Eine für beliebige Call-Prozesse geltende Verallgemeinerung der Palmschen Formeln. Math. Scand. 16, 121-47.

Fisher, L. (1972). Mathematical theory of multidimensional point processes. This volume.

Franken, P. & Richter, G. (1965). Über eine Klasse von Zufällige Punktfolgen. Wiss. Z. Friedrich-Schiller-Univ. Jena 14, 247-49.

Gel'fand, I.M. (1955). Generalized random processes. Dokl. Akad. Nauk SSSR 100, 853-56.

Gel'fand, I.M. & Vilenkin, N. Ya. (1964). *Generalized Functions*, Vol. 4. New York: Academic Press. (Translation by A. Feinstein from 1st Russian ed., Moscow, 1961.)

Gnedenko, B. V. & Kovalenko, I. N. (1966). *Introduction to Queueing Theory*. Moscow: Nauka. (Translation: (1968) Jerusalem: Israel Israel Program for Scientific Translations.)

Goldman, J. R. (1967). Stochastic point processes: limit theorems. Ann. Math. Statist. 38, 771-9.

Grigelionis, B. (1963). On the convergence of sums of random step processes.to a Poisson process. Teor. Veroyatnost. i Primenen. 8, 189-94. (English translation in Theory Prob. Appl. 8, 177-82.)

Gurland, J. (1957). Some interrelationships among compound and generalized distributions. Biometrika 44, 265-8.

Harris, T. E. (1963). *The Theory of Branching Processes*. Berlin: Springer.

Harris, T. E. (1971). Random motions and point processes. Z. Wahrscheinlichkeitstheorie verw. Geb. 18, 85-115.

Hawkes, A. G. (1971). Spectra of some self-exciting and mutually exciting point processes. Biometrika 58, 83-90.

Ito, K. (1954). Stationary random distributions. Mem. Coll. Sci. Univ. Kyoto Ser. A 28, 209-23.

Jagers, P. (1971). On the weak convergence of superpositions of point processes. Z. Wahrscheinlichkeitstheorie verw. Geb. 22, 1-7.

Jirina, M. (1964). Branching processes with measure-valued states. *Trans. 3rd Prague Conf. on Information Theory*, 333-57.

Jirina, M. (1966). Asymptotic behavior of measure-valued branching processes. Rozpravy Ceskoslovenske Akad. Ved. Rada Mat. Prirod. Ved. 76, No. 3.

Jowett, J. & Vere-Jones, D. (1972). Prediction of point processes. This volume.

Kaplan, E. L. (1955). Transformations of stationary random sequences. Math. Scand. 3, 127-49.

Kendall, D. G. (1949). Stochastic processes and population growth. J. R. Statist. Soc. B 11, 230-64.

Kennedy, D. P. (1970). Weak convergence for the superposition and thinning of point processes. Tech. Report No. 11, Operations Research Dept., Stanford Univ.

Kerstan, J. & Matthes, K. (1964). Stationäre zufällige Punktfolgen, II. Jber. Deutsch. Math.-Verein. 66, 106-18.

Kerstan, J. & Matthes, K. (1965). A generalization of the Palm-Khinchin theorem. Ukrain. Mat. Z. 17, 29-36.

Kerstan, J., Matthes, K. & Mecke, J. (1972). *Unbegrenst teilbare Punktprozesse*. Berlin: Akademie Verlag (to appear).

Khinchin, A. Ya. (1955). *Mathematical Methods in the Theory of Queueing*. Trudy Mat. Inst. Steklov 49. (English translation (1960). London: Griffin.)

Khinchin, A. Ya. (1956a). Sequences of chance events without after-effects. Teor. Veroyatnost. i Primenen. 1, 3-18. (English translation in Theory Prob. Appl. 1, 1-15.)

Khinchin, A. Ya. (1956b). On Poisson streams of random events. Teor. Veroyatnost. i Primenen 1, 320-7. (English translation in Theory Prob. Appl. 1, 291-7).

Kingman, J.F.C. (1967). Completely random measures. Pacific J. Math. 21, 59-78.

Kolmogorov, A.N. (1935). La transformation de Laplace dans les espaces linéaires. C. R. Acad. Sci. Paris 200, 1717-18.

Kolmogorov, A.N. (1940). Kurven im Hilbertschen Raum, die Gegenüber einer einparametrigen Gruppe von bewegungen invariant sind; and Wienersche Spiralen und einige andere interessant Kurven im Hilbertsche Raum. C.R. Acad. Sci. URSS 26, 6-9 and 115-8.

König, D. & Matthes, K. (1963). Verallgemeinerungen der Erlangsche Formeln, I. Math. Nachr. 26, 45-56.

Leadbetter, M.R. (1966). On streams of events and mixtures of streams. J. R. Statist. Soc. B 28, 218-27.

Leadbetter, M.R. (1968). On three basic results in the theory of stationary point processes. Proc. Amer. Math. Soc. 19, 115-7.

Leadbetter, M.R. (1972a). On basic results of point process theory. Proc. 6th Berkeley Symp. (to appear).

Leadbetter, M. R. (1972b). Point processes generated by level crossings. This volume.

Le Cam, L. (1947). Un instrument d'étude des fonctions aléatoires: la fonctionelle charactéristique. C. R. Acad. Sci. Paris 224, 710-1.

Le Cam, L. (1961). A stochastic theory of precipitation, Proc. 4th Berkeley Symp. 3, 165-86.

Lee, P. M. (1967). Infinitely divisible stochastic processes. Z. Wahrscheinlichkeitstheorie verw. Geb. 7, 147-60.

Lee, P. M. (1968). Some examples of infinitely divisible point processes. Studia Sci. Math. Hung. 3, 219-24.

Lewis, P.A.W. (1964a). A branching Poisson process model for the analysis of computer failure patterns. J. R. Statist. Soc. B 26, 398-456.

Lewis, P.A.W. (1964b). The implications of a failure model for the use and maintenance of computers. J. App. Prob. 1, 347-68.

Lewis, P.A.W. (1969). Asymptotic properties and equilibrium conditions for branching Poisson processes. J. App. Prob. 6, 355-71.

Lewis, P.A.W. (1970). Remarks on the theory, computation and application of the spectral analysis of series of events. J. Sound Vib. 12, 353-75.

Lewis, P.A. W. (1972). Asymptotic properties of branching renewal processes (to appear).

Lewis, T. (1961). The intervals between regular events displaced in time by independent random deviations of large dispersion. J. R. Statist. Soc. B 23, 476-83.

Liemant, A. (1969). Invariante zufällige Punktfolgen. Wiss. Z. Friedrich-Schiller-Univ. Jena 18, 361-72.

McFadden, J.A. (1962). On the lengths of intervals in a stationary point process. J. R. Statist. Soc. B 24, 364-82. Corrigenda, J. R. Statist. Soc. B 25, 500.

Matérn, B. (1960). Spatial variation. Meddelanden fran Statens Skogsforskninsinstitut 49, No. 5.

Matthes, K. (1963a). Stationäre zufällige Punktfolgen, I. Jber. Deutsch. Math.-Verein. 66, 66-79.

Matthes, K. (1963b). Unbeschränkt teilbare Verteilungsgesetze stationärer zufälliger Punktfolgen. Wiss. Z. Hochsch. Electro. Ilmenau 9, 235-8.

Matthes, K. (1972). Infinitely divisible point processes. This volume.

Mecke, J. (1967). Stätionare zufällige Masse auf lokalkompakten Abelschen Gruppen. Z. Wahrscheinlichkeitstheorie verw. Geb. 9, 36-58.

Mecke, J. (1968). Eine charakterische Eigenschaft der doppelt stochastischen Poissonschen Prozesse. Z. Wahrscheinlichkeitstheorie verw. Geb. 11, 74-81.

Miles, R.E. (1969). The asymptotic values of certain coverage probabilities. Biometrika 56, 661-80.

Miles, R.E. (1970). On the homogeneous planar Poisson point process. Math. Biosci. 6, 85-127.

Milne, R.K. (1971). *Stochastic Analysis of Multivariate Point Processes*. Canberra: Australian National Univ. Ph.D. thesis.

Milne, R.K. & Westcott, M. (1972a). Further results for Gauss-Poisson processes. Adv. Appl. Prob. (to appear).

Milne, R.K. & Westcott, M. (1972b). Gauss-Poisson processes. This volume.

Moyal, J.E. (1962a). The general theory of stochastic population processes. Acta Math. 108, 1-31.

Moyal, J.E. (1962b). Multiplicative population chains. Proc. Roy. Soc. A 266, 518-26.

Nawrotzki, K. (1962). Eine Grenzwertsatz für homogene zufällige Punktfolgen. Math. Nach. 24, 201-17.

Neveu, J. (1968). Sur la structure des processus ponctuels stationnaires. C. R. Acad. Sci. Paris Sér. A 267, A561-4.

Newman, D. S. (1970). A new family of point processes which are characterized by their second moment properties. J. App. Prob. 7, 338-58.

Neyman, J. (1963). On finiteness of the process of clustering. Sankhyā A 25, 69-74.

Neyman, J. & Scott, E. L. (1958). A statistical approach to problems of cosmology. J. R. Statist. Soc. B 20, 1-43.

Ososkov, G. A. (1956). A limit theorem for flows of homogeneous events. Teor. Veroyatnost. i Primenen 1, 274-82. (English translation in Theory Prob. Appl. 1, 248-55.)

Palm, C. (1943). Intensitätsschwankungen im Fernsprechverkehr. Ericsson Techniks 44, 1-189.

Papangelou, F. (1970). The Ambrose-Kakutani theorem and the Poisson process. *Springer-Verlag Lecture Notes Math.* 160, 234-40.

Papangelou, F. (1972). On the Palm probabilities of processes of points and processes of lines. In Rollo Davidson Memorial Volume. New York: Wiley (to appear).

Parthasarathy, K. R. (1967). *Probability Measures on Metric Spaces*. New York: Academic Press.

Prekopa, A. (1957a). On the compound Poisson distribution. Acta Sci. Math. Szeged. 18, 23-8.

Prekopa, A. (1957b). On Poisson and compound Poisson stochastic functiones. Studia Math. 16, 142-55.

Prohorov, Yu. V. (1961). The method of characteristic functionals. Proc. 4th Berkeley Symp. 2, 203-19.

Ramakrishnan, A. (1950). Stochastic processes relating to particles distributed in a continuous infinity of states. Proc. Camb. Phil. Soc. 46, 595-602.

Renyi, A. (1956). A characterization of Poisson processes. Magyar Tud. Akad. Mat. Kutato Int. Közl. 1. 519-27.

Rosenblatt, M. (1962). *Random Processes*. New York: Oxford Univ. Press.

Rozanov, Yu. A. (1959). On the extrapolation of generalized stationary random processes. Teor. Veroyatnost. i Primenen 4, 465-71 (English translation in Theory Prob. Appl. 4, 426-31).

Ryll-Nardzewski, C. (1961). Remarks on processes of calls. Proc. 4th Berkeley Symp. 2, 455-65.

Shiryaev, A.N. (1960). Some problems in the spectral theory of higher order moments, I. Teor. Veroyatnost. i Primenen 5, 293-313 (English translation in Theory Prob. Appl. 5, 265-84).

Slivnyak, I.M. (1962). Some properties of stationary flows of homogeneous random events. Teor. Veroyatnost. i Primenen 7, 347-52. (English translation in Theory Prob. Appl. 7, 336-41). Addendum 9, 190 (English translation in Theory Prob. Appl. 9, 168).

Slivnyak, I.M. (1966). Stationary streams of homogeneous random events. Vest. Harkov. Gos. Univ., Ser. Mech.-Math. 32, 73-116. (English translation: Statistics Dept., Australian National Univ.)

Smith, W.L. (1958). Renewal theory and its ramifications. J. R. Statist. Soc. B 20, 243-302.

Spitzer, F. (1970). Interaction of Markov processes. Adv. Math. 5 246-90.

Srinivasan, S.K. (1969). *Stochastic Theory and Cascade Processes.* New York: Elsevier.

Stone, C. (1963). Weak convergence of stochastic processes defined on semi-infinite time intervals. Proc. Amer. Math. Soc. 14, 694-6.

Stone, C. (1968). On a theorem of Dobrushin. Ann. Math. Statist. 39, 1391-1401.

Straf, M. (1971). Weak convergence of stochastic processes with several parameters. Proc. 6th Berk. Symp. (To appear).

Thedeen, T. (1964). A note on the Poisson tendency in traffic. Ann. Math. Statist. 35, 1823-4.

Thedeen, T. (1967). Convergence and invariance questions for point systems in R_1 under random motion. Arkiv. för Mat. 7, 211-39.

Vasil'ev, P.I. (1965). On the question of ordinariness of a stationary stream. Kisinev. Gos. Univ. Ucen. Zap. 82, 44-8. (English translation: Statistics Dept., Australian National Univ.).

Vere-Jones, D. (1968). Some applications of probability generating functionals to the study of input/output streams. J. R. Statist. Soc. B 30, 321-33.

Vere-Jones, D. (1970). Stochastic models for earthquake occurrences. J. R. Statist. Soc. B 32, 1-62.

Vere-Jones, D. (1971). Appendix to Daley (1971).

Vere-Jones, D. (1972). An elementary approach to the spectral theory of stationary random measures. (Rollo Davidson Memorial Volume) New York: Wiley (to appear).

Vere-Jones, D. & Davies, R.D. (1966). A statistical survey of earthquakes in the main seismic region of New Zealand. II: Time series analysis. N.Z.J. Geol. Geophys. 9, 251-84.

Volkonskii, V.A. (1960). An ergodic theorem on the distribution of duration of fades. Teor. Veroyatnost. i Primenen 5, 357-60 (English translation in Theory Prob. Appl. 5, 323-6).

Westcott, M. (1971). On existence and mixing results for cluster point processes. J. R. Statist. Soc. B 33, 290-30.

Westcott, M. (1972). The probability generating functional. J. Aust. Math. Soc. (to appear).

Whitt, W. (1971). Weak convergence of probability measures on the function space $D[0,\infty)$.

Whitt, W. (1972). Representation and convergence of point processes on the line (to appear).

Yaglom, A.M. (1955). Correlation theory of processes with random stationary nth order increments. Mat. Sb. (N.S.) 37, 141-96. (English translation in Amer. Math. Soc. Transl. Ser. 2 8, 87-141).

Zitek, F. (1957). On a theorem of Korolyuk. Czech. Math. J. 7, 318-19.

INFINITELY DIVISIBLE POINT PROCESSES

K. Matthes
Deutsche Akademie der Wissenschaften zu Berlin, DDR

SUMMARY

The present paper gives a supplement to the comprehensive paper of D. J. Daley and D. Vere-Jones (1972). It does not intend to render an exhaustive presentation of the theory of infinitely divisible point processes. Such a presentation will be given in the forthcoming book by Kerstan, Matthes and Mecke (1972). The paper is an attempt to clarify the role of cluster-fields within this theory and to deal with some functional equations.

1. BASIC DEFINITIONS

Let (A,ρ_A) be an arbtirary complete separable metric space, A the corresponding σ-algebra of Borel sets, B the ring of all bounded $X \in A$ and N the set of all measures ν in A, such that $\nu(X)$ is finite for all $X \in B$. An element Φ of N is integer-valued iff the representation

$$\Phi = \sum_{a\varepsilon A,\Phi(\{a\})>0} \Phi(\{a\})\ \delta_a$$

is valid. We denote by M the set of all integer-valued measures $\Phi \in N$. An element of M is said to be *simple*, if $\Phi(\{a\}) \leq 1$ for all $a \in A$.

Let M be the smallest σ-algebra of subsets of M, such that for all $X \in B$ the real-valued function

$$\Phi \longrightarrow \Phi(X) \qquad (\Phi \in M)$$

is measurable. Any probability field (M,M,P) is called a *point process* (with state space (A,ρ_A)). Let $\mathbb{P}$ denote the set of all probability measures P in M and $\mathcal{P}$ the smallest σ-algebra of subsets of $\mathbb{P}$, such that for all Y ε *M* the real-valued function

$$P \longrightarrow P(Y) \qquad (P \in \mathbb{P})$$

is measurable. For any measure H on *M* we introduce the corresponding *intensity-measure*

$$\rho_H(X) = \int \Phi(X)\ H(d\Phi) \qquad (X \in A) .$$

The mapping

$$\Phi \longrightarrow {}_X\Phi = \Phi\{(\cdot) \cap X\} \qquad (\Phi \in M)$$

is measurable for all X ε *A*. For any measure H in *M* we write

$${}_XH = H\{{}_X\Phi \in (\cdot)\} .$$

The mapping

$$\Phi \longrightarrow \{\Phi(X_1), \ldots , \Phi(X_m)\} \qquad (\Phi \in M)$$

is measurable for all $X_1, \ldots , X_m \in B$. For any measure H in *M* we write

$$H_{X_1,\ldots,X_m} = H[\{\psi(X_1), \ldots , \psi(X_m)\} \in (\cdot)] .$$

A probability measure P ε $\mathbb{P}$ is determined uniquely by its finite-dimensional distributions $P_{X_1,\ldots,X_m}$. Mönch (1971) has proved the following interesting theorem:

Theorem 1.1. Let P, Q be elements of $\mathbb{P}$, satisfying the conditions

a) P(Φ is simple) = 1,

b) $P_X = Q_X$ for all X ε *B*.

Then P = Q.

The superposition

$$(\Phi_1, \Phi_2) \longrightarrow \Phi_1 + \Phi_2 \qquad (\Phi_1, \Phi_2 \in M)$$

defines a measurable mapping of $M \times M$ onto M. For any σ-finite measure H_1, H_2 in M we write

$$H_1 * H_2 = (H_1 \times H_2)\{(\Phi_1 + \Phi_2) \in (\cdot)\} \quad .$$

The measure $H_1 * H_2$ is called the *convolution* of H_1 and H_2. For all H we have

$$H * \delta_0 = H \ ,$$

where 0 denotes the zero-measure on A. For any finite measure E in M and any nonnegative integer n we define E^n by

$$E^n = \underbrace{E * \cdots * E}_{n \text{ times}} \quad \text{for } n \geq 1, \qquad P^0 = \delta_0 \ .$$

Let $(A', \rho_{A'})$ be a second state space, A' the corresponding σ-algebra of Borel sets, M' the corresponding set of integer-valued measures in A' etc. Any measurable mapping

$$a' \longrightarrow \kappa_{(a')} \qquad (a' \in A') \ ,$$

of A' into $\mathbb{P}$ is called a *cluster-field* in $(A', \rho_{A'})$ with state space (A, ρ_A): To every point a' of A' corresponds the probability distribution $\kappa_{(a')}$ of a random cluster of points of A. We write

$$\kappa_{(\Phi)} = \mathop{*}_{a' \in A', \Phi(\{a'\})>0} (\kappa_{(a')})^{\Phi(\{a'\})}$$

if the infinite convolution exists. The measure $\kappa_{(\Phi)} \in \mathbb{P}$ describes the probability distribution of the superposition of all random clusters generated independently by the points of Φ. Let M'_κ denote the set of all $\Phi \in M'$ such that $\kappa_{(\Phi)}$ exists. If a measure H on M' satisfies the condition

$$H(\Phi \notin M'_\kappa) = 0 ,$$

we introduce a measure H_κ in M by means of

$$H_\kappa = \int \kappa_{(\Phi)} (\cdot)\ H(d\Phi) .$$

The operation of "clustering" commutes with convolution:

Theorem 1.2. Let κ be a cluster-field in $(A',\rho_{A'})$ with state space (A,ρ_A). Then for any at most denumerable family $(P_i)_{i\in I}$ of probability measures $P_i \in \mathbb{P}'$ such that $\underset{i\in I}{*} P_i$ exists,

$$\left(\underset{i\in I}{*} P_i\right)_\kappa = \underset{i\in I}{*} (P_i)_\kappa ,$$

where the existence of one side is equivalent to the existence of the other side.

Let $(A',\rho_{A'})$ and (A,ρ_A) coincide. Then to each $X \in A'$ corresponds the deterministic cluster-field

$$\gamma^X_{(a')} = \begin{cases} \delta_{\delta_a} & (a \in X) , \\ \delta_0 & (a \notin X) . \end{cases}$$

For all measures H in M

$$H_{\gamma^X} = {}_XH .$$

Theorem 1.2 yields

$$_{X}(P*Q) = (_{X}P) * (_{X}Q) \qquad (P, Q \in \mathbb{P}) \ . \tag{1.3}$$

Let (A,ρ_A) be the finite set $\{1, \ldots, m\}$. Then the mapping

$$\Phi \longrightarrow \{\Phi(\{1\}), \ldots, \Phi(\{m\})\}$$

of M onto $\{0,1,\ldots\}^m$ is one-to-one and $\mathbb{P}$ may be identified with the set of all probability distributions of m-dimensional random vectors with nonnegative integer components. For any finite sequence $X_1,\ldots, X_m$ of sets from B' we define a deterministic cluster-field

$$\sigma^{X_1,\ldots,X_m}$$

in $(A',\rho_{A'})$ with state space $\{1,\ldots,m\}$ by means of

$$\sigma^{X_1,\ldots,X_m}_{(a')} = \delta_{\sum_{i=1}^{m} h_{X_i}(a')\delta_i} \ ,$$

where h_{X_i} denotes the indicator-function of X_i. Then for every $P \in \mathbb{P}$ the probability distribution $P_{\sigma^{X_1,\ldots,X_m}}$ coincides with $P_{X_1,\ldots,X_m}$, and Theorem 1.2 yields

$$(P * Q)_{X_1,\ldots,X_m} = (P_{X_1,\ldots,X_m}) * (Q_{X_1,\ldots,X_m}) \qquad (P,Q \in \mathbb{P}) \ . \tag{1.4}$$

Let (A,ρ_A) be the Euclidean space R^n. Then we define

$$T_x\Phi = \sum_{a\in R^n, \Phi(\{a\})>0} \Phi(\{a\})\ \delta_{a-x} \qquad (x \in R^n,\ \Phi \in M) \ .$$

A measure H in M is called *stationary*, if

$$H = H\{(T_x\Phi) \in (\cdot)\} \qquad \text{for all } x \in R^n \ .$$

For all stationary H the value

$$i_H = \rho_H([0,1)^n)$$

is called the *intensity* of H. Let μ be the Lebesgue-measure in the σ-algebra R^n of all Borel sets of R^n. Then

$$\rho_H = i_H \mu$$

if $i_H < +\infty$.

2. QUEUE MEASURES

A probability measure $P \in \mathbb{P}$ is called *infinitely divisible*, if for any positive integer n the equation $Q^n = P$ is solvable by an element Q of $\mathbb{P}$. Let $\mathbb{T}$ denote the set of all infinitely divisible $P \in \mathbb{P}$. On account of Equation 1.4 all finite dimensional distributions $P_{X_1,\ldots,X_m}$ of any $P \in \mathbb{T}$ are infinitely divisible. Conversely, $P \in \mathbb{P}$ belongs to $\mathbb{T}$ if all $P_{X_1,\ldots,X_m}$ are infinitely divisible.

Let P_ν denote the Poisson-distribution in *M* with intensity-measure $\nu \in N$. For all $\nu_1, \nu_2 \in N$,

$$P_{\nu_1} * P_{\nu_2} = P_{\nu_1+\nu_2} \ ;$$

especially for all $\nu \in N$ and all positive integers n

$$\left(P_{\frac{1}{n}\nu}\right)^n = P_\nu.$$

Therefore, $P_\nu \in \mathbb{T}$ for all $\nu \in N$.

The following elementary model yields a comprehensive set of elements of $\mathbb{T}$. Let $\mathbb{E}^+$ denote the set of all finite measures in *M*. For every $E \in \mathbb{E}^+$ we define an element

$$U_E = e^{-E(M)} \sum_{n=0}^{\infty} \frac{1}{n!} E^n$$

of $\mathbb{P}$. The usual calculations lead us to

$$U_{E_1} * U_{E_2} = U_{E_1+E_2} \qquad (E_1, E_2 \in \mathbb{E}^+) \quad . \tag{2.1}$$

Especially for all $E \in \mathbb{E}^+$ and all positive integers n

$$\left(U_{\frac{1}{n}E}\right)^n = U_E \; .$$

Therefore, all U_E, $E \in \mathbb{E}^+$ belong to $\mathbb{T}$. On account of

$$U_E = U_{E((\cdot)\setminus\{0\})} \qquad (E \in \mathbb{E}^+) \tag{2.2}$$

we may assume without loss of generality $E(\{0\}) = 0$.

For all finite $\nu \in N$ we define $Q_\nu \in \mathbb{E}^+$ to be the image of ν with respect to the mapping $a \longrightarrow \delta_a$ of A into M. Then

$$P_\nu = U_{Q_\nu} \qquad \text{for all finite } \nu \in N \; . \tag{2.3}$$

If $\nu \in N$ is not finite, P_ν does not possess the form

$$U_E,\ E \in \mathbb{E}^+ : \quad P_\nu(\{0\}) = 0 \; ,$$

whereas

$$U_E(\{0\}) = e^{-E(M)} > 0 \qquad \text{for all} \quad E \in \mathbb{E}^+ \; !$$

Using a simple second limit operation we exhaust all $P \in \mathbb{T}$.

Theorem 2.4. Let $(E_i)_{i\varepsilon I}$ be an at most denumerable family of measures $E_i \varepsilon \mathbb{E}^+$, $E_i(\{0\}) = 0$. Then $\underset{i\varepsilon I}{*} U_{E_i}$ exists iff for all $X \varepsilon B$

$$(\sum_{i\varepsilon I} E_i) \; \{\psi(X) > 0\} < +\infty .$$

We denote by $\mathbb{W}$ the set of all measures W on M such that

a) $W(\{0\}) = 0$,

b) $W(\psi(X) > 0) < +\infty$ for all $X \varepsilon B$.

Any $W \varepsilon \mathbb{W}$ is representable as a sum $\Sigma_{i\varepsilon I} E_i$ of an at most denumerable family of measures $E_i \varepsilon \mathbb{E}^+$, $E_i(\{0\}) = 0$.

Theorem 2.5. Let $(E_i)_{i\,\varepsilon\, I}$, $(G_j)_{j\varepsilon J}$ be at most denumerable families of measures of $\mathbb{E}^+$. If

$$\sum_{i\varepsilon I} E_i = \sum_{j\varepsilon J} G_j \;\varepsilon\; W,$$

then

$$\underset{i\varepsilon I}{*} U_{E_i} = \underset{j\varepsilon J}{*} U_{G_j} .$$

On account of Theorem 2.5, to each $W \varepsilon \mathbb{W}$ there corresponds a well-defined probability measure $P = \underset{i\varepsilon I}{*} U_{E_i}$, for $\sum_{i\varepsilon I} E_i = W$. We write $\tilde{P} = W$. The measure W is called the queue-measure corresponding to P.

Theorem 2.6. $P \longrightarrow \tilde{P}$ is a one-to-one mapping of $\mathbb{T}$ onto $\mathbb{W}$.

Theorem 2.7. Let $(P_i)_{i\varepsilon I}$ be an at most denumerable family of probability measures $P_i \varepsilon \mathbb{T}$. Then $\underset{i\varepsilon I}{*} P_i$ exists iff $\sum_{i\varepsilon I} \tilde{P}_i$ belongs to $\mathbb{W}$ and

$$\widetilde{\underset{i\varepsilon I}{*} P_i} = \sum_{i\varepsilon I} \tilde{P}_i .$$

Let P be an element of $\mathbb{T}$. For all $t \geq 0$ the measure $t\tilde{P}$ belongs to $\mathbb{W}$. We denote by P(t) the corresponding element of $\mathbb{T}$. Then

$$(+) \quad P(1) = P, \quad P(t_1) * P(t_2) = P(t_1 + t_2) \qquad \text{for all} \quad t_1, t_2 \geq 0 .$$

Theorem 2.8. An element P of $\mathbb{P}$ is infinitely divisible iff there is a mapping $t \longrightarrow P(t)$ of $[0, +\infty)$ into $\mathbb{P}$ satisfying (+). This mapping is unique:

$$P(t) \in \mathbb{T}, \qquad \widetilde{P(t)} = t\tilde{P} \qquad \text{for all } t \geq 0 .$$

As an immediate consequence of Theorem 1.2 we get the following "cluster theorem" due to Franken, Liemant and Matthes (1965):

Theorem 2.9. Let P be an element of $\mathbb{T}'$ and κ be a cluster-field in $(A', \rho_{A'})$ with state space (A, ρ_A). Then P_κ exists iff $(\tilde{P})_\kappa$ exists and $(\tilde{P})_\kappa((\cdot)\setminus\{o\}) \in \mathbb{W}$. The probability measure P_κ is infinitely divisible

$$\widetilde{P_\kappa} = (\tilde{P})_\kappa((\cdot)\setminus\{o\}) .$$

Applying 2.8 to the special cluster-fields γ^X, $\sigma^{X_1,\ldots,X_m}$, we obtain

Theorem 2.10. Let P be an element of $\mathbb{T}$ and X be an element of $\mathbb{A}$. Then ${}_XP \in \mathbb{T}$ and

$$\widetilde{{}_XP} = {}_X(\tilde{P})((\cdot)\setminus\{o\}) .$$

Theorem 2.11. $\widetilde{P_{X_1,\ldots,X_m}} = (\tilde{P})_{X_1,\ldots,X_m}((\cdot)\setminus\{(0,\ldots,0)\})$ for all $P \in \mathbb{T}$; $X_1, \ldots, X_m \in B$.

For all $P \in \mathbb{T}$, $X \in B$ the measure $\widetilde{{}_XP} = {}_X(\tilde{P})((\cdot)\setminus\{0\})$ has total mass

$$\tilde{P}\{\psi(X) > 0\} < +\infty .$$

In this way we have proved

Theorem 2.12.

$$_X P = U_{_X(\widetilde{P})((\cdot)\setminus\{o\})} \qquad \text{for all } P \in \mathbb{T},\ X \in B\ .$$

For all $P \in \mathbb{T}$, $X \in B$

$$P\{\Phi(X) = 0\} = ({}_X P)(\Phi = o) = U_{_X(\widetilde{P})((\cdot)\setminus\{o\})}(\Phi = o)$$

$$= \exp\{-{}_X(\widetilde{P})(\psi \neq o)\} = \exp[-\widetilde{P}\{\psi(X) > 0\}] > 0\ .$$

Theorem 2.13. Let P be an element of $\mathbb{T}$, X be an element of B. Then $P\{(\cdot)|\Phi(X) = 0\}$ is infinitely divisible, and

$$\widetilde{P\{(\cdot)|\Phi(X) = 0\}} = \widetilde{P}\{\psi \in (\cdot),\ \psi(X) = 0\}\ .$$

By Theorem 2.7, for all $P \in \mathbb{T}$, $X \in B$ the functional equation

$$P = [P\{(.)|\Phi(X) = 0\}] * Q$$

possesses the infinitely divisible solution

$$Q = U_{\widetilde{P}\{\psi \in (\cdot),\ \psi(X) > 0\}}$$

The following theorem is due to Matthes (1969).

Theorem 2.14. Let P be an element of $\mathbb{P}$. If

$$P\{\Phi(\{a\}) > 0\} = 0 \qquad \text{for all } a \in A\ ,$$

then P is infinitely divisible iff for all $X \in B$

a) $P\{\Phi(X) = 0\} > 0$,

b) the functional equation $P = [P\{(\cdot)|\Phi(X) = 0\}] * Q$ possesses a solution $Q \in \mathbb{P}$.

If P is without after-effects and $P\{\Phi(\{a\}) > 0\} = 0$ for all $a \in A$, then conditions a) and b) can be verified easily.

The following theorems present translation of some fundamental properties of infinitely divisible probability measures into the language of the corresponding queue-measures.

Theorem 2.15. An element P of $\mathbb{T}$ is Poisson iff

$$\tilde{P}\{\psi(A) \neq 1\} = 0 .$$

Lee (1967) has proved

Theorem 2.16. An element P of $\mathbb{T}$ is without after-effects iff

$$\tilde{P}\{\psi(\{a\}) > 0 \text{ for more than one } a \in A\} = 0 .$$

A probability measure $P \in \mathbb{T}$ is called *regular infinitely divisible*, if for all $X \in B$ and all $\varepsilon > 0$ there exists a real number $r_{X,\varepsilon}$, such that

$$\sup_{X' \in B} |P\{\Phi(X) = 0|\Phi(X') = 0\} - P\{\Phi(X) = 0\}| \leq \varepsilon ,$$

$$\inf_{x \in X, x' \in X'} \rho_A(x,x') \geq r_{X,\varepsilon} .$$

Theorem 2.17. An element P of $\mathbb{T}$ is regular infinitely divisible iff

$$\tilde{P}\{\psi(A) = +\infty\} = 0.$$

The next theorem, valid in the case $(A,\rho_A) = R^n$, is due to Kerstan and Matthes (1965):

Theorem 2.18. A stationary $P \in \mathbb{T}$ with finite intensity i_P is ergodic iff

$$\tilde{P}\left\{\overline{\lim_{m\to\infty}} \frac{\{[-m,m)^n\}}{(2m)^n} > 0\right\} = 0 .$$

The preceding three theorems demonstrate a simple heuristic idea: The intensity of the after-effects with respect to $P \in \mathbb{T}$ is closely related to the thickness of the typical realisations ψ with resepct to the corresponding queue-measure $\tilde{P}$.

On account of Theorem 2.9, for all $\nu \in N'$ and all cluster-fields κ in $(A', \rho_{A'})$ with state space (A, ρ_A) the probability measure $(P_\nu)_\kappa$ exists if

$$\int \kappa_{(a')}\{\chi(X) > 0\}\ \nu(da') < +\infty$$

for all $X \in B$. Also $(P_\nu)_\kappa$ belongs to $\mathbb{T}$, and

$$\widetilde{(P_\nu)_\kappa} = \int \kappa_{(a')}((\cdot)\setminus\{o\})\ \nu(da') .$$

Any representation $P = (P_\nu)_\kappa$ of a $P \in \mathbb{P}$ is called a *cluster-representation* of P in $(A', \rho_{A'})$. All $P \in \mathbb{P}$ possessing a cluster-representation are infinitely divisible. Conversely, each $P \in \mathbb{T}$ possesses a cluster-representation in $\{1,2,\ldots\}$: Let $\tilde{P} = \Sigma_{n=1}^{\infty} E_n$ be a representation of $\tilde{P}$ as a sum of an infinite series of measures $E_n \in \mathbb{E}^+$. Then

$$P = \mathop{*}_{n=1}^{\infty} U_{E_n} = (P_\nu)_\kappa ,$$

where

$$\nu(\{n\}) = E_n(M) \qquad (n = 1,2,\ldots)$$

$$\kappa_{(n)} = \begin{cases} (E_n(M))^{-1} E_n & \text{for } E_n(M) \neq 0 , \\ \delta_0 & \text{for } E_n(M) = 0 . \end{cases}$$

An element P of $\mathbb{T}$ possesses a cluster-representation in A' = {1} iff $\tilde{P}$ is finite. Then ν is given by one value $\nu(\{1\})$ and κ by one probability measure $\kappa_{(1)}$, and $P = (P_\nu)_\kappa$ is equivalent to

$$P = U_{\nu(\{1\})\kappa_{(1)}} \; .$$

Theorem 2.19. A probability measure $P \in \mathbb{P}$ is infinitely divisible iff P possesses a cluster-representation in the state space (A, ρ_A).

This cluster-representation in general is not unique. An interesting special cluster-representation for arbitrary $P \in \mathbb{T}$ in the state space $A' = M\backslash\{o\}$ (furnished with a suitable metric) was given by Lee (1964, 1967).

3. CAMPBELL MEASURES

The following elementary assertion is often quoted as "Campbell's theorem":

Theorem 3.1. For all measures H in M and all nonnegative real-valued measurable functions f in A

$$\int \Big\{ \sum_{a\varepsilon A, \Phi(\{a\})>0} f(a)\; \Phi(\{a\}) \Big\}\; H(d\Phi) = \int f(a)\; \rho_H(da) \; .$$

It is possible to give a refinement of Theorem 3.1 by admitting functions f depending on a and Φ. For this purpose we introduce a refinement of the intensity-measure ρ_H. Let C denote the subset $\{(a,\Phi):(a,\Phi)\varepsilon\, A \times M, \Phi(\{a\}) > 0\}$ of A × M. This set C belongs to $A \otimes M$. We denote by C the σ-algebra of all subsets Z of C which belong to $A \otimes M$. For all measures H in M

$$C_H = \int \Big\{ \sum_{a\varepsilon A, \Phi(\{a\})>0} \Phi(\{a\})\; \delta_{(a,\Phi)}(\cdot) \Big\}\; H(d\Phi)$$

$$= \int (\Phi \times \delta_\Phi)(\cdot)\; H(d\Phi)$$

defines a measure C_H in *C*, which is called the *Campbell measure* corresponding to H.

Theorem 3.2. For all measures H in *M* and all nonnegative measurable real-valued function f in C

$$\int \{ \sum_{a\varepsilon A, \Phi(\{a\})>0} \Phi(\{a\}) \, f(a,\Phi)\} \, H(d\Phi) = \int f(c) \, C_H(dc) \; .$$

Theorem 3.3. For all measures H_1, H_2 in *M* the relation $C_{H_1} = C_{H_2}$ is equivalent to $H_1((\cdot)\setminus\{o\}) = H_2((\cdot)\setminus\{o\})$.

Theorem 3.4. A measure H in *M* satisfying the condition $H(\{o\}) = 0$ is σ-finite iff C_H is σ-finite.

For all σ-finite measures V resp. H in C resp. *M* the mapping

$$\{(a,\Phi), \psi\} \longrightarrow (a, \Phi + \psi)$$

transforms $V \times H$ into a measure $V \circledast H$ in *C*. The operations $\circledast$, $*$ are closely connected.

Theorem 3.5. Let V be a σ-finite measure in *C* and H_1, H_2 be σ-finite measures in *M*, such that the measures $V \circledast H_1$, $H_1 * H_2$ are σ-finite. Then

$$(V \circledast H_1) \circledast H_2 = V \circledast (H_1 * H_2) \; .$$

Theorem 3.6. Let $(P_i)_{i\varepsilon I}$ be an at most denumerable family of measures $P_i \varepsilon \mathbb{P}$ for which $\underset{i\varepsilon I}{*} P_i$ exists. Then

$$C_{\underset{i\varepsilon I}{*} P_i} = \sum_{i\varepsilon I} C_{P_i} \circledast \underset{j\varepsilon I\setminus\{i\}}{*} P_j$$

For all $E \in \mathbb{E}^+$

$$C_{U_E} = \sum_{n=0}^{\infty} \frac{e^{-E(M)}}{n!} C_{E^n} = \sum_{n=1}^{\infty} \frac{e^{-E(M)}}{n!} (nC_E \circledast E^{n-1})$$

$$= C_E \circledast \left(\sum_{m=0}^{\infty} \frac{e^{-E(M)}}{m!} E^m \right) = C_E \circledast U_E$$

Let P be an arbitrary element of $\mathbb{T}$ and $\tilde{P} = \Sigma_{i\in I} E_i$ a representation of $\tilde{P}$ as a sum of an at most denumerable family of elements of $\mathbb{E}^+$. Applying Theorem 3.5 and once more Theorem 3.6 we get

$$C_P = C_{\underset{i\in I}{*} U_{E_i}} = \sum_{i\in I} C_{U_{E_i}} \circledast \underset{j\in I\setminus\{i\}}{*} U_{E_j} = \sum_{i\in I} (C_{E_i} \circledast U_{E_i}) \circledast \underset{j\in I\setminus\{i\}}{*} U_{E_j}$$

$$= \sum_{i\in I} C_{E_i} \circledast \underset{j\in I}{*} U_{E_j} = \left(\sum_{i\in I} C_{E_i} \right) \circledast P = C_{\sum_{i\in I} E_i} \circledast P = C_{\tilde{P}} \circledast P \ .$$

The functional equation $C_P = C_{\tilde{P}} \circledast P$ is characteristic for elements of $\mathbb{T}$. Kummer and Matthes (1970) (see also Mecke, 1967, and Tortrat, 1969) have proved the following theorem:

Theorem 3.7. A measure $P \in \mathbb{P}$ is infinitely divisible iff the functional equation

$$C_P = V \circledast P$$

is solvable by a σ-finite measure V in *C*. The solution is unique: $V = C_{\tilde{P}}$.

In the case $(A,\rho_A) = R^n$, the following version of the Campbell measure is useful.

The one-to-one mapping

$$(a,\Phi) \longrightarrow (a,T_a\Phi)$$

transforms C onto $R^n \times M^0$, where M^0 denotes the set $\{\Phi : \Phi \in M, \Phi(\{0,\ldots,0)\}) > 0\} \in \mathcal{M}$. For every any measure V in C we write

$$V^T = V\{(a, T_a\Phi) \in (\cdot)\} \ .$$

The measure V^T is defined in $R^n \times M^0$, where $\mathcal{M}^0$ comprises each $Y \in \mathcal{M}$, $Y \subseteq M^0$.

Mecke (1967) has given an interesting characterization of stationarity:

Theorem 3.8. A σ-finite measure H in M is stationary iff C_H^T splits in a direct product

$$C_H^T = \mu \times Q \ .$$

The σ-finite measure Q in M^0 is unique. We write $Q = H^0$. The measure H^0 is called the *Palm measure* corresponding to H. The total mass $H^0(M^0)$ is equal to the intensity i_H of H. In the case $0 < i_H < +\infty$ the probability distribution

$$H_0 = (i_H)^{-1} H^0$$

is called the *Palm distribution* corresponding to H. For stationary $P \in \mathbb{T}$

$$\mu \times P^0 = C_P^T = (C_{\widetilde{P}} \circledast P)^T = C_{\widetilde{P}}^T \circledast P = \{\mu \times (\widetilde{P})^0\} \circledast P = \mu \times \{(\widetilde{P})^0 * P\}.$$

The original functional equation $C_{\widetilde{P}} = C_P \circledast P$ reduces to $P^0 = (\widetilde{P})^0 * P$, and instead of 3.7, we get (see Kerstan and Matthes, 1964; Ambarzumjan, 1966; Mecke, 1967, and Tortrat, 1969)

Theorem 3.9. A stationary $P \in \mathbb{P}$ is infinitely divisible iff the functional equation

$$P^0 = H * P$$

is solvable by a σ-finite measure H in M^0. The solution is unique: $H=(\tilde{P})^0$.

In the case $P = P_{\ell\mu}$, $\ell \geq 0$ the measure $(\tilde{P})^0$ coincides with $\ell\delta_{\delta_{(0,\ldots,0)}}$, and Theorem 3.9 yields (Slivnyak, 1962)

Theorem 3.10. A stationary $P \in \mathbb{P}$ is equal to $P_{\ell\mu}$, $\ell \geq 0$ iff

$$P^0 = \ell\delta_{\delta_{(0,\ldots,0)}} * P .$$

4. HOMOGENEOUS CLUSTER-FIELDS

Within this Section we assume (A,ρ_A) to coincide with R^n. A cluster-field (in R^n with state space R^n) is called *homogeneous* if

$$\kappa_{(x)} = \kappa_{\{(0,\ldots,0)\}}\{(T_{-x}\chi) \in (\cdot)\} \qquad (x \in R^n) \quad .$$

A homogeneous cluster-field κ is uniquely determined by the probability measure $\kappa_{\{(0,\ldots,)\}}$:

Theorem 4.1. The mapping

$$\kappa \longrightarrow \kappa_{\{(0,\ldots,0)\}}$$

of the set of all homogeneous cluster-fields into $\mathbb{P}$ is onto and one-to-one.

We denote by $[Q]$ the homogeneous cluster-field corresponding to $Q \in \mathbb{P}$.

Theorem 4.2. A $P \in \mathbb{P}$ possesses a cluster-representation

$$P = (P_{\ell\mu})_{[Q]} \qquad (\ell \geq 0,\ Q \in \mathbb{P})$$

iff P is stationary and regular infinitely divisible.

Let $\mathbb{D}$ be the set of all $D \in \mathbb{P}$ such that

$$\rho_D(R^n) = 1 .$$

For any stationary $P \in \mathbb{P}$ with finite intensity i_P the probability measure $P_{[D]}$ exists. $P_{[D]}$ is stationary and

$$i_{P_{[D]}} = i_P .$$

We define

$$D^{[0]} = \delta_0, \quad D^{[n]} = (D^{[n-1]})_{[D]} \qquad \text{for } n = 1,2,\ldots .$$

Then

$$\underbrace{\{\ldots(P_{[D]})\ldots\}_{[D]}}_{n \text{ times}} = P_{[D^{[n]}]} .$$

Any solution $P \in \mathbb{P}$ of the functional equation

$$P_{[D]} = P$$

describes a possible statistical equilibrium with respect to the clustering-operation $Q \longrightarrow Q_{[D]}$. Liemant (1969) has proved:

Theorem 4.3. For any $D \in \mathbb{D}$, such that ρ_D is a non-lattice distribution R^n, the functional equation

$$P_{[D]} = P$$

possesses a non-trivial (different from δ_0) stationary solution $P \in \mathbb{P}$ with finite intensity iff

$$\inf_{n=1,2,\ldots} \int D^{[n]}[\chi\{[0,1)^n + x\} > 0]\, \mu(dx) > 0 .$$

The following theorem is due to Debes, Kerstan, Liemant and Matthes (1970);

Theorem 4.4. Let D be an element of $\mathbb{D}$ such that ρ_D is a non-lattice distribution and

$$\inf_{n=1,2,\ldots} \int D^{[n]}[\chi\{[0,1)^n + x\} > 0]\ \mu(dx) > 0 .$$

Then for every $\ell \geq 0$ there exists exactly one ergodic stationary solution $]D[_\ell \in \mathbb{P}$ with intensity ℓ of the functional equation

$$P_{[D]} = P .$$

The probability measures $]D[_\ell$, $\ell \geq 0$ form a semi-group:

$$]D[_{\ell_1} *]D[_{\ell_2} =]D[_{\ell_1+\ell_2} \qquad (\ell_1, \ell_2 \geq 0) .$$

Every stationary solution $P \in \mathbb{P}$ with finite intensity of the functional equation is uniquely representable as a mixture

$$P = \int]D[_\ell(\cdot)\ \sigma(d\ell)$$

of the infinitely divisible probability measures $]D[_\ell$.

To any probability distribution γ in R^n there corresponds the probability distribution Q_γ in M. Clustering with respect to $[Q_\gamma]$ is equivalent to a random displacement of the points with respect to γ. It is due to Doob (1953) that

$$(P_{\ell\mu})_{[Q_\gamma]} = P_{\ell\mu} \qquad \text{for all } \ell \geq 0 .$$

Because $\rho_{Q_\gamma} = \gamma$, Theorem 4.4 yields (Dobrushin, 1956; see also Stone, 1968):

Theorem 4.5. Let γ be an arbitrary non-lattice probability distribution in R^n. Then for every $\ell \geq 0$ the stationary Poisson measure $P_{\ell\mu}$ is the unique ergodic solution with intensity ℓ of the function equation

$$P_{[\mathfrak{Q}_\gamma]} = P \quad .$$

Every stationary solution $P \in \mathbb{P}$ with finite intensity of this functional equation is uniquely representable as a mixture

$$P = \int P_{\ell\mu}(\cdot)\ \sigma(d\ell) \ .$$

REFERENCES

Ambarzumian, R. W. (1966). On an equation for stationary point processes. Akad. Nauk Armjan SSR Dokl. 42, 3, 141-7 (in Russian).

Debes, H., Kerstan, J., Liemant, A. & Matthes, K. (1970). Verallgemeinerungen eines Satzes von Dobruschin I. Math. Nachr. 47, 183-244.

Dobruschin, R. L. (1956). On Poisson laws for distributions of particles in space. Ukrain. Math. J. 8, 127-34.

Doob, J. L. (1953). *Stochastic Processes*. New York: Wiley.

Franken, P., Liemant, A. & Matthes, K. (1965). Stätionare zufällige Punktfolgen III. Jber. Deutsch Math-Verein 67, 4, 183-202.

Kerstan, J. & Matthes, K. (1964). Verallgemeinerung eines Satzes von Sliwnjak. Revue Roumaine de mathématiques pures et appliquées 9, 811-29.

Kerstan, J. & Matthes, K. (1965). Ergodische unbegrenzt teilbare stationäre zufällige Punktfolgen. *Trans. 4th Prague Conf. on Information Theory, Statistical Decision Functions, Random Processes*, 399-415.

Kerstan, J., Matthes, K. & Mecke, J. (1972). *Unbegrenzt teilbare Punktprozesse*. Berlin: Akademie-Verlag (in preparation).

Lee, P. M. (1964). A structure theorem for infinitely divisible point processes. Unpublished address to I.A.S.P.S., Berne.

Lee, P. M. (1967). Infinitely divisible stochastic processes. Z. Wahrscheinlichkeitstheorie verw. Geb. 7, 147-60.

Lee, P. M. (1968). Some examples of infinitely divisible point processes. Studia Sci. Math. Hung. 3, 219-24.

Kummer, G. & Matthes, K. (1970). Verallgemeinerung eines Satzes von Slivnyak II. Revue Roumaine de Mathématiques pures et appliquees 15, 845-70.

Liemant, A. (1969). Invariante zufällige Punktfolgen. Wiss. Z. Friedrich-Schiller-Universitat Jena, Math.-Nat. Reihe 38, 97-114.

Matthes, K. (1969). Eine Charakterisierung der kontinuierlichen unbegrenzt teilbaren Verteilungsgesetze zufälliger Punktfolgen. Revue Roumaine de Mathématiques pures et appliquées 14, 1121-7.

Mecke, J. (1967). Stationäre Masse auf lokalkompakten Abelschen Gruppen. Z. Wahrscheinlichkeitstheorie verw. Geb. 9, 36-58.

Mönch, G. (1971). Verallgemeinerung eines Satzes von A. Renyi. Studia Sci. Math. Hung. 6, 81-90.

Slivnyak, I. M. (1962). Some properties of stationary flows of homogeneous random events. Teor. Veroyatnost. i Primenen 7, 347-52 (in Russian).

Stone, C. (1968). On a theorem of Dobrushin. Ann. Math. Statist. 39, 1391-401.

Tortrat, A. (1969). Sur les mesures aléatoires dans les groupes non Abéliens. Ann. Inst. H. Poincaré B (N.S.) 5, 31-47.

THE PREDICTION OF STATIONARY POINT PROCESSES

J. Jowett
Victoria University, Wellington, New Zealand

D. Vere-Jones
Victoria University, Wellington, New Zealand

SUMMARY

Two approaches to the prediction theory of point processes are reviewed. The first is an extension of the second order linear least squares prediction theory of mean square continuous processes. It represents a special case of the more general extension of this theory to stationary random distributions. The second approach consists in an attempt to determine the conditional distribution of the time to the next occurring event, given some configuration of past events. In principle, such conditional distributions can be obtained by an extension of the theory of Palm functions; however, in order to obtain formulae that can be used in practice it is necessary to resort to ad hoc methods. The two approaches are illustrated by several examples.

1. INTRODUCTION

By and large, prediction problems for point processes fall into two classes. In the first, we may be interested in finding the distributions of the numbers of events in future time intervals; in the second, the distributions of the times to and between various configurations of events. The first type leads naturally to a consideration of the point process as a random measure $dN(t)$ and more generally as a random distribution or random generalized function; the second to a consideration of the process as a set of intervals between events. It is our aim in this paper to review some results which are relevant to these different approaches, and to illustrate their application by constructing some particular examples of prediction formulae.

To make a start on the first type of problem, we may consider a point process as a sequence of delta functions occurring at random points in time. The process then becomes a special case of the "stationary random distributions", studied by Ito (1954), Gel'fand and Vilenkin (1964), Rozanov (1959) and Balagangadharan (1960) among others. These authors have shown that the Wiener-Kolmogorov theory for the linear least squares prediction of a second-order stationary random process has an elegant and natural generalization to random distributions. In particular, the Wold decomposition of a stationary process into deterministic and non-deterministic parts, and the backward moving average representation of the non-deterministic part, have been generalized to random distributions by the above authors, and these results are immediately applicable to stationary point processes, or stationary random measures.

A virtually equivalent generalization of the Wiener-Kolmogorov theory, due mainly to Yaglom (1955), is to mean-square continuous processes with stationary nth order increments. Since a point process or random measure may be regarded as a process with stationary first order increments, these results also are directly applicable to point processes and random measures. It is mainly a matter of personal taste which approach is preferred. We have chosen to outline here the theory of stationary random distributions, because the results seem to us more transparent when applied to point processes, and closer to the spirit of the empirical spectral analysis of point processes developed by Bartlett (1963), Cox and Lewis (1966) and others. However, we have followed Yaglom by devoting particular attention to the case of processes with rational spectral density.

This material is outlined in Sections 2 and 3 of the present paper. Section 2 contains a summary of basic results on the spectral theory for stationary random distributions, and Section 3 a more detailed account of prediction theory for point processes with rational spectral density. The main result to emerge here is that for the latter class of processes, the "spectral characteristic for extrapolation", $\psi_\tau(\lambda)$, satisfies the inequality

$$\int |\psi_\tau(\lambda)|^2 f(\lambda)\, d\lambda < \infty,$$

where $f(\lambda)$ is the spectral density. In such cases the prediction formula provides an estimate $m(t + \tau|t)$ of a quantity which may be regarded as the mean rate of occurrence, or density, of the process at time $t + \tau$, conditional on the past of the process up to time t.

In a sense, this result only serves to indicate the limitations of this second-order theory in the point process context. An estimate of the mean rate of occurrence at some time in the future may be of interest in discussing some broad, general questions, or in planning for periods which are large in comparison with the time interval between events. More commonly, however, we will have chosen to analyse the process as a point process precisely because we are concerned with time intervals which are of the same order of magnitude as the intervals between events. In many cases the outstanding quantity of interest will be the time to the next event. In attempting to estimate quantities of this kind, a knowledge of the expected mean rate of occurrence is of very limited value. These considerations are related to the general fact that for processes which deviate substantially from the Gaussian model, as all point processes must when considered in the small, the second order properties characterize the full distributions only very partially.

If, therefore, one is looking for prediction techniques which are likely to be practically effective, the second approach appears to hold out the better prospects. The main difficulty here is that little exists in the way of a systematic theory, so that any prediction results are likely to depend on incidental features of the particular model which render the prediction problem tractable. At first glance it might be thought that useful results could be obtained by treating the intervals as a stationary sequence, and applying the Wiener-Kolmogorov prediction theory to these. However, the most vital piece of information concerning the past of the process is the time that has elapsed since the immediately preceding event, and we have not been able to discover any way in which this information can be combined with a second order analysis of the intervals. In effect we are concerned with quantities such as the conditional distribution of the time to the next event, given some or all of the past of the process, and it is the immediate past, and in particular the time which has elapsed since the preceding event, which is of most relevance.

From this point of view, a natural starting point for the second type of problem is the Palm-function theory for stationary point processes as outlined, for example, in Chapter 7 of the review paper by Daley and Vere-Jones (1972). (This paper will subsequently be referred to as D and V-J.) This theory provides conditional distributions of quantities such as the time to the next event, given that an event occurred "at the origin". More generally, we describe extensions of this theory which provide the conditional distributions of similar quantities, given more elaborate configuration of past events. The main difficulty in the face of such an extension does not lie in proving the existence of such conditional distributions, so much as in finding any sort of tractable mathematical formulae which can be used to compute the conditional distributions when more than one or two past events are included.

In the present paper we present in Section 4 a very brief discussion of the general situation, and then pass on in Section 5 to some explicit results for a simple particular model--a Neyman-Scott clustering process in which the distribution of distances (time intervals) of subsidiary events after cluster points is exponential. It is possible to analyse this process in terms of an associated Markov process, and thereby to solve in principle all prediction problems relating to the process. Even with this example, the formulae become very complex, but a computationally feasible algorithm can be obtained. Some possible extensions of this work are also indicated--for example, to the case where the parent process (for cluster events) is a renewal process rather than a Poisson process.

2. BASIC RESULTS IN THE SPECTRAL THEORY OF STATIONARY RANDOM DISTRIBUTIONS

As a preliminary to discussing the prediction formulae themselves, we shall briefly summarize the basic results pertaining to the spectral theory of stationary random distributions. We shall not enter into detailed proofs, as these are available in the original papers by Ito (1954), Rozanov (1959), and Balagangadharan (1960): a thorough introduction to the topic is contained in the monograph by Gel'fand and Vilenkin (1964). We shall, however, sketch a derivation of the principal results for the special case of a point process or random measure, based on one of the standard techniques in this area--smoothing the process until it becomes mean square continuous and the standard theory can be applied.

Let K denote the space of test functions which are infinitely differentiable, with bounded support, and suppose that K is given its usual topology (see, for example, Schwartz, 1950). Then by a random distribution we shall understand a linear, continuous mapping from K into the Hilbert space $L_2(\rho)$ of equivalence classes of random variables defined on a given probability space (Ω, B, ρ), and possessing finite second moments. Thus a random distribution is a continuous, linear random-variable valued functional on K. As a matter of convenience, we shall also require, and assume henceforth without further mention, that *the means are taken to be zero*: if X is a random distribution, then $E[X(\emptyset)] = 0$, (all $\emptyset \in K$).

As examples, any mean square continuous stochastic process x(t) with zero mean defines a random distribution X(.) by the equation

$$X(\emptyset) = \int x(t)\, \emptyset(t)\, dt ,$$

where the integral is to be understood as a mean square limit (Bochner integral). Similarly, a point process or random measure dN(t) with finite second moment measure defines a random distribution through the equation

$$X(\emptyset) = \int \emptyset(t)\, dN(t) - m \int \emptyset(t)\, dt ,$$

where $m\,dt = E\{dN(t)\}$, so that m is the intensity or mean rate of the process. In both cases the linearity is obvious, and the continuity condition required of a random distribution follows from the assumptions made on the second moments of the processes.

Now define a shift operator for functions by the equation $T_h\phi(h) = \phi(t + h)$, and for random distributions by the equation $T_h X(\phi) = X(T_{-h}\phi)$. Then we shall say that the random distribution $X(\phi)$ is *stationary* (actually second-order stationary, but in this and the following section we shall have no need for any stronger concept), if

$$E\{T_h X(\phi)\ \overline{T_h X(\psi)}\} = E\{X(\phi)\ \overline{X(\psi)}\}\ . \tag{2.1}$$

In the two illustrations above, the random distributions induced by the processes $x(t)$, $dN(t)$ are stationary in this sense if the processes which induce them are stationary in the usual sense for such processes. The boundedness properties of stationary processes (as illustrated, for example, by the mean square ergodic theorem) imply that a stationary process or random measure defines a random distribution not only on K, but also on the larger space S of test functions which are infinitely differentiable and together with all their derivatives converge to zero at infinity faster than any power of t.

Whenever the random distribution satisfies condition (2.1), it can be shown that there exists a unique distribution C(.), which we shall refer to as the *covariance distribution* of X, such that

$$E\{X(\phi)\ X(\psi)\} = C(\phi * \bar{\psi}_-)\ , \tag{2.2}$$

where $\psi_-(t) = \psi(-t)$ and $*$ denotes convolution. Since $E\{|X(\phi)|^2\} \geq 0$, this covariance distribution is *positive definite* in the sense that for all $\phi \in K$,

$$C(\phi * \bar{\phi}_-) \geq 0\ . \tag{2.3}$$

The first principal result in the theory of random distributions is then the following extension of Bochner's Theorem.

Theorem 1 (Bochner-Schwartz Theorem). Let C(.) be a positive definite distribution on K. Then there exists a Borel measure dF, which may not be totally finite, but which integrates $(1 + \lambda^2)^{-k}$ for some finite, positive k, such that for $\emptyset \in K$, and with $\tilde{\emptyset}(\lambda) = \int e^{i\lambda t} \emptyset(t)\, dt$,

$$C(\emptyset) = \int \tilde{\emptyset}(\lambda)\, dF(\lambda) . \tag{2.4}$$

A measure satisfying the boundedness condition of the theorem is usually referred to as a tempered measure. Equation (2.4) is of Parseval type, and hence asserts that any positive definite distribution can be represented as the Fourier transform of a tempered measure. Since a tempered measure defines a distribution on S as well as on K, and since S is self-dual under the Fourier transformation, it is implicit in the statement of the theorem (and an essential ingredient in its proof) that any positive definite distribution on K is in fact a distribution on S also. From this it can be deduced that any stationary random distribution on K can be extended to a stationary random distribution on S.

When C is the covariance distribution associated with a random distribution X, the measure dF will be referred to as the *spectral measure* of X. The value of the smallest integer k satisfying the condition of the Theorem is then a measure of the degree of local irregularity of the realizations of X. In particular, if X is derived from a mean-square continuous process, we can take k = 0, and Theorem 1 reduces to the familiar version of Bochner's Theorem. In the case of a point process or random measure, it follows from the arguments presented below that we can always take k = 1, and in fact it is true (see Section 4 of D and V-J) that the measure integrates $(1 + \lambda^2)^{-\alpha}$ for any $\alpha > 1/2$.

Equation (2.4) provides a Khinchin-type representation of the covariance distribution, and from it, by essentially identical arguments to those of the classical case, follows a Cramer-type representation for the process itself.

Theorem 2. *(Ito).* Let X(.) be a stationary random distribution with spectral measure dF. Then there exists a random measure $d\xi$, with orthogonal increments, such that $E|d\xi(\lambda)|^2 = dF(\lambda)$, and for $\phi \in K$,

$$X(\phi) = \int \tilde{\phi}(\lambda)\ d\xi(\lambda) \quad \text{a.s.} \tag{2.5}$$

As usual, the integral on the right of (2.5) is to be interpreted as an integral in mean square. Equation (2.5) again has the form of a Parseval equation and as such uniquely determines the correspondence between X and ξ; we omit any derivation of explicit inversion theorems.

The basis for the linear least squares prediction theory is the Wold decomposition of the process into its deterministic and non-deterministic parts, and the backward moving average representation of a purely non-deterministic process. To describe these results in the present context, let $L_2(X)$ denote the Hilbert space spanned by the random variables $X(\phi)$ $(\phi \in K)$, and let $L_2(X;a)$ denote the subspace spanned by those $X(\phi)$ for which ϕ has its support in $(-\infty,a]$. Thus $L_2(X;a)$ has its usual interpretation as the space spanned by linear combinations of past values of the process. The random distribution X is then said to be deterministic if $\bigcap_a L_2(X;a) = L_2(X)$, and purely non-deterministic if $\bigcap_a L_2(X;a) = \{0\}$. The following two theorems are slight rearrangements of results given by Rozanov (1959) and Balagangadharan (1960).

Theorem 3. Any stationary random distribution can be written uniquely in the form $X = X_1 + X_2$, where the random distributions X_1 and X_2 are mutually orthogonal, X_1 is deterministic, and X_2 is purely non-deterministic. Furthermore, the deterministic component is absent if and only if the spectral measure dF is absolutely continuous, and its density $f(\lambda)$ satisfies the condition

$$\left| \int \frac{\log f(\lambda)}{(1+\lambda^2)}\ d\lambda \right| < \infty\ . \tag{2.6}$$

Theorem 4. (i) For $X(\emptyset)$ to be purely non-deterministic, it is necessary and sufficient that the spectral measure should be expressible in the form

$$dF(\lambda) = |g(\lambda)|^2 \, d\lambda \,, \tag{2.7}$$

where $g(\lambda)$ is the product of a polynomial and a square-summable function, and is the Fourier transform of a distribution G (on S) with support in $(-\infty, 0]$; or equivalently that the process itself can be represented in either of the two forms

$$X(\emptyset) = \int g(\lambda) \, \tilde{\emptyset}(\lambda) \, d\beta(\lambda) \qquad \text{(frequency domain)}, \tag{2.8}$$

$$X(\emptyset) = \int (G * \emptyset)(u) \, dB(u) \qquad \text{(time domain)} \,, \tag{2.9}$$

where $d\beta$ and dB are random measures with stationary orthogonal increments:

$$E\{|d\beta(\lambda)|^2\} = d\lambda \,, \qquad E\{|dB(t)|^2\} = dt \,.$$

(ii) The representations (2.7)-(2.9) are not unique in general, but may be made so (up to a constant of unit modulus in $g(\lambda)$) by the requirement that the representations in (2.8) and (2.9) be invertible, for which it is necessary and sufficient that $g(\lambda)$ be zero-free (as well as analytic) in the lower half-plane $Im(\lambda) < 0$.

Note. The last condition for mean-square continuous processes is often framed in terms of a maximality condition on G(0). This condition cannot be directly applied here as G(0) is infinite, but an equivalent condition could be stated in terms of the maximality of the decomposition for smoothed versions of the process.

The particular representation (2.9) referred to in part (ii) of the theorem will be referred to as the *canonical representation* of X.

For the special case of a point process or random measure, very simple proofs of these results can be obtained by introducing the smoothed process

$$x(t) = \int_0^\infty e^{-u}\, dN_u(t - u) . \tag{2.10}$$

When dN is stationary with finite second moment measure, x(t) is stationary and mean square continuous, and so has spectral measure, say $dF_x(\lambda)$, which is totally finite. We then define the spectrum of the process as $dF(\lambda) = (1 + \lambda^2)\, dF_x(\lambda)$, and verify that the Parseval relation (2.4) is valid. Similarly, if the x(t) process has a Cramer representation $x(t) = \int e^{i\lambda t}\, dZ_x(\lambda)$, we define $dZ(\lambda) = (1 + i\lambda)\, dZ_x(\lambda)$, and verify that (2.5) holds. Finally, it is not difficult to show that the pasts of the x- and N-processes are equivalent, so that $L_2(N) = L_2(x)$, and $L_2(N;a) = L_2(x;a)$ for every a; thus the x- and N-processes are deterministic or purely non-deterministic together, and their Wold decompositions correspond. The remaining results in Theorems 3 and 4 now follow from the corresponding results for the x(t)-process, and we leave the details to the reader (see also Vere-Jones, 1972).

Example 1. Two-Point Cluster Process. We suppose that cluster centres occur at the instants of a Poisson process with parameter μ, and that each cluster contains just two members, one at the cluster centre and the other at a fixed distance h behind the cluster centre. Evidently, the process can be represented in the symbolic form

$$dN(t) = dP(t) + dP(t - h) \tag{2.11}$$

where dP(t) is the Poisson process of cluster centres. (Here, as elsewhere in the text, we shall use such differential formulae as a convenient and almost essential shorthand for writing more precise results which can be obtained by integrating both sides of the equation with a suitable test function; thus for example (2.11) is to be understood as an abbreviation for

$$\int \phi(t)\, dN(t) = \int \phi(t)\, dP(t) + \int \phi(t)\, dP(t-h) . \tag{2.11a}$$

Since the Poisson process itself has orthogonal increments, this is already in the form of a moving average representation (2.9), although we should more strictly write it as

$$dX(t) = \mu\{dB(t) + dB(t-h)\} , \tag{2.12}$$

where $dX(t) = dN(t) - 2\mu\, dt$, $dB(t) = \{dP(t) - \mu dt\}/\mu$, in conformity with our convention that all processes should have zero means. Thus we find $G(u) \doteq \mu\{\delta(u) + \delta(u+h)\}$, where $\delta(u)$ is the Dirac delta-function, and hence $g(\lambda) = \mu\{1 + e^{-i\lambda h}\}$, $f(\lambda) = \mu^2|1 + e^{-i\lambda h}|^2$. Furthermore, the representation is invertible, since $g(\lambda)$ is zero-free in the upper half-plane, and corresponding to (2.12) we have the inverse relationship

$$dB(t) = \frac{1}{\mu} \sum_{j=0}^{\infty} (-1)^j\, dX(t - jh) . \tag{2.13}$$

(This formula is meaningful since only a finite number of terms will enter into the integral with respect to a test function from K, and the integral will clearly converge if we take a test-function from S.) The interpretation of this formula is that $dB(t)$ will have an atom at t whenever $dX(t)$ has an atom at t that is not cancelled out by a preceding atom at $t - h$.

This example illustrates the degree of irregularity introduced by the fact that we are dealing with point processes rather than mean square continuous functions. The remaining two examples illustrate that the class of point processes with rational spectral density, which will be the main object of study in the next section, is wide enough to contain a variety of useful models.

Example 2. The Renewal Process with Gamma Density. The spectral density of a stationary renewal process may be written in the form (see for example Cox and Lewis, 1966, p. 79)

$$f(\lambda) = \frac{\mu}{\pi}\{1 + \frac{h(i\lambda)}{1 - h(i\lambda)} + \frac{h(-i\lambda)}{1 - h(-i\lambda)}\} ,$$

where $h(s)$ is the Laplace-Stieltjes transform of the distribution of the time interval between renewals and μ^{-1} is its mean. Clearly, this will be a rational function whenever $h(s) = (1 - \beta s)^{-n}$, corresponding to a gamma distribution between renewals, and more generally whenever the time between renewals is a sum or mixture of exponential distributions.

Example 3. Cluster Processes with Negative Exponential Distances. The spectral density of a general cluster process can be written in the form

$$f_X(\lambda) = \text{var } S(\lambda) + |\text{ave } S(\lambda)|^2 f_Y(\lambda) ,$$

where $f_Y(\lambda)$ is the spectrum of the parent (cluster centre) process, $S(\lambda)$ is the random variable $S(\lambda) = \Sigma_j e^{i\lambda t_j}$, the t_j are the points in a cluster with centre at zero and "ave" and "var" refer to expectations over the cluster distributions. We shall consider the special case where the cluster members are independently and exponentially distributed about the cluster centres, with mean cluster size μ_c and cluster size variance var_c. Then

$$\text{ave } S(\lambda) = \frac{\alpha^2}{\alpha^2 + \lambda^2} \mu_c ,$$

$$\text{var } S(\lambda) = |\text{var}_c - \mu_c| \frac{\alpha^2}{\alpha^2 + \lambda^2} + \mu_c ,$$

where α is the parameter of the negative exponential distribution for the distances.

If the process of cluster centres is Poisson, with rate μ, then

$$f_X(\lambda) = \mu \frac{\alpha^2(\text{var}_c + \mu_c) + \mu_c \lambda^2}{\alpha^2 + \lambda^2}$$

and we have a sufficient number of parameters at our control to construct a process of this type with any given spectral density of the form

$$f(\lambda) = a^2 \frac{b^2 + \lambda^2}{c^2 + \lambda^2}$$

where $b^2 > c^2$, and a^2, b^2, c^2 are positive constants.

Now taking this as the parent process, and generating a cluster process by the same technique, we may construct processes with spectra of the form

$$f(\lambda) = a^2 \frac{(b_1^2 + \lambda^2)(b_2^2 + \lambda^2)}{(c_1^2 + \lambda^2)(c_2^2 + \lambda^2)}$$

provided $b_1^2 > c_1^2$, $b_2^2 > c_2^2$, and so on. At each stage we introduce two new parameters, and thus have at our disposal quite a range of rational spectral densities. Further examples may be built up by combining this cluster process structure with the renewal structure of the preceding example for the process of cluster centres.

3. LINEAR LEAST SQUARES PREDICTORS

In principle, the linear least squares prediction problem is solved once the canonical representation (2.9) has been found. Indeed, (2.9) reduces the problem of predicting the original process X(.) to the problem of predicting the orthogonal random measure dB(.), and since the linear least squares predictor of $dB(t + \tau)$, given the past of B (or equivalently of X, since the canonical representation is invertible, and the past of either process determines that of the other) is just its unconditional expected value, namely zero, the latter problem is trivial. Developing this idea more formally, and specializing to the case of a point process or random measure, we can state the following theorem.

Theorem 5. Let dX(.) be a completely non-deterministic random measure, and let

$$dX(t) = \int_{-\infty}^{0} G(u)\, d_u B(t+u)\, dt; \qquad dB(t) = \int_{-\infty}^{0} H(v)\, d_v X(t+v)\, dt \qquad (3.1)$$

be the canonical representation and its inverse. Then the prediction formula for $dX(t+\tau)$, based on the past of the process up to time t, is given by

$$dX(t+\tau|t) = \int_{-\infty}^{0} G(u)\, d_u B(t+\tau+u)\, d\tau = \int_{-\infty}^{0} d_w X(t+w)\, K_\tau(w)\, d\tau , \quad (3.2)$$

where the distribution $K_\tau(x)$ is given by

$$K_\tau(w) = \int_{w}^{0} G(v-\tau)\, H(w-v)\, dv . \quad (3.3)$$

In the frequency domain this representation takes the form

$$dX(t+\tau|t) = \int e^{it\lambda}\, \psi_\tau(\lambda)\, d\xi(\lambda), \quad (3.4)$$

where the function $\psi_\tau(\lambda)$ is the Fourier transform of $K_\tau(w)$.

Following Yaglom (1962) we shall refer to $\psi_\tau(\lambda)$ as the *spectral characteristic for prediction*. Like $g(\lambda)$, it can be expressed as the product of a polynomial and a square summable function.

Example 5. The two-point cluster process (continued). Here we can write

$$dX(t+\tau|t) = dB(t+\tau|t) + dB(t+\tau-h|t) .$$

Now the first term on the right is zero for $\tau > 0$, and the second will be zero for $\tau > h$. For $0 < \tau \leq h$, however, the second term takes the form

$$dB(t+\tau-h) = \sum_{j=1}^{\infty} (-1)^{j-1}\, dX(t+\tau-jh);$$

this will in turn be zero unless there is an event at $t+\tau-h$ which is not matched by an earlier event distance h before it. In terms of the original point process $dN(t)$ we find

$$dN(t+\tau \mid t) = 2\mu \, d\tau \qquad (\tau > h)$$

$$= \mu \, d\tau + \sum_{1}^{\infty} (-1)^j \, dN(t+\tau-jh) \qquad (0 < \tau \leq h).$$

The spectral characteristic for prediction is here given by

$$\psi_\tau(\lambda) = 0 \qquad (\tau > h)$$

$$= -e^{-i(h-\tau)\lambda}/(1-e^{-ih\lambda}) \qquad (0 < \tau \leq h) .$$

It will be noted that for $0 < \tau \leq h$ the function $\psi_\tau(\lambda)$ is not square integrable with respect to the spectral density of the process, corresponding to the fact that, considered as a function of t, the predictor is not mean square continuous, but contains delta-function components. It may also be noted that in this example the distributions $G(u)$ and $K_\tau(u)$ are measures, a situation which may be regarded as typical for point processes and random measures. However, we have not been able to determine whether this is true in general; the open question is whether the conditions on the growth at infinity of the spectral density imposed by the fact that it corresponds to a random measure (see Section 4 of D and V-J) are sufficient to imply that $G(u)$ (the Fourier transform of its factor $g(\lambda)$) is of locally bounded variation.

In practice, both in the present situation and in the more familiar context of mean square continuous processes, the main difficulty in applying the preceding results is to convert them into a form suitable for numerical calculation. Effectively, the only general situation for which computationally feasible formulae can be obtained is where the process has rational spectral density. We therefore turn to a more detailed examination of this special case. Even for point processes, we have already seen that the class of processes with rational spectral density is wide enough to contain a variety of processes of practical importance, and hence also to provide useful approximations to processes of an even wider class.

First let us examine the canonical representation of a random measure with rational spectral density. The condition that the density $f(\lambda)$

should integrate $(1+\lambda^2)$ then implies that the denominator in $f(\lambda)$ is a polynomial of exact degree at least as great as the exact degree of the numerator. If it is strictly greater, the process has mean square continuous density, and the standard theory for mean square continuous processes can be applied to the density. Thus the case of interest is when the numerator and denominator have the same exact degree; in particular, this condition must obtain for any point process, in view of the delta function (atom) of the covariance measure at the origin. In this case the canonical factor $g(\lambda)$ can itself be taken to be a rational function with the same property, so that it can be written in the form

$$g(\lambda) = c + g_1(\lambda) \quad ,$$

where $g_1(\lambda)$ is rational with the denominator of strictly greater degree, and there is a similar equation for the inverse function $h(\lambda) = 1/g(\lambda)$. Now this implies that $g_1(\lambda)$ is an L_2-function, and hence that it is the Fourier transform of an L_2 function, $G_1(u)$, say. Then we can write the moving average representation for $dX(t)$ in the form

$$dX(t) = c\,dB(t) + \int_{-\infty}^{0} G_1(u)\, dB(t+u) \quad ,$$

with a similar inverse relation

$$dB(t) = c^{-1}\, dX(t) + \int_{-\infty}^{0} H_1(u)\, dX(t+u) \quad ,$$

where both G_1 and H_1 are square integrable functions. Now it is clear from these equations that the predictor $dX(t+\tau|t)$ has the form

$$dX(t+\tau|t) = \int_{-\infty}^{-\tau} G_1(u)\, dB(t+\tau+u) \quad . \tag{3.5}$$

Now $G_1(u)$ is square integrable, and hence (3.5) has the form of the moving average representation of a mean-square continuous process (in t). Consequently the spectral characteristic for prediction is square integrable with respect to the spectral density. Thus we reach the somewhat surprising conclusion that for a random measure with rational spectral density, even if the measure itself does not have a mean square continuous density, its predictor does. In this case (3.3) is more appropriately written as

$$x(t+\tau|t) = \int e^{it\lambda}\psi_\tau(\lambda)\, d\xi(\lambda) , \tag{3.6}$$

or in terms of the original process $dN(t)$, using $m(t+\tau|t)$ to denote the conditional density,

$$m(t+\tau|t) = m + \int e^{it\lambda}\psi_\tau(\lambda)\, d\xi(\lambda) . \tag{3.7}$$

From these formulae, the predictor for any integral $\int \phi(t-u)\, dN(u) = x_\phi(t)$ may be obtained by integrating with respect to τ, using the fact that the linear least squares predictor for a linear combination is the same linear combination of the least squares predictors.

The equation (3.6) is of exactly the same type, and involves exactly the same conditions on $\psi_\tau(\lambda)$ relative to the spectral density $f(\lambda)$, as the corresponding equation when the underlying process is mean square continuous. We can therefore take over without change the prescriptions for determining the function $\psi_\tau(\lambda)$ described by Yaglom (1962), and we obtain the following theorem.

Theorem 6. Let $dN(t) = m\, dt + dX(t)$ be a random measure with rational spectral density. Then the prediction of the process can be expressed in terms of a "prediction density" (3.7) which is, for fixed τ, a mean square continuous process in t. Correspondingly, the spectral characteristic for prediction is square integrable with respect to the spectral density, and may be identified by the following properties:

(i) $\psi_\tau(\lambda)$ *is analytic in the lower half-plane* $Im(\lambda) < 0$;

(ii) $[e^{-\tau\lambda} - \psi_\tau(\lambda)]f(\lambda)$ *is analytic in the upper half plane* $Im(\lambda) > 0$;

(iii) $\psi_\tau(\lambda)$ *has no worse than power growth as* $|\lambda| \to \infty$;

(iv) $\int |\psi_\tau(\lambda)|^2 f(\lambda)\, d\lambda < \infty$.

The application of these rules to derive the spectral characteristic for prediction in particular cases is discussed in detail by Yaglom (1962), and it is hardly necessary to go into further detail here. However, we may illustrate the preceding discussion by one of the simplest examples of a point process with rational spectral density, a Neyman-Scott cluster process with Poisson cluster centre process and negative exponential distances (Example 3 of the preceding section). Here we can write $f(\lambda)$ in the form

$$f(\lambda) = \left| a \frac{b + i\lambda}{c + i\lambda} \right|^2$$

where the latter is clearly the canonical factorization of Theorem 4. This corresponds to the moving average representation

$$dX(t) = a\{dB(t) + (b-c) \int_{-\infty}^{0} e^{cu} \, dB(u) \, dt\} ,$$

with the inverse relation

$$dB(t) = a^{-1}\{dX(t) - (b-c) \int_{-\infty}^{0} e^{bu} \, dX(u) \, dt\} .$$

Then the prediction density is

$$x(t+\tau | t) = a(b-c) \, e^{-c\tau} \int_{-\infty}^{0} e^{cu} \, dB(t+u)$$

$$= (b-c) \, e^{-c\tau} \int_{-\infty}^{0} e^{bu} \, dX(t+u) .$$

The spectral characteristic for prediction is

$$\psi_\tau(\lambda) = \frac{(b-c) \, e^{-c\tau}}{b+i\lambda} ,$$

and the reader may verify that it satisfies the condition (i)-(iv) of Theorem 6.

In terms of the original point process, $c = \alpha$, the parameter of negative exponential distribution of distances, $b = \alpha(1 + \mathrm{var}_c/\mu_c)^{1/2}$. Hence the prediction density can be written in the form

$$m(t+\tau|t)$$

$$= \frac{m}{(1+\text{var}_c/\mu_c)^{1/2}} + \alpha\{(1+\text{var}_c/\mu_c)^{1/2} - 1\}$$

$$\times e^{-\alpha\tau} \int_{-\infty}^{0} \exp\{u\alpha(1+\text{var}_c/\mu_c)^{1/2}\} dN(u) .$$

4. PREDICTORS BASED ON PALM FUNCTIONS

In this and the following section we take up the second approach fore-shadowed in the introduction, which may be briefly summarized as the attempt to obtain the distributions of times to future events, conditional on all or part of the past history of the process. It will be noted that this approach is more thorough-going than that of the preceding sections where, roughly speaking, we were content to characterize distributions in terms of their first and second moments, and our estimates were in the form of expectations of certain conditional distributions. Here, in addition to focussing attention on interval rather than counting properties, we shall attempt to determine the complete conditional distributions of our estimates.

We shall suppose that the process is (strictly) stationary and orderly (using the terms as in D and V-J), with finite rate m. We shall also suppose given the finite-dimensional distributions of the process; in the present context thse are most conveniently written in the form

$$P_{ijk\ldots}(x,y,z,\ldots) = \text{prob}\{i \text{ events in } (0,x),\ j \text{ events in } (x,x+y),$$
$$k \text{ events in } (x+y,\ x+y+z),\ \ldots\}.$$

The choice of zero as the first end-point is of course arbitrary, in view of the stationarity assumed.

In order to restrict the discussion to essentials, we shall suppose throughout that the quantity we wish to estimate is the time X to the next occurring event. Its distribution is most conveniently characterized in terms of the survivor function (one minus the corresponding distribution function). Thus, denoting the given past history of the process by H, our aim will be to find a meaningful interpretation for, and then to compute, the quantity

$$S(x;H) = \text{prob}(X > x \mid H) \ . \tag{4.1}$$

Let us start from the simplest cases. At the moment when we first observe a point process, we have no knowledge of its past history, and (4.1) reduces to the unconditional probability

$$S(x) = P_0(x) \ .$$

Similarly, if our only information is that we have observed the process for a time interval τ, without any events occurring, our conditional probability becomes

$$S_0(x;\tau) = S(x+\tau)/S(\tau) \ .$$

The situation becomes rapidly more complicated as soon as we start to observe any events, since we are then conditioning on infinitesimal probabilities, and at best we may hope that meaningful conditional probabilities can be written down in terms of derivatives of the $P_{ijk\ldots}(x,y,z,\ldots)$. In a few examples, such as the Renewal Process, or Wold's Markovian Process of Intervals (see Lawrence, 1972) we may have directly available the densities of the multivariate distributions of the time intervals between successive events. In such cases, of course, deriving the prediction formulae presents no great difficulties, and may (we hope) be safely left to the reader. In the general case, our problem is essentially to build up formulae for these interval densities from the given distributions $P_{ijk\ldots}(x,y,z,\ldots)$. Now, the fact that a well-defined set of finite-dimensional distributions for the intervals exists, and a formal algorithm for deriving them in terms of the original finite dimensional distributions, are two principal assertions of the theory of *Palm Functions* or *Palm measures*, for a discussion of which we may refer in the first instance to Section 7 of D and V-J, and to the references given therein. However, what we need in the present context, is not so much a proof of the existence of the interval distributions, as a set of formulae for computing their densities when the latter exist. This aspect of the theory does not seem to have been treated in any detail; what it calls for is an extension of the Palm-Khinchin equations (Section 7.3 of D and and V-J) to a multiple derivative situation. Without attempting to set down any formal theorems, let us indicate some of the points involved by deriving some more complicated examples of prediction formulae.

We first consider the situation where our information concerning the past of the process is restricted to the length of time τ back to the last occurring event. We might then attempt to write down the conditional probability that we require in the form

$$S_{10}(x;\tau) = \lim_{h\to 0} P_{10}(h, \tau+x)/P_{10}(h,\tau) .$$

Now the simplest example of the Palm-Khinchin formulae asserts the existence of the derivative

$$\psi_{10}(x) = -\frac{1}{m}\frac{dP_0(x)}{dx} = \lim_{h\to 0} P_{10}(h,x)/\{1 - P_0(h)\} ,$$

and interprets the limit as a kind of conditional probability that an interval of length x will follow an event at zero, given the event at zero. From this we obtain the required conditional probability in the form

$$S_{10}(x;\tau) = \psi_{10}(x+\tau)/\psi_{10}(\tau) .$$

We may note in passing that the formula above is exact for renewal processes, and may be expected to provide a reasonable approximation for more general processes when the intervals are approximately independent, as they often appear to be in real examples.

The next most complicated situation is to suppose that we have a past history of length $\tau_1+\tau_2$, comprising an initial interval of length τ_1 before an observed event, and then a subsequent interval, with no further events, of length τ_2. Here, we may treat the required conditional probability as the limit

$$S_{010}(x;\tau_1,\tau_2) = \lim_{h\to 0} P_{010}(\tau_1,h,\tau_2+x)/P_{010}(\tau,h,\tau_2).$$

But we can write

$$P_{010}(\tau_1,h,\tau_2) = P_{01}(\tau_1, \tau_2+h) - P_{01}(\tau_1+h, \tau_2) ,$$

which leads us to introduce the Palm-type function

$$\psi_{010}(\tau_1,\tau_2) = \lim_{h\to 0} P_{010}(\tau_1,h,\tau_2)/[1 - P_0(h)]$$

$$= \frac{1}{m} \left(\frac{\partial P_{01}}{\partial \tau_2} - \frac{\partial P_{01}}{\partial \tau_1}\right) ,$$

where we assume the existence of the partial derivatives here, as also in the formulae below. Then we obtain the required conditional distribution in the form

$$S_{010}(x;\tau_1,\tau_2) = \psi_{010}(\tau_1, \tau_2+x)/\psi_{010}(\tau_1,\tau_2) .$$

Similarly, if we are given that times τ_2, $\tau_1 + \tau_2$ have elapsed since the last two preceding events, we set

$$\psi_{1010}(\tau_1,\tau_2) = \frac{1}{m^2} \left(\frac{\partial^2 P_{01}}{\partial \tau_1 \partial \tau_2} - \frac{\partial^2 P_{01}}{\partial \tau_2^2}\right)$$

and find for the required conditional probability

$$S_{1010}(x;\tau_1,\tau_2) = \psi_{1010}(\tau_1, \tau_2+x)/\psi_{1010}(\tau_1;\tau_2) .$$

If we are given intervals τ_1 up to the first observed event, τ_2 from this to the second observed event, and a final interval τ_3 with no observed events, we set

$$\psi_{01010}(\tau_1,\tau_2,\tau_3) = \left(\frac{\partial}{\partial\tau_2} - \frac{\partial}{\partial\tau_1}\right)\left(\frac{\partial}{\partial\tau_3} - \frac{\partial}{\partial\tau_2}\right) P_{011}(\tau_1,\tau_2,\tau_3)$$

and derive the conditional probability in the form

$$S_{01010}(x;\tau_1,\tau_2,\tau_3) = \psi_{01010}(\tau_1,\tau_2,\tau_3+x)/\psi_{01010}(\tau_1,\tau_2,\tau_3) \, .$$

The reader will perceive the pattern upon which the above results are built up, and we forbear from further elaboration, not least from the suspicion that in practice there are few cases in which the higher order finite dimensional distributions are available explicitly, and that where they are, the prediction formulae may be obtained more easily in other ways.

5. PREDICTORS BASED ON REDUCTION TO A MARKOV PROCESS

In a few special cases it may be possible to relate the evolution of the point process to the evolution of an underlying Markov process. Since the transition probabilities of a Markov process are nothing other than the conditional distributions of a future state of the system, given its present state, one may hope that in such cases the formulae of the preceding section will reduce to a more manageable form. We shall analyze one example of this kind, the Neyman-Scott process with exponential distances from the cluster centre already described in Section 2 and Section 3.

In this example, the underlying Markov process may be described as the process of *unexpired secondary events*. If we associate with each cluster centre t_i the random variable M_i equal to the number of secondary events in the cluster it generates, and consider the number $U_{S,T}$ of such events which have been generated by cluster centres in the interval (S,T) and have not expired (been observed) up to time T, then $U_{S,T}$, as a function of T, has the Markovian property. Indeed, it is equivalent to an "immigration-death" process in which the "immigrants" arrive in batches of size M_i, and have life-times which are exponentially distributed with parameter α (keeping to the notation of example 3 of Section 2).

If the mean cluster size is finite, we can also show that the random variable $U_{S,T}$ increases monotonically as $S \to -\infty$, to a proper limit random variable, say U_T. To prove this, it is enough to show that the limit is a.s. finite. But $E(U_{T-1,T}) = \rho$, say, is clearly finite if the mean cluster size is finite, and then $E(U_{T-k,T}) = \Sigma_{r=0}^{k-1} \rho\, e^{-r\alpha}$ remains bounded as $k \to \infty$, so that $E(U_T) < \infty$, and hence U_T is a.s. finite.

From stationarity, the distribution of U_T may be identified with the limit distribution of $U_{S,T}$ as $T \to \infty$, and is therefore just the equilibrium distribution of the Markov process. Using standard techniques based on the forward Kolmogorov equations, we find that the generating function $P(z)$ of U_T (which is, of course, independent of T, in view of the stationarity) satisfies the differential equation

$$\alpha(1-z)\frac{dP}{dz} + \mu Q(z)P = -\mu P , \tag{5.1}$$

where as before μ is the parameter of the Poisson process of cluster centres, and $Q(z) = \Sigma_0^\infty q_n z^n$ is the generating function of the cluster size distribution.

The solution of (5.1) is straightforward, but in the present context it is perhaps more instructive to derive the generating function of U_T directly from the following more general lemma.

Lemma 5.1. Consider a cluster point process in which cluster members follow cluster centres at distances which are independently distributed with common d.f. $F(x)$, cluster sizes are independently distributed with finite mean and common p.g.f. $Q(z)$, and the cluster centre process has p.g.fl. $G[\xi]$. Then the process U_T, defined as above, has p.g.f.

$$\Pi_T(z) = G\{Q[1 - (1-z)\{1 - F^*(T-t)\}]\} , \tag{5.2}$$

where $F^*(x) = 1$ for $x < 0$ and $F^*(x) = F(x)$ for $x \geq 0$.

Of course, when the cluster centre process is stationary, the dependence on T is illusory.

To prove the lemma observe that conditional on a cluster centre at t_i, the number of unexpired cluster events at T has p.g.f. $Q(p+qz)$, with $p = F(T-t_i)$ and $q = 1-p$. In view of the independence of different clusters assumed in the definition of a cluster process, the overall number of unexpired cluster members at T, conditional on cluster centres at t_i, is

$$\prod_{t_i \leq T} Q[1 - (1-z)\{1 - F(T-t_i)\}] .$$

Now taking expectations with respect to the t_i, and referring back to the definition of the probability generating functional (D and V-J, Section 3) we obtain the formula (5.2). (For related results, see Westcott, 1970, and the discussion round Section 3.2 of Smith, 1958.)

In the special case of interest, with Poisson cluster centre process and negative exponential distances, $G[\xi] = \exp[\mu \int_{-\infty}^{\infty} \{\xi(t)-1\}\, dt]$, $1-F(x) = e^{-\alpha x}$, and we find that $\Pi_T(z)$ is independent of T and given by

$$P(z) = \exp\{(\mu/\alpha) \int_0^1 [Q\{1 - (1-z)y\} - 1] \frac{dy}{y}\} . \tag{5.3}$$

The reader may verify that (5.3) does provide the solution of (5.1).

Equation (5.3) provides a starting point from which we may proceed to the calculation of conditional probabilities as outlined in the previous section. Let us first derive the function $P_0(x)$ giving the probability of no events in the interval $(T,T+x)$. We can write this in the form prob{no events on (T, T+x)}

= prob(none of the unexpired events at T expire before T+x)

× prob(none of the events generated in (T, T+x) expire before T+x).

Now the first of the probabilities on the right of this equation is just $P(e^{-\alpha x})$. To derive the second probability, consider the generating function $Q_x(z)$ of the probabilities

$${}_0p_n(x) = \text{prob}\{n \text{ events are generated in } (T, T+x) \text{ and none expires before } T+x\}.$$

This generating function may be obtained by arguing as in the proof of the lemma. Conditional on cluster centres at t_i $(T < t_i < T+x)$, it may be written

$$\prod_{T<t_i<T+x} Q\{ze^{-\alpha(T+x-t_i)}\} .$$

Then taking expectations with respect to the t_i, we have, after simplification,

$$Q_x(z) = \exp\left[(\mu/\alpha) \int_{e^{-\alpha x}}^{1} \{Q(zy) - 1\} \frac{dy}{y}\right] . \tag{5.4}$$

Hence we can write finally

$$P_0(x) = P(e^{-\alpha x})\ Q_x(1) . \tag{5.5}$$

This yields an alternative derivation of the formula for the hazard function

$$j(x) = -\frac{d}{dx} \log P_0(x) = \mu \frac{1 - Q(e^{-\alpha x})}{1 - e^{-\alpha x}}$$

given by Vere-Jones (1970, equation 28; note that the constant μ was inadvertently omitted from this equation in the paper cited). We leave the details of the differentiation to the reader.

Further results may now be built up, corresponding to the successive formulae of the preceding section. The most effective approach is to start from the left-hand end of the interval for which observations are given, and to make use of Bayesian arguments to provide successive modifications of the basic formula (5.3). To run over the argument briefly, let us suppose that observations are given over the interval (S,T) and that the point S was chosen without regard to the development of the process; further, let $P_t(z)$ denote our estimate for the generating function of the number of unexpired events at time t $(S < t < T)$, so that in particular we may set $P_S(z) = P(z)$. Now suppose that t is followed by an interval h with no events. Then our new estimate $P_{t+h}(z)$ will be formed from $P_t(z)$ by conditioning on this extra information, the past history up to t having no further effect, in view of the Markovian property. We shall therefore have, by a similar argument to that leading to (5.5),

$$P_{t+h}(z) = P_t(ze^{-\alpha h})\, Q_h(z)/P_t(e^{-\alpha h})\, Q_h(1) = L_h\, P_t(z) \; . \tag{5.6}$$

In a similar way, if an event is observed at t, and $P_{t-h}(z)$ denotes our estimate of the generating function just prior to t, we can write

$$p_{t,n} = \alpha h(n+1)\, e^{-\alpha(n+1)h}\, p_{t-h,n+1} + o(h)$$

for the probability that n unexpired cluster events remain at t, and that one expires in the interval $(t-h,t)$. Conditioning on the event in $(t-h,t)$, taking generating functions, and letting $h \to 0$, we obtain

$$P_t(z) = P_{t-}'(z)/P_{t-}'(1) = EP_{t-} . \tag{5.7}$$

The limit operations involved in the above derivation will be valid if the distribution corresponding to $P_{t-}(z)$ has finite first moment; in fact we shall apply the operator E only after an operator L_x, and it is clear that for any P, $L_x P$ will have finite first moment.

Now suppose that the starting point S is followed by an initial period τ_1 without any events, so that the first observed event is at $S+\tau_1$, and that further events are observed after successive intervals of lengths τ_2, τ_3, ... , τ_n, with a final interval, containing no further events, of length τ_{n+1}. Thus $T-S = \Sigma_1^{n+1} \tau_k$. Then $P_T(z)$ is obtained from $P_S(z) = P(z)$ by successively operating with L_{τ_1}, E, L_{τ_2}, ... , E, $L_{\tau_{n+1}}$. Finally, the probability that no events will occur in a further period x following T is given, repeating the argument leading to (5.5), by $P_T(e^{-\alpha x})\ Q_x(1)$. Returning to a notation more consistent with that of the previous section, these results may be summarized as the following Theorem.

Theorem 7. Let $P_{0101...010}(z;\tau_1,\tau_2, \ldots ,\tau_n,\tau_{n+1})$ denote the generating function for the conditional probability distribution of U_T, given events at times $T-\tau_{n+1}$, $T-\tau_{n+1} - \tau_n$, ... , $T-\tau_{n+1} - \tau_n - \ldots - \tau_2$, and that observations on the process started at time $S=T-\tau_{n+1} - \tau_n - \cdots - \tau_1$. Then

$$P_{0101...010}(z;\ \tau_1,\ \tau_2,\ldots,\ \tau_{n+1}) = L_{\tau_{n+1}}\ E\ L_{\tau_n}\ \ldots\ E\ L_{\tau_1}\ P(z)\ ,$$

where P(z) is the equilibrium distribution (5.3), and the operations L_x, E are defined by (5.6), (5.7) respectively. Furthermore, if

$$S_{0101...010}(x;\ \tau_1,\ \tau_2,\ldots,\ \tau_{n+1})$$

has the same interpretation as in the previous section, then

$$S_{0101...010}(x;\ \tau_1,\ \tau_2,\ldots,\ \tau_{n+1})$$
$$= P_{0101...010}(e^{-\alpha x};\ \tau_1,\ \tau_2,\ldots\ ,\ \tau_{n+1})\ Q_x(1)\ ,$$

where $Q_x(z)$ is defined by (5.4).

Although the formulae given by the above equations soon become algebraically unmanageable, they are in a form which (when the generating functions are written out again in terms of the probabilities) is quite suitable for numerical computation. A particularly pleasing feature from this point of view is that new information merely upgrades the previous estimates, and does not require the whole calculation to be repeated from the outset.

In principle, similar techniques can be applied to the more general case in which the cluster centre process is a renewal process rather than a Poisson process. Provided the cluster members are still exponentially distributed about the cluster centres, the process of unexpired cluster members becomes Markovian if one adjoins to it, as a supplementary variable, the time that has elapsed since the last cluster centre. Hence we have a 2-dimensional underlying Markov process (U_t, c_t), where U_t again denotes the number of unexpired cluster members at time t, and c_t is the time back to the last cluster centre. The generating function $P_c(z) = \Sigma z^n \text{prob}(U_t = n, c_t < c)$ now satisfies the partial differential equation

$$\frac{\partial P}{\partial c} = \alpha(1-z) \frac{\partial P}{\partial z} - h(c)P ,$$

where h(c) is the hazard function for the distribution of the time between clusters. The solution is of the form

$$P_c(z) = \{1 - F(c)\} \phi\{(1-z) e^{-\alpha c}\} ,$$

where F(.) is the distribution corresponding to h(c), and ϕ is an arbitrary function to be determined from the boundary condition

$$P_0(z) = Q(z) \int_0^\infty P_c(z) h(c) dc \quad .$$

The latter equation yields the integral equation for ϕ

$$\phi(z) = Q(1-z) \int_0^\infty f(c) \phi(ze^{-\alpha c}) dc \quad .$$

However, we have not been able to solve this equation, or to determine the form of the operations corresponding to L_τ and E in this more general case.

Another generalization, which was pointed out to us by Dr. Snyder at the Conference, is to any doubly stochastic Poisson process with Markovian stochastic rate function. Dr. Snyder himself (Snyder, 1972) has treated filtering problems for such doubly stochastic processes by methods similar to those developed here.

ACKNOWLEDGMENTS

This paper had its genesis in the discussion to Vere-Jones (1970). We should like to take this opportunity of thanking the participants in that discussion, particularly Professor Brillinger, for their comments, references and suggestions, of which we have made free use in the present discussion.

REFERENCES

Balagangadharan, K. (1960). The prediction theory of stationary random distributions. Mem. Coll. Sci. Univ. Kyoto, Ser. A. Math 33, 253-56.

Bartlett, M.S. (1963). The spectral analysis of point processes. J. R. Statist. Soc. B 25, 264-96.

Cox, D.R. & Lewis, P. A. W. (1966). *The Statistical Analysis of Series of Events*. London: Methuen & New York: Barnes & Noble.

Daley, D.J. & Vere-Jones, D. (1972). A survey of the mathematical theory of point processes. This volume.

Gel'fand, I.M. and Vilenkin, N.Ya. (1964). *Generalized Functions*, 4, *Applications of Harmonic Analysis*. New York: Academic Press (trans.)

Ito, K. (1954). Stationary random distributions. Mem. Coll. Sci. Univ. Kyoto, Ser. A. Math. 28, 209-23.

Lawrance, A.J. (1972). Some models for series of stationary univariate events. This volume.

Schwartz, L. (1950). *Théorie des Distributions*. Paris: Herman & Cie.

Smith, W.L. (1958). Renewal theory and its ramifications. J.R. Statist. Soc. B 20, 243-302.

Snyder, D.L. (1972). Filtering and detection for doubly stochastic Poisson processes. To appear in I.E.E.E. Transactions on Information Theory.

Vere-Jones, D. (1970). Stochastic models for earthquake occurrence. J.R. Statist. Soc. B 32, 1-62.

Vere-Jones, D. (1972). An elementary approach to the spectral theory of stationary random measures. To appear in *Stochastic Analysis and Geometry* (Rollo Davidson Memorial Volume). New York: Wiley.

Westcott, M. (1970). *Some Applications of the Probability Generating Functional to Point Processes*. Ph.D. Thesis, Australian National University.

Yaglom, A.M. (1962). *The Theory of Stationary Random Functions*. New York: Prentice Hall (trans.).

POINT PROCESSES GENERATED BY LEVEL CROSSINGS

M. R. Leadbetter
University of North Carolina at Chapel Hill

SUMMARY

This paper consists primarily of a review of available literature (and especially more recent work) concerning the crossings of levels and curves by stochastic processes. Attention is particularly directed towards those properties for which it is most profitable to emphasize the point process nature of the crossings. The topics considered are summarized in the following list of contents:

Section 1 *Introduction*

Section 2 *The Point Process of Level Crossings. Counting Procedures for the Mean Number of Events*

Section 3 *Applications and Generalizations*

Section 4 *Moments of the Number of Crossings*

Section 5 *Distribution of the Time Between Crossings by a Stationary Process*

Section 6 *"Crossings" by Vector Processes and Fields*

Section 7 *Poisson Approximations for Level Crossings*

Section 8 *Local Extremes.*

1. INTRODUCTION

The problems concerning crossings of levels and curves by stochastic processes are very closely related to a number of areas of stochastic process theory. Such areas include local properties (continuity, etc.) of stochastic process sample functions, extreme value theory of stochastic processes, and the theory of point processes (cf. Cramér and Leadbetter, 1967). While other aspects must necessarily enter, it is the latter connection--the particular emphasis on purely point process properties--which we wish especially to develop in this paper. The selection of topics reflects this interest and has resulted in the exclusion of some problems which might have been treated.

Within this framework, the aim of the paper is to review the currently available literature on a spectrum of level- and curve-crossing problems, with particular reference to recent work. The historical references will be largely confined to this introductory section and the summary of the contents of the paper given below. A much fuller discussion of work up to 1967 may be found in Cramér and Leadbetter (1967).

The interest in crossing problems really dates back to the original papers of M. Kac (1943) and S. O. Rice (1945). In particular, Rice obtained the now familiar formula for the mean number E(C) of crossings of a level u by a stationary (zero mean) normal stochastic process $\xi(t)$ in an interval $0 \leq t \leq T$, viz.,

$$E(C) = \frac{T}{\pi}\left(\frac{\lambda_2}{\lambda_0}\right)^{1/2} \exp(-u^2/2\lambda_0) , \tag{1.1}$$

where, if $r(\tau)$ denotes the covariance function of $\xi(t)$, $\lambda_0 = r(0) =$ var $\xi(t)$ and $\lambda_2 = -r''(0)$, the second derivative of r at the origin. (λ_2 is also the "second spectral moment", $\lambda_2 = \int \lambda^2 \, dF(\lambda)$ if F denotes the spectral function for $\xi(t)$.)

The conditions under which (1.1) holds were successively weakened by various authors, including Ivanov (1960), Bulinskaya (1961), Ito (1964) and Ylvisaker (1965), the latter two authors giving minimal conditions. In particular, Ylvisaker used a counting procedure for the number of zeros which may be readily adapted to a general point process framework. This will be discussed in Section 2 following the definition of terms concerning crossings and a description of the previous counting procedures used (following Kac).

In Section 2 and throughout the paper, we shall mainly think in terms of upcrossings of levels, rather than all crossings, for reasons of "cleanliness" of exposition in the later sections. The modifications required to deal with downcrossings or all crossings will be evident. The notation N(a,b) will be used in two senses; first to denote the number of upcrossings of some fixed level (or curve) in the interval $a < t \leq b$, and also to denote the number of events of an arbitrary point process in that interval. It seems preferable to do this to avoid duplicating notation, since it will always be clear which use is intended, and the first use is, of course, a special case of the second. In later sections, we will need also to use the notation $N_u(a,b)$ where the level u considered will be changing.

Section 3 contains applications of the general results of Section 2. The standard results for crossings of levels and curves by normal (stationary and non-stationary) processes are given, and a general result for non-normal processes, including, in particular, application to the envelope of a stationary normal process. This section is concluded with some recent general results of Fieger concerning the generalization of the crossing framework to include possibly discontinuous curves and sample paths.

The higher (factorial) moments of the number N(0,T) of upcrossings are considered in Section 4. Again a general point process framework is used in which the kth factorial moment of N(0,T) is exhibited as the mean number of events of a process, the "kth derived process", formed from the original one, but in k-dimensional Euclidean space R^k. This follows essentially the treatment of the problem by Belayev (1969). The specific methods used for obtaining the moments are indicated from this basic viewpoint. Questions of *finiteness* of the moments are briefly mentioned.

One of the main uses of the factorial moments of the number of upcrossings of a level in a given time, is to describe the distribution of the time between successive upcrossings by a stationary normal process. Series expressions, of which one was given by Rice (1945) and many more, later, by Longuet-Higgins (1962a), may be given for such distributions. It turns out that the terms in the series are in fact (derivatives of) the factorial moments.

Again these are special cases of results which hold in a much wider framework--that of stationary point processes. This is described in Section 5, from which the specific application to the intervals between, say, upcrossings may be obtained. The concept of a "mixture" of point processes is also mentioned in that section--leading in particular to the distribution of the length of, say, an upwards excursion of a level u (i.e., the time from an upcrossing to the next downcrossing).

In Section 6, we consider generalizations of crossing problems to include vector processes and fields. Specific results are described, and it is shown that these are typically special cases of a crossing of a fixed vector by a "vector field", or equivalently the solutions of n random equations in n unknowns.

The behavior of a sequence of stationary point processes on the line, with intensities tending to zero, tends to take on more and more a Poisson character in certain cases. The important application for our purposes here concerns the upcrossings of an increasingly high level by a normal stationary process. (In particular, this leads to the asymptotic distribution of the maximum of such a process.) These questions are discussed in Section 7.

The final topic of the paper--in Section 8--concerns the occurrence of local extremes of a stationary normal process. In particular, we indicate some interesting recent work of G. Lindgren (1970) on the behavior of such a process near a local maximum and the asymptotic properties of crest-trough times and heights.

2. THE POINT PROCESS OF LEVEL CROSSINGS. COUNTING PROCEDURES FOR THE MEAN NUMBER OF EVENTS

In this section, we shall review the basic theory for calculation of the mean number of crossings of a level by a stochastic process, including the original "counting procedure" due to Kac (1943), and the now more often used method due to Ylvisaker (1965). This latter procedure will be cast here within a point process framework. Specific applications will be given in the next section.

It will be assumed throughout this section that the process $\xi(t)$ considered has continuous sample functions (see e.g. Cramér and Leadbetter,

1967, for a variety of sufficient conditions for this to hold). The relaxation of the continuity condition will be considered in the next section.

Following Cramér and Leadbetter (1967, Section 10.2), we say that $\xi(t)$ has an *upcrossing* of the level u at t_0 if for some $\varepsilon > 0$, $\xi(t) \le u$ for $t_0-\varepsilon < t < t_0$, and $\xi(t) \ge u$ for $t_0 < t < t_0+\varepsilon$. Downcrossings are defined by simply reversing the inequalities $\xi(t) \le u$, $\xi(t) \ge u$, whereas $\xi(t)$ has a *crossing* of the level u at t_0 if in each neighborhood of t_0 there are points t_1, t_2 with $\{\xi(t_1)-u\}\{\xi(t_2)-u\} < 0$. (For a discussion of these definitions, we refer to Cramér and Leadbetter, 1967, Section 10.2.)

To obtain the mean number of, say upcrossings, of u in an interval I, it is necessary to approximate the number of upcrossings as the limit of a sequence of quantities whose means can be calculated from the process distributions. (This point will be developed further in relation to a general point process later in this section.) For example, Kac (1943) used a counting procedure of the form

$$W_n = \int_I \delta_n\{\xi(t)-u\}\, \xi_+'(t)\, dt$$

in which $\xi_+'(t)$ is the derivative $\xi'(t)$ of ξ where this is non-negative, and zero otherwise, and $\delta_n(\cdot)$ is a sequence of functions "becoming more like a δ-function" as n increases (e.g., $\delta_n(x) = n$ for $|x| \le 1/2n$, $\delta_n(x) = 0$ otherwise). That W_n approximates the number $N(I)$ of upcrossings of u by $\xi(t)$ in I may be seen intuitively by regarding I as being composed of disjoint intervals in each of which $\xi'(t)$ has constant sign. The integrand vanishes on the intervals on which $\xi'(t) \le 0$ and on each of the other intervals may be transformed to give a contribution $\int_\alpha^\beta \delta_n(v)\, dv$ where $\alpha < 0 < \beta$ if and only if that interval contains an upcrossing. For such intervals $\int_\alpha^\beta \delta_n(v)\, dv \to 1$ whereas for the others (with $\alpha < \beta < 0$ or $0 < \alpha < \beta$) $\int_\alpha^\beta \delta_n(v)\, dv \to 0$. Thus in the limit W_n "counts" the number of upcrossings of u by $\xi(t)$ in I. The mean number of upcrossings is then obtained as $\lim_{n\to\infty} E(W_n)$.

Later authors (e.g., Bulinskaya, 1961; Ivanov, 1960; Leadbetter, 1965) followed Kac in using similar counting procedures. More recently, it has become customary to use a very simple and in some ways more natural procedure (due to Ylvisaker, 1965) which does not (at least explicitly) make any differentiability assumption necessary. It does, however, appear that the "Kac counting procedure" may provide a rather useful and natural method of dealing with certain crossing problems for vector processes (see Section 6).

Instead of describing the counting procedure used by Ylvisaker directly, we give a modified general version within a purely point process framework. This, we feel, provides an illuminating viewpoint and is certainly relevant to the aims of the present paper. Specifically, the following result holds. (We shall consider the interval (0,1] for simplicity--the required alterations for any other interval will be obvious.)

Theorem 2.1. Consider a point process without multiple events and let $N(s,t) = N(I)$ denote the number of events in the semiclosed interval $(s,t] = I$. Divide the interval (0,1] into n sub-intervals

$$I_{ni} = ((i-1)/n,\ i/n], \qquad i = 1,2,\dots, n; \qquad n = 1,2,\dots .$$

Then

$$E\{N(0,1)\} = \lim_{n\to\infty} \sum_{i=1}^{n} \text{prob}\{N(I_{ni}) \geq 1\} \quad . \tag{2.1}$$

The proof of (2.1) is obtained very simply by writing $\chi_{ni} = 1$ if $N(I_{ni}) \geq 1$, $\chi_{ni} = 0$ otherwise and $N_n = \Sigma_{i=1}^n \chi_{ni}$. Then it is clear that $N_n \to N(0,1)$ with probability one, and hence by Fatou's lemma, since $N_n \leq N(0,1)$, it follows that $E(N_n) \to E\{N(0,1)\} \leq \infty$, from which (2.1) follows.

It may be noticed that this argument can be regarded also as the heart of simple proofs of (a) the equivalence of the "principal" and "parametric" measures of a point process on the line ("Korolyuk's Theorem"--cf. Belayev, 1969) and (b) the expression of $E\{N(0,1)\}$ as the "Burkill integral" $\int_0^1 P\{N(\cdot) \geq 1\}$ (cf. Fieger, 1971).

Equation (2.1) expresses the mean number of events in the interval (0,1] in terms of distributions of numbers of events in (small) intervals. If the events are, say, upcrossings of a level by a process $\xi(t)$, we wish to express $E\{N(0,1)\}$ in terms of the finite-dimensional distributions of $\xi(t)$. It is intuitively clear that we should be able to replace $\text{prob}\{N(I_{ni}) \geq 1\}$ in (2.1) by $\text{prob}[\xi\{(i-1)/n\} < u < \xi(i/n)]$ in such a case. Specifically, we have the following result.

Theorem 2.2. Let $\xi(t)$ be a stochastic process whose sample functions are continuous with probability one, and such that for the given level u, $\text{prob}\{\xi(t) = u\} = 0$ for all t. Then if $N(s,t) = N(I)$ denote the number of upcrossings of u by $\xi(t)$ in $I = (s,t]$, it follows that

$$E\{N(0,1)\} = \lim_{n\to\infty} \sum_{i=1}^{n} \text{prob}\{\xi(\tfrac{i-1}{n}) < u < \xi(\tfrac{i}{n})\} . \qquad (2.2)$$

This result may be proved from Theorem (2.1) by defining χ_{ni} as in the proof of that theorem and $\chi'_{ni} = 1$ if $\xi\{(i-1)/n\} < u < \xi(i/n)$, $\chi'_{ni} = 0$ otherwise. Clearly $N'_n = \Sigma_{i=1}^n \chi'_{ni} \leq \Sigma_{i=1}^n \chi_{ni}$ and hence $\lim\sup E(N'_n) \leq E\{N(0,1)\}$ by Theorem 2.1. But it is apparent that with probability one for n fixed, $N_n \leq N'_m$ when m is sufficiently large, since then an upcrossing of u in $((i-1)/n,\ i/n]$ implies $\xi\{(j-1)/m\} < u < \xi(j/m)$ for some subinterval I_{mj} of I_{ni}. Thus, by Fatou's lemma $\liminf_m E(N'_m) \geq E(N_n)$ and hence

$$\liminf_m E(N'_m) \geq \liminf_n E(N_n) = E\{N(0,1)\}$$

by Theorem 2.1, showing that $E(N'_n) \to E\{N(0,1)\}$, which yields (2.2).

Thus the calculation of the mean number of upcrossings in an interval is better carried out by substituting the simpler event "$\xi\{(i-1)/n\} < u < \xi(i/n)$" for "$N\{(i-1)/n,\ i/n\} \geq 1$". This is rather obvious and indeed it is easy to proceed *directly* to Theorem 2.2 without using Theorem 2.1.

The mean number of *downcrossings* of the level u in (0,1) is, of course, obtained under the same conditions by reversing the inequalities in (2.2), whereas the mean number of crossings is obtained by adding the results for upcrossings and downcrossings.

These results may be applied at once to give the standard formulae for normal and other processes. Such applications are noted in the next section. To conclude this section, we mention that the above method for calculation the mean number of events from a simpler result such as (2.2) rather than (2.1) may be given considerable generality. Such very general considerations have been given by Fieger (1971) in a related, but differently organized framework. We shall use a somewhat less general and simplified version of what Fieger terms an "ausschöpfend" set function (Fieger, 1971, Section 3), designed to handle standard applications.

Specifically, let (Ω, F, P) denote the basic probability space on which a given point process is defined. For each $n = 1,2,\ldots$, $i = 0,\pm 1,\pm 2,\ldots$ let $S_{ni} = S(I_{ni})$ be an F-measurable set determined by the interval $I_{ni} = ((i-1)/n, i/n]$. Then S_{ni} will be called a *sufficient set function* for the point process if for each n, i

(a) $S_{ni} \subset \{\omega : N(I_{ni}) \geq 1\}$

and

(b) with probability one, if $N(I_{ni}) \geq 1$, there exists an integer $m_0 = m_0(\omega)$ such that for any $m \geq m_0$ an integer j may be found with $\omega \in S_{mj}$ and $I_{mj} \subset I_{ni}$.

Theorem 2.2 may then be generalized as follows.

Theorem 2.3. Let S_{ni} be a sufficient set function for the point process considered in Theorem 2.1. Then

$$E\{N(0,1)\} = \lim_{n\to\infty} \sum_{i=1}^{n} P(S_{ni}) ,$$

(cf. Fieger, 1971).

3. APPLICATIONS AND GENERALIZATIONS

For a *stationary* process $\xi(t)$ satisfying the conditions of Theorem 2.2, we have at once, for the intensity $E\{N(0,1)\}$ of upcrossings of the level u;

$$E\{N(0,1)\} = \lim_{n\to\infty} \text{prob}\{\xi(0) < u < \xi(n^{-1})\}/n^{-1} \quad . \tag{3.1}$$

In the case of a stationary normal process (zero mean, unit variance, covariance function $r(t)$), this easily yields the familiar formula of Rice (1945) viz.,

$$E\{N(0,1)\} = (\lambda_2^{1/2}/2\pi)\ \exp(-u^2/2), \tag{3.2}$$

where λ_2 is the "second spectral moment" taking the value $-r''(0)$ if r is twice differentiable, and $+\infty$ otherwise. The mean number of downcrossings of u in $0 \leq t \leq 1$ has the same value and the mean number of crossings, twice the value given by (3.2).

For a non-stationary normal process $\xi(t)$ the mean number of upcrossings of the *zero* level in $0 \leq t \leq 1$ may be similarly obtained and has the expression

$$E\{N(0,1)\} = \int_0^1 \gamma\sigma^{-1}(1-\mu^2)^{1/2}\ \phi(m/\sigma)[\phi(\eta) + \eta\{\Phi(\eta) - 1/2\}]\ dt\ , \tag{3.3}$$

in which (if $\xi(t)$ has mean $m = m(t)$ and covariance function $r(t,s)$),

$$\sigma^2 = r(t,t); \qquad\qquad \gamma = \partial^2 r(t,s)/\partial t\partial s\big|_{s=t};$$

$$\mu = \partial r(t,s)/\partial s\big|_{s=t}/(\gamma\sigma);$$

$$\eta = (\frac{dm}{dt} - \gamma\mu m/\sigma)/\{\gamma(1-\mu^2)^{1/2}\}\ ,$$

it being assumed that dm/dt and $\partial^2 r(t,s)/\partial r\partial s$ are continuous functions and that $\sigma > 0$, $|\mu| < 1$ for each t. Here ϕ is the standard normal density (with d.f. Φ). A discussion of this result, and details of proof, are contained in Cramér and Leadbetter (1967, Section 13.2). While it applies to upcrossings of *zero*, it can be modified to apply to any level u by simply writing m(t)-u for m(t) (since $\xi(t)-u$ is a normal process with mean m(t)-u and covariance function r(t,s) which crosses zero whenever $\xi(t)$ crosses u). Indeed, we may clearly consider upcrossings of a (continuously differentiable) *curve* u(t) (defined to be upcrossings of zero by $\xi(t) - u(t)$) by simply writing m(t) - u(t) in place of m(t) in (3.3).

Another way of writing (3.3) is

$$E\{N(0,1)\} = \int_0^1 \int_0^\infty y p_t(0,y)\, dy\, dt \tag{3.4}$$

where $p_t(x,y)$ denotes the joint density for the normal process $\xi(t)$ and its q.m. derivative $\xi'(t)$. In fact, this result holds for a wide variety of stochastic processes $\xi(t)$--not necessarily stationary or normal. A variety of conditions are possible to ensure (3.3) for an arbirary process. Those in the following theorem--given in Leadbetter (1966a)--are sometimes useful.

Theorem 3.1. Let $\xi(t)$ be a stochastic process with continuous sample functions and joint density $f_{t,s}(x,y)$ for $\xi(t)$, $\xi(s)$, $t \neq s$. Write $g_{t,\tau}(x,y) = \tau f_{t,t+\tau}(x,x+\tau y)$ (i.e., the joint density for $\xi(t)$ and $\{\xi(t+\tau) - \xi(t)\}/\tau$). Assume that $\xi(t)$ and its q.m. derivative $\xi'(t)$ have a joint density $p_t(x,y)$. Then the following three conditions imply the truth of Equation (3.4):

(i) $g_{t,\tau}(x,y)$ is continuous in (t,x) for each y, τ;

(ii) $g_{t,\tau}(x,y) \to p_t(x,y)$ as $\tau \to 0$ uniformly in (t,x) for each y;

(iii) $g_{t,\tau}(x,y) \leq h(y)$ for all t, τ, x, where $\int_0^\infty yh(y)\, dy < \infty$.

Equation (3.3) for normal processes, may be obtained from Theorem 3.1. As another application we mention the upcrossings of a level u by the *envelope* R(t) of a stationary normal (zero mean, unit variance) stochastic process $\xi(t)$, defined (Cramér and Leadbetter, 1967) to be

$$R(t) = \{\xi^2(t) + \hat{\xi}^2(t)\}^{1/2} ,$$

where, if $\xi(t)$ has the spectral representation

$$\xi(t) = \int_0^\infty \cos \lambda t \, du(\lambda) + \int_0^\infty \sin \lambda t \, dv(\lambda) ,$$

$\hat{\xi}(t)$ is its Hilbert transform, defined by

$$\hat{\xi}(t) = \int_0^\infty \sin \lambda t \, du(\lambda) - \int_0^\infty \cos \lambda t \, dv(\lambda) .$$

Application of Theorem 3.1 yields, after calculation,

$$E\{N(0,1)\} = (\Delta/2\pi)^{1/2} \, u \exp(-u^2/2) , \tag{3.5}$$

in which Δ is a constant determined by the first two spectral moments λ_1, λ_2 of ξ, $\Delta = \lambda_2 - \lambda_1^2$ ($\lambda_i = \int_0^\infty \lambda^i \, dF(\lambda)$ if F denotes the "real form" of the spectrum of ξ). The details of this calculation may be found in Leadbetter (1966a). We will describe another method for obtaining this result in Section 6.

Our final topic in this section concerns the possibility of relaxing the requirement of sample function continuity. As far as *stationary normal* processes are concerned, the question is of little importance since it is known (Belayev, 1971) that for a stationary normal process if the sample functions are not continuous with probability one, then they are unbounded (above and below) in every finite interval. Thus in this case, one should certainly regard the number of crossings of a level as being infinite with probability one. This combined with Equation (3.2) (with a value ∞ if $\lambda_2 = \infty$) describes the mean number of (up)crossings for a stationary normal process in all circumstances.

On the other hand, for other types of discontinuous processes, it may be of interest to obtain the mean number of crossings of a level. A result of this type could then be used also to calculate the mean number of crossings of a possibly discontinuous curve by a process with continuous or discontinuous sample paths.

Fieger (1971) has recently investigated this problem in considerable generality. He uses a slightly different definition of crossings and upcrossings from ours--especially relative to the endpoints of the interval considered, and obtains an equation like (2.2), expressed as a Burkill integral. For a normal process $\xi(t)$ (not necessarily stationary), Fieger is then able to give necessary and sufficient conditions for finiteness of the mean number of crossings of a level in a given interval. Temporarily using his notation, in which m(a,b;g) denotes the number of crossings (computed according to his definition) of the curve $g(\cdot)$ in the interval with endpoints a, b, Fieger obtains the following result.

Theorem 3.2 (Fieger). If $\xi(t) - g(t)$ is separable, $\xi(t)$ being normal with zero mean, unit variance and covariance function r(t,s), then E{m(a,b;g)} exists if and only if

(a) g(t) is of bounded variation on [a,b]

and

(b) $\int_a^b \{1 - r(\cdot,\cdot)\}^{1/2}$ exists as a Burkill integral.

The formula for the mean number of crossings also appears compactly in this framework. Specifically, Fieger (1971) defines a function S(t) as the Burkill integral $\int_a^t \{1 - r(.,.)\}^{1/2}$ and gives a formula for the mean number $E\{m(a,b;\gamma)\}$ of crossings of a level γ by the (same) normal process $\xi(t)$ in the interval with endpoints a, b. When $S(\cdot)$ is a continuous function this becomes

$$E\{m(a,b;\gamma)\} = (2^{1/2}/\pi)\exp(-\gamma^2/2)\int_a^b \{1 - r(\cdot,\cdot)\}^{1/2}\,. \tag{3.6}$$

For a *stationary* normal process for which $1 - r(t,s) \sim \frac{1}{2}\lambda_2(t-s)^2$ (i.e., with second spectral moment λ_2) the Burkill integral $\int_a^b \{1 - r(\cdot,\cdot)\}^{1/2}$ clearly becomes $(\lambda_2/2)^{1/2}$ (b-a), which gives

$$E\{m(a,b;\gamma)\} = (\lambda_2^{1/2}/\pi)\ (b-a)\ \exp(-\gamma^2/2)$$

(consistently with (3.2) in view of the difference between *crcssings* and *upcrossings*). If $\lambda_2 = \infty$ this again gives $E\{m(a,b;\gamma)\} = \infty$.

Equation (3.6) is a pleasant and very compact way of expressing the mean number of crossings of a level. Of course, in particular cases one would still need to make assumptions about the behavior of $r(t,s)$ (such as those leading to (3.3)) in order to evaluate the Burkill integral $\int_a^b \{1 - r(\cdot,\cdot)\}^{1/2}$.

4. MOMENTS OF THE NUMBER OF CROSSINGS

Again, to be definite, we shall consider *upcrossings* of a level u by the stochastic process $\xi(t)$ and write $N(s,t)$ for the number in $(s,t]$. One is interested in the moments of the random variables $N(s,t)$ for at least two reasons. First, since the distributions of these random variables are usually unknown, the moments are helpful quantities to calculate, particularly if limiting results of some kind are of interest. Second--as discussed further in the next section--knowledge of the moments permits a calculation (even though it may be complicated) of the distribution of the time from an "arbitrary" upcrossing to the next or of the length of an "arbitrary excursion above the level u".

Both for neatness of expression and for usefulness of application, it is more convenient to calculate the *factorial* moments $M_k(T) = E[N(0,T)\{N(0,T) - 1\} \cdots \{N(0,T) - k + 1\}]$ of $N(0,T)$ rather than the moments themselves. A history of the calculation of such moments is given in Cramér and Leadbetter (1967) for the case when $\xi(t)$ is a stationary normal process. These calculations are summarized by the following result from that reference.

Theorem 4.1. Let $\xi(t)$ be a stationary normal process whose spectrum possesses a continuous component, and finite second (spectral) moment λ_2. Then for any positive integer k

$$M_k(T) = \int_0^T \cdots \int dt_1 \ldots dt_k \int_0^\infty \cdots \int y_1 \ldots y_k \, p_{\underline{t}}(u,\underline{y}) \, dy_1 \ldots dy_k, \qquad (4.1)$$

in which $p_{\underline{t}}(u,\underline{y})$ denotes the joint density of the random variables $\xi(t_1)\ldots\xi(t_k)$ and the q.m. derivatives $\xi'(t_1)\ldots\xi'(t_k)$, evaluated at $(u,\ldots u, y_1 \ldots y_k)$. Equation (4.1) holds whether both sides are finite or infinite.

Equation (4.1) can be written in more specific forms for k = 1 (Equation 3.2), 2 and 3 (cf. Cramér and Leadbetter, 1967, and Longuet-Higgins, 1962b), whereas for larger values of k such results have not been obtained, to our knowledge.

For non-stationary and non-normal processes, conditions can be given under which Equation (4.1) still holds. For example sufficient conditions --analogous to those of Theorem 3.1--are given in Leadbetter (1967). Before quoting this result, we shall indicate briefly the "counting procedure" which we feel is simplest for the calculation of the moments. While not being simplified in any way, the calculation is again illuminated by focussing on the purely point-process properties of the upcrossings, this time in a way developed in some generality by Belayev (1969). For simplicity, we shall again consider T = 1.

Consider, then, an arbitrary point process without multiple events on the real line R. Given a fixed integer $k > 1$, define a new point process, the "kth derived process", on R^k (k-dimensional Euclidean space) to consist of all points $(t_1 \ldots t_k) \in R^k$ such that $t_1 \ldots t_k$ are times of occurrence of (distinct) events of the original point process. (Note that even if the original point process is stationary, the new one will not be.) Write N for the number of events of the original point process in the unit interval and $N^{(k)}$ for the number in the new process which lie in the unit cube. Clearly

$$N^{(k)} = N(N-1) \cdots (N-k+1)$$

and thus the kth factorial moment of N is

$$M_k = E\{N^{(k)}\} . \tag{4.2}$$

That is, the kth factorial moment of the number of events in the unit interval is simply the mean number of events of the kth derived process in the unit cube of R^k.

Now the method for obtaining the mean number of events of a point process in the unit interval outlined in Section 2, involved splitting the interval (0,1] into subintervals of length 1/n, for n = 1,2,... . Similarly, here we may calculate $E\{N^{(k)}\}$ by splitting the unit cube $\{0 < t_i \leq 1:i = 1...k\}$ into smaller and smaller cubes. (This method may be generalized and set in an abstract framework--cf. Belayev, 1969, or Leadbetter, 1972).

Specifically, for each n let $I_{n\underline{i}}$ denote the small cube consisting of points $(t_1...t_k)$ with $(i_j-1)/n < t_j \leq i_j/n$ where $\underline{i} = (i_1...i_k)$, $1 < i_j \leq n$. Then it follows as for (2.1) that

$$E\{N^{(k)}\} = \lim_{n\to\infty} \sum_{\underline{i}} \text{prob}\{N^{(k)}(I_{n\underline{i}}) \geq 1\} , \tag{4.3}$$

where the sum is over all the small cubes (i.e., all vectors $\underline{i}$) and $N^{(k)}(I_{n\underline{i}})$ denotes the number of events of the kth derived process in $I_{n\underline{i}}$.

Looking now at the case where the original events are upcrossings of a level u by a stochastic process $\xi(t)$ (with continuous sample functions and such that $\text{prob}\{\xi(t) = u\} = 0$), we may slavishly proceed to the analogue of Equation (2.2). To avoid degeneracies of the joint distribution of $\xi(t)$ at points close together, it is more convenient to consider only those small cubes, the coordinates of whose points do not lie too close together. To be specific, let $\varepsilon > 0$ and for fixed n let $C_{n\varepsilon}$ index the cubes with the following property: $\underline{i} = (i_1...i_k) \in C_{n\varepsilon}$ when, if $\underline{t} = (t_1...t_k) \in I_{n\underline{i}}$, then $|t_r-t_s| > \varepsilon$ for all $r \neq s$.

For $\underline{i} = (i_1...i_k)$, write $\chi_{n\underline{i}} = 1$ if $N^{(k)}(I_{n\underline{i}}) \geq 1$, ($\chi_{n\underline{i}} = 0$ otherwise) and $\chi'_{n\underline{i}} = 1$ if $\xi\{(i_j-1)/n\} < u < \xi(i_j/n)$ all $j = 1...k$ ($\chi'_{n\underline{i}} = 0$ otherwise). Instead of approximating $\Sigma\chi_{n\underline{i}}$ by $\Sigma\chi'_{n\underline{i}}$ we use just $\Sigma^*\chi_{n\underline{i}}$ where the * denotes that the summation is only taken over $\underline{i} \in C_{n\varepsilon}$. As before of course $\chi'_{n\underline{i}} = 1$ implies that $\chi_{n\underline{i}} = 1$ and hence $\Sigma^*\chi'_{n\underline{i}} \leq \Sigma\chi_{n\underline{i}}$ (the latter sum being over all $\underline{i} = (i_1...i_k)$, $i \leq i_j \leq n$), and it follows from (4.3) that

$$E\{N^{(k)}\} \geq \liminf_{\varepsilon \to 0} \liminf_{n \to \infty} \Sigma^*E(\chi'_{n\underline{i}}) . \tag{4.4}$$

But if ε is sufficiently small and n is fixed, it is not difficult to see that

$$\Sigma\chi_{n\underline{i}} \leq \Sigma^*\chi'_{m\underline{i}} \qquad \text{(summing over } C_{m\varepsilon}\text{)}$$

when m is sufficiently large, which gives

$$\Sigma\chi_{ni} \leq \liminf_{\varepsilon \to 0} \liminf_{m \to \infty} \Sigma^*\chi'_{mi}$$

and thus from (4.3) and two applications of Fatou's lemma, the reverse inequality to (4.4) holds. We thus have

$$E\{N^{(k)}\} = \liminf_{\varepsilon\to 0} \liminf_{n \to \infty} \sum_{\underline{i}\varepsilon C_{n\varepsilon}} \text{prob}\left\{\xi\left(\frac{i_j-1}{n}\right) < u < \xi\left(\frac{i_j}{n}\right),\ j = 1\ldots k\right\}. \tag{4.5}$$

This formula may be used to obtain Theorem 4.1, or to give the conditions referred to above for the validity of Equation (4.1) for an arbitrary process $\xi(t)$. In these conditions, it is assumed first of all that the 2k-dimensional distributions of $\xi(t)$ are defined by densities $f_{t_1\ldots t_k\ s_1\ldots s_k}(x_1\ldots x_k,\ y_1\ldots y_k)$ from which a function $g_{\underline{t}\tau}(\underline{x},\underline{y})$ is defined for $\underline{t} = (t_1\ldots t_k)$, $\underline{x} = (x_1\ldots x_k)$, $\underline{y} = (y_1\ldots y_k)$ by

$$g_{\underline{t}\tau}(x,y) = \tau^k f_{t_1\ldots t_k,\ t_1+\tau\ldots t_k+\tau}(x_1\ldots x_k,\ y_1\ldots y_k)$$

($g_{\underline{t},\tau}$ is the joint density for $\xi(t_1)\ldots\xi(t_k)$ and the increments $\{\xi(t_i+\tau) - \xi(t_i)\}/\tau$). Now $g_{\underline{t},\tau}$ is assumed to satisfy the following three conditions:

(i) $g_{\underline{t}\tau}(\underline{x},\underline{y})$ is continuous in $(\underline{t},\underline{x})$ for each $\underline{y}$, τ;

(ii) For each $\varepsilon > 0$, $g(\underline{t},\tau)(x,y) \to p_{\underline{t}}(\underline{x},\underline{y})$ as $\tau \to 0$ uniformly in the region $D(\varepsilon) = \{\underline{t} = (t_1\ldots t_k): |t_i-t_j| \geq \varepsilon$ for all $i \neq j\}$ and each y

(iii) For each $\varepsilon > 0$ there is a function $h_\varepsilon(\underline{y})$ such that for $\underline{t} \in D(\varepsilon)$, $g_{\underline{t},\tau}(\underline{x},\underline{y}) \leq h_\varepsilon(\underline{y})$ and

$$\int_0^\infty \cdots \int y_1 \cdots y_k \, h_\varepsilon(\underline{y}) \, d\underline{y} < \infty .$$

Under these conditions (4.1) holds for the process $\xi(t)$.

Finally, we note that simple modifications of Equation (4.1) yield the factorial moments of the number of downcrossings and of the total number of crossings. In the former case $y_1 \cdots y_k$ are replaced by $|y_1| \cdots |y_k|$ and the y-integrals are each taken from $-\infty$ to 0. For the total number of crossings $|y_1| \cdots |y_k|$ again appear and the integrals are each taken over the entire range $(-\infty,\infty)$.

Our final comments of this section concern *finiteness* of these factorial moments. Under the conditions stated, Equation (4.1) gives the factorial moments for the number of upcrossings of a level u by a process $\xi(t)$, whether or not that moment is finite. Even for stationary normal processes, it is not known in general precisely when the $M_k(T)$ are finite. For $k = 1$, the situation is clear. For $k = 2$, a sufficient condition for finiteness given in Cramér and Leadbetter (1967) is that the covariance function $r(\tau)$ should satisfy

$$\int_0^\delta \tau^{-1}\{\lambda_2 + r''(\tau)\} \, d\tau < \infty$$

for some $\delta > 0$. We understand that it has been shown by D. Geman that this condition is, in fact, also necessary for finiteness of the second moment. Belayev (1966) has shown that a sufficient condition for finiteness (still in the normal case) of $M_k(T)$ is that $\xi(t)$ should possess k q.m. derivatives. (This condition is, as noted by Belayev, clearly not a necessary one.) Aside from that, the results are scattered. For example, Longuet-Higgins (1962b) does some calculations which seem to imply finiteness of moments under certain assumptions, and some criteria (for particular kinds of covariance behavior) are given by Piterbarg (1968).

5. DISTRIBUTION OF THE TIMES BETWEEN CROSSINGS BY A STATIONARY PROCESS

The statistical properties of the intervals between zero crossings or between crossings of some level by a stationary stochastic process, have important engineering applications--for example, with reference to the "limiting" of stochastic waveforms. Further, it is often of importance to consider intervals such as that between an upcrossing of a level and the next downcrossing--or vice versa. For example, the interval that the envelope of a radio signal spends below some level may represent fading and consequent communication loss.

The interest in this problem again dates back to the pioneering work of S. O. Rice (1945) who gives a series expression and a very simple approximation for the case of successive zero crossings by a stationary normal process. More generally, one may consider the distribution of times between an axis crossing and the nth subsequent axis crossing or between an upcrossing of zero and the nth subsequent downcrossing, and so on. Problems of this latter type have been discussed by Longuet-Higgins (1962b), with particular reference to the normal case. In particular, Longuet-Higgins calculates (by somewhat heuristic methods) a series for the probability density of the time between an "arbitrary" upcrossing of zero and the (r+1)st subsequent upcrossing. (Precise definitions of what is meant by an "arbitrary" upcrossing are not usually given in the relevant literature, but this will be commented on below.) The series just referred to may be written as

$$f(\tau) = \sum_{i=0}^{\infty} (-)^i \binom{r+1}{i} \int_{0<t_2<\cdots<t_{n-1}<\tau} \cdots \int W(0,t_2 \ldots t_{n-1},\tau)/W(0)\, dt_2 \ldots dt_{n-} \tag{5.1}$$

where $W(t_1 \ldots t_n)\, dt_1 \ldots dt_n$ represents the "probability of an upcrossing of the axis in each of the intervals (t_i, t_i+dt_i)". (These W-functions were used originally by Ramakrishnan (cf. Bartlett, 1956) and termed "product densities".)

Again, in these situations it is useful to think in terms of a purely (now stationary) point process framework, and ask for the distribution of such quantities as the time from an "arbitrary event" to say the nth subsequent event. Precisely, this means the distribution of the time to the nth event after that at zero under the "Palm distribution". For a more pedestrian, but more intuitive viewpoint it is the conditional distribution of the time to the nth event after time zero, given an event "at" zero in the limiting sense:

$$F_n(t) = \lim_{\delta \downarrow 0} \text{prob}\{N(0,t) \geq n | N(-\delta,0) \geq 1\} \tag{5.2}$$

(using $N(s,t)$ for the number of events in $(s,t]$ again). Now $N(0,t) \geq n$, of course, if and only if the nth event after zero occurs before time t. We note that $F_n(t)$ as defined is a distribution function, but is *not* the same in general as distribution functions "defined with reference to a fixed time point" such as that for the interval from the first event prior to zero to the nth event after zero. The reason for the importance of $F_n(t)$ as opposed to these other possible distribution functions is its frequency interpretation under ergodic assumptions. (These points are discussed further in Leadbetter, 1966a.)

It turns out that bounds, and exact series expressions for $F_n(t)$ may be obtained in terms of the factorial moments

$$\beta_k(t) = E\{N(N-1) \ldots (N-k+1)\} \qquad (k = 1,2,\ldots)$$

of $N = N(0,t)$, under general conditions. When the point process consists of, say, upcrossings of a level by a stochastic process, the factorial moments may be obtained from the expressions of the previous section, leading to equations such as (5.1) of Longuet-Higgins (1962b).

The specific results obtainable for a stationary point process are as follows (from Cramér, et al., 1971).

Theorem 5.1 (bounds for $F_n(t)$): Suppose that for given positive integers n, k, $E\{N^{n+k+1}(0,t)\} < \infty$ for some (and hence all) $t > 0$. Then if k is even

$$F_n(t) \leq \lambda^{-1} \sum_{j=n+1}^{n+k+1} (-)^{j-n-1} \binom{j-2}{n-1} \beta_j'/j! \, , \tag{5.3}$$

where β_j' is written for the right hand derivative of $\beta_j(t)$ and λ is the intensity, $E\{N(0,1)\}$, of the process. If k is odd, the inequality is reversed.

Theorem 5.2 (series for $F_n(t)$): If $E\{N^k(0,t)\} < \infty$ for all $k = 1,2,...$ then

$$F_n(t) = \lambda^{-1} \sum_{j=n+1}^{\infty} (-)^{j-n-1} \binom{j-2}{n-1} \beta_j'/j! \quad , \tag{5.4}$$

provided the terms of the series tend to zero.

The proof of these results is quite straightforward and will not be given in detail here. It basically relies on easily obtained bounds for the distribution of any discrete random variable in terms of its factorial moments. Using this result, the F_n's may be bounded by expressions involving "conditional factorial moments" $\alpha_k(t)$ which are like the $\beta_k(t)$ but defined conditional on the occurrence of an event at the origin, in the same way as the F_n were defined by (5.2). The quantity α_k is then related to β_{k+1} by means of "Palm's formulae" to give Theorem 5.1. Theorem 5.2 follows from Theorem 5.1.

It should be noted that Theorem 5.1 may be applied to give bounds for $F_n(t)$ whenever a sufficient (but finite) number of moments of $N(0,t)$ are finite. For "crossing problems", the difficulties of determining whether the moments are finite were mentioned in the last section. Bounds may, of course, be obtained in both directions by altering k. Theorem 5.2 gives an exact expression for $F_n(t)$, but requires knowledge and finiteness of all the moments. The series (5.4) incidentally converges, but not necessarily absolutely, under the stated assumptions.

Using these results--together with the expressions of the previous section for the factorial moments, we may obtain (at least in principle), the distributions of the time from one upcrossing of zero (or a level) by a stationary process to the nth subsequent upcrossing, or from a downcrossing to the nth later downcrossing, or indeed from an arbitrary crossing (unidentified as to direction) to the nth later one. The point is that we deal with a single point process--upcrossings, downcrossings, or all crossings. The problem of determining the distribution of the time from say an *upcrossing* to the nth later *downcrossing* is somewhat different. There we are really dealing with a point process composed of two point processes consisting of events of "type 1" (e.g., upcrossings) and "type 2" (e.g., downcrossings) which occur alternately. Previously, we considered the point process of each type of event separately, or else together without identifying the types. Now it is important to consider them to be "mixed" in this way but to retain their identities. (Or we may regard them as "marked" in the sense of Matthes, 1963.)

Similar results to those in Theorem 5.1, 5.2 are possible here. One may define, for example, the distribution function $F_n^{(1)}(t)$ for the time from an arbitrary event of type 1 to the nth subsequent event of type 2. The factorial moments $v_k(t)$ of a random variable M now enter, M being defined so that M-1 is the number of those type 1 events in (0,t) for which the following type 2 event is also in (0,t). We then obtain, analogously to (5.4),

$$F_n^{(1)}(t) = \lambda_1^{-1} \sum_{j=n+1}^{\infty} (-)^{j-n-1} \binom{j-2}{n-1} \nu_j'/j!\ ,$$

where λ_1 is the mean number of type 1 events in (0,1), provided all moments ν_j are finite (cf. also Leadbetter, 1969a), and the terms tend to zero.

In view of the difficulty of calculating factorial moments of numbers of level crossings, or zeros, of a stochastic process (because of the multiple integrations involved), we feel that bounds of the type given in Theorem 5.1, for small values of n, k are likely to be the most useful of these results in practice. Series such as (5.4) are, of course, exact, and widely applicable in principle. When these series fail to converge, "Euler's q-transform" may be used to give other double series which may converge. The form of this result corresponding to (5.4) is

$$F_n(t) = \lambda^{-1} \sum_{k=n+1}^{\infty} (1+q)^{-k-1} \sum_{s=n+1}^{k} (-)^{s-k-1} \binom{k}{s} \binom{s-2}{n-1} q^{k-s}\ \beta_s'/s!\ .$$

While this result (which applies when q is large enough) is theoretically interesting, it is not likely to be very helpful in practice (cf. Leadbetter, 1969b).

6. "CROSSINGS" BY VECTOR PROCESSES AND FIELDS

One obvious generalization of the problem of obtaining the mean (or moments) of the number of zeros of a stochastic process, is to consider a *vector* process $\underline{\xi}(t) = (\xi_1(t)\ldots\xi_n(t))$. This is a curve in n-dimensional space and it is sensible to consider the (one-dimensional) point process consisting of the instants at which this curve intersects a surface, say $\phi(x_1\ldots x_n) = 0$.

There is no difficulty in principle in solving this problem since if we define a stochastic process

$$\zeta(t) = \phi\{\xi_1(t)\ldots\xi_n(t)\}$$

we are simply concerned with the zeros of $\zeta(t)$--and the results of Section 4 may be applied. That is, the vector case really reduces to the familiar scalar situation. Of course, the vector process $\xi(t)$ and the function ϕ must be suitably restricted in order to use these results.

Belayev (1968) treats this topic and gives the following version for the kth factorial moment M_k of the number of intersections of the vector process with the subset Γ of the surface for t in the time interval Δ

$$M_k = \int\cdots\int_{t_i \in \Delta} \int\cdots\int_{x_i \in \Gamma} E\{\prod_{i=1}^{k} |\underline{n}(\underline{x}_i)' \dot{\underline{\xi}}_{t_i}| \,|\, \underline{\xi}_{t_i} = \underline{x}_i,\ i=1\ldots k\}$$

$$\times\, p_{t_1\ldots t_k}(\underline{x}_1\ldots\underline{x}_k)\, dS(\underline{x}_1)\ldots dS(\underline{x}_k)\, dt_1\ldots dt_k\ , \qquad (6.1)$$

in which $n(\underline{x})$ denotes the normal to the surface at $\underline{x}$, the dot denotes time differentiation, $p_{t_1\ldots t_k}$ is the joint density for $\underline{\xi}(t_1)\ldots\underline{\xi}(t_k)$ and $dS(\underline{x})$ is the element of surface area at $\underline{x}$.

The conditions for the validity of (6.1) are naturally fairly complicated and we refer to Belayev (1968) for the details. Belayev applies (a modified version of) this result in an interesting way to obtain the mean number of upcrossings of a level by the envelope of a stationary normal process--thus providing an alternative derivation to that in Section 3. Specifically, with the notation leading to Equation (3.5) it can be seen that the envelope R(t) crosses the level u when the vector process $\{\xi(t), \hat{\xi}(t)\}$ crosses the circle $x^2 + y^2 = u^2$. Application of the version of (6.1) applicable to "outward crossings", with k = 1, yields Equation (3.5) again. Of course, while the point of view is illuminating, no different mathematics is really involved since, as noted, the vector case can be reduced to a scalar one.

Crossing problems for (scalar) stochastic processes may also be regarded as the discussion of the solutions to a random equation $\xi(t) = 0$, $\xi(t) - u = 0$, etc. We typically ask, "How many solutions are there (for fixed "sample point" ω) to the single equation ($\xi(t) = 0$) in one unknown (t) in some interval, and what are the mean and moments of this number?" As we saw, the case of a vector process really belongs within this framework.

A more general question is to consider the point process (in R^n) consisting of the solutions to n (random) equation in n unknowns. That is, we consider a vector stochastic process $\underline{\xi}(\underline{t}) = \{\xi_1(\underline{t}) \ldots \xi_n(\underline{t})\}$ where the parameter $\underline{t} = (t_1 \ldots t_n)$ is also a member of n-dimensional Euclidean space R^n. (We will call such a process a "vector field".) One may, of course, even more generally let $\underline{\xi}$ have n components and $\underline{t}$ m-components, and this may lead to interesting results (e.g., in terms of Hausdorff dimension and measure of the set of zeros) when $n < m$. We will not need this extra generality here, however.

It is not our intention to pursue any general theory for the zeros of vector-fields, but merely to point out that it is this framework which is really appropriate for a number of applications. Particularly (in addition to all the cases considered with n = 1), we would refer to work by Longuet-Higgins (1962a) and especially Belayev (1967b) concerning certain point processes associated with (ordinary) random fields.

Specifically Belayev (1967b) considers a real random field $\{\zeta_{\underline{s}}: \underline{s} = (s_1 \ldots s_n) \in R^n\}$. He deals with three point processes in R^n formed from such a field, which he terms "bursts", "shines" (alias "flashes", "splashes" according to translational preferences) and saddle points. A "burst" occurs at the point $\underline{s}$ if $\zeta_{\underline{s}}$ has a local maximum of height greater than (some fixed) u at that point. A "shine" occurs at $\underline{s}$ if $\zeta_{\underline{s}} > u$ and the normal to the ζ-surface $s_{n+1} = \zeta(s_1 \ldots s_n)$ has fixed direction--which means that a light ray along this direction should be reflected straight back by the element of surface there. A saddle point occurs at $\underline{s}$ if the normal to the surface there is parallel to the (n+1)-axis and the principal curvatures have opposite sign.

Thus, bursts, shines and saddle points are defined primarily by conditions on the direction of the normal to the surface--which immediately leads to n equations to be satisfied in each case. (Of course, in working out, say, the mean number of "bursts" $\underline{s}$ in a set, we must use the added condition that $\xi_{\underline{s}} > u$.)

The point processes of bursts, shines and saddle points are homogenous with respect to the group of parallel shifts if the field ζ_s is homogeneous. Belayev gives conditions--based on the covariance function for ζ_s--for these point processes to have finite intensities. He further obtains explicit formulae (generalizing (4.1)) for the factorial moments of each of the number of bursts, shines and saddle points in a given set. We refer to Belayev (1967b) for the details.

7. POISSON APPROXIMATIONS FOR LEVEL CROSSINGS

If a stationary point process on the line is "thinned out" by making events rarer in some way, it is natural to ask whether the remaining events tend (in some sense) to behave approximately like a Poisson process. Clearly the "less dependency" there is in the process between distinct time points, the "more likely" such a result will be. Such questions were investigated first (to our knowledge) by Volkonski and Rozanov (1959, 1961) who used *strong mixing* assumptions to describe the "dependence decay".

One most useful situation in which this has been thoroughly investigated is that for the upcrossings of a high level u, by a stationary normal process. As the level increases, the upcrossings become rarer, thus raising the possibility of Poisson-like behavior. Let us now write $N_u(0,T)$ for the number of upcrossings of the level u in (0,T) by a stationary normal process $\xi(t)$ (with zero mean, unit variance and covariance function $r(\tau)$). As u increases, $N_u(0,T)$ will tend to become smaller. Hence we consider a normalized version of $N_u(0,T)$ formed by letting $T \to \infty$ as $u \to \infty$, in a suitably coordinated way. Specifically, writing $\mu = E\{N_u(0,1)\} = (\lambda_2^{1/2}/2\pi)\exp(-u^2/2)$, and fixing $\tau > 0$ we wish to have

$$\lim_{u\to\infty} \text{prob}\{N_u(0,T) = k\} = \exp(-\tau)\tau^k/k! \qquad (7.1)$$

with corresponding independent Poisson limits for the joint distributions of $N(a_i b_i)$ in disjoint (normalized) intervals.

Volkonski and Rozanov (1959, 1961) obtained this result under strong mixing conditions on the normal process, as an application of their general result referred to above. For normal processes, however, conditions on the covariance function are much preferable to mixing conditions and the Poisson limit was obtained first under such assumptions by Cramér (1966). Specifically, Cramér's basic assumptions were that the process possessed a finite fourth spectral moment (i.e., $r(\tau)$ is four times differentiable at $\tau = 0$) and that $r(t) = 0(t^{-\alpha})$ for some $\alpha > 0$, as $t \to \infty$.

These conditions were subsequently weakened by Belayev (1967a), Qualls (1968) and finally recently by Berman (1971). We quote Berman's result.

Theorem 7.1 (Berman). With the above notation, let the stationary normal process $\xi(t)$ have finite second spectral moment $\lambda_2 = -r''(0)$ and suppose that either (i) $r(t)\log t \to 0$ as $t \to \infty$, or (ii) $\int_0^\infty r^2(s)\,ds < \infty$. Then the upcrossings asymptotically approximate a Poisson process in the sense that (7.1), and the corresponding limits for the joint distributions, hold.

In fact Berman proves more than this. He shows that the upcrossings of a high level and downcrossings of a low level are independently distributed in Poisson fashion asymptotically as the levels tend to $\pm\infty$.

An important use for Theorem 7.1 concerns the evaluation of the asymptotic distribution for the maximum $M(T)$ of the stationary normal process $\xi(t)$ in $0 \leq t \leq T$. This connection is easily seen since

$$\text{prob}\{N_u(0,T) = 0\} = \text{prob}\{\xi(0) > u,\ N_u(0,T) = 0\} + \text{prob}\{M(T) \leq u\}.$$

The first term on the right is dominated by $\text{prob}\{\xi(0) > u\}$ which tends to zero as $u \to \infty$. Hence $\text{prob}\{M(T) \leq u\}$ can be approximated by $\text{prob}\{N_u(0,T) = 0\}$ as $u, T \to \infty$ in a coordinated manner. This leads in fact to

$$\text{prob}[(2 \log T)^{1/2} \{M(T) - (2 \log T)^{1/2}\} - \log(\lambda_2^{1/2}/2\pi) \leq z]$$
$$\to \exp(-e^{-z})$$

as $T \to \infty$ (cf. Cramér and Leadbetter, 1967, Section 12.3).

Other topics related to the Poisson nature of the upcrossings of a high level include the distribution of the total time spent above a high level in an interval, and of the length of a single excursion above a high level. This latter quantity is again the distribution of time from an "arbitrary" upcrossing to the next downcrossing, for which the series expression holds. However, if the level is very high, it may be shown that such an excursion--suitably scaled--has an asymptotic distribution with density $\frac{1}{2}\pi t \exp(-(\pi/4)t^2)$. A discussion of the history of results of this kind appears in Cramér and Leadbetter (1967, 12.4, 12.5). More recent work appears in Belayev and Nosko (1969) and closely related work in Berman (1972).

The basic assumption made in considering the Poisson nature of the upcrossings of a very high level was that the stationary normal process $\xi(t)$ had a finite second spectral moment λ_2. If $\lambda_2 = \infty$, then the mean number of upcrossings of any level u is infinite and indeed there may be positive probability that the number of upcrossings itself is infinite. This of course does not provide a point process with any of the usual properties of good behavior. However, some interesting properties may still be obtained under useful assumptions, in particular when the covariance function $r(\tau)$ satisfies

$$1 - r(\tau) \sim C|t|^{\alpha} \qquad \text{as } |t| \to 0, \tag{7.2}$$

where C is a finite non-zero constant and $0 < \alpha < 2$. The case $\alpha = 2$ is that previously considered ($\lambda_2 < \infty$). The case $\alpha = 1$ includes the Ornstein-Uhlenbeck process with $r(\tau) = \exp(-a|\tau|)$.

It has been shown by S. Orey (1970) that when $r(\tau)$ satisfies (7.2) (and certain other assumptions) then the set of crossings has Hausdorff dimension $1 - \alpha/2$. Thus even though infinitely many crossing events occur in finite intervals, their "density" may be described via their Hausdorff dimension.

There is, of course, no hope of obtaining a Poisson result for upcrossings of a high level. However, a device of Pickands (1969a) for "thinning" the upcrossings to get a well-behaved point process does allow a Poisson result to be derived, and this result is useful in discussing extreme value distributions for such processes.

To be precise, Pickands (1969a) discussed "ε-upcrossings" which are defined for $\varepsilon > 0$ to be those upcrossing points t for which $(t-\varepsilon,t)$ contains no other upcrossing. The number of ε-upcrossings of a level u in any finite interval is finite--indeed a bounded random variable since it cannot exceed (length of interval)/ε. Pickands (1969a) shows that if, in addition to (7.2), either $r(t) \log t \to 0$ as $t \to \infty$ or $\int_0^\infty r^2(t)\, dt < \infty$, then an asymptotic Poisson property holds for the ε-upcrossings. This result may again be used to obtain the asymptotic distribution of the maximum Pickands (1969b).

Finally, we note that other forms of limiting result are possible (for, say, the zeros of a stationary normal process with $\lambda_2 < \infty$). In the case considered the crossings were made rare and hence approximately

Poisson by letting the level $u \to \infty$. For a fixed level, it may be shown under certain conditions that the number of zeros (suitably normalized) has an asymptotic normal distribution (cf. Malevich, 1969) in long time intervals.

8. LOCAL EXTREMES

In this final section, we shall briefly consider some aspects of the occurrence of local extremes of a stationary normal process $\xi(t)$ (again with zero mean, unit variance and covariance function $r(\tau)$).

A local maximum or minimum may be defined in terms of "continuity properties" of the sample path (e.g., a local maximum may be said to occur at t_0 if $\xi(t) < \xi(t_0)$ for all t in some neighborhood of t_0). However, there is little point in so doing, since the conditions one wishes to assume will virtually always imply the existence of a continuous sample derivative. This being so, the local maxima (minima) of $\xi(t)$ are simply the down crossings (upcrossings) of zero by $\xi'(t)$. Since $\xi'(t)$ has covariance function $-r''(\tau)$, it follows that the mean number of local maxima and minima per unit time are finite if and only if $\lambda_4 = -r^{(4)}(0) < \infty$ and are each given by $(1/2\pi)(\lambda_4/\lambda_2)^{1/2}$. These and other properties of the local extremes are developed in Leadbetter (1966). (See also Fieger (1969) for a development via "continuity properties".)

A problem of some interest--which has recently been studied in considerable detail by G. Lindgren (1970)--concerns the behavior of a stationary normal process $\xi(t)$ in the neighborhood of a local maximum. In particular, what Lindgren calls the "wavelength" (time to the next minimum or "crest-trough time") and the "wave-height" (drop in height from the maximum to the minimum or "crest-trough height") have had some attention in the past (cf. Cartwright and Longuet-Higgins, 1956).

Lindgren (1970) shows that given a local maximum of height u at $t=0$, $\xi(t)$ has the same finite dimensional distributions as a process $uA(t) + \Delta_1(t) - \Delta_2(t)$, where $A(t)$ is a certain deterministic function, Δ_1 is a non-stationary zero mean normal process with a certain covariance function, and Δ_2 is an independent deterministic process of the form $\zeta B(t)$, where ζ is a random variable and B a known function. (The condition "given an event at t = 0" must be appropriately interpreted, of course--cf. Lindgren (1970) for the details of this, the specific functional forms of A, B and the distribution of ξ.) This representation is, incidentally, motivated by one of Slepian in a different context.

Lindgren studies the behavior of $\xi(t)$ in the neighborhood of a very low maximum (Lindgren, 1970) and uses the results in Lindgren (1971a) to obtain the limiting distribution of the wavelength and wave-height as $u \to -\infty$. Similar questions are studied in Lindgren (1971b) for the (perhaps most interesting) case when $u \to +\infty$.

ACKNOWLEDGMENTS

This research was supported by the Office of Naval Research under contract N00014-67-A-0321-0002.

REFERENCES

Bartlett, M.S. (1956). *An Introduction to Stochastic Processes*. Cambridge University Press.

Belayev, Yu. K. (1961). Continuity and Holder's conditions for sample functions of stationary Gaussian processes. Proc. 4th Berkeley Symp. 2, 23-33.

Belayev, Yu. K. (1966). On the number of intersections of a level by a Gaussian stochastic process. Teor. Veroyatnost i Primenen 11, 120-8.

Belayev, Yu. K. (1967a). On the number of intersections of a level by a Gaussian stochastic process II. Teor. Veroyatnost i Primenen 12, 444-57.

Belayev, Yu. K. (1967b). Bursts and shines of random fields. Doklady Akad. Nauk SSSR 176, 495-7.

Belayev, Yu. K. (1968). On the number of exists across a boundary of a region by a vector stochastic process. Teor. Veroyatnost i Primenen 13, 333-7.

Belayev, Yu. K. (1969). Elements of the general theory of random streams. Appendix to Russian edition of *Stationary and Related Stochastic Processes* by H. Cramér and M. R. Leadbetter, MIR Moscow 1969 (English translation in University of North Carolina Mimeo Series No. 703).

Belayev, Yu. K. & Nosko, V.P. (1969). Characteristics of excursions above a high level for a Gaussian stochastic process and its envelope. Teor. Veroyatnost i Primenen 14, 302-14.

Berman, S. M. (1971). Asymptotic independence of the numbers of high and low level crossings of a stationary Gaussian process. Ann. Math. Statist. 42, 927-45.

Berman, S. M. (1972). Excursions above high levels for stationary Gaussian processes. Pacific J. Math. to appear.

Bulinskaya, E. V. (1961). On the mean number of crossings of a level by a stationary Gaussian process. Teor. Veroyatnost i Primenen 6, 474-7.

Cartwright, D. E. & Longuet-Higgins, M. S. (1956). The statistical distribution of the maxima of a random function. Proc. Roy. Soc. A 237, 212-32.

Cramér, H. (1966). On the intersections between the trajectories of a normal stationary process and a high level. Ark. Mat. 6, 337-49.

Cramér, H. and Leadbetter, M. R. (1967). *Stationary and Related Stochastic Processes*. New York: Wiley.

Cramér, H., Leadbetter, M. R. & Serfling, R. J. (1971). On distribution function-moment relationships in a stationary point process. Z. Wahrscheinlichkeitstheorie verw. Geb. 18, 1-8.

Fieger, W. (1969). The number of local maxima of a stationary normal process. Math. Scand. 25, 218-26.

Fieger, W. (1971). The number of γ-level crossing points of a stochastic process. Z. Wahrscheinlichkeitstheorie verw. Geb. 18, 227-60.

Ito, K. (1964). The expected number of zeros of continuous stationary Gaussian processes. J. Math. Kyoto Univ. 3-2, 207-16.

Ivanov, V. A. (1960). On the average number of crossings of a level by sample functions of a stochastic process. Teor. Veroyatnost i Primenen 5, 219-23.

Kac, M. (1943). On the average number of real roots of a random algebraic equation. Bull. Amer. Math. Soc. 49, 314-20.

Leadbetter, M. R. (1965). On crossing of arbitrary curves by certain Gaussian processes. Proc. Amer. Math. Soc. 16, 60-8.

Leadbetter, M. R. (1966a). On crossings of levels and curves by a wide class of stochastic processes. Ann. Math. Statist. 37, 260-7.

Leadbetter, M. R. (1966b). On streams of events and mixtures of streams. J. R. Statist. Soc. B 28, 218-27.

Leadbetter, M. R. (1966c). Local maxima of stationary processes. Proc. Camb. Phil. Soc. 62, 263-8.

Leadbetter, M. R. (1967). A note on the moments of the number of axis crossings by a stochastic process. Bull. Amer. Math. Soc. 73, 129-32.

Leadbetter, M. R. (1969a). On the distributions of the times between events in a stationary stream of events. J. R. Statist. Soc. B 31, 295-302.

Leadbetter, M. R. (1969b). On certain results for stationary point processes and their application. Bull. I.S.I. 43, 309-20.

Leadbetter, M. R. (1972). On basic results of point process theory. Proc. 6th Berkeley Symp., to appear.

Lindgren, G. (1970). Some properties of a normal process near a local maximum. Ann. Math. Statist. 41, 1870-83.

Lindgren, G. (1971a). Extreme values of stationary normal processes. Z. Wahrscheinlichkeitstheorie verw. Geb. 17, 39-47.

Lindgren, G. (1971b). Wavelength and amplitude for a stationary process after a high maximum-decreasing covariance function. University of North Carolina Institute of Statistics Mimeo Series No. 745.

Longuet-Higgins, M. S. (1962a). The statistical geometry of random surfaces. Hydrodynamic Instability. Proc. Symp. Appl. Math. 13, Amer. Math. Soc., 105-43.

Longuet-Higgins, M. S. (1962b). The distribution of intervals between zeros of a stationary random function. Phil. Trans. Roy. Soc. 254, 557-99.

Malevich, T. L. (1969). Asymptotic normality of the number of crossings of level zero by a Gaussian process. Teor. Veroyatnost i Primenen 14, 287-309.

Matthes, K. (1963). Stationäre Zufällige Punktfolgen I. Jber. Deutsh. Math.-Verein 66, 66-79.

Orey, S. (1970). Gaussian sample functions and the Hausdorff dimension of level crossings. Z. Wahrscheinlichkeitstheorie verw. Geb. 15, 249-56.

Pickands, J. (1969a). Upcrossing probabilities for stationary Gaussian processes. Trans. Amer. Math. Soc. 145, 51-73.

Pickands, J. (1969b). Asymptotic properties of the maximum in a stationary Gaussian process. Trans. Amer. Math. Soc. 145, 75-86.

Piterbarg, V. I. (1968). The existence of moments for the number of level crossings by a stationary Gaussian process. Doklady Akad. Nauk. SSSR 182, 46-8.

Qualls, C. (1968). On a limit distribution of high level crossings of a stationary Gaussian process. Ann. Math. Statist. 39, 2108-13.

Rice, S. O. (1945). Mathematical analysis of random noise. Bell Syst. Tech. J. 24, 46-156.

Volkonski, V. A. & Rozanov, Yu. A. (1959). Some limit theorems for random functions I. Teor. Veroyatnost i Primenen 4, 186-207.

Volkonski, V. A. & Rozanov, Yu. A. (1961). Some limit theorems for random functions II. Teor. Veroyatnost i Primenen 6, 202-15.

Ylvisaker, N. D. (1965). The expected number of zeros of a stationary Gaussian process. Ann. Math. Statist. 36, 1043-6.

A SURVEY OF THE MATHEMATICAL THEORY OF MULTIDIMENSIONAL POINT PROCESSES

L. Fisher
University of Washington

1. INTRODUCTION AND BASIC DEFINITIONS

This paper is to be a survey of the mathematical theory of multidimensional point processes (m.p.p.). By a m.p.p. we shall mean an entity whose sample value is a finite or countable collection of points in R^k (k-dimensional Euclidean space) without a finite limit point. Most of the work done to date is either a conscious generalization of results for point processes on R^1 or were developed specifically for the statistical analysis of m.p.p.'s. Since several of the papers at this conference are in these areas (including Daley and Vere-Jones, 1972, Holgate, 1972, and Neyman and Scott, 1972), I will apologize in advance for the great overlap that must inevitably appear. We shall only consider point processes static in time in this paper. That is, we shall not consider the evolutionary behavior of point processes except as it affects the consideration of the point processes at a fixed moment in time. For these other very interesting ideas pertaining to the evolution in time the reader is referred to Bartlett (1955, 1960), Harris (1963) and Moyal (1957, 1962a).

We now mention a few areas in which m.p.p.'s give appropriate models. The references included will allow the interested reader to pursue the subject in more detail and gain acquaintance with the subject.

a) Blood cell counts in hematology are discussed in Turner and Eadie (1957) who reference a history going back at least to Abbe (1837) and Student (1907).

b) Models with a base point giving rise to a cluster were considered in Neyman (1939) and these ideas extended in Neyman and Scott (1957, 1958, 1959 and 1972).

c) Bombing or coverage problems fall into the m.p.p. area (see Eikler, 1969 and Miles, 1970).

d) M.p.p.'s occur as models of space-time clustering or interaction (see Gani, 1972, for a discussion in this area).

e) Of late there has been much interest in taking multidimensional data (and other types of data as well) and trying to group it into natural groups or clusters or into a hierarchical tree of clusters (see Sokal and Sneath, 1963, and Cormack, 1971, which has an extensive list of references). Such procedures lie in an area called cluster analysis and so far there has been relatively little distribution theory associated with the subject. (For an exception see Scott and Symons, 1971.) It would seem reasonable that clustering models such as Neyman and Scott's would give reasonable models with which to evaluate such procedures.

f) M.p.p.'s form one of the main theoretical tools of statistical ecology. The interested reader might refer to books by Greig-Smith (1964) and Pielou (1969). A good place to look for more recent work is in the three volume series edited by Patil, Pielou and Waters (1971). An excellent source of references related to all of the topics touched on so far is "A Dictionary and Bibliography of Discrete Distributions" by Patil and Joshi (1968). Another extremely pertinent reference is Matern (1960).

g) Percolation processes (see Frisch and Hammersley, 1963) have continuous analogs (Gilbert, 1961, and Roberts, 1967) that may be considered.

With this background we move on to a more formal study of m.p.p.'s and a survey of some of the more important results in the field. We shall not try for great generality in the presentation, but shall present the theory on a level appropriate to most of the applied work that has so far been done on m.p.p.'s. Much of the material presented (particularly in Sections 1, 2, 4 and 5) will be from Moyal (1962a, 1962b). We shall not present the material in the generality of Moyal's 1962a paper, but will restrict ourselves to point processes on R^k which have no duplicate points (except for Milne's work given in Section 7). Similarly, in Section 2 we present product density functions with respect to Lebesgue measure and not an arbitrary product measure.

Definition. X will denote the set of all finite or countable subsets of k-dimensional Euclidean space, R^k, which do not have a finite limit point. For any subset A of R^k let N(A) be the function on X such that N(A)(x) = Cardinality of $A \cap \{x\}$ for $x \in X$. Let F be the smallest σ-field of subsets of X which contains all subsets of the form; $\{x: N(A_i)(x) = k_i,\ i = 1,2,\ldots, m\}$ where m is a positive integer, $A_1, \ldots, A_m$ are Borel sets of R^k and $k_i \in \{0,1,2,\ldots;\infty\}$, $i = 1,2,\ldots, m$.

Thus, the σ-field F is the smallest σ-field such that all N(A) are measurable functions for Borel sets A.

Definition. *A multidimensional point process* (m.p.p.) is a probability space (X,F,P). If R^k is the underlying Euclidean space we shall talk about the m.p.p. P on R^k. Since X and F are fixed (for a given k) we will only refer to the measure P.

Definition. A m.p.p. on R^k is finite if $P\{N(R^k) < \infty\} = 1$.

Definition. A m.p.p. is *stationary* if for every finite collection of Borel subsets of R^k; $A_1, \ldots, A_m$, vector $r \in R^k$ and k_i as above;

$$P\{N(A_i) = k_i,\ i = 1,2,\ldots, m\} = P\{N(A_i + r) = k_i,\ i = 1,2,\ldots, m\}.$$

When we have a finite m.p.p. it is convenient to decompose the measure P into the measures when the number of points is 0,1,2,... . Let us note that if A is a bounded Borel set of R^k and P a m.p.p. then we can define a new finite m.p.p. P_A as follows;

$$P_A(Q) = P\{(x: y = x \cap A,\ y \in Q)\}$$

for each $Q \in F$. More intuitively, P_A is achieved by throwing away all points that do not land in the set A. Since X contains only point sets in R^k without finite limit points it follows that P_A is a finite m.p.p.

In decomposing a finite m.p.p. according to how many points appear one runs into slight technical difficulty since on the one hand if a sample realization has n-points, say $\{x_1,\ldots, x_n\}$, it is natural to think of them as an unordered set of n-elements. On the other hand when setting up a probability measure for the distribution of n points in R^k it is natural to use the product space of R^k with itself n-times. In this case each n-tuple of distinct points $\{x_1,\ldots, x_n\}$ is represented by its n! different permutations. This accounts for the factorials appearing in the following formulation.

Let $R^{k*n} = R^k \times \cdots \times R^k$ (n-times) with the usual Borel σ-field, F^{k*n}. Let $X' = \bigcup_{n=0}^{\infty} R^{k*n}$ (where $R^{k*0} = \{0\}$). Let F' be the σ-field generated by $\bigcup_{n=0}^{\infty} F^{k*n}$. A probability measure P' on (X',F') will be called symmetric if for any Borel sets $A_1,\ldots,A_n$ of R^k and permutation $\{\pi(1), \ldots, \pi(n)\}$ of $(1,2,\ldots, n)$,

$$P'(A_1 \times \cdots \times A_n) = P'\{A_{\pi(1)}\times\cdots\times A_{\pi(n)}\} .$$

(Recall that $A_1 \times \cdots \times A_n \subseteq R^{k*n} \subseteq X'$.) Let

$$D_n = \{x:x \in R^{k*n},\ x = (x_1,\ldots, x_n),\ x_i \in R^k,\ \prod_{j=1}^{n-1} \prod_{m=j}^{n} |x_i-x_j| = 0\}.$$

We say that P' puts mass zero on the diagonal if $P'(\bigcup_{n=2}^{\infty} D_n) = 0$. We then have the following theorem.

Theorem 1.1. There is a 1-1 correspondence between finite m.p.p. (X,F,P) and symmetric probability measures P' on (X',F') which put mass zero on the diagonal. The correspondence is determined by relations of the form: For any disjoint Borel sets $A_1,\ldots,A_n$ of R^k,

$$P\{N(R^k) = n,\ N(A_i) = 1,\ i = 1,2,\ldots, n\}$$

$$= P'\{\bigcup_{\pi} \pi(A_1\times \cdots \times A_k)\}, \qquad (1.1)$$

where $\bigcup_{\pi} \pi(A_1 \times \cdots \times A_k)$ denotes the union over all permutations of $A_1 \times \cdots \times A_k$, that is $A_{\pi(1)}\times \cdots \times A_{\pi(n)}$ for a particular permutation $\{\pi(1),\ldots ,\pi(n)\}$ of $(1,\ldots, n)$.

Let us note several things about this result. If we do not require mass zero on the diagonal, but have symmetry then we have point processes where several points of the process may be the same point in R^k. If we abandon the symmetry requirement then we have distinguishable points, i.e., $A_1 \times \cdots \times A_n$ corresponds to the first point in A_1, the second in A_2, etc. We will let $P^{(n)}$ be the restriction to (R^{k*n}, F^{k*n}) of P'. In terms of the $P^{(n)}$ many quantities have an appealing form (see Moyal, 1962a) of which we only mention a few.

1) $P\{N(R^k) = n\} = P^{(n)}(R^{k*n})$.

2) If $A_1, \ldots, A_n$ is a partition of R^k and $\{k_1, \ldots, k_n\} \subseteq Z_+ = \{0,1,2,\ldots\}$, then

$$P\{N(A_i) = k_i,\ i = 1,\ldots, n\} = \binom{k_1+\cdots+k_n}{k_1,\ldots,k_n} P^{(k_1+\cdots+k_n)}(A_1^{k_1}\times\cdots\times A_n^{k_n}),$$

where $A_i^{k_i} = A_i \times \cdots \times A_i$ (k_i-times) is a set in R^{k*k_i}.

3) For any Borel set $A \subseteq R^k$,

$$P\{N(A) = n\} = \sum_{m=0}^{\infty} \binom{n+m}{n} P^{(n+m)}\{A^n \times (R^k - A)^m\} .$$

Note that if A is a bounded Borel subset of R^k and $A \subseteq B$ is a bounded Borel set then if we have a m.p.p. and form P_B (or equivalent P_B') that there are obvious consistencies involving the calculations of joint probabilities of $N(A_i)$ where the A_i are subsets of A regardless of which B we choose.

In the next section we consider product density functions in some detail since the usage of product densities is quite prevalent in the m.p.p. literature and with the exception of the exposition embedded in Bhabha (1950) and Moyal (1962a) the presentation is on a somewhat intuitive level, e.g., Bartlett (1954, 1955). In the following it should also be kept in mind that x often denotes a point in a higher dimensional Euclidean space and single integral signs denote integration over higher dimensional sets.

2. PRODUCT DENSITY FUNCTIONS

Product density functions go back at least to Yvon in the 1940's. Bhabha (1950) and Ramakrishnan (1950) used the product density for the study of cosmic radiation cascades. We shall develop the subject here in a manner analogous to Bhabha (1950) as contrasted with the exposition in Ramakrishnan (1950) and Bartlett (1954, 1955).

Suppose that we have a stochastic point process on R^n such that for every bounded Borel set $S \subset R^n$ we have density functions with respect to Lebesgue measure describing the distribution of sample points in S, that is, we have functions $g_i(x_1,\ldots, x_i)$ such that for disjoint Borel subsets $A_1,\ldots, A_i$ of S

$$P\{N(S) = i,\ N(A_j) = 1,\ j = 1,\ldots, i\}$$

$$= \int_{A_1}\int_{A_2}\cdots\int_{A_i} g_i(x_1,\ldots, x_i)\, dx_1\cdots dx_i \ .$$

Let r be a positive integer and assume that for every $x \in R^n$ there is a neighborhood of x, say $\sigma(x)$, such that $E[N\{\sigma(x)\}^r] < \infty$. This implies that for any bounded Borel set A, $E\{N(A)^r\} < \infty$ since we may cover each point in $\bar{A}$ (the closure of A) by an appropriate neighbohood $\sigma(x)$. Since $\bar{A}$ is compact we may pick a finite subcovering $\sigma(x_1),\ldots, \sigma(x_k)$ and $E\{N(A)^r\} \leq E([\sum_{i=1}^{k} N\{\sigma(x_i)\}]^r) \leq 2^r \sum_{i=1}^{k} E[N\{\sigma(x_i)\}^r] < \infty$. If $S \subset R^n$ we denote by S^k the cartesian product of S with itself. If $g(x_1,\ldots, x_m)$, $m \geq k$ is a measurable function we will denote by $\int_{S^k} g(x_1,\ldots, x_m)$ the integral with respect to be last k arguments and Lebesgue measure (assuming the function is integrable). If we write an iterated interval $\int_{B^j}\int_{A^k} g(x_1,\ldots, x_m)$ we mean that first the last k arguments are integrated out and then the next j arguments. Noting that g_i is symmetric in its arguments (or at least one version is) it is easy to see that

$$P\{N(S) = i\} = \int_{S^i} g_i(x_1,\ldots, x_i)/(i!)\ . \tag{2.1}$$

The i! results from the fact that each way of placing the i points in S is added in each of its i! permutations.

We now want to define the density function of order r associated with the probability density of having points at $x_1, \ldots, x_r$ regardless of what happens elsewhere.

Definition. Let $\{x_1, \ldots, x_r\} \subseteq S$, S be bounded, and the assumptions of this section hold. Then we define the rth order probability density functions as

$$f_r(x_1, \ldots, x_r) = \sum_{k=0}^{\infty} \int_{S^k} g_{r+k}(x_1, \ldots, x_{r+k})/(k!) \ . \tag{2.2}$$

We note several points about this definition.

1) Interpretation. $\int_{S^k} g_{r+k}(x_1, \ldots, x_{r+k})/k!$ gives the contribution to the probability density of points at $x_1, \ldots, x_r$ when there are a total of r+k points in S. The k! is needed since the remaining k positions are counted in each of their k! permutations.

2) $f_r(x_1, \ldots, x_r)$ is finite a.s. (with respect to Lebesgue measure on R^r). This follows since

$$\begin{aligned} \int_{S^r} f_r(x_1, \ldots, x_r) &= \sum_{k=0}^{\infty} \frac{(r+k)!}{k!} \left\{\int_{S^{r+k}} g_i(x_1, \ldots, x_{r+k})/(r+k)!\right\} \\ &= \sum_{k=0}^{\infty} (r+k)(r+k-1) \cdots r \cdot P\{N(S) = r+k\} \\ &= E[N(S)\ \{N(S) - 1\} \cdots \{N(S) - r + 1\}] \\ &\leq E\{N(S)^r\} < \infty \ . \end{aligned}$$

Thus we have the following lemma.

Lemma 2.1. Let every $x \in R^n$ have a neighborhood $\sigma(x)$ with $E[N\{\sigma(x)\}^r] < \infty$. Then the rth order product density function $f_r(x_1, \ldots, x_r)$ exists and for any bounded set S,

$$E[N(S)\{N(S)-1\}\cdots\{N(S)-r+1\}] = \int_{S^r} f_r(x_1,\ldots,x_r). \quad (2.3)$$

From these factorial moments one may determine the usual moments by using coefficients $C_j^{(r)}$ (see W. L. Stevens, 1937) such that $n^r = \sum_{j=1}^{r} C_s^{(r)} n(n-1)\cdots(n-j+1)$. The $C_s^{(r)}$ are the Stirling numbers of the second kind (see Douglas, 1971). Then

$$E[N(S)^r] = \sum_{j=1}^{r} C_j^{(r)} \int_{S^j} f_j(x_1,\ldots,x_j) .$$

3) The definition is independent of the set S. Let A be a set disjoint from S then we have (a.s.) if $\{x_1,\ldots,x_r\} \subseteq S$,

$$g_i^S(x_1,\ldots,x_i) = \sum_{k=0}^{\infty} \int_{A^k} g_{i+k}^{S\cup A}(x_1,\ldots,x_{i+k})/k! . \quad (2.4)$$

(If we have i points in S at $x_1,\ldots,x_i$ then we have i+k points in $S\cup A$, $k = 0,1,2,\ldots$.) Now consider f_r defined by "using" $S\cup A$ then using the symmetry of the g_i's, (2.4) and Fubini's theorem and we have

$$\begin{aligned} f_r(x_1,\ldots,x_r) &= \sum_{k=0}^{\infty} \int_{(S\cup A)^k} g_{r+k}^{S\cup A}(x_1,\ldots,x_{r+k})/k! \\ &= \sum_{k=0}^{\infty}\sum_{j=0}^{k} \binom{k}{j} \int_{S^j}\int_{A^{k-j}} g_{r+k}^{S\cup A}(x_1,\ldots,x_{r+k})/k! \\ &= \sum_{j=0}^{\infty} \{\int_{S^j} \sum_{k=j}^{\infty} \int_{A^{k-j}} g_{r+j+(k-j)}^{S\cup A}(x_1,\ldots,x_{r+k})/(k-j)!\}/j! \\ &= \sum_{j=0}^{\infty} \int_{S^j} g_{r+j}^{S}(x_1,\ldots,x_{r+j})/j! . \end{aligned}$$

This last expression is the one used to define f_r when using the set S. Thus, except for a slight quibble over the fact that the functions are defined only a.s. we have the same result. In applications there is usually a "natural" choice for the f_r. For example, f_r may be chosen to be continuous in most cases.

4) *Lemma 2.2.* Let S be a set such that $\sum_{r=0}^{\infty} E\{N(S)^r\}/r! < \infty$. Then

$$g_i^S(x_1,\ldots, x_i) = \sum_{r=0}^{\infty} \frac{(-1)^r}{r!} \int_{S^r} f_{i+r}(x_1,\ldots, x_{i+r}) .$$

Proof. The right hand side is equal to

$$\sum_{r=0}^{\infty} \frac{(-1)^r}{r!} \int_{S^r} \{ \sum_{k=0}^{\infty} \int_{S^k} g_{i+r+k}^S(x_1,\ldots, x_{i+r+k})/k!\} .$$

Using the absolute convergence and rearranging in terms of m = r+k this is equal to

$$g_i^S(x_1,\ldots, x_r) + \sum_{m=1}^{\infty} \int_{S^m} g_{i+m}^S(x_1,\ldots, x_{i+m}) \frac{1}{m!} \sum_{k=0}^{m} \frac{m!}{k!(m-k)!} (-1)^{m-k}$$

$$= g_i^S(x_1,\ldots, x_r) .$$

Let A and B be bounded sets. We now calculate cov{N(A), N(B)} in terms of the product density functions. Suppose that f_1 and f_2 are defined. We will use several steps.

1) Let A and B be disjoint. Then

$$E\{N(A)\ N(B)\} = \sum_{j=1}^{\infty} \sum_{k=1}^{\infty} kj\ P\{ N(A) = j,\ N(B) = k\}$$

$$= \sum_{j=1}^{\infty} \sum_{k=1}^{\infty} kj \int_{A^j} \int_{B^k} g_{j+k}^{A\cup B}(x_1,\ldots, x_{j+k})/(j!k!)$$

$$= \int_A \int_B \sum_{j=1}^{\infty} \sum_{k=1}^{\infty} \int_{A^{j-1}} \int_{B^{k-1}} g_{j+k}^{A\cup B}(x_1,\ldots,x_{j+k})/(j-1)!(k-1)!,$$

using symmetry. Putting $m = j+k-2$ we then have

$$E\{N(A)\ N(B)\} = \int_A \int_B \sum_{m=0}^{\infty} \frac{1}{m!} \sum_{n=0}^{m} \binom{m}{n} \int_{A^n} \int_{B^{m-n}} g_{m+2}^{A\cup B}(x_1,\ldots,x_{m+2})$$

$$= \int_A \int_B \sum_{m=0}^{\infty} \frac{1}{m!} \int_{(A\cup B)^m} g_{m+2}(x_1,\ldots,x_{m+2})$$

$$= \int_A \int_B f_2(x_1,x_2) \ .$$

This result is clearly intuitively correct from the interpretation of $f_2(x_1,x_2)$.

2) Let A = B. Then using (2.3)

$$E\{N(A)\ N(B)\} = \int_A f_1(x_1) + \int_{A^2} f_2(x_1,x_2) \ .$$

3) For general A and B let $C = A \cap B^c$, $D = A \cap B$ and $E = A^c \cap B$. Then

$$E\{N(A)\ N(B)\} = E\{(N(C) + N(D)\}\{N(D) + N(E)\}$$

$$= \int_C \int_D f_2(x_1,x_2) + \int_{D^2} f_2(x_1,x_2) + \int_D f_1(x_1)$$

$$+ \int_C \int_E f_2(x_1,x_2) + \int_D \int_E f_2(x_1,x_2)$$

$$= \int_A \int_B f_2(x_1,x_2) + \int_{A\cap B} f_1(x_1) \ .$$

Thus

$$\mathrm{cov}\{N(A),\ N(B)\} = \int_A \int_B \{f_2(x_1,x_2) - f_1(x_1)\ f_1(x_2)\} + \int_{A\cap B} f_1(x_1)$$

$$= \int_A \int_B \{f_2(x_1,x_2) - f_1(x_1)\ f_1(x_2) + \delta(x_1-x_2)\ f_1(x_1)\} ,$$

where $\delta(x_1-x_2)$ is the Dirac δ-function.

Definition. Let f_2 exist. Then the *covariance density function* of the process is

$$h(x_1,x_2) = f_2(x_1,x_2) - f_1(x_1)\, f_1(x_2) + \delta(x_1-x_2)\, f_1(x) \ .$$

It is convenient to use differential notation for these relationships, such as

$$E\{dN(x_1)\, dN(x_2)\} = \{f_2(x_1,x_2) + \delta(x_1-x_2)\, f(x_1)\}\, dx_1 dx_2 .$$

3. SPECTRAL THEORY.

It is reasonable to think of multidimensional point processes as a limiting case of the usual real-valued multi-dimensional stochastic processes where at each point of a realization we have a δ-function. Consider a stationary m.p.p. on R^k and suppose that the m.p.p. has second order product densities. From the m.p.p. we form a multidimensional parameter stochastic process by replacing each point $x \in R^k$ by a constant density $1/\delta^k$ over a cube of side δ, denoted by $C(x,\delta)$, with x as center. Let $X_\delta(x)$ denote the resulting process. Let $N(x,\delta)$ be the number of points of the m.p.p. in this cube. Then $X_\delta(x) = N(x,\delta)/\delta^k$. Thus,

$$\begin{aligned} \operatorname{cov}\{X_\delta(x),\, X_\delta(y)\} &= \operatorname{cov}\{N(x,\delta)/\delta^k,\, N(y,\delta)/\delta^k\} \\ &= \operatorname{cov}\{N(x,\delta),\, N(y,\delta)\}/\delta^{2k} \\ &= \int_{C(x,\delta)} \int_{C(y,\delta)} h(x,y)\, dx\, dy/\delta^{2k} \ . \end{aligned}$$

This converges a.s. (Lebesgue measure) to $h(x,y)$ as $\delta \to 0$.

In the case of a stationary m.p.p. the corresponding X_δ will also be stationary and passing to the limit in

$$\frac{1}{(2\pi)^k} \int_{R^k} e^{-iwx} \operatorname{cov}\{X_\delta(x),\, X_\delta(0)\}\, dx$$

we have

Definition 3.1. The *complete spectral function* or *spectral function* of a stationary m.p.p. which has second-order product densities with h(x,0) integrable over R^k is defined to be:

$$f(w) = \frac{1}{(2\pi)^k} \int_{R^k} e^{-iwx} h(x,0)\, dx ,$$

or, equivalently,

$$g(w) = \int_{R^k} e^{-iwx} h(x,0)\, dx ,$$

where 0 is the origin in R^k and w is the usual dot-product of vectors in R^k.

Spectral theory for m.p.p. was first developed for m.p.p.'s by Gelfand within the framework of generalized random fields, see Gelfand and Vilenkin (1964, Chap. III) and the notes and references to the literature at the end of the book. In this context a spectral measure exists under much more general circumstances and a spectral decomposition also holds. See again Gelfand and Vilenkin (1964, Chap. III) and also Daley (1971). In line with the philosophy (expressed in Section 9) of making the exposition as close to applications as possible we restrict ourselves to this more concrete case. The spectral function in the form considered here was first introduced by Bartlett (1963a, 1963b, 1964). Bartlett was the first to investigate the statistical properties of estimation. Anticipating the terminology of Section 5, Bartlett (1963b) showed that if the m.p.p. arose from a doubly stochastic Poisson process whose rate function was an appropriate stationary linear process, then with smoothing the usual theory of spectral analysis went through. Cox and Lewis (1966), Section 5.5 consider such matters, as does Hawkes (1971). Unfortunately, there is not (to my knowledge) a good foundational article outlining a fairly large class of m.p.p. for which the analysis proceeds. Such an article, similar to Brillinger (1970), would serve a very useful purpose in the literature at this time.

Such spectral analysis would seem to be quite useful, not only for hypothesis testing because of the usual sampling properties, but particularly in R^2 where a print out of the estimated spectral density function allows visual inspection with possible interpretations. Although Professor D. R. Cox notes in the discussion to Bartlett (1963a) that he cannot see any physical meaning to such a spectral decomposition in R^1, it seems that periodicities might result in ecological applications in R^2 due to the way organisms distribute themselves and on occasion might have some physical explanation.

4. CHARACTERISTIC FUNCTIONALS AND PROBABILITY GENERATING FUNCTIONALS

When Paul Levy introduced the characteristic function with such great success it was natural that people looked for generalizations in more complex situations. In Kolmogorov (1935) the characteristic functional was introduced for probability distributions in a Banach space. This line of development has been followed up by many authors, in particular, LeCam (1957) and Prohorov (1961). Perhaps the most general extension of the characteristic functionals has been in embedding the theory of stochastic processes in the theory of generalized functions or Schwartz distributions. The reader is referred to Chapter III of Gelfand and Vilenkin (1964). In our context the characteristic functional becomes

$$\Phi(\phi) = E[\exp\{i \int_{R^k} \phi(x)\, dN(x)\}] ,$$

where ϕ is a real infinitely differentiable function with compact support in R^k. As a function of $Z \in X$, $\Phi(\phi) = \int_{R^k} \phi(x)\, dN(x)$ is defined to be $\sum_{i=1}^{\infty} \phi(x_i)$, where $Z = \{x_1, x_2, \ldots\}$. Since ϕ has compact support all but a finite number of terms are zero. If we let B be a bounded, Borel set containing the support of ϕ then using the notation of Section 1,

$$\Phi(\phi) = \sum_{n=0}^{\infty} \int_{R^{k*n}} \exp\{i \sum_{i=1}^{n} \phi(x_i)\}\, P_B^{(n)}(dx_1 \cdots dx_n) .$$

In Gelfand and Vilenkin (1964), analogs of Bochner's theorem and results on stationary processes will be found. For our purposes it will be convenient to enlarge the allowable class of functions ϕ beyond the testing functions of distribution theory.

Definition. The *characteristic functional* (c.fl.) of a m.p.p. on R^k is

$$\Phi(\phi) = E[\exp\{ i \int_{R^k} \phi(x)\, dN(x)\}]$$

defined for all bounded, measurable functions ϕ with compact support.

It is immediately found that the c.fl. uniquely determines the distribution of the m.p.p. Let $A_1, \ldots, A_n$ be bounded, Borel sets and let χ_i be the indicator function of the set A_i, that is

$$\chi_i(x) = \begin{cases} 0 & \text{if } x \notin A_i \\ 1 & \text{if } x \in A_i \end{cases} \qquad (x \in R^k) .$$

Let $\phi(t_1, \ldots, t_n, x) = \sum_{i=1}^{n} t_i \chi_i$, the t_i being real numbers. Then

$$\Psi(t_1, \ldots, t_n) = E[\exp\{i \int_{R^k} \phi(t_1, \ldots, t_n, x)\, dN(x)\}] \tag{4.1}$$

$$= E[\exp\{i \sum_{j=1}^{n} t_i N(A_i)\}]$$

is the characteristic function of $\{N(A_1), \ldots, N(A_n)\}$ which uniquely determines the joint distribution of $\{N(A_1), \ldots, N(A_n)\}$ and it is easily seen that such distributions uniquely determine the distribution of the m.p.p.

Characteristic functionals are a special case of the probability generating functionals (p.g.fl.) as developed systematically in Moyal (1962a). We follow his treatment below with slight modifications. For each complex-valued, bounded, measurable function ϕ on R^k let $\|\phi\| = \sup_{x \in R^k} |\phi(x)|$. Let S be the set of functions ϕ such that ϕ is complex-valued, measurable with $\phi(x) = 1$ on the complement of a bounded set and $\|\phi\| = 1$.

Definition. Let $\phi \in S$ and $\phi(x) = 1$ for x in the complement of the bounded measurable set B. Then the *probability generating functional* is defined to be

$$G(\phi) = \sum_{n=0}^{\infty} \int_{R^{k*n}} \phi(x_1) \cdots \phi(x_n) \, P_B^{(n)}(dx_1 \cdots dx_n) ,$$

where the term for n = 0 is defined to be $P_B^{(0)}(R^{k*0}) \equiv P\{N(B) = 0\}$.

It is easy to see that the definition is independent of B. If $\phi(x) = e^{i\theta(x)}$ where the support of θ is bounded it follows that $G(\phi) = \Phi(\theta)$. Thus, the c.fl. may be considered a special case of the p.g.fl.

If we take $\log \phi$ to be the principle branch of the logarithm we may write

$$\begin{aligned} G(\phi) &= \sum_{n=0}^{\infty} \int_{R^{k*n}} \phi(x_1) \cdots \phi(x_n) \, P_B^{(n)}(dx_1 \cdots dx_n) \\ &= \sum_{n=0}^{\infty} \int_{R^{k*n}} \exp\{ \sum_{i=1}^{n} \log \phi(x_i) \} \, P_B^{(n)}(dx_1 \cdots dx_n) \\ &= \sum_{n=0}^{\infty} E[\exp\{ \int \log \phi(x) \, dN(x) \} | N(B) = n] \, P\{N(B) = n\} \\ &= E[\exp\{ \int \log \phi(x) \, dN(x) \}] , \end{aligned}$$

where $e^{-\infty} = 0$. Thus we have

Lemma 4.1. If $\phi \in S$,

$$G(\phi) = E[\exp\{ \int \log \phi(x) \, dN(x) \}] .$$

The following relations are quite easy to see.

Proposition 4.2. Let $A_1, \ldots, A_n$ be disjoint bounded Borel sets, and under some m.p.p. let $P\{N(A_i) = k_i,\ i = 1,2,\ldots, n\} = P_{k_1,\ldots,k_n}$ for $k_i \in Z_+$.

a) The probability generating function

$$\Psi(t_1,\ldots, t_n) = \sum_{k_1,\ldots, k_n} P_{k_1,\ldots, k_n} t_1^{k_1} \cdots t_n^{k_n}$$

is equal to $G\{\phi(t_1,\ldots, t_n; \cdot)\}$, where $\phi = \sum_{i=1}^{n} t_i \chi_i + \chi_{n+1}$, where χ_i is the indicator function of A_i and χ_{n+1} is the indicator function of the complement of $\bigcup_{i=1}^{n} A_i$.

b) Let $\mu_{k_1,\ldots,k_n} = \sum_{j_1,\ldots, j_n} j_1^{k_1} \cdots j_n^{k_n} P_{j_1,\ldots, j_n}$ be such that the moment generating function

$$\chi(t_1,\ldots, t_n) = \sum_{k_1,\ldots, k_n} \frac{\mu_{k_1,\ldots,k_n}}{k_1! \cdots k_n!} t_1^{k_1} \cdots t_n^{k_n}$$

converges for $(t_1,\ldots, t_n)$ such that $\sum_{i=1}^{n} t_i^2 < C$. Then for $\sum_{i=1}^{n} t_i^2 < C$,

$$\chi(t_1,\ldots , t_n) = G\{\phi(e^{t_1},\ldots, e^{t_n};\cdot)\} .$$

The analogous results hold under appropriate conditions for the factorial generating function, cumulant generating function and factorial cumulant generating function. We may also take Gateaux derivatives (see Kantoravich and Akilov, 1964, Chapt. XVII, for example or Hille and Phillips, 1957, Sections 3.16 through 3.19) in certain cases.

Let η_i, $i = 1,\ldots, n$ be bounded, measurable functions with bounded supports S_i, $i = 1,\ldots, n$. Let $C = \bigcup_{i=1}^{n} S_i$. Let $\sup_{x\in C}|\eta(x)| < 1-\delta$ and $\|\eta\| = 1$. Then $G(\eta + \sum_{i=1}^{n} \lambda_i\eta_i)$, where the λ_i are complex numbers, is an analytic function of the λ_i for $\sum_{i=1}^{n} |\lambda_i| < \delta/(\max\{\|\eta_1\|,\ldots, \|\eta_n\|\})$. Using dominated convergence the nth order variation of G is defined to be

$$\delta^n(\eta_1,\ldots,\eta_n;\eta) = \frac{\partial^n}{\partial\lambda_1\cdots\partial\lambda_n} G(\eta + \sum_{i=1}^{n} \lambda_i\eta_i)\Big|_{\lambda_1=\cdots=\lambda_n=0}$$

$$= \sum_{m=n}^{\infty} n!\binom{m}{n} \int_{B^m} \eta_1(x_1)\cdots\eta_n(x_n)\,\eta(x_{n+1})\cdots\eta(x_m)\,P_B^{(m)}(dx_1\cdots dx_n),$$

where $B \supseteq C$.

In particular if η is the indicator function of C complement, η_i is the indicator function of S_i. Then the above expression reduces to

$$n!\,P_C^{(n)}(A_1 \times \cdots \times A_n).$$

Some further points are mentioned in Moyal (1962a).

Paloheimo (1971a) pointed out the uses of the p.g.fl. in computing nearest neighbor distributions. For example, fix x and let χ_r be the indicator function of $\{y:|x-y| \leq r\}$, and let $\phi_r(t,x) = t\chi_r(x) + \{1-\chi_r(x)\}$. Then by Proposition 4.2,

$$P[N(\{y:|x-y| \leq r\}] = 0) = G\{\phi_r(0,\cdot)\}\ .$$

Consider a stationary m.p.p. Then the distance from a point to its nearest neighbor has a distribution function given as follows:

$$P(\text{Nearest neighbor} > r)$$
$$= \lim_{\varepsilon\to 0} \frac{P[N(\{y:|y| \leq \varepsilon\} > 0,\ N(\{y:\varepsilon < |y| \leq r\}) = 0]}{P[N(\{y:|y| \leq \varepsilon\} > 0]}\ .$$

Let χ_ε be the indicator function of $\{y:|y| \leq \varepsilon\}$, χ_r the indicator function of $\{y:\varepsilon < |y| \leq r\}$ and χ_c the indicator function of $\{y:|y| > r\}$. Let $\phi_{\varepsilon,r}(t_1,t_2;x) = t_1\chi_\varepsilon(x) + t_2\chi_r(x) + \chi_c(x)$. Then the distribution function of the nearest neighbor distribution is given by

$$F(r) = 1 - \lim_{\varepsilon\to 0} \{\frac{\partial}{\partial t_1} G\{\phi_{\varepsilon,r}(t_1,0;\cdot)\}\Big|_{t_1=0} / [1 - G\{\phi_{\varepsilon,r}(0,1;\cdot)\}]\}\ .$$

Finally we note that p.g.fls. may naturally be developed in terms of factorial moment measures. See Moyal (1962a) and Vere-Jones (1970) for more discussion and details.

5. SPECIFIC MODELS AND DISTRIBUTIONS

Several of the models below deal with cluster models which may be thought of as being generated in two stages. Initially a m.p.p., say P with p.g.fl. G, is generated giving centers for clusters. Associated with each center x_i the cluster is formed according to a m.p.p., say Q_{x_i} with a p.g.fl. G_{x_i}. The points of the different clusters are independent given the location of the centers $\{x_1, x_2, \ldots\}$. Probability generating functionals are natural for this type of work as is usual in branching processes or population processes (see for example Bartlett and Kendall, 1951; Moyal, 1962a, 1962b; and Harris, 1963). We call P the base process and Q_{x_i} the cluster structure process(es). If Q_{x_j} is independent of x_i when translated by $-x_i$ to the origin of R^k we shall denote such a cluster process by $P_{P,Q}$ where $Q \equiv Q_0$. If P is a Poisson process with constant rate λ we will denote the cluster process by $P_{\lambda,Q}$.

We note that some care must be taken in forming cluster processes since a given P and Q may be such that $P_{P,Q}$ in a bounded region has probability one of having an infinite number of points, i.e., we do not have a m.p.p. For example, let P put mass one on the point x consisting of all integer valued lattice points. Label the lattice points as $x_1, x_2, \ldots$. Let $Q \equiv Q_0$ be developed as follows. With probability P_n $(n = 1,2,\ldots)$, Q has n points which are independently and identically distributed. With probability r_i a point of Q is distributed at random on the unit square with center at x_i. Let A_i be the event that one or more sample points from the cluster with center $-x_i$ fall into the unit square with center at the origin. Then

$$\begin{aligned} P(A_i) &= \sum_{n=1}^{\infty} P(\text{Point}|n \text{ points in cluster})\, P(n \text{ points}) \\ &= \sum_{n=1}^{\infty} P_n\{1 - (1 - r_i)^n\}. \end{aligned}$$

Thus, if

$$\sum_{i=1}^{\infty} P(A_i) = \sum_{i=1}^{\infty} \sum_{n=1}^{\infty} P_n\{1 - (1 - r_i)^n\} = \infty,$$

by the Borel-Cantelli lemma each unit square will, with probability one, contain an infinite number of points and thus will not give a m.p.p. This may be achieved, e.g., $P_n = (6/\pi n^2)$, $r_i = 1 - (6/\pi i^2)$.

In this context we have the following theorem by Goldman (1967a).

Theorem 5.1. The Poisson cluster process $P_{\lambda,Q}$ $(\lambda > 0)$ exists if for every bounded rectangle I,

$$\int_{R^k} Q\{N(I + x) \neq 0\}\, dx < \infty \, , \tag{5.1}$$

where $I + x = \{y; y - x \in I\}$. Further if (5.1) holds then

$$Q\{N(R^k) < \infty\} = 1 \, . \tag{5.2}$$

Due to the heavy usage of clustering processes in the literature we proceed to a somewhat detailed mathematical analysis of the basic constructions and results. In the following discussion $\tilde{x}$ denotes points in X; **x** (with no tilda) denotes a point in R^k, and $\tilde{x}(i)$, i = 1, 2, ... , denotes the point in R^k where $\tilde{x} = \{\tilde{x}(1), \tilde{x}(2), \ldots\}$. Label the $\tilde{x}(1)$, $\tilde{x}(2)$, ... by looking at the coordinates in the following manner. For each n partition R^k up into half-open cubes with sides of length $1/2^n$ which successively partition the larger cubes. For each n number the cubes A(n,i), i = 1,2,... , first numbering the cubes near the origin in some orderly manner. We say $x < y$ $(x,y \in R^k)$ if for the first n such x and y are in separate cubes, $x \in A(n,i)$ and $y \in A(n,j)$ with $i < j$. We use the following assumption.

(A). Let P be a m.p.p. and let $Q_{\tilde{x}}$ $(\tilde{x} \in X)$ also be m.p.p.'s. Assume that for each $A \in \mathcal{F}$, $Q_{\tilde{x}(i)}(A)$ is a measurable function of $\tilde{x}$ $((X,\mathcal{F}) \to (R^1, B^1))$.

Lemma 5.2. Let $\tilde{x} \in X$ and (A) hold. On $\chi = (X \times X \times \cdots)$, $\tilde{F} = (F \times F \times \cdots)$ construct the product probability space with measure $Q_{\tilde{x}(1)} \times Q_{\tilde{x}(2)} \times \cdots$ (see Loeve, 1963, p. 91). Then for each $A \in F \times F \times F \times \cdots$, $Q_{\tilde{x}}(A)$ is a measurable function of $\tilde{x}$ (as a function on (X, F)).

Proof. (Note if $\tilde{x} = \{\tilde{x}(1), \ldots, \tilde{x}(n)\}$ then $Q_{\tilde{x}(i)}$, $i = n+1, 2, \ldots$ puts mass one on ϕ (the empty set of points in R^k).) It is enough to show $Q_{\tilde{x}}$ measurable on $A = A_1 \times \cdots \times A_n \times \cdots$. But $Q_{\tilde{x}}(A) = \prod_{i=1}^{\infty} Q_{\tilde{x}(i)}(A_i)$ where the infinite product converges monotonely and each finite product is measurable as it is the product of measurable functions.

On $(X, F) \times (\chi, \tilde{F})$ we now form the probability measure Π such that $\Pi(A \times A) = \int_A Q_{\tilde{x}}(A)\, P(d\tilde{x})$.

Consider now the function T on $X \times \chi$. If $\bar{x} = \{\tilde{x}_0, (\tilde{x}_1, \tilde{x}_2, \ldots)\}$ we define $T(\bar{x}) = \bigcup_{n=1}^{\infty} \tilde{x}_i$. X is a subset of the range of T. For $A \in F$ we now see that $T^{-1}(A) \in F \times \tilde{F}$ (the product σ-field). To show this it is enough to show it for $A = \bigcap_{i=1}^{\infty} \{N(A_i) = k_i\}$, where the A_i are disjoint bounded Borel subsets which partition R^k and $k_i \in Z_+$. Note that $A^n = \bigcap_{i=1}^{n} \{N_{\tilde{x}_j}(A_i) = \ell_i\} \in F \times \tilde{F}$ so that the event $\bigcap_{j=1}^{\infty} \bigcap_{i=1}^{n} \{N_{\tilde{x}_j}(A_i) = \ell_i\}$ $\in F \times \tilde{F}$. Now $T^{-1}(A^n)$ is the union of a countable number of such possibilities (since all but at most $k_1 + \cdots + k_n$ of the ℓ_i must be zero). Now intersect over all n.

Definition. The clustering measure associated with P and Q_x $(x \in X)$ under assumptions (A) is the measure space (X, F, P_c) where $P_c(A) = \Pi(T^{-1}(A))$. P_c is a subprobability measure. If $P_c(X) = 1$ we call P_i the cluster process with *base process* P and *structure processes* Q_x.

The usual structure processes Q_x, are such that $Q_x(A) = Q(A-x)$ where $A - x = \{\tilde{x}: (\tilde{x}(1) + x, \tilde{x}(2) + x, \ldots) \in A\}$. In this case we say Q_x is generated by translation.

Lemma 5.3. If $Q_x(A) = Q(A - x)$ for a m.p.p. Q then $Q_{\tilde{x}(i)}(A)$ is a measurable function of $\tilde{x}$.

Proof. Fix a positive integer i. We first show that $f_i(\tilde{x}) = \tilde{x}(i)$ is measurable $(X,F) \to (R^k,B^k)$ where B^k is the Borel sigma-field. It is enough to show that for a half-closed, bounded rectangle A used in the definition of the ordering of points in R^k, $B = \{\tilde{x}:f_i(\tilde{x}) \in A\} \in F$. Since each bounded set has a finite number of points, $\tilde{x} \in B$ iff for the first n such that $N\{A(n,j)\} = 0$ or 1 for all $j \leq m$ (where $\bigcup_{j=1}^{m} A(n,j) \supseteq A = A(r,\ell)$, $r \leq n$ and m is a minimum) the ith set $A(n,j)$ such that $N\{A(n,j)\} = 1$ is such that $A(n,j) \subseteq A$. Since $Q_{\tilde{x}(i)}(A) = Q_{f_i(\tilde{x})}(A)$ it is enough to complete the proof if we show that for $A \in F$, $g(x) = Q_x(A)$ is measurable $(R^k,B^k) \to (R^1,B^1)$. It is sufficient to show this for A of the form $A = \{\tilde{x}:N(A_i) = k_i,\ i = 1,2,\ldots,n\}$, A_i bounded, and Borel as usual. Let $\tilde{x}|_n = \{\tilde{x}(i),\ldots,\tilde{x}(n)\}$. We can show this for $Q|_n$, Q restricted to the $\tilde{x}|_n$ space, since $Q|_n(A + x)$ converges monotonically to $Q(A + x)$. But $Q|_n$ may be decomposed into Borel measures on $\bigcup_{j=0}^{n} R^{k*n}$ as in Section 1, in which case the measurability is the measurability of translates of Borel sub-probability measures on Euclidean space, a fact that is well known (for its use in taking convolutions for sums of independent random vectors).

In order to prove the fundamental result relating the p.g. fls. of the cluster process and the base and structure processes we shall characterize the p.g.fl. in a slightly different manner (which is clear from Section 4).

Lemma 5.4. If $\|\phi\| \leq 1$ and P is a m.p.p. with p.g.fl. G, then $G(\phi) = E_{\tilde{x}\in X}[\prod_{\tilde{x}(i)} \phi\{\tilde{x}(i)\}]$.

Proof. Each $\prod_{i=1}^{n} \phi(x_i)$ is measurable and converges monotonically to the possibly infinite product. If $|\phi| < 1$ on B and $\phi = 1$ on B complement, then

$$\prod_{i=1}^{\infty} \phi(x_i) = \prod_{x_i \varepsilon B} \phi(x_i)$$

and

$$E\{\prod_{i=1}^{\infty} \phi(x_i)\} = \sum_{n=0}^{\infty} \int_{R^{k*n}} \phi(x_1) \cdots \phi(x_n)\, P_B^{(n)}(dx_1 \cdots dx_n)$$

$$= G(\phi) \ .$$

More generally, if $\phi \not\in S$ we enlarge the domain of the p.g.fl. to ϕ with $||\phi|| \leq 1$ and define it to be as given in this lemma.

Theorem 5.5. Let $\tilde{P}$ be a cluster process with base P and cluster processes Q_x (under assumption (A)) such that for $x_1 \neq x_2$, $Q_{x_1} \times Q_{x_2}(\{(\tilde{x}_1,\tilde{x}_2):\tilde{x}_1(j) = \tilde{x}_2(k)$ for some j and k$\}) = 0$.

If $\phi \in S$ and $\tilde{G}$, G_1 and $G_2(\cdot|x)$ are the p.g.fls. of $\tilde{P}$, P and Q_x respectively, then

$$\tilde{G}(\phi) = G_1\{G_2(\phi|\cdot)\}.$$

Proof. With the mechanism at hand this drops out easily. We have

$$E^{\tilde{P}}_{\tilde{x}\varepsilon X}\{\prod_i \phi(x_i)\} = E^{\pi}_{x*\varepsilon\chi}[\prod_{i=1}^{\infty}\prod_{j=1}^{\infty} \phi\{\tilde{x}_i(j)\}]$$

$$= E^{P}_{\tilde{x}_0 \varepsilon X}(E^{Q_{\tilde{x}_0}}_{(\tilde{x}_1,\tilde{x}_2,\ldots)}[\prod_{i=1}^{\infty}\prod_{j=1}^{\infty} \phi\{\tilde{x}_i(j)\}])$$

$$= E^{P}_{\tilde{x}_0 \varepsilon X}(\prod_{i=1}^{\infty} E^{Q_{\tilde{x}_0}(i)}_{\tilde{x}_i \varepsilon X}[\prod_{j=1}^{\infty} \phi\{x_i(j)\}])$$

$$= E^{P}_{\tilde{x}_0 \varepsilon X}[\prod_{i=1}^{\infty} G_2\{\phi|\ \tilde{x}_0(i)\}] = G_1\{G_2(\phi|\cdot)\},$$

where $E^{\tilde{T}}_{x\varepsilon X}$ means we take the expectation with respect to the random quantity x whose distribution is determined by the probability measure $\tilde{T}$. Step one follows since $\tilde{P}$ is defined in terms of the mapping T and probability measure π; step two breaks π up into its conditional definition;

step three uses the independence of $Q_{\tilde{x}_0(1)}, Q_{\tilde{x}_0(2)}, \ldots$, and step four uses the definition of the p.g.fl.

(The result of Theorem 5.5 holds in general provided that we allow multiple points at the same position in R^k. In this case if x' occurs r-times, $\int \log \phi(x)\, dN(x)$ would have a contribution of $r \log \phi(x')$.)

Proposition 5.6. A m.p.p. P with p.g.fl. G is stationary iff for every ϕ, $\|\phi\| \leq 1$, $x \in R^k$,

$$G(\phi) = G(\phi_x) \quad , \tag{5.3}$$

where

$$\phi_x(t) = \phi(t - x) \ .$$

Proof. The if direction follows from the definition of stationarity and Proposition 4.2a. The only if direction follows from approximation of ϕ by simple functions.

Corollary 5.7. Let $\tilde{P}$ be a cluster process with base process P and structure processes Q_x. Then $\tilde{P}$ is a stationary m.p.p. if either;

a) P is stationary and Q_x is generated by translation;

b) Q_x is stationary for each x.

Proof. a) If Q_x is generated by translation it is easy to see that

$$G_2(\phi_x|\cdot) = G_2(\phi|x+\cdot) = G_2(\phi|\cdot)_x \ ,$$

so that

$$G(\phi_x) = G_1\{G_2(\phi|\cdot)_x\} = G_1\{G_2(\phi|\cdot)\} = G(\phi) \ .$$

b) $G_1\{G_2(\phi_x|\cdot)\} = G_1\{G_2(\phi|\cdot)\}$.

These are certainly not the only methods of generating stationary m.p.p. as will be shown by a later example.

We now turn to the standard specific examples of m.p.p. For each example, where known, we give a) the p.g.fl. b) the first two product densities, c) the covariance density function and d) the spectral function.

Example 5.1. Simple Poisson Process. The cornerstone of m.p.p.'s is the simple Poisson m.p.p. It is characterized by a positive constant λ and the fact that for any set of disjoint bounded measurable Borel sets of R $\{A_1,\ldots,A_n\}$ with Lebesgue measure $V(A_1),\ldots V(A_n)$ the random variables $N(A_1),\ldots, N(A_n)$ are independent Poisson random variables with parameters $\lambda V(A_1),\ldots, \lambda V(A_n)$ respectively. The importance of the simple Poisson m.p.p. is analogous to that of the normal distribution and uniform distribution combined in the usual study of random variables. The simple Poisson m.p.p. is the limiting m.p.p. in many limit theorems (see Section 7). It is the natural model for the expression that "points are distributed at random." Indeed, in statistical ecology (see Greig-Smith, 1964, Chap. 3) the term random refers specifically to the simple Poisson m.p.p. and non-random refers to any other distribution (as opposed to many fields where random refers to some stochastic variability and non-random to something that is deterministic).

The p.g.fl. is $G(\phi) = \exp[\lambda\int\{\phi(x) - 1\}\,dx]$. The first two product densities are $f_1(x) = \lambda$, $f_2(x,y) = \lambda^2$. The covariance density function is $h(x,y) = \delta(x-y)\lambda$ and the spectral function is $g(w) = \lambda$.

Example 5.2. Poisson m.p.p. An immediate generalization of the simple Poisson m.p.p. is to let the density λ vary with x. Let $\lambda(x)$ be a non-negative measurable function which is integrable on compact subsets of R^k. The Poisson m.p.p. with density function $\lambda(x)$ is characterized by assigning to $N(A_i)$, $i = 1,2,\ldots, n$ (as above) independent Poisson random variables with parameters $\int_{A_i}\lambda(x)\,dx$, $i = 1,2,\ldots, n$. (More, generally one may use a measure Λ with the parameters being $\Lambda(A_i)$. If Λ has an atomic part then duplicate points need to be allowed.) If $\lambda(x) \equiv \lambda$ this reduces to the simple Poisson m.p.p.

The p.g.fl. is $G(\phi) = \exp[\int \lambda(x)\{\phi(x) - 1\}\, dx]$. The product densities and the covariance density function are given by

$$f_1(x) = \lambda(x), \qquad f_2(x,y) = \lambda(x)\lambda(y)$$

and

$$h(x,y) = \delta(x-y)\lambda(x) \quad .$$

As the process is not stationary unless $\lambda(x) = \lambda$ a.s. (Lebesgue measure) the spectral function is not appropriate.

Example 5.3. Neyman-Scott clustering m.p.p. We shall abbreviate this to N-S m.p.p. Professors Neyman and Scott developed a clustering model as outlined at the beginning of this section in their work on the clustering of galaxies. They postulate a stationary m.p.p. for the base process (Neyman and Scott, 1958). The structure processes are generated by translation. Conditionally upon the cluster center at 0 they postulate a nonnegative integer valued random variable ν giving the number of points in the cluster. Conditionally upon ν the cluster points are independently and identically distributed with a density function $f(x)$ with respect to Lebesgue measure. The p.g.fl. for the cluster process is thus given by

$$\begin{aligned} G_2(\phi|0) &= E\{\pi\phi(x_i)\} = \sum_{n=0}^{\infty} E\{\prod_{i=1}^{n} \phi(x_i)\,|\,\nu = n\}\, P(\nu = n) \\ &= \sum_{n=0}^{\infty} \prod_{i=1}^{n} E\{\phi(x_i)\}\, P(\nu = n) \\ &= \sum_{n=0}^{\infty} \{\int \phi(x)\, f(x)\, dx\}^n\, P(\nu = n) \\ &= g\{\int \phi(x)\, f(x)\, dx\}, \end{aligned}$$

where

$$g(z) = \sum_{n=0}^{\infty} z^n\, P(\nu = n)$$

is the probability generating function for ν. Thus, the p.g.fl. for the cluster process is $G_1(G_2(\phi|\cdot)) = G_1[g\{\int \phi(x + \cdot)\ f(x)\ dx\}]$, where G_1 is the base process p.g.fl. In particular, Neyman and Scott consider the case where the base process is a Poisson process, so that the generating function is

$$G(\phi) = \exp[\lambda\int\{g\{\int\phi(x + y)\ f(x)\ dx\} - 1\}\ dy] .$$

To evaluate the product densities we use the rth order variation as described in Section 4. We also use the following

Proposition 5.8. Let S be a bounded, Borel set with indicator function χ_S and m a positive integer. Then $E[N(S)\ \{N(S) - 1\} \cdots \{N(S) - m + 1\}]$ is finite iff

$$\lim_{\varepsilon\to 0} \delta^m(\underbrace{\chi_S,\ldots,\chi_S}_{\text{m times}},\ 1 - \varepsilon\chi_S) < \infty \quad ,$$

in which case the two are equal.

Proof. $E[N(S) \cdots \{N(S) - m + 1\}]$

$$= \lim_{\eta\uparrow 1} \sum_{n=m}^{\infty} \frac{n!}{(n-m)!} \int_{S^m} \eta^{n-m}\ P_S^{(n)}(dx_1 \cdots dx_n)$$

$$= \lim_{\varepsilon\to 0} \sum_{n=m}^{\infty} \frac{n!}{(n-m)!} \int_{S^m} \chi_S(x_1)\cdots\chi_S(x_m)\{1-\varepsilon\chi_S(x_{m+1})\}$$

$$\cdots \{1-\varepsilon\chi_S(x_n)\}\ P_S^{(n)}(dx_1 \cdots dx_n)$$

$$= \lim_{\varepsilon\to 0} \delta^m(\chi_S,\ldots,\chi_S;\ 1 - \varepsilon\chi_S).$$

Another equivalent method of getting at the moments is to use the characteristic function as given in Section 4 and to differentiate that in the usual manner.

Using the monotone convergence theorem we find that

$$\lim_{\varepsilon\to 0} \frac{\partial}{\partial\lambda_1} G(1 - \varepsilon\chi_S + \lambda_1\chi_x)\Big|_{\lambda_1=0} = \lim_{z\to 1} g'(z)\ \lambda V(S) \ .$$

But $g'(1) = E(\nu)$ so that the first order product density exists iff $E(\nu) < \infty$, in which case $f_1(x) = \lambda E(\nu)$. Similarly,

$$\lim_{\varepsilon\to 0} \frac{\partial^2}{\partial\lambda_1\partial\lambda_2} G(1 - \varepsilon\chi_S + \lambda_1\chi_S + \lambda_2\chi_S)\Big|_{\lambda_1=\lambda_2=0}$$

$$= \lambda^2\{\lim_{z\to 1} g'(z)\}^2\ v(S)^2 + \lambda \lim_{z\to 1} g''(z) \int_S\int_S \{\int_y f(w-y)\ f(v-y)\ dy\} dw\ dr.$$

Thus, we need $E\{\nu(\nu - 1)\} < \infty$ for the second order product density function to exist, in which case

$$f_2(w,v) = \lambda^2 E(\nu)^2 + \lambda E\{\nu(\nu - 1)\} \int f(w-y)\ f(v-y)\ dy$$

$$= \lambda^2\ E(\nu)^2 + \lambda E\{\nu(\nu - 1)\} \int f(z)\ f\{(v-w) + z\}\ dz\ .$$

From this the covariance density function is

$$h(x,y) = \lambda E\{\nu(\nu - 1)\} \int f(z)\ f\{(x-y) + z\}\ dz + \delta(x-y)\ \lambda E(\nu)\ .$$

The spectral function is thus

$$g(w) = \lambda E(\nu) + \lambda E\{\nu(\nu - 1)\} \Big| \int e^{iwz}\ f(z)\ dz\Big|^2 \ .$$

In proceeding thus we assumed that we actually have a m.p.p., that is, the condition of Theorem 5.1 is satisfied. We note that this is always satisfied if $E(\nu) < \infty$. To see this let

$$Q^* = \int Q\{N(I + x) \neq 0\}\ dx$$

$$= \int \sum_{n=1}^{\infty} P(\nu = n)\ [1 - \{1 - \int_I f(y-x)\ dx\}^n]\ dy\ .$$

Now $1 - (1 - p)^n \leq np$ if $0 \leq p \leq 1$ so that

$$Q^* \leq \int \sum_{n=1}^{\infty} P(\nu = n) n \int_I f(y - x)\, dx\, dy = E(\nu)\, V(I) < \infty$$

for a bounded rectangle I.

If in the foregoing a parent point is included at the cluster center the above may easily be modified to obtain the appropriate formulas.

Neyman and Scott call the clustering process outlined above a first order clustering process or a simple clustering process. One may consider the possibilities of clusters of clusters, etc. They define inductively n + 1st order clustering by choosing a base process which is an nth order clustering process and using a structure process of the type used above (with p.g.fl. $g\{\int \phi(x)\, f(x)\, dx\}$). Thus, the p.g.fl. of an n + 1st order clustering process is $G(\phi) = G_n[g\{\int \phi(x + \cdot)\, f(x)\, dx\}]$ where G_n is the p.g.fl. of an nth order clustering process. It is an open question to describe (in a reasonable sense) the closure of the class of m.p.p. generated in this manner.

It is typical of the field that many m.p.p. may be generated in a variety of ways. Consider the following example

Example 5.4. Consider a cluster process on R^2 whose base process is a Poisson process with $\lambda(x) = (1 + \sin x_1)(1 + \sin x_2)$, where $x = (x_1, x_2)$. The cluster processes are formed as follows. Given a center at z = (z_1, z_2) there is one point in the cluster which is distributed uniformly on the square $I_z = (z_1 \pm \delta_1(z),\ z_2 \pm \delta_2(z))$ where

$$\delta_1(z) \begin{cases} = 2\pi & \text{if } [|z_1|] \equiv \text{least integer} > |z_1| \text{ is even,} \\ = 4\pi & \text{if } [|z_1|] \text{ is odd.} \end{cases}$$

The function δ_2 is defined analogously for $[|z_2|]$. Then the cluster process has p.g.fl.

$$G(\phi) = \exp(\int \lambda(x)\ [\int \{\phi(x + z)\ \chi_{I(x)}(z)/4\delta_1(x)\ \delta_2(x)\}\, dz - 1]\, dx)\ .$$

Changing variables x + z = t, x = u,

$$G(\phi) = \exp(\int\{\phi(t) - 1\}[\int\{\lambda(u)\ \chi_{I(u)}(t-u)/4\delta_1(u)\ \delta_2(u)\}\ du]\ dt) .$$

But the term in brackets integrates to one, so that $G(\phi) = \exp[\int\{\phi(t)-1\}dt]$ and the process is the simple Poisson m.p.p. with $\lambda = 1$. This example also shows that a cluster process may be stationary when the base process is not stationary and the structure processes are neither stationary nor generated by translation. For other more "realistic" examples, see Bartlett (1964) and Vere-Jones (1970). Thus, in m.p.p. if identifiability is required in some class of m.p.p. care must be taken to check that it holds.

Example 5.5. Doubly stochastic Poisson m.p.p. In this case the intensity $\lambda(x)$ of Poisson process is chosen itself as a realization of a stochastic process with nonnegative locally integrable sample functions (or more generally, one could choose a random measure Λ). Intuitively a sample $x \in X$ is chosen in two steps.

1) We observe the process giving $\lambda(x)$.

2) Having $\lambda(x)$ we then sample a Poisson process with intensity $\lambda(x)$ to give our m.p.p. The term doubly stochastic refers to those two conceptual stochastic operations.

A formal construction of such processes would involve roughly the following steps (compare to our construction of cluster processes). Let P' be a probability measure on the sample path space of the $\lambda(x)$'s, say, Λ. On $\Lambda \times X$ (the product space Λ with X) construct measures by $\tilde{P}(A \times B) = \int_A P_\lambda(B)\ P'(d\lambda)$ where P_λ is the Poisson m.p.p. with intensity λ. Use the map $T:\Lambda \times X \to X$ defined by $T(\lambda,x) = x$ and define the resulting doubly stochastic Poisson m.p.p. by $P(A) = \tilde{P}(T^{-1}(A))$ for $A \in \mathcal{F}$. We shall not go into the details of such a construction here.

Such processes are considered by Cox (1955) and are discussed in Bartlett (1964a, 1964b), Cox and Lewis (1966), and Matern (1960, 1971).

The calculation of the p.g.fl. is best done conditionally.

$$G(\phi) = E\{\pi\phi(x_i)\} = E_\lambda[E\{\pi\phi(x_i)|\lambda\}]$$

$$= E_\lambda(\exp[\int\lambda(x)\{\phi(x) - 1\}\, dx]) .$$

Assuming that sufficient regularity for the following operations exists and that the first two product density functions exist we find that;

$$\lim_{\varepsilon\to 0} \frac{\partial}{\partial\lambda_1} E_\lambda(\exp[\int\lambda(x)\ \{-\varepsilon\chi_{S^1}(x) + \lambda_1\chi_2(x)\}\ dx])\Big|_{\lambda_1=0}$$

$$= \lim_{\varepsilon\to 0} E\{\frac{\partial}{\partial\lambda_1}\ (-)\}\Big|_{\lambda_1=0} = \int_S E\{\lambda(x)\},$$

so $f_1(x) = E\{\lambda(x)\}$.

Similarly,

$$\lim_{\varepsilon\to 0} \delta^2(\chi_S,\ \chi_S,\ 1 - \varepsilon\chi_S) = E_\lambda[\{\int_S\ \lambda(x)\ dx\}^2]$$

$$= \int_S \int_S E_\lambda\{\lambda(x)\ \lambda(y)\}\ dx\ dy\ ,$$

so that

$$f_2(x,y) = E_\lambda\{\lambda(x)\ \lambda(y)\},$$

$$h(x,y) = E_\lambda\{\lambda(x)\ \lambda(y)\} - E_\lambda\{\lambda(x)\}\ E_\lambda\{\lambda(y)\} + \delta(x-y)\ E\{\lambda(x)\},$$

which is the covariance function for the underlying process $\lambda(\cdot)$. Assuming stationarity one can write down $g(\omega)$ the spectral function.

If $\lambda(x)$ is a constant λ a.s., then the process is called a mixed Poisson process (we use this in Section 7). In that case λ is a random variable and $f_1(x) = E(\lambda)$, $f_2(x,y) = E(\lambda^2)$ and $h(x,y) = \text{var}(\lambda) + \delta(x-y)E(\lambda$ It is clear that mixed Poisson processes are highly nonergodic in the sense that any one realization can only give (in fact determine with probability one) information about the value of λ for that realization, but not the distribution of λ. In this case if F is the distribution function of λ the p.g.fl. reduces to;

$$G(\phi) = \int_{R^1} \exp[\lambda \int_{R^k} \{\phi(x) - 1\} dx] \, dF(\lambda) \; .$$

From the p.g.fl. it follows immediately that if $\lambda(x)$ is a stationary process, then the corresponding doubly stochastic Poisson m.p.p. is stationary.

Example 5.6. Compound Poisson m.p.p. We present an intuitive picture. Let N be a Poisson random variable with parameter λ. Let Q be a m.p.p. If N = n then we take n independent samples of Q and take the union of the points in all n samples as our realization. Let the distribution of such a sum be denoted by Q^{*n}. The compound Poisson process has measure $P = e^{-\lambda} \sum_{m=0}^{\infty} \frac{\lambda^m}{m!} Q^{*m}$ where Q^{*0} puts mass one on the empty set of points.

(A note on terminology: Matern (1960) has used the name compound Poisson process for what we have called a doubly stochastic m.p.p. In Matern (1971) he uses compound Poisson process for what we have called a mixed Poisson process. The terminology used here seems preferable, since the construction is analogous to the usual compound Poisson distribution for random variables (e.g., Lamperti, 1966, p. 89; Feller, 1966, p. 177) and plays the same role in infinite divisibility in both situations (cp. Matthes, 1972 and Lamperti, 1966, p. 89, Theorem 2). It is interesting to note how terminology changes. In Feller (1943) what is now a compound Poisson distribution was called the generalized Poisson distribution and the term compound Poisson distribution was applied to what is most naturally now called a mixture of Poisson distributions.)

Let G_Q be the p.g.fl. for Q. Then the p.g.fl. for the compound Poisson process P is

$$\begin{aligned} G(\phi) &= E_N[E\{\pi\phi(x_i)|N\}] \\ &= e^{-\lambda} \sum_{m=0}^{\infty} \frac{\lambda^m}{m!} G_Q(\phi)^m \\ &= \exp[\lambda\{G_Q(\phi) - 1\}] \; . \end{aligned}$$

(Strictly speaking this is the p.g.fl. where a point may be duplicated; if Q has a first order density, for example, such events have probability zero.) Suppose that Q has first and second order product densities, f_1^Q and f_2^Q. Then from the p.g.fl. we see that for our compound Poisson m.p.p.

$$f_1(x) = \lambda f_1^Q(x) ,$$

$$f_2(x,y) = \lambda^2 f_1^Q(x)\ f_1^Q(y) + \lambda f_2^Q(x,y) ,$$

$$h(x,y) = \lambda\{f_2^Q(x,y) + \delta(x-y)\ f_1^Q(x)\}.$$

6. INFINITE DIVISIBILITY

This has been one of the most developed topics in the mathematical theory of m.p.p.'s. We refer the reader to Matthes (1972) who discusses this in detail.

7. LIMIT THEOREMS

The first result we describe is due to Goldman (1967b) and Tulya-Muhika (1970). We will need some definitions before stating the theorem.

Definition. A m.p.p. P on R^k is a well-distributed or a proper G-process if there is a finite measurable function μ on (X,F) which satisfies the following:

Consider any sequence I_i of rectangles aligned along the coordinate axes in R^k for which

a) the ratios of the lengths of the sides in any two fixed directions is a positive constant for all i (the constant may vary with the directions),

b) the origin is in $\bar{I}_i$ (the closure of I_i) for each i and all I_i lie in the same quadrant and

c) if the diameter expands to ∞. Then $N(I_i)/V(I_i) \to \mu$ as $i \to \infty$ a.s. (P) where $V(I_i)$ the Lebesgue measure of I_i.

Definition. A doubly stochastic Poisson process is a mixed Poisson process if $P\{\lambda(x) = \text{a constant}\} = 1$. (The constant may vary, but each realization is from the usual Poisson process with constant density λ.)

Definition. Let P, P_n, $n = 1,2,\ldots$ be m.p.p. on R^k. We write $P_n \xrightarrow{\text{rect}} P$ if for any finite set of bounded rectangles $I_1,\ldots, I_m$ and non-negative integers $i_1,\ldots, i_m$

$$\lim_n P_n\{N(I_j) = i_j,\ j = 1,2,\ldots, m\} = P\{N(I_j) = i_j,\ j = 1,2,\ldots, m\}$$

The following theorem is based upon the idea of introducing variability into a point process by deleting points independently with a probability p where $p \to 1$. As p converges to one in any bounded set S the number of points will then converge to zero in probability. To rectify this each set is associated with an appropriately enlarged set for the original process.

Definition. Let $0 \le p \le 1$ and P be a m.p.p. on R^k. Then P_p is a m.p.p. such that for any disjoint bounded Borel sets $S_1,\ldots, S_m$ and non-negative integers $k_1,\ldots, k_m$

$$P_p\{N(S_i) = k_i,\ i = 1,2,\ldots, m\}$$
$$= \sum_{n_1=0}^{\infty} \cdots \sum_{n_m=0}^{\infty} [P\{N(S_i/(1-p)^{1/k}) = k_i+n_i, i=1,\ldots,m\} \prod_{j=1}^{m} \binom{k_i+n_i}{k_i}$$
$$\times (1-p)^{k_i} p^{n_i}]$$

where $S_i/(1-p)^{1/k} = \{x : x(1-p)^{1/k} \in S_i\}$.

It may be verified by the extension theorem that these P_p uniquely define a probability measure P_p on (X,F). The above expression results from the fact that if k_i points are in $S_i/(1-p)^{1/k}$ after deletion there must have been k_i+n_i for some non-negative integer n_i before deletion.

Theorem 7.1. Let P be a proper G-process on R^k and p_n a sequence of numbers in (0,1), $p_n \to 1$. Then $P_{p_n} \xrightarrow{rect} P'$, where P' is a mixed Poisson process whose parameter has the same distribution as the μ in the definition of a proper G-process.

Goldman (1967b) gave a different definition of a G-process than Tulya-Muhika. For any increasing sequence of rectangles I_i such that at least one dimension became infinite he required that a.s. (P) $N(I_i)/V(I_i) \to \mu$. To see that the definitions differ consider a m.p.p. P on R^2 such that $P(\{\omega\}) = 1$, where ω is the set of points in the plane with both coordinates integer-valued. This is a proper G-process such that $P_{p_n} \xrightarrow{rect}$ the Poisson process with parameter $\lambda = 1$. However let $I_i = \{(x,y): 0 \leq x < i,\ 0 \leq y < i\}$. Then $N(I_i)/V(I_i) = 1$, but if we let $I_i = \{(x,y); 0 \leq x \leq 1/2,\ 0 \leq y < i\}$, then $N(I_i)/V(I_i) = 2$, so that P is not a G-process in Goldman's sense. We also note from this example that we do not get convergence of N(S), for every bounded Borel set S, to the correct Poisson distribution. Let $p_n = 1 - (1/n)^2$ and S be the set of all points in the unit square with irrational coordinates. Then $P_{p_n}\{N(S) = 0\} = 1$, whereas the limiting m.p.p. assigns probability e^{-1} to this event. This points up the difference between $\to$ and $\xrightarrow{rect}$.

For each $\omega \in X$ let P^{ω} be the m.p.p. that puts mass one on ω. Tulya-Muhika (1970) gives the following

Theorem 7.2. Let P be a m.p.p. Then $P^{\omega}_{p_n} \xrightarrow{rect}$ a.s. (P) iff P is a proper G-process. If P is a proper G-process with parameter function $\mu(\omega)$ then $P^{\omega}_{p_n} \xrightarrow{rect}$ Poisson$(\mu(\omega))$ a.s. (P) where Poisson $(\mu(\omega))$ is the mixed Poisson process with parameter $\lambda = \mu(\omega)$.

Various other situations in which Poisson random variables result as a limit occur in ecological situations, for example the number of points in a quadrat. As examples see Neyman (1939, p. 44); Skellam (1952. p. 350) who is essentially stating Theorem 7.1 and Neyman and Scott (1957).

The next result we describe is due to Milne (1971) using a technique due to Leadbetter (1968). It is somewhat unfair to present these results since the proofs are so simple and clean that the reader is best advised to refer to Milne's paper. In this case the m.p.p. may duplicate points so the appropriate space will be X', say, where any $x \in X'$ is $x = \{(x_1,i_1), (x_2,i_2),\ldots;\ x_j \in R^k,\ i_j \in Z^+\}$. Here i_j denotes the multiplicity associated with the point x_i. As before the set of the x_i is assumed to have no finite limit point. We denote these processes by n.o.m.p.p. (non-orderly multidimensional point processes). (Strictly speaking if all $i_j = 1$ a.s. we have our old m.p.p. and should reserve the term non-orderly for a process with $P\{\text{at least one } i_j > 1\} > 0$.) Let C_t be the cube of side t at the origin, $C_t = \{x:x = (x_1,\ldots,x_k),\ 0 \leq x_i < t\}$ and N_j be the random variable giving the number of groups of size j within C_1, that is, the number of (x_n,i_n) where $x_n \in C_1$, and $i_n = j$. Let $N^* = \sum_{j=1}^{\infty} N_j$ and $N = \sum_{k=1}^{\infty} kN_k$. Let $P_j(t) = P\{N(C_t) = j\}$, and $m = E\{N(C_1)\}$. A m.p.p. (in this generality) will be called one-stationary if for every bounded Borel set S, vector $x \in R^k$ and integer ℓ, we have $P\{N(S) = \ell\} = P\{N(S + x) = \ell\}$. Here N(S) counts the multiplicities associated with each point. Milne's generalizations of the one-dimensional results follow:

Theorem 7.3. a) Let P be a one-stationary n.o.m.p.p. Then $\lambda \lim_{t\downarrow 0} t^{-k}[1 - P_0(t)] = E(N^*)$. If $E(N^*) < \infty$, then for $j = 1,2,\ldots$, $E(N_j) = \lim_{t\downarrow 0} t^{-k} P_j(t)$.

b) Let $E(N^*) < \infty$. Then for $j = 1,2,\ldots$ the following exist;

$$\pi_j = \lim_{t\downarrow 0} P_j(t)[1 - P_0(t)]^{-1}$$

and these form a set of probabilities ($\sum_{j=1}^{\infty} \pi_j = 1$). Also

$$E(N_k) = \lambda\pi_k ,$$

$$E(N^*) = \lambda ,$$

$$m = E(N) = \lambda \sum_{j=1}^{\infty} j\pi_j .$$

c) If $\alpha \geq 1$, $E(N^\alpha) < \infty$ then $\lim_{t \to 0} t^{-k} \sum_{j=1}^{\infty} j^\alpha P_j(t) = \lambda \sum_{j=1}^{\infty} j^\alpha \pi_j$.

The last equation of b) has the interpretation that the intensity with which points appear is equal to the intensity with which groups appear times the average size of a group. For a work and references on weak convergence for point processes see Jagers (1971) and Grigelionis (1972).

8. RESULTS OF A MORE GEOMETRIC NATURE

In this section we consider results which are of a more geometric nature. Since the majority of the results pertain somewhat directly to statistical analysis we shall only briefly point out some of the appropriate literature and results and defer to the more detailed discussion in Holgate (1972).

The geometrical analysis of a random set of points in Euclidean space has a long history. A good place to begin reading in the area is Kendall and Moran (1963). In particular a sample of size n distributed on a set in Euclidean space has drawn much attention. More recent references to such topics are Geffroy (1958a,b), Renyi and Sulanke (1963, 1964), Efron (1965) and Fisher (1969, 1971). In particular, such papers have a bearing on the pattern in which members of a cluster are dispersed. Related topics are coverage problems as mentioned in Section 1.

Many applications of m.p.p.'s have considered the distance between a random point in space or a random point of the realization of the m.p.p. and the nearest, second nearest,... point of the sample. We refer to points from the m.p.p. as sites and a random point refers to an auxiliary point chosen for sampling purposes. See Matern (1960), Persson (1964), Persson (1965) and Holgate (1965a). In the case of a simple Poisson m.p.p. the distributions are particularly easy to calculate since the radial distance of sites from a fixed point or from sites follows a one-dimensional Poisson process of integrated rate πr^2 in a circular area about the fixed point. In applications if we choose a random site of the sample, care must be taken with the meaning of the term random.

Persson (1964) considers such distances in both the Poisson case and when the point is randomly selected but the sites fall on a square lattice. He also considers the distribution of the number of seedlings in a randomly located circular plot. The case where the sites consist of the vertices of an equilateral triangular lattice has been considered by Holgate (1965b) and Persson (1965). This is a natural object for study as the centers of tightly packed circles in the plane form such a lattice. Warren (1971) considers distance in the Neyman-Scott clustering model following procedures analogous to those discussed in Sections 4 and 5, but specializing the models even further in order to obtain more specific results. Warren uses the term center-satellite structure for what we have called the Neyman-Scott clustering m.p.p.

Another topic of much applied interest is in sampling a m.p.p. by the quadrat method (i.e., counting the number of points in an area). Probability g.fls. may be useful in this context in certain situations. See Neyman and Scott (1958).

Another application of clustering is given in Thompson (1955). Greig-Smith (1952) has discussed using the analysis of variance by nesting quadrat sampling in m.p.p.'s. For the simple Poisson process this goes through in a fairly straightforward manner due to the independence of the counts in disjoint quadrants. Thompson (1955) considers the modifications necessary when a N-S clustering m.p.p. occurs rather than a simple Poisson m.p.p. He looks at specific moments by using product density functions to arrive at his results.

The closest m.p.p. analog of the interval is perhaps the Dirichlet cells. The Dirichlet cell of a site consists of those points closest to the given site; see Gilbert (1962), and Matern (1960, section 3.5) who states the known results. Roberts (1969) considers the probability that a Poisson site is the nearest neighbor to n other points by getting bounds and by a Monte Carlo study. Naus (1965) gives bounds on the probabilities of getting n of N random points in a square in a smaller subsquare. Such ideas may provide tests for clustering or nonclustering.

There is a considerable literature on the distribution of the number of points to be found in a quadrat. We previously discussed Neyman

(1939). Feller (1943) showed that Neyman's type A distribution results as both a mixture of Poisson variables and a compound Poisson random variable (and other results along this line). Such lack of identifiability springs up in many places in the theory. Various types of distributions including the Poisson, the negative binomial and logarithmic series have been fitted to such data. Recent papers on such topics are Rao (1971), and Gurland and Hinz (1971) which contains good background material.

Warren (1971) indicates that the contagious distributions of Beall and Rescia (1953) and Gurland (1958) may be obtained as quadrat samples from N-S clustering m.p.p.'s.

9. CONCLUDING REMARKS

The mathematical theory of m.p.p. as opposed to point processes on R^1 is quite small. A small fraction of the R^1 results have been extended to R^k but certainly not much at this time. The crux of the matter, of course, is the presence or lack of an ordering. In R^1 the natural ordering gives the interplay between points and intervals as well as the generation of many natural kinds of processes, e.g., renewal processes. There is no natural analog in R^k. Perhaps the appropriate analog is x_t, where each x_t is a m.p.p.; population processes fall into this area. As mentioned in the introduction we have not discussed this topic. The reader might refer to Moyal (1957, 1962a), Harris (1963), Bartlett (1955, 1960), Skellam (1951) and Bailey (1964). There is a great deal of literature on limit theorems in branching processes or population processes. Here, once again, there is a natural ordering to be considered.

As we look over the theory of m.p.p.'s as delimited in this paper one must conclude that the mathematical theory has not been too extensively developed. This is in sharp contrast to the statistical development which has been flourishing at an ever expanding rate and to the model building work which has also been fairly substantial. In this latter context it should be pointed out that there appears to be a need for a period of consolidation while more experience is gained with some specific parametric models. For example, is it possible that we have so much freedom that many of the models will give a reasonable fit of the same data? As a recent example of these models in a somewhat applied context we might mention Paloheimo (1971b).

Further mathematical progress with m.p.p. certainly seems possible, particularly in proving ergodic and/or mixing sorts of results. Dr. P. A. W. Lewis in the discussion to Vere-Jones (1970) indicates reservations as to the usefulness of p.g.fls. in specific calculations. He notes that N-S m.p.p.'s are a special case of generalized branching Poisson processes (Lewis, 1969, 1970). It would appear that an analogous approach may have possibilities in the multidimensional case, but unfortunately I have not found enough time to explore this possibility in the preparation of this paper.

ACKNOWLEDGMENTS

I should like to thank Mary Ann Byhre, Laura Pratt and Sonja Ogle for assistance during the preparation of this paper and to express my apologies to my mother, Mrs. Clare Fisher, who had the misfortune to have her yearly visit during a period of extensive work on the paper.

The writing of this paper was supported in part by NSF grants.

REFERENCES

Abbe, E. (1878). Ueber Blutkörper-Zahlung. S.B. Jena. Ges. Med. Naturiv. No. 48, XCViii-CV.

Bailey, N. T. J. (1964). *The Elements of Stochastic Processes with Applications to the Natural Sciences*. New York: Wiley.

Bartlett, M. S. (1954). Processes stochastiques ponctuels. Annals Inst. Henri Poincare 14, 35-60.

Bartlett, M. S. (1955). *An Introduction to Stochastic Processes*. Cambridge: Cambridge University Press.

Bartlett, M. S. (1960). *Stochastic Population Models in Ecology and Epidemiology*. London: Methuen.

Bartlett, M.S. (1963a). The spectral analysis of point processes. J. R. Statist. Soc. B. 25, 264-96.

Bartlett, M.S. (1963b). Statistical estimation of density functions. Sankhyā A 25, 245-54.

Bartlett, M.S. (1964). The spectral analysis of two-dimensional point processes. Biometrika 51, 299-311.

Bartlett, M.S. & Kendall, D.G. (1951). On the use of the characteristic functional in the analysis of some stochastic processes occurring in physics and biology. Proc. Camb. Phil. Soc. 47, 65-76.

Beall, G. & Rescia, R.R. (1953). A generalization of Neyman's contagious distributions. Biometrics 9, 354-85.

Bhabha, H.J. (1950). On the stochastic theory of continuous parametric systems and its application to electron cascades. Proc. R. Soc. A. 202, 301-22.

Brillinger, D.R. (1970). The frequency analysis of relations between stationary spatial series. In *Proc. 12th Canadian Math. Congress*, R. Pyke, ed. Montreal: Canadian Math. Congress, 39-81.

Cormack, R.M. (1971). A review of classification. J. R. Statist. Soc. A 134, 321-67.

Cox, D.R. (1955). Some statistical methods connected with series of events. J. R. Statist. Soc. B 17, 129-57.

Cox, D.R. & Lewis, P.A.W. (1966). *The Statistical Analysis of Series of Events*. London: Methuen and New York: Barnes & Noble.

Daley, D.J. (1971). Weakly stationary point processes and random measures J. R. Statist. Soc. B, to appear.

Daley, D.J. & Vere-Jones, D. (1972). A summary of the mathematical theory of point processes. This volume.

Douglas, J. (1971). Stirling numbers in discrete distributions. In *Statistical Biology, Volume 1*, Patil, Pielou and Waters, Ed., University Park: Penn. State Univ. Press, 69-98.

Efron, B. (1965). The convex hull of a random set of points. Biometrika 52, 331-43.

Eikler, A.R. (1969). A survey of coverage problems associated with point and target areas. Technometrics 11, 561-89.

Feller, W. (1943). On a general class of "contagious" distributions. Ann. Math. Statist. 14, 389-400.

Feller, W. (1966). *An Introduction to Probability Theory and Its Applications*. Vol. II, New York; Wiley.

Fisher, L. (1969). Limiting sets and convex hulls of samples from product measures. Ann. Math. Statist. 40, 1824-32.

Fisher, L. (1971). Limiting convex hulls of samples; theory and function space examples. Z. Wahrscheinlichkeitstheorie verw. Geb. 18, 281-97.

Frisch, H. L. & Hammersley, J. M. (1963). Percolation processes and related topics. J. Soc. Ind. Appl. Math. 11, 894-918.

Gani, J. (1972). Point processes in epidemiology. This volume.

Geffroy, J. (1958). Contribution a la theorie des valeuis extremes. Publs. Inst. Statist. Univ. Paris 7, 36-123 and 8, 3-52.

Gel'fand, J. M. & Vilenkin, N. Ya. (1964). *Generalized Functions*. Vol. 4 (English translation). New York: Academic Press.

Gilbert, E. N. (1961). Random plane networks. J. Soc. Ind. Appl. Math. 9, 533-43.

Gilbert, E. N. (1962). Random subdivisions of space into crystals. Ann. Math. Statist. 33, 958-72.

Goldman, J. R. (1967a). Stochastic point processes: Limit theorems. Ann. Math. Statist. 38, 771-9.

Goldman, J. R. (1967b). Infinitely divisible point processes in R^n. J. Math. Analysis Applic. 17, 133-46.

Greig-Smith, P. (1952). The use of random and contiguous quadrats in the study of the structure of plant communities. Ann. Bot. 16, 293-316.

Greig-Smith, P. (1964). *Quantitative Plant Ecology*. 2nd Ed. London: Butterworths.

Grigelionis, B. (1972). On weak convergence of the sums of multidimensional stochastic point processes. This volume.

Gurland, J. (1958). A generalized class of contagious distributions. Biometrics 14, 229-49.

Gurland, J. & Hinz, P. (1971). Estimating parameters, testing fit, and analyzing untransformed data pertaining to the negative binomial and other distributions. In *Statistical Ecology, Vol. 1*, Patil, Pielou and Waters, eds. University Park: Penn. State Univ. Press, 1943-78.

Harris, T. E. (1963). *The Theory of Branching Processes*. Berlin: Springer.

Hawkes, A. G. (1971). Spectra of some self-exciting and mutually exciting point processes. Biometrika 58, 83-104.

Hille, E. & Phillips, R. S. (1957). *Functional Analysis and Semi-groups* (Revised Edition). American Mathematical Society.

Holgate, P. (1965a). Tests of randomness based on distance methods. Biometrika 52, 345-53.

Holgate, P. (1965b). The distance from a random point to the nearest point of a closely packed lattice. Biometrika 52, 261-3.

Holgate, P. (1972). Statistical analysis of multidmensional point processes. This volume.

Jagers, P. (1971). On the weak convergence of the superposition of point processes. Stanford Technical Report.

Kantorovich, L. V. & Akilov, G. P. (1964). *Functional Analysis in Normed Spaces*. (Translated from Russian). New York: Macmillan.

Kendall, M. G. & Moran, P. A. P. (1963). *Geometrical Probability*. London: Griffen.

Kersten, J. & Matthes, K. (1963). Stationare Zufallige Punktfolgen. II. Jber. Deutsch. Math.-Verein. 66, 106-18.

Kolmogorov, A. N. (1935). La transformation de Laplace dans les espaces lineaires. C. R. Lebd. Seanc. Aca. Sci. Paris 200, 1717-18.

Lamperti, J. (1966). *Probability: A Survey of the Mathematical Theory*. New York: Wiley.

Leadbetter, M. R. (1968). On three basic results in the theory of stationary point processes. Proc. Amer. Math. Soc. 19, 115-17.

LeCam, L. (1957). Convergence in distribution of stochastic processes. Univ. Calif. Publ. Statist. 2 (11), 207-36.

Lewis, P. A. W. (1969). Asymptotic properties and equilibrium conditions for branching Poisson processes. J. Appl. Prob. 6, 355-71.

Lewis, P. A. W. (1970). Asymptotic properties of branching renewal processes, to appear.

Loeve, M. (1963). *Probability Theory, 3rd Edition.* Princeton: Van Nostrand.

Matérn, B. (1960). Spatial variation. Stochastic models and their application to some problems in forest surveys and other sampling investigations. Meddn. St. Skogsforskninst. 44 (5), 1-144.

Matérn, B. (1971). Doubly stochastic Poisson processes in the plane. In *Statistical Ecology, Vol. 1,* Patil, Pielou and Waters, eds. University Park: Penn. State Univ. Press, 195-214.

Matthes, K. (1963a). Stationare Zufallige Punktfolgen I. Jber. Deutsch Math. Verein. 66, 66-79.

Matthes, K. (1963b). Unberschrankt Neilbare Verteilungsgesetze Stationarer Zufallinger Punktfolgen, Wirs. Zeitschrift der Hochschule fun Electrotechnik Ilmenare 9, 235-8.

Matthes, K. (1972). Infinitely divisible point processes. This volume.

Miles, R. E. (1970). On the homogenous planar Poisson process. Math. Biosciences 6, 85-127.

Milne, R. K. (1971). Simple proofs of some theorems on point processes. Ann. Math. Statist. 42, 368-72.

Moyal, J. E. (1957). Discontinuous Markoff processes. Acta Math. Stockh. 98, 221-64.

Moyal, J. E. (1962a). The general theory of stochastic population processes. Acta. Math. Stockh. 108, 1-31.

Moyal, J. E. (1962b). Multiplicative population chains. J. R. Soc. (A) 266, 518-26.

Naus, J. I. (1965). Clustering of random points in two dimensions. Biometrika 52, 263-9.

Neyman, J. (1939). On a new class of "contagious" distributions, applicable in entomology and bacteriology. Ann. Math. Statist. 10, 35-57.

Neyman, J. & Scott, E.L. (1957). On a mathematical theory of populations conceived as conglomerations of clusters. Cold Spring Harbor Symp. Quant. Biol. 27, 109-20.

Neyman, J. & Scott, E. L. (1958). Statistical approach to problems of cosmology (with discussion). J. R. Statist. Soc. B 20, 1-43.

Neyman, J. & Scott, E.L. (1959). Stochastic models of population dynamics. Science 130, 303-8.

Neyman, J. & Scott, E.L. (1972). Processes of clustering and applications. This volume.

Paloheimo, J. E. (1971a). Discussion in *Statistical Ecology, Vol. 1*, Patil, Pielou and Waters, eds. University Park: Penn. State Univ. Press, 210-12.

Paloheimo, J. E. (1971b). On a theory of search. Biometrika 58, 61-75.

Patil, G. P. and Joshi, S. W. (1968). *A Dictionary and Bibliography of Discrete Distributions.* Edinburgh: Oliver & Boyd.

Patil, G. P., Pielou, E. C. & Waters, W. E., Eds. (1971). *Statistical Ecology, Volume 1: Spatial Patterns and Statistical Distributions. Volume 2: Sampling and Modeling Biological Populations and Population Dynamics. Volume 3: Many Species Populations, Ecosystems, and Systems Analysis.* University Park: Penn. State Univ. Press.

Persson, O. (1964). *Distance Methods. The Use of Distance Measurements in the Estimation of Seeding Density and Open Space Frequency.* Studia Forestalia Suecica, Nr. 15, Stockholm: Skogshogskolan.

Persson, O. (1965). Distance methods II. Distributions of distance from a randomly located sample point to plants in an equilateral triangular lattice. Institutionen for Shoglig Matematisk Statistik, Research notes, Nr. 6, Stockholm.

Pielou, E. C. (1969). *An Introduction to Mathematical Ecology.* New York: Wiley.

Prohorov, Y. V. (1961). The method of characteristic functionals. *Proc. Fourth Berkeley Symp.* Vol. 1, 403-19.

Ramakrishnan, A. (1950). Stochastic processes relating to particles in a continuous infinity of states. Proc. Camb. Phil. Soc. 46, 595-602.

Rao, C. R. (1971). Some comments on the logarithmic series distribution in the analysis of insect trap data. In *Statistical Ecology, Vol. 1*, Patil, Pielou and Waters, eds. University Park: Penn. State Univ. Press, 131-42.

Renyi, A. & Sulanke, R. (1963, 1964). Uber die konvexe hulle von n zufallig gewahlten, Punkten I. and II. Z. Wahrscheinlichkeitstheorie verw. Geb. 2, 75-84 and 3, 138-48.

Roberts, F. D. K. (1967). A Monte Carlo solution of a two-dimensional cluster problem. Biometrika 54, 625-8.

Roberts, F. D. K. (1969). Nearest neighbors in a Poisson ensemble. Biometrika 56, 401-6.

Roberts, F. D. K. & Storey, S. H. (1968). A three-dimensional cluster problem. Biometrika 55, 258-60.

Scott, E. L. (1963). Subclustering. In *Classical and Contagious Discrete Distributions*. London: Pergamon Press.

Scott, A. J. & Symons, M. J. (1971). Clustering methods based on the likelihood ratio criteria. Biometrics, 27, 387-97.

Skellam, J. G. (1951). Random dispersal in theoretical populations. Biometrika 38, 196-218.

Skellam, J. G. (1952). Studies in statistical ecology. I. Spatial Pattern. Biometrika 39, 346-62.

Skellam, J. G. (1958). On the derivation and applicability of Neyman's type A distribution. Biometrika 45, 32-6.

Sokal, R. R. & Sneath, P. H. (1963). *Principles of Numerical Taxonomy.* San Francisco: Freeman.

Stevens, W. L. (1937). Significance of grouping. Ann. Eugen. 8, 57-73.

Stiteler, W. M. & Patil, G. P. (1971). Variance-to-mean ratio and Morisita's index as measures of spatial patterns in scological populations. In *Statistical Ecology, Vol. 1,* Patil, Pielou and Waters, eds. University Park: Penn. State Univ. Press, 423-60.

Student. (1907). On the error of counting with a haemocytometer. Biometrika 5, 351-60.

Thompson, H. R. (1955). Spatial point processes, with applications to ecology. Biometrika 42, 102-15.

Tulya-Munika, S. (1970). A characterization of G-processes and Poisson processes in R^n. Report 71/STM 2, The Manchester-Sheffield School of Probability and Statistics.

Turner, M. E. & Eadie, G. S. (1957). The distribution of red blood cells in the hemacytometer. Biometrics 13, 485-95.

Vere-Jones, D. (1968). Some applications of probability generating functionals to the study of input-output systems. J. R. Statist. Soc. B 30, 321-33.

Vere-Jones, D. (1970). Stochastic models for earthquake occurrence (with discussion). J. R. Statist. Soc. B 32, 1-62.

Warren, W. G. (1971). The center-satellite concept as a basis of ecological sampling. In *Statistical Ecology, Vol. 2,* Patil, Pielou and Waters, eds. University Park: Penn. State Univ. Press.

Warren, W. G. (1972). Point processes in forestry. This volume.

THEORY OF HYPERPLANE PROCESSES

K. Krickeberg
Fakultät für Mathematik, Universität Bielefeld, West Germany

SUMMARY

Let $S = S_{n-1} = \{\eta : \eta \in R^n, |\eta| = 1\}$ be the unit hypersphere in R^n, and describe a hyperplane x in R^n by its equation $x = \{\xi : \xi \in R^n, (\xi|\eta) + p = 0\}$, $p \in R$, $\eta \in S$, where η, the "direction" of x, is determined by the orientation of x. In this way, $X = R \times S$ becomes a natural parameter space for the space of all oriented hyperplanes in R^n. Denote by G the group induced in X by the euclidean motions (rotations and translations) of R^n. The present paper is a survey on properties of point processes in X which depend heavily on the structure of X and *G*.

1. GENERAL PRELIMINARIES

Let *M* be the space of all positive Radon measures in X. Endowed with the vague topology, *M* is a Polish space. By a random measure in X we mean a probability measure in *M*; the set of all random measures will be denoted by $\mathbb{P}$.

A point measure in X is a measure carried by countably many points, each of them having measure 1. The set *S* of all point measures is borelian in *M*; a random measure carried by *S* will be called a point process in X, or hyperplane process. Similarly, a diffuse process in X is a random measure carried by the diffuse measures in X.

The group *G* induces a group of transformations in *M* via the concept "image of a measure in X under a transformation of X"; a random measure is called strictly stationary if it is invariant under this group.

Let k be a positive integer. The map $\mu \to \mu^{Ⓚ} = \mu \otimes \cdots \otimes \mu$ of *M* into the set of all positive Radon measures in X^k is vaguely continuous.

The image $P^{(k)}$ of a random measure P in X under this map is a random measure in X^k, and it is a point process or diffuse process in X^k, respectively, if P is. In the former case P describes, intuitively speaking a random distribution of countably many hyperplanes in R^n, whereas $P^{(k)}$ describes the corresponding random distribution of all k-tuples of such hyperplanes.

For any borelian $C \subseteq X^k$ and any $P \in \mathcal{P}$ we set

$$\nu_P^k(C) = \int_{\mathcal{M}} \mu^{(k)}(C)\ P(d\mu) = \nu^1_{P^{(k)}}(C) \ . \tag{1}$$

P is said to be of k-th order if $\nu_P^k(C) < \infty$ for every relatively compact Borel set C. In this case, ν_P^k is a Radon measure in X^k which we call the k-th moment measure of P. If P is a hyperplane process, ν_P^k determines all ν_P^ℓ with $\ell \leq k$ (see Equation 6 below), and $\nu_P^k(C)$ represents the "mean number of k-tuples of hyperplanes in C".

The random measure P is said to be k-th order stationary if it is of k-th order, and if ν_P^k is invariant under the "diagonal" group G_k of all transformations of X^k of the form

$$(x_1, \ldots , x_k) \longmapsto (Tx_1, \ldots , Tx_k) , \qquad \text{where } T \in G.$$

Clearly, a strictly stationary random measure of k-th order is k-th order stationary.

So far, the main problems of the theory of hyperplane processes have been those of the construction and structural analysis of stricly or k-th order stationary hyperplane processes.

2. CONSTRUCTION OF STATIONARY HYPERPLANE PROCESSES

Apart from more or less obvious elementary constructions, the prominent tool has been that of constructing Poisson processes or mixtures of them, the so-called Cox processes (see Krickeberg, 1972a, 1972b). Given any diffuse $\rho \in \mathcal{M}$, there is one and only one hyperplane process Q_ρ such that $\nu^1_{Q_\rho} = \rho$, and such that the functions $\mu \longmapsto \mu(A_1), \ldots , \mu \longmapsto \mu(A_\ell)$

considered as random variables on the probability space (M, Q_ρ), are independent whenever $A_1, \ldots, A_\ell$ are disjoint. Q_ρ is called the Poisson process with intensity measure, or mean number of hyperplanes, ρ.

Next, let V be a diffuse process. Then, the Cox process, or doubly stochastic Poisson process, Q_V given by V is defined by

$$Q_V = \int_M Q_\rho V(d\rho) \quad . \tag{2}$$

The hyperplane process Q_V determines V uniquely, and Q_V is strictly stationary if and only if V is. The moment measures of V up to the order k can be found from the moment measures of Q_V up to the order k and vice versa; in fact

$$\nu^k_{Q_V}(A_1 \times \cdots \times A_k) = \sum_{m=1}^{k} \sum_{\{J_1,\ldots,J_m\}} \nu^m_V(\mathop{\times}_{\ell=1}^{m} \bigcap_{j\in J_\ell} A_j), \tag{3}$$

where $\{J_1, \ldots, J_m\}$ runs through all partitions of $\{1, \ldots, k\}$ into m disjoint and non-empty sets. Therefore, Q_V is stationary up to the order k if and only if V is.

Also, necessary and sufficient conditions on V can be given in order that Q_V be k-fold non-degenerate, that is, Q_V-almost surely any k distinct hyperplanes have linearly independent directions. These conditions may be formulated in terms of the decomposition (13) of ν^k_V given below.

3. STRUCTURAL ANALYSIS OF STATIONARY HYPERPLANE PROCESSES

Let P be a strictly stationary hyperplane process, and $1 \leq k \leq n$. Then, if P is not k-fold non-degenerate, P-almost surely there are infinitely many k-tuples of hyperplanes with linearly dependent directions (see Davidson, 1968, for the case $k = n = 2$).

One of the big problems of the theory, posed by Davidson (1968) in the case $n = 2$, the case of line processes, is unsolved: is every strictly stationary n-fold non-degenerate hyperplane process of the n-th order a Cox process? It is only under additional assumptions that the answer is positive: see Papangelou (1972).

The corresponding problem for the moment measures, however, has a solution, and in order to find it, we start with an analysis of the moment measures themselves. In the following, k will be fixed, $1 \leq k \leq n$, and P a k-th order stationary hyperplane process.

Consider a partition $\mathcal{J} = \{J_1, \ldots, J_m\}$ of $\{1, \ldots, k\}$ into $m = \#\mathcal{J}$ disjoint and non-empty sets. By the $\mathcal{J}$-diagonal $D_{\mathcal{J}}$ of X^k we mean the set of all $(x_1, \ldots, x_k) \in X^k$ such that $x_i = x_{i'}$ if i and i' belong to the same J_ℓ. By the projection $\Pi_{\mathcal{J}}: (x_1, \ldots, x_k) \longmapsto (y_1, \ldots, y_m)$ with $x_i = y_\ell$ for all $i \in J_\ell$, the diagonal $D_{\mathcal{J}}$ is mapped in a one-to-one fashion onto R^m. We set $\mathcal{J}' \subset \mathcal{J}$ if $\mathcal{J}$ is a subpartition of $\mathcal{J}'$, that is, $D_{\mathcal{J}'} \subseteq D_{\mathcal{J}}$, and

$$E_{\mathcal{J}} = D_{\mathcal{J}} - \bigcup_{\mathcal{J}':\mathcal{J}'\subset\mathcal{J};\mathcal{J}'\neq\mathcal{J}} D_{\mathcal{J}'} .$$

Then $E_{\mathcal{J}}$ is the set of all $(x_1, \ldots, x_k)$ such that $x_i = x_{i'}$ if and only if i and i' belong to the same J_ℓ. The sets $E_{\mathcal{J}}$ are disjoint, and

$$\nu_P^k(C) = \sum_{\mathcal{J}} \nu_P^k(E_{\mathcal{J}} \cap C) \tag{5}$$

for every borelian $C \subseteq R^k$ where the part of ν_P^k carried by $E_{\mathcal{J}}$ is, in fact, given by

$$\nu_P^k(E_{\mathcal{J}} \cap C) = \nu_P^{\#\mathcal{J}}\{\Pi_{\mathcal{J}}(E_{\mathcal{J}} \cap C)\} . \tag{6}$$

Assuming henceforth P to be k-fold non-degenerate, we have

$$\nu_P^k(E_{\mathcal{J}} \cap C) = \nu_P^k(E_{\mathcal{J}}^o \cap C) , \tag{7}$$

where $E_{\mathcal{J}}^o$ is the set of all $(x_1, \ldots, x_k)$ of $D_{\mathcal{J}}$ such that the directions of $x_{i_1}, \ldots, x_{i_m}$ with $m = \#\mathcal{J}$ are linearly independent if $i_\ell \in J_\ell$, $\ell = 1, \ldots, m$. Clearly, $E_{\mathcal{J}}^o \subseteq E_{\mathcal{J}}$.

Next we study the action of G_k within E^o_J, and decompose E^o_J into equivalence classes with respect to the group G_k restricted to E^o_J. Let Γ^o_J be the set of these equivalence classes, endowed with the quotient topology which makes it locally compact with a countable base. Given any $\gamma \in \Gamma^o_J$, there is one and, up to a factor, only one positive Radon measure τ_γ in X^k carried by γ and invariant under G_k. Upon proper normalization, the theory of disintegration of measures yields

$$\nu^k_P(E^o_J \cap C) = \int_{\Gamma^o_J} \tau_\gamma(C)\ \kappa(d\gamma)\ , \tag{8}$$

where κ is a certain positive Radon measure in Γ^o_J, uniquely determined by ν^k_P. In fact, the explicit knowledge of τ_γ from integral geometry gives us a natural normalization and a certain factorization of ν^k_P derived from formula (8). To describe this factorization conveniently, we permute the factors of $X^k = (R \times S)^k$ and write $X^k = R^k \times S^k$ instead. We define also the J-diagonals $D_J(R)$ and $D_J(S)$ of R^k and S^k as in X^k, and accordingly the map Π_J. Then we have

$$\nu^k_P\{E^o_J \cap (M \times T)\}$$

$$= \lambda^{(m)}[\Pi_J\{D_J(R) \cap M\}] \int_{\Gamma^o_J} \sigma_\gamma[\Pi_J\{D_J(S) \cap T\}]\ \kappa(d\gamma) \tag{9}$$

for any Borel sets $M \subseteq R^k$ and $T \subseteq S^k$ where $\lambda^{(m)}$ is the m-dimensional Lebesgue measure and σ_γ a measure in S^m studied in integral geometry, in fact a generalization of the so-called Poincaré kinematic measure.

We now combine (5), (7), and (9) and use the resulting representation of ν^k_P to derive, via an averaging procedure, the existence of a k-fold non-degenerate Cox process $\tilde{P} = Q_V$ such that

$$\nu^\ell_P = \nu^\ell_{\tilde{P}} \qquad (\ell = 1, \ldots, k)\ . \tag{10}$$

(See Davidson, 1970, and Krickeberg, 1972a.) Moreover, it turns out that, in fact,

$$\nu_P^k(E_J^o \cap C) = \nu_V^{\#J}(\Pi_J C) \tag{11}$$

for all Borel sets C in X^k, hence the representations (3) and (5) of $\nu_P^k = \nu_{Q_V}^k$ are identical term by term, and the measure $C \longmapsto \nu_V^{\#J}\{\Pi_J(D_J \cap C)\}$ admits the factorization (9). In particular,

$$\nu_V^k(M \times T) = \lambda^{\textcircled{k}}(M) \int_{\Gamma^o_{J_{max}}} \sigma_\gamma(T)\ \kappa(d\gamma) \qquad (M \subseteq R^k,\ T \subseteq S^k)\ , \tag{12}$$

where $J_{max} = \{\{1\},\ldots,\{k\}\}$. Note that $E^o_{J_{max}}$ is the set of all k-tuples $(x_1,\ldots,x_m)$ with linear independent directions.

By a Hilbert space argument (Davidson, 1970), the factorization (12) of ν_V^k implies a factorization of the random measure V itself: we have V-almost surely $\mu = \lambda^{\textcircled{1}} \otimes \alpha_\mu$ where α_μ is a diffuse measure in S. This fact admits interesting geometrical and statistical interpretations.

Generalizing (6) and (8) we may write the moment measure ν_V^k of any k-th order stationary random measure V in the form

$$\nu_V^k = \int_\Gamma \tau_\gamma\ \kappa(d\gamma)\ , \tag{13}$$

where Γ is the set of equivalence classes of X^k under G_k and τ_{γ_k} an "integral geometric measure" carried by γ which does not depend on ν_V^k. In this case, however, the quotient topology of Γ need not be separated, and we have to use a "natural" topology instead which makes Γ locally compact with a countable base. The formula (13) may be used, e.g., to study invariance properties of ν_V^k under maps not in G_k.

4. CONCLUDING REMARKS

Particular properties of random lines in the plane, and more generally random hyperplanes, have been studied by many authors. A typical problem would be to find the average number of edges of a certain polygon formed by random lines. The usual set-up is that of a finite and fixed number of random lines that meet a fixed compact set K, the probability distribution being derived from an integral geometric measure. Infinitely many lines distributed randomly in the entire plane R^2 appear only by a limit procedure whereby K expands to absorb the plane. The resulting limit line processes are usually of the Poisson type. A survey on various special problems is given in Kendall & Moran (1962), Moran (1966), and Moran (1969). A more systematic account is given in Miles (1969, 1971). In addition, interesting results not quoted in these surveys are to be found in Ambartsumian (1970), Kingman (1969), Vinogradov and Zaregradskii (1970) and Ziezold (1970).

In the preceding notes we have tried to look at once at the entire line process, imposing only invariance restrictions. However, the resulting representation by a Cox process described in Section 3 will often allow to carry over some of the classical theorems to this general situation. An excellent survey on the earlier theory of line processes as defined here is contained in the monograph by Bartlett (1966). This book also gives an account of the statistical analysis of line processes.

REFERENCES

Ambartsumian, R. V. (1970). The method of invariant imbedding in the theory of random lines. Izv. Akad. Nauk Armjan. SSR Ser. Mat. 5, 167-206.

Bartlett, M.S. (1966). *An Introduction to Stochastic Processes*. 2nd Ed. Cambridge: Cambridge University Press.

Davidson, R. (1968). *Stochastic processes of flats*. Thesis, University of Cambridge.

Davidson, R. (1970). Construction of line processes: second order properties. Izv. Akad. Nauk Armjan. SSR Ser. Mat. 5, 219-34.

Kendall, M.G. & Moran, P.A.P. (1962). *Geometrical probability*. London: Griffin.

Kingman, J.F.C. (1969). Random secants of a convex body. J. Appl. Prob. 6, 660-72.

Krickeberg, K. (1970). Invariance properties of the correlation measure of line processes. Izv. Akad. Nauk Armjan. SSR Ser. Mat. 5, 251-62.

Krickeberg, K. (1972a). Moments of point processes. Lecture Notes in Mathematics. To appear.

Krickeberg, K. (1972b). The Cox process. To appear.

Miles, R. E. (1969). Poisson flats in Euclidean space. Part I. Adv. Appl. Prob. 1, 211-37.

Miles, R. E. (1971). Poisson flats in Euclidean space. Part II. Adv. Appl. Prob. 3, 1-43.

Moran, P.A.P. (1966). A note on recent research in geometrical probability. J. Appl. Prob. 3, 453-63.

Moran, P.A.P. (1969). A second note on recent research in geometrical probability. Adv. Appl. Prob. 1, 73-89.

Papangelou, F. (1972). Summary of some results on point and line processes. This volume.

Vinogradov, O.P. & Zaregradskii, I.P. (1970). Geometrical probabilities in the theory of optimal search. Izv. Akad. Nauk Armjan. SSR Ser. Mat. 5, 207-18.

Ziezold, H. (1970). Über die Eckenanzahl zufälliger konvexer Polygone. Izv. Akad. Nauk Armjan. SSR Ser. Mat. 5, 296-312.

SUMMARY OF SOME RESULTS ON POINT AND LINE PROCESSES

F. Papangelou
Department of Mathematics, Ohio State University and
Institut für math. Statistik, Universität Göttingen

SUMMARY

In the present note we give a brief description of some results on point and line processes, the details of which will appear in Papangelou (1972a, 1972b). We also make some comments on the connection between our work and K. Murali Rao's proof of the Meyer decomposition theorem.

1. POINT PROCESSES

Beginning with point processes let R be the real line and $\mathcal{B}$ the σ-field of Borel sets in R. Following Ryll-Nardzewski (1961) we denote by Ω the space of all countable subsets of R which are unbounded on both left and right and are locally finite (i.e., have no finite accumulation point). For $\omega \in \Omega$ and $Q \in \mathcal{B}$ we denote by $N(\omega,Q)$ the number of points of ω in Q and we let $\mathcal{F}$ be the minimal σ-field relative to which all functions $N(\omega,Q)$ (bounded $Q \in \mathcal{B}$) are measurable.

A point process on R is defined by a probability measure P on $\mathcal{F}$. In the sequel, if no explicit statement to the contrary is made, we assume for all processes under consideration that

(i) they are stationary, i.e.,

$$P\{N(Q_1) = j_1,\ldots, N(Q_n) = j_n\}=P\{N(Q_1+t) = j_1,\ldots, N(Q_n+t) = j_n\}$$

for any bounded Borel sets $Q_1, Q_2,\ldots, Q_n$, any non-negative integers $j_1,\ldots, j_n$ and any real t, and

(ii) $\alpha = \int_\Omega N(\omega, [0,1])\, P(d\omega) < \infty$.

For each real t, F_t will denote the σ-field of events that happen left of t, i.e., the σ-field generated by the random variables $N(\omega,Q)$ with $Q \subseteq (-\infty,t)$ (bounded $Q \in B$). As is customary for random variables, we shall occasionally write $N(Q)$ for $N(\omega,Q)$, $N[x,y)$ for $N\{\omega,[x,y)\}$ etc. We let $E(X)$ denote the expectation of X and $E(X|F_t)$ its conditional expectation given F_t. We shall write $E(X|F_t)(\omega)$ whenever we want to exhibit the latter as a random variable in ω.

If $\Delta = \{b = \xi_0 < \xi_1 < \cdots < \xi_{n+1} = c\}$ is a partition of a given interval [b,c] we set

$$S_\Delta = \sum_{\nu=0}^{n} E\{N[\xi_\nu,\xi_{\nu+1})|F_{\xi_\nu}\} . \tag{1}$$

Theorems 1 and 3 below assert the almost sure and mean convergence of these random sums when the mesh $\|\Delta\| = \max_{0\le\nu\le n} (\xi_{\nu+1} - \xi_\nu)$ of the partition goes to zero. Recall that an interval function $F[x,y)$, $x < y$, is said to be Burkill integrable over [b,c) if

$$\lim_{\|\Delta\|\to 0} \sum_{\nu=0}^{n} F[\xi_\nu, \xi_{\nu+1}) = W$$

exists, i.e., if for every $\varepsilon > 0$ there is $\delta > 0$ such that $\|\Delta\| < \delta$ implies

$$\left| \sum_{\nu=0}^{n} F[\xi_\nu, \xi_{\nu+1}) - W \right| < \varepsilon .$$

W is called the Burkill integral of F over [b,c).

Theorem 1. There is a version $Z\{\omega,[x,y)\}$, $\omega \in \Omega$, $x < y$, of the random interval function $E\{N[x,y)|F_x\}$, which is left-continuous in x and y and such that for all $\omega \in \Omega$ the sample interval function $Z\{\omega,[x,y)\}$, $x < y$, is Burkill integrable over any finite interval.

We denote the Burkill integral of $Z\{\omega, [x,y)\}$ over [b,c) by $W\{\omega,[b,c)\}$. From Theorem 1 we obtain:

Theorem 2. The limit

$$Y = \lim_{\varepsilon \downarrow 0} \frac{Z\{\omega, [0,\varepsilon)\}}{\varepsilon}$$

exists almost surely and $E(Y) \leq \alpha$.

In fact the more general limit

$$Y(\omega) = \lim_{[x,y] \downarrow \{0\}} \frac{Z\{\omega, [x,y)\}}{y-x}$$

exists a.s., where $[x,y] \downarrow \{0\}$ means $x \leq 0$, $y \geq 0$, $y-x > 0$, $x \to 0$, $y \to 0$.

Theorem 3. Let $[b,c]$ be a finite interval. With S_Δ as defined in (1), we have

$$\lim_{\|\Delta\| \to 0} \int_\Omega \left| S_\Delta(\omega) - W\{\omega, [b,c)\} \right| P(d\omega) = 0 \ .$$

Corollary. The process $U(t,\omega) = N\{\omega,[0,t)\} - W\{\omega,[0,t)\}$, $t > 0$, is a martingale. (I owe this observation to H. Rost.)

The next theorem asserts that if the random transformation $t \longrightarrow W\{\omega,[0,t)\}$ of $[0,\infty)$ onto itself is continuous (i.e., has no jumps), then it transforms the non-negative points of the given process into a Poisson process in $[0,\infty)$ with rate 1 and independent of F_0. First note that the random interval function $W\{\omega,[b,c)\}$ has a unique extension to a random measure $W(\omega,Q)$, $Q \in B$, which we call the (random) *conditional intensity measure* of the process. Its defining property is, roughly, $W[t,t+dt) = E\{N[t,t+dt) | F_t\}$.

Theorem 4. Suppose W is a.s. atomless, let

$$0 \leq t_1(\omega) < t_2(\omega) < t_3(\omega) < \cdots \qquad (2)$$

be the non-negative points of ω and define the sequence

$$0 \leq \tau_1(\omega) \leq \tau_2(\omega) \leq \tau_3(\omega) \leq \cdots \qquad (3)$$

by $\tau_n(\omega) = W\{\omega, [0, t_n(\omega))\}$. Under the probability P the sequence (3) is almost surely strictly increasing and is, in fact, a homogeneous Poisson process with rate 1 and independent of the σ-field F_0.

The proof is based on the Palm probabilities (see Section 3) and the following characterization of the Poisson process given in Papangelou (1972a).

Definition. A (not necessarily stationary) point process is said to have conditional rate 1, if for every interval $[s,r]$, $E(N[s,r] | F_s) = r-s$ a.s.

Theorem 5. If a (not necessarily stationary) point process has conditional rate 1, then it is a Poisson process.

In this connection we may remark that the a.s. limit Y in Theorem 2 is a good concept of "conditional rate" for the process only if it is also a mean limit, and a necessary and sufficient condition for this is given in Theorem 7 below. In the general case the correct approach is through W.

Theorem 4 implies that if the conditional intensity measure of a process is a.s. the Lebesgue measure then the process is Poisson with rate 1.

If $\Delta = \{0 = \xi_0 < \xi_1 < \xi_2 < \cdots\}$ is a partition of $[0,\infty)$, we define a random transformation $t \longrightarrow T^\Delta(t,\omega)$ of $[0,\infty)$ onto itself by setting first

$$T^\Delta(\xi_n,\omega) = \sum_{\nu=0}^{n-1} Z\{\omega, [\xi_\nu, \xi_{\nu+1})\} \qquad (n = 1,2,\ldots),$$

and then interpolating $T^\Delta(t,\omega)$ linearly for the other t's. Theorems 1 and 4 imply:

Theorem 6. Suppose W is a.s. atomless, let (2) be the sequence of non-negative elements of ω and for each partition Δ of $[0,\infty)$, define the sequence

$$0 \leq \tau_1^\Delta(\omega) \leq \tau_2^\Delta(\omega) \leq \cdots \tag{4}$$

by $\tau_n^\Delta(\omega) = T^\Delta\{t_n(\omega),\omega\}$. When the mesh of Δ in each finite interval $[0,K]$ goes to zero, the process (4) converges P-almost surely to a Poisson process with rate 1 and independent of F_0.

This almost sure convergence can be taken to mean $\tau_n^\Delta(\omega) \to \tau_n(\omega)$ a.s. for each n, where $\tau_n(\omega)$ is as defined in (3), but can also be defined in terms of a natural topology in the space of sequences (3) (see Papangelou, 1972a).

The intuitive interpretation of Theorems 4 and 6 is this: Suppose an observer, who at each instant is able to see only the past, travels on $[0,\infty)$ with variable "motion" which (given up-to-the-moment information on the past) he tries to regulate so as to meet expected future points at an instantaneous rate of one per time unit. Then he will meet the actual points of the given process at the times of a Poisson process with rate 1.

Let now P_t be the Palm probability measure of the given point process, i.e., the conditional probability measure, given that the process has a point at t (see Ryll-Nardzewski, 1961, for a rigorous definition). P_0 is carried by the set $\Omega_0 = \{\omega \in \Omega : 0 \in \omega\}$ and is therefore singular relative to P on the σ-field F. However, P_0 can be absolutely continuous relative to P on the σ-subfield F_0. (Below we use the symbol $<<$ to denote absolute continuity.) If this is so, let $X(\omega)$ be the corresponding F_0-measurable Radon-Nikodym derivative

$$P_0(A) = \int_A X(\omega)\, P(d\omega) \qquad (A \in F_0)$$

and define the stationary process $X(t,\omega)$ by $X(t,\omega) = X(\omega-t)$, where $\omega-t = \{x-t : x \in \omega\}$. Note that this process is progressively measurable relative to the σ-fields F_t, and that $P_t(A) = \int_A X(t,\omega)\, P(d\omega)$, $A \in F_t$.

Theorem 7. The following conditions are equivalent:

(i) $P_0 << P$ on F_0.

(ii) The limit $Y = \lim_{\varepsilon \downarrow 0} \frac{1}{\varepsilon} E\{N[0,\varepsilon) | F_0\}$ exists in the mean.

(iii) The function $\phi_\omega(t) = W\{\omega,[0,t)\}$ is a.s. absolutely continuous.

In this case $Y(\omega) = \alpha X(\omega)$ and $W(\omega,Q) = \alpha\int_Q X(t,\omega)\,dt$, for $Q \in \mathcal{B}$.

Thus for fixed t, $X(t,\omega)$ is the Radon-Nikodym derivative of the Palm probability P_t relative to P on F_t, while for fixed ω, $\alpha X(t,\omega)$ is the Radon-Nikodym derivative of $W(\omega,\cdot)$ relative to the Lebesgue measure (i.e., $W[t, t+dt) = \alpha X(t)\,dt$).

We will conclude this section with criteria (expressed in terms of P_0 alone) for the atomlessness and the absolute continuity of W. For $\omega \in \Omega_0$ let $\eta_0(\omega)$ be the smallest positive element of ω, let $\eta_0(\omega) + \eta_1(\omega)$ be the next smallest and so on, and in the same way let $-\eta_{-1}(\omega)$ be the greatest negative point etc. The process $\ldots, \eta_{-1}, \eta_0, \eta_1, \ldots$ of interpoint distances defined on Ω_0 is shift-stationary under P_0 and in fact determines the given point process (Ryll-Nardzewski, 1961).

Theorem 8. W is a.s. atomless or absolutely continuous (relative to the Lebesgue measure) if and only if the conditional distribution $F(x) = P_0(\eta_0 < x | \eta_{-1} = x_1, \eta_{-2} = x_2, \ldots)$ is continuous or absolutely continuous, respectively, for almost all $(x_1, x_2, \ldots)$ ("almost" in the sense of the distribution of $(\eta_{-1}, \eta_{-2}, \ldots)$ under P_0).

2. LINE PROCESSES

The origin of the present research was a problem on line processes in the plane, posed by R. Davidson (1968, Chap. 6). The processes we have in mind are those with finite intensity that are stationary under the Euclidean group (of translations and rotations). The problem is this. If such a process has a.s. no parallel lines, is it necessarily a doubly stochastic Poisson process, i.e., a Poisson process built on a random measure whose distribution is stationary under the Euclidean group?

We describe a partial solution; it was announced in Papangelou (1970) and the proof will appear in Papangelou (1972a). Let L be the space of all oriented lines in the plane. We choose a fixed line $\ell_0 \in L$ and a fixed origin 0 on ℓ_0. The normal coordinates of an oriented line are (θ,p), where θ is the angle it makes with ℓ_0 and p its signed distance from 0 $(0 \leq \theta < 2\pi, -\infty < p < \infty)$. The line processes we are considering are thus essentially point processes on the cylinder $C \times R$, where $C = \{\theta : 0 \leq \theta < 2\pi\}$.

The Palm probabilities P_ℓ, $\ell \in L$, for a stationary line process can be introduced by the method of Ryll-Nardzewski (1961); see Papangelou (1972a). Suppose the process satisfies the following condition:

$$P_\ell << P \text{ on the } \sigma\text{-field of events describable in terms of what happens in } L - \{\ell\} .$$

Let $X(\ell,\omega)$ be the corresponding Radon-Nikodym derivative. Using a lemma of R. Davidson one can prove that if two lines ℓ_1, ℓ_2 are parallel with the same orientation, then $X(\ell_1) = X(\ell_2)$ a.s., so that $X(\ell,\omega)$ is essentially a process $X(\theta,\omega)$ parametrized by the first coordinate θ of $\ell = (\theta,p)$, i.e., a stochastic process on the circle $0 \leq \theta < 2\pi$, stationary under rotations. Then the process is doubly stochastic Poisson built on the random measure $\alpha(\nu \otimes \lambda)$ on $C \times R$, where $\nu(\omega)$ is the (stationary under rotations) measure on the circle with density $X(\theta,\omega)$, $0 \leq \theta < 2\pi$, and λ is the Lebesgue measure on R.

3. THE METHOD

Our main tool in proving the above results is the Palm probability. There are half a dozen approaches to this concept (more on this in Papangelou, 1972a). One approach that is simple and direct enough to yield our results with chiefly measure-theoretic manipulations is outlined in Papangelou (1972a). One starts by defining for each $Q \in \mathcal{B}$ and each $A \in \mathcal{F}$ the set-function

$$m(Q \times A) = \int_A N(\omega,Q)\ P(d\omega) . \tag{5}$$

This m can be extended to a unique measure on $\mathcal{B} \otimes \mathcal{F}$, which is the "integral" over the space $(\Omega,\mathcal{F}, P)$ of the measures $N(\omega,\cdot)$, each of which lives on a section $R \times \{\omega\}$ of the product space $R \times \Omega$. (G. Kummer and K. Matthes, 1970, have also considered such a measure recently and called it the Campbell measure.)

However m can also be expressed as an integral over $(R,\mathcal{B},\lambda)$ of a family of probability measures P_t, $-\infty < t < \infty$, each living on a section

$\{t\} \times \Omega$. The existence of these probabilities follows from the Radon-Nikodym theorem as Ryll-Nardzewski (1961) has indicated. By the Fubini theorem, if $F(t,\omega)$ is a non-negative $\mathcal{B} \otimes \mathcal{F}$-measurable function, then

$$\int F \, dm = \int_{\Omega} \int_{R} F(t,\omega) \, N(\omega,dt) \, P(d\omega) = \int_{R} \int_{\Omega} F(t,\omega) \, P_t(d\omega) \alpha \, dt . \qquad (6)$$

This is the magic formula that is at the root of most of the results. It is capable of yielding the well-known identities connecting P_0 with P without use of, for instance, the differentiation and continuity techniques of Ryll-Nardzewski (1961). To indicate the wide applicability of the method we mention here that the Palm-Khintchine formulae

$$P\{N(0,c) > j\} = \alpha \int_0^c P_0\{N(0,x) = j\} \, dx ,$$

which are proved in Khintchine (1960) by means of differential equations, follow purely measure-theoretically from (6) with the proper choice of F. We refer the reader to Papangelou (1972a) for details.

If in analogy to (5) we define

$$\bar{m}(Q \times A) = \int_A W(\omega,Q) \, P(d\omega)$$

and extend $\bar{m}$ to a measure on $\mathcal{B} \otimes \mathcal{F}$ as before, then it can be proved (see Papangelou, 1972b, Theorem 4) that

$$m(D) = \bar{m}(D)$$

for every set $D \in \mathcal{B} \otimes \mathcal{F}$ which is "adapted" to $\mathcal{F}_t$, $-\infty < t < \infty$, i.e., such that $D_t = \{\omega : (t,\omega) \in D\} \in \mathcal{F}_t$ for every t. This is one of the crucial facts in the proof of Theorem 4 in Section 1.

4. MURALI RAO'S PROOF OF THE MEYER DECOMPOSITION THEOREM.

K. Murali Rao (1969) has used sums such as those in (1) to prove the Meyer decomposition theorem for continuous parameter supermartingales. I learned of his work after the above results were obtained and I wish to indicate here some connections between the present work and his.

P. A. Meyer [1966, p. 128] has developed a classification of stopping times that throws much light on his decomposition theorem. In a complete probability space $(\Omega,\bar{F},P)$ let $\bar{F}_t$, $t \geq 0$, be a right-continuous increasing family of σ-fields each of which contains all P-null sets; let T be a stopping time relative to this family and $\bar{F}_T$ the σ-field of events that happen before time T, i.e., $A \in \bar{F}_T$ if and only if $A \cap \{T \leq t\} \in \bar{F}_t$ for all $t \geq 0$. Then there is an (essentially unique) partition of the sample space Ω into three disjoint events A, B, C with the following properties:

(i) $A, B, C \in \bar{F}_T$;

(ii) $\{T = +\infty\} = C$;

(iii) On B, T is "inaccessible", i.e., if S_n, $n = 1,2,\ldots$, is a non-decreasing sequence of stopping times, such that $S_n \leq T$ for all n then $P[\{\omega \in B: \lim_{n\to\infty} S_n(\omega) = T(\omega) \text{ and } S_n(\omega) < T(\omega) \text{ for all } n\}] = 0$;

(iv) On A, T is "accessible", i.e., A can be written as $A = \bigcup_{k=1}^{\infty} A_k$, where for each k there is a non-decreasing sequence S_n^k, $n = 1,2,\ldots$, of stopping times such that $S_n^k \leq T$ and $A_k \subseteq \{\omega: \lim_{n\to\infty} S_n^k(\omega) = T(\omega),\ S_n^k(\omega) < T(\omega) \text{ for all } n\}$.

We call the restrictions of T to A and B, the accessible and inaccessible parts of T, respectively.

Suppose X_t, $t \geq 0$, is a non-decreasing right-continuous process, adapted to the family $\bar{F}_t$, $t \geq 0$. We say that (X_t) charges a stopping time T (or its accessible or inaccessible parts) if $P(X_T \neq X_{T-}) > 0$ or $P(A \cap \{X_T \neq X_{T-}\}) > 0$ or $P(B \cap \{X_T \neq X_{T-}\}) > 0$ respectively.

In our present case of a point process we define $\bar{N}_t = N[0,t]$ and let $\bar{F}$ be the completion of F and $\bar{F}_t$, $t \in R$, the σ-field of events happening left of or at t, augmented by all null events. (In Section 1 we used left-continuous versions for an obvious reason: the condition "$P_0 \ll P$ on F_0" of Theorem 7. The necessary modifications are easy to see.)

Using dyadic partitions of the interval [0,t] we define for each $t > 0$ and each n

$$\bar{S}_n(t) = \sum_{i=0}^{k_n(t)-1} E\left\{ N\left(\frac{i}{2^n}, \frac{i+1}{2^n}\right] \Big| \bar{\mathcal{F}}_{(i/2^n)} \right\} + E\left\{ N\left(\frac{k_n(t)}{2^n}, t\right] \Big| \bar{\mathcal{F}}_{(k_n(t)/2^n)} \right\},$$

where $k_n(t)$ is the unique integer such that $\{k_n(t)/2^n\} < t \leq \{(k_n(t)+1)/2^n\}$. Easy adaptations of arguments due to P. A. Meyer and K. Murali Rao yield partial results on the convergence of the sums $\bar{S}_n$. Their methods were designed to handle processes (supermartingales) much more general than point processes and this is why their convergence results are weaker than Theorems 1 and 3 in Section 1. We summarize what can be done.

For each $t > 0$, the random variables $\bar{S}_n(t)$, $n = 1,2,\ldots$, are uniformly integrable, hence every subsequence of $\{\bar{S}_n(t)\}$ contains a sub-subsequence which converges in the weak $(\sigma(L_1, L_\infty))$ topology to a limit in L_1. Next one proves the uniqueness of such a limit and this implies that the entire sequence $\{\bar{S}_n(t)\}$ converges in the $\sigma(L_1, L_\infty)$-topology to a random variable $\bar{W}(t)$. There is a right-continuous (non-decreasing) version of the process $\bar{W}(t)$, $t \geq 0$, and such a version does not charge the inaccessible parts of stopping times (in other words the jumps of $\bar{W}$ are accessible and the process contains no "surprises"; indeed it is the "expected" part of the point process).

If in addition the given point process $\bar{N}(t)$ does not charge the accessible parts of stopping times, then the same holds for $\bar{W}(t)$ and thus the latter charges no stopping time, i.e., it is a.s. continuous. In fact this is a necessary and sufficient condition for continuity of $\bar{W}(t)$. Further, the methods of Meyer and Murali Rao establish that in this case $\bar{S}_n(t)$, $n = 1, 2,\ldots$, converges to $\bar{W}(t)$ in the mean. (As proved in Section 1, mean as well as a.s. convergence hold in *all* cases.) Comparison with Theorem 7 in Section 1 shows that $\bar{N}(t)$ does not charge the accessible parts of stopping times if and only if no point of the process can be "predicted" with positive probability.

ACKNOWLEDGMENTS.

I owe much to my late friend R. Davidson, whose thesis was so stimulating, to D. G. Kendall for early encouragement and to H. Rost who made me aware of Murali Rao's work. This work was written with support from the National Science Foundation (Grant GP-21339).

REFERENCES

Davidson, R. (1968). *Stochastic processes of flats.* Chapter 6 of Thesis, University of Cambridge.

Khintchine, A.Y. (1960). *Mathematical methods in the theory of queueing.* London: Griffin.

Kummer, G. & Matthes, K. (1970). Verallgemeinerung eines Satzes von Sliwnjak II. Rev. Roumaine Math. Pures Appl. 15, 845-70.

Meyer, P.A. (1966). *Probability and Potentials.* Waltham, Mass: Blaisdell.

Murali Rao, K. (1969). On decomposition theorems of Meyer. Math. Scand. 24, 66-78.

Papangelou, F. (1970). The Ambrose-Kakutani theorem and the Poisson process. In *Contributions to Ergodic Theory and Probability*, Lecture Notes in Math 160, Berlin-Heidelberg: Springer-Verlag, 234-40.

Papangelou, F. (1972a). On the Palm probabilities of processes of points and processes of lines. In *Stochastic Analysis and Geometry.* New York: Wiley, to appear.

Papangelou, F. (1972b). Integrability of expected increments of point processes and a related random change of scale. Trans. Amer. Math. Soc., to appear.

Ryll-Nardzewski, C. (1961). Remarks on processes of calls. Proc. 4th Berkeley Symp. 2, 455-65.

STOCHASTIC THEORY OF EVOLUTIONARY PROCESSES (1937-71)

A. Ramakrishnan
The Institute of Mathematical Sciences, Madras-20, India

SUMMARY

We describe several new mathematical techniques developed by the author to deal with evolutionary point processes with particular reference to cascades. The method of product densities was introduced in 1950 to circumvent the pathological problem of describing the distribution of an enumerable number of particles in continuous infinity of states. Further refinements and developments in the method leading to a new approach to cascade theory are summarized.

Among other concepts involved in the theory of stochastic processes, attention is drawn to integrals of point processes, inverse probability and ergodicity.

1. HISTORICAL INTRODUCTION

It is a puzzling fact of scientific history that though probability theory is an old and well established branch of mathematics, its application to dynamical processes is of comparatively recent origin. It was Furry (1937) who first wrote down the equation for the simplest of evolutionary multiplicative processes and in late 1930's differential equations for the probability generating functions of the pure birth process, pure immigration process (Poisson process) and pure death process (binomial) were available.

In the above processes the events are represented by points on the t axis and a discrete (integral number) number of points are distributed in random in a continuous time interval. Kendall (1949) combined these into the birth (immigration) - death processes governing the growth of a population and studied the solutions under various conditions.

Two important generalizations were necessary for further development. The first involved branching processes (Kolmogrov & Dimitriev, 1947) with individuals of different types.

The second arose from the work of Bhabha and Heitler (1937) when they perceived that the electromagnetic cascade showers are in fact stochastic processes in every sense of the term. They noted that not only the numbers of particles of various types are random but also that they were distributed in the energy space in a random manner. Hence they had to deal with a stochastic variable representing the number of particles distributed in a continuous infinity of states characterized by the energy parameter E. If the E space were discrete and the states represented by E_1, $E_2, \ldots,$ then it would be possible to define a function $\pi(\nu_1, E_1, \nu_2, E_2, \ldots; t)$ representing the probability at t that there are ν_i particles in $E_i (i=1,2,3,\ldots)$. When the E space is continuous, no such function *can be defined as we are led to a pathological problem of dealing with a continuous infinity of stochastic variables representing the number in the elementary range* dE. The concept of correlation has also to be generalized.

Similar problems were realized by Kendall (1949) in connection with age dependent population process.

Scott and Uhlenbeck (1942) divided the continuous E into N intervals and allowed N to tend to infinity. The analytical solution to the problem came with the independent introduction of product density functions by Bhabha (1950) and Ramakrishnan (1950) for cosmic ray cascades and cumulant densities for the population process by Kendall (1949). While Bhabha and Kendall confined themselves to second order densities, Ramakrishnan defined product densities of general order n. Bhabha's treatment followed closely that of Scott and Uhlenbeck and involved a transition from a discrete E-space to continuous E, while that of the author dealt with continuous E space ab initio.

2. PRODUCT DENSITIES

If M(E) is the number of particles with energy less than E, dM(E) is the number of particles in the range dE. We shall assume that the probability

$$\text{prob}\{dM(E) = n\} = p(n) = o(dE^n) \qquad (n = 1,2,\ldots).$$

Let

$$f_1(E)dE = E\{dM(E)\}, \tag{2.1}$$

where $E\{dN(E)\}$ represents the average number of particles in dE. Thus

$$p(1) = f_1(E)dE + o(dE^2) = E\{dN(E)\} + o(dE^2),$$

$$p(0) = 1 - f_1(E)dE - o(dE^2),$$

$$p(n) = o(dE^n) \qquad (n > 1). \tag{2.2}$$

Hence

$$E(n^2) = E[\{dM(E)\}^r] = E(n) = E\{dM(E)\}. \tag{2.3}$$

Thus all the moments of the stochastic variable dM(E) are equal to the probability that the stochastic variable assumes value 1. But it must be noted that while $f_1(E)dE$ is a probability, its integral over E yields only the mean number of particle in the range of integration as the integration does not correspond to the addition of infinitesimal probabilities relating to mutually exclusive events. Hence

$$E\{M(E_u) - M(E_\ell)\} = \int_{E_\ell}^{E_u} dE \times E\{dM(E)\} = \int_{E_\ell}^{E_u} f_1(E)dE. \tag{2.4}$$

We can now introduce the concept of product density. We form the product of the stochastic variables $dM(E_1)$ and $dM(E_2)$ and define

$$f_1(E_1,E_2)dE_1dE_2 = E\{dM(E_1)dM(E_2)\}, \tag{2.5}$$

which is also equal to the joint probability that a particle lies in dE_1 and a particle lies in dE_2 where dE_1 and dE_2 do not overlap. When they do, a degeneracy occurs and

$$E[\{dM(E_1)\}^2] = E\{dM(E_1)\} = f_1(E)dE. \tag{2.6}$$

We call f_2 a product density of degree 2. In view of the degeneracy it follows directly that

$$E[\{M(E_u) - M(E_\ell)\}^2] = \int_{E_\ell}^{E_u}\int_{E_\ell}^{E_u} E\{dM(E_1)dM(E_2)\}dE_1 dE_2$$

$$= \int_{E_\ell}^{E_u} f_1(E)dE + \int_{E_\ell}^{E_u}\int_{E_\ell}^{E_u} f_2(E_1 E_2)dE_1 dE_2. \tag{2.7}$$

Similarly, we can define product densities f_n of higher order n:

$$f_n(E_1,E_2,\ldots,E_n)dE_1 dE_2\ldots,dE_n = E\{dM(E_1)dM(E_2)\ldots dM(E_n)\}, \tag{2.8}$$

where intervals dE_1, $dE_2,\ldots,dE_n$ do not overlap. If $dE_n = dE_{n-1}$ then

$$E[dM(E_1)\ldots\{dM(E_{n-1})\}^2] = f_{n-1}(E_1,E_2,\ldots,E_{n-1})dE_1 dE_2\ldots dE_{n-1}. \tag{2.9}$$

In view of the degeneracies mentioned we can write the rth moment as

$$E[\{M(E_u) - M(E_\ell)\}^r] = \sum_{s=1}^{r} C_s^r \int_{E_\ell}^{E_u}\int_{E_\ell}^{E_u}\ldots\int_{E_\ell}^{E_u} f_s(E_1,E_2,\ldots,E_s)dE_1 dE_2\ldots dE_s \tag{2.10}$$

The coefficients C_s^r are functions of r and s and in no way dependent on the function f_s. The coefficients can be obtained by considering the particular case of the distribution of N particles in E-space. We have

$$N^r = \sum_{s=1}^{r} C_s^r N(N-1)\ldots(N-s+1). \tag{2.11}$$

The Markovian character of cascade processes is preserved because the product density functions depend upon the energies of the *particles under consideration and not on other particles present in the E-space.*

The product densities for the Poisson, Binomial and Furry processes are given below for the sake of comparison.

Poisson Process. The product density of degree n in the t-space is given by

$$f_n(t_1,t_2,\ldots,t_n) = f_1(t_1)f_1(t_2)\ldots f_1(t_n), \tag{2.12}$$

which indicates that an absence of correlation between the entities distributed in different parametric intervals.

Binomial Process. The product density of degree n is

$$f_n(t_1,t_2,\ldots t_n) = 1(1-\frac{1}{N})(1-\frac{2}{N})\ldots(1-\frac{n-1}{N})\ f_1(t_1)\ldots f_1(t_n). \tag{2.13}$$

Furry Process. The product density of degree n is

$$f_n(t_1,t_2,\ldots,t_n) = n!\ f_1(t_1)f_1(t_2)\ldots f_1(t_n). \tag{2.14}$$

3. METHOD OF REGENERATION POINTS

Janossy (1950), in his treatment of Cascade showers, defined a function $\pi^i(n,E;E_0;t)$ representing the probability that there are n electrons above the energy E at t, given that at t = 0 there was a primary with energy E_0, with i = 1,2, referring to an electron or photon respectively. This obviates the introduction of a continuous infinity of stochastic variables.

He observed that we can deal with the function $\pi^i(n,E;E_0;t)$ directly, if we consider what happens after the first event occurs at some τ between 0 and t, since the initial conditions are well defined and that the future development of the process in the interval $t - \tau$ can be considered in terms of the first event which occurred at τ. This approach was recognized by Bartlett and Kendall (1951) to be the regeneration point method of Bellman and Harris (1948).

Adopting the regeneration point method let us look at the equations governing some simple stochastic processes studied by the author (1951a). For the Furry process the generating function G(u;t) satisfied the equation:

$$\frac{\partial G(u;t)}{\partial t} = -\lambda[G(u;t) + \{G(u;t)\}^2]. \tag{3.1}$$

If we confine ourselves to the generating functions for the total number of electrons in electromagnetic cascades, we obtain

$$\frac{\partial G}{\partial t} = (GF-G)\alpha \qquad \frac{\partial F}{\partial t} = (G^2-F)\beta\ , \tag{3.2}$$

where α and β are related respectively to the probability that an electron of given energy radiates a photon in the desired energy range and a photon creates an electron position pair. Here G and F refer to the generating function of electron distribution in electron and photon initiated showers respectively.

We can express $\pi(n,E;E_0;t)$ as

$$\pi(n,E;E_0;t) = \int_E^\infty \int_E^\infty \ldots \int_E^\infty J_n(E_1,\ldots,E_n;t)\,dE_1 dE_2 \ldots dE_n, \tag{3.3}$$

where the Janossy density function $J_n(E_1,\ldots,E_n;t)dE_1dE_2\ldots dE_n$ denotes the probability that there is one particle in dE_1, one in dE_2,..., one in dE_n and *none in other states*. Note that J_n is not a probability density.

On further analysis it was shown by the author (1952) that Janossy functions when differentiated yield product density functions, i.e.,

$$f_n(E_1,E_2,\ldots,E_n;t) = \sum_n \frac{1}{(n-h)!} \int_{E_n} \int_{E_{n-1}} \ldots \int_{E_{h+1}} J_n(E_1,E_2,\ldots,E_n) \times$$

$$\times\, dE_{h+1}\ldots dE_n.$$

Bartlett and Kendall (1951) have defined a characteristic functional for a continuous infinity of random variables. If N(E) represents the number of particles with energy below E, then dN(E) is a stochastic variable which represents the number in dE and the characteristic functional is defined as

$$C\{\theta(E)\} = E[\exp\{i\int_E \theta(E)dN(E)\}], \tag{3.4}$$

where $\theta(E)$ is an arbitrary chosen function of E. The characteristic functional, when determined, contains in portmanteau form all the probability relationships associated with the state of 'population' of particles in the energy space. We can recover the product densities from the characteristic functional as

$$\frac{(-i)^h \partial^h C\{\theta(E)\}}{\partial\theta(E_1)\partial\theta(E_2)\ldots\partial\theta(E_n)}\Bigg|_{\theta(E)=0} = f_n(E_1,E_2,\ldots,E_h;t)$$

4. FURTHER DEVELOPMENTS IN PRODUCT DENSITIES

The power of the product density method consists essentially in the fact that we associate a single entity with an infinitesimal interval in a continuous space. This may even refer to a twin or a triplet or a multiplet. Thus we can introduce multiple product densities (Ramakrishnan and Srinivasan, 1958) and this is useful in dealing with the problem of age distribution in population growth involving twins and higher multiplets.

Exactly as in the case of product densities, if we denote by $N_i(E_i;t)$ the number of i-tuple entities having parametric values less than or equal to E, we can introduce a function $f_1^i(E;t)$, $1 = 1,2,\ldots,m$;

$$f_1^i(E;t)dE = \mathcal{E}\{dN_i(E;t)\}, \tag{4.1}$$

$$dM(E;t) = \sum_i i dN_i(E;t) \tag{4.2}$$

and we have

$$\mathcal{E}\{M(b;t) - M(a;t)\} = \sum_i \int_a^b f_1^i(E;t)dE. \tag{4.3}$$

It is possible to calculate the higher order moments of M(E;t) though certain new combinatorial features are encountered.

i) Sequent Product Densities.

One of the striking analytic features of product density is that when one computes the r-th moment of a stochastic variable, it not only involves the product density of degree r but also of all the lower orders. However, we can define a new type of n-th order correlation function (Ramakrishnan and Radha, 1961)

$$\pi(E_1,E_2,\ldots,E_n;t_1,t_2,\ldots,t_n)dE_1dE_2dE_3,\ldots,dE_n \ , \tag{4.4}$$

which denotes the existence of a particle in dE_1 at t_1, a particle in dE_2 at t_2, etc., with $t_1 < t_2 < t_3 < \ldots < t_n$. No degeneracies can arise here but as the t_i's tend to one another we should expect again overlap of energy intervals. Such correlation functions are called sequent product densities. An n-th order sequent density corresponding to different thicknesses is connected to all the instant densities of order m less than equal to n.

ii) New Approach to Cascade Theory

In applications of product density to cascades there is one serious defect which makes all the limiting distributions tend to zero because of the attenuation of the energy of the primary particle. It is worthwhile defining a function which has non-trivial limits as $t \to \infty$ (Ramakrishnan and Srinivasan, 1956). Such a function is suggested when we study cascades in nuclear emulsion where it is more convenient to estimate the energies of particles at the point of their product rather than the energies of all particles at a particular thickness. In this approach the energy of the particle will refer to that at the time of production and hence will be called the primitive energy. We can define

$$F_k(E_1,E_2,\dots,E_k;\ t_1,t_2,\dots,t_k)dE_1,\dots,dE_k dt_1,\dots dt_k$$

as the new product density of degree k which denotes the joint probability that an electron of energy between E_1 and $E_1 + dE_1$ is produced between t_1 and $t_1 + dt_1$, etc. The function is a product density over E space as well as t space. If F_k is integrated over $E_1,E_2,\dots,E_k$ over the range E_c to ∞ it is still a product density of particles produced between and t_i and $t_i + dt$, $i = 1,2,\dots,k$, the primitive energy of each particle being greater than E_c. The moments can be obtained from this by integration over the proper range of t. The r-th moment of the number of electrons produced between 0 and t, the primitive energy each of the electrons being greater than E_c, is given by

$$\mathcal{E}[\{N(E_c;t)\}^r] = \sum_i C_i^r \int_{E_c}^{\infty} dE_1 \int_{E_c}^{\infty} dE_2 \dots \int_{E_c}^{\infty} dE_i \times$$
$$\times \int_0^t dt_1 \int_0^t dt_2 \dots \int_0^t dt_i F_i(E_1,E_2,\dots,E_i;t_1,t_2,\dots,t_i). \qquad (4.5)$$

We can express the new product densities in terms of the old product densities e.g.,

$$F_1(E|E_0;t) = 2\int_E^{E_0}\int_0^t g_1(E',\ E_0;t)R(E|E')dE'e^{-\beta(t-\tau)}d\tau, \qquad (4.6)$$

where g_1 is a product density of degree 1 for photons.

iii) Numerical Calculations

The product density treatment of cascades becomes useful only if we can numerically evaluate the mean and the fluctuations in the number of electrons and photons both at a thickness t and in the number produced in a thickness t. Bhabha and Chakrabarty (1943) have employed the saddle point approach to perform numerical calculations on the number of electrons above a certain energy at a thickness t. Bhabha and Ramakrishnan (1950) obtained for the first time using product densities an analytical solution for the fluctuation in the number of electrons produced which lends itself for numerical evaluation by the saddle point method as was done by the author and Mathews (1954). Mean number of electrons produced in a thickness t have been evaluated by Srinivasan and Ranganthan (1957) for both small and large thickness and the numerical results confirm the asymptotic behaviour expected on theoretical grounds. In all these discussions we have omitted the ionization loss suffered by charged particles traversing a material medium but this can be taken into account though it complicates the treatment.

5, INTEGRALS OF POINT PROCESSES

It turns out that a point process is a particular example of a *Basic Random Process* (B.R.P.) in which the number of jumps are finite. A typical trajectory of the B.R.P in an interval 0 to t will be characterized by finite jumps at finite number of points and in between the transitions, the realized values of the stochastic variable remain constant. The interesting feature of this process is that it is quite easy to assign a probability measure to a typical trajectory.

If a process is characterized by infinitesimal random changes in infinitesimal periods of time, the author has shown (1955a,1956) how to view the process as an integral of another process and so on till we reach a process which is a random process characterized by finite random changes. It is reasonable to suppose this is a B.R.P. This observation has led the author to develop a phenomenological interpretation of integrals of random functions using only the concept of realized trajectory of a stochastic process.

When the B.R.P is of a very simple type with certain symmetric properties, we can introduce the notion of an 'equivalent' process. Two processes are said to be equivalent to each other if the distribution function of the two stochastic variables $y(t)$ and $Y(t)^*$ are identical at all t. We are not stating that the joint distribution of $Y(t)$ at more than one point, say t_1 and t_2, is identical with the joint distribution of $y(t)^*$ at the same points t_1 and t_2. Since a stochastic process $y(t)$ represents a continuous infinite set of random variables, the distribution function deals with one of them 'at a time' and hence any information only about the distribution function cannot give us a complete picture of the stochastic process.

We shall illustrate this by considering processes associated with a Poisson process. We can study a different type of integrals connected with random functions, in consequence of the simple fact that the duration of the process to which integral refers, can be increased in two ways i.e., by changing either of the end points of the interval. Let us consider random points distributed according to Poisson's law along a line AB. Starting with A, we can associate with every point P on the line AB a number $n(\tau)$ which represents the number of random points lying in AP = τ. Let $n^R(\tau)$ be the realized curve in the interval $(0,t)$ and the area enclosed by the realized curve $n^R(\tau)$ is the realized value $y^R(t)$ of a random variable defined by

$$y(t) = \int_0^t n(\tau)\, d\tau. \tag{5.1}$$

If we consider the stochastic variable $n(t-\tau)$ in which the events are counted from B, its realized value or trajector is $n^R(t-\tau)$, which we shall call the *inverse trajectory* $n^R(\tau)$. The area enclosed by this inverse trajectory is the realized value $y^R(t)^*$ of another stochastic variable defined by

$$y(t)^* = \int_0^t n(t-\tau)d\tau. \tag{5.2}$$

It is clear that even though the realized values of $y(t)^*$ and $y(t)$

will in general be different, the stochastic variables $y(t)^*$ and $y(t)$ will have *identical distribution functions for all values of t*. This remark allows us to obtain differential equations for the probability frequency function of $y(t)$ in two ways. We have for $\pi(y;t)$ of $y(t)^*$ the equation

$$\frac{\partial\pi(y;t)}{\partial t} = -\lambda\pi(y;t) + \lambda\pi(y-t;t), \tag{5.3}$$

with $\pi(y;0) = \delta(y)$.

The same $\pi(y;t)$ also holds for $y(t)$. However, we may start from $y(t)$ directly and deduce a differential equation for $\pi(y,n;t)$, the joint probability that $y(t)$ lies between y and $y + dy$ and $b(t) = n$. We have

$$\frac{\partial\pi(y,n;t)}{\partial t} = -\lambda\pi(y,n;t) + \lambda\pi(y,n-1;t) - n\frac{\partial\pi(y,n;t)}{\partial y} .$$

It is an important question whether the equation (5.4) can be derived from (5.5) but this is a difficult one and hence the *advantage of the inverse trajectory and the concept of equivalent* process.

The author and his collaborators have successfully applied these concepts involving integrals of random function to several physical problems like the energy depletion on collisions where the total energy loss is $y(t)$ and energy loss in each collision is given by $n(\tau)$ and thus is $y(t)$ $y(t) = \int_0^t n(\tau)d\tau$. This phenomenological interpretation of random integrals have opened out a new chapter in solution of stochastic differential equations and as such is capable of wide applications (Ramakrishnan, 1959a).

6. OUTLOOK FOR FUTURE

In the opinion of the author there exist three aspects which look promising for future developments in the theory of stochastic processes and their applications to physical problems. Even the current methods can find new applications and a beginning has been made by the author (1961). We shall discuss in particular below

a) Stochastic methods and Quantum Mechanics;

b) Concepts of inverse probability, direction of stochastic process and ergodicity;

c) New mathematical methods.

i) Applications to Quantum Mechanics

The relationship between the theory of probability and quantum mechanics can be traced even to the beginnings of quantum mechanics when the Born interpretation of the wave function for a particle was adopted as a fundamental postulate. Feynman (1957) has remarked that the nature seems to have devised a law of superposition of probability amplitude in quantum mechanics different from that of superposition of probabilities for the events related to a roulette wheel. We however wish to modify his statement in the following manner. In quantum mechanics, to every event we assign a complex probability amplitude and the probabilistic interpretation for the square of the modulus of complex amplitude is applicable only to a certain class of events.

As long as we remember the principle of superposition of probability amplitudes in quantum mechanics for a class of events and take care to introduce the probability interpretation at a correct stage for suitably superposed amplitudes, there is no reason why methods of stochastic theory should not find application in the study of quantum mechanical problems. Ramakrishnan and Ranganathan (1960) have suggested the possible utility of the following methods in quantum mechanics:

a) the superposition principle;

b) the concept of realization;

c) the concept of inverse probability;

d) product densities and many particle system;

e) the regeneration point method;

f) the ambigenous stochastic process.

We shall be merely content to state our results of this approach. Methods (a), (b) and (c) lead us to an alternative view of Feynman formalism in quantum electrodynamics and also to a simple and direct proof of the equivalence of field theoretic formalism with that of Feyman. The proof makes use of a new prescription by the author (1961) for ordering the operators in the interaction terms in quantum electrodynamics which follows from suitable adaptation of the concepts of inverse probability. The concept of product densities is found quite useful in dealing with many particle systems and the method of regeneration point method in the study of scattering problems.

We suggest that it is possible to develop a non-perturbative approach using Feynman formalism by adopting the methods used in the study of the ambigeneous stochastic processes (1961) which have a new and interesting feature, in that they involve back scattering in relation to parameter with respect to which the process unfolds, a feature which disturbs the otherwise Markovian character of the process. The paths traced in such a process bears close relation to the paths traced by the electrons in the electrons in the Feynman formalism of quantum electrodynamics. Thus there exists a possibility that a comparison between the two may suggest a new method in the study of quantum electrodynamics. Ambigenous stochastic process are studies by switching off back-scattering probabilities outside a finite interval. The question now arises whether meaningful results can be obtained in quantum electrodynamics by considering the process in a finite duration t and later letting $t \to \infty$. This is possible since in quantum mechanical scattering we already have the concepts of 'switch on' and 'switch off' of the interactions. But there is one important difference between an ambigenous stochastic process and Feynman scattering. The stochastic process is symmetric with respect to the directions of velocity while in Feynman formalism the direction of time is important in determining the electron or positron character of the particle.

ii) Inverse Probability and Ergodicity

There is still one aspect of cascade theory which has not claimed much attention and this relates to the use of inverse probability in the solution of cascade equations.

The basic problem of inverse probability in a Markovian stochastic process involves the question: Can we make statements about the past given the present state of the system? In other words does a stochastic process have a sense of direction? It has been shown by the author (1955b) that a stochastic process can also be studied for $t < 0$ just as for $t > 0$ but the process can be *traced back only up to its primitive origin.* The spectrum at this origin cannot be the result of the evolution of a process started earlier. This result can be established for any stochastic process and therefore applicable to point processes.

We wish to suggest that the concept of inverse probability can be applied to yield new information in the theory of cascade showers. In particular we can answer the question: given a distribution $\phi(E)$ in E space of the cascade particles at a particular point on t (material thickness) what is thickness of matter traversed? We can given an unambiguous answer by solving the cascade equations for $\pi(E,t)$ with the initial condition $\pi(E|E_0;t) = \phi(E)$ and then find the negative value of $t = -\tau$ at which $\pi(E|E_0 t) = 0$ for some E in the permissible energy range. Then τ is the thickness of matter traversed.

iii) Ergodic Properties

One of the most important problems in regard to dynamical system is the so called ergodicity of the system. The author has succeeded in studying the ergodic properties of some simple stochastic processes by adopting a phenomenological approach (1957).

The ergodic property in case of stochastic systems can be established by considering the fraction of time spent in a state, which is random variable. The author has shown that this variable converges in probability to the stationary distribution.

REFERENCES

Bhabha, H. J. (1950). On the stochastic theory of continuous parametric systems and its applications to electron cascades. Proc. R. Soc. A 202, 301-22.

Bhabha, H. J. & Heitler, W. (1937). The Theory of Cosmis Ray Showers. Proc. R. Soc. A 159, 432-58.

Bhabha, H. J. & Chakrabarty, S. K. (1943). The cascade theory with collision loss. Proc. R. Soc. A 181, 267-303.

Bhabha, H. J. & Ramakrishnan, A. (1950). The mean square deviation of the number of electrons and quanta in the cascade theory. Proc. Ind. Acad. Sci. A 32, 141-53.

Bartlett, M. S. & Kendall, D. G. (1951). On the use of the characteristic functional in the analysis of some stochastic processes occurring in physics and biology. Proc. Camb. Phil. Soc. 47, 65-76.

Bellman, R. E. & Harris, T. E. (1948). On the theory of age dependent stochastic branching processes. Proc. Nat. Acad. Sci. (USA) 34, 601-4.

Feyman, R. P. (1951). The concept of probability in quantum mechanics. Proc. 2nd Berkeley Symp. Berkeley: University of California Press.

Furry, W. H. (1937). On fluctuation phenomena in the passage of high energy electrons through lead. Phys. Rev. 52, 569-81.

Janossy, L. (1950). On the absorption of a nucleon cascade. Proc. R. Irish. Acad. Sci. A 53, 181-8.

Kendall, D. G. (1949). Stochastic processes and population growth. J. R. Statist. Soc. B 11, 230-64.

Kolomogorov, A. N. & Dmitriev, N. A. (1947). Branching stochastic processes. Com. Rendus Acad. Sci. URSS (N. S.)(Doklady) 56, 5-8.

Ramakrishnan, A. (1950). Stochastic processes relating to particles distributed in continuous infinity of states. Proc. Camb. Phil. Soc. 46, 595-602.

Ramakrishnan, A. (1951a). Some simple stochastic processes. J. R. Statist. Soc. R 13, 131-40.

Ramakrishnan, A. (1951). Stochastic processes and their applications to physical problems. Ph.D. Thesis. University of Manchester.

Ramakrishnan, A. (1952). A note on Janossy's mathematical model of nucleon cascade. Proc. Camb . Phil. Soc. 48, 451-6.

Ramakrishnan, A. (1955a). Phenomenological interpretation of the integrals of a class of random functions. Proc. Kon. Nederl. Akad. Wentensch, A 58, 470-82 and 634-45.

Ramakrishnan, A. (1955b). Inverse probability and evolutionary Markov stochastic processes. Proc. Ind. Acad. Sci. A 41, 145-53.

Ramakrishnan, A. (1956). Process represented as integrals of a class of random functions. Proc. Kond. Nederl. Akad. Wentensch, A 59, 120-7.

Ramakrishnan, A. (1957). Ergodic properties of some simple stochastic process. Zeit. Ang. Math. Mech. 37, 336-44.

Ramakrishnan, A. (1959a). Ambigeneous stochastic process. Zeit. Ang. Math. U. Mech. 39, 388-90.

Ramakrishnan, A. (1959b). Probability and Stochastic Processes. *Handbuch der Physik*. Springer Verlag Vol. III/2, 524-651.

Ramakrishnan, A. (1961). In *Elementary Particles and Cosmic Rays*. London: Pergamon Press, 522-30.

Ramakrishnan, A. & Mathews, P.M. (1954). Studies in the stochastic problem of electron photon cascade. Proc. Theor. Phys. 11, 95-117.

Ramakrishnan, A. & Radha, T. K. (1961). Correlation problems in evolutionary stochastic process. Proc. Camb. Phil. Soc. 57, 843-7.

Ramakrishnan, A. & Ranganathan, N. R. (1962). Stochastic methods in quantum mechanics. Jour. Math. Anal. Appl. 3, 261-94.

Ramakrishnan, A. & Srinivasan, S. K. (1956). A new approach to cascade theory. Proc. Ind. Acad. Sci. A 44, 263-73.

Ramakrishnan, A. & Srinivasan, S. K. (1958). On age distribution in population growth. Bull. Math. Biophysics. 20, 289-303.

Scott, W. T. & Uhlenbeck, G. E. (1942). On the theory of cosmic ray showers. Phys. Rev. 62, 497-508.

Srinivasan, S. K. & Ranganathan, N. R. (1957). Numerical calculations on the new approach to the cascade theory. Proc. Indi. Acad. Sci. 45, 69-73.

SUPERPOSITION OF POINT PROCESSES

E. Çinlar
Northwestern University, Evanston, Illinois

SUMMARY

If two independent recurrent processes are superposed to obtain a recurrent process, then in general, each process involved is a Poisson process. A review of such results is the first of six problems discussed in this paper.

If a large number of uniformly sparse point processes are superposed, the result is approximately a Poisson process. A review of the most general such theorems is given along with various specializations to stationary and to recurrent processes. If the component processes are marked point processes, then the convergence is to a marked Poisson process. If the components are all clustered point processes, then the convergence is to a clustered Poisson process. In all these results convergence is in the sense of the weak convergence of the finite dimensional distributions.

If only a finite number of processes are superposed, then the finite dimensional distributions differ from the appropriate Poisson distribution. The error involved in using the Poisson distribution to approximate such distributions is considered and various bounds are given along with references to further results. Some asymptotic expansions using the Charlier B-series are given and some preliminary results are stated.

If one can observe the superposition process only, then we have the statistical problem of determining the number of components and estimating the probability law of each from the known data on the superposition process. The paper concludes with a review of known results on this problem.

1. INTRODUCTION

A point process is pictured best as a collection of random points; the superposition of two point processes is then obtained by taking the union of the points of both. The simplest and best known result is that the superposition of two independent Poisson processes is again a Poisson process. That is, the class of Poisson processes is invariant under the superposition of its independent members. This raises the question of whether there are other such classes of processes. Most of Section 2 is a survey of results in this direction. The answer obtained is, roughly, that Poisson processes are unique in this regard. This section also contains an unpublished very clever proof due to Professor K. L. Chung; we thank him for allowing it to appear here.

Poisson processes appear in practical experience with unreasonable frequency and an explanation of this phenomenon is put in Section 3. The results surveyed show that the superposition of a large number of uniformly sparse point processes is a (not necessarily stationary) Poisson process. A many dimensional generalization of the same result is put in Section 4.

Somewhat more generally, the points of a Poisson process may attract (or cause) a secondary series of points. In geography, central places have secondary population centers attached to them; in reliability, the physical failure of a component (in a complex machine such as a computer) causes secondary malfunctionings. In such situations one obtains a clustered Poisson process. In Section 5 we give an explanation of the phenomenon by showing that, if a large number of processes, each sparsely clustered, are superposed, the result is a clustered Poisson process.

Knowing the limiting results mentioned above we next ask, if we have a moderate number of point processes being superposed, how well can the superposition be approximated by a Poisson process? In Section 6 we give bounds for the "distance" between the finite dimensional distributions of the superposition process and the corresponding ones of a suitably chosen Poisson process. In Section 7 we give asymptotic expansions in terms of Charlier B-series which can be used to approximate the one-dimensional distributions of the superposition process. These two sections draw heavily upon the results of Franken (1963a, 1963b, 1964) which seem not to have received the attention they deserve.

The final section is devoted to statistical problems which arise when one is able to observe only the superposition process and wants to draw inferences concerning the number and nature of the component processes. Some results in this direction can be found in the well known book of Cox and Lewis (1966); we have therefore concentrated on the work of Ambartzumian (see the references).

The approximation results of the type reported in Sections 6 and 7 and statistical ones given in Section 8 are of great interest from both the standpoint of theory and that of applications. The available results on the other hand are either rudimentary or very complicated and cumbersome. We hope that our survey here generates interest in these problems.

In the remainder of this section we introduce the notations and definitions to be used throughout this paper.

Throughout we write $\mathbb{R} = (-\infty,+\infty)$, $\mathbb{R}_+ = [0, +\infty)$, and R, R_+ for their respective Borel subsets. We will have a fixed probability space (Ω,F,P) which will be assumed to be rich enough to hold all the processes we will be considering. By a point process we mean a stochastic process $\{N(t), t \geq 0\}$ defined over (Ω,F,P) so that, for almost all $\omega \in \Omega$, the mapping $t \to N(\omega,t)$ is non-decreasing, right continuous, non-negative, finite, integer valued, and has $N(\omega,0) = 0$. Such a process increases by jumps only, and the times of jumps will be called the *points* of N. In particular, a point process N with points $T_1, T_2,\ldots$ is called a *renewal process* if $T_1, T_2-T_1,\ldots$ are independent and identically distributed; it is called a stationary renewal process if $T_2-T_1, T_3-T_2,\ldots$ are independent and have a common distribution function F, T_1 is independent of the other intervals but its distribution has the density $\{1 - F(x)\}/\mu$, where $\mu = \int_0^\infty x\, F(dx)$. This is also sometimes called the *equilibrium* renewal process.

If (E,E) is an arbitrary measurable space, by a *random counting measure* we mean a stochastic process $\{N(A);\ A \in E\}$ taking non-negative integer values so that $N(A \cup B) = N(A) + N(B)$ whenever A and B are disjoint. A counting measure is purely atomic, and we will call the atoms *points* (if an atom has weight greater than one, the corresponding point is repeated as many times as the weight). Then, N(A) is the number of points of N in the set A. Any point process can be looked upon as a random measure in $\mathbb{R}$ by identifying $N(t) - N(s)$ with $N((s,t])$ for $s < t$. We will do this for all the time without further mention. If $\{N(t); t \geq 0\}$ is a point process, we write N(A) for the number of points of N in the set A; if $\{N(A); A \in \mathcal{R}\}$ is a random counting measure, we write N(t) for $N((0,t])$.

A random counting measure N on $(E,\mathcal{E})$ is called a Poisson random measure with mean measure μ provided that $N(A_1),\ldots, N(A_m)$ are independent whenever $A_1,\ldots, A_m \in \mathcal{E}$ are disjoint and

$$P\{N(A) = k\} = e^{-\mu(A)} \mu(A)^k/k! \qquad (k = 0,1,\ldots) ,$$

whenever $\mu(A) < \infty$. A Poisson process with parameter function $\mu(t)$ is a Poisson random measure on $(\mathbb{R}_+,\mathcal{R}_+)$ with mean measure element $\mu(dx)$. A Poisson process is stationary if its mean measure is a constant multiple of the Lebesgue measure.

A point process is stationary if the joint distribution of $N(t + A_1)$, $\ldots$, $N(t + A_m)$ is independent of t for all $m \geq 1$ and $A_1,\ldots, A_m \in \mathcal{R}_+$. A stationary Poisson process is stationary in this sense, and similarly for a stationary renewal process.

The *superposition* of two point processes N_1 and N_2 is the point process N defined by

$$N(t) = N_1(t) + N_2(t) \qquad (t \geq 0) .$$

Then we simply write $N = N_1 + N_2$. If N_1 and N_2 are random counting measures, then their superposition N is similarly defined by $N(A) = N_1(A) + N_2(A)$. One can think of the superposition as the union of the points of both components in the case where there are no multiplicities. Finally, by weak convergence we mean the weak convergence of the finite dimensional distributions.

2. INVARIANCE UNDER SUPERPOSITION

The superposition of two independent Poisson processes is again a Poisson process, and it is natural to ask if there are other classes of point processes which are invariant under superposition. Many authors have considered the problem and the answer is that, amongst renewal processes, Poisson processes are the only ones with this property. We shall give below an account of the results in this direction. We will give proofs of some of these in order to illustrate the diversity of techniques employed.

Throughout the following we limit ourselves to point processes N whose sample paths are right-continuous non-decreasing with $N(0) = 0$, and $N(t) - N(t - 0) \leq 1$ for all t. The more general case where multiple jumps are allowed can be studied by first considering only the number of jumps by the techniques we will describe, and then separately discuss the magnitudes of a given jump.

The earliest result in this direction is due to McFadden (1962, p. 379).

Proposition (2.1). Suppose N_1 is a stationary renewal process with finite variance for the inter-renewal times and let N_2 be a Poisson process independent of N_1. If $N_3 = N_1 + N_2$ is a stationary renewal process, then N_1 is a Poisson process.

Remark (2.2). If multiple jumps are allowed then the jump times of N_1 form a Poisson process and the magnitudes of jumps are independent of each other and of the jump times and have a common geometric distribution.

Proof. Let $\phi_j(k,t) = P\{N_j(t) = k\}$, $j = 1,2,3$. Since N_1, N_3 are stationary renewal processes,

$$\phi_j(1,t) = \frac{1}{\mu_j}\int_o^t \{1 - F_j(x)\}\{1 - F_j(t - x)\}\,dx \tag{2.3}$$

$$= \mu_j\phi_j'(0,t) * \phi_j'(0,t) \qquad (j = 1,3) ,$$

where prime denotes derivative and "*" convolution. By the independence of N_1, N_2 one can write, letting $a = 1/\mu_1$,

$$\phi_3(0,t) = e^{-at} \phi_1(0,t) ,$$

$$\phi_3(1,t) = e^{-at}[\phi_1(1,t) + at\, \phi_1(0,t)] .$$

Substituting these in (2.3) for $j = 3$ and then using (2.3) again for $j = 1$ and factoring out a term e^{-at} one obtains an equation involving $\phi_1(0,t)$ only. Taking Laplace transforms, one obtains an equation with $y = \int_0^\infty e^{-\lambda t} \phi_1(0,t)\, dt$,

$$a^2 y' = \{(\mu_2-\mu_3)\lambda^2 + 2a\mu_3\lambda - a^2\mu_3\}y^2 - 2\{(\mu_2-\mu_3)\lambda + a\mu_3\}y + \mu_2 - \mu_3 .$$

Solving this and putting $b = 1/\mu_2$, $c = 1/\mu_3$ one gets

$$y = \frac{\lambda + b - bcD}{(\lambda + b)^2 - bcD(\lambda + b) + bc}, \tag{2.4}$$

where D is some constant. Since $\phi_1'(0,t) = b\{1 - F_1(t)\}$, one can get from this the distribution F_1 and show that it would have a negative derivativ (from the right) at zero unless $|D| = +\infty$. Letting D go to infinity, and then inverting the Laplace transform one obtains from (2.4) the desired result that $\phi_1(0,t) = e^{-bt}$.

McFadden and Weissblum (1963) have generalized the preceding result to the following

Theorem (2.5). Suppose N_1, N_2 are independent stationary renewal processes with finite variances for their intervals. If $N_3 = N_1 + N_2$ is a stationary renewal process, then N_1, N_3 are both Poisson processes.

Their proof uses tools similar to that of the preceding proposition. There are some minor errors (cf. paragraphs 2, 3, 4 on p. 430) which are easy to clear; also in p. 427, the case $F_1 = F_2$ needs to be distinguishe when computing p and their arguments need to be replaced by a symmetry argument.

This theorem has been generalized in two directions. Ambartzumian (1967a) has shown under some regularity conditions that the same conclusion holds if, instead of assuming N_3 to be a stationary renewal process, it is only assumed that the lengths of intervals of N_3 form a Markov chain of finite order. He had further extended the result to point processes on arbitrary spaces (rather than the line) in Ambartzumian (1967b). In another direction, Mecke (1967) showed that the assumption about N_2 can be relaxed. The following is his result.

Theorem (2.6). Suppose N_1 is a stationary renewal process, N_2 is an arbitrary stationary point process independent of N_1, and suppose $N_3 = N_1 + N_2$ is a stationary renewal process. Then, under the further assumption that the intervals of N_1, N_3 possess probability densities which are continuous on $(0,\infty)$ and right continuous at 0, the process N_1 must be Poisson.

We sketch the main lines of Mecke's proof. If N is a stationary renewal process with a probability density function f (for its interrenewal times) which is continuous on $(0,\infty)$ and right continuous at 0, and if

$$P(t,j,k) = P\{N(0,t) = j,\ N(\tau + t,\ \tau + 2t) = k\} \tag{2.7}$$

for some fixed $\tau > 0$, then

$$\lim_{t\to 0} P(t,0,0) = 1 \tag{2.8}$$

$$\lim_{t\to 0} P(t,k,0)/t^k = \lim_{t\to 0} P(t,0,k)/t^k = \lambda f(0)^{k-1}/k! \tag{2.9}$$

$$\lim_{t\to 0} P(t,j,k)/t^{j+k} = \lambda h(\tau)\ f(0)^{j+k-2}/j!k!\ , \tag{2.10}$$

where λ is the intensity of N and h its renewal density. These equations are obtained by straightforward computations by writing P(t,j,k) out completely; and then obtaining lower and upper bounds by replacing the $f(t)$ by $f_1(x) = \inf_{0\le t\le x} f(t)$, $f_2(x) = \sup_{0\le t\le x} f(t)$ and h by h_1 and h_2 in a similar fashion; and finally taking limits and using the continuity of f and h.

Next let P_i be defined by (2.7) when N is replaced by N_i. Since $N_3 = N_1 + N_2$ and N_1, N_2 are independent one can write P_3 as the convolution of P_1 and P_2 and then separating one term from the sum write

$$P_2(t,j,k)\ P_1(t,0,0)$$
$$= P_3(t,j,k) - \sum_{m+n>0} P_1(t,m,n)\ P_2(t,j-m,k-n)\ . \tag{2.11}$$

By (2.8)-(2.10) the limits

$$p_i(j,k) = \lim_{t\to 0} P_i(t,j,k)/t^{j+k} \tag{2.12}$$

exist for i = 1,3; this implies by an induction on j+k in (2.11) that the limits (2.12) exist for i = 2 also. Letting

$$Q_i(x,y) = \sum_{j,k} p_i(j,k)\ x^j y^k \qquad (|x|,\ |y| < 1)\ , \tag{2.13}$$

one has

$$Q_3(x,y) = Q_1(x,y)\ Q_2(x,y)\ . \tag{2.14}$$

It follows from (2.8)-(2.10) that Q_1 and Q_3 are entire functions; since, obviously,

$$0 \le p_i(j,k) \le p_3(j,k) \qquad (i = 1,2) \tag{2.15}$$

for all j, k $\ge$ 1, Q_2 is also entire and (2.14) holds for all complex x, y (not only for x, y with $|x|$, $|y| < 1$). This implies that every root (x_0,y_0) of the function Q_1 must also be a root of Q_3. There remains to show that these demands can be met only when N_1 is a Poisson process.

First it is shown that $f_3(0) > 0$; for, if $f_3(0) = 0$ then by (2.9), (2.10), (2.15) one has

$$Q_3(x,0) = 1 + \lambda_3 x, \qquad Q_1(x,0) = 1 + \lambda_1 x, \qquad Q_2(x,0) = 1 + cx$$

so that (2.14) is impossible to satisfy no matter what c is.

Writing out Q_1 and Q_3 by using (2.8)-(2.10) and examining in detail the zeros of Q_1 (which must also be of Q_3) one concludes that we must have $f_1(0) > 0$ also. Using these facts and again arguing about the zeros of the Q_i one shows that only the values 0, λ_1, $f_1(0)$ are possible for $h(\tau)$. Since the renewal density h_1 is continuous by the hypothesis of continuity about f_1, we must have

$$h_1(\tau) = \lambda_1 = f_1(0).$$

which can hold only if N_1 is Poisson.

This last result is also applicable to the superposition of any number of stationary renewal processes:

Corollary (2.16). Suppose $N_1, \ldots, N_n$ and $N = N_1 + \cdots + N_n$ are stationary renewal processes each with a density function which is continuous on $(0,\infty)$ and right continuous at 0. Assume $N_1, \ldots, N_n$ are independent. Then, $N_1, \ldots, N_n$ are all Poisson processes.

Proof is immediate from the grouping $N = N_1 + (N_2 + \cdots + N_n)$ to which Theorem (2.6) applies to show that N_1 is Poisson.

The following result and the short proof is due to Chung (1971).

Proposition (2.17). Suppose N_1, N_2 are two independent renewal processes with a common density function f for their intervals, and assume f is continuous on $(0,\infty)$ and right continuous at 0. If $N = N_1 + N_2$ is a renewal process, then both N_1 and N_2 are Poisson processes.

Proof. Consider the joint density function $g(x,y)$ of the first two intervals in the superposition process. By elementary reasoning (one can make the following much more rigorous) we have

$$g(x,y) = 2f(x)\ f(y)\ \{1 - F(x+y)\} + 2f(x)\ f(x+y)\ \{1 - F(y)\}\ ,$$

where F is the distribution function whose density is f. Since N is a renewal process, $g(x,y) = g(y,x)$ which implies that

$$f(x)\ f(x+y)\ \{1 - F(y)\} = f(y)\ f(y+x)\ \{1 - F(x)\}.$$

Taking limits as $y \to 0$, we obtain

$$f(x)^2 = f(0)\ f(x)\ \{1 - F(x)\}\ .$$

Remark (2.18). One can use the same idea, now by considering the joint distribution of the first three intervals in the superposition process, to prove the more general case where N_1, N_2 do not necessarily have the same distribution.

So far it has always been assumed that the component processes are independent. Çinlar and Agnew (1968) have considered the superposition of non-independent processes. The main idea there is to use an indicator process which labels the points of the superposition process according to the components to which they belong. Let N_1, N_2 be two point processes and let $N = N_1 + N_2$. If $T_1, T_2, \ldots$ are the points of N, we define

$$X_n = k \tag{2.19}$$

if and only if T_n belongs to N_k, $k = 1,2$; (since by our hypothesis, N has no multiple points there is no ambiguity here). The process $X = \{X_1, X_2, \ldots\}$ is called the indicator process for (N_1, N_2). It is clear that the sample paths of N_1 and N_2 are determined by those of X and N and vice-versa. The following is a converse to the well known result that if N is Poisson, X Bernoulli, and N and X are independent, then N_1, N_2 are independent Poisson processes.

Proposition (2.20). Suppose X is a Bernoulli process independent of N. Then, N_1 and N_2 are independent only if N is a (possibly non-stationary) Poisson process.

Proof. Let A be a finite union of disjoint bounded intervals and let g denote the generating function of N(A). Since X is Bernoulli independent of N, the generating function of $N_1(A)$ is $g(p + qz)$ and that of $N_2(A)$ is $g(q + pz)$. Since N_1, N_2 are independent

$$g(z) = g(p + qz)\, g(q + pz) \qquad |z| \leq 1 .$$

Then one can show that $(d/dz) \log g(z)$ must be constant; this implies that N(A) has a Poisson distribution with some parameter $\lambda(A)$. Since A is an arbitrary finite union of disjoint intervals, this is possible only if N is a Poisson process (cf. Renyi, 1967).

By the same type of computational methods one can show the following

Proposition (2.21). Suppose X is a Markov chain independent of N, and suppose the intensities of N_1 and N_2 are not the same. Then, N is a renewal process if and only if N_1 and N_2 are independent.

If X is a two state Markov chain, it is well known that times of visits to any one state form a recurrent process. The converse also holds and can be used to prove the following

Theorem (2.22). Suppose X is independent of N. If N_1, N_2, N are renewal processes, then X is a Markov chain.

Proof. Let $\tau(1), \tau(2), \ldots$ be the succesive $n \geq 1$ for which $X_n = 1$. If $T_1, T_2, \ldots$ are the renewal times for N, then $T_{\tau(1)}, T_{\tau(2)}, \ldots$ are the renewal times for N_1. For $\lambda_i \geq 0$ $(i = 1, \ldots, n)$ we have by conditioning on the τ_i

$$E[\exp\{-\sum_{i=1}^{n} \lambda_i (T_{\tau(i+1)} - T_{\tau(i)})\}] = E\{\prod_{i=1}^{n} z_i^{\tau(i+1)-\tau(i)}\} ,$$

where $z_i = E\{\exp(-\lambda_i T_1)\}$. Since N_1 is a renewal process, the left hand side is also equal to

$$\prod_{i=1}^{n} E\{z_i^{\tau(i+1)-\tau(i)}\} \quad .$$

Hence, since z_i takes on at least every value in (0,1) as λ_i varies, we must have that $\tau(i+1) - \tau(i)$, $i = 1,2,\ldots$, are independent and identically distributed. One can in fact show that $\tau(1)$ is also independent of all the rest.

The next step is to show that this implies the assertion about $\{X_n\}$. Consider the event $A = \{X_1 = j_1, \ldots, X_{n=1} = j_{n-1}, X_n = 1\}$; let $n_1, \ldots, n_k$ be the successive m for which $j_m = j_n$; and note that $n_k = n$ and we have

$$A = \{\tau(1) = n_1, \ldots, \tau(k-1) = n_{k-1}, \tau(k) = n\} \ .$$

Therefore

$$P(X_{n+1} = j|A) = \begin{cases} P\{\tau(k+1) = n+1|\tau(k) = n\} & \text{if } j = 1 \ , \\ 1 - P\{\tau(k+1) = n+1|\tau(k) = n\} & \text{if } j = 2, \end{cases}$$

by the fact that $\tau(k+1) - \tau(k)$ is independent of $\tau(1), \ldots, \tau(k)$. But

$$P\{\tau(k+1) = n+1|\tau(k) = n\} = P(X_{n+1} = 1|X_n = 1) \ ,$$

and hence we have that

$$P(X_{n+1} = j|X_1 = j_1, \ldots, X_n = j_n) = P(X_{n+1} = j|X_n = j_n) \qquad (2.23)$$

if $j_n = 1$. By interchanging the roles of N_1 and N_2 throughout we get (2.23) for $j_n = 2$ also. This completes the proof.

Most of the difficulty in proving theorems of the type (2.1), (2.5), (2.6) is circumvented by assuming X is independent of N. The following result due to Störmer (1969) illustrates this (Störmer actually assumes only that N_1, N_2, N are *delayed* renewal processes, but the added generality does not add any difficulties).

Proposition (2.24). Suppose N_1, N_2, $N = N_1 + N_2$ are all renewal processes. Assume X is independent of N, and N_1 is independent of N_2, and N_2 has a finite mean interval length. Then, N_1 and N_2 are both Poisson.

Proof. Let U_n be the time of the nth renewal in the process N_1. Then since N_1, N_2 are independent,

$$P\{N(U_n, U_n+t] = 0 | U_n\} = \{1 - F_1(t)\} \int_0^{U_n} R_2(du) \{1 - F_2(U_n+t-u)\},$$

where F_i in the interval distribution for N_i and $R_i = \sum_0^\infty F_i^{(n)}$. On the other hand, since X is independent of N, the distribution of the interval starting at U_n is the same as that of an arbitrary interval: that is,

$$P\{N(U_n, U_n+t] = 0 | U_n\} = 1 - F(t) = \{1 - F_1(t)\}\{1 - F_2(t)\} . \quad (2.25)$$

Equating the two equations obtained and letting $n \to \infty$, we get

$$\begin{aligned} 1 - F_2(t) &= \lim_{n\to\infty} \int_0^{U_n} R_2(du) \{1 - F_2(U_n+t-u)\} \\ &= \frac{1}{\mu_2} \int_t^\infty \{1 - F_2(y)\} dy , \end{aligned}$$

by the key renewal theorem since $U_n \uparrow \infty$ as $n \uparrow \infty$. But this last equation is possible only if F_2 is the exponential distribution with parameter $1/\mu_2$.

In general, however, one cannot write (2.25) without some argument about the independence of the length of the next interval from the indicator of the point involved. This is in fact true (cf. Proposition (2.17) and remark (2.18) following it), but we do not know of a direct proof.

3. CONVERGENCE TO POISSON PROCESSES

In this section we show that the superposition of a large number of independent and uniformly sparse point processes is approximately a Poisson process. The essential intuitive basis for the theorems to follow can be explained best in the case of stationary component processes. The conditions we impose are such that, in the limit, the number of processes being superposed increases while each individual component contributes less and less so as to keep the intensity of the superposition process constant. This way, no one component can contribute more than one point to the superposition in any finite interval [0,t]. Suppose, there are n_t points in the interval [0,t]; then each belongs to a different component process; the positions of these n_t points are independent of each other by the independence of the components; and each position has the uniform distribution over [0,t] by the stationarity of the component it belongs to. The number n_t must be so that n_t/t is the intensity of the superposition as $t \to \infty$. But this is precisely the description of the Poisson process as "the random distribution of points on the line",(cf. Doob, 1953, p. 405). We will shortly give a precise and general treatment of results of this nature.

The convergence theorems involved are generalizations of the now-classical results for the convergence of sums of random variables to the case of sums of stochastic point processes. The earliest such result is due to Gnedenko (1938) and Marcinkiewicz (1938) who gave, independently and almost simultaneously, necessary and sufficient conditions for the convergence of the distribution of the sum of infinitesimal independent random variables to the Poisson distribution.

An attempt to explain the unreasonably frequent occurrence of the Poisson process in practical experience led Palm (1943) to consider the superposition of a large number of stationary point processes each with a small intensity. He showed that the intervals between the points of the superposed process have an exponential distribution. Led by applications in neurophysiology, Cox and Smith (1953, 1954) considered the superpositions of strictly periodic processes and renewal processes.

In the case of a large number of summands, they showed that the intervals in the superposition process have exponential distributions.

Both Palm (1943) and Cox and Smith (1953, 1954) point that the superposition in the limit is a Poisson process. The first proof of this result was given by Khintchine (1960) (which appeared in Russian in 1955). It is limited to stationary point processes and gives only sufficient conditions, but is very direct and original. Using the methods of Khintchine, Ososkov (1956) gave equivalent sufficient criteria for convergence and further showed their necessity. Franken (1963a) and Grigelionis (1963) extended these results by considering arbitrary point processes as components and gave necessary and sufficient conditions for convergence to a (possibly non-stationary) Poisson process. Franken (1963a) and Grigelionis (1966a) contain equivalent and simpler conditions in the case of renewal processes as summands. Generalizations to random counting measures were considered in Goldman (1967) and Jagers (1971).

In this section we will give a fairly comprehensive survey of these results. As the main result we choose to give the theorem due to Franken (1963a) and Grigelionis (1963) whose necessity we prove by appealing to the Gnedenko-Marcinkiewicz theorem (as did Grigelionis, 1963) and whose sufficiency we prove by the method of Franken (1963a) (which seems to us to be the simplest). The proof is such that it holds, word for word, also for the convergence of random measures, and we chose to use the language of point processes only because it is more familiar. Then, specializing to stationary processes we obtain the results of Ososkov (1956) and Khintchin (1960), and by specializing to renewal processes those of Franken (1963a) and Grigelionis (1963).

Before formulating the problem we will describe a distance function on the space of probability distributions and list some results which will be of use in the proof of our main theorem.

If ϕ and Ψ are two probability distributions on a *discrete* space E, we introduce the *distance* between them as

$$d(\phi,\Psi) = \sum_{x \in E} \left|\phi(x) - \Psi(x)\right| , \tag{3.1}$$

(that is, the total variation of the signed measure $\phi - \Psi$). It is clear that $d(\phi,\Psi) = 0$ if and only if $\phi = \Psi$, and that the triangle inequality holds;

$$d(\phi,\Psi) \leq d(\phi,\eta) + d(\eta,\Psi) \tag{3.2}$$

for any distribution η on E.

If ϕ_1, ϕ_2 are two probability distributions on E, F (both discrete spaces), then $\phi_1 \times \phi_2(x,y) = \phi_1(x)\ \phi_2(y)$ is a probability distribution on E × F. It follows from (3.2) with $\eta = \phi_2 \times \Psi_1$ that

$$d(\phi_1 \times \phi_2, \Psi_1 \times \Psi_2) \leq d(\phi_1,\Psi_1) + d(\phi_2,\Psi_2) \quad . \tag{3.3}$$

From the definition $\phi_1 * \phi_2(x) = \sum_{y+z=x} \phi_1 \times \phi_2(y,z)$ of convolution we have, through (3.3),

$$d(\phi_1 * \phi_2, \Psi_1 * \Psi_2) \leq d(\phi_1,\Psi_1) + d(\phi_2,\Psi_2) \quad . \tag{3.4}$$

Example (3.5). Let Ψ_a denote the Poisson distribution on $\{0,1,2,...\}$ with parameter a; and consider the distance between Ψ_a and Ψ_b for $a > b$; i.e., consider

$$d(\Psi_a,\Psi_b) = \sum_{i=0}^{\infty} |e^{-a}a^i - e^{-b}b^i|/i! \quad .$$

Since we can write $\Psi_a = \Psi_b * \Psi_{a-b}$, $\Psi_b = \Psi_b * \Psi_0$, we have from (3.4) that

$$d(\Psi_a,\Psi_b) \leq d(\Psi_{a-b},\Psi_0) = |e^{-(a-b)}-1| + 1 - e^{-(a-b)} \leq 2(a-b) \ .$$

Example (3.6). More generally, if $a = \{a(1),..., a(m)\}$ and $b = \{b(1),..., b(m)\}$, and if Ψ_a and Ψ_b are the m-dimensional Poisson distributions with parameters a and b, then we get from (3.3) and (3.5) that

$$d(\Psi_a,\Psi_b) \le 2 \sum_{j=1}^{m} |a(j) - b(j)| .$$

The following is a useful approximation result.

Proposition (3.7). Let M be a point process and B the union of disjoint bounded sets $B_1,\ldots, B_m \in R$. Let ϕ be the distribution of the random vector $\{M(B_1),\ldots, M(B_m)\}$ and Ψ the m-dimensional Poisson distribution with parameter $\{\mu(B_1),\ldots, \mu(B_m)\}$ where

$$\mu(B_j) = P\{M(B_j) = 1,\ M(B - B_j) = 0\} .$$

Then,

$$d(\phi,\Psi) \le 2P\{M(B) \ge 2\} + 2[P\{M(B) = 1\}]^2 .$$

Proof. Let $p = P\{M(B) = 1\}$ and $q = P\{M(B) \ge 2\}$; then from the definition of the distance,

$$\begin{aligned} d(\phi,\Psi) &= |(1-p-q) - e^{-p}| + \sum_{j=1}^{m} |\mu(B_j) - \mu(B_j)\, e^{-p}| + q + 1 - e^{-p} - pe^{-p} \\ &= e^{-p} - 1 + p + q + p - pe^{-p} + q + 1 - e^{-p} - pe^{-p} \\ &= 2q + 2p(1-e^{-p}) \le 2q + 2p^2 \end{aligned}$$

as claimed.

We next formulate the main limit theorem. We consider a double sequence

$$\begin{matrix} M_{11}, & \ldots, & M_{1k_1} & & \\ \vdots & & & \ddots & \\ M_{n1}, & \ldots, & M_{ni}, & \ldots, & M_{nk_n} \end{matrix}$$

of point processes where $k_n \to \infty$ as $n \to \infty$. We assume that all the processes on the same row are mutually independent, and let M_n be the superposition of the processes on the nth row, i.e.,

$$M_n = M_{n1} + \cdots + M_{nk_n} \qquad (n = 1,2,\ldots) \quad . \tag{3.8}$$

The processes M_{ni} are called *infinitesimal* if

$$\lim_{n\to\infty} \sup_{1\le i\le k_n} P\{M_{ni}(B) \ge 1\} = 0 \tag{3.9}$$

for any bounded interval $B \in \mathcal{R}$ (or in particular for $B = [0,t]$).

The following is the main limit theorem which is due to Franken (1963a) and Grigelionis (1963); our proof uses Franken's method.

Theorem (3.10). Suppose the M_{ni} are infinitesimal, that is

$$\lim_{n\to\infty} \sup_{1\le i\le k_n} P\{M_{ni}(B) \ge 1\} = 0 \tag{3.9'}$$

for any bounded interval $B \in \mathcal{R}$. Then, the superposition process M_n converges weakly to a Poisson process with mean measure μ if and only if

$$\lim_{n\to\infty} \sum_{i=1}^{k_n} P\{M_{ni}(B) = 1\} = \mu(B) \tag{3.11}$$

and

$$\lim_{n\to\infty} \sum_{i=1}^{k_n} P\{M_{ni}(B) \ge 2\} = 0 \tag{3.12}$$

for any finite interval $B \in \mathcal{R}$.

Remark. It is clear that when (3.12) holds, in condition (3.11) the event $\{M_{ni}(B) = 1\}$ can be replaced by the event $\{M_{ni}(B) \ge 1\}$.

Remark. For point processes, (3.9) and (3.12) hold for any bounded $B \in \mathcal{R}_+$ if and only if they hold for $B = (0,t]$ for arbitrary $t > 0$.

Proof. We need to show that the joint distribution ϕ_n of $M_n(B_1)$, ..., $M_n(B_m)$ converge as $n \to \infty$ to the Poisson distribution Ψ with parameter $\mu = \{\mu(B_1), \ldots, \mu(B_m)\}$ if and only if (3.11) and (3.12) hold for any disjoint collection $B_1, \ldots, B_m \in R$. It is enough to do this for intervals $B_1, \ldots, B_m$ whose union B is again an interval. We take them so and fix them throughout this proof.

Necessity follows on taking m = 1 from the classical Gnedenko-Marcinkiewicz theorem whose conditions are precisely the ones we list except for the last two which are satisfied automatically since we allow only integer values (cf. Gnedenko and Kolmogorov, 1968, p. 132).

To prove the sufficiency, first note that the convergence of ϕ_n to Ψ is implied if we show that $d(\phi_n, \Psi) \to 0$. We introduce the notations

$$\mu_{ni}(B_j) = P\{M_{ni}(B_j) = 1,\ M_{ni}(B - B_j) = 0\}\ , \tag{3.13}$$

$$\mu_{ni} = \{\mu_{ni}(B_1), \ldots, \mu_{ni}(B_m)\}, \tag{3.14}$$

$$\mu_n = \mu_{n1} + \cdots + \mu_{nk_n}\ , \tag{3.15}$$

and let Ψ_n and Ψ denote the m-dimensional Poisson distributions with parameter vectors μ_n and μ respectively. Then, by the triangle inequality (3.2), we have

$$d(\phi_n, \Psi) \leq d(\phi_n, \Psi_n) + d(\Psi_n, \Psi)\ . \tag{3.16}$$

By the theorem of total probability

$$\sum_{i=1}^{k_n} P\{M_{ni}(B_j) = 1\}$$

$$= \sum_{i=1}^{k_n} P\{M_{ni}(B_j)=1,\ M_{ni}(B)=1\} + \sum_{i=1}^{k_n} P\{M_{ni}(B_j)=1,\ M_{ni}(B) \geq 2\}\ .$$

Observing that the second term is smaller than $\sum_{i=1}^{k_n} P\{M_{ni}(B) \geq 2\}$ and taking limits on both sides we have

$$\mu(B_j) = \lim_{n\to\infty} \mu_n(B_j)$$

by the conditions (3.11), (3.12) and the expressions (3.13)-(3.15). Thus, using the result of example (3.6),

$$0 \le d(\Psi_n, \Psi) \le 2 \sum_{j=1}^{k_n} \left|\mu_n(B_j) - \mu(B_j)\right| \to 0 \,, \qquad (n \to \infty) \,.$$

On the other hand, using the independence of $M_{n1}, \ldots, M_{nk_n}$ and (3.15), we can apply (3.4) to $d(\phi_n, \Psi_n)$ and then apply Proposition (3.7) to each term M_{ni} and thus obtain

$$d(\phi_n, \Psi_n) \le 2 \sum_{i=1}^{k_n} [P\{M_{ni}(B) \ge 2\} + (P\{M_{ni}(B) = 1\})^2]$$

$$\le 2 \sum_{i=1}^{k_n} P\{M_{ni}(B) \ge 2\}$$

$$+ 2[\sup_{1\le i\le k_n} P\{M_{ni}(B) = 1\}] [\sum_{i=1}^{k_n} P\{M_{ni}(B) = 1\}] \,.$$

The hypothesis (3.9') together with the conditions (3.11) and (3.12) imply that the right side approaches zero; that is, $d(\phi_n, \Psi_n) \to 0$ as $n \to \infty$. Together with (3.16) and (3.17), this implies that $d(\phi_n, \Psi) \to 0$ as $n \to \infty$, thus completing the proof.

Remark (3.18). The fact that the processes M_{ni} were all point processes defined on the real line played no role in our proof. The theorem holds also for point processes defined on $(\mathbb{R}^n, \mathcal{R}^n)$ or, more generally, on measurable spaces $(E, \mathcal{E})$ where E is any locally compact space with a countable base and $\mathcal{E}$ is the σ-algebra generated by the open subsets of E. Both in the wording and the proof we then need to replace "bounded intervals" by "bounded Borel sets". In this sense, (3.10) is slightly stronger than that of Goldman (1967) who proved that (3.11) is necessary and sufficient for the convergence assuming that (3.9) and (3.12) hold.

On the other hand, the condition (3.11) can be replaced by the slightly weaker condition that, if

$$\lambda_n(A,B) = \sum_{i=1}^{k_n} P\{M_{ni}(A \cap B) = 1,\ M_n(B) = 1\}$$

for B compact and A ε *E*, the measures $\lambda_n(\cdot,B)$ be weakly convergent. Then, the resulting convergence is also slightly weaker (cf. Jagers, 1971, for details).

Next we consider the special case of stationary point processes as components.

Corollary (3.19). Suppose each M_{ni} is stationary and let λ_{ni} be its intensity, ie., $\lambda_{ni} = E\{M_{ni}(1)\}$. If

$$\lim_{n\to\infty} \sup_{1\le i\le k_n} \lambda_{ni} = 0 \tag{3.20}$$

holds, then the conditions

$$\lim_{n\to\infty} \sum_{i=1}^{k_n} \lambda_{ni} = \lambda \tag{3.21}$$

and

$$\lim_{n\to\infty} \sum_{i=1}^{k_n} P\{M_{ni}(t) \ge 2\} = 0 \qquad (t > 0) \tag{3.22}$$

are necessary and sufficient for the superposition $M_n = \sum_{i=1}^{k_n} M_{ni}$ to converge weakly to a stationary Poisson process with intensity λ.

Remark. This result captures the essence of these convergence theorems in its most basic form. In the limit, because of the hypotheses (3.20), the component processes become uniformly sparse and no one component, by (3.22), can contribute more than one point to the superposition. We had already mentioned the manner in which these lead to a Poisson process.

Proof. We have

$$\lambda_{ni}t = E[M_{ni}(t)] = \sum_{m=1}^{\infty} P\{M_{ni}(t)\ge m\} \ge P\{M_{ni}(t)\ge 1\} + P\{M_{ni}(t)\ge 2\} . \tag{3.23}$$

Thus, (3.21) and (3.22) imply that

$$\lim_{n\to\infty} \sum_{i=1}^{k_n} P\{M_{ni}(t) \geq 1\} = \lambda t .$$

In view of (3.22) this is equivalent to (3.11). The proof follows now from Theorem (3.10) since, by (3.23), the assumption (3.20) implies the infinitesimality assumption (3.9).

Remark (3.24). It is clear from the proof that the conditions (3.21) and (3.22) together are equivalent to the conditions (3.21) and

$$\lim_{n\to\infty} \sum_{i=1}^{k_n} P\{M_{ni}(t) \geq 1\} = \lambda t \qquad (t \geq 0) .$$

Remark (3.25). Both Khintchine (1960) and Ososkov (1956) assumed that $\Sigma_i \lambda_{ni} = \lambda$ for all n. Khintchine showed that (3.20) and (3.22) are sufficient, and Ososkov showed that given (3.20), (3.22) is also necessary.

Suppose next that each M_{ni} is a renewal process with inter-renewal time distribution F_{ni}. Then, the conditions (3.9), (3.11), (3.12) are equivalent to

$$\lim_{n\to\infty} \sup_{1\leq i\leq k_n} F_{ni}(t) = 0 , \tag{3.26}$$

$$\lim_{n\to\infty} \sum_{i=1}^{k_n} \int_0^s R_{ni}(dx) \{F_{ni}(t-x) - F_{ni}(s-x)\} = \mu(s,t] , \tag{3.27}$$

$$\lim_{n\to\infty} \sum_{i=1}^{k_n} F_{ni} * F_{ni}(t) = 0 , \tag{3.28}$$

for all s, t $\geq$ 0 with s < t (here R_{ni} is the renewal function $\Sigma_{k=0}^{\infty} F_{ni}^{(k)}$ corresponding to F_{ni}). Since $F * F \leq F^2$, (3.26) implies (3.28); further, (3.26) implies that $\lim_n \sup_i R_{ni}(t) = 1$ for all $t \geq 0$ so that the integral in (3.27) becomes $F_{ni}(t) - F_{ni}(s)$. Thus we have

Corollary (3.29). Suppose each M_{ni} is a renewal process with distribution function F_{ni} for its intervals. Suppose

$$\lim_{n\to\infty} \sup_{1\le i\le k_n} F_{ni}(t) = 0 \qquad (t \ge 0) . \tag{3.26'}$$

Then, M_n converges weakly to a Poisson process with parameter function $\mu(t)$ if and only if

$$\lim_{n\to\infty} \sum_{i=1}^{k_n} F_{ni}(t) = \mu(t) . \tag{3.30}$$

For the case of stationary renewal processes we get the following simple result.

Corollary (3.31). Suppose each M_{ni} is a stationary renewal process with mean interval length $1/\lambda_{ni}$. If

$$\lim_{n\to\infty} \sup_{1\le i\le k_n} F_{ni}(t) = 0 \qquad (t \ge 0) , \tag{3.32}$$

then the condition

$$\lim_{n\to\infty} \sum_{i=1}^{k_n} \lambda_{ni} = \lambda \tag{3.33}$$

is necessary and sufficient for M_n to converge weakly to a stationary Poisson process with intensity λ.

Proof. The hypothesis (3.32) implies (3.20); and (3.33) is the same as (3.21). Necessity part therefore follows immediately from Corollary (3.19). To prove the sufficiency, again using Corollary (3.19), we need only to show that (3.32) and (3.33) imply (3.22). But now,

$$\begin{aligned}\sum_i P\{M_{ni}(t) \ge 2\} &= \sum_i \lambda_{ni} \int_0^t \{1 - F_{ni}(y)\} F_{ni}(t-y)\, dy \\ &\le \sum_i \lambda_{ni} t F_{ni}(t) \\ &\le \{\sup_i F_{ni}(t)\} \{\sum_i \lambda_{ni} t\} \to 0 \qquad (n \to \infty) .\end{aligned}$$

4. CONVERGENCE TO MARKED POISSON PROCESSES

Theorem (3.10) strengthened by the remark (3.18) following it can be used to prove further results which are more general in appearance. In this section we will give three such results. First is a generalization of a theorem in Çinlar (1968) on the convergence of m-variate point processes to an m-variate Poisson process. By an m-variate point process is meant a vector valued stochastic process $\underline{N}(t) = \{N^1(t), \ldots, N^m(t)\}$ where $N(t) = N^1(t) + \cdots + N^m(t)$ is a point process. One may think of it as a sequence of points on the line each one having one of the colors (or types of events) $1, \ldots, m$. By an m-variate Poisson process with mean measure $\mu(j,B)$ is meant an m-variate point process $\underline{N} = (N^1, \ldots, N^m)$ where $N^1, \ldots, N^m$ are independent Poisson processes with mean measures $\mu(1,A), \ldots, \mu(m,A)$ respectively.

We now consider a double sequence $\{\underline{N}_{ni};\ n = 1,2,\ldots;\ i = 1,\ldots, k_n\}$ of m-variate point processes, define $N_{ni} = N^1_{ni} + \cdots + N^m_{ni}$, and let

$$\underline{N}_n = \underline{N}_{n1} + \cdots + \underline{N}_{nk_n} .$$

We assume that, for each n, $\underline{N}_{n_1}, \ldots, \underline{N}_{nk_n}$ are independent.

Proposition (4.1). Suppose

$$\lim_{n\to\infty} \sup_{1\leq i\leq k_n} P\{N_{ni}(t) \geq 1\} = 0 \tag{4.2}$$

for every $t \to 0$. Then, the superposition $\underline{N}_n$ converges weakly to an m-variate Poisson process with mean measure $\mu(j,A)$ if and only if

$$\lim_{n\to\infty} \sum_{i=1}^{k_n} P\{N^j_{ni}(A) \geq 1\} = \mu(j,A) \qquad (j = 1,\ldots, m) \tag{4.3}$$

and

$$\lim_{n\to\infty} \sum_{i=1}^{k_n} P\{N_{ni}(t) \geq 2\} = 0 . \tag{4.4}$$

Proof. Necessity follows from Theorem (3.10) upon noting that the jth component N_n^j of the vector $\underline{N}_n$ must converge to a Poisson process with mean measure $\mu(j,B)$ and that $N_n = N_n^1 + \cdots + N_n^m$ must converge to a Poisson random measure with mean measure $\mu(1,B) + \cdots + \mu(m,B)$.

To prove the sufficiency let $K = \{1,2,\ldots, m\}$, $\mathcal{K}$ be the set of all subsets of K, and $E = K \times \mathbb{R}$, $\mathcal{E} = \mathcal{K} \times \mathcal{R}$. For each $j \in K$ and $B \in \mathcal{R}$, define

$$M_{ni}(\{j\} \times B) = N_{ni}^j(B) \qquad (n=1,2,\ldots,\ i=1,\ldots,\ k_n) \quad .$$

We need to show that if (4.2), (4.3), (4.4) hold, the process $M_n = M_{n1} + \cdots + M_{nk_n}$ converges weakly to a Poisson random measure on $\mathcal{K} \times \mathcal{R}$ with mean measure $\mu(A \times B) = \Sigma_{j\in A}\ \mu(j,B)$ for every $A \in \mathcal{K}$, $B \in \mathcal{R}$. But this follows from Theorem (3.10) (strengthened by our remark (3.18)) once we note that for any subset A of K and bounded set $B \in \mathcal{R}$, we have

$$P\{M_{ni}(A \times B) \geq 1\} \leq P\{N_{ni}(B) \geq 1\}$$

$$P\{M_{ni}(A \times B) \geq 2\} \leq P\{N_{ni}(B) \geq 2\} \ .$$

If all the processes are taken to be stationary, then the resulting limit process is also stationary. Let λ_{ni}^j, λ_{ni} be the intensities of N_{ni}^j and N_{ni} respectively (then, $\lambda_{ni} = \lambda_{ni}^j + \cdots + \lambda_{ni}^m$). In view of (3.19) we can see that the following holds.

Corollary (4.5). Suppose each $\underline{N}_{ni}$ is stationary and assume

$$\lim_{n\to\infty}\ \sup_{1\leq i\leq k_n}\ \lambda_{ni} = 0 \ . \tag{4.6}$$

Then, $\underline{N}_n$ converges to an m-variate Poisson process with mean rates $\lambda^1,\ldots,\ \lambda^m$ if and only if

$$\lim_{n\to\infty} \sum_{i=1}^{k_n} \lambda_{ni}^j = \lambda^j \qquad (j = 1,\ldots, m) \tag{4.7}$$

and

$$\lim_{n\to\infty} \sum_{i=1}^{k_n} P\{N_{ni}(t) \geq 2\} = 0 . \tag{4.8}$$

An application of this result to the superposition of Markov renewal processes was given in Cinlar (1968).

The technique we have used in the proof of Proposition (4.1) can be applied also to the case of infinite dimensional processes (in which case one needs the further condition that $\Sigma_{j=1}^{\infty} \mu(j,A) = \mu(A) < \infty$). Indeed, it is possible to think of further generalizations for which the proof will go through with minor, obvious and mainly notational changes. We describe the general case next.

Let K be an arbitrary set and $\mathcal{K}$ a σ-algebra on K. Following Matthes (1963) we call a process $\underline{N} = \{N(A,t);\ A \in \mathcal{K},\ t \geq 0\}$ a *marked point process* provided that $N^A = \{N(A,t);\ t \geq 0\}$ is a point process for any $A \in \mathcal{K}$. For any such process $\underline{N}$ we denote N^K simply by N. Such processes are studied extensively in a sequence of papers by Matthes (1963), Kerstan and Matthes (1964), and Franken, Liemant, Matthes (1965). In the special case where K = {1,..., m} we get the m-variate point processes. For purposes of visualization, we think of the point process formed by the times electrons are emitted at the cathode of an electron tube and with each such instant we associate a "mark" giving the speed of the electron; then $N\{(a,b),t\}$ would be the number of electrons emitted in the time interval (0,t] whose speeds were greater than *a* but less than *b*.

A marked point process $\underline{N}$ is called a marked Poisson process with mean measure μ_A if $N^{A_1},\ldots, N^{A_m}$ are independent Poisson processes with mean measures $\mu_{A_1},\ldots, \mu_{A_m}$ whenever the sets $A_1,\ldots, A_m \in \mathcal{K}$ are disjoint.

The result we are putting next can be proved along the lines of the proof of (4.1). The result was first proved by Grigelionis (1966b) and includes as special cases both (4.1) and the results of Cinlar (1968) and Sobel (1971) on this problem.

Theorem (4.9). Let $\underline{N}_{ni}$ $(n = 1,2,\ldots;\ i = 1,\ldots,k_n)$ be a double sequence of marked point processes with marks in the space $(K,\mathcal{K})$. Suppose

$$\lim_{n\to\infty} \sup_{1\le i\le k_n} P\{N_{ni}(t) \ge 1\} = 0 \qquad (t \ge 0)\ . \tag{4.10}$$

In order for $\underline{N}_n = \Sigma_{i=1}^{k_n} \underline{N}_{ni}$ to converge weakly to a marked Poisson process with mean measure $\mu_A(B)$ it is necessary and sufficient that

$$\lim_{n\to\infty} \sum_{n=1}^{k_n} P\{N_{ni}(A \times B) \ge 1\} = \mu_A(B)\ , \tag{4.11}$$

$$\lim_{n\to\infty} \sum_{i=1}^{k_n} P\{N_{ni}(t) \ge 2\} = 0\ . \tag{4.12}$$

5. CONVERGENCE TO CLUSTERED POISSON PROCESSES

A clustered Poisson process is a collection of random points of two types, usually called primary and secondary; primary points form a Poisson process; and each primary point serves as "center" to a "cluster" of secondary points. Clusters are taken to be independent of each other and the law describing the positions of points belonging to the same cluster, in general, depends only on the position of the primary point serving as its center. Neyman and Scott (1958) have proposed the model to describe stellar distributions; it is also applicable in geography (for clusters of communities around a central place), in bacteriology (for bacterial colonies), etc. When the points are all on the real line, we get models for traffic flow (leader-follower models: where the instants the "leaders" pass a given spot form the sequence of primary points and their followers are the secondary points forming clusters behind them). In analyzing some data on times of computer failures, Lewis (1964) has proposed a similar model. In his model, the primary points form a stationary Poisson process and each such point serves as the origin for a "cluster" of secondary points forming a terminating renewal process (Lewis calls the resulting process a "branching Poisson process"). This problem serves as the best example as far as motivating our convergence result in this section, and therefore we give a fairly detailed description of it.

A computer is made up of a large number of components. The computer is said to fail whenever any one component fails to perform its prescribed function. A computer does not use all of its components all the time and thus can go on working after the initial failure of the particular component even if it had not yet been repaired. In such a case, the computer will fail again the next time the failed component is needed for its correct operation. This process continues on until the failed component is located and removed.

Consider now a fixed component. Its first failure time $T(1)$ is followed by a number of secondary failures (due to repeated attempts to use it) at $T(2), T(3), \ldots, T(N_1)$; at $T(N_1)$ the component is finally removed and replaced, so it works for some time again without failure and then fails again at $T(N_1+1)$, a number of attempts to use it is made at $T(N_1+2), T(N_1+3), \ldots$ and a replacement at $T(N_2)$ occurs, and so on. It is reasonable to assume that the intervals between failures are independent but they cannot all have the same distribution: presumably, lifetime of the component is "long" when compared with the times between successive uses of it.

Let $X(n) = 1$ or 2 according as the nth failure due to that fixed component is a primary or a secondary failure. Then, $\{X(n), T(n)-T(n-1); n = 1,2,\ldots\}$ is a sequence of independent pairs so that, for $n \geq 2$,

$$P\{X(n) = 2,\ T(n) - T(n-1) \leq t\} = G(t) ,$$

$$P\{X(n) = 1,\ T(n) - T(n-1) \leq t\} = \{1 - G(\infty)\} F(t) , \tag{5.1}$$

and for $n = 1$,

$$P\{X(1) = 1,\ T(1) \leq t\} = F(t) .$$

Here F is the lifetime distribution for the component and $G(t)/G(\infty)$ is the distribution of the time between successive instants that component is needed; the defect $1 - G(\infty)$ then can be interpreted as the probability of repair. We assume

$$F(+\infty) = 1, \qquad G(+\infty) \leq 1 .$$

A process $\{X(n),\ T(n);\ n \geq 1\}$ of the type described will be called an (F,G)-process. The point process N whose points are the T(n) of an (F,G)-process $\{X(n),\ T(n);\ n \geq 1\}$ will also be called an (F,G)-process.

Failures due to ith component form an (F_i, G_i)-process N_i, $1 \leq i \leq k$, and the failure times for the computer is the superposition $N = N_1 + \cdots + N_k$ of these processes. We will show below that for large k, under reasonable conditions, N is approximately a clustered Poisson process. This, then, provides a theoretic justification for the agreement between the data and model of Lewis (1964) and many authors in traffic theory. A further generalization leading to clustered Poisson process on arbitrary spaces will be reported on separately.

An explanation of facts of the kind we are considering was undertaken by Kovalenko (1965). Though his component processes N_i are of simpler structure than ours, because of the workings of the limit theorem, our limiting results are the same. Our proof here takes advantage of the results of the preceding section.

Before we list the main result we now describe the limiting clustered Poisson process in some detail. Let $\mathbb{K}$ be the set of all (possibly defective) distribution functions G defined on $[0,\infty)$; that is, each $G \in \mathbb{K}$ is a non-decreasing right continuous function satisfying

$$0 \leq G(0) \leq G(+\infty) \leq 1 \quad . \tag{5.2}$$

A sequence G_n in $\mathbb{K}$ is said to converge weakly to $G \in \mathbb{K}$ if $G_n(x) \to G(x)$ for all continuity points x of G. This introduces a topology on $\mathbb{K}$ and we let $\mathcal{K}$ denote its Borel sets (that is, $\mathcal{K}$ is the σ-algebra generated by the open sets of $\mathbb{K}$). Finally, let $Q(t,A)$ be a transition probability from $(\mathbb{R},\mathcal{R})$ into $(\mathbb{K},\mathcal{K})$, that is,

a) for each $t \in \mathbb{R}$, the mapping $A \to Q(t,A)$ is a probability measure on $\mathcal{K}$;

b) for each $A \in \mathcal{K}$, the mapping $t \to Q(t,A)$ is $\mathcal{R}$-measurable.

Next let $\ldots, T_{-1}, T_0, T_1, \ldots$ be the points of a Poisson process with mean measure μ. For each n, let G_n be "sampled from $\mathbb{K}$ according to the distribution $A \to Q(T_n,A)$" independent of all the other G_i and T_i, that is

$$P\{G_n \in A | T_i, G_j; i, j = 0, \pm 1, \ldots; j \neq n\} = Q(T_n,A) \tag{5.3}$$

for each $n = 0, \pm 1, \ldots$ and $A \in \mathcal{K}$. For each n, let a renewal process $U_{n0}, U_{n1}, U_{n2}, \ldots$ be constructed with G_n as the interval distribution, that is,

$$0 = U_{n0} \le U_{n1} \le U_{n2} \le \cdots .$$

This renewal process is possibly terminating since $G_n(+\infty)$ might be less than one, and the successive differences between the U_{ni} are independent and identically distributed with the common distribution G_n.

Definition (5.4). The point process N whose points are $\{T_n + U_{ni};$ $n = 0, \pm 1, \ldots, i = 0,1,\ldots\}$ is called a *C*-process with primary measure μ and secondary kernel Q.

Example (5.5). Suppose there are k distribution functions $G_1, G_2, \ldots, G_k$ so that $\Sigma_{i=1}^{k} Q(t,\{G_i\}) = 1$ for all t, we then write $q(t,i)$ for $Q(t,\{G_i\})$. Then we have the following interpretation (in the language of the computer failure problem). There are k components; given that there is a primary failure at time t, the conditional probability that it is due to the failure of the ith component is $q(t,i)$. Then, the probability that the failure is repaired immediately is $1 - G_i(+\infty)$ and in that case there are no secondary failures. Otherwise, with probability $G_i(\infty)$ the failure goes uncorrected, and causes a secondary failure. $G_i(x)/G_i(+\infty)$

is the distribution of the time between the successive needs for that component, and $G_i(x)$ is therefore the probability that a secondary failure occurs before $t + x$. Thus, starting at the primary failure time t, we have a renewal process of secondary failure times with distribution function G_i.

As usual we consider a double sequence of point processes N_{ni}, $n = 1,2,\ldots$; $i = 1,\ldots, k_n$, where $k_n \to \infty$ as $n \to \infty$, and for each n the processes $N_{n1},\ldots, N_{nk_n}$ are independent. We write

$$N_n = N_{n1} + \cdots + N_{nk_n} \qquad (n = 1,2,\ldots) .$$

We now list the promised convergence theorem.

Theorem (5.6). Let N_{ni} be an (F_{ni},G_{ni})-process and suppose

$$\lim_{n\to\infty} \sup_{1\le i\le k_n} F_{ni}(t) = 0 \qquad (t = 0) . \tag{5.7}$$

Then, the superposition N_n converges weakly to a C-process with primary measure μ and secondary kernel Q if and only if

$$\lim_{n\to\infty} \sum_{i=1}^{k_n} F_{ni}(t)\, I_A(G_{ni}) = \int_0^t \mu(ds)\, Q(s,A) \tag{5.8}$$

for every set $A \in K$ of the form

$$A = \{G \in \mathbb{K}\colon G(x_i) \le y_i,\ i = 1,\ldots, m\}$$

for some m (here I_A is the indicator function of the set A).

Proof. Let $E = \mathbb{R} \times \mathbb{K}$, $E = R \times K$ and let M_{ni} be the random measure defined on (E,E) by the points $\{(T_{ni}(m), G_{ni});\ m = 1,2,\ldots\}$ where the $T_{ni}(1), T_{ni}(2),\ldots$ are the positions of the primary points in the N_{ni} process. For any $A \in K$

$$P[M_{ni}\{(0,t] \times A\} \geq 1] = F_{ni}(t)\ I_A(G_{ni})$$

$$P[M_{ni}\{(0,t] \times A\} \geq 2] \leq F_{ni} * [\sum_{m=0}^{\infty} \{1 - G(\infty)\} G_{ni}^{(m)}] * F_{ni}(t)$$

$$\leq \{F_{ni}(t)\}^2 .$$

Thus, (5.7) is equivalent to

$$\lim_{n\to\infty} \sup_{1\leq i\leq k_n} P[M_{ni}\{(0,t] \times A\} \geq 1] = \lim_{n\to\infty} \sup_{i} F_{ni}(t) = 0 \tag{5.10}$$

and when (5.7) (and therefore (5.10)) holds the condition (5.8) is equivalent to the following two conditions: for any $A \in K$ and $t > 0$,

$$\lim_{n\to\infty} \sum_{i=1}^{k_n} P[M_{ni}\{(0,t] \times A\} \geq 1] = \int_0^t \mu(ds)\ Q(s,A) \tag{5.11}$$

and

$$\lim_{n\to\infty} \sum_{i=1}^{k_n} P[M_{ni}\{(0,t] \times A\} \geq 2] = 0 . \tag{5.12}$$

To prove the necessity of (5.8) we note that if the superposition N_n converges to a C-process, the primary points of clusters with distributions belonging to A will have to form a Poisson process with mean measure given by

$$\mu_A(B) = \int_B \mu(ds)\ Q(s,A) , \qquad B \in R . \tag{5.13}$$

By Theorem (4.9), (5.11) and (5.12) are necessary for this. But we just showed that (5.8) is equivalent to them.

To prove the sufficiency suppose (5.7) and (5.8) hold; then (5.10), (5.11), (5.12) hold; which, by Theorem (4.9) imply that the process

$$M_n = \sum_{i=1}^{k_n} M_{ni}$$

converges weakly to a Poisson random measure M on $(E,\mathcal{E})$ with a mean measure λ which satisfies

$$\lambda\{(s,t] \times A\} = \int_s^t \mu(du)\ Q(u,A) \ .$$

Let the points of M be $\{(T_n,G_n);\ i = 1,2,\ldots\}$. For each n, let $U_{n0} = 0$, U_{n1}, $U_{n2}-U_{n1},\ldots$ be independent and identically distributed random variables with distribution function G_n. Let N be the point process whose points are $\{T_n + U_{nm};\ n = 1,2,\ldots;\ m = 0,1,\ldots\}$. Then, N is a *C*-process with primary measure μ and secondary kernel Q and it is clear that N_n converges weakly to N.

6. APPROXIMATION OF THE SUPERPOSITION PROCESS BY MEANS OF POISSON PROCESSES

By now we have seen a sufficient number of theorems testifying to the rough statement that the superposition of a large number of uniformly sparse point processes approaches a Poisson process. In applied problems, however, we have only a certain number of processes to superpose. It is therefore of great importance to know how rapid the convergence to a Poisson process is, and to have some measure of the error involved in using a Poisson process to approximate the superposition of a number of point processes.

In this section we will give upper bounds for the "error" involved in such approximations. In the next section we will consider asymptotic expansions and give estimates for the speed of convergence to Poisson processes. The most comprehensive such study we are aware of is by Franken (1963b, 1964) whose work, for some reason, has gone quite unnoticed in the literature of point processes. Apparently, the problem was suggested to Franken by Kolmogorov; the 1964 paper was a dissertation and in fact precedes the 1963 paper. Franken's results extend those of Kerstan (1964) which in turn is a generalization of those of Prokhorov (1953) and Le Cam (1960) who were amongst the earliest to consider the error involved in approximating the distribution of sums of independent integer valued random variables by means of Poisson distributions.

Grigelionis (1962a, 1962b, 1962c) contain results in this direction which, however, are very complicated and cumbersome.

In Franken (1964) the problem considered is the "approximation of the distribution of sums of independent non-negative integer valued random variables by means of Poisson distributions," (which is also the title). So the results apply to the distribution of the number of points of the superposition process in an interval.

Let $X_1, X_2, \ldots, X_n$ be independent non-negative integer valued random variables and let

$$X = X_1 + X_2 + \cdots + X_n . \tag{6.1}$$

Let

$$\begin{aligned} \phi(k) &= P(X = k) \qquad (k = 0,1,\ldots) , \\ a &= E(X) \end{aligned} \tag{6.2}$$

and let Ψ be the Poisson distribution with parameter a.

In the case where X_j can take only the values 1 and 0 with probabilities p and 1-p respectively, Prokhorov (1953) has shown that

$$d(\phi,\Psi) = \sum_{k=0}^{\infty} |\phi(k) - \Psi(k)| \leq Kp , \tag{6.3}$$

where K is some constant independent of n and p. The results of Prokhorov were generalized by Le Cam (1960) to the case where X_j takes the values 1 and 0 with probabilities p_j and $1-p_j$ respectively. He shows that

$$d(\phi,\Psi) \leq 2 \sum_{j=1}^{n} p_j^2 , \tag{6.4}$$

$$d(\phi,\Psi) \leq D_1 \sup_{1\leq j\leq n} p_j , \tag{6.5}$$

and if $\sup_j p_j \leq 1/4$,

$$d(\phi,\Psi) \leq D_2(\Sigma_j p_j^2)/(\Sigma_j p_j) , \tag{6.6}$$

where D_1 and D_2 are two deterministic constants and we have

$$D_1 \leq 9, \qquad D_2 \leq 16 .$$

The proofs of (6.5) and (6.6) are not elementary and use operator methods which in fact yield analogous results for random variables on locally compact Abelian groups. These results were generalized to the case where X_j are any non-negative integer valued random variables by Kerstan (1964) who proved that, if $\sup_j a_j \leq 1/4$,

$$d(\phi,\Psi) \leq 5.4\ \Sigma_j c_j/(\Sigma_j a_j) \tag{6.7}$$

and

$$d(\phi,\Psi) \leq 2\Sigma_j c_j , \tag{6.8}$$

where $a_j = E(X_j)$, $b_j = E\{X_j(X_j-1)\}$ and, with $q_j = P(X_j=0)$,

$$c_j = b_j + a_j^2 + 4 - 4a_j - 2e^{-a_j} - 2q_j + 2|e^{-a_j} - q_j| .$$

Note that though more general, (6.7) and (6.8) are not as good as (6.6) and (6.4) in the special case of 0-1 summands.

For a different distance,

$$d_0(\phi,\Psi) = \sup_{0\leq x<\infty} \left| \sum_{k\leq x} \phi(k) - \sum_{k\leq x} \Psi(k) \right| , \tag{6.9}$$

Hodges and Le Cam (1960) obtained, by elementary methods, the bounds

$$d_0(\phi,\Psi) \leq 2\ \Sigma_j p_j^2 , \tag{6.10}$$

$$d_0(\phi,\Psi) \leq 3(\sup_j p_j)^{1/3} \tag{6.11}$$

for the case of 0-1 summands.

Considering the general case we introduce

$$a_j = E(X_j),\ b_j = a_j^2 + E\{X_j(X_j-1)\}, \qquad a = \Sigma a_j, \qquad b = \Sigma b_j. \tag{6.12}$$

The following theorem is due to Franken (1964).

Theorem (6.13). Supposing that the independent random variables X_j have finite second order moments we have

$$d_0(\phi,\Psi) \le 2b/\pi \tag{6.14}$$

and, if $a_j > b_j$ for every j,

$$d_0(\phi,\Psi) \le Qb\{1 - e^{-2(a-b)}\}/\{\pi(a-b)\}\ , \tag{6.15}$$

where

$$Q = \begin{cases} \sup_j\{\exp(a_j^2/2b_j)\} & \text{if } a_j^2/2b_j \le 1\ , \\ \sup_j\{\exp(2a_j-2b_j)\} & \text{otherwise}\ . \end{cases}$$

Remark. For the case of 0-1 summands (6.14) becomes the same as (6.10) whereas from (6.15) one obtains

$$d_0(\phi,\Psi) \le 0.6(\Sigma p_j^2)/(\Sigma p_j) \le 0.6 \sup_j p_j \tag{6.16}$$

for the case $p_j \le 1/4$.

Proof. Let $f_j(z) = E(z^{X_j})$ and $f(z) = E(z^X)$ for $|z| \le 1$. Below we will show that

$$|f(e^{i\lambda}) - e^{a(e^{i\lambda}-1)}| \le 2b \sin^2 \frac{\lambda}{2} \tag{6.17}$$

and that

$$|f(e^{i\lambda}) - e^{a(e^{i\lambda}-1)}| \leq 2Qb \sin^2 \frac{\lambda}{2} \exp\{-2(a-b) \sin^2 \frac{\lambda}{2}\} , \qquad (6.18)$$

from which the assertions (6.14) and (6.15) follow immediately with the help of the result

$$d_0(\phi,\Psi) \leq \frac{1}{4\pi} \int_{-\pi}^{\pi} |\{f(e^{i\lambda}) - e^{a(e^{i\lambda}-1)}\}/\sin \frac{\lambda}{2}| \, d\lambda ,$$

which is a special case of a theorem due to Tsaregradsky (1958).

Let $g(z) = f(z)\, e^{a(1-z)}$. Since $g(1) = 1$ we have the identity

$$g(e^{i\lambda}) - 1 = \int_0^{e^{i\lambda}} g'(z) \, dz$$

for every integration path lying entirely in $|z| \leq 1$. By substituting $z = 1 + \tau(e^{i\lambda} - 1)$ in here we get

$$\begin{aligned} &|f(e^{i\lambda}) - e^{a(e^{i\lambda}-1)}| \\ &= |e^{a(e^{i\lambda}-1)}\{g(e^{i\lambda})-1\}| \\ &= |e^{a(e^{i\lambda}-1)}|\,|e^{i\lambda}-1|\cdot|\int_0^1 g'\{1 + \tau(e^{i\lambda}-1)\} \, d\tau| \\ &\leq \exp(-2a \sin^2 \frac{\lambda}{2}) \; |2 \sin \frac{\lambda}{2}| \int_0^1 |g'\{1 + \tau(e^{i\lambda}-1)\}| d\tau, \end{aligned} \qquad (6.19)$$

since

$$Re(e^{i\lambda}-1) = -2 \sin^2 \frac{\lambda}{2} \qquad \text{and} \qquad |e^{i\lambda}-1| = |2 \sin \frac{\lambda}{2}| .$$

Noting that $f(z) = f_1(z)\times\ldots\times f_n(z)$ we obtain

$$g'(z) = e^{a(1-z)} \sum_j \{f_j'(z) - a_j f_j(z)\} \prod_{i\neq j} f_i(z) ,$$

which in turn yields, for $z_0 = 1 + \tau(e^{i\lambda}-1)$,

$$|g'(z_0)| \le \exp(2a\tau \sin^2 \frac{\lambda}{2}) \sum_j |f_j'(z_0) - a_j f_j(z_0)| \prod_{i\neq j} |f_i(z_0)|. \tag{6.20}$$

By hypothesis a_j, b_j are finite, so, for $|z| \le 1$ we have

$$f_j(z) = 1 + \theta_j(z-1) \qquad (\theta_j \le a_j) ,$$

$$f_j'(z) = a_j + \gamma_j(z-1) \qquad (\gamma_j \le b_j - a_j^2) ,$$

so that

$$\sum_j |f_j'(z_0)-a_j f_j(z_0)| \le b\tau|\sin \frac{\lambda}{2}| , \tag{6.21}$$

where $z_0 = 1 + \tau(e^{i\lambda}-1)$.

Replacing the $|f_i(z_0)|$ in (6.20) by 1 we get from (6.19), (6.20), (6.21)

$$|f(e^{i\lambda}) - e^{a(e^{i\lambda}-1)}|$$

$$\le 4b \sin^2 \frac{\lambda}{2} \exp(-2a \sin^2 \frac{\lambda}{2}) \int_0^1 \tau \exp(2a\tau \sin^2 \frac{\lambda}{2})\, d\tau, \tag{6.22}$$

which becomes (6.17) once we notice that

$$\exp(2a\tau \sin^2 \frac{\lambda}{2}) \le \exp(2a \sin^2 \frac{\lambda}{2}) .$$

If we estimate $|f_j(z_0)|$ in (6.20) by using

$$f_j(z_0) \le |1 + a_j(z_0-1)| + \frac{1}{2} (b_j-a_j^2)||z_0-1|^2$$

$$\le \{1 - 4(\sin^2 \frac{\lambda}{2})(a_j\tau)(1-a_j\tau)\}^{1/2} + 2\tau^2(b_j-a_j^2) \sin^2 \frac{\lambda}{2}$$

$$\le 1 - 2 \sin^2 \frac{\lambda}{2} (a_j\tau-a_j^2\tau^2) + 2\tau^2(b_j-a_j^2) \sin^2 \frac{\lambda}{2}$$

$$\le \exp\{-2(a_j\tau-b_j\tau^2) \sin^2 \frac{\lambda}{2}\}$$

we get, making use of (6.20) and (6.21), the bound

$$|g'(z_0)| \le \left|2 \sin \frac{\lambda}{2}\right| \tau b \exp(2b\tau^2 \sin^2 \frac{\lambda}{2})$$
$$\times \sup_j [\exp\{2(a_j\tau - b_j\tau^2) \sin^2 \frac{\lambda}{2}\}]$$
$$\le Qb\tau \left|2 \sin \frac{\lambda}{2}\right| \exp(2b\tau^2 \sin^2 \frac{\lambda}{2}) .$$

The bound (6.18) now follows from (6.23) and (6.19). This completes the proof.

In addition to these Franken gives upper bounds also for the distances

$$\sup_k |\phi(k) - \Psi(k)| ,$$

$$\sum_k |\phi(k) - \Psi(k)| ,$$

$$\sum_k |\phi(k) - \Psi(k)|^2 .$$

The methods used are illustrated quite well by the preceding proof.

So far we have considered the problem of approximating the distributions of $X = X_1 + \cdots + X_n$ by a Poisson distribution only in the case where $X_1, \ldots, X_n$ are integer valued univariate random variables. In considering the superposition M of point processes $M_1, \ldots, M_n$ we would like to be able to approximate the distribution of the m-variate vector

$$\underline{X} = \{M(A_1), \ldots, M(A_m)\}$$

by the corresponding m-variate Poisson distribution.

In this direction, we put below the results due to Grigelions (1962b, 1962c) and to Franken (1963b).

Let

$$\underline{X}_1 = \{X_1(1),\ldots, X_1(m)\},\ldots\ldots\ldots, \underline{X}_n = \{X_n(1),\ldots, X_n(m)\} \tag{6.24}$$

be independent random vectors with non-negative integer valued components and let

$$\underline{X} = \underline{X}_1 + \cdots + \underline{X}_n . \tag{6.25}$$

Further, we write $X_j = \Sigma_{k=1}^m X_j(k)$, $1 \le j \le n$, and $X = \Sigma_{k=1}^m X(k)$. Define

$$\mu_j(k) = P\{X_j(k) = 1, X_j = 1\} \tag{6.26}$$

and let

$$\underline{\mu}_j = \{\mu_j(1),\ldots, \mu_j(m)\}, \tag{6.27}$$

$$\underline{\mu} = \underline{\mu}_1 + \cdots + \underline{\mu}_n . \tag{6.28}$$

The following was essentially proved within the sufficiency part of the proof of Theorem (3.10), and thus we omit a second proof.

Theorem (6.29). Let ϕ be the distribution of $\underline{X}$ and ψ the m-variate Poisson distribution with parameter vector μ. Then,

$$d(\phi,\Psi) = \sum_x |\phi(x) - \Psi(x)| \le 2 \sum_{j=1}^n (p_j^2 + q_j) ,$$

where

$$p_j = P(X_j=1), \qquad q_j = P(X_j \ge 2) .$$

Next we consider a different distance function which we define by

$$d_1(\phi,\Psi) = \sup_x |\phi(x) - \Psi(x)| . \tag{6.30}$$

Throughout the following we will assume that $\underline{X}_j$ have finite moments of second order, and define

$$
\begin{aligned}
a_j(k) &= E\{X_j(k)\}; \qquad \mu_j(k,\ell) = E\{X_j(k)\ X_j(\ell)\}\ , \\
b_j(k) &= a_j(k)^2 + \mu_j(k,k) - a_j(k)\ , \\
c_j(k,\ell) &= \mu_j(k,\ell) + a_j(k)\ a_j(\ell)\ , \\
a(k) &= a_1(k) + \cdots + a_n(k) = E\{X(k)\}\ .
\end{aligned}
\tag{6.31}
$$

Then we have the following approximation theorem.

Theorem (6.32). Let ϕ be the distribution of the random vector $\underline{X}$ and let Ψ denote the m-variate Poisson distribution with parameter a = {a(1),..., a(m)}. Then,

$$d_1(\phi,\Psi) \leq \sum_{j=1}^{n} \left\{ \sum_{k=1}^{m} b_j(k) + \frac{16}{\pi^2} \sum_{k>\ell} c_j(k,\ell) \right\}, \tag{6.33}$$

$$d_1(\phi,\Psi) \leq 2 \sum_{j=1}^{n} \left[\sum_{k=1}^{m} \frac{b_j(k)}{a(k)} + \sum_{k<\ell} c_j(k,\ell) \left\{ \frac{1}{a(k)} + \frac{1}{a(\ell)} \right\} \right]. \tag{6.34}$$

Remark. By a further estimation we obtain from (6.34) the following coarser bound,

$$d_1(\phi,\Psi) \leq \sum_{k=1}^{m} \sup_j \frac{b_j(k)}{a_j(k)} + \sum_{k<\ell} \left\{ \sup_j \frac{c_j(k,\ell)}{a_j(k)} + \sup_j \frac{c_j(k,\ell)}{a_j(\ell)} \right\} \tag{6.35}$$

which is uniform with respect to n.

Proof of this theorem follows along the lines of that of (6.13) and we omit it here; the interested reader (that elusive animal!) is referred to Franken (1963b, Theorem 2). There one can also find further special results in the case where $\underline{X}_j = \{M_j(A_1),\ldots, M_j(A_m)\}$ for some renewal process M. Grigelionis (1962b, 1962c) has put his work in the framework of double series schema of our sections 3, 4, 5 and always assumed that convergence to Poisson conditions are fulfilled.

7. ASYMPTOTIC EXPANSIONS OF THE ONE-DIMENSIONAL DISTRIBUTIONS OF THE SUPERPOSITION PROCESS

As before we let $X_1, \ldots, X_n$ be independent non-negative integer valued random variables, define

$$X = X_1 + \cdots + X_n ,$$

and let ϕ be the probability distribution of X. It is of interest, both from the theoretic and applied standpoints, to be able to find a decomposition

$$\phi(k) = \Psi(k) + \sum_{i=1}^{\infty} \beta_i(k) \tag{7.1}$$

so that the error estimate

$$d_m = d_0(\phi, \Psi + \sum_{i=0}^{m-1} \beta_i) \qquad (m = 1,2,\ldots) \tag{7.2}$$

is of a small order of magnitude; here, as before,

$$d_0(\phi,\eta) = \sup_{0\leq x<\infty} \left| \sum_{k\leq x} \phi(x) - \sum_{k\leq x} \eta(k) \right| . \tag{7.3}$$

This, then, is a refinement of the problem of the last section.

Grigelionis and Franken have considered this problem. Grigelionis (1962a) has put the problem in the framework of double sequences $\{X_{nk};\ n = 1,2,\ldots;\ i = 1,\ldots, k_n\}$ and their sums. Under the hypothesis that the characteristic functions of the X_{ni} vanish nowhere, and that the distribution of $X_{n1} + \cdots + X_{nk_n}$ converges to a Poisson distribution, and, moreover, that there exists a constant K independent of n such that

$$E(\log \hat{X}_{n1} + \cdots + \log \hat{X}_{nk_n}) \leq K ,$$

where $\hat{X}_{ni} = \max(1, X_{ni})$, he considered the asymptotic behavior of the remainder term. The computations and the expressions which precede them are however very complicated and require full knowledge of the distribution of the X_{ni}; and furthermore, the estimate obtained contains only an assertion on the asymptotic behavior without the exact specification of parameters.

The results we are giving below are due to Franken (1964). Unfortunately, these are also very complicated (it is possible to find theorems whose statements are over a page long) and the proofs rather non-instructive. We will therefore sketch the results and only try to capture the flavor of the methods involved.

Let ϕ be a probability distribution on $\{0,1,\ldots\}$ and let $m(i)$ be its ith factorial moment, i.e.,

$$m(i) = \sum_k k(k-1)\cdots(k-i+1)\,\phi(k)\ . \tag{7.4}$$

Let Ψ be the Poisson distribution with parameter a; the rth finite difference of Ψ is defined by

$$\Psi_r(k) = \sum_{j=0}^{r} \binom{r}{j} (-1)^j\, \Psi(k-j)\ . \tag{7.5}$$

The so-called B-series of Charlier for a discrete distribution ϕ is given by

$$\phi(k) \sim \sum_{r=0}^{\infty} \frac{(-1)^r}{r!} c_r \Psi_r(k)\ . \tag{7.6}$$

The coefficients c_r are called the Charlier coefficients and they are given by

$$c_r = \sum_{j=0}^{r} (-1)^j \binom{r}{j} a^j\, m(r-j)\ ; \tag{7.7}$$

note that, when $a = m(1)$,

$$c_0 = 1, \qquad c_1 = 0.$$

We will now express these coefficients by means of the factorial semi-invariants $\mu(i)$ defined by

$$\mu(i) = \frac{d^i}{dz^i} \log f(z) \bigg|_{z=1} , \tag{7.8}$$

where f is the generating function $\Sigma z^k \phi(k)$. From the easily obtained result

$$\sum_{k=0}^{\infty} \Psi_r(k) z^k = (1-z)^r e^{-a(1-z)} ,$$

and the series (7.6) we obtain

$$g(z) = f(z) e^{a(1-z)} = \sum_{r=0}^{\infty} \frac{c_r}{r!} (z-1)^r . \tag{7.9}$$

Thus the Charlier coefficients for ϕ are, formally, nothing other than the derivatives of the function g at the point z=1. In terms of the semi-invariants, it follows from (7.7) and (7.9), and noting that we put $a = m(1)$,

$$\exp\{\sum_2^{\infty} \frac{\mu(r)}{r!} (z-1)^r\} = 1 + \sum_2^{\infty} \frac{c_r}{r!} (z-1)^r ,$$

from which we have

$$c_r = \sum_{s=1}^{[r/2]} c_r(s) , \tag{7.10}$$

where [x] is the greatest integer less than or equal to x and

$$c_r(s) = \frac{r!}{s!} \sum_{r_1+\cdots+r_s=r} \cdots \sum \frac{\mu(r_1) \cdots \mu(r_s)}{r_1! \cdots r_s!} . \tag{7.11}$$

Letting Ψ be the Poisson distribution with parameter $a = m(1) = E(X)$, separating the initial term of the series (7.6), noting that then $c_1=0$, and using (7.10) and (7.11) we get the expansion

$$\phi(k) \sim \Psi(k) + \sum_{i=1}^{\infty} \beta_i(k) , \tag{7.12}$$

where

$$\beta_i(k) = \sum_{I_i} \frac{(-1)^r}{r!} c_r(s) \Psi_r(k) , \tag{7.13}$$

where the summation is over the index set

$$I_i = \{(r,s): r-s=i, 1 \leq s \leq [\tfrac{r}{2}]\} .$$

The series (7.12) is considered to be the asymptotic series for $\phi(k)$. To justify the term asymptotic we now show that, in the case where $X_1, \ldots, X_n$ are identically distributed, it indeed is an asymptotic series. We first note that for any distribution η with factorial moments $m^*(r)$, the distribution

$$\eta_n(k) = \sum_{i=k}^{\infty} \binom{i}{k} \eta(i) \left(\frac{1}{n}\right)^k \left(1 - \frac{1}{n}\right)^{i-k}$$

has factorial moments $m^*(r)/n^r$. Thus, in the case $X_1, \ldots, X_n$ are identically distributed with distribution ϕ_n each, the factorial moments of the X_j are

$$m_j(r) = \frac{m^*(r)}{n^r} \qquad (r = 0,1,\ldots) , \tag{7.14}$$

where $m^*(r)$ are independent of n. In this case, the factorial semi-invariants $\mu_j(r)$ are also of the form $\mu^*(r)/n^r$ so that, from the definition of $c_r(s)$ we have

$$c_r(s) + O\left(\frac{1}{n^{r-s}}\right) . \tag{7.15}$$

Since the parameter of the Poisson distribution Ψ is $a = m^*(1)$, independent of n, the finite differences $\Psi_r(k)$ are independent of n also. Hence, from the definition (7.13),

$$\beta_i(k) = O\left(\frac{1}{n^i}\right). \tag{7.16}$$

The following result due to Franken (1963a) is now self-evident.

Proposition (7.17). Let $M_1, \ldots, M_n$ be independent renewal processes all having the same distribution function F for their intervals. Suppose $H(t) = E\{M_i(t)\}$ is such that $nH(t) = E\{N(t)\}$ for some renewal process N. If $F(0) < 1/2$, then for any interval A we can expand the (cumulative) probability distribution of $M_1(A) + \cdots + M_n(A)$ as

$$\sum_{k \le x} \phi(k) = \sum_{k \le x} \left\{ \Psi(k) + \sum_{i=1}^{m} \frac{\gamma_i(k)}{n^i} \right\} O\left(\frac{1}{n^{m+1}}\right)$$

where Ψ is the Poisson distribution with parameter H(A) and the γ_i are certain polynomials in k.

Remark (7.18). For the convergence of the Charlier B-series (7.6) it is necessary and sufficient that $\lim_k \phi(k)\, 2^k r^k = 0$ for $r = 0,1,\ldots,$ so that we must have $f(z) = \Sigma z^k \phi(k)$ analytic in $|z| < 2$ and continuous, with all its derivatives, in $|z| \le 2$. This explains the need for $F_j(0) < 1/2$. In fact, if $F_j(0) < 1/3$, the estimates obtained above are uniform in x.

In general for any distribution ϕ we have the following estimate for the mth remainder term of its Charlier B-series.

Proposition (7.19). If ϕ has finite moments of rth order, then

$$\left|\phi(k) - \Psi(k) - \sum_{2}^{r-1} \frac{(-1)^i}{i!} c_i \Psi_i(k)\right|$$
$$\le \frac{2^{r+1}}{r!\,\pi} \int_0^{\pi/2} \sin^r t \, dt \sum_{i=0}^{r} \binom{r}{i} m(1)^i\, m(r-i)\,. \tag{7.20}$$

The bound is obtained by considering the inversion formula for the Fourier transforms which give in this case that the left hand side of (7.20) is

$$\left|\frac{1}{2\pi} \int_{-\pi}^{\pi} e^{i\lambda k} \{f(e^{i\lambda}) - g(e^{i\lambda})\} \, d\lambda\right| ,$$

where f, g are the generating functions of ϕ and $\Sigma_0^{r-1} \frac{(-1)^i}{r!} c_i \Psi_i(k)$ respectively. Then, one uses the fact that Charlier coefficients are the derivatives at z = 1 of the function $f(z)\, e^{a(1-z)}$ to further estimate the integral.

From the preceding, which is of independent interest, through very complicated and cumbersome reasonings one first obtains an estimate (cf. Franken, 1964, Theorem 5) for the Fourier transform of the remainder term

$$\rho_m(k) = \phi(k) - \Psi(k) - \sum_{i=0}^{m-1} \beta_i(k)$$

and then using this obtains an upper bound for

$$d_m = \sup_{0 \le x < \infty} \left| \sum_{k \le x} \phi(k) - \sum_{k \le x} \{\Psi(k) + \sum_{i=1}^{m-1} \beta_i(k)\}\right| .$$

For complete details we refer to Franken (1964, Theorem 6). To give an idea about the result we restate the simpler looking half of the theorem mentioned in the present notation.

Theorem (7.22). Suppose the independent random variables $X_1, \ldots, X_n$ have finite moments of (m+1)th order, and further suppose

$$\delta_j = \inf_{-\pi<\lambda<\pi} |f_j(e^{i\lambda})| > 0$$

for all j. Then,

$$d_m \le \frac{1}{\pi} (e^{A_m} - 1 + e^{B_m} - 1 - D_m) \int_0^{\pi/2} \sin^m \lambda \, d\lambda ,$$

where

$$B_m = \sum_2^m \mu(k) 2^k/k!; \qquad \mu(k) = \sum_j \mu_j(k) ;$$

$$D_m = \sum_{k=1}^{m-1} \sum_I \frac{2^r}{s!} \sum_J \frac{\mu(i_1)\cdots\mu(i_s)}{i_1! \cdots i_s!} ,$$

$$I = \{(r,s): r-s=k,\ 1 \le s \le [k/2]\} ,$$

$$J = \{(i_1,\ldots,\ i_s): i_1+\cdots+i_s=r,\ 2 \le i_s \le m\} ;$$

and

$$A_m = \sum_{j=1}^n \frac{1}{\delta_j} \left\{ \frac{2^{m+1}}{(m+1)!} \sum_{k=0}^{m+1} \binom{m+1}{k} \{m_j(1)\}^k m_j(m+1-k) \right.$$

$$\left. + \sum_{I\cap(r>m)} \frac{2^r}{s!} \sum_J \frac{\mu_j(i_1) \cdots \mu_j(i_s)}{i_1! \cdots i_s!} \right\} .$$

There is a further, sharper, estimate in the case where certain quantities are positive. We omit the result because of its extreme length.

In the case where $X_1,\ldots,\ X_n$ are identically distributed and

$$\mu_j(k) = \mu^*(k)/n^k$$

for all k, we have $\delta_j = \delta$ for all j, and

$$\mu(k) = \mu^*(k)/n^{k-1} ,$$

so that the estimate of (7.22) becomes

$$d_m \le K_m/n^m , \tag{7.23}$$

where

$$K_m \leq \frac{1}{\pi} \left(\int_0^{\pi/2} \sin^m \lambda \, d\lambda \right) \left(e^{A^*_m} - 1 + e^{B^*_m} - 1 - D^*_m \right)$$

and A^*_m, B^*_m, D^*_m are obtained from the formulas for A_m, B_m, D_m by replacing μ in them throughout by μ^*. We see here that the asymptotic series (7.12) approximates the distribtuion ϕ to any desired precision level.

8. CONVERSE PROBLEMS AND CORRELATION PROPERTIES

So far, in considering the superposition problem, we always assumed that the number and properties of the component processes are known. In this section we do the opposite; supposing we are able to observe only the superposition process, we would like to infer from the data available, the number and type of the components being superposed.

Suppose we observe a point process N which we know is the superposition

$$N = N_1 + \cdots + N_k$$

of stationary renewal processes $N_1, \ldots, N_k$ which are independent and have the same distribution F for their intervals. Our first problem is to find the number k and the distribution function F from the observed data on N.

Suppose $N = N_1 + N_2$ and that N_1 is a stationary Poisson process and N_2 a stationary renewal process with interval distribution function F. Our second problem is to find the intensity a of N_1 and the distribution function F describing N_2 from the observed data on N.

More generally, if $N = N_1 + \cdots + N_k$ where $N_1, \ldots, N_k$ are known to be stationary renewal processes with interval distribution functions $F_1, \ldots, F_k$, the problem is to find k and $F_1, \ldots, F_k$ from the known data on N. One general method, due to Ambartzumian (1969), is to consider the generating function $\phi(z) = \Sigma \; \rho^n z^n$ where ρ_n is the correlation coefficient between the mth and (m+n)th intervals of the superposition N. Quite an amount of the information concerning $F_1, \ldots, F_k$ is contained in the behavior of ϕ on the circle $|z| = 1$. Ambartzumian (1969) shows that ϕ is analytic everywhere in the disc $|z| \leq 1$, and that it has derivatives of kth order at $z = 1$ if and only if at least one of the F_i has moments of order k+1.

Another problem of the same nature, due again to Ambartsumian (1965), is concerned with clustered point processes. In observing such a process, one does not know in advance the cluster to which a point belongs. The problem then is to obtain information concerning the structure of the clusters from the observed properties of the entire process. We will not give any details here on this important problem; further, the well known book by Cox and Lewis (1966) contain a chapter on the statistical questions raised by the superposition problem. Considering the easy availability of the book we will not give any details here. Below we describe the results of Ambartzumian on the first and second problems.

We now consider the first problem. Suppose we are observing $N = N_1 + \cdots + N_k$ where $N_1, \ldots, N_k$ are independent stationary renewal processes with the common distribution function F for their intervals. We will, without loss of generality, take the intensity of N to be unity so that the expected length of an interval in N_j is k. We denote by H the "renewal function" corresponding to F (that is, $H = \Sigma_1^\infty F^{(n)}$, where $F^{(n)}$ is the n-fold convolution of F with itself), and we assume that $\eta = H'(0)$ exists.

Let

$$M_i(t) = \lim_{s \downarrow 0} E[N(s,s+t) \,|\, N(0,s) \geq i] \; ; \tag{8.1}$$

then one can show that (cf. Ambartzumian, 1965a, p. 450)

$$M_1(t) = H(t) + \frac{k-1}{k} t \tag{8.2}$$

and

$$M_2(t) = p\{H(t) + \frac{k-1}{k} t\} + (1-p) \{2H(t) + \frac{k-2}{k} t\} , \tag{8.3}$$

where

$$p = \frac{\eta/2}{\frac{\eta}{2} + \binom{k}{2}/k^2} = \frac{\eta}{\eta + \frac{k-1}{k}} .$$

Taking the derivatives a, b of M_1, M_2 at t = 0 we get

$$a = \eta + \frac{k-1}{k}, \tag{8.4}$$

$$b = pa + (1-p)(2a-1) = \eta + \frac{k-1}{k}\left(2 - \frac{1}{a}\right), \tag{8.5}$$

so that

$$k = \frac{1 - a}{1 - a - a^2 + ab}. \tag{8.6}$$

Once k is known, the distribution function F is easy to determine: either one can use the formula for M_1 to solve for H and then get F, or one can use the relation

$$\int_t^\infty \Phi(x)\, dx = \left[\frac{1}{k} \int_t^\infty \{1 - F(x)\}\, dx\right]^k$$

to get

$$1 - F(x) = \Phi(x) \left\{\int_x^\infty \Phi(y)\, dy\right\}^{-1+1/k}, \tag{8.7}$$

where $1-\Phi$ is the distribution function of an interval in N (or Φ is the Palm function of zeroth order).

The same method can be used also to solve the second problem: knowing $N = N_1 + N_2$ where N_1 is a Poisson process with intensity a and N_2 is a stationary renewal process independent of N_1 and has an interval distribution F, find the intensity a and the distribution function F. Here we will consider this problem by a different method.

For easy reference we now put some easy results concerning the superposition N. Let T_0, T_1,... be its points, put $X_n = T_{n+1}-T_n$ for $n \leq 0$, and introduce

$$\Phi_m(t) = P\{T_{n+m+1} - T_{n+m} > t\} \qquad (n \geq 1,\ m \geq 0), \tag{8.8}$$

which by stationarity is independent of n. Writing

$$c_m = E(X_0 X_m) \qquad (m = 0,1,2,\ldots), \tag{8.9}$$

we have by stationarity

$$E\{(T_{m+1}-T_0)^2\} = \begin{cases} c_0 & \text{if } m = 0, \\ (m+1)c_0+2mc_1+2(m-1)c_2+\cdots+2c_m & \text{otherwise}. \end{cases} \tag{8.10}$$

We will next compute the generating function

$$G(z) = \frac{1}{2}c_0 + \sum_{m=1}^{\infty} c_m z^m . \tag{8.11}$$

Suppose that some of the points of N are deleted by using independent coin flippings so that each point, independent of what happens to other points, has probability p of being deleted. The resulting process N^p is still the superposition of two independent point processes: namely, N_1^p and N_2^p which are obtained by the same deletion rules from N_1 and N_2 respectively. Then we have, with $q = 1-p$,

$$\begin{aligned} P\{N^p(t) = 0\} &= \sum_{m=0}^{\infty} qp^m P(T_m > t) \\ &= \sum_{m=0}^{\infty} qp^m \int_t^{\infty} \{\Phi_m(x) - \Phi_{m-1}(x)\}\, dx \\ &= q^2 \int_t^{\infty} \sum_{m=0}^{\infty} p^m \Phi_m(x)\, dx . \end{aligned} \tag{8.12}$$

(We are assuming, without loss of generality, that the intensities *a* and *b* of N_1 and N_2 sum to unity.) Now using the independence of the summands in $N^p(t) = N_1^p(t) + N_2^p(t)$, and using the same derivation as in

(8.12) we get

$$q^2 \int_t^\infty \sum_{m=0}^\infty p^m \Phi_m(x)\ dx = e^{-aqt} q^2 b \int_t^\infty \sum_{m=0}^\infty p^m F_m(x)\ dx\ , \tag{8.13}$$

where $1-F_m$ is the (m+1)-fold convolution of F with itself.

Noting that

$$2\int_0^\infty dt \int_t^\infty \Phi_m(x)\ dx = E\{(T_{m+1}-T_0)^2\}, \tag{8.14}$$

and using the notations (8.9), (8.11) with result (8.10) we see that

$$q^2 \int_0^\infty dt \int_t^\infty \sum_{m=0}^\infty p^m \Phi_m(x)\ dx = G(p)$$

by a change of the order of integration and summation. Now from integrating the right-hand-side of (8.13) by using the relation

$$\int_0^\infty e^{-\lambda t}\ dt \int_t^\infty F_m(x)\ dx = \frac{m+1}{\lambda b} - \frac{1}{\lambda^2} + \frac{f(\lambda)^{m+1}}{\lambda^2}\ , \tag{8.15}$$

where

$$f(\lambda) = \int_0^\infty e^{-\lambda t}\ F(dt),$$

we obtain the result

$$G(p) = \frac{a-b}{a^2 q} + \frac{bf(aq)}{a^2\{1 - pf(aq)\}} \tag{8.16}$$

We see that the distribution function F is determined by the intensity *a* of the Poisson component and the covariance generating function G:

Proposition (8.17). We have, for $\lambda < a$,

$$f(\lambda) = \left\{a(\lambda-a) + \frac{a(1-a)\lambda}{1 - 2a + \lambda a\ G(1 - \frac{\lambda}{2})}\right\}^{-1} .$$

To complete the solution of our problem we need to compute the intensity a. This we do by considering the coefficients of p^0 and p^1 in the two power series on both sides of (8.13). We obtain

$$\int_t^\infty \Phi_0(x)\,dx = e^{-at}\, b \int_t^\infty F(x)\,dx \ , \tag{8.18}$$

$$\int_t^\infty \Phi_1(x)\,dx = e^{-at}\, b \int_t^\infty F_1(x)\,dx + ate^{-at}\, b \int_t^\infty F_0(x)\,dx \ . \tag{8.19}$$

Taking integrals over $(0,\infty)$, using (8.14) and (8.15) and the fact that

$$6 \int_0^\infty t\,dt \int_t^\infty \Phi_0(x)\,dx = E(X_0^3) \equiv d \ , \tag{8.20}$$

the formulas (8.18) and (8.19) yield

$$\frac{1}{2}\, c_0 = b[1/ab - 1/a^2 + f(a)/a^2] \ , \tag{8.21}$$

$$c_0 + c_1 = b\{2/ab - 1/a^2 + f(a)^2/a^2\} + ad/6 \ . \tag{8.22}$$

Putting $b = 1-a$, solving for $f(a)$ in (8.21), putting the result in (8.22), and rearranging we obtain

$$(3c_0^2 - 2d)a^2 + 2(6c_1 - 6c_0 + d)a + 12(1 - c_1) = 0 \ . \tag{8.23}$$

Hence a is that root of (8.23) which lies in $(0,1)$.

REFERENCES

Ambartzumian, R. V. (1965a). Two inverse problems concerning the superposition of recurrent point processes. J. Appl. Prob. 2, 449-54.

Ambartzumian, R. V. (1965b). The Poisson superpositions of clusters. Akad. Nauk Armjan SSR Dokl. 41, 73-80 (In Russian.)

Ambartzumian, R. V. (1967a). On the characterization of Poisson processes in terms of expandability of a point process in independent recurrent components. Akad. Nauk Armjan SSR Dokl. 44, 8-13. (In Russian.)

Ambartzumian, R. V. (1967b). On a characterization of Poisson point fields in terms of decomposability of a point field into a bounded number of independent components. Izv. Akad. Nauk. Armjan. SSR Ser. Mat. 2, 57-65. (In Russian.)

Ambartzumian, R. V. (1969). Correlation properties of the intervals in the superpositions of independent stationary recurrent point processes. Studia Sci. Math. Hungar. 4, 161-70.

Chung, K. L. (1971). Private communication.

Çinlar, E. (1968). On the superposition of m-dimensional point processes. J. Appl. Prob. 5, 169-76.

Çinlar, E. & Agnew, R. A. (1968). On the superposition of point processes. J. R. Statist. Soc. B 30, 576-81.

Cox, D. R. & Lewis, P. A. W. (1966). *The Statistical Analysis of Series of Events*. Methuen: London and Barnes and Noble: New York.

Cox, D. R. & Smith, W. L. (1953). The superposition of several strictly periodic sequences of events. Biometrika 40, 1-11.

Cox, D. R. & Smith, W. L. (1954). On the superposition of renewal processes. Biometrika 41, 91-9.

Doob, J. L. (1953). *Stochastic Processes*. Wiley: New York.

Franken, P. (1963a). A refinement of the limit theorem for the superposition of independent renewal processes. Teor. Veroyatnost. i Primenen 8, 320-28.

Franken, P. (1963b). Approximation durch Poissonsche Prozesse. Math. Nachr. 26, 101-14.

Franken, P. (1964). Approximation der Verteilungen von Summen unabhängiger nichtnegativer ganzzahliger Zufallsgrossen durch Poissonsche Verteilungen. Math. Nachr. 27, 303-340.

Franken, P., Liemant, A. & Matthes, K. (1965). Stationäre zufällige Punktfolgen, III. Jber. Deutsch. Math.-Verein. 67, 183-202.

Gnedenko, B. V. (1938). Über die Konvergenz der Verteilungsgesetze von Summen voneinander unabhängiger Summanden. C. R. (Dokl.) Acad. Sci. URSS (N.S.) 18, 231-4.

Gnedenko, B. V. & Kolmogorov, A. N. (1968). *Limit Distributions for Sums of Independent Random Variables*, 2nd Ed. Reading, Mass.: Addison-Wesley.

Gnedenko, B. V. & Kovalenko, I. N. (1968). *Introduction to Queueing Theory*. Israel Program for Scientific Translations, Jerusalem.

Goldman, J. R. (1967). Stochastic point processes: limit theorems. Ann. Math. Statist. 38, 771-9.

Grigelionis, B. (1962a). On the asymptotic expansion of the remainder term in the case of convergence to the Poisson law. Litovsky. Mat. Sb. 2, 35-48. (In Russian.)

Grigelionis, B. (1962b). Sharpening of a higher-dimensional limit theorem on convergence to the Poisson law. Litovsk. Mat. Sb. 2, 127-33. (In Russian.)

Grigelionis, B. (1962c). On the degree of approximation of the composition of renewal processes by a Poisson process. Litovsk. Mat. Sb. 2, 135-43. (In Russian.)

Grigelionis, B. (1963). On the convergence of sums of random step processes to a Poisson process. Teor. Veroyatnost. i Primenen 8, 189-94. (English translation in Theory Prob. Appl. 8, 172-82.)

Grigelionis, B. (1964). A central limit theorem for sums of renewal processes. Litovsk. Mat. Sb. 4, 197-201. (In Russian.)

Grigelionis, B. (1966a). On the question of convergence of sums of step-like random processes to a Poisson process. Litovsk. Mat. Sb. 6, 241-4. (In Russian.)

Grigelionis, B. (1966b). Composition of integer valued random measures. Litovsk. Mat. Sb. 6, 359-63. (In Russian.)

Hodges, J. L. & Le Cam, L. (1960). The Poisson approximation to the Poisson binomial distribution. Ann. Math. Statist. 31, 737-40.

Jagers, P. (1971). On the weak convergence of superpositions of point processes. Technical Report No. 20, Department of Statistics, Stanford University.

Kerstan, J. (1964). Verallgemeinerung eines Satzes von Prochorov und Le Cam. Z. Wahrscheinlichkeitstheorie Verw. Geb. 2, 173-9.

Kerstan, J. & Matthes, K. (1964). Stationäre zufällige Punktfolgen, II. Jber. Deutsch. Math.-Verein. 66, 106-18.

Khintchine, A. Y. (1960). *Mathematical Methods in the Theory of Queueing*. London: Griffin.

Kovalenko, I. N. (1965). On the class of limit distributions for a sequence of series of sums of independent renewal processes. Litovsk. Mat. Sb. 5, 561-8.

Le Cam, L. (1960). An approximation theorem for the Poisson binomial distribution. Pacific J. Math. 10, 1181-97.

Lewis, P. A. W. (1964). A branching Poisson process model for the analysis of computer failure patterns. J. R. Statist. Soc. B 26, 398-456.

Marcinkiewicz, J. (1938). Sur les fonctions indépendants, II. Fund. Math. 30, 349-64.

Matthes, K. (1963). Stationäre zufällige Punktfolgen, I. Jber. Deutsch. Math.-Verein 66. 66-79.

McFadden, J. A. (1962). On the lengths of intervals in a stationary point process. J. R. Statist. Soc. Ser. B 24, 364-82.

McFadden, J. A. & Weissblum, W. (1963). Higher-order properties of a stationary point process. J. R. Statist. Soc. B 25, 413-31.

Mecke, J. (1967). Zum Problem der Zerlegbarkeit stationärer rekurrenter zufälliger Punktfolgen. Math. Nachr. 35, 311-21.

Neyman, J. & Scott, E. L. (1958). Statistical approach to problems of cosmology. J. R. Statist. Soc. B 20, 1-43.

Ososkov, G.A. (1956). A limit theorem for flows of homogeneous events. Teor. Veroyatnost. i Primenen 1, 274-82. (English translation in Theory Prob. Appl. 1, 248-55.)

Palm, C. (1943). Intensitätsschwankungen im Fernsprechverkehr. Ericsson Techniks 44, 1-189.

Prokhorov, Iu. V. (1953). Asymptotic behavior of the binomial distribution. Uspehi Mat. Nauk 8, 135-42. (In Russian.)

Renyi, A. (1967). Remarks on the Poisson process. Studia Sci. Math. Hungar. 2, 119-23.

Sobel, M. J. (1971). On the aggregation of point processes. CORE discussion paper No. 7103, Université Catholique de Louvain.

Störmer, H. (1969). Zur Überlagerung von Erneuerungsprozessen. Z. Wahrscheinlichkeitstheorie Verw. Geb. 13, 9-24.

Tsaregradsky, I. P. (1958). On uniform approximation to a binomial distribution by infinitely divisible laws. Teor. Veroyatnost. i Primenen 3, 470-4.

ON THE CONVERGENCE OF SUMS OF POINT PROCESSES WITH INTEGER MARKS

D. Szász
Mathematical Institute of the Hungarian Academy of Sciences, Budapest

SUMMARY

Necessary and sufficient conditions are given for the convergence of sums of independent point processes with integer marks to the compound Poisson processes.

1. RESULTS

Khintchin (1960), Ososkov (1956) and Grigelionis (1963, 1966) proved theorems which explain the fact that in applications point processes often can be approximated by Poisson processes. In particular Grigelionis (1963) gave necessary and sufficient conditions for the convergence of the sums of independent, uniformly asymptotically negligible point processes to the Poisson process. Starting from practical problems B. V. Gnedenko posed the question: what can we assert about the convergence of the sums of independent integer-valued step processes? Integer-valued step processes can be interpreted as marked point processes, where the marks can be positive or negative integers. Processes of this type can be observed in insurance-theory, where incomes are positive, outlays are negative integers.

Let $X(t)$, $Y(t)$, $Z(t)$, denote integer-valued step processes with a finite number of steps in every finite interval and satisfying the condition $X(0) = Y(0) = Z(0) = 0$. We always denote a finite interval by I, and for $I = [a,b]$ we write $X(I) = X(b) - X(a-0)$, etc. Assume that for every n the processes $X_{nk}(t), k=1,2,\ldots,k_n$, are independent and for every interval I

$$\lim_{n\to\infty} \max_{1\le k\le k_n} \text{prob}\{X_{nk}(I) \ne 0\} = 0.$$

Write

$$Y_n(t) = \sum_{k=1}^{k_n} X_{nk}(t) \qquad (n=1,2,\ldots). \tag{1}$$

From general results of Lee (1967) it follows that the limit process of $Y_n(t)$ is infinitely divisible (*convergence* is to be understood in the sense of weak convergence of finite dimensional distributions). In the same work necessary and sufficient conditions for the convergence of the sums (1) are also given. The case where the limit process has independent increments (for example the Poisson process), is of special interest. Our results are concerned with this case and can be proved independently of Lee's results on the basis of a simple lemma, which is a generalization of the theorem on the convergence to the Poisson law (Gnedenko & Kolmogorov 1968).

It is known that if $Y(t)$ has independent increments then for any I

$$E\{e^{iuY(I)}\} = \exp\{\sum_{s\neq 0} \lambda_s(I)(e^{ius}-1)\}, \tag{2}$$

where for every non-zero integer s, $\lambda_s(I)$ is a measure and $\sum_{s\neq 0} |s|\lambda_s(I) < \infty$ see Szász & Wojczyński (1970). Such processes are called *compound Poisson processes*. If the process $Y(t)$ has only positive steps, then in (2) for every negative s we have $\lambda_s(I) = 0$ and so

$$E\{e^{iuY(I)}\} = \exp\{\sum_{s=1}^{\infty} \lambda_s(I)(e^{ius}-1)\}.$$

In this case the necessary and sufficient conditions for convergence are simpler, so we shall start by assuming that all processes in question have only positive steps.

Let us introduce the following conditions:

A_1. For arbitrary disjoint intervals I_1 and I_2

$$\lim_{n\to\infty} \sum_{k=1}^{k_n} \text{prob}\{X_{nk}(I_1) \neq 0,\ X_{nk}(I_2) \neq 0\} = 0.$$

A_2. For any interval I, it is true that if J_{nk} denotes the smallest interval containing all steps of the process $X_{nk}(t)$ lying in the interval I, then for every $\varepsilon > 0$

$$\lim_{n\to\infty} \sum_{k=1}^{k_n} \text{prob}(|J_{nk}| > \varepsilon) = 0,$$

where $|J|$ denotes the length of the interval J.

A_3. For any interval I and $\varepsilon > 0$,

$$\lim_{n\to\infty} \text{prob}(\max_{1\le k\le k_n} |J_{nk}| > \varepsilon) = 0.$$

B. For any interval I and every $s \neq 0$

$$\lim_{n\to\infty} \sum_{k=1}^{k_n} \text{prob}\{X_{nk}(I) = s\} = \lambda_s(I) < \infty.$$

Theorem 1. Suppose that the processes $X_{nk}(t)$ have only positive steps. Then the processes $Y_n(t)$, defined by (1), converge to a compound Poisson process, iff the condition B and one of the conditions A_1, A_2, A_3 are satisfied. In the case of convergence the limit process Y(t) is given by (2).

Theorem 2. Suppose that the processes $X_{nk}(t)$ have only positive steps and the processes $Y_n(t)$ converge to a process Y(t). The process Y(t) is a compound Poisson process of form (2), iff the condition B is satisfied with the additional requirement that for every $s > 0$ the interval-function $\lambda_s(I)$ is additive.

In our general theorem we shall need one more condition.

A_4. For any interval I and $\varepsilon > 0$ there exist processes $Z_{nk}^{(\varepsilon)}(t)$ such that for every n and k the steps of the processes $Z_{nk}^{(\varepsilon)}(t)$ are contained in an interval of length ε, and if

$$X_{nk}^{(\varepsilon)}(t) = X_{nk}(t) - Z_{nk}^{(\varepsilon)}(t),$$

then for every sub-interval J of I,

$$\lim_{n\to\infty} \sum_{k=1}^{k_n} \text{prob}\{X_{nk}^{(\varepsilon)}(J) \neq 0\} = 0.$$

Theorem 3. The processes $Y_n(t)$, defined by (1) converge to a compound Poisson process Y(t) iff the condition B and one of the conditions A_1 or A_4 are satisfied. In the case of convergence the limit process Y(t) is given by (2).

We remark that condition A_1 is a special case of Lee's condition, while conditions A_2, A_3 and A_4 are new.

2. PROOFS

We consider for every n independent random vectors

$$\underline{\xi}_{nk} = (\xi_{nk}^{(1)},\ldots,\xi_{nk}^{(r)}) \in E^r \qquad (1 \leq k \leq k_n),$$

for which $\xi_{nk}^{(i)}$ takes on only integer values $(1 \leq i \leq r)$. We suppose that

$$\lim_{n\to\infty} \max_{1\leq k\leq k_n} \text{prob}(\underline{\xi}_{nk} \neq \underline{0}) = 0$$

and we denote

$$\underline{\eta}_n = \sum_{k=1}^{k_n} \underline{\xi}_{nk}.$$

Lemma 1. The distribution of random vectors $\underline{\eta}_n$ tends, as $n\to\infty$, to a limit distribution $F(\underline{x})$ in E^r iff

(i) for every $\underline{s} \neq \underline{0}$ with integer components

$$\lim_{n\to\infty} \sum_{k=1}^{k_n} \text{prob}(\underline{\xi}_{nk} = \underline{s}) = \lambda_{\underline{s}}$$

and

(ii)

$$\lim_{n\to\infty} \sum_{\underline{s}\neq\underline{0}} \left| \sum_{k=1}^{k_n} \text{prob}(\underline{\xi}_{nk} = \underline{s}) - \lambda_{\underline{s}} \right| = 0.$$

In the case of convergence

$$\int_{E^r} e^{i(\underline{u},\underline{x})}\, dF(\underline{x}) = \exp\{\sum_{\underline{s}\neq\underline{0}} \lambda_{\underline{s}} (e^{i(\underline{u},\underline{s})}-1)\}. \tag{3}$$

Proof. The lemma can be proved in the same way as the theorem on the convergence to the Poissonian law (Gnedenko & Kolmogorov, 1968, Sect. 26).

Denote the ith unit vector in E^r by $\underline{e}_i$ and let

$$B_{nk} = \bigcup_{1\leq i<j\leq r} \{\xi_{nk}^{(i)} \neq 0,\ \xi_{nk}^{(j)} \neq 0\}.$$

Lemma 2. For the convergence of the distributions of the random vectors $\underline{\eta}_n$ to a limit distrubution with independent components it is necessary that

$$\lim_{n\to\infty} \sum_{k=1}^{k_n} \text{prob}(B_{nk}) = 0 \tag{4}$$

and it is sufficient that both (4) and the condition that for every $s_i \neq 0$ $(1\leq i\leq r)$

$$\lim_{n\to\infty} \sum_{k=1}^{k_n} \text{prob}(\underline{\xi}_{nk} = s_i\,\underline{e}_i) = \lambda_{s_i\underline{e}_i}$$

be satisfied.

Proof. It is easy to see that the distribution with characteristic function (3) has independent components if and only if $\lambda_{\underline{s}}$ always vanishes when at least two components of the vector $\underline{s}$ differ from 0. This remark and the assertion of lemma 1 prove the lemma.

Proof of Theorem 1. Define

$$A_{nk} = \bigcup_{1\leq i<j\leq r} \{X_{nk}(I_i) \neq 0,\ X_{nk}(I_j) \neq 0\},$$

where $I_1,\ldots,I_r$ are intervals. It is easy to see that condition A_1 is equivalent to the following condition: for any disjoint system $I_1,\ldots,I_r$, the $\lim_{n\to\infty} \sum_{k=1}^{k_n} \text{prob}(A_{nk}) = 0$. So the fact that conditions A_1 and B are necessary and sufficient, follows immediately from Lemma 2. We shall prove that conditions A_1, A_2 and A_3 are equivalent.

Let I be an arbitrary interval and $\varepsilon > 0$. Let us partition I for disjoint intervals $I_1,\ldots,I_r$ with lengths no larger than $\varepsilon/2$. Then

$$\text{prob}(|J_{nk}| \geq \varepsilon) \leq \text{prob}(A_{nk}),$$

and so A_2 follows from A_1.

Let $I_1,\dots,I_r$ be disjoint intervals, each pair of which has neither common point nor common end-point. Suppose that the interval I contains all intervals $I_1,\dots,I_r$. If $\varepsilon > 0$ is less than any distance between intervals $I_1,\dots,I_r$, then

$$\text{prob}(A_{nk}) \leq \text{prob}(|J_{nk}| \geq \varepsilon).$$

Thus A_1 follows from A_2.

The equivalency of conditions A_2 and A_3 follows from the inequality

$$1-\exp \sum_{k=1}^{k_n} \text{prob}(|J_{nk}| \geq \varepsilon) \leq \text{prob}(\max_{1\leq k\leq k_n} |J_{nk}| \geq \varepsilon) \leq \sum_{k=1}^{k_n} (\text{prob}(|J_{nk}| \geq \varepsilon)$$

The theorem is proved.

Proof of Theorem 2. The necessity of the additional condition is obvious. We shall prove the sufficiency. Let $I_1,\dots,I_r$ be a system of disjoint intervals such that $\bigcup_{i=1}^{r} I_i = I$, where I is an interval. On the basis of lemma 1 it follows that the limit distribution of the vector $\{Y_n(I_1),\dots,Y_n(I_r)\}$ has a characteristic funtion of form

$$\exp[\sum_{\substack{s_1,\dots,s_r=0\\(s_1,\dots,s_r)\neq\underline{0}}}^{\infty} \lambda_{s_1,\dots,s_r}\{\exp(i\sum_{\ell=1}^{r} u_\ell s_\ell)-1\}].$$

It is easy to see that for $1 \leq i \leq r$, and $s > 0$

$$\lambda_s(I_i) = \sum_{s_1=0}^{\infty} \cdots \sum_{s_{i-1}=0}^{\infty} \sum_{s_{i+1}=0}^{\infty} \cdots \sum_{s_r=0}^{\infty} \lambda_{s_1\dots,s_{i-1},s,s_{i+1},\dots,s_r}.$$

Substituting $s = 1$ we get that $\lambda_{s_1,\dots,s_r}$ vanishes always when at least two of $s_1,\dots,s_r$ differ from 0 and one of them equals to 1. Similarly, substitute $s = 2,3,\dots$. We get that $\lambda_{s_1,\dots,s_r}$ always vanishes when at least two of $s_1,\dots,s_r$ differ from 0. This means that the components of the limit distribution of the random vectors $\{Y_n(I_1),\dots,Y_n(I_r)\}$ are independent. Consequently the limit process is of compound Poisson type.

Proof of Theorem 3. The necessity and sufficiency of the conditions A_1 and B follow from our lemmas. We shall prove that the conditions A_1 and A_4 are equivalent.

Let us divide the interval I as in the proof of Theorem 1. For fixed n and k we define the events

$$C_1 = \{X_{nk}(I_i) = 0,\ i = 1,2,\ldots,r\},$$

$$C_2 = \{X_{nk}(I_i) \neq 0 \text{ for exactly one index } i\},$$

$$C_3 = \{X_{nk}(I_i) \neq 0 \text{ for two or more indices } i\}.$$

Under condition C_1 or C_3 let $Z_{nk}^{(\varepsilon)}(t) \equiv 0$. Under condition C_2 there exists an I_i for which $X_{nk}(I_i) \neq 0$. In this case we define $Z_{nk}^{(\varepsilon)}(t)$ in such a way that for $J \subset I_i$

$$Z_{nk}^{(\varepsilon)}(J) = X_{nk}(J),$$

and for $J \cap I_i = 0$

$$Z_{nk}^{(\varepsilon)}(J) = 0.$$

Let $J \subset I$. Suppose that the endpoints of the interval J belong to the intervals I_{i_1} and I_{i_2} $(i_1 < i_2)$ and that they divide these intervals into the subintervals I'_{i_1}, I''_{i_1} and I'_{i_2}, I''_{i_2} respectively. The case $i_1 = i_2$ is treated analogously.

Clearly

$$\text{prob}\{X_{nk}^{(\varepsilon)}(J) \neq 0\} = \sum_{\ell=1}^{3} \text{prob}\{X_{nk}^{(\varepsilon)}(J) \neq 0;\ C_\ell\}. \tag{5}$$

Now

$$\text{prob}\{X_{nk}^{(\varepsilon)}(J) \neq 0;\ C_1\} = \text{prob}\{X_{nk}(J) \neq 0;\ C_1\} \leq$$
$$\text{prob}\{X_{nk}(I'_{i_1}) \neq 0;\ X_{nk}(I''_{i_1}) \neq 0\} + \text{prob}\{X_{nk}(I'_{i_2}) \neq 0;\ X_{nk}(I''_{i_2}) \neq 0\} \tag{6}$$

and

$$\text{prob}\{X_{nk}^{(\varepsilon)}(J)\neq 0; C_3\} = \text{prob}\{X_{nk}(J)\neq 0; C_3\} \leq \text{prob}(C_3) \leq \text{prob}(A_{nk}). \tag{7}$$

Lastly

$$\text{prob}\{X_{nk}^{(\varepsilon)}(J)\neq 0; C_2\} \leq \sum_{i\neq i_1} \text{prob}\{X_{nk}(I_i)\neq 0;\ X_{nk}(I''_{i_1})\neq 0\} + \sum_{i=i_2} \text{prob}\{X_{nk}(I_i)\neq 0;\ X_{nk}(I'_{i_2})\neq 0\}. \tag{8}$$

From (5), (6), (7) and (8) it follows that

$$\sum_{k=1}^{k_n} \text{prob}\{X_{nk}^{(\varepsilon)}(J)\neq 0\} \leq \sum_{k=1}^{k_n} [\sum_{j=1}^{2} \text{prob}\{X_{nk}(I'_{i_j})\neq 0; X_{nk}(I''_{i_j})\neq 0\} + \text{prob}\{A_{nk}\}$$

$$+ \sum_{i\neq i_1} \text{prob}\{X_{nk}(I_i)\neq 0; X_{nk}(I''_{i_1})\neq 0\} + \sum_{i\neq i_2} \text{prob}\{X_{nk}(I_i)\neq 0; X_{nk}(I'_{i_2})\neq 0\}]$$

and accordingly condition A_4 is a consequence of A_1.

Let $I_1,\ldots,I_r$ be disjoint intervals without common endpoints and ε be less than the distance between any two of them. Let us choose I so that $\bigcup_{i=1}^{r} I_i \subset I$ and define $X_{nk}^{(\varepsilon)}(t)$ in accordance with A_4. Then

$$\text{prob}\{A_{nk}\} \leq \text{prob}[\bigcup_{i=1}^{r} \{X_{nk}^{(\varepsilon)}(I_i) \neq 0\}] \leq \sum_{i=1}^{r} \text{prob}\ X_{nk}^{(\varepsilon)}(I_i) \neq 0\}$$

and

$$\sum_{k=1}^{k_n} \text{prob}(A_{nk}) \leq \sum_{i=1}^{r} \sum_{k=1}^{k_n} \text{prob}\{X_{nk}^{(\varepsilon)}(I_i) \neq 0\};$$

consequently condition A_1 follows from A_4. The theorem is proved.

ACKNOWLEDGMENTS

This work was done in the Department of Probability of Lomonosov University, Moscow. I express my sincere thanks to B. V. Gnedenko for the posed question and discussing results.

REFERENCES

Gnedenko, B. V. & Kolmogorov, A. N. (1968). *Limit distributions for sums of independent random variables,* 2nd ed.. Reading, Mass.: Addison Wesley.

Grigelionis, B. I. (1963). On the convergence of sums of random step processes to a Poisson process. Teor. Veroyatnost. i Priminen. 8, 189-94. (English translation in Theory Prob. Appl. 8, 177-82.)

Grigelionis, B. I. (1966). On the question of convergence of sums of step-like random processes to a Poisson process. Litovsk. Mat. Sb. 6, 241-4. (In Russian).

Khintchine, A. Y. (1960). *Mathematical methods in the theory of queues.* London: Griffin.

Lee, P. M. (1967). Infinitely divisible stochastic processes. Z. Wahrscheinlichkeitstheorie verw. Geb. 7, 147-60.

Ososkov, G. A. (1956). A limit theorem for processes of homogeneous events. Teor Veroyatnost. i Priminen. 1, 274-82 (English translation in Theory Prob. Appl. 1, 248-55).

Szász, D. & Woyczyński, W. A. (1970). Poissonion random measures and linear processes with independent increments. Bull. de l'Académie Polonaise des Sciences 18, 475-82.

ON WEAK CONVERGENCE OF THE SUMS OF MULTIVARIATE STOCHASTIC POINT PROCESSES

B. Grigelionis
Institute of Physics and Mathematics, Vilinus, U.S.S.R.

SUMMARY

The author earlier gave necessary and sufficient conditions for the convergence of superpositions of multivariate point processes, $\underline{N}_n(t)$, to multivariate, time-dependent, Poisson processes. In this paper we give conditions under which the distributions of a wide class of functionals of $\underline{N}_n(t)$ converge to the distributions of functionals of the limiting Poisson process.

1. INTRODUCTION

Let $\underline{N}_{nr}(t) = \{N^1_{nr}(t), \ldots, N^m_{nr}(t)\}$, $t \geq 0$, $r = 1,\ldots, k_n$, be independent, right continuous, m-variate stochastic processes such that the increments of the components $N^j_{nr}(t)$ take only non-negative integer values and $\text{prob}\{N^j_{nr}(0) = 0\} = 1$, $j = 1,\ldots, m$. Let

$$\underline{N}_n(t) = \{\sum_{r=1}^{k_n} N^1_{nr}(t),\ldots, \sum_{r=1}^{k_n} N^m_{nr}(t)\}, \tag{1}$$

and denote by $\underline{N}_0(t) = \{N^1_0(t),\ldots, N^m_0(t)\}$, $t \geq 0$, a right continuous m-variate Poisson process with independent components and with a mean function $\underline{\lambda}(t) = [E\{N^1_0(t)\}, \ldots, E\{N^m_0(t)\}] = \{\lambda^1(t), \ldots, \lambda^m(t)\}$.

Because a great majority of investigations in queueing theory and in reliability theory are based on the assumption that the arrival processes are Poisson processes and in many practical cases they can be represented as the sums of a large number of independent summands of small intensity, it is an important problem to investigate conditions when the sums (1) are near to some Poisson process.

Under natural assumption of smallness of summands, i.e., that for every fixed $t > 0$

$$\lim_{n\to\infty} \max_{1\le r\le k_n} \text{prob}\{\sum_{k=1}^{m} N_{nr}^{k}(t) > 0\} = 0 \qquad (2)$$

we found (Grigelionis, 1970) necessary and sufficient conditions for convergence of sums (1) to the Poisson process in the sense of weak convergence of the finite dimensional distributions of $\underline{N}_n(t)$ to the corresponding finite dimensional distributions of the Poisson process. The result is the following

Theorem 1. Under assumption (2) the finite dimensional distributions of $\underline{N}_n(t)$ converge weakly to the finite dimensional distributions of the Poisson process $\underline{N}_0(t)$ with the mean function $\underline{\lambda}(t) = \{\lambda^1(t),\ldots, \lambda^m(t)\}$ if and only if for every fixed $t > 0$

$$\lim_{n\to\infty} \sum_{r=1}^{k_n} \text{prob}\{N_{nr}^{k}(t) > 0\} = \lambda^k(t) \qquad (k = 1,\ldots, m) , \qquad (3)$$

and

$$\lim_{n\to\infty} \sum_{r=1}^{k_n} \text{prob}\{\sum_{k=1}^{m} N_{nr}^{k}(t) > 1\} = 0 . \qquad (4)$$

Under additional assumption of stationarity and ordinariness (see Daley and Vere-Jones, 1972, Sect. 6, this volume, for discussion of this terminology) some general results were obtained earlier by Palm (1943), Khintchine (1960), Ososkov (1956), Grigelionis (1963), Goldman (1967) for one-dimensional point processes and by Kerstan and Matthes (1965), Grigelionis (1966), Cinlar (1968) for multivariate point processes.

Let us suppose that the point processes $\underline{N}_{nr}(t) = \{N_{nr}^1(t), \ldots , N_{nr}^m(t)\}$, $r = 1, \ldots , k_n$, have stationary increments and are ordinary, i.e., for all $\ell \ge 1$, $0 \le t_0 < \cdots < t_\ell$, $t > 0$, the distribution of the probability of increments $\underline{N}_{nr}(t_k+t) - \underline{N}_{nr}(t_{k-1}+t)$, $k = 1,\ldots, \ell$, is independent of t, and

$$\lim_{t\to 0+} \text{prob}\{\tilde{N}_{nr}(t) > 1\}/t = 0 ,$$

where

$$\tilde{N}_{nr}(t) = \sum_{k=1}^{m} N_{nr}^{k}(t) .$$

If $\tilde{\lambda}_{nr} = E\{\tilde{N}_{nr}(1)\} < \infty$, then there exist Palm-Khintchine functions

$$\phi_{nr}^{k}(j,t) = \lim_{s\to 0+} \text{prob}\{N_{nr}^{k}(t+s) - N_{nr}^{k}(s) = j \mid N_{nr}^{k}(s) > 0\}$$

and

$$\tilde{\phi}_{nr}(j,t) = \lim_{s\to 0+} \text{prob}\{\tilde{N}_{nr}(t+s) - \tilde{N}_{nr}(s) = j \mid \tilde{N}_{nr}(s) > 0\}$$

$$(j = 0,1,\ldots) .$$

If we assume that

$$\lim_{n\to\infty} \max_{1\le r\le k_n} \tilde{\lambda}_{nr} = 0 , \tag{5}$$

then the condition (2) is obviously fulfilled.

Corollary 1. Under assumption (5) the finite dimensional distributions of $\underline{N}_n(t)$ converge weakly to the finite dimensional distributions of the stationary Poisson process with the parameter $\underline{\lambda} = (\lambda^1, \ldots , \lambda^m)$ if and only if for every fixed $t > 0$

$$\lim_{n\to\infty} \sum_{r=1}^{k_n} \lambda_{nr}^{k} \int_0^t \phi_{nr}^{k}(0,u)\, du = \lambda^k t \qquad (k = 1,\ldots, m) \tag{6}$$

and

$$\lim_{n\to\infty} \sum_{r=1}^{k_n} \tilde{\lambda}_{nr} \int_0^t \tilde{\phi}_{nr}(1,u)\, du = 0 , \tag{7}$$

where

$$\lambda_{nr}^{k} = E\{N_{nr}^{k}(1)\} .$$

2. GENERALIZATIONS

In some situations it is important to find general conditions not only when the finite dimensional distributions of the sums $\underline{N}_n(t)$ converge to the finite dimensional distributions of the Poisson process $N_0(t)$, but also to find conditions when the distributions of wide class of functionals from $\underline{N}_n(t)$ converge to the distributions of functionals from the limit Poisson process. For this purpose we shall use some general results of Skorohod (1956) and Stone (1963).

Following Stone's paper (1963) we denote by K the space of all m-dimensional vector valued functions $\underline{x}(t)$, $0 \leq t < \infty$, which at every point have a limit from the left and are continuous from the right. We define on K the Skorohod topology J_1: a sequence $\underline{x}_n(t)$ is said to be J_1-convergent to $\underline{x}(t)$ if there exists a sequence of continuous one-to-one mappings $\lambda_n(t)$ of the interval $[0,\infty)$ onto itself such that for each $t > 0$

$$\sup_{0 \leq s \leq t} |\lambda_n(s) - s| \to 0$$

and

$$\sup_{0 \leq s \leq t} \|\underline{x}_n(s) - \underline{x}\{\lambda_n(s)\}\| \to 0$$

as $n \to \infty$, where $\|\underline{x}\| = (\Sigma_{k=1}^{m} x_k^2)^{1/2}$ for $\underline{x} = (x_1, \ldots, x_m)$.

We denote by C the class of all functionals f on K such that $f\{X_n(\cdot)\}$ are random variables and f is continuous in the J_1 topology almost everywhere with respect to the measure on K corresponding to the process $\underline{N}_0(t)$.

The sequence $\underline{N}_n(t)$ is said to be weakly convergent to $\underline{N}_0(t)$ if for all $f \in C$ the distribution of $f\{X_n(\cdot)\}$ weakly converges to the distribution of $f\{X_0(\cdot)\}$ as $n \to \infty$.

From the theorem proved by Stone (1963) we obtain the following assertion.

Theorem 2. The sequence $\underline{N}_n(t)$ is weakly convergent to $\underline{N}_0(t)$ if the finite dimensional distributions of $\underline{N}_n(t)$ converge weakly to the finite dimensional distributions of $\underline{N}_0(t)$ as $n \to \infty$ and for every fixed $T > 0$

$$\lim_{\substack{n\to\infty \\ c\to 0}} \text{prob}\left[\sup_{\substack{t-c<t_1<t<t_2<t+c; \\ 0\le t_1<t_2\le T}} \min\{\|\underline{N}_n(t_1) - \underline{N}_n(t)\|, \|\underline{N}_n(t) - \underline{N}_n(t_2)\|\} > 0\right] = 0 .$$

We can easily prove that

$$\text{prob}\left[\sup_{\substack{t-c<t_1<t<t_2<t+c; \\ 0\le t_1<t_2\le T}} \min\{\|\underline{N}_n(t_1) - \underline{N}_n(t)\|, \|N_n(t) - N_n(t_2)\|\} > 0\right]$$

$$\le \sum_{r=1}^{k_n} \text{prob}\{\sum_{k=1}^{m} N_{nr}^k(T) > 1\} + \int_0^T \{\tilde{\Lambda}_n(t+c) - \tilde{\Lambda}_n(t-c)\}\, d\tilde{\Lambda}_n(t) \quad , (8)$$

where

$$\tilde{\Lambda}_n(t) = \sum_{r=1}^{k_n} \sum_{k=1}^{m} E\{N_{nr}^k(t)\} \qquad (t \ge 0) ,$$

$$\tilde{\Lambda}_n(t) = 0 \qquad (t < 0) .$$

Combining Theorem 1 and (8) we can obtain general conditions for the weak convergence of the sequence $\underline{N}_n(t)$ to the Poisson process $\underline{N}_0(t)$. Some results for one-dimensional point processes have recently been obtained by R. Banys.

Corollary 2. The sequence $\underline{N}_n(t)$ is weakly convergent to the Poisson process with the mean function $\underline{\lambda}(t) = \{\lambda^1(t), \ldots , \lambda^m(t)\}$, if conditions (2)-(4) are fulfilled and for every fixed $T > 0$

$$\lim_{\substack{n\to\infty \\ c\to 0}} \int_0^T \{\tilde{\Lambda}_n(t+c) - \tilde{\Lambda}_n(t-c)\}\, d\tilde{\Lambda}_n(t) = 0 .$$

Let us suppose now that the point processes $\underline{N}_{nr}(t) = \{N_{nr}^1(t), \ldots , N_{nr}^m(t)\}$, $r = 1, \ldots , k_n$, are stationary and ordinary, and

$$\lim_{n\to\infty} \sum_{r=1}^{k_n} \lambda_{nr}^k = \lambda^k \qquad (k = 1,\dots, m) \tag{9}$$

exists. From corollary 1, theorem 2, (8) and (9) we can derive the following assertion.

Theorem 3. Under assumptions (7) and (9) the sequence $\underline{N}_n(t)$ is weakly convergent to the stationary Poisson process with the parameter $\underline{\lambda} = (\lambda^1, \dots, \lambda^m)$ if for every fixed $t > 0$

$$\lim_{n\to\infty} \sum_{r=1}^{k_n} \lambda_{nr}^k \phi_{nr}^k(0,t) = \lambda^k \qquad (k = 1, \dots, m)$$

and

$$\lim_{n\to\infty} \sum_{r=1}^{k_n} \tilde{\lambda}_{nr} \tilde{\phi}_{nr}(0,t) = \tilde{\lambda},$$

where

$$\tilde{\lambda} = \sum_{k=1}^{m} \lambda^k .$$

3. AN EXAMPLE-MARKOV RENEWAL PROCESSES

As an example we investigate the case of Markov renewal processes.

Let $I_j^{n,r}$, $j \geq 0$, $r = 1, \dots, k_n$, be a finite Markov chain with state space $E = \{1, \dots, m\}$, initial distributions $a_i^{n,r} = \text{prob}\{I_0^{n,r} - i\}$ and transition probability matrices $\{p_{ij}^{n,r}\}$, $1 \leq i, j \leq m$. Let $Z^{n,r}(t) = \max\{j: \Sigma_{k=0}^{j} T_k^{n,r} < t\}$ where $T_0^{n,r}, \dots, T_k^{n,r}$ are sequences of non-negative random variables independent for different indexes r and such that

$$\text{prob}\{T_0^{n,r} = 0\} = 1,$$

$$\text{prob}\{I_1^{n,r} = j,\ T_1^{n,r} < t \mid I_0^{n,r} = i\} = p_{ij}^{n,r} \hat{F}_{ij}(t) = \hat{Q}_{ij}^{n,r}(t) ,$$

$$\text{prob}\{I_{k+1}^{n,r} = j,\ T_{k+1}^{n,r} < t \mid I_0^{n,r}, \ldots, I_{k-1}^{n,r}, T_1^{n,r}, \ldots, T_k^{n,r}, I_k^{n,r} = i\}$$

$$= p_{ij}^{n,r} F_{ij}^{n,r}(t) = Q_{ij}^{n,r}(t) \qquad (k \geq 1),$$

for all $t \geq 0$, $i, j \in E$; where $\hat{F}_{ij}^{n,r}(t)$ and $F_{ij}^{n,r}(t)$ are given distribution functions on $[0,\infty)$.

We define Markov renewal processes as

$$\underline{N}_{nr}(t) = \{N_{nr}^{1}(t), \ldots, N_{nr}^{m}(t)\},$$

where

$$N_{nr}^{j}(t) = \sum_{k=1}^{Z^{(n,r)}(t)} \delta_{jI_k^{n,r}}$$

and $\delta_{ij} = 1$ when $i = j$, and $\delta_{ij} = 0$ when $i \neq j$.

Assumption (2) in the case of Markov renewal processes is equivalent to the condition:

$$\lim_{n\to\infty} \max_{1\leq r\leq k_n} \sum_{i,j=1}^{m} a_i^{n,r} \hat{Q}_{ij}^{n,r}(t) = 0 \qquad (10)$$

for every fixed $t > 0$.

Using the Theorem 1, Sapagovas (1969) proved that under assumption (10) the finite dimensional distributions of the sums $\underline{N}_n(t)$ of Markov renewal processes converge weakly to the finite dimensional distributions of the Poisson process with the mean function $\underline{\lambda}(t) = \{\lambda^1(t), \ldots, \lambda^m(t)\}$ if and only if for every fixed $t > 0$

$$\lim_{n\to\infty} \sum_{r=1}^{k_n} \sum_{i=1}^{m} a_i^{n,r} \hat{Q}_{ik}^{n,r}(t) = \lambda^k(t) \qquad (k = 1, \ldots, m),$$

and

$$\lim_{n\to\infty} \sum_{r=1}^{k_n} \sum_{i,j,k=1}^{m} a_i^{n,r} \hat{Q}_{ij}^{n,r}(t) * Q_{jk}^{n,r}(t) = 0,$$

where * is a symbol of convolution.

Let

$$H_i^{n,r}(t) = \sum_{j=1}^{m} Q_{ij}^{n,r}(t) ,$$

$$0 < \eta_i^{n,r} = \int_0^\infty t \, dH_i^{n,r}(t) < \infty \qquad (i = 1, \ldots , m).$$

Let the system of equations

$$\sum_{j=1}^{m} \pi_j^{n,r} p_{j,k}^{n,r} = \pi_k^{n,r}, \qquad \sum_{k=1}^{m} \pi_k^{n,r} = 1 \qquad (r = 1, \ldots , k_n)$$

have a unique solution.

In the case when

$$a_i^{n,r} = \frac{\pi_i^{n,r} \eta_i^{n,r}}{\sum_{j=1}^{m} \pi_j^{n,r} \eta_j^{n,r}}$$

and

$$\hat{Q}_{ij}^{n,r}(t) = \frac{1}{\eta_i^{n,r}} \int_0^t \{p_{ij}^{n,r} - Q_{ij}^{n,r}(u)\} \, du \qquad (i,j=1,\ldots,m; r=1,\ldots,k_n)$$

it is well known (Pyke, 1961, and Pyke and Schaufele, 1966), that the point processes $\underline{N}_{nr}(t)$, $r = 1, \ldots , k_n$, are stationary.

Using Theorem 2, Equation (8) and the result of Sapagovas (1969), we can prove the following statement.

Theorem 4. If for every fixed $t > 0$

$$\lim_{n\to\infty} \max_{\substack{1\le r\le k_n; \\ 1\le i,j\le m}} Q_{ij}^{n,r}(t) = 0 ,$$

then the sequence of the sums $\underline{N}_n(t)$ of stationary Markov renewal processes is weakly convergent to the stationary Poisson process with parameter $\underline{\lambda} = \{\lambda^1, \ldots, \lambda^m\}$ if

$$\lim_{n\to\infty} \sum_{r=1}^{k_n} \frac{\pi_k^{n,r}}{\sum_{j=1}^{m} \pi_j^{n,r} \eta_j^{n,r}} = \lambda^k \qquad (k = 1, \ldots, m) .$$

We remark that in Theorem 4 summands $\underline{N}_{nr}(t)$, $r = 1, \ldots, k_n$, are not necessarily ordinary. They are ordinary if and only if

$$Q_{ij}^{n,r}(0+) = 0 \qquad (i, j = 1, \ldots, m, r = 1, \ldots, k_n).$$

REFERENCES

Cinlar, E. (1968). On the superposition of m-dimensional point processes. J. Appl. Prob. 5, 160-76.

Goldman, J. R. (1967). Stochastic point processes: limit theorems. Ann. Math. Statist. 38, 771-9.

Grigelionis, B. (1963). On convergence of sums of random step processes to a Poisson process. Teor. Veroyatnost. i Primenen 8, 189-94. (English translation in Theory Probl. Appl. 8, 177-82.)

Grigelionis, B. (1966). Composition of integer-valued random measures. Litovsk. Mat. Sb. 6, 359-63 (In Russian).

Grigelionis, B. (1970). Limit theorems for the sums of multidimensional stochastic step processes. Litovsk. Mat. Sb. 10, 29-49 (In Russian).

Kerstan, J. and Matthes, K. (1965). Generalization of the theorem of Palm-Khintchine. Ukrainian J. Math. 17, 29-36 (In Russian).

Khintchine, A. Y. (1960). *Mathematical Methods in the Theory of Queueing.* London: Griffin.

Ososkov, G. A. (1956). A limit theorem for flows of homogeneous events. Teor. Veroyatnost. i Primenen 1, 274-82 (English translation in Theory Prob. Appl. 1, 248-55).

Palm, C. (1943). Intensitätsschwankungen in Fernsprechverkehr. Ericsson Techniks 44, 1-189.

Pyke, R. (1961). Markov renewal processes with finitely many states. Ann. Math. Statist. 32, 1243-59.

Pyke, R. and Schaufele, R. (1966). The existence and uniqueness of stationary measures for Markov renewal processes. Ann. Math. Statist. 37, 1439-62.

Sapagovas, J. (1969). On the convergence of the sums of the Markov renewal processes to a many-dimensional Poisson process. Litovsk. Mat. Sb. 9, 817-26 (Russian).

Stone, C. (1963). Weak convergence of stochastic processes defined on semi-infinite time intervals. Proc. Amer. Math. Soc. 14, 694-6.

Skorohod, A. V. (1956). Limit theorems for stochastic processes. Teor. Veroyatnost. i Priminen 1, 289-319(English translation in Theory Prob. Appl. 1, 261-90).

PALM DISTRIBUTIONS AND SUPERPOSITIONS OF INDEPENDENT POINT PROCESSES IN R^n

R. V. Ambartzumian
Armenian Academy of Sciences, Erevan, U.S.S.R.

SUMMARY

Although much work has been done in the field of superpositions of independent homogeneous (= translation invariant) point processes in R^n, it seems that the role of Palm distributions in this theory has remained underestimated. The present paper is aimed at exposing this role, together with an attempt at a more rigorous introduction of Palm distributions by elementary integral geometry methods. In doing this we use a somewhat unusual definition of a point process in R^n; the latter may be modified appropriately for the study of homogeneous and isotropic (= invariant with respect to all Euclidean motions) point processes. This in turn gives rise to a problem which is described in the concluding section. New examples of infinitely divisible point processes in R^n are constructed.

1. DEFINITION OF RANDOM POINT PROCESSES IN R^n

Let T_n denote the set of all parallel rigid shifts (translations) of n-dimensional Euclidean space R^n. Each $t \in T_n$ may be identified with the point $t0 \in R^n$ where $0 \in R^n$ is the origin and $t0$ is the image of 0 under t. There will be no misunderstanding if we describe subsets of T_n as being the corresponding subsets of R^n. Let B denote the σ-algebra of subsets of T_n coinciding with the class of Borel subsets of R^n. Denote by dt the element of n-dimensional Lebesgue measure $|.|$ on T_n.

Let M_n be the class of all countable subsets of R^n satisfying the condition that any $m \in M_n$ has no point of accumulation in any finite part of R^n.

For any set $S \subset R^n$ let tS be its image under $t \in T_n$.

Define $N(D,m)$ as the number of points in $D \cap m$, where D is a domain in R^n and $m \in M_n$.

Let (Ω, A, μ) be a probability space. Given a mapping $m(\omega): \Omega \to M_n$, $N\{D, tm(\omega)\}$ is then a function defined on $T_n \times \Omega$. Introduce the σ-algebra $B \times A$ of subsets of $T_n \times \Omega$.

Definition. A random point process (r.p.p.) in R^n *is a mapping* $m(\omega)$: $\Omega \to M_n$ *such that for every sphere* $S \subset R^n$, $N\{S, tm(\omega)\}$ *is a function measurable on* $T_n \times \Omega$. (Here and in the sequel, spheres in R^n are assumed to be open.)

Denote by C the minimal σ-algebra of subsets of M_n containing all subsets of the form

$$\{m:N(S,m) = k\} \tag{1.1}$$

for every sphere $S \subset R^n$ and every integer $k \geq 0$. Those members of C which can be represented as finite intersections of sets of the type (1.1) are called *cylindrical*.

The mapping $m(\omega)$ induces a measure P on C which is called the *distribution* of the r.p.p. $m(\omega)$.

For any fixed sphere $S \subset R^n$, $N\{S, m(\omega)\}$ is then a random variable. Note that the distribution P on C is uniquely determined by the joint distributions of the random vectors

$$\underline{N}(\underline{S}) = \{N(S_1), \ldots, N(S_s)\},$$

where $S_1, \ldots, S_s$ is a set of non-overlapping spheres in R^n (in other words, P is uniquely determined by its values on cylindrical events).

The r.p.p. $m(\omega)$ is said to be *homogeneous* (h.r.p.p.) if for any $C \in C$ the value $P(tC)$ does not depend on $t \in T_n$, where $tC = \{m:t^{-1}m \in C\}$. In other words, for an h.r.p.p. the distribution of $tm(\omega)$ coincides with that of $m(\omega)$ for any $t \in T_n$.

2. PALM DISTRIBUTION

Let $S(r)$ denote the sphere of radius r centered at the origin in R^n, and for $h > 0$ set $\Delta(h,r) = S(r + h) - S(r)$. In particular then, $\Delta(h,0) = S(h)$. Introduce the events

$$H_k(r) = \{m:N\{\Delta(h,r),\ m\} = k\} .$$

Denote by $\partial S(r)$ the boundary of the sphere $S(r)$, and by dt^* an element of the surface $\partial S(r)$. Denote by $E(.)$ expectation with respect to P (equivalently, integration on Ω with respect to the measure μ), $\|D\| = \int_D dt^*$ for $D \subset \partial S(r)$ such that the integral is well-defined.

Lemma 1. Let P be the distribution of an h.r.p.p. $m(\omega)$ in R^n such that

$$E[N\{S(r),\ m(\omega)\}] = \lambda|S(r)| \qquad (\lambda < \infty) , \tag{2.1}$$

(i.e., $m(\omega)$ is an h.r.p.p. of finite intensity λ). Then for $h \downarrow 0$,

$$P\{H_1(0)\} = \lambda|S(h)| + o\{|S(h)|\} , \tag{2.2}$$

$$\sum_{k=2}^{\infty} kP\{H_k(0)\} = o\{|S(h)|\}, \tag{2.3}$$

and for every cylindrical $C \in \mathcal{C}$ there exists the limit

$$\pi^{(0)}(C) = \lim_{h\downarrow 0} P\{C \cap H_1(0)\}/|S(h)| . \tag{2.4}$$

Proof. Introduce the indicator of the set C

$$I_C(m) = \begin{cases} 1 & \text{if} \quad m \in C , \\ 0 & \text{if} \quad m \notin C , \end{cases}$$

and consider the integrals $\int_{S(1)} I_{H_k(0)}\{tm(\omega)\}\,dt$ for small h and fixed $\omega \in \Omega$. Since for every h.r.p.p. $P\{m(\omega) \cap \partial S(1) = \emptyset\} = 1$, there exists with probability one $h(\omega) > 0$ such that for all $h < h(\omega)$,

$$\int_{S(1)} I_{H_1(0)}\{tm(\omega)\}dt = N\{S(1),\, m(\omega)\}\cdot|S(h)| \ , \qquad (2.5)$$

$$\int_{S(1)} I_{H_k(0)}\{tm(\omega)\}\,dt = 0 \qquad (k = 2,3,\ldots) \ . \qquad (2.6)$$

Thus with probability one,

$$\lim_{h\downarrow 0} |S(h)|^{-1}\int_{S(1)} I_{H_1(0)}\{tm(\omega)\}\,dt = N\{S(1),\, m(\omega)\} \ , \qquad (2.7)$$

$$\lim_{h\downarrow 0} |S(h)|^{-1}\int_{S(1)} \sum_{k=2}^{\infty} kI_{H_k(0)}\{tm(\omega)\}\,dt = 0 \ . \qquad (2.8)$$

Choose cylindrical $C \in \mathcal{C}$ and let S_k, $k = 1,\ldots, s$, be the spheres associated with C. For fixed $\omega \in \Omega$ such that $m(\omega) \cap \partial S(1) = \emptyset$, put

$$F_\omega(C) = \{\rho_i : \rho_i \in m(\omega) \cap S(1),\ \rho_i^{-1}\, m(\omega) \cap (\bigcup_k \partial S_k) = \emptyset\} \ ,$$

$$G_\omega(C) = \{\rho_i : \rho_i \in m(\omega) \cap S(1),\ \rho_i^{-1}\, m(\omega) \cap (\bigcup_k \partial S_k) \neq \emptyset\} \ .$$

Then for $h < h(\omega)$,

$$\int_{S(1)} I_{CH_1(0)}\{tm(\omega)\}\,dt = \sum_{\rho_i \in m(\omega)\cap S(1)} \int_{\rho_i^{-1}S(h)} I_C\{tm(\omega)\}\,dt, \qquad (2.9)$$

since for $t \in \rho_i^{-1}S(h)$ and $h < h(\omega)$, $I_{CH_1(0)}(tm) = I_C(tm)$.

If $\rho_i \in F_\omega(C)$, then there exists $h_1(\omega)$ such that identically for all $t \in \rho_i^{-1}S(h)$ with $h < h_1(\omega)$, $I_C\{tm(\omega)\} = I_C\{\rho_i^{-1}\, m(\omega)\}$, so for $\rho_i \in F_\omega(C)$,

$$\lim_{h\downarrow 0} |S(h)|^{-1} \int_{\rho_i^{-1}S(h)} I_C\{tm(\omega)\}\, dt = I_C\{\rho_i^{-1}m(\omega)\} \quad . \tag{2.10}$$

In general (2.10) is not true for $\rho_i \in G_\omega(C)$ because then $\rho_i^{-1} \in \rho_i^{-1}S(h)$ for all $h > 0$ and ρ_i^{-1} is a point of discontinuity of $I_C\{tm(\omega)\}$ as a function of t. Nevertheless the limit

$$\lim_{h\downarrow 0} |S(h)|^{-1} \int_{\rho_i^{-1}S(h)} I_C\{tm(\omega)\}\, dt = W_\omega(C,\rho_i) \tag{2.11}$$

will still exist, its value depending on the position of the points belonging to $\rho_i^{-1} m(\omega) \cap (\bigcup_k \partial S_k)$. It follows from (2.9), (2.10) and (2.11) that

$$\lim_{h\downarrow 0} |S(h)|^{-1} \int_{S(1)} I_{CH_1(0)}\{tm(\omega)\}\, dt$$

$$= \sum_{\rho_i \in F_\omega(C)} I_C\{\rho_i^{-1} m(\omega)\} + \sum_{\rho_i \in G_\omega(C)} W_\omega(C,\rho_i) \quad . \tag{2.12}$$

On the other hand, for every $\omega \in \Omega$ and $h < 1$,

$$\int_{S(1)} N\{S(h),\ tm(\omega)\}\, dt \leq |S(h)|.N\{S(2),\ m(\omega)\} \ ,$$

so since $I_{CH_1(0)}\{tm(\omega)\} \leq I_{H_1(0)}\{tm(\omega)\} \leq N\{S(h),\ tm(\omega)\}$ and

$$\sum_{k=2}^{\infty} kI_{H_k(0)}\{tm(\omega)\} \leq N\{S(h),\ tm(\omega)\} \ ,$$

we can apply the dominated convergence theorem to (2.7), (2.8) and (2.12) and deduce that

$$\lim_{h\downarrow 0} |S(h)|^{-1} E[\int_{S(1)} I_{H_1(0)}\{tm(\omega)\}dt] = E[N\{S(1), m(\omega)\}] ,$$

$$\lim_{h\downarrow 0} |S(h)|^{-1} E[\int_{S(1)} \sum_{k=2}^{\infty} kI_{H_k(0)}\{tm(\omega)\} dt] = 0 ,$$

$$\lim_{h\downarrow 0} |S(h)|^{-1} E[\int_{S(1)} I_{CH_1(0)}\{tm(\omega)\} dt] \tag{2.13}$$

$$= E[\sum_{\rho_i \varepsilon F_\omega(C)} I_C\{\rho_i^{-1} m(\omega)\}] + E\{\sum_{\rho_i \varepsilon G_\omega(C)} W_\omega(C,\rho_i)\} .$$

Now for an h.r.p.p., $E[I_C\{tm(\omega)\}] = P(C)$ for every $C \varepsilon \mathcal{C}$ and all $t \varepsilon T_n$. Using this identity in conjunction with Fubini's theorem in (2.13), equations (2.2) and (2.3) follow, together with the existence of $\pi^{(0)}(C)$ as at (2.4) and the expression

$$|S(1)| \; \pi^{(0)}(C) = E[\sum_{\rho_i \varepsilon F_\omega(C)} I_C\{\rho_i^{-1} m(\omega)\}] + E\{\sum_{\rho_i \varepsilon G_\omega(C)} W_\omega(C,\rho_i)\}. \tag{2.14}$$

This completes the proof of Lemma 1.

Lemma 2. Let P be the distribution of an h.r.p.p. $m(\omega)$ in R^n of finite intensity λ. Then for $h \downarrow 0$ and every $r > 0$,

$$P\{H_1(r)\} = \lambda h\|\partial S(r)\| + o(h) , \tag{2.15}$$

$$\sum_{k=2}^{\infty} kP\{H_k(r)\} = o(h) , \tag{2.16}$$

and for every cylindrical $C \varepsilon \mathcal{C}$ there exists the limit (with $r > 0$)

$$\pi^{(r)}(C) = \lim_{h\downarrow 0} P\{C \cap H_1(r)\}/h . \tag{2.17}$$

Proof. We shall use the fact that for every $m \varepsilon M_n$ consisting of two distinct points, ρ_1 and ρ_2, say, $\int I_{H_2(r)}(tm)\, dt = o(h)$, $h \downarrow 0$, for the integral equals the volume common to the two shells $\Delta(h,r)$ and $\rho_1\rho_2^{-1}\Delta(h,r)$

On account of this fact, we have for every fixed $\omega \in \Omega$ that as $h \downarrow 0$,

$$\int_{S(1)} I_{CH_1(r)}\{tm(\omega)\}\, dt = \sum_{\rho_i \in m(\omega)} \int_{t\in S(1),\, t\rho_i \in \Delta(h,r)} I_C\{tm(\omega)\}\, dt + o(h) , \tag{2.18}$$

$$\int_{S(1)} \sum_{k=2}^{\infty} kI_{H_k(r)}\{tm(\omega)\}\, dt = o(h) , \tag{2.19}$$

noting that in fact the sum in (2.19) is a finite sum for every $\omega \in \Omega$.

For every cylindrical $C \in \mathcal{C}$, every $\rho_i \in m(\omega)$ with $\omega \in \Omega$, we now show that

$$\int_{t\in S(1),\, t\rho_i \in \Delta(h,r)} I_C\{tm(\omega)\}\, dt = h \int_{Z(r,\rho_i)} I_C\{tm(\omega)\}\, dt^* + o(h) , \tag{2.20}$$

where $Z(r,\rho_i)$ denotes the (n-1)-dimensional surface in T_n defined by

$$Z(r,\rho_i) = \{t \in T_n : t\rho_i \in \partial S(r),\ t \in S(1)\} .$$

To prove (2.20) consider the set

$$D_i = \{t' \in T_n : t' \in \bar{S}(1),\ t'\rho_i \in \bar{\Delta}(h,r),\ I_C\{tm(\omega)\} \text{ is not a constant function of } t \text{ in the domain } t'\rho_i S(h') \cap \bar{\Delta}(h,r) \text{ for any } h' > 0\},$$

where the bar denotes closure of the set and $\rho_i \in m(\omega)$. By definition of a cylindrical event C,

$$D_i \subset \bigcup_{\rho_j \in m(\omega)} \bigcup_k \{t' \in \bar{S}(1) : t'\rho_i \in \bar{\Delta}(h,r),\ t'\rho_j \in \partial S_k\} , \tag{2.21}$$

where the S_k, $k = 1,\ldots, s$, are the spheres associated with C, and consequently D_i is a subset of a finite number of parts of (n-1)-dimensional spherical surfaces. Thus (2.20) will follow on showing that $\|D_i \cap Z(r,\rho_i)\| = 0$. Indeed, $\|\{t':t'\rho_i \in \partial S(r),\ t'\rho_j \in \partial S_k\}\| = 0$ unless $\rho_i^{-1}\ \partial S(r) = \rho_j^{-1}\ \partial S_k$, which case can be excluded from the right-hand side of (2.21) because $t\rho_j$ can never then cross ∂S_k as t varies in $\bar{\Delta}(h,r)$.

Equations (2.18), (2.20), and (2.19) enable us to assert that with probability one,

$$\lim_{h\downarrow 0} h^{-1} \int_{S(1)} I_{CH_1(r)}\{tm(\omega)\}\, dt = \sum_{\rho_i \in m(\omega)} \int_{Z(r,\rho_i)} I_C\{tm(\omega)\}\, dt^* , \tag{2.22}$$

$$\lim_{h\downarrow 0} h^{-1} \int_{S(1)} \sum_{k=2}^{\infty} kI_{H_k(r)}\{tm(\omega)\}\, dt = 0 . \tag{2.23}$$

Now for every $\omega \in \Omega$,

$$\int_{S(1)} N\{\Delta(h,r),\ tm(\omega)\}\, dt \leq |\Delta(h,r)|.N\{S(r+1+h),\ m(\omega)\} , \tag{2.24}$$

so for h < 1 and bounded r, the right-hand side of (2.24) is bounded above by $h.K(r).N\{S(r+2),\ m(\omega)\}$ for appropriate K(r). Also

$$I_{CH_1(r)}\{tm(\omega)\} \leq I_{H_1(r)}\{tm(\omega)\} \leq N\{\Delta(h,r),\ tm(\omega)\} ,$$

and

$$\sum_{k=2}^{\infty} kI_{CH_k(r)}\{tm(\omega)\} \leq N\{\Delta(h,r),\ tm(\omega)\} ,$$

so we can take expectations in (2.22) and (2.23) and use the dominated convergence theorem and Fubini's theorem as in the proof of Lemma 1 to yield (2.16) and (2.17). Further, we find that

$$|S(1)|\ \pi^{(r)}(C) = E[\sum_{\rho_i \in m(\omega)} \int_{Z(r,\rho_i)} I_C\{tm(\omega)\}\, dt^*] . \tag{2.25}$$

Substituting $t\rho_i^{-1}$ for t, interchanging the order of summation and integration, and integration and expectation, yields

$$|S(1)|\ \pi^{(r)}(C) = \int_{\partial S(r)} \left\{ E\left[\sum_{\rho_i \in m(\omega),\, t\rho_i^{-1} \in S(1)} I_C\{t\rho_i^{-1}\ m(\omega)\}\right] \right\} dt^* . \tag{2.26}$$

Setting $m'(\omega) = t^{-1}\ m(\omega)$, it is simple to verify that

$$\sum_{\rho_i \in m(\omega),\, t\rho_i^{-1} \in S(1)} I_C\{t\rho_i^{-1}\ m(\omega)\} = \sum_{\rho_i \in m'(\omega),\, \rho_i \in S(1)} I_{t^{-1}C}\{\rho_i^{-1}\ m'(\omega)\}.$$

The expectations with respect to $m(\omega)$ and $m'(\omega)$ coincide, so (2.26) yields

$$|S(1)|\ \pi^{(r)}(C) = \int_{\partial S(r)} \left\{ E\left[\sum_{\rho_i \in m(\omega) \cap S(1)} I_{t^{-1}C}\{\rho_i^{-1}\ m(\omega)\}\right] \right\} dt^* . \tag{2.27}$$

Equation (2.15) follows from (2.27) when $C = M_n$, and this completes the proof of Lemma 2.

Comparing (2.27) with (2.14), observe that if for dt*-almost every point $t \in \partial S(r)$ the set $G_\omega(tC)$ is empty, then

$$\pi^{(r)}(C) = \int_{\partial S(r)} \pi^{(0)}(tC)\ dt^* . \tag{2.28}$$

We shall see in Lemma 4 that this may be so when S(r) is not one of the spheres $S_1, \dots, S_s$ associated with C. In the contrary case, clearly $G_\omega(tC)$ is with positive probability not empty for every $t \in \partial S(r)$, and (2.28) does not hold. In order to rectify this situation, we define a new set function Π on cylindrical members of *C*.

If $0 \in \partial S_1$, define the event $\tilde{H}_k$ by $\tilde{H}_k = \{m : N\{S_1^0 \cap S(h), m\} = k\}$, where S_1^0 is the complement of S_1, and set

$$\Pi(C) = \lim_{h \downarrow 0} P(C \cap \tilde{H}_1)/\{\lambda|S_1^0 \cap S(h)|\} , \tag{2.29}$$

where of course $S_1, \ldots, S_s$ are the spheres associated with C. In the case $0 \notin \bigcup_k \partial S_k$, set

$$\Pi(C) = \pi^{(0)}(C)/\lambda . \tag{2.29'}$$

Lemma 3. Let P be the distribution of an h.r.p.p. $m(\omega)$ of finite intensity λ, and $C \in \mathcal{C}$ a cylindrical event such that $0 \in \partial S_1$, $0 \notin \partial S_k$, $k = 2,\ldots, s$. Then $\Pi(C)$ defined by (2.29) exists and

$$\lambda|S(1)|.\Pi(C) = E[\sum_{\rho_i \varepsilon F'_\omega(C)} I_C\{\rho_i^{-1}m(\omega)\}] + E\{\sum_{\rho_i \varepsilon G'_\omega(C)} W_\omega(C,\rho_i)\} , \tag{2.30}$$

where $F'_\omega(C) = m(\omega) \cap S(1) - G'_\omega(C)$ and

$$G'_\omega(C) = \{\rho_i : \rho_i \in m(\omega) \cap S(1),\ \rho_i^{-1}\{m(\omega) - \rho_i\} \cap (\bigcup_k \partial S_k) \neq \emptyset\} .$$

Proof. The argument of Lemma 1 applies, except for the difference in definition of F' and G', which difference arises because in the analogue of (2.9) the point $t\rho_i$ never crosses ∂S_1 when $t \in \rho_i^{-1}\overline{\{S(h) \cap S_1^0\}}$.

$\Pi(C)$ is called the *Palm distribution* of the h.r.p.p. $m(\omega)$, and was first introduced for one dimension by quite different methods in Khintchine (1960), and developed further in Kerstan & Matthes (1964) and Ambartzumian (1966).

Lemma 4. For an h.r.p.p. $m(\omega)$ of finite intensity λ and every cylindrical $C \in \mathcal{C}$ such that $r \neq r_k$, $k = 1,\ldots, s$, where the r_k are the radii of the spheres S_k associated with C ,

$$\pi^{(r)}(C) = \lambda \int_{\partial S(r)} \Pi(tC)\, dt^* . \tag{2.31}$$

If S(r) coincides with one of the S_k, then (2.31) holds without any restrictions on the radii.

Proof. Denote by $D_\omega(C)$ the set of all $t \in \partial S(r)$ for which $G'_\omega(tC) \neq \emptyset$. Because of (2.27), (2.30), (2.14) and Fubini's theorem, it is enough to show that

$$\|\{t:t \in \partial S(r),\ P\{m(\omega):G'_\omega(tC) \neq \emptyset\} > 0\}\| = 0 \ ,$$

noting that $G'_\omega(tC) = G_\omega(tC)$ when $S(r)$ does not coincide with any S_k. Suppose the contrary is true. Then

$$\int_{\partial S(r)} P\{m(\omega):G'_\omega(tC) \neq \emptyset\}\, dt^* > 0 \ ,$$

and by Fubini's theorem it follows that $P\{m(\omega):\|D_\omega(C)\| > 0\} > 0$. For fixed $\omega \in \Omega$,

$$D_\omega(C) \subset \bigcup_{\rho_i,\rho_j \in m(\omega),\rho_i \neq \rho_j} \ \bigcup_k \{t \in \partial S(r):\rho_i^{-1}\rho_j \in t\,\partial S_k\} \ .$$

Denoting the sets under the union by A_{ijk}, it is easily seen that for each i, j, k such that $r_k \neq r$, $\|A_{ijk}\| = 0$, and hence under the restrictions on the r_k, $\|D_\omega(C)\| = 0$ for each $\omega \in \Omega$. This contradiction proves the first assertion of Lemma 4.

The proof of the second assertion is omitted here as the first is all that is used in the sequel.

3. GENERALIZED PALM FORMULAE

Generalized Palm formulae enable us to recover P from Π for cylindrical members of $\mathcal{C}$. In their simplest form (one-dimensional case, one interval) they appear in Khintchine (1955), while in Ambartzumian (1966) they appear as here though the arguments there are less detailed.

Let $C \in \mathcal{C}$ be a cylindrical event and let $S_1, \ldots, S_s$ be the system of spheres in R^n associated with C. Since we shall be concerned with P(C), for which $P(C) = P(tC)$ by homogeneity, we can and do assume without loss of generality that S_1 is centered at 0. We write

$$C = C(r_1,k_1) = \{m:N\{S_1(r_1)\} = k_1,\ N(S_2) = k_2,\ \dots\ ,\ N(S_s) = k_s\}\ ,$$

and emphasize that in writing $C = C(r_1,k_1)$, we are showing explicitly the dependence of C on the radius r_1 of S_1 and on $k_1 = N\{S_1(r_1),\ m(\omega)\}$.

On account of Lemma 2 we may write for small $h > 0$,

$$P\{C(r_1+h,\ k_1)\} = P\{C(r_1,k_1) \cap H_0(r_1)\} + P\{C(r_1,k_1-1) \cap H_1(r_1)\}+o(h), \tag{3.1}$$

$$P\{C(r_1,k_1)\} = P\{C(r_1,k_1) \cap H_0(r_1)\} + P\{C(r_1,k_1) \cap H_1(r_1)\} + o(h). \tag{3.2}$$

Subtracting (3.2) from (3.1) and dividing by h we find

$$h^{-1}[P\{C(r_1+h,k_1)\} - P\{C(r_1,k_1)\}]$$
$$= h^{-1}\ P\{C(r_1,k_1-1) \cap H_1(r_1)\} - h^{-1}\ P\{C(r_1,k_1) \cap H_1(r_1)\} + o(1)\ . \tag{3.3}$$

Lemma 2 assures us of the existence of the limit as $h \downarrow 0$ of the right-hand side of (3.3), and hence we obtain the equation

$$D^+_{r_i}\ P\{C(r_1,k_1)\} = \pi^{(r_1)}\{C(r_1,k_1-1)\} - \pi^{(r_1)}\{C(r_1,k_1)\}\ , \tag{3.4}$$

which may be rewritten in the integral form

$$P\{C(r_1,k_1)\} = P\{C(0,k_1)\} + \int_0^{r_1} [\pi^{(u)}\{C(u,k_1-1)\} - \pi^{(u)}\{C(u,k_1)\}]\ du. \tag{3.5}$$

Here, $P\{C(0,k_1)\} = 0$ if $k_1 > 0$, while for $k_1 = 0$ we set

$$P\{C(0,0)\} = P[\{m:N(S_2,m) = k_2,\ \dots\ ,\ N(S_s,m) = k_s\}]\ \ .$$

This $P\{C(0,0)\}$ thus depends on s-1 spheres, and may be treated in a similar fashion to $P(C)$, yielding at the sth stage $P[\{m:N\{S(0),m\} = k\}]$, which equals 0 if $k > 0$ and equals 1 for $k = 0$. In this way $P\{C(r_1,k_1)\}$ is determined completely in terms of the functions $\pi^{(r)}$.

Continuing by substituting the expressions (2.31) in (3.5), we obtain *generalized Palm formulae*

$$P\{C(r_1,k_1)\} =$$

$$= P\{C(0,k_1)\} + \lambda \int_{S(r_1)} [\Pi\{tC(\|t\|,k_1-1)\} - \Pi\{tC(\|t\|,k_1)\}]\, dt, \quad (3.6)$$

in which $\|t\|$ is here used to denote the length of the shift t. Hence the result

Theorem. For every h.r.p.p. $m(\omega)$ *in* R^n *of finite intensity* λ, *its Palm distribution together with the value of* λ *uniquely determines the distribution* P *of* $m(\omega)$.

The procedure based on (3.6) gives the solution of the problem of finding the values of P for cylindrical events using the values of Π. Note that the result of this procedure is independent of the order in which the "first" sphere is chosen at each step, and hence there are implied necessary conditions for a set function defined on the cylindrical members of C to be the Palm distribution of some h.r.p.p.

By means of the generating functions

$$G(\bar{z},\bar{S}_{r_1}) = \Sigma\, \bar{z}^{\bar{k}}\, P\{C(r_1,k_1)\} = E\left\{\bar{z}^{\bar{N}(\bar{S}_{r_1})}\right\},$$

$$g(\bar{z},\bar{S}_{r_1}) = \Sigma\, \bar{z}^{\bar{k}}\, \pi^{(r_1)}\{C(r_1,k_1)\}\ ,$$

where $\bar{z}^{\bar{k}} = z_1^{k_1} \cdots z_s^{k_s}$, $\bar{S}_{r_1} = \{S(r_1), S_2, \ldots, S_s\}$, (3.4) can be rewritten as

$$D^+_{r_1}\, G(\bar{z},\bar{S}_{r_1}) = (z_1 - 1)\, g(\bar{z},\bar{S}_{r_1})\ . \quad (3.7)$$

Note that according to Lemma 4, and writing

$$J(\bar{z},\bar{S}_{r_1}) = \Sigma\ \bar{z}^{\bar{k}}\ \Pi\{C(r_1,k_1)\}\ ,$$
$$g(\bar{z},\bar{S}_{r_1}) = \lambda \int_{\partial S(r_1)} J(\bar{z},\ t\bar{S}_{r_1})\ dt^*\ , \tag{3.8}$$

and finally then

$$G(\bar{z},\bar{S}_{r_1}) = G(\bar{z},\bar{S}_{r_1=0}) + \lambda(z_1 - 1) \int_{S(r_1)} J(\bar{z},t\bar{S}_{\|t\|})\ dt. \tag{3.9}$$

4. SUPERPOSITIONS OF INDEPENDENT h.r.p.p.'s

Given a number of r.p.p.'s $m_i(\omega)$, $i = 1,\ldots,\ell$, defined on the same probability space $(\Omega,\mathcal{A},\mu)$, their superposition is defined to be

$$m(\omega) = \bigcup_{i=1}^{\ell} m_i(\omega)\ .$$

We consider the case where the $m_i(\omega)$ are independent, i.e.,

$$\mu\{\omega:m_i(\omega)\ \varepsilon\ C_i,\ i = 1,\ldots,\ell\} = \prod_{i=1}^{\ell} \mu\{\omega:m_i(\omega)\ \varepsilon\ C_i\}$$

for every ℓ-tuple $C_1,\ldots,C_\ell$, $C_i\ \varepsilon\ \mathcal{C}$. If P_i is the distribution of $m_i(\omega)$ and the $m_i(\omega)$ are independent, then the distribution P of $m(\omega)$ is the convolution of the P_i's and we write

$$P = P_1 * \cdots * P_\ell.$$

If $P_1 = \cdots = P_\ell$, then we write $P = (P_1)^{*\ell}$.

Denote by Π and Π_i the Palm distributions of the h.r.p.p.'s $m(\omega)$ and $m_i(\omega)$, $i = 1,\ldots,\ell$.

Lemma 5. Let $m_i(\omega)$, $i = 1,\ldots,\ell$, be identically distributed independent h.r.p.p.'s of finite intensity λ/ℓ. Then for cylindrical events

$$\Pi = \Pi_1 * (P_1)^{*(\ell-1)} . \tag{4.1}$$

Proof. For simplicity assume that $C \in \mathcal{C}$ is the event $C = C_k = \{m: N(S,m) = k\}$ where S is a sphere in R^n. Writing H_0 and H_1 for $H_0(0)$ and $H_1(0)$, clearly

$$P(C_k \cap H_1) = \sum_{i=1}^{\ell} \sum_{k_1+\cdots+k_\ell=k} P_1(C_{k_1} \cap H_0)\ldots P_i(C_{k_i} \cap H_1)\ldots P(C_{k_\ell} \cap H_0). \tag{4.2}$$

Since we are assuming that the $m_i(\omega)$'s are distributed identically, (4.2) may be rewritten as

$$P(C_k \cap H_1) = \ell \sum_{k_1+\cdots+k_\ell=k} P_1(C_{k_1} \cap H_1)\, P_1(C_{k_2} \cap H_0)\ldots P_1(C_{k_\ell} \cap H_0) . \tag{4.3}$$

Dividing (4.3) by $\lambda|S(h)|$ and observing that for $0 \notin \partial S$ (cf. (2.29')), $\lim_{h\downarrow 0} \ell P_1(C_{k_1} \cap H_1)/\lambda|S(h)| = \Pi_1(C_{k_1})$, $\lim_{h\downarrow 0} P(C_k \cap H_1)/\lambda|S(h)| = \Pi(C_k)$, and $\lim_{h\downarrow 0} P_1(C_{k_1} \cap H_0) = P_1(C_{k_i})$, it follows that

$$\Pi(C_k) = \sum_{k_1+\cdots+k_\ell=k} \Pi_1(C_{k_1})\, P_1(C_{k_2})\ldots P_1(C_{k_\ell}),$$

which is just (4.1) for $C = C_k \in \mathcal{C}$. (If $0 \in \partial S$, then replace H_0 and H_1 by $\tilde{H}_0$ and $\tilde{H}_1$, $|S(h)|$ by $|S_1^0 \cap S(h)|$, and recall (2.29).) For more general cylindrical events (4.1) is proved similarly, and hence Lemma 5.

Now apply the formulae of (3.7) and (3.8) to the result of Lemma 5. First, by Lemma 5 for every $t \in T_n$ we have

$$J(\bar{z},t\bar{S}_{r_1}) = J_1(\bar{z},t\bar{S}_{r_1})\, G_1^{\ell-1}(\bar{z},t\bar{S}_{r_1}) \,, \tag{4.4}$$

where the index 1 refers to Π_1. In fact, $G_1(\bar{z},t\bar{S}_{r_1})$ does not depend on $t \in T_n$, and integration of (4.4) over $\partial S(r_1)$ gives, by virtue of (3.8),

$$g(\bar{z},\bar{S}_{r_1}) = \ell g_1(\bar{z},\bar{S}_{r_1})\, G_1^{\ell-1}(\bar{z},\bar{S}_{r_1}) \,. \tag{4.5}$$

Multiplying (4.5) by (z_1-1), using (3.7) and $G_1^{\ell} = G$, we derive the differential equation

$$D^{+}_{r_1}\, G(\bar{z},\bar{S}_{r_1}) = \ell(z_1 - 1)\, \frac{g_1(\bar{z},\bar{S}_{r_1})}{G_1(\bar{z},\bar{S}_{r_1})}\, G(\bar{z},\bar{S}_{r_1}) \,, \tag{4.6}$$

which on integration yields

$$G(\bar{z},\bar{S}_{r_1}) = \exp\left\{\ell(z_1 - 1) \int_0^{r_1} \frac{g_1(\bar{z},\bar{S}_{u_1})}{G_1(\bar{z},\bar{S}_{u_1})}\, du_1\right\} \cdot G(\bar{z}.\bar{S}_{r_1=0}) \,,$$

where $\bar{S}_{r_1=0} = \{S(0), S_2, \ldots, S_s\}$. We may use (3.8) to rewrite this in the form

$$G(\bar{z},\bar{S}_{r_1}) = \exp\left\{\lambda(z_1 - 1) \int_{S(r_1)} \frac{J_1(\bar{z},t\bar{S}_{\|t\|})}{G_1(\bar{z},t\bar{S}_{\|t\|})}\, dt\right\} \cdot G(\bar{z},\bar{S}_{r_1=0}) \,. \tag{4.7}$$

Note that $G(\bar{z},\bar{S}_{r_1=0})$ depends only on s-1 spheres in R^n, and that in the case $s = 1$, $G(\bar{z},\bar{S}_{r_1=0}) = 1$.

One obvious application of (4.7) lies in obtaining limiting results when ℓ, the number of independent h.r.p.p.'s (each of intensity λ/ℓ) being superposed, tends to ∞. We state here only the simplest result which is readily seen from (4.7).

Proposition. For each $\ell = 1,2,\ldots$ let $m_i^{(\ell)}$, $i = 1,\ldots,\ell$, be independent identically distributed h.r.p.p.'s of finite intensity λ/ℓ, and let $P_1^{(\ell)}$ and $\Pi_1^{(\ell)}$ be the distribution and Palm distribution respectively of each $m_i^{(\ell)}$, $i = 1,\ldots,\ell$. If $\Pi_1^{(\ell)} \to \Pi_1$, $\ell \to \infty$, for each cylindrical member of C, then the value of $(P_1^{(\ell)})^{*\ell}$ tends to a limit P for all cylindrical members of C. The Palm distribution Π corresponding to P satisfies the equation

$$\Pi = \Pi_1 * P . \tag{4.8}$$

The proof, based on (4.7) and the fact that $\lambda/\ell \to 0$ implies that $G_1(\bar{z},\bar{S}) \to 1$, is carried out by induction with respect to the number s of spheres associated with the cylindrical event. Equation (4.8) follows from the limiting form of (4.7),

$$G(\bar{z},\bar{S}_{r_1}) = \exp\{\lambda(z_1 - 1) \int_{S(r_1)} J_1(\bar{z},t\bar{S}_{\|t\|})\, dt\}\cdot G(\bar{z},\bar{S}_{r_1=0}). \tag{4.9}$$

Note that the theorem gives as a special case sufficient conditions for convergence to the Poisson distribution, namely when $\Pi_1^{(\ell)} \to \nabla$, where ∇ is the distribution of a point process which with probability one has only one point which is placed at the origin.

5. INFINITELY DIVISIBLE POINT PROCESSES IN R^n

We pass over the general questions connected with infinite divisibility that arise obviously in the present context of r.p.p.'s in R^n and proceed to construct two classes of infinitely divisible h.r.p.p.'s of finite intensity.

An r.p.p. $m(\omega)$ is called an s-*stripe* $(0 \leq s \leq n)$ *in* R^n if

(i) there exists an s-dimensional subspace $T_s \subset T_n$ such that the distribution of $tm(\omega)$ does not depend on t for $t \in T_s$; and

(ii) the random variable $N\{F,m(\omega)\}$ is finite with probability one for every "cylindrical" subset F which is the direct sum $F = F_1 \oplus F_2$ of a subspace F_1 of the orthogonal complement X_{n-s} of T_s and a subset F_2 of T_s of finite s-dimensional volume.

An s-stripe is said to be of finite intensity when $E[N\{F,m(\omega)\}] < \infty$. with F as in (ii).

Let X_{n-s} be the orthogonal complement to T_s, let $\{x_i\}$ be the set of points of a Poisson h.r.p.p. in X_{n-s}, and let $m_1(\omega)$, $m_2(\omega)$, ... be a sequence of independent identically distributed s-stripes of finite intensity. Putting

$$m(\omega) = \bigcup_i x_i m_i(\omega) \tag{5.1}$$

we obtain an r.p.p. that is obviously homogeneous and infinitely divisible. In the case $s = 0$ this construction leads to the familiar "Poisson clustering" h.r.p.p. in R^n (see Neyman and Scott, 1958, 1972). If $s=n$ then we obtain the superposition of a random number ν of independent h.r.p.p.'s of finite intensity, ν being independent of these h.r.p.p.'s and having a Poisson distribution.

Another construction for infinitely divisible h.r.p.p.'s satisfying (4.8) is obtained by superposing s-stripes $m_i(\omega)$, $i = 1,2,\ldots,$which are subjected not only to parallel shifts as in (5.1) but also rotations.

For example, let $n = 2$ (i.e., the planar case), and let $m_i(\omega)$, $i = 1,2,\ldots,$ be defined as mutually independent 1-stripes of finite intensity with the x-axis as a subspace of invariance and with a common distribution. Let M be the motion of R^2 which is the product of rotation (around the origin) through an angle $\phi - \frac{1}{2}\pi$, and a parallel shift of R^2 by the vector (p,ϕ). Let $\{M_i\}$ be a random set of motions of R^2 corresponding to a homogeneous Poisson process in the strip $0 \leq \phi < 2\pi$, $p > 0$. Then the h.r.p.p.

$$m(\omega) = \bigcup_i M_i m_i(\omega) \tag{5.2}$$

is infinitely divisible and is isotropic (i.e., its distribution is invariant with respect to all Euclidean motions of the plane).

Distributions of processes of the classes (5.1) and (5.2) may be obtained by limiting processes satisfying the conditions of the proposition and thus they have then to satisfy (4.8).

6. CONCLUDING REMARKS

First we note that the statement

$$P[N\{D_k, m(\omega)\} > 1] = o(|D_k|) \tag{6.1}$$

is not true for any arbitrary sequence of domains $D_k \subset R^n$ for which $|D_k| \downarrow 0$, when we assume merely that the point process $m(\omega)$ is homogeneous and has finite intensity.

For example, restricting ourselves to the planar case we readily find that for the sequence of domains

$$D_h = \{(x,y): 0 < x < 1,\ 0 < y < h\} \qquad (h \downarrow 0)\ , \tag{6.2}$$

and for the h.r.p.p. $m(\omega)$ defined by (5.1) where each $m_i(\omega)$ with probability one consists of the two points (0,0) and $(\frac{1}{2}, 0)$, equation (6.1) is not true.

However, the same sequence D_h of (6.2) may be used in (6.1) when the additional assumption of isotropy is imposed on $m(\omega)$, for then the condition

$$\int_{M} I_{\rho_1\rho_2}(M)\ dM = o(|\partial D_k|) \tag{6.3}$$

analogous to that used in the proof of Lemma 1 is valid, where M is the set of all Euclidean motions of R^2, dM is an element of "kinematic measure" on M (see Santalo, 1953), $|\partial D_k|$ = surface area of D_k, and for every two points $\rho_1, \rho_2 \in R^2$,

$$I_{\rho_1\rho_2}(M) = \begin{cases} 1 & \text{if } \rho_1, \rho_2 \in MD_k,\ M \in M, \\ 0 & \text{otherwise}\ . \end{cases}$$

As far as we can see, (6.3) leads to (6.1) when we assume a new definition of an r.p.p. in R^2 requiring measurability of $N\{S, Mm(\omega)\}$ on the product space $\Omega \times M$.

Finally we come to the following measure-theoretic problem. Denote by α the set of all distributions of r.p.p.'s (in the sense of Section 1, say), and by β the set of all probability measures on B. Obviously, $\beta \supseteq \alpha$; is it true or false that $\alpha = \beta$?

REFERENCES

Ambartzumian, R. V. (1966). On an equation for stationary point processes. Akad. Nauk. Armjan. SSR Dokl. 42 (In Russian), 141-7.

Kerstan, J. & Matthes, K. (1964). Stationäre zufällige Punktfolgen, II. Jber. Deutsch. Math-Verein. 66, 106-18.

Khintchine, A. Ya (1955). *Mathematical methods in the theory of queueing*. Trudy Mat. Inst. Steklov 49. (English translation (1960) London: Griffin.)

Neyman, J. & Scott, E. L. (1958). A statistical approach to problems of cosmology. J. R. Statist. Soc. B 20, 1-43.

Neyman, J. & Scott, E. L. (1972). Processes of clustering and applications. This volume.

Santalo, L. A. (1953). *Introduction to integral geometry*. Paris: Hermann. Act. Sci. Indust. 1198.

PROCESSES OF CLUSTERING AND APPLICATIONS

J. Neyman and E. L. Scott
University of California, Berkeley

SUMMARY

Definition of the stochastic process of clustering, a specialization of the general branching process, is followed by a list of applied work concerned with properties of natural phenomena. These include: (i) the plot distribution of larvae on an experimental field, (ii) some tactical problems of formation bombing, (iii) "contagion" of accidents, (iv) "within-family" contagion of a disease, (v) spatial distribution of galaxies, (vi) population dynamics, and (vii) theory of epidemics.

1. INTRODUCTION

The terms *mechanism of clustering* and *process of clustering* will be used to designate two different classes of entities both concerned with the dispersal of some observable particles to be denoted by Π, and of some other possibly unobservable particles Γ. The term mechanism of clustering will denote an empirical phenomenon either developing spontaneously or man made. The term process of clustering will mean a mathematical concept invented with the hope of approximating the properties of a given mechanism of clustering, a stochastic model of the latter.

We say that the phenomenon of scattering of particles Π in some empirical space E involves a mechanism of clustering if it consists in the following:

(i) First, particles Γ, termed cluster centers, observable or not, are dispersed in E in an arbitrary manner.

(ii) Second, for each cluster center Γ located at a point u, there exists a non-negative integer number $\nu(u)$, whether constant or variable described as *the size of the cluster at u*.

(iii) Finally, $\nu(u)$ observable particles Π are dispersed in some manner about the point u.

The particular mechanisms of clustering differ from each other in the mechanism of dispersing the cluster centers Γ, in the possible variability in the sizes $\nu(u)$ of clusters and in the manner in which the $\nu(u)$ cluster members are scattered about the point u.

The following sections of this paper describe briefly several phenomena involving the mechanism of clustering that came to our attention, and also the processes of clustering invented to approximate these phenomena. In all these processes, the empirical space E is represented by a Euclidean space R_s of s = 1, 2, 3 dimensions. Also, invariably, the sizes $\nu(u)$ of clusters are treated as mutually independent random variables, with distributions possibly dependent on u. These distributions are characterized by the probability generating functions (pgf) denoted by

$$g(t|u) = Et^{\nu(u)}. \tag{1}$$

The third predominantly common feature of the processes considered is that, given the location u of a cluster center and given $\nu(u) = n > 0$, the coordinates $X_i = (X_{i1}, X_{i2}, \ldots, X_{is})$ of the ith cluster member are random, independent of X_j, with $j \neq i$, and having the same probability density $f(x|u)$ possibly dependent upon u. This density is described as the structure of the cluster or as the dispersal function. In addition, given the location of all cluster centers, the locations of all the particles Π are assumed to be mutually independent.

The problems treated in the several cases vary somewhat, depending upon the nature of the phenomenon considered. However, they have the following common feature. Let $Q_1, Q_2, \ldots Q_r$ be for some $r > 1$, disjoint regions (really, measurable sets) in the space R_s and let N_i denote the number of particles Π either all those falling in Q_i or only of those among them that are in a defined sense "successful". The numbers N_i are observable random variables and the studies reported below are concerned with their joint pgf

$$G_{\underline{N}}(\underline{t}) = E[\prod_{i=1}^{r} t_i^{N_i}]. \qquad (2)$$

The definition of "success" of a particle Π located at a point x depends upon the nature of the problem studied. In some cases, it may be survival of an organism. In other cases, it may be the "visibility" of the particle, etc. Generally, the success of a particle at x is assumed to be a random event, independent of all other elements of the system, with probability $\theta(x)$ depending upon the location x of the particular particle, but otherwise left unspecified.

As a result of all those assumptions, the general form of the function (2) must depend upon the following elements of the datum of each particular problem:

(i) the distribution of the cluster centers,

(ii) the pgf denoted by $g(t|u)$, of cluster sizes,

(iii) the dispersal function $f(x|u)$, and

(iv) the probabilities $\theta(x)$ of success.

2. CLUSTERING OF LARVAE ON AN EXPERIMENTAL FIELD

One of our earliest studies of clustering goes back to pre-war years (Neyman, 1939). It originated from conversations with Geoffrey Beall who was then interested in field experiments with pesticides. The particles Π, whose clustering was to be studied, were larvae counted in small plots of an experimental field. The cluster centers Γ, unobservable in this case, were "egg masses" laid by moths more or less at the same time, at spots u on the field selected by themselves. The cluster size $\nu(u)$ was the number of those larvae hatched from an egg mass located at u and later crawling around in search of food, that survived up to the time of counting.

The basic assumptions of the process of clustering constructed were that the egg masses Γ are Poissonwise distributed over an indefinite plane R_2, with a uniform density per unit area, that the cluster sizes $\nu(u)$ are identically distributed for all $u \in R_2$, and that the dispersal functions $f(x|u)$ depend only upon the differences $x_1 - u_1$ and $x_2 - u_2$. All surviving larvae were assumed "successful" so that $\theta(x) \equiv 1$.

One further assumption was adopted regarding the dispersal functions. This is that, for sufficiently large $|x_1 - u_1|$ and $|x_2 - u_2|$, this dispersal function is identically zero. This corresponds to the obvious fact that the distances that larvae can travel after hatching must be rather limited.

The object of study was the distribution of the number N of surviving larvae counted in a single unit area plot Q on the experimental field, so that r = 1. It happens that the general formula for the pgf of N does not depend explicitly on the dispersal function f but only on the integral, say

$$p(u) = \int_Q f(x|u)dx \qquad (3)$$

representing the probability that a surviving larva from an egg mass at u will be found in Q. The assumptions about f imply that Q must be surrounded by an area A described as the area of accessibility, such that (3) is positive for all $u \in A$ but vanishes otherwise.

In the attempt to fit the model to actual counts of larvae, it was assumed that the cluster sizes ν are Poisson distributed. On the other hand, the functional form of the dispersal function was not specified. Instead, two different methods were used to achieve an arbitrarily close approximation to the probability p(u) considered as a function of $u = (u_1, u_2)$. These two methods resulted in two infinite families of "contagious" distributions labeled Type A and Type B. Particular members of each family differ in the number of adjustable parameters 2,3,...

The study was motivated by the hope of reaching a better understanding of the treatment effects exhibited in an appropriately designed experiment: did the treatment diminish $E(\nu)$, the average number of survivors per egg mass? Did the treatment alter the dispersal function?

To our knowledge, no such analysis of insecticide experiments was ever attempted. On the other hand, the Type A distribution with two parameters has been used as an interpolation formula to fit the counts of insects per unit area. Frequently, the fit was quite good.

In this connection, it may be interesting to mention a particular case where the use of the same Type A distribution with two parameters was singularly unsuccessful in approximating the counts of insects "per unit area". The case in question was the counts of fleas found on single wild rats. Here, then, a rat was treated as a "unit area". This particular incident indicates that the two families Type A and Type B should not be expected to approximate the distribution of the number of insects per unit of observation, whether it be a plot in the field or an animal, but rather a particular mechanism that governs this distribution. Obviously, the mechanism governing the number of fleas per rat, probably depending upon the age of the rat, is very different from that underlying the crawling of larvae over a field.

3. PROBLEM OF LAND MINES ON THE LANDING BEACHES IN NORMANDY

A modification of the process of clustering sketched above found application to problems of formation bombing practiced during World War II. Here, the cluster centers Γ were the so-called centers of impact of bombs released from a formation of planes. Particles Π were the points of impact of individual bombs. The distribution of cluster centers was determined by the choice of the aiming point and by the distribution of aiming errors in range and in deflection supposed known from earlier studies. The number ν of cluster members was the known number of bombs to be released. The dispersal function was determined by the selected shape of the formation of planes and by the dispersion of bombs released from a given altitude, supposed known. Most of the work was classified. However, one relevant paper by Bronowski and Neyman (1945) appeared in an open journal. One of the studies, originally classified but later declassified, was concerned with the selection of optimal combination of the shape of formation and of the type of bombs to be used in an effort to clear land mines from the landing beaches in Normandy. Next, the problem was to calculate how many bombing attacks should be made using the optimal formation-bomb type combination so as to reduce the probability of an unexploded mine on the beach to a prescribed low level.

The problems contemplated in the two preceding sections might be summarized as follows:

(i) To fit a process of clustering believed to be roughly equivalent to the mechanism of a natural phenomenon with the hope that the estimated parameters will provide some not immediately obvious information on the details of the mechanism studied.

(ii) With reference to a man-made mechanism of clustering intended to parallel a preconceived process involving several parameters subject to human control, to find a defined optimal combination of these parameters and to calculate the number of independent replications which would ensure the preassigned level of probability of a specified outcome.

The problems treated in the following sections are of an entirely different type. With reference to two specified mechanisms of scattering certain particles Π the question asked is whether these particles are clustered in a particular way. On the hypothesis tested, asserting the absence of clustering, the process of scattering the particles Π is a special case of the Bartlett-Lewis process (Bartlett, 1963; Lewis, 1964; Vere-Jones, 1970).

4. TRUE OR FALSE CONTAGION OF ACCIDENTS

As is well known, the empirical distribution of the number of accidents per unit of time, per person of a specified occupation frequently resembles a negative binomial distribution. It is also known that this distribution is implied by two sharply different processes. One, described by Greenwood and Yule (1920) is an appropriate mixture of Poisson variables each with its own expectation λ. With reference to accidents, the value of λ, supposed to characterize a given individual, is labeled this individual's accident proneness. This proneness λ is supposed to remain unchanged in time and the occurrence of one or more accidents in the past does not influence the chance of more accidents in the future.

The alternative process, implying the same distribution of accidents per unit of time is a particular case of the urn scheme of Pólya (1930).

With reference to the mechanism of recurrence of accidents among individuals of a population, the process of Pólya presupposes that initially all the individuals concerned are equally prone. However, each accident sustained by an individual is supposed to alter his proneness to more accidents. Here, then, the first accident incurred by an individual during the period of observation plays the role of a cluster center Γ and the particular scheme of contagion implies both the number of cluster members Π (subsequent accidents during the same period of observation) and their distribution. However, in this case, the coordinates (times) of these particles Π are not mutually independent.

It is easy to visualize practical situations in which the possibility of distinguishing between the two mechanisms, of "true" and "false" contagion, is quite important.

The study by Bates and Neyman (1952) of the above two schemes with reference to several consecutive periods of observations was culminated by a paper by Bates (1955) who deduced uniformly most powerful tests for the presence of Pólya contagion specified to be positive or negative (earlier accidents increase or decrease the probability of more accidents). Also, a uniformly most powerful unbiased test was deduced against the alternative specifying only the presence of Pólya contagion which may be positive or negative.

Still another aspect of the problem occurs to us which appears interesting. It seems likely that populations for which the accident proneness is being studied may be mixtures of individuals some of whom may be subjected to a positive and some others to a negative Pólya type contagion. Here it will be realized that the negative contagion, representing the ability to learn by experience how to avoid accidents, is a desirable attribute of a candidate for particular kinds of employment. Thus, a proper statistical test for the presence of a randomly distributed contagion, occasionally positive and occasionally negative, might be important to deduce. Here the technique described by Neyman and Scott (1967) of optimal $C(\alpha)$ tests may yield the desired result.

5. PROBLEM OF WITHIN-FAMILY CONTAGION

Briefly, the problem, labeled the problem of within-family contagion, arises in those cases where, for a disease D known to be communicable, the mechanism of its spread is not known. One possibility is that D is spread among humans by some vectors, perhaps flies or mosquitoes, but cannot be directly transmitted from an individual suffering from D to anyone else. Here, then, we may speak of "outside exposure". The other possibility is that, in addition to the outside exposure, the members of small communities, such as families, have another kind of exposure. This depends on the presence within the given family of one or more members who already contracted the disease D. If the presence in a family of persons ill with D increases (or decreases, perhaps through immunization) the probability of more cases of D in the same family, then we speak of positive (or negative) within family contagion.

The problem of within-family contagion is the subject of a study by William R. Gaffey (paper in preparation). This problem attracted the attention of epidemiologists long ago and Table 1, reproduced from Puffer (1950) illustrates the classical approach.

The first three columns of Table 1 refer to the general population while the subsequent columns are restricted to families in which there was at least one case of poliomyelitis during the period studied. Here, then, "persons exposed" mean family members minus one person who was attacked first. "Subsequent cases" are cases that occurred among those persons described as persons exposed.

In Puffer (1950), Table 1 is accompanied by a comment that while the secondary cases might be due to the same outside source of the disease as the first, "the chances are great, however, that they were contracted at home." Unfortunately, as Gaffey shows, the differences between the secondary rates (last column in Table 1) and the overall rates (in tabular column 3) may be entirely due to differences in the exposure of particular families to the outside sources of the disease. For example, the intensity of the outside source may well depend on the distance from a city dump, etc.

TABLE 1. *Incidence of Poliomyelitis in General Population and in Families with a Case, New York City, 1916*

Age group	General population			Persons in families with a case		
	Estimated population	Cases in each group	Incidence per 10,000	Persons exposed*	Subsequent cases	Incidence per 10,000
Total	5,087,332	9,345	18.4	37,300	407	109
0 to 5 years	629,710	7,838	124.5	10,540	335	318
6 to 10 years	460,700	956	20.8	4,575	58	127
11 to 20 years	995,020	228	2.3	4,994	10	20
Over 20 years	3,001,902	109	0.4	17,191	4	2
Age unknown		214				

*Initial cases excluded.

The term within-family contagion is already conceptual and, in order to deduce a test for its presence, it is unavoidable to push the conceptualization a little further. Gaffey's model on which his test is based is as follows.

(i) The intensity of the outside exposure of n members of a given family is measured by a constant $\lambda \geq 0$ which may be different from that of any other family. (This hypothesis seems entirely realistic.)

(ii) Given that prior to time $t > 0$ there were exactly $m < n$ cases in the family, each of the remaining $n-m$ individuals may contract the disease within the time interval $(t, t+\tau)$, with $\tau > 0$, with the same probability

$$(\lambda + m\mu)\tau + o(\tau), \tag{4}$$

where $\mu \geq 0$ characterizes the intensity of within family contagion. It is assumed that μ has the same value for all families considered, which is likely to be an oversimplification.

The observable variables refer to those families, N of them, in which during the period of observation T, say T = 1, there were $k > 0$ cases of the disease. For a family of n members, the observable random variables are k = total number of cases, and the times

$$0 < t_1 < t_2 < \ldots < t_k < 1 \tag{5}$$

when the consecutive cases occurred. The hypothesis H_0 to test asserts $\mu = 0$.

It will be seen that, on the hypothesis H_0, the variables t_i referring to a given family represent a segment of the Bartlett-Lewis process. Thus, Gaffey's test for within family contagion amounts to a test that N sequences of numbers (5) represent independent realizations of as many segments of particular Bartlett-Lewis processes. Gaffey's model of within-family contagion amounts to the selection of hypotheses against which the test to be deduced is desired to be particularly powerful. One possibility, not treated by Gaffey, is not to insist that μ has the same value for all the families considered but to allow it to vary from one family to the next, with the possible restriction of being non-negative or nonpositive. As in the similar case of accident proneness, the technique of optimal $C(\alpha)$ tests might prove useful.

In deducing his test, Gaffey used the fact that, on the hypothesis tested, the sequence of statistics $\{S_i\}$

$$S_i = \sum_{j=1}^{k_i} t_{ij}, \tag{6}$$

where t_{ij} and k_i have the defined meaning for the i-th family, $i = 1,2,\ldots,N$ is a minimal sufficient statistic for the N nuisance parameters λ_i, each representing the intensity of exposure of the given family to the outside source of the disease.

The test criterion deduced by Gaffey is, say

$$T = \sum_{i=1}^{N} \sum_{j=1}^{k_i} j t_{ij}. \tag{7}$$

Depending on whether the alternative hypotheses are "one sided" or not the test deduced is a uniformly most powerful test or an unbiased uniformly most powerful test of H_0.

6. DISTRIBUTION OF GALAXIES IN SPACE

By far the strongest and the most sustained effort expended by us in studying natural phenomena through appropriately selected aspects of the process of clustering referred to astronomy, specifically to galaxies. As in the case of larvae, the stimulus came from the substantive scientists, that is from astronomers. The first stimulus came in about 1950 from C. D. Shane, then Director of the Lick Observatory, who just completed a survey of the part of the sky visible from the Lick Observatory and was engaged in counting images of galaxies in little squares on his photographic plates. In this process Shane was impressed by the fact that, even in the parts of the sky near the galactic pole, apparently free from interference by the patchy dust clouds within the Milky Way System, and even after some corrections for various errors ("plate error", "east-west effect" due to the pecularity of the telescope's lens etc.), the square-to-square distribution of images of galaxies had no resemblance to the expected Poisson distribution. Shane's conclusion was that the space distribution of galaxies must be somehow "not random" and he was anxious to bring out the underlying mechanism.

The contemporary ideas found in the astronomical literature were somewhat indistinct. On the one hand, a few large local concentrations of galaxies, labeled clusters, had been noticed and began to be studied per se. These may be exemplified by the Coma and the Virgo Clusters. On the other hand, whenever the large scale distribution of galaxies was discussed, almost invariably the discussion was based on the assumptions of "randomness", that is, of Poissonness, with a uniform density per unit volume ("cosmological principle" and "perfect cosmological principle"). However, as the years went by, there was apparent a marked evolution of ideas. More and more clusters, some large, some small, were identified and studied as separate entities. A large catalogue of clusters of galaxies was recently compiled by Zwicky (1961-68).

A classification of clusters begins to emerge, some compact, some diffuse, some large and some small, etc. Our own ideas about an appropriate stochastic model participated in this evolution. The account given in this section refers predominantly to the beginning period of work, done in the early 1950's. The basic assumptions adopted by Neyman & Scott (1952), Neyman, Scott & Shane (1953) and Scott, Shane & Swanson (1954) were simple:

(i) Galaxies (treated as particles Π) occur only in clusters of varying sizes.

(ii) The cluster centers Γ are unobservable; their distribution in space is "quasi-uniform." This term is used to mean: (a) for any two congruent regions (Borel sets) in space R_3, the numbers of cluster centers in these two regions are identically distributed, and (b) the numbers of clusters in any two disjoint regions in R_3 are mutually independent random variables.

(iii) The number $\nu(u)$ of galaxies (particles Π) in a cluster centered at any point u in R_3 has a fixed distribution with pgf g(t) independent of u. Also, the cluster sizes ν of all the clusters are mutually independent and independent of all other variables of the system.

(iv) Given the locations of all the cluster centers, the locations of particular galaxies are mutually independent and independent of all other variables of the system.

(v) The dispersal function $f(x|u)$ of galaxies from a cluster centered at u depends only upon the distance

$$\{(x_1-u_1)^2 + (x_2-u_2)^2 + (x_3-u_3)^2\}^{1/2}. \tag{8}$$

(vi) A galaxy located at a point x in R_3 is called successful if its image on the photographic plate is distinct enough to be counted. The probability of this, denoted by $\theta(\xi)$ is supposed to depend only upon the distance ξ from the observer. The success of any galaxy is assumed independent of that of any other galaxy.

The above six postulates determine the structure of the original process of clustering proposed to approximate the large scale organization of the distribution of galaxies in space. The efforts to fit this model to the observations required additional hypotheses of an interpolatory character. In particular such hypotheses were needed to specify the function $\theta(\xi)$.

Let Q be an arbitrary region in R_3 of finite measure V and let $M(Q)$ stand for the number of cluster centers in Q. Clearly, postulate (ii) implies that the distribution of $M(Q)$ is infinitely divisible with its pgf having the form

$$G_{M(Q)}(t) = \exp\{Vh(t)\}. \tag{9}$$

where the function h vanishes at $t = 1$ and has the form

$$h(t) = -h_0 + \sum_{k=1}^{\infty} h_k t^k, \tag{10}$$

with $h_0 > 0$ and with nonnegative coefficients of higher powers of t. Simple calculations imply then that, whatever be the disjoint regions $Q_1, Q_2, \ldots, Q_r$ in R_3, the joint pgf of the numbers $N_1, N_2, \ldots, N_r$ of successful galaxies in these regions has the form, say,

$$G_N(t) = \exp\{\int_{R_3} h[g\{1 - \sum_{k=1}^{r} (1-t_i)p_i(u)\}]du\}, \tag{11}$$

where

$$p_i(u) = \int_{Q_i} \theta(x)\ f(x|u)dx \tag{12}$$

stands for the probability that a galaxy from a cluster centered at $u = (u_1, u_2, u_3)$ will be located in Q_i and that it will be successful.

The problem of fitting the model considered to any kind of observational data is greatly simplified by the following result.

THEOREM 1. *Whatever be the function* h *in (10), whatever be the* pgf g *and whatever the other details of the model considered, there exists a positive number* λ *and a pgf* $g^*(t)$ *with* $g^*(0) = 0$, *such that the distribution determined by (11) is identical with that determined by the same formula with the substitutions of* $-\lambda(1-t)$ *for* $h(t)$ *and of* $g^*(t)$ *for* $g(t)$, *where both* λ *and* $g^*(t)$ *are uniquely determined.*

The meaning of Theorem 1 is that no generality is lost by assuming that the cluster centers Γ are Poisson distributed in space with constant density λ per unit volume and that each of the hypothetical clusters contains at least one galaxy. As a result, and dropping the now unnecessary asterisk, formula (11) can be rewritten in the following form:

$$G_N(t) = \exp\{-\lambda\int_{R_3} [1 - g(1 - \Sigma(1-t_i)p_i(u))]du\}. \tag{13}$$

Formula (13) refers to numbers N_i of "successful" galaxies falling within an arbitrary number r of arbitrarily selected disjoint regions Q_i. In order to use this formula for analysis of counts of images of galaxies on photographic plates, particularly Shane's counts, the regions Q_i must be appropriately specialized.

In order to count the images of galaxies in a survey, the photographic plates are divided into squares of equal size, perhaps 10' x 10' or $1^\circ \times 1^\circ$, regularly placed in rows and columns, parallel to the edges of the plate. In our studies, formula (13) was used with reference to a pair of disjoint squares both in the same row or in the same column, with known distance between their centers, measured in terms of the dimension of each square taken as a unit. Let w_1 and w_2 be such a pair, k units of distance apart, k = 1,2,... . It will be realized that to each of these squares w on the photographic plate there corresponds a solid angle Q in R_3 subtended by w and extending to infinity.

From now on, formula (13) will be understood to refer to a pair of such solid angles Q_1 and Q_2 (which implies r = 2). Because these solid angles are unbounded and because the integral, say $I(t_1,t_2)$ in the exponent of (13) extends over the whole space R_3, the first question to consider is whether, under the various possible specifications of g,f and θ, this integral converges. As things stand in general, the only information available is that, for fixed t_i each between zero and unity, the integrand cannot be negative. Thus, if $I(t_1,t_2)$ is divergent, it must diverge to $+\infty$, with the result that the pgf of N_1 and N_2 is identically zero. The interpretation of this is that N_1 and N_2 are degenerate and, with probability one, must be infinite.

The question of convergence of $I(t_1, t_2)$ has been considered under two sets of assumptions. Both sets include the hypotheses that g is at least twice differentiable at t = 1, which is not essential to the proof. Also, both sets presuppose that the universe is static. One set of assumptions specialized the probability of success $\theta(x)$ in accordance with the contemporary views on the distribution of absolute brightness of galaxies in space, about their dimming due to distance from the observer, and about the details of the mechanism of counting on the photographic plates. Some of these original views are presently considered antiquated; the details will not be reported here. The second set of assumptions, also referring to $\theta(x)$, is frankly unrealistic but its consequences are interesting because they indicate the importance of some phenomena that one is tempted to ignore. The relevant theorem is given in the next section. It refers to just one solid angle Q.

7. OLBER'S PARADOX

Commonly, the term Olber's paradox refers to the fact that each photograph of the sky contains only a finite number of images of galaxies. On the other hand, according to traditional thinking, the space around us is infinite in all directions and is filled with galaxies, perhaps clustered, but with a large scale uniform density. It is true that the light from a very distant galaxy takes a long time to reach the photographic plate. *However, in a stationary universe this delay plays no role.* In consequence, one will expect an infinity of images of galaxies on the photographic plate. Furthermore, if the above musings were realistic, the energy of the light arriving from all the galaxies would be infinite. What is the solution?

Let us assume the following simplifying hypotheses.

(i) Each galaxy is of the same absolute brightness and emits photons Poissonwise with a uniform density $\mu > 0$ per unit time, uniformly in all directions. Each emitted photon has a fixed energy E.

(ii) All these photons arriving at the telescope carry the same energy E independently of the distance ξ between the emitting galaxy and the observer.

(iii) The photograph of the sky is taken by an ideally constructed telescope (perfect focusing) with light gathering mirror of area A; the exposure time is a preassigned period of T hours.

(iv) The photographic plate used is also ideal, with sensitivity of its emulsion characterized by a number $s \geq 1$ as follows: in order that the plate produce an image of a point source of light, it is both necessary and sufficient that during the period of exposure at least s photons emitted by this source reach the mirror of the telescope and hence the plate.

As formerly, let N stand for the number of "successful" galaxies in the solid angle Q subtended by the photographic plate, the success being determined by the appearance of the image of the galaxy on the photographic plate. The probability $\theta(\xi)$ is the probability that, during the exposure T, the mirror of the telescope receives at least s photons from a galaxy at distance ξ.

Obviously, the distribution of N must depend upon the area A of the telescope mirror, on the time T of the exposure and on the sensitivity s of the emulsion. A priori it seems plausible that a change in one of these parameters can be compensated by changes in the other two, so that the essential properties of the distribution of N remain unaltered. The somewhat unexpected result (Neyman, 1963) is summarized in the following theorem.

THEOREM 2. *With probability one, the number* N *of images of galaxies on the photographic plate will be finite or infinite according to whether the sensitivity* s *is at least equal to* 2 *or is* s = 1, *respectively.*

Out of the four frankly oversimplifying assumptions underlying Theorem 2, assumption (ii) is the most unrealistic. It ignores the phenomennon known as the redshift-distance relation, the farther a galaxy is from the observer, the more its whole spectrum is shifted towards the red. Thus, each photon, emitted by a galaxy at distance ξ with some energy E, arrives at the observer with an energy $E(\xi)$ which is negatively correlated with the distance travelled ξ. The exact nature of this effect is a subject of discussion, one of the possibilities being the Doppler effect due to the recession of galaxies from each other with velocities proportional to distance.

8. QUASI-CORRELATIONS AS THE BASIS FOR ESTIMATING THE DISPERSAL FUNCTIONS

The procedurally easiest, but probably not very efficient, method of estimating, or interpolating the dispersal function of a process of clustering is based on moments and, more particularly on "quasi-correlations" between counts of particles Π in regularly spaced equal regions in the space considered. This method is described below with particular reference to Shane's counts of galaxies in small squares on his photographic plates.

Turning to formula (13) with $r = 2$, we consider two solid angles Q_1 and Q_2 subtended by two equal squares where counts are made, in the same row or in the same column, k units of distance apart, $k = 1,2,3,\ldots$. Using the assumption that the number ν of galaxies per cluster has at least two finite moments, ν_1 and ν_2, we differentiate the logarithm of (13) with respect to t_1 and t_2 and substitute $t_1 = t_2 = 1$. Easy calculations yield in turn

$$E(N_i) = \lambda\nu_1 J_{10} = \Lambda J_{10}, \tag{14}$$

$$\text{var}\,(N_i) - E(N_i) = \lambda(\nu_2-\nu_1)J_{20} = \Lambda(c-1)J_{20}, \tag{15}$$

$$\text{cov}(N_1,N_2) = \lambda(\nu_2-\nu_1)J_{11} = \Lambda(c-1)J_{11}, \tag{16}$$

where the symbol $J_{m,n}$ denotes the integral

$$J_{m,n} = \int_{R_3} p_1^m(u)p_2^n(u)\,du, \tag{17}$$

with $p_i(u)$ representing the probability that a galaxy from a cluster centered at $u = (u_1,u_2,u_3)$ will be located in Q_i and that it will be successful.

The detail of these formulas that may appear regrettable is that they depend upon the space density λ of cluster centers and on the moments

ν_k of the cluster sizes in a manner that makes it difficult to have these constants estimated separately. This difficulty is partly avoided by noticing that the product, say $\Lambda = \lambda\nu_1$, has a physical meaning of its own, namely, the expected number of galaxies per unit volume of space. Letting $\nu_2 = c\nu_1$, the above formulas can be written in a more convenient form, in terms of Λ and in terms of $c > 1$, which can be accepted as a compromise characterization of the distribution of cluster sizes ν.

Dividing formula (16) by formula (15) we obtain a quotient, say

$$q_k = \frac{\text{cov}(N_1, N_2)}{\text{var}(N_i) - E(N_i)} = \frac{\int_{R_3} p_1(u)p_2(u)du}{\int_{R_3} p_i^2(u)du}, \tag{18}$$

which is labeled the quasi-correlation between counts of galaxies in equal squares on the photographic plate, both in the same row or in the same column, k units of distance apart. Since N_1 and N_2 are identically distributed, q_k would have been the correlation coefficient between these two variables, were it not for the negative term in the denominator.

Examination of the right side of (18) indicates that q_k cannot be negative and that, by Schwarz inequality, it cannot exceed unity. The occurrence of negative q_k is an indicator that the process of clustering does not represent the actual mechanism of the counts considered. With reference to counts of galaxies, negative empirical q_k may indicate the presence of patchy light absorbing clouds in the photographed part of the sky.

The procedure of fitting the model to the actual counts of galaxies consisted of the following steps. First, using interpolatory formulas chosen to approximate the unknown dispersal function f and the unknown success probability θ, with the combined number of adjustable parameters equal to 3, a system of combinations of estimates was determined to fit formula (18) to empirical values of the q_k computed for $k = 1,2,\ldots$. Next, the tentative combinations of f and θ were substituted in the right side of (14) to calculate the corresponding Λ. Finally the empirical value of the left side of (15) yielded the corresponding c.

The final choice among the several estimates of the quadruplet (f,θ,Λ and c) was made by comparing the consequences of each of them with certain qualitative characteristics of the photographic plates. These included the numbers of tentatively identified clusters of varying apparent sizes, etc.

The particular interpolatory assumptions were:

(i) The dispersal functions were assumed to be spherical normal densities centered at the center of the cluster (thus far, there is no objection to this approximation) with standard deviation for each coordinate σ the same for all clusters. (This latter assumption is now believed unrealistic, but with no distinct indication as to how the value of σ might vary).

(ii) It was assumed that, to be successful, a galaxy must be brighter than the limiting apparent magnitude m_1, left unspecified. This assumption is now believed unrealistic.

(iii) It was assumed that the distribution of the absolute magnitude M in the population of all galaxies, irrespective of their morphological types (whether ellipticals, or spirals, etc.), is normal with an unspecified mean M_0 and standard deviation σ_M. Now, this assumption (iii) appears most unrealistic (see next section).

As a result of these assumptions, the tentative formula for q appeared to depend only on three adjustable parameters σ, $m_1 - M_0$ and σ_M. The process of fitting is illustrated in Figure 1, where the continuous curves correspond to the indicated trial values of the three parameters and filled circles represent the empirical values of quasi-correlation q_k computed for Shane's counts in $1^o \times 1^o$ regularly spaced squares. It is seen that quite a few combinations of parameter values fit the data well, which encouraged further efforts of estimation. In particular it was found that the assumption of ν having a negative binomial distribution was quite acceptable. However, it is obvious that, with a large number of adjustable parameters at one's disposal, a great variety of interpolatory formulas can be fitted to the same data, without any of these formulas representing a realistic picture of the actual chance mechanism governing the observations. Thus, an effort was made to

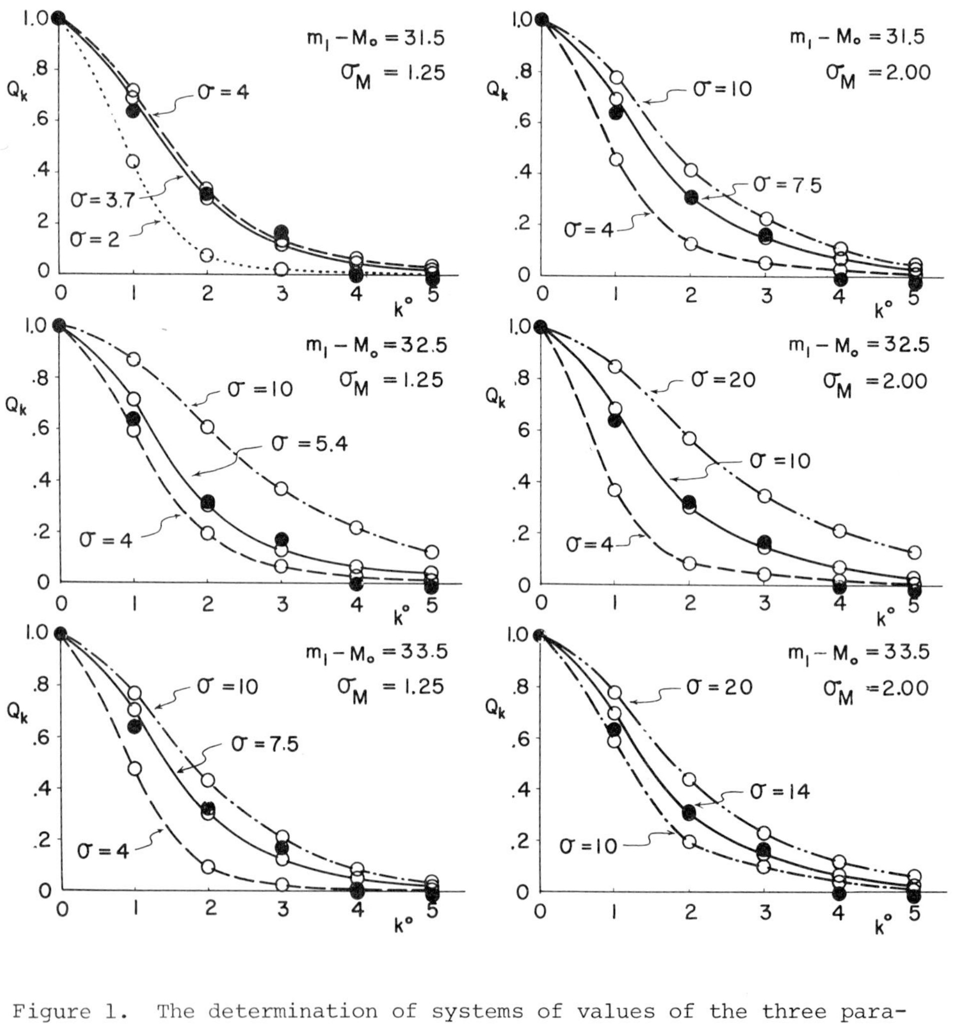

Figure 1. The determination of systems of values of the three parameters m_1-M_0, σ_M and σ from the quasi-correlation.

substantiate the model. The basic consideration was that, if the model is realistic, then the estimates of the parameters, obtained by fitting the counts of images of galaxies in $1^o \times 1^o$ squares, should be in reasonable agreement with those fitting the counts in 10' x 10' squares. When the effort was made at this verification (Neyman & Scott, 1959) there resulted a shocking disappointment. Both sequences of empirical quasi-correlations could be fitted by assuming appropriate values of the adjustable parameters. However, the values that fitted one sequence, would not fit the other. The unavoidable conclusion is that the model as a whole is not realistic.

In these circumstances, it was natural to probe each of the originally adopted postulates (see next section) and look for the one most likely to require modifications. Thus far, no agreement was reached in this respect. The particular suggestions include the following:

(a) The assumption of Poisson distribution of cluster centers may be seriously wrong. In fact, studies are published indicating that cluster centers may be clustered themselves [superclustering or subclustering (Abell, Neyman, Scott, 1964)].

(b) The frequent descriptions of some clusters as compact and some others as diffuse suggest that the assumed identity of the dispersal functions for all clusters is seriously wrong. In fact, from certain points of view it seems plausible that the dispersal function of a cluster must be dependent on the number of members in the cluster.

While these and some other suggestions may be valid, thus far no specific ideas seems to have emerged that could be used for a reconstruction of the model.

REMARK: This is in reply to the question, more or less, why introduce quasi-correlations which are a novelty, rather than deal with a familiar correlation coefficient?

The answer is implied by formulas (14), (15), and (16). It will be seen that the correlation coefficient of N_1 and N_2 depends not only on the dispersal function f and on probabilities of success θ, but also on the properties of the distribution of the cluster sizes ν. Specifically, the correlation depends on the quotient c. Thus, the information that might be provided by the empirical serial correlation coefficients between the N_1 and N_2 will be less specific than that stemming from quasi-correlations. Furthermore, the correlation coefficients of N_1 and N_2 must be influenced in an undesirable manner by the sizes of the squares in which the counts of galaxies are made. It will be seen that, if these squares are diminished, tending to zero, then the probabilities p_1 and p_2 become infinitesimals of the same order of magnitude. The consequence of this is that, with a decrease in the squares, the correlation coefficient of N_1 and N_2 must tend to zero, more or less irrespective of c and of the properties of f and θ. This is not the case with quasi-correlations.

A particular difficulty of astronomical studies of clustering is connected with the success probability θ. As already mentioned, the nature of these probabilities is subject to considerable doubts. On the other hand, in certain ecological studies, perhaps concerned with wanderings of some larvae over a field as described in section 2, it may be appropriate to assume $\theta \equiv 1$. In this case, depending upon the choice of the interpolatory formula for the dispersal function f, the integrals J_{mn} would be evaluated explicitly and a comparison between the empirical and the theoretical quasi-correlation would provide direct estimates of the relevant adjustable parameters. For example, this would be the case if f could be approximated by the product of two normal densities in terms of differences x_1-u_1 and x_2-u_2, each with its appropriate standard deviation, treated as an adjustable parameter. Actually, this method was used and revealed that on a given field, certain insects have a preferred direction of travel: in this direction they travel somewhat faster than in the perpendicular direction. Figure 2 reproduced from Neyman and Scott (1957) illustrates the procedure.

9. AN APPARENTLY SUBSTANTIATED SUBSTANTIVE FINDING

In the foregoing section it was briefly mentioned that the ideas about the distribution of absolute magnitudes of all the galaxies

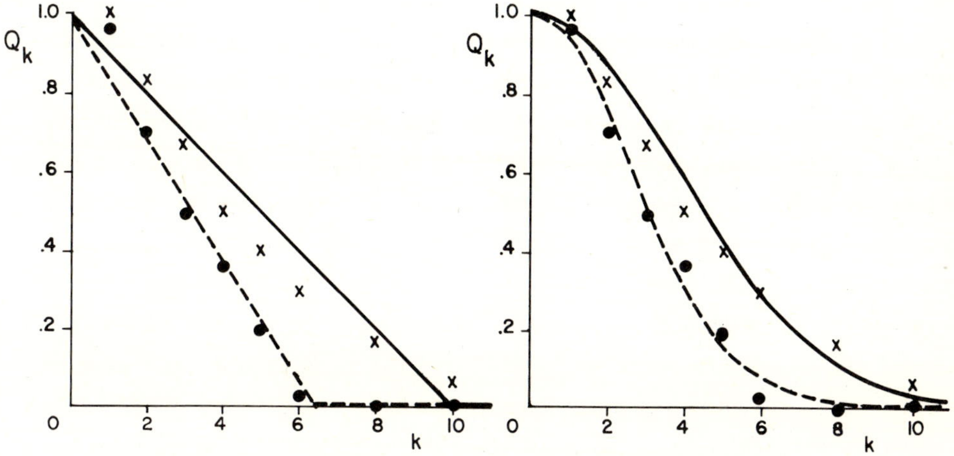

Figure 2. Search for a realistic dispersal function for larvae of Japanese beetles counted in equal regularly arranged square plots on a field. Solid lines and crosses indicate the theoretical and the empirical quasi-correlations for rows of plots in the East-West direction. Filled circles and dashed lines refer similarly to the North-South direction. The left panel refers to the dispersal function tentatively visualized for type A contagious distributions with two parameters (see section 2). The right panel refers to the assumption that the dispersal function is a product of two normal densities. The fit on the right is better than on the left. Both panels indicate that the dispersal in the E-W direction is broader than in the perpendicular direction.

combined, the ideas prevailing in 1950's, were later found unrealistic. Because every sample of galaxies for which the brightness and distance could be measured (with unavoidable errors!) had to be a biased sample, with preference for those objects that were brighter than the others, an appropriate statistical method had to be devised to correct for the bias.

The basic concept introduced (Neyman & Scott, 1961) is a "homogeneous class" of galaxies defined as follows. A class C of galaxies is called homogeneous if the inclusion of any of its members into a catalogue is a chance event, with the probability of inclusion, say $\Phi(m)$ depending only upon the apparent magnitude m of this galaxy and on nothing else. Another theoretical concept is that a list of apparent magnitudes m and a list of absolute magnitudes M compiled for those galaxies of a homogeneous class C that are actually included in the catalogue are random samples from respective distributions labeled catalogue distributions of m and M. The basic theoretical result is that, once the catalogue distributions of m and M are adequately approximated by some method of curve fitting, direct estimates of $\Phi(m)$ and also of the "space distribution" of the absolute magnitudes M for the given class can be unambiguously and easily estimated. Of course, all this applies to a homogeneous class as defined above and, a priori, there was no guarantee that such an entity exists.

To solve this problem, a test for homogeneity of a given class of galaxy was developed. Its application to the totality of all galaxies in the HMS catalogue (Humason, Mayall, Sandage, 1956) gave a definitive answer in the negative. The next step was to try the same test on galaxies in the same catalogue characterized by one or another of the several "morphological types", known as ellipticals E0-E3, ellipticals E4-E7, spirals, etc. Here, the test deduced failed to detect the presence of heterogeneity in several catagories. This encouraged further work to estimate the selection probabilities $\Phi(m)$, separately for each presumed homogeneous class, and also the corresponding space distribution of the absolute magnitude (so-called luminosity function). In turn, these results permitted the estimation of the space frequencies of the several homogeneous classes of galaxies. All this work was done with reference to galaxies in the HMS catalogue that were labeled "field galaxies", presumably each being a member of a cluster composed of just one galaxy, and also of a few members of "near by" and of "intermediate groups".

The question arose whether the theory behind these calculations could be somehow substantiated. Could this theory, and the numerical results cbtained, be used to calculate something to be compared with some direct observations not used in obtaining the original estimates?

Table 2 summarizes the results (Neyman, Scott, Zonn, 1962) of the verification performed. The column labeled $\lambda\%$ gives the estimates of the percentages of all field galaxies in space that belong to the

TABLE 2. *Percentage of galaxies of different morphological types*

Morpho-logical type	Field galaxies HMS		Near groups n=32		Virgo n=80		Intermediate groups n=72		Coma N=46		"Far" clusters n=48	
	n	Space λ%	Exp	Obs	Exp	Obs	Exp	Obs	Exp	Obs	Exp	Obs
EO-E3	83	3.7	7.0	6.2	10.6	12.5	18.1	22.2	34.8	34.8	44.1	47.9
E4-E7,Ep	28	7.6	5.3	9.4	5.0	15.0	6.2	9.7	8.8	19.6	12.3	8.3
SB0,SBa	21	2.2	4.1	9.4	5.6	10.0	4.6	13.9	2.3	0.0	1.1	2.1
SBb	26	1.9	3.4	3.1	5.0	2.5	6.3	4.2	7.8	2.2	5.8	0.0
S0,S0p	66	11.5	12.5	6.2	13.7	15.0	16.1	23.6	20.5	28.3	21.0	14.6
Sa,Sap,Sab	51	3.3	7.1	12.5	11.2	8.8	11.6	11.1	8.8	8.7	4.3	18.7
Sb,Sbc	77	20.8	21.8	12.5	20.6	12.5	17.5	8.3	11.0	2.2	8.0	6.2
Sc,Scp,SBc	94	49.0	38.8	40.6	28.5	23.8	19.5	6.9	5.9	4.3	3.3	2.1
Correlation coefficient			0.90		0.73		0.32		0.88		0.90	

morphological types (and some of their combinations) symbolized in the left margin. These are presumed to form homogeneous classes. It is seen that in space the morphological types E0-E3 are estimated to represent only about 3.7% of the total. On the other hand, the combined class of spirals Sc, SCp, SBc, appears very frequent, with 49% of the total.

The subsequent double columns of the table compare the estimates of percentages of the same presumed homogeneous classes as estimated from the theory and as provided by direct observations in the HMS catalogue. The most relevant double columns are those relating to the Virgo, to the Coma and to the "far clusters". None of the galaxies in these systems were used in the process of estimating the selection probabilities and of the luminosity functions and the indicated expected percentages are pure predictions.

It is seen that, on the whole, the theoretical expectations in Table 2 agree quite well with the observed percentages of the several presumed homogeneous classes of galaxies. It may be taken, thus, that the theory of homogeneous classes and also the various estimates optained through it have been reasonably substantiated. In the light of these results, it will be realized that the assumptions regarding the luminosity function of galaxies at large, that were adopted in the treatment of Shane's counts on the photographic plates, were much too primitive to promise success.

The findings in Table 2 have another interesting aspect. In order to appreciate its significance, it must be remembered that we see galaxies and study their properties not as they are now, but as they were in the epoch when the light reaching our instruments now was emitted by the galaxies. In the case of the relatively near Virgo Cluster, the difference between the two epochs is about 22 million years. Thus, we see the Virgo Cluster as it was 22 million years ago. For other clusters, the estimated distance into the past runs into several hundred million and, occasionally, into more than a billion years.

The agreement between the predictions, based on near galaxies, on the one hand, and the observations for distant clusters, on the other, indicates, then, that in the far past the morphological type composition of the population of galaxies and also the luminosity functions of these types must have been similar to those in the more recent epoch to which the observations on near galaxies refer. Thus, Table 2 implies a support for the highly disputed theory of continuous creation-steady state of the universe proposed some two decades ago by Bondi, Gold and Hoyle (Hoyle, 1956).

10. RELATION BETWEEN THE PROCESS OF CLUSTERING AND THE GENERAL BRANCHING PROCESSES. POPULATION DYNAMICS.

The general branching processes (Harris, 1963) are a set of generalizations and extensions of the Galton-Watson process constructed to represent the mechanism of extinction of human family trees. At a given moment, a family has so many male members. Each of them produces a certain number of male descendants, considered as a random variable with some fixed distribution. Each male of this first generation produces male descendants of the second generation, subject to the same probability law, etc. The basic problem of the theory is concerned with the sequence of generations so formed and, in particular, with the probability of ultimate extinction.

The branching processes are concerned with sequences of generations of particles which replace the human males of the Galton-Watson process. The particles considered simultaneously are of several different kinds and each is capable of creating a family of descendants that may be of a variety of kinds. The "kind" of a particle may be characterized by a variety of physical properties, including location in some space, etc.

The processes of clustering considered in the earlier sections of this paper appear as segments of a branching process, segments composed of only two generations, the cluster centers Γ and their "descendant" particles Π. Roughly speaking, the problems studied were mostly limited to the mechanisms of the relationship between the "parents" Γ and the "descendants" Π, which are not the typical problems of the theory of branching processes.

The purpose of the present section, and that of the next, is to sketch briefly two applications of the process of clustering to biological phenomena involving problems closely related to those of general branching processes. One of these applications refers to the development of a biological population, perhaps of weeds, insects or bacteria, conceived as a conglomeration of clusters. An attempt in this direction has been already mentioned (Neyman & Scott, 1957).

Here, the successive generations of the population studied, perhaps started by a few "immigrants," were assumed to multiply in a fixed manner, just as is assumed in the Galton-Watson process. Also, the dispersal functions of clusters were assumed to be fixed, independent of the generation. Formulas were deduced to characterize recursively the distribution of the organisms of the successive generations, separately males and females. Also, the possibility is considered that in actual counts (perhaps of mice) it may be impossible to separate the organisms according to the generations to which they belong.

While the results obtained have some interest as the first step in the indicated direction, it is well to be aware of their important weaknesses. One is the omnipresent assumption that the distribution of the number ν of "descendants" of one particular member Γ of the n-th generation is assumed independent of n, irrespective of anything. This particular assumption excludes the possibility of a more or less stable equilibrium which is actually observed in many laboratory experiments. Thus, in order to make the theory mentioned closer to reality, it appears essential to modify this particular assumption. One way of doing so is to utilize the set of theoretical assumptions (Neyman, Park & Scott, 1956) which proved quite effective in explaining some curious phenomena observed by Thomas Park and by the generation of his students in their careful experiments with flour beetles. Briefly and roughly the phenomena are as follows.

Each of the two not interbreeding species, say A and B, of the beetles, kept separately in vials filled with a fixed amount of flour, gradually multiplies and, in due course, reaches an apparent level of equilibrium, say E(A) and E(B), respectively. The two levels differ considerably, perhaps one of them being about 100 and the other 150.

With unavoidable fluctuations from one census to the next, the populations appear to maintain these levels more or less indefinitely. This applies to the two populations kept separately. What might be expected if one starts a mixed population, beginning with equal numbers of both species A and B?

Intuitively, one might expect an equilibrium, perhaps at a level E(A,B) intermediate between E(A) and E(B), probably with a preponderance of the species found capable of maintaining the higher numerical level if kept in isolation. The interesting point is that no such combined equilibrium was ever observed. Invariably, in all the experiments, one of the two species died out completely, not always the same, leaving the field to the other, which eventually reached its own equilibrium level.

According to the ideas then prevailing among the biologists, the explanation of the phenomena described had to depend on some biological changes in the individual beetles caused by factors of crowding and of cohabitation. The statistical treatment showed that no assumptions of biological changes are necessary and that the explanation of all observed facts can be reached by inserting a stochastic element into the known mechanism governing the successive generations of flour beetles.

The life cycle of both species A and B is known to involve two inert forms (egg and pupa) and also two active forms (larva and adult). Also, it is known that the active forms do not fight with each other, but like to feed on their own and also on the other species' inactive forms. Depending upon the mean number of inactive forms (eggs) produced by a female, upon the duration of the inactive forms and depending upon the voracity of the active form, a given species kept in isolation was found to have its own equilibrium level. Also, it was found that the presence or absence of non-zero equilibria for the two species kept together depends upon the relationship between their characteristics enumerated above and, in addition, on whether they have a preference to eat their own eggs or the eggs of the other species.

While the above stochastic scheme must oversimplify the phenomena in many respects, subsequent experiments by Park and by his students tend to indicate that, in the main, it does reflect the actual mechanism of reproduction and of competition. It seems likely that the incorporation of some such hypotheses into the original model of population dynamics might make it substantially more realistic.

11. PROCESS OF CLUSTERING AS A TOOL IN MODELING THE PHENOMENA OF EPIDEMICS

The problems of the classical stochastic theory of epidemics are well illustrated in the book by Bailey (1957) and by more recent papers of Gani and Daniels, published in the Proceedings of the Fifth Berkeley Symposium on Mathematical Statistics and Probability (1967), which utilize the works of the other principal authors on the subject, M. S. Bartlett (1956), D. G. Kendall (1965) and P. Whittle (1955).

Briefly, the studies refer to "closed populations" initially composed of n uninfected susceptibles and of a infectious. For any time $t > 0$, the random variables considered are $r(t)$, the number of uninfected susceptibles, and $s(t)$, the number of infectious. Suitable assumptions are made about the probabilities of changes in the values of these variables due to a susceptible becoming infected and due to the disappearance of an infectious, either through death or otherwise. The joint probability distribution of r and s is hard to obtain and the literature contains a number of efforts in this direction, some very ingeneous.

Our own work on the theory of epidemics was stimulated by the question about the kind of epidemic and about the kind of population to which a theory based on the above assumptions might possibly apply. While there is some latitude in this respect, our most confident answer to this question is an epidemic developing in a well guarded boarding school for girls in the Victorian era. One reason for this diagnosis is the implication of the classical theory that an additional infectious increases the danger of contracting the contagious disease *equally* for every single susceptible in the population. Contrary to this, if one considers epidemics as they develop in a country or even in a city, a new infectious, say A_0, who spends the period of his infectiousness in New York City, may infect a number of individuals in his immediate surroundings but not directly anyone, say B, who spends this same period of time in San Francisco. On the other hand B may well be infected indirectly. For this to happen one or more intermediary individuals are necessary, say $A_1, A_2, \ldots, A_n$, with A_1 being infected by A_0 and B being infected by A_n. In addition, a hypothesis is necessary about the intermediary A_i traveling from New York to some other places and, ultimately to San Francisco.

One other hypothesis is necessary for which the classical theory has no place. This is that the number, say $\nu(u)$ of susceptibles whom a single infectious located at a point u may infect is not the same and, if considered as a random variable, the distribution of $\nu(u)$ is very likely to depend upon u. For example, consider the possibility of u being somewhere in slums or, alternatively, in some comfortable suburb.

Considerations of this kind led us to believe that, in order to be anywhere realistic in modern times, a theory of epidemics must involve considerations of geography and of geographical diversification. The first attempts in this direction, from a point of view different from ours, seem to be due to D. G. Kendall (in Bartlett, 1957). See also Kendall (1965). Interesting Monte Carlo studies are due to Bartlett (1961). Our own effort (Neyman & Scott, 1964) represents an application of the process of clustering as follows.

The epidemic is supposed to develop in a habitat H identified with R_2. To each point u in H there correspond two functions $g(t|u)$ and $f(x|u)$. The first is the probability generating function of the number $\nu(u)$ of susceptibles who would be infected by a single infectious at u. The second function $f(x|u)$ is the dispersal function. It determines, for each point x in H the probability density of the random location X at which a susceptible infected at u will become infectious himself. Apart from certain conditions of regularity including the measurability of the two functions in terms of u, both g and f are left unspecified. The additional omnipresent hypothesis is that, before becoming infectious, all the infected individuals travel independently of each other.

It is believed that the above hypotheses bring the theory quite close to the actual mechanism of the spread of epidemics. This is not the case with the following two assumptions. One is that the incubation period, that is the period between the moment a susceptible is infected and the moment he becomes infectious, is an absolute constant which may be taken as a unit of time. This assumption, then, led to considering an epidemic as a sequence of disjoint generations. Our second unrealistic assumption is that the duration of infectiousness of each infected individual is equal to zero. An effort to remove these assumptions is due to Bartoszyński (1967).

A degree of unrealisticness is attached also to the third assumption, namely that the numbers, say $\nu(u')$ and $\nu(u'')$ of susceptibles infected by any two infectious, one at u' and the other at u", are mutually independent. Here, the possibility is that u' and u" may be close together both in space and in time and that the two infectious at these points might "compete" with each other for susceptibles to infect. Here, then, the phenomenon of partial exhaustion of susceptibles, nicely taken care of by the classical theory, is ignored in our process of clustering approach.

The problems treated were of two kinds. First we were interested in the development of epidemics, either in the whole habitat or in a particular region R, of epidemics started by a single individual ("immigrant") infectious at a specified point u. The second category of problems refers to epidemics continuously recurring due to mutations of some bacteria carried by members of the population studied. The mutations are assumed to occur Poissonwise in H with unspecified local densities $\lambda(u)$, possibly connected with the local densities of the population.

One particular question interested us referring to the first category of problems. This is the probability of ultimate extinction of an epidemic started by a single infectious located at u_0. The following two theorems appear interesting.

THEOREM 3. *If for every point* u *in the habitat the probability* $P\{\nu(u) > 0\} > 0$ *and if the dispersal functions* $f(x|u) > 0$ *for all* x *and* u, *then the probability of the ultimate extinction of an epidemic in* H, *started by a single infectious at* u_0, *will be equal to unity either for all* u_0 *in* H *or for none.*

In order to formulate the next theorem, we consider a region R and two epidemics initiated by a single infectious at u_0. One of these epidemics develops in the whole habitat H. The other epidemic develops in R and is partial to that in H.

THEOREM 4. *If there exists a positive number* α *such that, for all* u *in* H, *the probability* $P(R|u)$, *that an individual infected at* u *will become infectious in* R, *is at least equal to* α, *then, in order that an epidemic started at* u_0 *and developing in* R *have a unity probability of extinction, it is both necessary and sufficient that the same hold for the epidemic developing in the whole habitat* H.

The two theorems 3 and 4 support the tendency of the public health authorities to keep all the localities in the country on a reasonable level of sanitary conditions.

Some refinements and further developments of the theory are due to Bühler (1967) and to Bartoszyński, Loś and Wycech-Loś (1965).

ACKNOWLEDGMENTS

Prepared with the partial support of NIH Grant USPHS GM 10525-08 and US Army Grant DA-ARO-D-31-124-G1135.

REFERENCES

Abell, G. O., Neyman, J. & Scott, E. L. (1964). Subclustering of galaxies. Astronomical J. 69, 529.

Bailey, N.T.J. (1957). *The Mathematical Theory of Epidemics*. London: Griffin.

Bailey, N. T. J. (1967). The simulation of stochastic epidemics in two dimensions. *Proc. Fifth Berkeley Symp. Math. Stat. and Prob.* 4 237-257, ed. by L. LeCam and J. Neyman. Berkeley: Univ. of Calif. Press.

Bartlett, M. S. (1956). Deterministic and stochastic models for recurrent epidemics. *Proc. Third Berkeley Symp. Math. Stat. and Prob.* 4, 81-109, ed. by J. Neyman. Berkeley: Univ. of Calif. Press.

Bartlett, M. S. (1963). The spectral analysis of point processes. J. R. Statist. Soc. B 25, 264-96.

Bartoszyński, R., Łoś, J. & Wycech-Łoś, M. (1965). Contribution to the theory of epidemics. *Bernoulli, Bayes, Laplace* 1-8, ed. by J. Neyman and L. LeCam. Berlin-Heidelberg: Springer-Verlag.

Bartoszyński, R. (1967). Branching processes and the theory of epidemics. *Proc. Fifth Berkeley Symp. Math. Stat. and Prob.* 4, 250-69, ed. by L. LeCam and J. Neyman. Berkeley: Univ. of Calif. Press.

Bates, G. E. & Neyman, J. (1952). Contribution to the theory of accident proneness, II. True or false contagion. *Univ. of Calif. Publ. in Statist.* 1, 255-76.

Bates, G. E. (1955). Joint distributions of time intervals for the occurrence of successive accidents in a generalized Pólya scheme. Ann. Math. Statist. 26, 705-20.

Bronowski, J. & Neyman, J. (1945). The variance of the measure of a two-dimensional random set. Ann. Math. Statist. 16, 330-41.

Bühler, W. J. (1966). A theorem concerning the extinction of epidemics. Biometrische Zeitschrift 7, 10-14.

Daniels, H. E. (1967). The distribution of the total size of an epidemic. *Proc. Fifth Berkeley Symp. Math. Stat. and Prob.* 4, 281-93, ed. by L. LeCam and J. Neyman. Berkeley: Univ. of Calif. Press.

Gani, J. (1967). On the general stochastic epidemic. *Proc. Fifth Berkeley Symp. Math. Stat. and Prob.* 4, 271-79, ed. by L. LeCam and J. Neyman. Berkeley: Univ. of Calif. Press.

Greenwood, M. & Yule, G. U. (1920). An inquiry into the nature of frequency distributions representative of multiple happenings with particular reference to the occurrence of multiple attacks of disease as repeated accidents. J. R. Statist. Soc. 83, 255-79.

Harris, T. E. (1963). *The Theory of Branching Processes*. Berlin: Springer.

Hoyle, F. (1956). The steady-state universe. *The Universe*, 77-86. New York: Simon & Schuster.

Humason, M. L., Mayall, N. U. & Sandage, A. R. (1956). Redshifts and magnitudes of extragalactic nebulae. Astronomical J. 61, 97-162.

Kendall, D. G. (1965). Mathematical models of the spread of infection. *Mathematics and Computer Science in Biology and Medicine*, 213-25. London: Her Majesty's Stationery Office.

Lewis, P. A. W. (1964). A branching Poisson process model for the analysis of computer failure patterns. J. R. Statist. Soc. B. 26, 398-456.

Neyman, J. (1939). On a new class of contagious distributions, applicable in entomology and bacteriology. Ann. Math. Statist. 10, 35-57.

Neyman, J. (1963). On finiteness of the process of clustering. Sankhyā A 25, 69-74.

Neyman, J., Park, T. & Scott, E. L. (1956). Struggle for existence; the tribolium model. *Proc. Third Berkeley Symp. Math. Stat. and Prob.* 4, 41-79, ed. by J. Neyman. Berkeley: Univ. of Calif. Press.

Neyman, J. & Scott, E. L. (1952). A theory of spatial distribution of galaxies. Astrophysical J. 116, 144-63.

Neyman, J. & Scott, E. L. (1953). Frequency of separation and of interlocking of clusters of galaxies. Proc. Nat. Acad. Sci. 39, 737-43.

Neyman, J. & Scott, E. L. (1957). On a mathematical theory of populations conceived as conglomerations of clusters. Proc. Cold Spring Harbor Symposia on Quantitative Biology 22, 109-20.

Neyman, J. & Scott, E. L. (1959). Large scale organization of the distribution of galaxies. *Handbuch der Physik* 53, 416-44. Berlin: Springer-Verlag.

Neyman, J. & Scott, E. L. (1964). A stochastic model of epidemics. *Stochastic Models in Medicine and Biology,* ed. by J. Gurland. Madison, Wis.: Univ. of Wisc. Press.

Neyman, J. & Scott, E. L. (1967). On the use of optimal $C(\alpha)$ optimal tests of composite hypotheses. Bull de l'Institut International de Statistique 41, 477-96.

Neyman, J., Scott, E. L. & Shane, C. E. (1953). On the spatial distribution of galaxies: A specific model. Astrophysical J. 117, 92-133.

Neyman, J., Scott, E. L. & Zonn, W. (1962). Abundances of morphological types among galaxies in clusters and in the field. Astronomical J. 67, 583.

Pólya, G. (1930). Sur quelques points de la théorie des probabilités. Ann. Inst. Henri Poincaré 1, 117-61.

Puffer, R. R. (1950). *Practical Statistics*. New York: McGraw-Hill.

Scott, E. L., Shane, C. D. & Swanson, M. D. (1954). Comparison of the synthetic and actual distribution of galaxies on a photographic plate. Astrophysical J. 119, 91-112.

Vere-Jones, D. (1970). Stochastic models for earthquake occurrence. J. R. Statist. Soc. B 32, 1-62.

Whittle. P. (1955). The outcome of a stochastic epidemic — a note on Bailey's paper. Biometrika 42, 116-22.

Zwicky, F. & Herzog, E. (1961-68). *Catalogue of Galaxies and of Clusters of Galaxies*, 1-5, Pasadena Calif.: California Institute of Technology.

SOME POINT PROCESS MODELS OF LUNAR AND PLANETARY SURFACES

A. Marcus
John Hopkins University, Baltimore, Maryland

SUMMARY

In Part I a Markovian multivariate immigration and multiple death process is developed which gives the moments of the numbers of craters of different sizes in different regions on the surface. The spatial cross-covariance density of crater numbers is derived. For an initial crater diameter distribution which is Paretian with any exponent $\gamma > 2$, the effect of obliteration is to produce an equilibrium crater expected number density which is proportional to $(\text{diameter})^{-3}$. In Part II a moving average model, generated by a multi-dimensional Poisson process, is developed for lunar elevations. It is shown that when crater depth is proportional to crater diameter, and $2 < \gamma < 3$, then the elevation distribution is approximately "stable" with exponent $\gamma - 2$, and the slope distribution has power-law ("stable" or Paretian) tails with characteristic exponent $\gamma - 1$. A simple result is obtained for the spectral density of elevations.

1. DISTRIBUTION OF CRATERS AND CRATER COUNTS

1.1. Introduction

The surface of the Moon is covered by roughly circular *craters* with raised rims; these range in size from about a thousand kilometers in diameter (about 30% of the lunar diameter — probably close to the size of theoretically largest crater that could be formed without completely shattering the planet) down to tiny pits of a few microns in diameter found in the samples returned by the Apollo expeditions. Mars is somewhat less densely covered by similar objects, at least to diameters of a kilometer and more. Radar observations of Venus have found large circular objects beneath the clouds, which may be similar to the large lunar and Martian craters and circular seas *(maria)*. Presumably the planet Mercury, the asteroids, and the other moons of the solar system are also covered by craters.

On the Earth there are dozens — possibly hundreds — of circular craters which are almost certainly the result of impacts of meteoroids on the Earth's surface. There are other numerous craters - *morphologically indistinguishable from these* — which are of internal origin, e.g. explosive maars and calderas. The crater morphology does not, therefore, give unambiguous evidence of its origin except (perhaps!) in a few extreme cases. The origin of most of the craters observed on the surfaces of the other planets of the solar system must, therefore, be deduced indirectly and in the main rather than individually.

It is here that probabilistic models and their associated statistical analyses are of great use. Because craters are circular, we can regard them as *points* (the crater centers), each with an associated random variable, the crater diameter. Physical hypotheses about the origin, interaction and evolution of craters may then lead to testable hypotheses about the spatial and size distributions of the craters.

1.2. Meteoridal Impact Hypothesis

It is now commonly believed that most lunar craters larger than 2 km in diameter, and most smaller than about 50 meters in diameter, are of primary meteoroidal impact origin (most of the others may also be, but the proportion of craters of secondary impact or internal origin is, at the very least, somewhat greater at diameters of 50 meters - 2 kilometers than at other sizes). The meteoroid flux even now is nearly uniform on the lunar surface (except for small variations due to "gravitational focusing" by the Earth) and was probably even more nearly isotropic in the past. We thus suppose that *craters are formed Poissonwise in space and time on the surface*. The probability density function p(x) of diameter of newly formed craters is usually assumed to be an inverse-power law, with an arbitrary minimum crater diameter x_0 determined by the limits of observational completeness, and with characteristic exponent γ;

$$p(x) = \gamma x_0^{\gamma} x^{-\gamma-1} \qquad (x > x_0 > 0). \tag{1}$$

We will let f(t) denote the mean number of craters formed per unit area per unit time at time t, of diameter $\geq x_0$.

The lunar surface is composed of large regions obviously not of the same age. The light-colored, heavily cratered highlands or terrae are generally considered quite old, and the darker maria are relatively young; even within these respective regions there are smaller regions which have apparently been more or less effectively rejuvenated by lava flows and other internal processes. It is however usually possible to distinguish homogeneous regions by physical and morphological evidence, although with increasing frequency the variations of crater numbers and distribution are themselves being used to discriminate among apparently similar regions (see e.g. Fielder, 1970).

The theory developed here can be generalized, e.g. to first-order clustering processes, but only with some difficulty (Marcus, 1964).

1.3. An Approximate Model

The following model of crater formation and survival (Marcus, 1967) has adequately explained many of the observed properties of crater distribution. We will start first with a discrete model. Suppose that a population consists of m types of individuals, and that an individual of type j can destroy an individual of type i (i < j) when the j enters the population. We further suppose that i's are destroyed independently of each other (this is not quite correct for craters), hence the probability that a newly arrived j destroys k_1 out of the n_1 type 1's in the population when he arrives, k_2 out of the previously existing type 2's, etc., is (for some q_{ij}, $0 \leq q_{ij} \leq 1$)

$$\prod_{i=1}^{j-1} \binom{n_i}{k_i} q_{ij}^{k_i} (1-q_{ij})^{n_i-k_i}. \tag{2}$$

Suppose that individuals of all types arrive independently of each other,

Poissonwise at rate λp_j for type j, where $\sum_{j=1}^{m} p_j = 1$. Let $N_j(t)$ be the number of type j's at time t, and assume $N_j(0) = 0$. We can easily show that the probability generating function

$$G(\underline{z};t) = \sum_{n_1,\ldots,n_m} \text{prob}\{N_i(t) = n_1,\ldots,N_m(t) = n_m\} \prod_{i=1}^{m} z_i^{n_i}$$

satisfied the forward Kolmogorov equation

$$\frac{1}{\lambda}\frac{\partial}{\partial t} G(\underline{z};t) + G(\underline{z};t) = \sum_{j=1}^{m} G[\underline{q}(j) + \{\underline{1}-\underline{q}(j)\} \otimes \underline{z};t]p_j z_j, \tag{3}$$

where the following are m-vectors:

$$\underline{z} = (z_1,\ldots,z_m) \qquad (0 \leq z_j \leq 1)$$

$$\underline{q}(j) = (q_{1j},\ldots,q_{j-1,j},0,\ldots,0) = (q_{ij}, i = 1,\ldots,m);$$

$$\underline{1} = (1,\ldots,1);$$

$$\underline{a}\otimes\underline{b} = (a_1b_1,\ldots,a_mb_m).$$

Stationary moments are easily obtained:

$$E\{N_i(\infty)\} = p_i/\mu_i \qquad (i < m); \tag{4}$$

$$\text{var}\{N_i(\infty)\} = \{1 + \frac{p_i\mu_{ii}}{\mu_i(2\mu_i-\mu_{ii})}\}\frac{p_i}{\mu_i} \qquad (i < m); \tag{5}$$

$$\text{cov}\{N_i(\infty),N_j(\infty)\} = \frac{\mu_{ij}-\mu_j q_{ij}}{\mu_i+\mu_j-\mu_{ij}}\,\frac{p_i p_j}{\mu_i\mu_j} \qquad (i < j < m), \tag{6}$$

where

$$\mu_i = \sum_{k=i+1}^{m} p_k q_{ik}, \tag{7}$$

$$\mu_{ij} = \sum_{k=j+1}^{m} p_k q_{ik} q_{jk} \qquad (i \leq j). \tag{8}$$

Of course, $N_m(t)$ is a Poisson $(\lambda p_m t)$ random variable. Note that the index of dispersion $var\{N_i(\infty)\}/E\{N_i(\infty) > 1$.

The extension to a real spatial counting process with a continuum of crater sizes is easily made. Suppose that the basic region of interest is finite; denote it by R(0). In order to obtain a completely Markovian process, we must count all craters which may affect other craters. Hence, we will count all craters of diameter x whose centers are within a region R(x) consisting of R(0) plus a strip of width x/2 surrounding R(0). This correction is not usually important when the region is much larger than the largest craters, but has substantially decreased the estimated density of craters from Mariner 4 photos of Mars (Marcus, 1968). Denote the area of R(x) by A(x). We will also suppose that a crater of diameter x_i can be destroyed by a crater of diamter x_k if the crater of diameter x_k forms within a region of area $C(x_i,x_k)$ centered on the crater of diameter x_i; we assume that $x_k > \theta x_i (0 < \theta \leq 1)$. Our model is now applicable, with

$$\lambda = \bar{A}\, f(t),$$

$$\bar{A} = \int_{x_0}^{\infty} A(y/\theta - x_0)p(y)dy, \tag{9}$$

$$p_i = A[x_i/\theta - x_0)p(x_i)dx_i/\bar{A}] + 0(dx_0)dy; \tag{10}$$

$$q_{ik} = C(x_i,x_k)/A(x_k/\theta - x_0); \tag{11}$$

$$\Lambda(x) = \bar{A}\mu(x) = \int_{\theta x}^{\infty} C(x,y)p(y)dy. \tag{12}$$

The moments for counts of craters in an interval of diameters can be expressed in terms of moment densities $f_1(x;t), f_2(x_i,x_j;t)$ etc. (Bartlett, 1966). Let $N(x_1,x_2;t)$ be the number of craters whose diameters are between x_1 and X_2 that are counted at time t. Then

$$E\{N(x_1,x_2;t)\} = \int_{x_1}^{x_2} f_1(y;t)A(y/\theta - x_0)dy; \tag{13}$$

$$\mathrm{var}\{N(x_1,x_2;t)\} = \int_{x_1}^{x_2} f_1(y;t)A(y/\theta-x_0)dy$$
$$+\int_{x_1}^{x_2}\int_{x_1}^{x_2} f_2(y,z;t)A(y/\theta-x_0)A(z/\theta-x_0)dydz; \quad (14)$$

$$\mathrm{cov}\{N(x_1x_2;t),N(x_3,x_4;t)\}=\int_{x_1}^{x_2}\int_{x_3}^{x_4} f_2(y,z;t)A(y/\theta-x_0)A(z/\theta-x_0)dydz \quad (15)$$

for $x_1 < x_2 \leq x_3 < x_4$. We find

$$f_1(y;t) = f_1(y)[1-\exp\{-\Lambda(y)\int_0^t f(u)du\}], \quad (16)$$

$$f_1(y) = p(y)/\Lambda(y) \quad (17)$$

and, if

$$\int_0^\infty f(u)du = \infty.$$

then for $y \leq z$

$$f_2(y,z;\infty) = \frac{[\Lambda_2(y,z)-\{\Lambda(z)C(y,z)/A(z/\theta-x_0)\}]f_1(y)f_1(z)}{\Lambda(y) + \Lambda(z) - \Lambda_2(y,z)}, \quad (18)$$

where

$$\Lambda_2(y,z) = \int_{\theta z}^\infty \frac{C(y,v)C(z,v)}{A(v/\theta-x_0)} p(v)dv. \quad (19)$$

1.4. Spatial Structure of Crater Counts; Approximate Model

We generalize the preceding model to obtain the joint p.g.f. and moments of crater counts in separate but interacting regions $R_1(0)$ and $R_2(0)$. Let $R_n(x)$ be the enlargement of region $n(n=1,2)$ and let $A_n(x)$ be its area. Let

$$A_{1+2}(x) = \text{area of } R_1(x) \cup R_2(x). \quad (20)$$

The spatial interactions are defined by

$$q_{ij}^{(n)}(\underline{r}) = C_n(x_i, x_j, \underline{r})/A_{1+2}(x_j - x_0), \tag{21}$$

where $C_n(x_i, x_j, \underline{r})$ is the area of the intersection of $R_1(x_i - x_0) \cup R_2(x_i - x_0)$ with the region in which the center of a crater of diameter x_i must lie in order to be destroyed by a crater of diameter x_j whose center has coordinate $\underline{r}$. (In this section we take $\theta = 1$). The approximation employed in Section 1.3 that craters are destroyed independently of each other, is even less defensible than before; but a convenient analysis does not otherwise seem possible. We obtain in a straightforward way the p.g.f. of the number of j-types in region n, namely $N_j(t)$;

$$G(\underline{y}, \underline{z}; t)$$

$$= \sum_{k_1, \ldots, k_m, n_1, \ldots, n_m} \text{prob}\{N_{11}(t) = k_1, \ldots, N_{1m}(t) = k_m, N_{21}(t) = n_1, \ldots, N_{2m}(t) = n_m\} \prod_{i=1} y_i^{k_i} z_i^{n_i}$$

as

$$\frac{1}{\lambda}\frac{\partial}{\partial t} G(\underline{y}, \underline{z}; t) + G(\underline{y}, \underline{z}; t) = \sum_{j=1} \frac{p_j}{A_{1+2}(x_j - x_0)} \Big[y_j z_j \int_{R_1(x_j - x_0) \cap R_2(x_j - x_0)}$$

$$+ y_j \int_{R_1(x_j - x_0) - R_2(x_j - x_0)} + z_j \int_{R_2(x_j - x_0) - R_1(x_j - x_0)} \tag{22}$$

$$G[\underline{q}^{(1)}(j, \underline{r}) + \{\underline{1} - \underline{q}^{(1)}(j, \underline{r})\} \otimes \underline{y}, \underline{q}^{(2)}(j, \underline{r}) + \{\underline{1} - \underline{q}^{(2)}(j, \underline{r}\} \otimes \underline{z}; t] dr \Big]$$

where

$$\underline{q}^{(n)}(y, r) = \{q_{1j}^{(n)}(\underline{r}), q_{2j}^{(n)}(\underline{r}), \ldots, q_{j-1,j}^{(n)}(\underline{r}), 0, \ldots 0\}.$$

The moment densities are

$$f_1^{(n)}(x) = p(x)/\mu^{(n)}(x) A_{1+2}(x - x_0), \tag{23}$$

where

$$\mu^{(n)}(x) = \int_x^\infty \frac{p(y)dy}{A_{1+2}^2(y-x_0)} \int_{R_1(y-x_0)\cup R_2(y-x_0)} C_n(x,y,\underline{r})d\underline{r} \tag{24}$$

and

$$f_2(y,z) = \left[\mu_2(y,z) - \frac{\mu^{(1)}(y)\int_{R_1(y-x_0)} C_1(z,y,\underline{r})d\underline{r}}{A_1(y-x_0)\, A_{1+2}(y-x_0)}\right.$$

$$\left. - \frac{\mu^{(2)}(z)\int_{R_2(z-x_0)} C_2(y,z,\underline{r})dr}{A_2(z-x_0)A_{1+2}(z-x_0)}\right] \frac{f_1^{(1)}(y)f_1^{(2)}(z)}{\mu^{(1)}(y)+\mu^{(2)}(z)-\mu_2(y,z)}, \tag{25}$$

where for $y \leq z$

$$\mu_2(y,z) = \int_z^\infty \frac{p(v)dv}{A_{1+2}^3(v-x_0)} \int_{R_1(z-x_0)\cup R_2(z-x_0)} C_1(y,v,\underline{r})C_2(z,v,\underline{r})d\underline{r}\ . \tag{26}$$

The covariance density of the counting process can be obtained as a function of the distance r_{12} between regions 1 and 2 by taking $A_1(0) = 0$, $A_2(0) = 0$, in the above equations, which then simplify considerably. The spectral density of the counting process is simply the Hankel transform of the covariance density with respect to r_{12}. The same is true of course for cross-covariances and cross-spectra for craters of different diameters in the two regions.

1.5. Applications

In order to use these results we must know when a crater is destroyed. This is a problem of some complexity, both physical and perceptual (Marcus, 1970a). Two possibilities are the following

MODEL 1. A crater of diameter x is destroyed when and only when a crater of diameter y, $y \geq \theta x$, is formed whose center is within a distance $y - \theta x$ of the center of the original crater, Thus

$$C(x,y) = \frac{\pi}{4}(y-\theta x)^2. \tag{27}$$

MODEL 2. A crater of diameter x is destroyed when and only when a crater of diameter y, $y \geq \theta x$, is formed whose perimeter or rim edge touches that of the original crater. Thus

$$C(x,y) = \begin{cases} \pi xy & (y \leq x), \\ \frac{\pi}{4}(x+y)^2 & (y > x). \end{cases} \tag{28}$$

Using the probability density (1), one finds for $\gamma > 2$

$$f_1(x) = \{2\gamma(\gamma-1)(\gamma-2)\theta^{\gamma-2}/\pi\}x^{-3} \qquad \text{(Model 1)}, \tag{29}$$

$$f_1(x) = [4\gamma(\gamma-1)(\gamma-2)/\{2 + \frac{4\gamma(\gamma-2)}{\theta^{\gamma-1}}\}\pi]x^{-3} \qquad \text{(Model 2)}.$$

Empirical evidence and arguments too lengthy to reproduce here suggest that for large craters (x > 5 km) Model 2 is more plausible than Model 1, with a very large range of possible values of γ and θ ($\gamma \sim 2.65$ if $\theta = 1$, $\gamma \sim 3$ if $\theta = 0.833$).

The fact that the index of dispersion of the counts is greater than unity suggests that the hypothesis of Poissonness of crater centers will be rejected by the usual tests. This is in fact observed (Fielder, 1966; Marcus, 1966) but, for craters larger than 5 km, only to an extent consistent with an originally Poisson distribution. At smaller sizes there are some real clusters of craters. (Fielder & Marcus, 1967).

There are interesting spatial variations in crater counts. Available techniques in the spectral analysis of point processes (Bartlett, 1964) should make it possible to test the model of Section 1.4, but the actual analysis has not yet been carried out.

2. DISTRIBUTION OF ELEVATIONS AND SLOPES ON A CRATERED PLANETARY SURFACE

2.1. Introduction

A statistical description of some geometric properties of planetary surfaces (elevations, slopes, curvature) is of interest in radar astronomy and in the design of vehicles for planetary exploration. The usual first choice of a model is that the surface is Gaussian; but analysis of lunar photographs taken by the Ranger and Surveyor space craft have shown that this is not so. In any case, it is attractive to try to derive a surface model from fundamental hypotheses about the nature of the lunar surface.

Our basic hypothesis is that the major factor affecting the small-scale (~1 meter) surface roughness is cratering due to meteoroidal impacts. A basic mathematical assumption is that the number of craters $dN(x,\underline{R})$ of diameter x to x + dx in a small region $d\underline{R}$ surrounding the d-dimensional point $\underline{R}$ is a Poisson random variable with mean $\xi(x)dxd\underline{R}$. A second basic assumption is that the elevation change at $\underline{R}$ due to the formation of a crater of diameter x at a point $\underline{R} + \underline{r}$ whose distance from $\underline{R}$ depends only on r = (length of $\underline{r}$); denote it by $\zeta(x,r)$. Finally, we assume that elevation changes add linearly, that is the elevation $Z(\underline{R})$ at $\underline{R}$ is given by

$$Z(\underline{R}) = \int \zeta(x,r)dN(x,\underline{R} + \underline{r}). \tag{31}$$

This last assumption was introduced by Chernov (1967) and has been discussed at length by Marcus (1970b). It is approximately correct on the lightly cratered maria, but may not be useful when the surface is crater-saturated (see Section 1). This d-dimensional generalization of shot noise is well known (Takacs, 1957; Feller, 1966); $Z(\underline{R})$ has a compound Poisson distribution with characteristic function

$$\phi_{Z(\underline{R})}(u) = E[\exp\{iu\, Z(\underline{R})\}]$$

$$= \exp[\int_{-\infty}^{\infty} \xi(x)dx \int_{0}^{\infty} [\exp\{iu\, \zeta(x,r)\}-1]s_d(r)dr] \quad , \tag{32}$$

where

$$s_d(r) = d\pi^{d/2}\, r^{d-1}/\Gamma(1 + \tfrac{1}{2}d). \tag{33}$$

Regularity conditions for the random field $Z(\underline{R})$ seem to have escaped the attention of most authors until quite recently (Daley, 1970). In our notation, $Z(\underline{R})$ is

(i) almost surely finite if for sufficiently large $a > 0$,

$$\int_0^\infty \xi(x)\, dx \int_a^\infty \zeta r^{d-1} \zeta(x,r)\, dr < \infty \; ; \tag{34}$$

(ii) stationary if

$$\int_0^\infty \xi(x)\, dx \lim_{R\to\infty} \int_R^{R+a} r^{d-1} \zeta(x,r)\, dr = 0. \tag{35}$$

The problem of the simultaneous distribution of $Z(\underline{R})$ and $Z(\underline{R}+\underline{r})$ is more difficult because a single crater can affect elevations at both points. The second-order properties of $Z(\underline{R})$ are well known (e.g., Matérn, 1960). If

$$\int_0^\infty \xi(x)\, dx \int_0^\infty r^{d-1} \zeta^2(x,r)\, dr < \infty \, , \tag{36}$$

then the covariance function

$$c(r) = E\{Z(\underline{R})\; Z(\underline{R}+\underline{r})\} - E^2\{Z(\underline{R})\}$$

exists and is finite, with

$$c(r) = \int_0^\infty \xi(x)\, dx \int_0^\infty \int_0^\infty \zeta(x,r) \zeta(x,r_2) \nu_d(r_1,r_2;r)\, dr_1 dr_2 , \tag{37}$$

where (Marcus, 1969a)

$$\nu_d(r_1,r_2;r) = \frac{2\pi^{(d-1)/2}}{\Gamma(\frac{d-1}{2})} \frac{r_1 r_2 r_3^{d-3}}{r} \qquad (|r_1 - r_2| < r_1, r_1 + r_2 \geq r), \tag{38}$$

$$= 0 \qquad\qquad (\text{otherwise}),$$

and

$$r_3 = \frac{1}{2r} \{4r_1^2\, r_2^2 - (r_1^2 + r_2^2 - r^2)^2\}^{\frac{1}{2}}. \tag{39}$$

It is then possible to obtain the spectral density function $\sigma(\omega)$ as the d-dimensional Fourier (Hankel) transform of $c(r)$ (Takacs, 1957; Matern, 1960);

$$\sigma(\omega) = \int\ldots\int \exp(i\underline{\omega}\cdot\underline{r})c(r)d\underline{r}$$

$$= (2\pi)^{-d/2}\omega^{-(d-2)/2}\int_0^\infty c(r)r^{d/2}J_{d-2)/2}(\omega r)dr, \quad (40)$$

where $J_k(\mathbf{x})$ is a standard Bessel function of the first kind of degree k. Note that $\sigma(\omega)$ is *not* the radial spectral density function $G'(\omega)$, given by

$$G'(\omega) = \frac{2\pi^{d/2}}{\Gamma(r/2)}\omega^{d-1}\sigma(\omega)$$

$$= \frac{2}{\Gamma(d/2)}\int_0^\infty c(r)(\omega r/2)^{d/2}J_{(d-2)/2}(\omega r)dr. \quad (41)$$

It is sometimes easier to first calculate $\sigma(\omega)$ from (Matérn, 1960)

$$\sigma(\omega) = \int_0^\infty \xi(x)\,|\tilde{\zeta}(x,\omega)|^2 dx, \quad (42)$$

where the Fourier transform of $\zeta(x,r)$ is

$$\tilde{\zeta}(x,\omega) = (2\pi)^{-d/2}\omega^{-(d-2)/2}\int_0^\infty \zeta(x,r)r^{d/2}J_{(d-2)/2}(\omega r)dr. \quad (43)$$

We can then calculate $c(r)$ as the Hankel transform of $\sigma(\omega)$.

The techniques developed here can readily be extended to first-order clustering processes.

2.2 Joint Distribution of Elevations: Slope Distribution

As noted earlier, a crater of diameter x changes the elevations $Z(\underline{R})$, $Z(\underline{R}+\underline{r})$ at points a distance r apart by amounts $\zeta(x,r_1)$ and $\zeta(x,r_2)$ respectively, where r_1 is the distance from $\underline{R}$ and r_2 the distance from $\underline{R} + \underline{r}$. In order for a crater center to be distant from $\underline{R}$ by r_1 to $r_1 + dr_1$, and distant from $\underline{R} + \underline{r}$ by r_2 to dr_2, it must lie within a region of volume

$$\nu_d(r_1,r_2;r)dr_1dr_2,$$

where ν_d is given by (38). Hence, after elementary calculations the joint characteristic function is

$$\phi_{Z(\underline{R})Z(\underline{R+r})}(u,v) = E[\exp\{iu\ Z(\underline{R}) + iv\ Z(\underline{R+r})\}]$$

$$= \exp[\int_0^\infty \xi(x)dx \iint [\exp\{iu\zeta(x,r_1) + iv\ \zeta(x,r_2)\}-1]\nu_d(r_1,r_2;r)dr_1dr_2]$$

$$= \exp(\int_0^\infty \xi(x)dx \int_0^\infty dw_1 \int_{-r}^{r} dw_2[\exp[iu\zeta\{x,(w_1+w_2)/2\}$$

$$+ iv\zeta\{x,(w_1-w_2)/2\}]-1]\bar{\nu}_d(w_1,w_2;r)dw_1dw_2, \tag{44}$$

where (Marcus, 1969a)

$$\bar{\nu}_d(w_1,w_2;r) = \frac{(\frac{\pi}{4})^{(d-1)/2}}{\Gamma(\frac{d-1}{2})} \frac{(w_1^2-w_2^2)(r^2-w_2^2)^{(d-3)/2}(w_1^2-r^2)^{(d-3)/2}}{r^{d-2}} . \tag{45}$$

The slope across a finite span of length r,

$$S(r) = [Z(\underline{R+r}) - Z(\underline{R})]/r, \tag{46}$$

is of great interest. Once (44) is calculated we can find

$$\phi_{S(r)}(u) = E[\exp\{iu\ S(r)\}]$$

$$=\phi_{Z(R),Z(\underline{R+r})}(-u/r,\ u/r). \tag{47}$$

In practice, the joint distribution is not readily obtained, hence we must content ourselves with moments. Obviously

$$E\{S(r)\} = 0,$$

$$\text{var}\{S(r)\} = 2\{c(0) - c(r)\}/r^2. \tag{48}$$

2.3. Applications to Elevation Distribution

We suppose that regions of interest on the moon can be modelled as infinite planes (d=2). From (1) we set, if $x_0 < x < x_m$,

$$\xi(x) = \int_0^T f(t)dt\gamma x_0^\gamma x^{-\gamma-1}\{1-(x_0/x_m)^\gamma\}^{-1}, \tag{49}$$

implicitly a function of the present epoch T. A useful approximation to crater shape is that of a paraboidal bowl with an inverse-power external rim

$$\zeta(x,r) = R_0 x^h + C_0 x^\delta\{(2r/x)^2-1\} \qquad (x < 2r),$$

$$\zeta(x,r) = R_0 x^h (2r/x)^{-k} \qquad (x > 2r), \tag{50}$$

where $C_0 > R_0 > 0$, k, h and δ are positive constants with approximate values $h = \delta$ (usually = 1), $k \sim 4$, $C_0 \sim 0.25$, $R_0 \sim 0.085$. It can be shown (Marcus, 1970b) that if

$$h = \delta,\ 2/k < \alpha = (\gamma-2)/\delta < 1,$$

and

$$\lim_{x_0 \to 0} x_0^\gamma \int_0^T f(t)dt = C < \infty, \tag{51}$$

then for $x_0 \to 0$, $x_m \to \infty$,

$$\phi_Z(u) = \exp[-\lambda|u|^\alpha\{1-i\beta \operatorname{sgn}(u) \tan(\pi\alpha/2)\}], \tag{52}$$

i.e., the surface elevations Z are distributed as a *stable law with* characteristic exponent $\alpha = (\gamma-2)/\delta$, scale parameter

$$\lambda = \frac{\pi\gamma C\Gamma(1-\alpha)\cos(\pi\alpha/2)}{2\alpha\delta}\left[\frac{\{R_0^{\alpha+1}+(C_0-R_0)^{\alpha+1}\}}{2C_0(1+\alpha)} + \frac{R_0^\alpha}{(\alpha-\{2/k\})}\right] \tag{53}$$

and skewness parameter

$$\beta = \left[\frac{\{R_0^{\alpha+1}-(C_0-R_0)^{\alpha+1})\}}{2C_0(1-\alpha)} - \frac{R_0^\alpha}{\{\alpha-(2/k)\}}\right] \Big/ \left[\frac{\{R_0^{\alpha+1}+(C_0-R_0)^{\alpha+1}\}}{2C_0(1+\alpha)} + \frac{R_0^\alpha}{\{\alpha-(2/k)\}}\right]. \tag{54}$$

Although this result does not hold for $\alpha \geq 1$, we can in that case compute the mean value of Z. It is, however, necessary to hold $x_0 > 0$. The result is, assuming $h = \delta$, $\gamma > 2+\delta$ and $x_m = \infty$, (Marcus, 1970b)

$$E(Z) = \frac{\pi\gamma x_0^{2+h}\int_0^T f(t)dt}{4(\gamma-2-h)}\left\{\frac{2(k-1)R_0}{k-2} - C_0\right\}. \tag{55}$$

This model has considerable support from observations of surface elevations and of the thickness of the fragmental surface layer on the lunar maria.

It has also been possible to calculate the theoretical correlation function and the power-spectral density function of lunar elevations (Marcus, 1970). The theory agrees well with observations. It is interesting to note that over a wide range of spatial frequencies, i.e., wave numbers ω, the two-dimensional spectral density $\sigma(\omega)$ is approximately an inverse power law

$$\sigma(\omega) \sim (\text{const})\ \omega^{-\varepsilon-1}, \tag{56}$$

with

$$\varepsilon = 3 + 2\delta - \gamma. \tag{57}$$

There is independent support for these results from the good agreement of theoretical and observed mean square slopes (Marcus, 1969b).

If the crater external rims are ignored as contributing significantly less to roughness than do crater bowls, we set $R_0 = 0$. A fairly simple expression for $\sigma(\omega)$ now results. Observe that in terms of Legendre polynomials of degree k, $P_k(\cdot)$ say, we can express $\zeta(x,r)$ as

$$\zeta(x,r) = -C_0 x[P_0\{1-2(2r/x)^2\} + P_1\{1-2(2r/x)^2\}]/4 \tag{58}$$

if $2r < x$, hence, using (42) and (43) with d = 2, and (Erdelyi et. al., 1954, 8.31), we find

$$\sigma(\omega) = [C_0^2 \int_0^T f(t)dt\{1-(x_0/x_m)^\gamma\}^{-1} \gamma x_0^\gamma/4]\omega^{-(6-\gamma)} \int_{x_0\omega/2}^{x_m\omega/2} y^{3-\gamma}\{J_1(y)+J_3(y)\}^2 dy. \tag{59}$$

2.4. Application to Distribution of Slopes

The model function (50) has proven to be too complicated to handle easily. Certain arguments can be made (Marcus, 1969bc, 1970b) that for many purposes a useful order-of-magnitude approximation can be made, which is in most cases qualitatively correct, that craters are rimless cylinders

$$\zeta(x,r) \begin{cases} = -H_0 x & (2r < x), \\ = 0 & (2r > x), \end{cases}$$

for $H_0 > 0$. Then from (47)

$$\phi_{S(r)}(u) = \exp\left[\frac{\gamma x_0^{\gamma}\int_0^T f(t)\,dt}{1-(x_0/x_m)^{\gamma}}\, r^{2-\gamma}\int_{x_0/r}^{x_m/r} H(y)\{\cos(H_0 y)-1\}y^{-\gamma-1}dy\right], \tag{60}$$

where

$$H(y) = \int_1^{y+1} [(w^2-\tfrac{1}{2})\arccos(w-y)-\tfrac{1}{2}(w-y)\{1-(w-y)^2\}^{\frac{1}{2}}]\,dw/\{w^2-1\}$$

if $0 < y \leq 1$;

$$H(y) - \int_{\max(1,y-1)}^{y} (w^2-\tfrac{1}{2})\arccos(w-y)-\tfrac{1}{2}(w-y)\{1-(y-w)^2\}^{\frac{1}{2}}dw/(w^2-1)^{\frac{1}{2}}$$

$$+\int_y^{y+1} [(w^2-\tfrac{1}{2})\arccos(y-w)-\tfrac{1}{2}(y-w)\{1-(y-w)^2\}^{\frac{1}{2}}dw(w^2-1)^{\frac{1}{2}} \tag{61}$$

if $1 \leq y$.

It is easy to show that

$$H(y) \sim \frac{\pi}{2}\,y^2 \qquad \text{as } y \to 0 \qquad (y << 1),$$

$$H(y) \sim 2y \qquad \text{as } y \to \infty \qquad (y >> 2)$$

and in view of (62), if $x_0 \to 0$, $x_m \to \infty$, then as $u \to 0$

$$\phi_{S(r)}(u) \cong 1 - \frac{2\gamma\Gamma(\gamma-3)}{(\gamma-2)(\gamma-1)}\cos\{\tfrac{\pi}{2}(3-\gamma)\}\,CH_0^{\gamma-1}\; r^{2-\gamma}|u|^{\gamma-1}, \tag{63}$$

i.e., the distribution of $S(r)$ has heavy power law tails with characteristic exponent γ-1. This also accords well with observations (Marcus, 1969b) and makes possible a neat and elegant solution of an important problem in planetary radar astronomy (Marcus, 1969c).

REFERENCES

Bartlett, M. S. (1964). The spectral analysis of two-dimensional point processes. Biometrika 51, 299-311.

Bartlett, M. S. (1966). *An Introduction to Stochastic Processes,* 2nd ed.. Cambridge: Cambridge University Press.

Chernov, Yu. A. (1967). The meteoroid mass distribution and autocorrelation function for a planetary surface eroded by a meteor stream. Sov. Astron. - A. J. 10, 853-9.

Daley, D. J. (1970). The non-existence of stationary infinite Newtonian universes and a multi-dimensional model of shot noise. Nature 227, 935. Also J. Applied Probability 8 (1971), 127-35.

Erdelyi, A., Magnus, W., Oberhettinger, F. & Tricomi, F. G. (1954). *Tables of Integral Transforms,* Vol. II. New York: McGraw-Hill.

Feller, W. (1966). *An Introduction to Probability Theory and its Applications, Vol. II.* New York: Wiley.

Fielder, G. (1966). Tests of randomness in the distribution of lunar craters. Mon. Not. R. Astr. Soc. 132, 413-22.

Fielder, G. (1970). On the distribution of lunar craters. Paper presented at NATO Advanced Study Institute on the Moon and Planets, Newcastle, April 1970.

Fielder, G. & Marcus, A. H. (1967). Further tests for randomness of lunar craters. Mon. Not. R. Astro. Soc. 136, 1-10.

Marcus, A. H. (1964). A stochastic model of the formation and survival of lunar craters, Part I. Icarus 3, 460-72.

Marcus, A. H. (1966). A stochastic model of the formation and survival of lunar craters, Part IV. Icarus 5, 190-200.

Marcus, A. H. (1967). A multivariate immigration with multiple death process and applications to lunar craters. Biometrika 54, 251-61.

Marcus, A. H. (1968). Martian craters: Number density. Science 160, 1333-5.

Marcus, A. H. (1969a). Compound Poisson vector fields, with applications to astronomy. Tech. Rept. 103, Dept. of Statistics, Johns Hopkins Univ.,Baltimore, Md.

Marcus, A. H. (1969b). Distribution of slopes on a cratered planetary surface: Theory and preliminary applications. J. Geophys. Res. 74, 5253-67.

Marcus, A. H. (1969c). Applications of a statistical surface model to planetary radar astronomy. J. Geophys. Res. 74, 4958-62.

Marcus, A. H. (1970a). Comparison of equilibrium size distributions for lunar craters. J. Geophys. Res. 75, 4977-84.

Marcus, A. H. (1970b). Distribution and covariance function of elevation on a cratered planetary surface. Part I. The Moon 1, 297-337.

Matern, B. (1960). *Spatial Variation.* Medd. fr. Statens Skogforskingsinstitut 49 (5).

Takacs, L. (1957). On secondary stochastic processes generated by a multidimensional Poisson process. A Magyar Tudo. Akad. Mat. Kut. Int. Köaleményi 2, 71-80.

THE STOCHASTIC PROPERTIES OF SPIKE TRAINS RECORDED FROM NERVE CELLS

R. B. Stein
Department of Physiology, University of Alberta, Canada

SUMMARY

The aim of this paper is to introduce mathematicians to the types of problems that are involved in analyzing the activity of single nerve cells and systems of nerve cells. No proofs of mathematical assertions have been included, but the text together with the equations should suggest the type of mathematical problems involved. References to the original proofs and to a considerable number of other related papers in neurobiology should help in pursuing particular aspects further. Some of the major results of recent work on the nervous system discussed in this paper are:

1) The output from single nerve cells can be considered as a stochastic point process whenever the fibers from these cells are long enough and thin enough that the signal would decrement and need to be boosted several times in traveling along the fibers. This criterion is fulfilled for all major fiber tracts in higher animals.

2) Under these conditions, transmission is via nerve impulses which have a stereotyped waveform. The form can be quantitatively described by a set of partial and ordinary differential equations describing the ionic currents crossing the membrane of the nerve cell.

3) The pattern of nerve impulses can be modeled using much simpler equations, and the behaviour of the models can be compared with experimental data using methods in both the time and frequency domain.

4) Analysis of experimental data in the time domain illustrates some of the important non-linearities which can be incorporated into neuronal models (e.g. saturation, rectification, phase-locking).

5) Analysis in the frequency domain permits efficient determination of the linear frequency response function of nerve cells or groups of cells in a way which can be easily extended to consideration of larger systems. A coherence function can also be computed which measures the extent to which the linear frequency response function characterizes the behaviour of the cell or system. The coherence function is closely related to measures of information transmission, and offers a new approach to quantifying the amount of information transferred by nerve cells.

6) The mean rate of discharge (the inverse of the mean interval between nerve impulses) appears to act as a carrier rate which can be modulated by inputs to the nerve cell. The variability of the discharge with constant inputs (measured for example by the coefficient of variation for intervals between impulses) plays an important role in reducing the distortion resulting, for example, from the phase-locking of nerve impulses to cyclic stimuli, although it does limit the information capacity of a nerve cell.

1. INTRODUCTION

I will begin by defining a number of important terms, as it is often unfamiliar words, not uninteresting mathematics that deters mathematicians from tackling biological problems. Readers already familiar with the terminology of neurophysiology, may wish to go directly to Section 2.

1) The Mammalian Spinal Cord

Figure 1 shows a schematic cross section through the mammalian spinal cord, together with a schematic representation of a few nerve cells (*neurons*) and their connections (*synapses*). Sensory nerve fibers or *axons* coming from the skin and muscles are gathered together on the upper (dorsal) surface into what is known as a dorsal root. Nerve fibers carrying information *to* the central nervous system are referred to as *afferent* fibers, while those conveying information out *from* the central nervous system are referred to as *efferent* fibers. The cell bodies, or *somas*, of the afferent fibers are also gathered together into a *ganglion*,

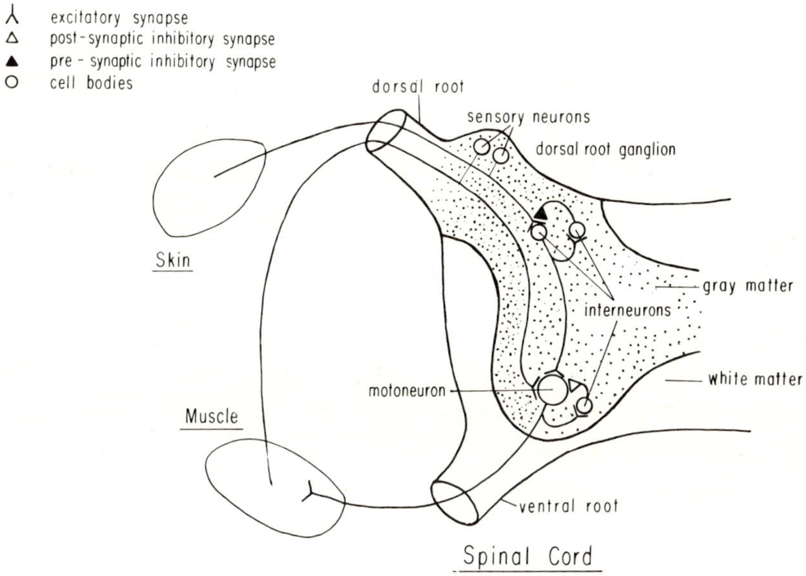

Figure 1. Schematic cross-section through the mammallan spinal cord, indicating a few nerve cells and their connections.

which is simply a collection of nerve cells. Generally, there are no synaptic connections between the cell bodies in the dorsal root ganglion. Connections are made in the *gray matter* of the spinal cord, directly from sensory neurons either to neurons which innervate muscles (*motoneurons)* or more commonly to neurons contained completely within the central nervous system *(interneurons)*. Connections from the interneurons may also feed forward to motoneurons, but feedback pathways are known from motoneurons to interneurons via particular interneurons first described by Renshaw (Renshaw cells), and from interneurons to the synaptic connections of sensory neurons. Synaptic connections may be either *excitatory* or *inhibitory,* depending on whether activity in one (*presynaptic)* cell tends to produce or prevent activity in the second (*postsynaptic)* cell. The connection may be to the postsynaptic cell directly or it may be indirect, acting via the terminals of the presynaptic axon. Thus, the inhibitory pathway shown from an interneuron back onto the terminals of a sensory axon would be referred to as a presynaptic inhibitory pathway. It constitutes a negative feedback pathway which can be used to regulate the transmission of a specific type of sensory information to the *central nervous system* (brain plus spinal cord). The schematic representation shown in Fig. 1 has two major deficiencies. First, for simplicity, only a few of the tens of thousands of cells in a spinal segment have been shown (for further information, see Henneman, 1968). Secondly, connections have been shown in a one-to-one fashion, whereas connections in the nervous system are almost always on a many-to-many basis. Individual motoneurons may receive thousands of synaptic connections, not only from cells in their vicinity, but also from higher centers in the brain. Similarly, interneurons may send axons to a large number of other cells, which may be in the vicinity, or in distant centers in the brain. This redundancy is an important element in living systems where individual cells may die, and are unreliable to some extent. The problem of constructing a reliable organism from basically unreliable components was considered already in the fifties (von Neumann, (1956).

Groups of axons are gathered together as *tracts* in the *white matter* surrounding the gray matter of the spinal cord. The difference in coloration arises from the presence of a thin sheath of an insulating material known as *myelin* surrounding axons which extend some distance and are thicker than a particular diameter (1-2 μm). The purpose of myelination will be considered in Section 2. Werner (1970) has recently discussed some of the topological considerations which arise in representing a three-dimensional object (the body) onto a two-dimensional surface (the brain) and which arise from the pattern of many-to-many connections found in the brain. More detailed discussions of connections will be found in standard neuroanatomy texts (e.g. Ranson and Clark, (1959). Realizing the complexity of the nervous system, many early invesitgators were inclined to use greatly simplified models of single neurons (McCulloch and Pitts, 1943), and assume rather arbitrarily a random pattern of connections (Hebb, 1949). Cowan (1971) has recently reviewed some mathematical methods for analyzing large neural systems. Such studies have also led to an interesting branch of artificial intelligence, which may have little relevance to the function of the brain. The individual neuron is a complex element which must be modeled carefully, and the pattern of connections between cells, though diverse, is nonetheless organized in a specific and structured way (Werner, 1970).

2) Single Nerve Cells:

With this brief introduction to the organization of the nervous system, let us consider single nerve cells or neurons in more detail. The electrical activity of single cells can be recorded in the central nervous system through the use of fine *microelectrodes*, with tips of the order of 1 millionth of a meter (1 μm) or less. The smallest microelectrodes may be used to record from inside the larger nerve cells without damaging the cell membrane too much, and can also be used to pass current or dyes directly into the cell. Figure 2 shows a reconstruction of a single motoneuron. As well as an axon and a cell body or soma,

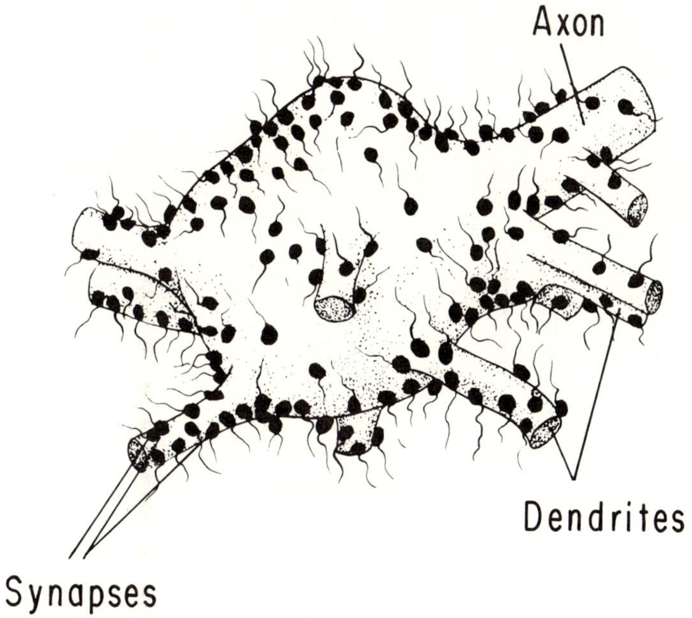

Figure 2. *Reconstruction of a motoneuron*. Modified from Hagger and Barr (1950).

there are a large number of thin processes or *dendrites* which provide a large surface area for synaptic contacts. The dendrites and the somas of motoneurons are covered to a large extent with synaptic connections, and these connections constitute the input to the cell. In addition, there is an *integrative* or *impulse-generating region* of the cell, which in a motoneuron is located near the beginning of the axon. The stereotyped *nerve impulses* which are produced, conduct along the length of the axon. They are the *output* of the cell and are an aspect of the nervous activity which may be treated as a stochastic point process. The validity of the assumption that the series of nerve impulses constitute a stochastic point process has been discussed elsewhere (Rushton, 1961; Stein, 1970) but will be summarized briefly in the next Section.

2. NERVE IMPULSES AS A POINT PROCESS

The axon can be considered as a cylinder as illustrated in Fig. 3.

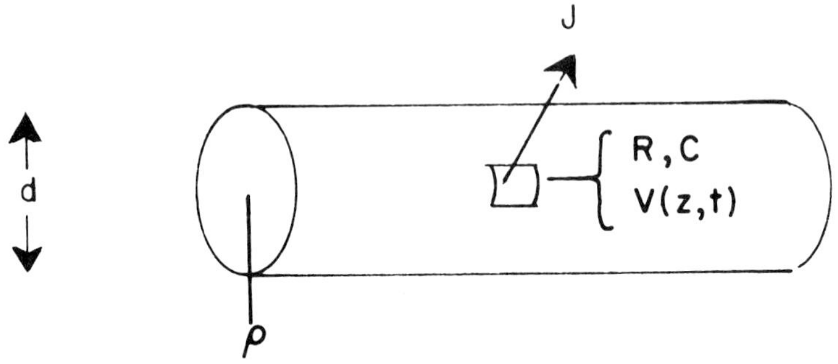

Figure 3. Representation of a nerve fiber as a cylindrical cable.

The core of the cylinder represents an Ohmic fluid of resistivity ρ. The membrane of the cell by virtue of its composition represents a capacitance C and also contains some pores through which ions can flow. The flow of charged ions through a unit area of membrane represents a current density J and it is easy to show (Hodgkin and Rushton, 1946; Noble, 1966; Noble and Stein, 1966) that the following partial differential equation applies to the voltage V inside a cylinder having the constants indicated in Fig. 3:

$$\frac{d}{4\rho}\frac{\partial^2 V}{\partial z^2} = C\frac{V}{\partial t} + J. \tag{1}$$

This equation is well known in the theory of heat flow (Carslaw and Jaeger, 1963). Its solution depends on the assumption made about the form of the ionic current density, J. The simplest assumption is Ohm's Law, namely that the current density is proportional to the voltage across the membrane, $J = V/R$ where R is the membrane resistance. Then, with a constant input at one point, say $z = 0$, the steady-state voltage inside the cable will decline exponentially with a space constant λ, where

$$V(z) = V(0)\ e^{-z/\lambda} \tag{2}$$

and

$$\lambda = \{(Rd)/(4\rho)\}^{\frac{1}{2}}.$$

Because the fluid in a nerve axon must inevitably contain a dilute salt solution, ρ will be large and hence λ must be short. In a long length of nerve the signal must be boosted repeatedly. Vertebrates have developed the device of myelination, in which a particular type of cell, first described by Schwann, wraps itself many times around the nerve, thereby increasing the membrane resistance and reducing its capacity. Rushton (1951) showed that the thickness of the myelin sheath formed by the Schwann cell was such that it maximized the speed of transmission along the nerve fiber for a given total diameter (fiber + Schwann cell). Nonetheless, the signal will still decrement markedly within a millimeter or so in a fiber of typical diameter (2-20 μm). Thus, it needs to be boosted hundreds of times to cover the meter or more which may separate a sense organ or muscle from the spinal cord. The booster stations are small gaps in the myelin sheath, known, again after the person who first described them, as nodes of Ranvier. At these nodes the current density is a grossly non-linear function of voltage and of time, which can be described by a series of equations (Frankenhaeuser and Huxley, 1964).

A similar set of equations was developed earlier for invertebrate nerve fibers (Hodgkin and Huxley, 1952) which are unmyelinated and show the non-linear current-voltage properties all along their axons. The current density can be decomposed into components carried by individual ionic species, principally sodium (Na) and potassium (K). The effect of other ions are lumped together as a leakage current for which the letter L is used:

$$J = J_{Na} + J_K + J_L . \tag{3}$$

The current for any ion depends on the permeability P of the membrane to that ion, the voltage V across the membrane, the electrochemical equilibrium potential of the ion E and other known constants k_1, k_2, according to an equation first derived by Goldman (1943). For example:

$$J_{Na} = P_{Na}\, k_2 \frac{\exp\{k_1(V - E_{Na})\} -1}{\exp\{k_1 V\} -1} . \tag{4}$$

The permeability has a maximum value $\bar{P}$ which is multiplied by normalized variables that vary with time according to first order equations

$$P_{Na} = \bar{P}_{Na}\, m^2 h, \tag{5}$$

where, for example,

$$dm/dt = \alpha_m(1-m) - \beta_m m . \tag{6}$$

The coefficients of equation (6), α_m and β_m, depend on voltage but not on time, and are amenable to direct measurement in various tissues by using feedback to control the membrane voltage (*voltage clamp*). The entire set of equations which includes equations (3) - (6) plus a cable equation such as equation (1) for the region between the nodes can be solved numerically (Goldman and Albus, 1968; Hutchinson, Smith and Koles, 1970). Whenever the computed voltage reaches a *threshold* value a stereotyped response about 0.1V in height and 1 msec in duration is generated which conducts with a constant delay from one node to the next. Any transients die away within a few nodes, so the only detail of any possible significance remaining at the end of a length of axon is the time of occurrence of each sterotyped response.

Rushton (1961) argued in general that since there may be hundreds of nodes between a sense organ and the central nervous system,accurate transmission of any analog signal would be extremely difficult. The boosters would have to be made of very high tolerance components if the signal at the end is to bear any detailed relation to that at the beginning. Experimental evidence for sufficiently high tolerance components was lacking, so Rushton concluded that little information could be transmitted in this way. Rushton's arguments apply with less force to axons of large diameter, because it follows directly from the cable equations (e.g. equation (2); see also Rushton, 1951, and Goldman and Albus, 1968) that the space constant will increase with diameter. Recent experimental work indicates that some large sensory fibers in invertebrates which are not too long (Bush and Roberts, 1968) also do not use nerve impulses, i.e., their output cannot be consided as stochastic point processes.

One final qualification should be mentioned. Even if nerve impulses occur and are sterotyped in form, there is no guarantee that they will have identical effects at a synaptic junction. The effect of one impulse in potentiating or depressing the effect of a subsequent impulse is well know (Eccles, 1964) and is presumably very important in the adaptive behaviour of the nervous system. Such a situation is, however, basically different from that of a signal based on analog parameters of the membrane voltage. It merely indicates that the decoding device (i.e., the synapse) weights nerve impulses more or less strongly when they occur at certain intervals. This weighting clearly affects the input-output relationships, but the response of the next cell still depends entirely on the times of occurrence of nerve impulses. Thus, the discharge pattern of a nerve cell represents a stochastic point process, and the information conveyed by the nerve cell must be related to the parameters of this process. To determine which parameters are relevant, let us consider models of the processes which generate the pattern of nerve impulses. The approach is similar to that of Stein and French (1970), but more recent results have been added.

3. MODELS FOR THE INPUT AND INTEGRATIVE PROCESSES IN NERVE CELLS

A simple model for these processes was suggested by McCulloch and Pitts (1943). Time was divided into discrete periods and if the number of input fibers which were active in the nth period exceeded the threshhold number v_0 in that period, an output pulse was produced during the next period. Such a model can be represented by the following equation:

$$y(n+1) \begin{cases} = 1 \\ = 0 \end{cases} \text{if} \quad \sum_{i=1}^{k} x_i(n) \begin{cases} \geq v_0 \\ < v_0 \end{cases},$$

where x_i is the ith input and y is the output. The McCulloch-Pitts model does have serious deficiencies. Perhaps the most serious is that by dividing time into small discrete periods and not allowing effects to carry over from one period to another, the important neural property of integrating events over time has been lost. Several authors (Gerstein and Mandelbrot, 1964; Baly, 1968; Calvin and Stevens, 1968; Johannesma, 1969; Knight, 1969; Knox, 1969; Stein et al., 1972a) have considered the opposite extreme, a model which ingetrates inputs without loss up to a threshold level. Beyond threshold an output is produced and the integrator is reset. Various terms have been suggested such as *perfect integrator* or *integrate and fire model*. The defining equation for the time t to reach a constant voltage threshold v_0 when the last impulse occurred at t = 0, is:

$$v_0 = \int_0^t x(u)\, du, \tag{8}$$

where x(u) is the input at time u. The input may be a continuous sensory stimulus applied to a sensory nerve cell, or a sum of discrete inputs from a number of other nerve cells. The number of synaptic inputs to an interneuron or a motoneuron can be very large, and each individual input may have a small effect relative to the voltage change required to reach threshold. Then, the voltage changes can be approximated by a diffusion process,and Gerstein and Mandelbrot (1964) showed that this model is formally equivalent to the well-known Weiner process (Cox and Miller, 1965). The voltage across the membrane will show a constant drift c produced by the average value of x(t) and superimposed upon this drift will be random fluctuations with zero mean and an amplitude a which depends on the fluctuations of x(t). The time taken to reach a constant threshold is formally equivalent to the first passage time to an absorbing barrier. The probability density function of the first passage times f(t) is (Gerstein and Mandelbrot, 1964):

$$f(t) = \frac{v_0}{a\sqrt{(2\pi t^3)}} \exp\{-\frac{1}{2t}(\frac{v_0 - ct}{a})^2\}. \tag{9}$$

Note that successive passage times (the intervals between impulses) will be independent, i.e., a renewal process. Thus, equation (9) completely specifies the statistical properties of the output. The function f(t) will be referred to as the *interval density function* (Stein, 1967c). The mean interval μ for equation (9) is:

$$\mu = \int_0^\infty t\ f(t)\ dt = v_0/c, \tag{10}$$

which is independent of the magnitude of the random fluctuations. The rate of generation of nerve impulses ν will then be

$$\nu = 1/\mu = c/v_0, \tag{11}$$

which is directly proportional to c, and hence to the average value of the input voltage. This property mimics the linear relation often found in nerve cells between frequency and a measured *generator potential* (Mountcastle, 1968; cf, Stein, 1967a). This model also behaves linearly for all deterministic time-varying signals (Knight, 1969; Stein et al., 1972a) as long as the inputs produce a monotonically increasing voltage. A decrease in voltage cannot generate nerve impulses (i.e. negative rates are not possible), so the response shows many of the properties of an electrical rectifier (Stein et al., 1972a). When there are random fluctuations in voltage having a mean amplitude a, the model behaves as a low-pass filter, i.e., it responds less well to high frequency signals. The frequency at which the response begins to decline varies inversely with the amplitude a (Stein et al., 1972a).

The model can be modified in several ways to reproduce more accurately properties found experimentally in nerve cells.

1) Refractoriness

During the production of a nerve impulse, further inputs cannot be integrated, and this is referred to as the *absolute refractory period.* The effect of including an absolute refractory period of duration t_0 is simply to produce the non-linearity of saturation with a maximum rate equal to $1/t_0$ for strong inputs (Stein, 1967a). After the production of one nerve impulse, another impulse is more difficult to generate for a time which is referred to as the *relative refractory period.* Both absolute and relative refractory periods can be demonstrated by numerical solutions of the full system of equations given earlier [(1), (3) - (6)], but it is generally sufficient simply to make the threshold a simple, time-dependent function such as an exponential (Fuortes and Mantegazzini, 1962; Geisler and Goldberg, 1966; Stein, 1967a; Roy and Smith, 1969) or a hyperbolic function (Eysel, 1971).

2) Leaky Integrators

Real neurons will not be able to behave as perfect integrators over indefinite periods of time. The presence of a resistance and capacitance in parallel in the membrane (Fig. 3) means that inputs will leak away exponentially with a time constant $\tau = RC$. Then, the time to reach a threshold v_0 (which, according to paragraph 1 above, may be time-varying) will be:

$$v_0(t) = \int_0^t x(u)\, e^{-(t-u)/\tau} du. \quad (12)$$

The effect will be to introduce a non-linearity at low rates of discharge. If the inputs contain stochastic as well as deterministic components, it has been shown (Stein, 1967a; Johannesma, 1969; Roy and Smith, 1969) that the direction of the non-linearity will depend on the amplitude of the stochastic inputs. The input-output relation has been fitted by a variety of equations (Stein, 1967a; Johannesma, 1969; Cowan, 1971). At high rates of discharge, the input-output relation is much less dependent on the presence of stochastic inputs, but these inputs will still have important effects. However, consideration of their effects is best delayed until later sections.

3) Dendritic Properties

The dendrites also contain many synapses and must be considered as distributed cables, since their length is often long compared to their length constant (Rall, 1962). By taking the Laplace transform of equation (1) and solving for a point input to a linear cable, the transfer function G(s) for a linear cable of infinite length is proportional to (Stein, 1967c):

$$G(s) \propto \frac{\exp\{-\sqrt{(1+s\tau)}\,x/\lambda\}}{\sqrt{(1+s\tau)}} \ . \tag{13}$$

Thus, the decline as a function of distance for signals with a frequency component s is much sharper than that of a steady signal. Comparing equation (13) with equation (2), the steady-state response $e^{-x/\lambda}$ is raised to the $(1+s\tau)^{\frac{1}{2}}$ power and divided by $(1+s\tau)^{\frac{1}{2}}$. This very sharp frequency cutoff means that only slowly changing signals can be transmitted considerable distances in a linear cable (see also Rall, 1962). This conclusion may not apply to non-linear cables. The finite length of the dendrites may also alter the form of equation (13), as will the nature of the ending (i.e., whether it is electrically closed or open). These topics are dealt with by Rall (1969), Hubbard, Llinás and Quastel (1969) and Jack and Redman (1971). There is now considerable evidence for non-linearity in dendrites, and in some cases, nerve impulses may actually arise in dendritic as well as axonal structures (Eccles, 1960; Kandel and Spencer, 1961; Llinás and Nicholson, 1969). The complexity introduced by these considerations has rarely been considered in functional models of neural systems. To determine which of these effects are important in the functioning of particular neurons requires analysis of experimental data from these neurons and comparison of it with model systems.

4. ANALYZING THE STOCHASTIC PROPERTIES OF NEURONS IN THE TIME DOMAIN

When studying the discharge of a nerve cell under steady-state conditions, one obvious measurement is the interval density function mentioned in the previous section, e.g. equation (9). Even in one species (e.g. the cat), different types of nerve cells show strikingly different patterns. In the neurophysiological literature, the data is often presented as a histogram, giving the number of intervals found of various durations in a particular period of recording. Fibers in the auditory nerve (Fig. 4A) give a nearly exponential interval histogram characteristic of a Poisson process, except for a lack of short intervals due to refractoriness. Muscle spindle receptors can discharge regularly and give a histogram characteristic of a normal distribution or they may show a skewed interval histogram (Fig. 4C), while fibers in the visual system may have multimodal histograms (Fig. 4B). One parameter which is of considerable importance in signalling sensory information is the mean rate ν or the mean interval μ, as recognized by Adrian forty years ago (Adrian, 1932). Yet the role of other parameters such as the variance of the interval density function remains controversial, and will be considered later.

Unless the discharge constitutes a renewal process, as in the simple linear integrator model of Section 3 (i.e., successive intervals are statistically independent), higher order probability density functions for the duration of two or more intervals are required to characterize the discharge. Typical effects which will prevent the discharge of a nerve cell from being a renewal process are

1) the accumulation of refractoriness from interval to interval; or

2) patterns in the input which extend over more than one interval.

Accumulation of refractoriness has been shown to arise in equations such as equations (3) - (6) (Stein, 1967c), while the cause of the multimodal distribution of Fig. 5c is thought to arise from periodic processes in the input to these cells (Ten Hoopen, 1966).

Another function which is often measured is the *intensity function*

$$h(t) = \sum_{n=1}^{\infty} f_n(t), \tag{14}$$

where $f_n(t)$ is the nth order interval density function mentioned above.

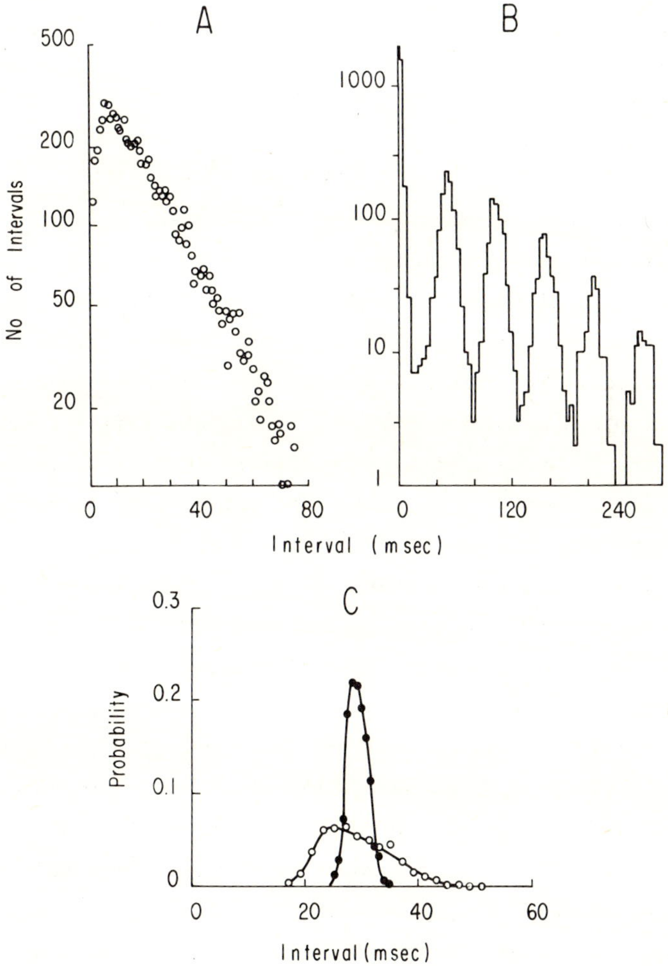

Figure 4. Interval histograms for (A) an auditory nerve fiber (Kiang et al., (1965), (B) a nerve cell in the visual system (lateral geniculate body) (Bishop, Levick and Williams, 1964), and (C) sensory fibers from muscle spindles (Mathews and Stein, 1969b). In (A) and (B) the number of intervals of given durations have been plotted on a logarithmic scale, while in (C) the number has been divided by the total number of intervals to give a probability and plotted on a linear scale.

The intensity function has been called the renewal density function, the expectation density function and other names in the neurophysiological literature (Moore, Perkel and Segundo, 1966).

To determine the synaptic relationship between two or more cells whose discharge can be recorded in the nervous system, *cross-intensity functions* can be computed (Cox and Lewis, 1966). These functions, which are also referred to as cross-correlation functions (Moore et al., 1970), are defined by:

$$c(t)dt = \text{prob(event in 2 between t and t+dt} \mid \text{event in 1 at 0)}. \quad (15)$$

The events in one process may, in fact, be stimulus pulses, in which case the cross-intensity function is referred to in neurophysiology as a *post-stimulus time histogram* or they may be marker pulses from an applied sinusoidal waveform, in which case the function is referred to as a *cycle histogram*. The term histogram is used loosely and may refer to the raw data collected, data converted to probabilities (i.e., $c(t)\ \Delta t$, where Δt is the accuracy of the measurements) or data converted to probabilities per unit time (i.e., $c(t)$ in equation (15) above). This last form is perhaps the most useful since the units of the function are then impulses/sec. The quantity measured is then a rate which is a biologically important quantity as mentioned above, and the function measures the effect of one process on the rate of the other process. In Fig. 5, cycle histograms for two mechanoreceptors are shown during the application of sinusoidal vibrations at 1 and at 40 Hz. The discharge of the muscle spindle receptor (Fig. 5A) is smoothly and approximately sinusoidally modulated, while that of a skin receptor (Fig. 5B) occurs at one and only one location in the sinsoidal cycle. This phenomenon is referred to as *phase-locking* and occurs to some extent in all sensory systems. If the vibratory stimulus applied to the muscle spindle of Fig. 5A were increased sufficiently in frequency on in amplitude, it too would show phase-locking.

Phase-locking has been investigated using neuronal models and this phenomenon is not shown by the perfect integrator model mentioned previously. Cycle histograms are always sinusoidally modulated until the amplitude is large enough that the discharge stops during part of the sinusoidal cycle (the phenomenon of rectification mentioned earlier). Furthermore, the modulation amplitude produced in this model is independent of the sinusoidal frequency applied, and the response is always

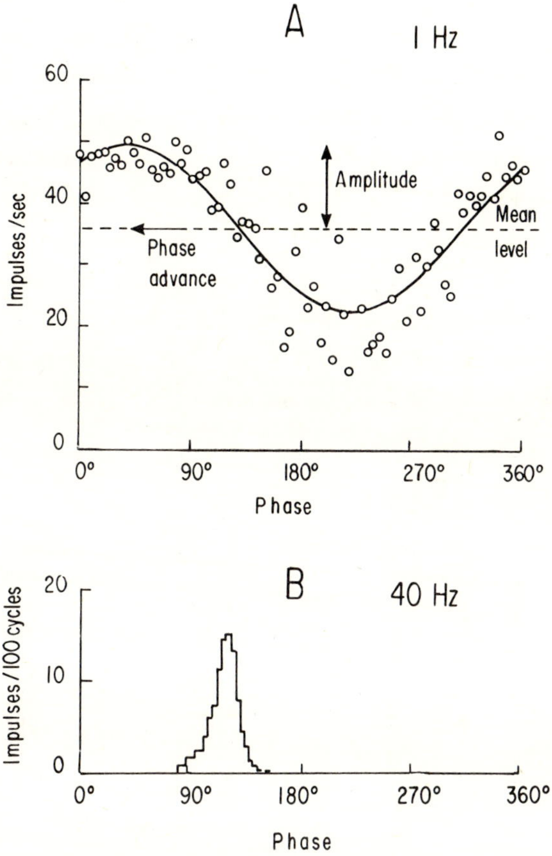

Figure 5. Cycle histograms for sensory fibers from (A) a muscle spindle (Matthews and Stein, 1969), (B) the skin (Talbot et al., 1968). Sinusoidal stimuli were applied with the frequency indicated.

precisely in phase with the input if no sources of variability are included. This remarkable property can be demonstrated mathematically for the perfect integrator model and has been discussed elsewhere (Knight, 1969; Stein et al., 1972a). However, almost any modification of the perfect integrator model will destroy this property.

For example, if the time constant of integration τ is finite [the leaky integrator of equation (12)], Rescigno et al., (1970) showed, again in the absence of variability, that there will always be a phase-locked pattern with a finite number m impulses discharged in a number n cycles. The numbers m and n tend to be smaller integers, the shorter the time constant of integration relative to the mean interval in the absence of modulation (Stein et al., 1972a). The skin receptor shown in Fig. 5B is an extreme example in that it does not discharge at all ($\mu \to \infty$) unless stimulated with a time-varying stimulus. As a result, m and n are generally 1 as in Fig. 5B, i.e., there is one-to-one phase-locking. Another important factor is the intrinsic variability of the process, which can most easily be measured from the standard deviation σ or the coefficient of variation σ/μ of the interval density function in the absence of time-varying stimuli. A high level of variability tends to disrupt the phase-locked patterns, as shown in Fig. 6 using an electronic

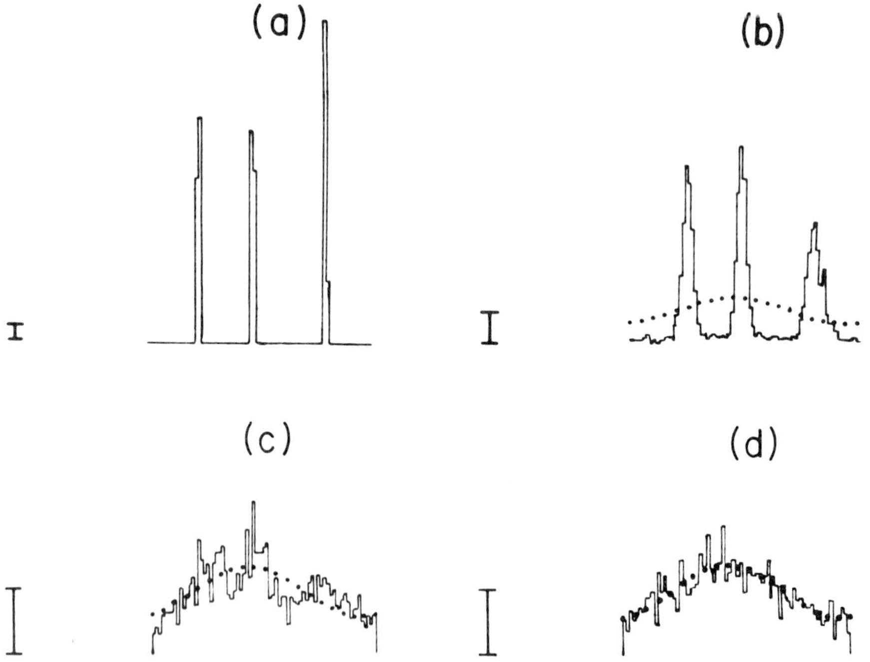

Figure 6. Cycle histograms for a neuronal model to which a sinusoidal stimulus plus increasing amounts of noise were applied (Stein, 1970). The vertical bars in each part represent a modulation of 100 impulses/sec, and the dots give the best fitting sine wave for the response. Note the increasingly sinusoidal response obtained by the addition of noise with little change in the parameters of the best fitting sinusoid.

neural analog of a leaky integrator constructed from integrated circuits (French and Stein, 1970). With almost no variability, the discharge showed a three-to-two phase-locking (i.e., $m = 3$ impulses in $n = 2$ cycles). The cycle histogram (Fig. 6A) had three sharp peaks which were initially broadened by addition of variability (Fig. 6B) and eventually at high levels disappeared altogether (Fig. 6D) leaving an approximately sinusoidally modulated discharge.

The phase-locked peaks represent a form of distortion and the presence of variability reduces this form of distortion and also that due to rectification (Spekreijse and Oosting, 1970). Thus, the presence of intrinsic variability can reduce the average distortion produced in transmitting the details of a cyclic stimulus as a rate of impulses in a single nerve cell or a population of nerve cells. In other words, if the steady discharge of a nerve cell is viewed as a stochastic point process, the mean rate ν (the inverse of the mean interval μ) represents a *carrier* rate which is modulated by sensory or synaptic inputs, while the variability of the stochastic point process determines the *distortion* produced on average in transmitting signals by this method. Further insight into these ideas results from analysis in the frequency domain.

5. ANALYZING THE STOCHASTIC PROPERTIES OF NEURONS IN THE FREQUENCY DOMAIN

The traditional methods of going from the time domain into the frequency domain follow directly from the functions already defined. In Fig. 5 and Fig. 6, responses were shown at one frequency, and by applying a number of frequencies, the frequency response of the system can be measured. Also, the Fourier transforms of the normalized intensity function and the cross-intensity function mentioned earlier, give, after slight modifications, the auto- and cross-spectral density as a function of frequency. The development of the fast Fourier transform algorithm by Cooley and Tukey (1965) offers a much more rapid and direct method of obtaining the spectral densities. The method can be easily applied to the time record of a continuous signal such as a sensory stimulus or the tension produced by the motoneurons to a muscle. The only requirements are that

1) the signal must be sampled regularly,

2) the number of samples should preferably be an integral power of 2,

3) the sampling interval Δt should be small enough that the *Nyquist frequency* $f_N = 1/(2\Delta t)$ is above the highest frequency component of the continuous signal.

If the third requirement is not met, high frequency components will be mistaken for lower frequency components, a phenomenon known as *aliasing*. The problems of using the fast Fourier transform method with point processes are more substantial, but have recently been overcome in an elegant way by my colleagues, French and Holden (1971b). The time of each event is recorded accurately, and then convolved with a sampling function (sin $[\theta]/\theta$) multiplied by the Nyquist frequency. If the ith impulse occurs at time t_i and is treated as a Dirac δ-function, then the contribution from it to the output at time t is:

$$y_i(t) = \frac{\sin\{2\pi f_N(t-t_i)\}}{2\pi(t-t_i)}, \tag{16}$$

where $\theta = 2\pi f_N(t-t_i)$.

In the frequency domain, the transforms of the functions which were convolved are multiplied and the transform of the sin $(\theta)/\theta$ function has a value of 1 for all frequencies up to the Nyquist frequency and a value of 0 for all higher frequencies. Thus, this method represents an ideal digital filter, and it should be useful whenever the spectrum of a point process is required. *Furthermore, because of the relation between the Nyquist frequency and the sampling interval, it turns out that only one value of sin* θ *must be calculated per event* since

$$\sin\{2\pi f_N(t+\Delta t-t_i)\} = \sin\{2\pi f_N(t-t_i) + \pi\}$$

$$= -\sin 2\pi f_N(t-t_i)\}.$$

Most of the computation required by equation (16) can be done virtually on-line between taking analog samples, even using a small computer

(Digital Equipment Corp., PDP-8). The entire spectral analysis package developed by French and Holden (1971a) permits rapid bias-free and alias-free estimates to be obtained and averaged for the *spectral densities* $S_{xx}(f)$ and $S_{yy}(f)$ of any two signals x(t) and y(t) together with the *cross-spectral density* between the two signals $S_{xy}(f)$.

From these spectra, one can easily calculate the two important functions, a *frequency response function*

$$G(f) = S_{xy}(f)/S_{xx}(f) \tag{17}$$

and a *coherence function*

$$\gamma^2(f) = \frac{|S_{xy}(f)|^2}{S_{xx}(f)\ S_{yy}(f)} . \tag{18}$$

The magnitude of the frequency response function gives the amplitude for the part of the response y which is linearly related to each unit of the input x while the argument of the frequency response function gives the phase change produced as a function of frequency. The coherence function indicates the fraction (it varies between 0 and 1) of the data at a given frequency which is accounted for by the linear frequency response function. The frequency response function can be an important stage in characterizing the behaviour of a nerve cell or system of cells, and in being able to predict its response to a variety of stimuli. As mentioned earlier, it can be computed in other ways, for example by applying a sinusoid of one frequency after another and measuring the response to each. The method of sinusoidal analysis has been applied to many neuronal systems since the initial study by Pringle and Wilson (1952). However, determining the frequency response function by spectral methods has several advantages:

1) One application of a randomly varying input can be used rather than requiring application of a number of sinusoidal inputs. This saving is important in biological systems which rarely remain completely stationary. It also makes studies feasible on time-varying systems in which adaptation, learning or fatigue are important. The resolution in frequency

which can be obtained by spectral methods is also much greater than is practical by applying large numbers of sinusoids.

2) The frequency response function obtained by spectral methods is the best-fitting linear input-output relation in the sense that it minimizes mean square deviations (Bendat and Piersol, 1966). In addition, the coherence function provides a measure of the extent to which the frequency response function completely describes the system. With sinusoidal inputs, one can only measure harmonic distortion, while the non-harmonic distortion (Bayly, 1968) and any variability inherent in the processes generating the nerve impulses are "averaged out" by the sinusoidal method.

3) If the coherence is low, and hence the linear frequency response function does not fully describe the system, the spectral methods can be extended to the computation of non-linear Wiener kernels (Wiener, 1958). In principle, these kernels offer a method for predicting the response of a non-linear system to any input, but the computational difficulty increases rapidly with the number of kernels required to characterize the system. Second order kernels (the first order kernel is the frequency response function) have been computed for one biological system (Stark, 1969).

On the other hand, sinusoidal methods may be more helpful in suggesting the nature of important non-linearities such as the rectification and phase-locking mentioned earlier. Also, for non-linear biological systems, the frequency response functions computed may sometimes differ in interesting ways using different methods, so a varied approach is useful. What should be clear is that the methods offer a means for building up a quantitative picture of nervous systems. The stochastic point process produced by one group of cells represents the input to other groups. The process will again be filtered by the synaptic junctions, integrated and modified to produce a new stochastic point process. At each stage the functional picture that emerges can help in deciding which of the possible complications mentioned in Section 3 are required to model accurately the activity of the nerve cells considered. Already the methods have proven useful in analyzing the early stages of pattern recognition in the visual system (Kirschfeld and Reichardt, 1964; Hughes and Maffei, 1966; Maffei and Rizzolatti, 1967; Knight, Toyoda and Dodge, 1970), the auditory system (Kiang et al., 1965), the vestibular system (Melvill Jones and Milsum, 1970) and the somatosensory system (Talbot et al., 1968; Mountcastle, et al., 1969).

I would like to illustrate the methods using work in progress in my lab on the systems involved in motor control. As indicated in Fig. 1, applying a small stretch to a muscle will excite receptors in the muscle, in particular, fibers coming from the *muscle spindles*. The larger sensory fibers from the muscle spindle, the so-called primary afferents, are the ones which directly excite motoneurons via a single synapse, and so produce a muscular contraction (*stretch reflex)*. This shortening in response to stretch represents a negative feedback pathway which will tend to maintain a constant position in the face of external disturbances. To study the properties of the muscle spindles, a sinusoidally varying length signal can be applied or a randomly varying one, and the frequency response computed. This is shown in Fig. 7 taken from the work of Matthews and Stein (1969c) applying sinusoidal stimuli to soleus muscle in the cat. Similar results have been obtained using other muscles (Poppele and Bowman, 1970) and random stimulation (Koles, 1970). The sensitivity of the primary afferents as measured by the modulation in impulses/sec produced per mm of stretch is greater than that of the smaller or secondary afferents. However, both increase sharply above a "corner" frequency of 1-3 Hz, and the sensitivity, particularly of the primaries at high frequencies, becomes very large (note that the sensitivity is plotted on a logarithmic scale).

The converse experiment can also be done. Stimuli can be applied to the motor nerve with rates which are modulated in a sinusoidal (Rosenthal et al., 1970) or a random manner (Stein et al., 1972b). The saving in time using random stimulation has particular advantages on the motor side because muscles fatigue with prolonged stimulation. The magnitude of the frequency response is shown in Fig. 8 using random stimulation to the same muscle mentioned previously, soleus of the cat. The force in grams produced per impulse/sec modulation is a declining

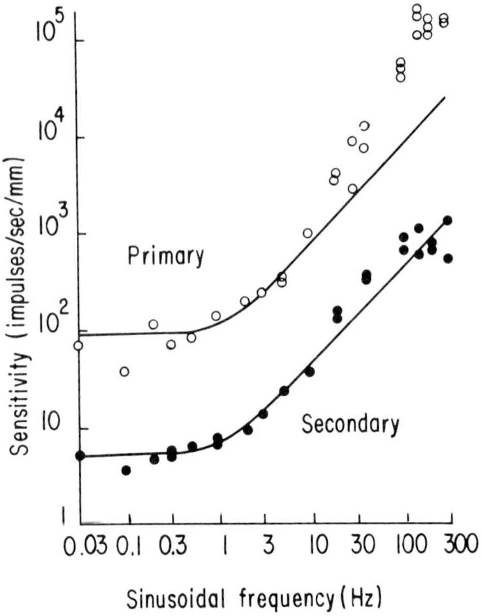

Figure 7. *Frequency response of sensory fibers from muscle spindles* (Mathews and Stein, 1969a). The sensitivity of the fiber in terms of the response in impulses/sec per mm of stretch is plotted on a logarithmic scale against the logarithm of frequency.

function of frequency with a somewhat higher corner frequency (4-10 Hz). Analysis of the corresponding phase changes indicates that the response of the muscle spindles can lead the sensory stimulus by more than 90° at some frequencies, whereas the response of the muscle always shows a phase lag which increases monotonically with frequency. Similarly, the frequency response is measurable for the synaptic transfer (the impulses/sec change in motoneurons produced per impulse/sec modulation of muscle receptors) and for the peripheral part of the feedback loop (the contraction in mm produced per gram of force generated with a given load). Thus, detailed mathematical analysis of the system should permit answers to questions such as the following:

1) To what extent does the increased sensitivity of the muscle spindles compensate for the decreased ability of the muscles to follow high frequency fluctuations?

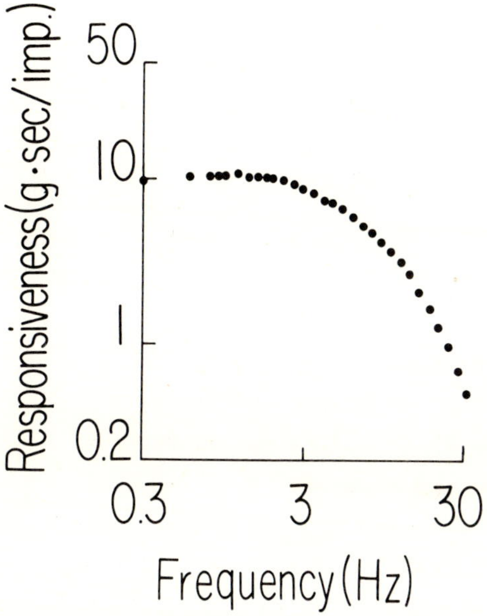

Figure 8. *Frequency response of a muscle in the cat measured using random stimulation of the motor nerve.* Both the ordinate and the abscissa are again logarithmic scales.

2) Over what range of frequencies can the phase advance of the muscle spindles compensate for the phase lag of the muscle and the delays around the feedback loop?

3) Do the characteristics of the muscle spindles tend to stabilize the loop and so prevent the tendency for oscillations which is present in any feedback system, or do they tend to make the loop unstable and so promote the rhythmic oscillations involved in normal tremor or in locomotion (walking, running, etc.)?

4) How can the central nervous system modify the properties of the feedback system through its synaptic connections onto

a) the α motoneurons which go to the normal muscle fibers,

b) the smaller γ motoneurons which go specifically to the muscle fibers inside muscle spindles, or

c) the synaptic terminals of the primary muscle spindle afferent fibers?

Similarly, basic questions should be answerable in other neural systems with these methods in the next few years. *The answers to these questions are all being determined from knowledge of the linear frequency response function, and make the large assumption that the biological systems can be treated as linear systems over a reasonable range of behaviour.* The validity of this assumption can probably be tested at each stage by use of the coherence function mentioned earlier. The coherence function will be reduced from its maximum value of unity by

1) non-linearities,
2) variability inherent in the system, and,
3) the non-stationarity of biological systems.

The role of the inherent variability in linearizing the system was mentioned previously, so variability in neural systems can increase coherence in some instances (see Stein, et al., 1972a), although it generally reduces coherence. This is of interest since measures of coherence are closely related to measures of information transfer first introduced by Shannon (1948). Derivation of the relationship between coherence measures and information capacity (Stein and French, 1970) makes the assumption that the noise producing the variability is approximately Gaussian and adds linearly with the input signal. This relation has been used to measure the information capacity of neural models (Stein et al., 1972a) and the formulae may be applied to measure information transmission in nerve cells whenever these assumptions are justified. Where non-linear processing and non-Gaussian sources of variability are important, the formulae will generally underestimate the ability of nerve cells to transmit information. Thus, the method offers a new approach to measuring information transmission by nerve cells, a problem which has been fraught with difficulties in the past (Stein, 1967b; Stein, 1970; Kenton and Kruger, 1971).

ACKNOWLEDGMENTS

I wish to thank Professor M. N. Oguztöreli and Doctor E. Butz for

for their helpful suggestions on this paper. The work described in this paper was supported by grants to the author from The Medical Research Council of Canada and the Muscular Dystrophy Association of Canada.

REFERENCES

Adrian, E. D. (1932). *The Mechanism of Nervous Action.* Philadelphia: Saunders.

Bayly, E. J. (1968). Spectral analysis of pulse frequency modulation in the nervous system. IEEE Trans. Bio-Med. Engng. BME 15, 257-65.

Bendat, J. S. and Piersol, A. G. (1966). *Measurement and Analysis of Random Data.* New York: Wiley.

Bishop, P. O., Levick, W. R. and Williams, W. O. (1964). Statistical analysis of the dark discharge of lateral geniculate neurons. J. Physiol. (London) 170, 598-612.

Bush, B. M. H. and Roberts, A. (1968). Resistance reflexes from a crab muscle receptor without impulses. Nature (London) 218, 1171-3.

Calvin, W. H. and Stevens, C. F. (1968). Synaptic noise and other sources of randomness in motoneuron interspike intervals. J. Neuro-Physiol 31, 574-87.

Carslaw, H. S. and Jaeger, J. C. (1963). *Operational Methods in Applied Mathematics.* New York: Dover.

Cooley, J. W. and Tukey, J. W. (1965). An algorithm for the machine calculation of complex Fourier series. Math. Comput. 19, 297-301.

Cox, D. R. and Lewis, P. A. W. (1966). *The Statistical Analysis of Series of Events.* London: Methuen and New York: Barnes and Noble.

Cox, D. R. and Miller, H. D. (1965). *The Theory of Stochastic Processes.* London: Methuen.

Cowan, J. D. (1971). Stochastic models of neuroelectric activity. In *Proc. I.U.P.A.P. Conference on Statistical Mechanics,* Rice, S., Light, J., Freed, K., eds. Chicago: University of Chicago Press.

Eccles, J. C. (1960). The properties of dendrites. In *Structure and Function of the Cerebral Cortex,* Tower, D. B., Schadé, J. P., eds. Amsterdam: Elsevier, 192-202.

Eccles, J. C. (1964). *The Physiology of Synapses.* Berlin: Springer.

Eysel, U. T. (1971). Computer simulation of the impulse pattern of muscle spindle afferents under static and dynamic conditions. Kybernetik 5, 171-8.

Frankenhaeuser, B. and Huxley, A. F. (1964). The action potential in the myelinated nerve fiber of *Xenopus laevis* as computed on the basis of voltage clamp data. J. Physiol. (London) 171, 302-15.

French, A. S. and Holden, A. V. (1971a). Frequency domain analysis of neurophysiological data. Comp. Prog. Biomed. 1, 219-34.

French, A. S. and Holden, A. V. (1971b). Alias-free sampling of neuronal spike trains. Kybernetik 8, 165-71.

French, A. S. and Stein, R. B. (1970). A flexible neural analog using integrated circuits. IEEE Trans. Bio-Med. Engng, BME 17, 248-53.

Fuortes, M. G. F. and Mantegazzini, F. (1962). Interpretation of the repetitive firing of nerve cells. J. Gen. Physiol. 45, 1163-79.

Geisler, C. D. and Goldberg, J. M. (1966). A stochastic model of the repetitive activity of neurons. Biophys. J. 6, 53-69.

Gerstein, G. L. and Mandelbrot, B. (1964). Random walk models for the spike activity of a single neuron. Biophys. J. 4, 41-68.

Goldman, D. E. (1943). Potential, impedance and rectification in membranes. J. Gen. Physiol. 27, 37-60.

Goldman, L. and Albus, J. S. (1968). Computation of impulse conduction in myelinated fibers: theoretical basis of the velocity-diameter relation. Biophys. J. 8, 596-607.

Hagger, R. A. and Barr, M. L. (1950). Quantitative data on the size of synaptic end-bulbs in the cat's spinal cord. J. Comp. Neurol. 93, 17-35.

Hebb, D. O. (1949). *The Organization of Behaviour*. New York: Wiley.

Henneman, E.(1968). Organization of the spinal cord. In *Medical Physiology*, Mountcastle, V. B., ed. St. Louis: Mosby, 1717-32.

Hodgkin, A. L. and Huxley, A. F. (1952). A quantitative description of membrane current and its application to conduction and excitation in nerve. J. Physiol. (London) 117, 500-44.

Hodgkin, A. L. and Rushton, W. A. H. (1946). The electrical constants of a crustacean nerve fiber. Proc. Roy, Soc. (London) B 133, 444-79.

Hubbard, J. I., Llinas, R. and Quastel, D. M. J. (1969). *Electrophysiological Analysis of Synaptic Transmission.* London: Arnold.

Hughes, G. W. and Maffei, L. (1966). Retinal ganglion cell response to sinusoidal light stimulation. J. Neurophysiol. 29, 333-52.

Hutchinson, N. A., Koles, Z. J. and Smith, R. S. (1970). Conduction velocity in myelinated nerve fibers of *Xenopus laevis*. J. Physiol. (London) 208, 279-89.

Jack, J. J. B. and Redman, S. J. (1971). The propagation of transient potentials in some linear cable structures. J. Physiol. (London) 215, 283-320.

Johannesma, P. I. M. (1969). *Stochastic Neural Activity - A Theoretical Investigation.* Ph. D. Thesis, Catholic University, Nijmegen, Netherlands.

Kandel, E. R. and Spencer, W. A. (1961). Electrophysiology of the hippocampal neurons. II. After-potentials and repetitive firing. J. Neurophysiol. 24, 253-9.

Kiang, N. Y., Watanabe, T., Thomas, E. C. and Clark, L. F. (1965). *Discharge Patterns of Single Fibers in the Cat's Auditory Nerve.* M.I.T. Res. Monograph, No. 35. Cambridge: M.I.T. Press.

Kirschfeld, K. and Reichardt, W. (1964). Die Verarbeitung stationärer optischen Nachrichten im Komplexauge von *Limulus*. Kybernetik 2, 43-61.

Knight, B. W. (1969). Frequency response for sampling integrator and for voltage to frequency convertor. In *Systems Analysis in Neurophysiology,* Terzuolo, C. A., ed. Minneapolis: University of Minnesota, 61-72.

Knight, B. W., Toyoda, J. I. and Dodge, F. A. (1970). A quantitative description of the dynamics of excitation and inhibition in the eye of *Limulus*. J. Gen. Physiol. 56, 421-37.

Kenton, B. and Kruger, L. (1971). Information transmission in slowly adapting mechanoreceptor fibers. Exp. Neurol. 31, 114-39.

Koles, Z. J. (1970). *A Study of the Sensory Dynamics of a Muscle Spindle.* Ph.D. Thesis, University of Alberta, Edmonton, Canada.

Knox, C. K. (1969). The power spectral density of random spike trains. in *Systems Analysis in Neurophysiology,* Terzuolo, C. A. ed. Minneapolis: University of Minnesota, 73-85.

Llinás, R. and Nicholson, C. (1969). Electrophysiological analysis of alligator cerebellar cortex: A study of dendritic spikes. In *Neurobiology of Cerebellar Evolution and Development*, Llinás, R., ed. Chicago: A.M.A. Education and Research Foundation, 421-66.

Maffei, L. and Rizzolatti, G. (1967). Transfer properties of the lateral geniculate body. J. Neurophysiol. 30, 333-340.

Matthews, P. B. C. and Stein, R. B. (1969a). The sensitivity of muscle spindle afferents to small sinusoidal changes of length. J. Physiol. (London) 200, 732-743.

Matthews, P. B. C. and Stein, R. B. (1969b). The regularity of primary and secondary muscle spindle afferent discharges and its functional implications. J. Physiol. (London) 202, 59-82.

McCulloch, W. S. and Pitts, W. H. (1943). A logical calculus of the ideas immanent in nervous activity. Bull. Math. Biophys. 5, 115-33.

Melvill Jones, G. and Milsum, G. H. (1970). Characteristics of neural transmission from the semicircular canal to the vestibular nuclei of cats. J. Physiol. (London) 209, 295-316.

Moore, G. P., Perkel, D. H. and Segundo, P. (1966). Statistical analysis and functional interpretation of neuronal spike data. Ann. Rev. Physiol. 28, 493-522.

Moore, G. P., Segundo, J. P., Perkel, D. H. and Levitan, H. (1970). Statistical signs of synaptic interaction in neurons. Biophys. J. 10, 876-900.

Mountcastle, V. B.(1968). *Medical Physiology*. St. Louis: Mosby.

Mountcastle, V. B., Talbot, W. H., Sakata, H. and Hyvärinen, J. (1969). Cortical neuronal mechanisms in flutter-vibration studied in unanaesthetized monkeys, neuronal periodicity and frequency discrimination. J. Neurophysiol. 32, 452-84.

Noble, D. (1966). Applications of Hodgkin-Huxley equations to excitable tissues. Physiol. Rev. 46, 1-50.

Noble, D. and Stein, R. B. (1966). The threshold conditions for initiation of action potentials by excitable cells. J. Physiol. (London) 187, 129-66.

Poppele, R. E. and Bowman, R. J. (1970). Quantitative description of linear behavior of mammalian muscle spindles. J. Neurophysiol. 33, 59-72

Pringle, J. W. S. and Wilson, V. J. (1952). The response of a sense organ to a harmonic stimulus. J. Exp. Biol. 29, 220-34.

Rall, W. (1962). Electrophysiology of a dendritic neuron model. Biophys. J. 2, (Suppl.) 145-67.

Rall, W. (1969). Time constants and electronic length of membrane cylinders and neurons. Biophys, J. 9, 1483-508.

Ranson, S. W. and Clark, S. L. (1959). *The Anatomy of the Nervous System*. Philadelphia: Saunders.

Rescigno, A., Stein, R. B., Purple, R. L. and Poppele, R. E. (1970). A neuronal model for the discharge patterns produced by cylic inputs. Bull. Math. Biophys. 31, 341-57.

Rosenthal, N. P., McKean, T. A., Roberts, W. J. and Terzuolo, C. A. (1970). Frequency analysis of stretch reflex and its main subsystems in triceps *surae* muscles of the cat. J. Neurophysiol. 33, 713-49.

Roy, B. K. and Smith, D. R. (1969). Analysis of the exponential decay model of the neuron showing frequency threshold effects. Bull. Math. Biophys. 31, 341-57.

Rushton, W. A. H. (1951). A theory of the effects of fiber size in medullated nerve. J. Physiol. (London) 115, 101-122.

Rushton, W. A. H. (1961). Peripheral coding in the nervous system. In *Sensory Communication*, Rosenblith, W. A., ed. Cambridge: M.I.T. Press, 169-81.

Shannon, C. E. (1948). A mathematical theory of communication. Bell System Tech. J. 27, 379-423, 623-656. Reprinted in Shannon, C. E. and Weaver, W. (1962), *The Mathematical Theory of Communication*. Urbana: University of Illinois Press.

Shepherd (1970). The olfactory bulb as a simple cortical system: experimental analysis and functional implications. In *The Neurosciences: Second Study Program*, Schmitt, F. O., ed., New York: Rockefeller University Press, 539-52.

Spekreijse, H. and Oosting, H. (1970). Linearizing: A method for analyzing and synthesizing non-linear systems. Kybernetik 7, 22-31.

Stark, L. (1969). The pupillary control system: Its non-linear adaptive and stochastic engineering design characteristics. Automatica 5 655-76.

Stein, R. B. (1967a). The frequency of nerve action potentials generated by applied currents. Proc. Roy. Soc. (London) B 167, 64-86.

Stein, R. B. (1967b). The information capacity of nerve cells using a frequency code. Biophys. J. 7, 797-826.

Stein, R. B. (1967c). Some models of neuronal variability. Biophys. J. 7, 37-68.

Stein, R. B. (1970). The role of spike trains in transmitting and distorting sensory signals. In *The Neurosciences: Second Study Program*, Schmitt, F. O., ed.. New York: Rockefeller University Press, 597-604.

Stein, R. B. and French, A. S. (1970). Models for the transmission of information by nerve cells. In *Excitatory Synaptic Mechanisms*, Anderson, P., Jansen, J. K. S., eds.. Oslo: Oslo University Press, 247-57.

Stein, R. B., French, A. S. and Holden, A. V. (1972a). The frequency response, coherence and information capacity of two neuronal models. Biophys. J., 12, 295-322.

Stein, R. B., French, A. S., Mannard, A. and Yemm, R. (1972b). New methods for analyzing motor function in man and animals. Brain Res., in press.

Talbot, W. H., Darian-Smith, I., Kornhuber, H. H. and Mountcastle, V. B. (1968). The sense of flutter vibration: Comparison of the human capacity with response patterns of mechanoreceptive afferents from the monkey hand. J. Neurophysiol. 31, 301-34.

Ten Hoopen, M. (1966). Multimodal interval distributions. Kybernetik 3, 17-24.

Von Neumann, J. (1956). Probabilistic logics and the synthesis of reliable organisms from unreliable components. In *Automata Studies*, Shannon, C. E., McCarthy, J., eds.. Princeton, N. J.: Princeton University Press, 43-98.

Werblin, F. S. and Dowling, J. E. (1969). Organization of the retina of the mudpuppy, *Necturus maculosus*. II. Interacellular recording. J. Neurophysiol. 32, 339-55.

Werner, G. (1970). The topology of the body representation in the somatic afferent pathway. In *The Neurosciences: Second Study Program*, Schmitt, F. O., ed.. New York. Rockefeller University Press, 605-17.

Wiener, N. (1958). *Non-linear Problems in Random Theory*. Cambridge: M.I.T. Press.

SOME POINT PROCESSES IN MOTOR AND SENSORY NEUROPHYSIOLOGY

B. W. Knight
Rockefeller University, New York

SUMMARY

Illustrative examples are given of the interplay between point processes and continuous processes in the nervous system. The first example is the neuromuscular junction, at which a point process gives rise to a continuous process. A similar but more elaborate example is an invertebrate photoreceptor. The invertebrate photoreceptor also illustrates how a continuous process gives rise to a point process in the nervous system. These examples are examined in a quantitative way.

Evolution exploits a few basic principles and mechanisms again and again. For this reason a few themes occur repeatedly in the study of biology. A good example is the way point processes are used by the nervous system as it receives, manipulates, and acts upon information from the outside world.

We find that as a signal passes through the nervous system its form changes back and forth, from the continuous to the discrete — from a continuous process to a point process — and back again. The material discussed in this paper will illustrate this, and can mostly be found in Katz (1966), Dodge, Knight and Toyoda (1968), Knight, Toyoda and Dodge (1970), and Shapley (1971). The first two of these references examine the transformation of a discrete point process into a continuous signal. The last two study the conversion of a continuous signal back into a discrete point process.

I must apoligize for picking two publications in which I was personally involved. I have done this because I am particularly familiar with that research, and because they are fairly representative of the mainstream of thinking and research in this area.

The underlying basis of nervous excitation is illustrated in Fig. 1.

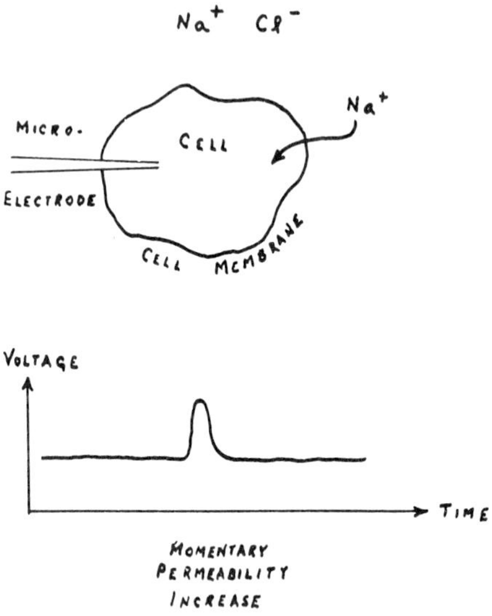

Figure 1. Ion movement and resulting voltage change within a cell.

Materials such as ordinary salt (NaCl) when they dissolve in water separate naturally into constituents (Na^+ and Cl^-) which are electrically charged. Under ordinary circumstances the Na^+ is prevented from diffusing into the inside of a cell, by an impermeable cell membrane. Now if part of the cell membrane becomes permeable to sodium, the sodium will flow into the inside of the cell, carrying its positive charge in with it. The voltage within the cell will swing in the positive direction, and this may be observed in the laboratory by the insertion of a microelectrode.

Figure 2 illustrates one way in which the change in membrane permeability may take place. This is an electron micrograph of a neuromuscular junction. The striped structure is a muscle cell. Sitting on it from above, the smaller structure with a granular texture is the ending of a

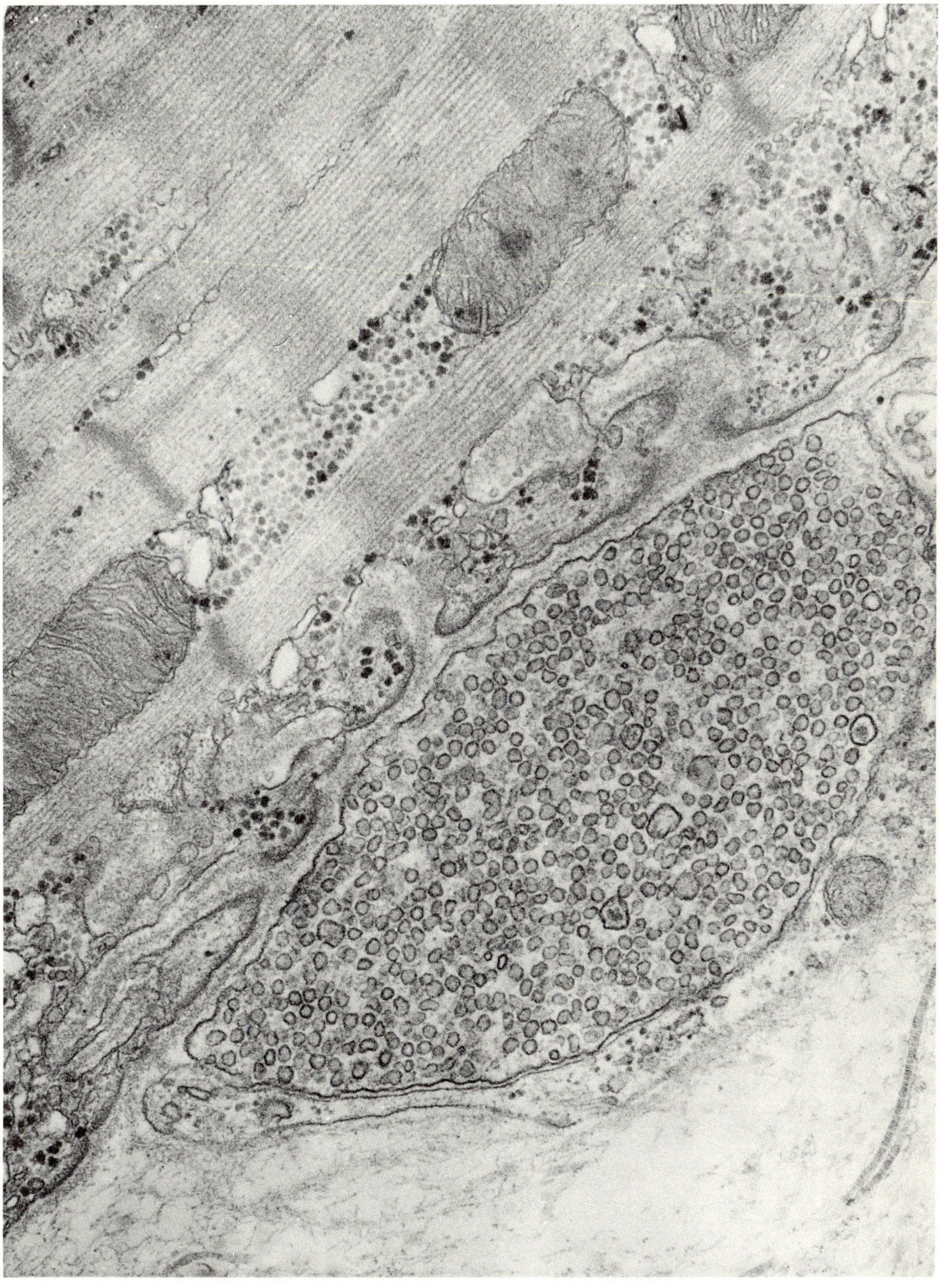

Figure 2. A neuromuscular junction (Cecarelli, personal communication)

nerve cell, which will change the permeability of the muscle cell membrane. The whole complex is called a synapse. The granules within the nerve cell are the so-called synaptic vessicles. Each one is a tiny packet of the chemical acetylcholine. Acetylcholine has the property that it makes the muscle cell membrane more permeable to sodium.

Figure 3 shows the essential features of the synapse. Acetylcholine is released from the nerve cell to the membrane of the muscle cell.

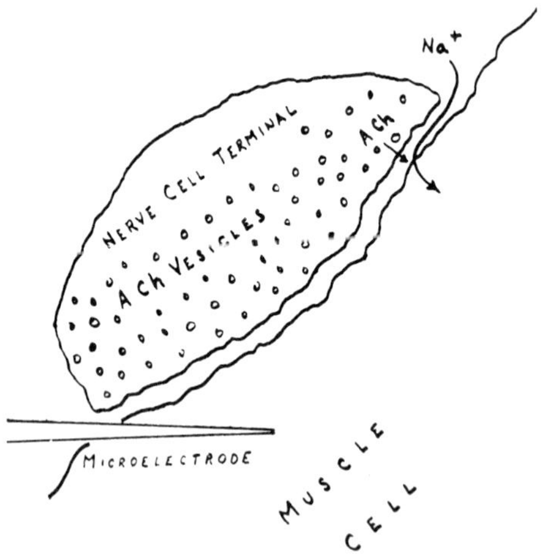

Figure 3. Schematic of the neuromuscular junction.

There it increases the membrane permeability to sodium. The sodium, with its positive charge, moves into the muscle cell, driving its internal voltage positive. This voltage change may be observed with a microelectrode, which is shown schematically.

Figure 4 shows what we observe with the aid of the microelectrode, in the absence of nerve activity. There is a spontaneous occurrence of little positive fluctuations in voltage within the muscle cell. Each one results from the release of a single packet of acetylcholine from the associated nerve cell. I will call these fluctuations quantal potentials. The arrivals form a Poisson point process. Moreover, the

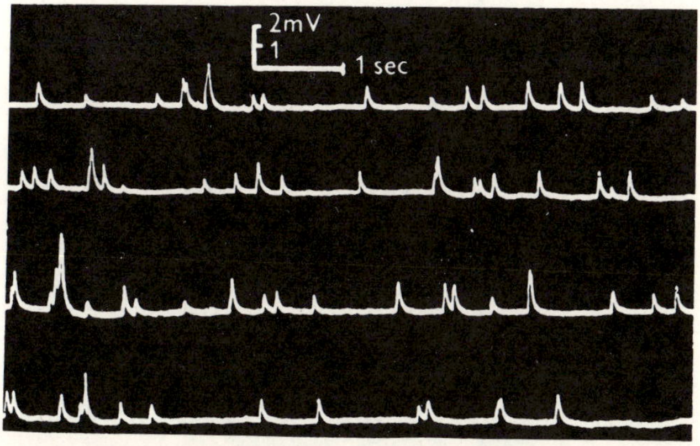

Figure 4. Quantal potentials in the muscle cell (del Castillo & Katz, 1954).

mean arrival rate depends on the voltage level within the associated nerve cell. There is some variation in the heights of the quantal potentials; however, these heights form a fairly compact distribution about a mean value as Fig. 5 shows.

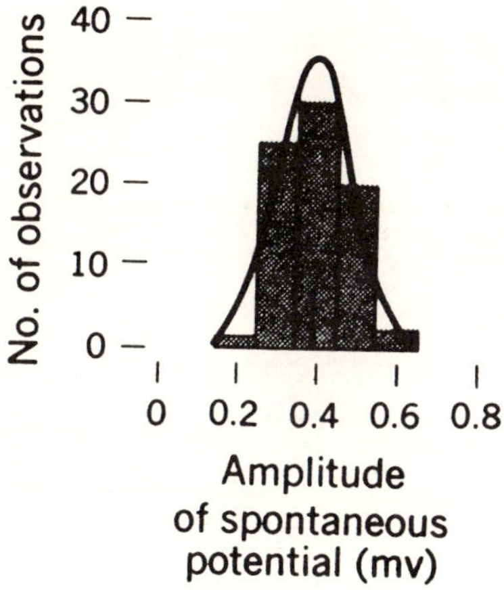

Figure 5. Distribution of amplitudes of quantal potentials (Katz, 1966).

We will return to this distribution of quantal heights in a moment.

The result of a single nerve impulse in the nerve cell is shown in Fig. 6.

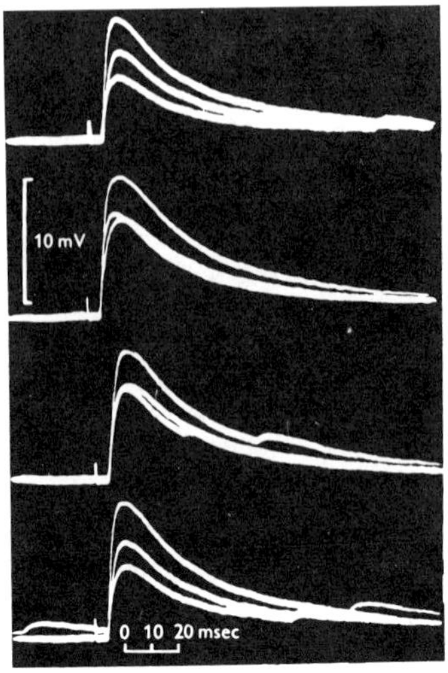

Figure 6. The muscle cell's voltage response to a nerve impulse (del Castillo & Katz, 1954).

Each frame shows three superimposed muscle cell voltage responses to single nerve impulses. The scales for both voltage and time are different from the previous record. In fact, we can relate to the previous record by noting a single quantal potential in the second frame from the bottom. These voltage responses to nerve impulses are called end-plate potentials. Note that they are distributed in size. It is reasonable to suspect that they are formed from several quantal potentials superimposed. We may speculate further that they result from the same Poisson point process, but enormously speeded up during the brief passage of the nerve impulse.

This conjecture is formulated quantitatively as follows. Let the mean quantal release number, m, be defined by m = (mean end-plate potential)/(mean quantal potential). The Poisson release hypothesis then is that the number of quanta has distribution $P_n = \frac{m^n}{n!} e^{-m}$. In particular, the probability of complete failure (no quanta) is

$$P_o = e^{-m}.$$

The mean size of both the quantal potential and the end-plate potential are measurable directly. If our Poisson process conjecture is correct, we should be able to predict the proportion of failures.

Actually, we should be able to do a great deal more as well.

The Poisson hypothesis leads to a well defined prediction of what the entire distribution of peak voltage should be:

For 1 quantum: $Q(v)$,

For 2 quanta: $Q^{(2)}(v) = Q * Q$,

For 3 quanta: $Q^{(3)}(v) = Q * Q^{(2)}$, etc.,

where * denotes convolution. The distribution for end-plate potential is thus

$$S(v) = \sum_{n=1}^{\infty} P_n Q^{(n)}(v) = e^{-m} \sum_{n=1}^{\infty} \frac{m^n}{n!} Q^{(n)}(v).$$

Now, let

$$C(\alpha) = E(e^{-\alpha v}).$$

Then, including failures, we have the generating function

$$C_S(\alpha) = e^{-m\{1-C_Q(\alpha)\}}.$$

There are various ways, plain and fancy, for evaluating the distribution. (Note that the characteristic function method immediately relates the moments of the two distributions, and even more immediately relates the moments of the quantal distribution to the semi-invariants of the end plate potential distribution.)

In Fig. 7, the prediction is compared with experiment. Only the measured quantal distribution in the inset, and the measured mean content m, have been used to make this prediction. The predicted and observed failures are shown at the left.

This beautiful work is due mostly to Bernard Katz and his collaborators over the past 20 years.

Let us now move from muscle to an organ of a very different sort. Fig. 8 shows a micrograph of a light receptor organ, the eye of the horseshoe crab, *Limulus*. At the top we see the light receptors, and below them the optic nerve fibers on their way to the horseshoe crab's brain. It is feasible to insert a microelectrode into a light-sensitive photoreceptor cell. What we discover in darkness or dim light is shown in Fig. 9.

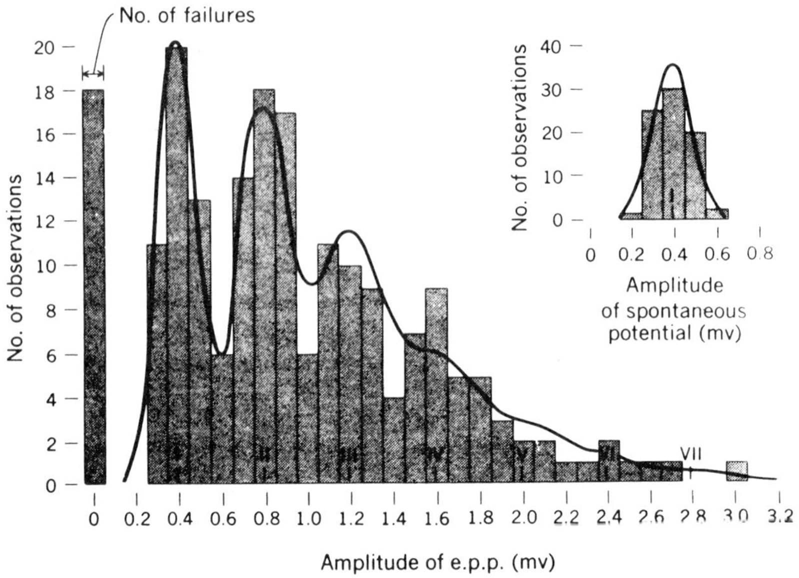

Figure 7. Amplitude distribution of the responses (Katz, 1966).

Figure 8. Micrograph of a section through a Limulus visual cell (micrograph by W. H. Miller, taken from Hartline, Wagner and Ratliff, 1956).

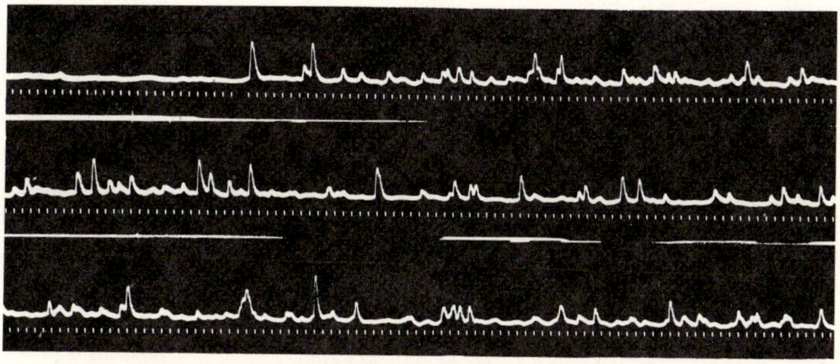

Figure 9, Quantal potentials in a *Limulus* visual cell (Adolph, 1963).

Something is going on here which is very much akin to the spontaneous voltage fluctuations in the neuromuscular junction. But the analogy cannot be perfect.

Figure 10 shows the voltage records we observe with increasing intensities of light. Notice that as the response increases, the fluctuations

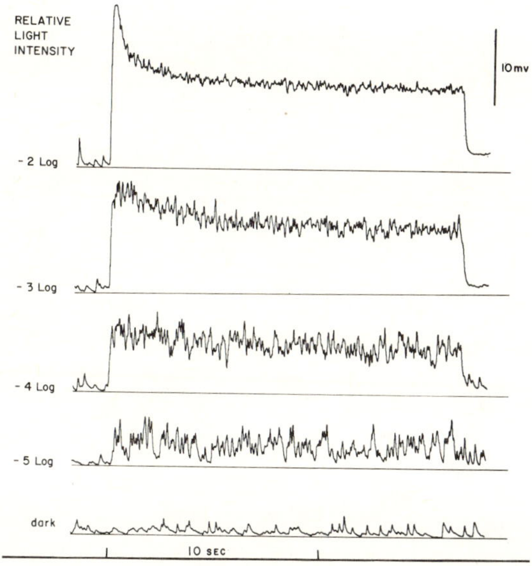

Figure 10. The cell's voltage response to light (Dodge, Knight & Toyoda, 1968).

in the response decrease, not only relatively but also absolutely. A random superposition of quantal events which had a fixed size distribution would lead to fluctuations increasing proportionally to the square root of the response.

We notice that response to dim light is followed by spontaneous quantal responses in the dark, but no visible quantal responses follow bright light. It seems that the normal effect of light is to reduce the sizes of the quantal events, and this leads us to propose the model shown in Fig. 11. We have noted that the chemical which changes membrane conductance comes in discrete packages. In this model for the *Limulus* photoreceptor we assume that any single package refills gradually after

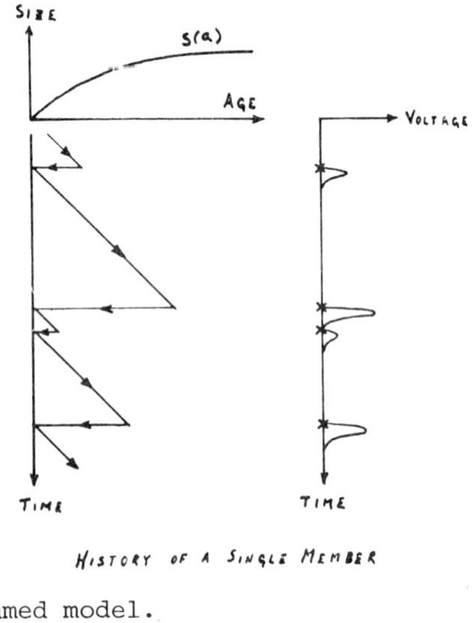

Figure 11. The assumed model.

discharge, so that the size of its chemical load increases with its age since previous discharge. We note two things: first, if the average discharge rate were increased, as with added light intensity, the average size would decrease (which is consistent with observation); and second, a discharge which follows its predecessor in unusally quick succession, as the third event shown here, will contribute a quantal response of less than average size.

Figure 12 shows schematically the distribution over ages of a whole population of packages. The size of the stipple marks is intended to indicate the size of a package at any age. It is easy to see that an increase in light intensity should lead to a surge in chemical release, followed by a decline as the constant population readjust to a smaller average age.

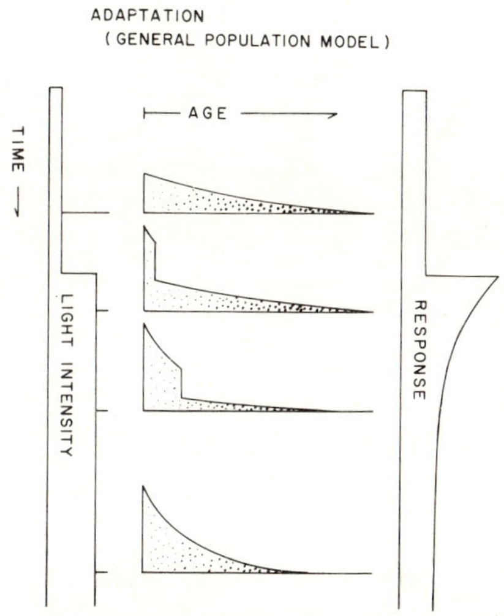

Figure 12. The model's theoretical response to light.

The model can be expressed formally, of course, in terms of dynamical equations. The age distribution equation for a single member is given by

$$\frac{\partial P}{\partial t} = -\frac{\partial P}{\partial a} - H(a,t)P + \delta(a)\int_0^\infty da' H(a',t)P(a',t),$$

where H(a,t) is hazard/(sec^{-1}). The voltage contributed by a single member is

$$V(t) = \sum_{\ell=-\infty}^{\infty} S(t_\ell - t_{\ell-1})g(t-t_\ell),$$

where S(a) is the size and g(t) the shot shape.

The differential equation expresses the evolution of an age distribution as time progresses. The first term on the right simply says that age advances linearly with time. The second term says that dropouts are due to a hazard; and the third term says each dropout gives a rebirth at age zero.

These equations completely specify the dynamics of the model. In principle we can answer any question about the model by being technically clever in manipulating the equations. For example, we can predict the autocorrelation, which is defined as

$$C(\tau) = E[\{V(t) - \bar{V}\}\{V(t+\tau) - \bar{V}\}].$$

This is interesting because the autocorrelation of the noisy voltage response to steady light also may be measured experimentally.

Another thing we may predict from the model is the frequency response to a flickering light. If the light time-course

$$L(t) = L_0 + L_1 \cos(\omega t)$$

leads to the expected voltage response

$$E\{V(t)\} = \langle V(t)\rangle = V_0 + V_1 \cos(\omega t+\phi),$$

then

$$R(\omega) = (V_1/L_1)\, e^{i\phi(\omega)}$$

is the frequency response or "transfer function." This voltage frequency response is also measurable in the laboratory, as shown in Fig. 13. (We won't discuss the lowest frame yet; it shows the point process we will consider later.) The top frame of Fig. 13 shows the stimulating light, sinusoidally modulated. The middle frame shows the voltage response: The noise jiggles about a sinusoidal voltage modulation. This experimental run contributes one point to the frequency response shown in Fig. 14. These are two frequency response curves taken from the same visual receptor cell at two mean light intensities differing by a factor of one hundred thousand. The appreciable scatter in the data points for low intensity is because voltage records from dim light are very irregular, as we have seen.

According to our theoretical model, we find that the autocorrelation should be predicted from the frequency response. Let $H(t)=H$, i.e., be independent of age, and $S(a) = S_0(1-e^{-\gamma a})$. Then, we may show that the

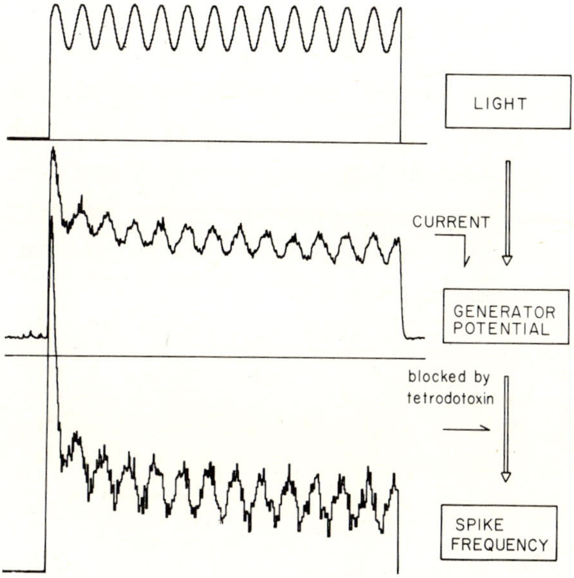

Figure 13. The intracellular voltage's frequency response (middle frame) to modulated light (top frame).

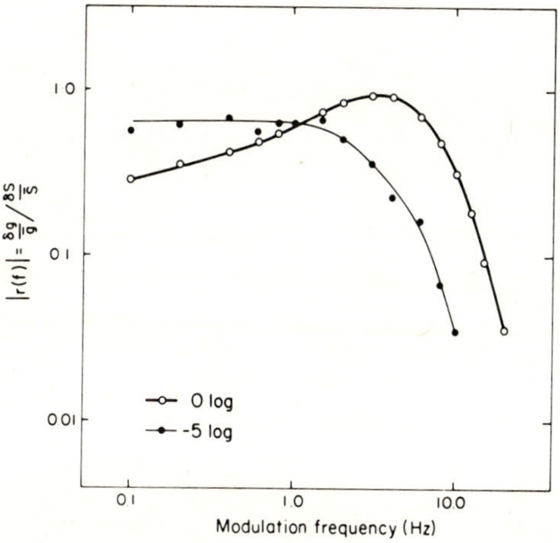

Figure 14. Amplitude of the frequency response, measured at two light intensities (Dodge, Knight & Toyoda, 1968).

autocorrelation and frequency response are related by

$$C(\tau) = A \int_{-\infty}^{\infty} d\omega |R(\omega)|^2 e^{i\omega\tau}.$$

Thus the autocorrelation should be proportional to the Fourier transform of the squared frequency response. We have done this operation on the frequency responses shown in Fig. 14, and the result is compared to autocorrelations measured on the same cell, as shown in Fig. 15. The points show the autocorrelations directly measured; the lines show what is predicted from the frequency responses.

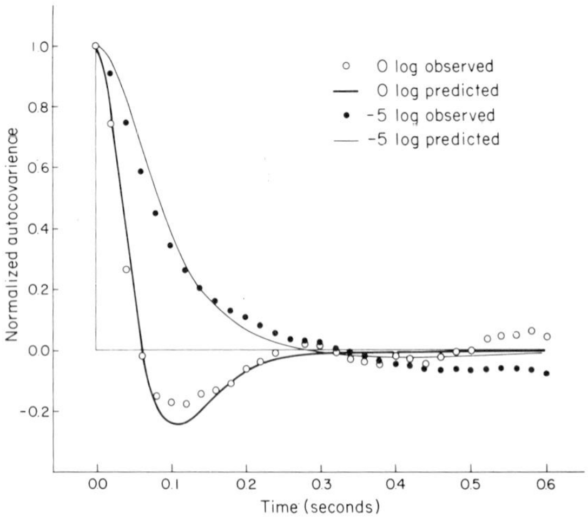

Figure 15. Observed and predicted autocorrelations (Dodge, Knight & Toyoda, 1968).

One interesting feature of the data for the bright light is that the autocorrelation swings negative at short times. According to the theoretical model this is because a quantal event which follows its predecessor quickly, will make a less-than-average contribution.

We now show how a quantal event may be characterized by an effective amplitude and duration.

Define the "effective" amplitude α and duration T of the average shot $\bar{S}g(t)$ by

$$\alpha = \frac{\int_0^\infty \{\bar{S}g(t)\}^2 dt}{\int_0^\infty \bar{S}g(t)\,dt} = \text{effective amplitude;}$$

$$T = \frac{\{\int_0^\infty \bar{S}g(t)\,dt\}^2}{\int_0^\infty \{\bar{S}g(t)\}^2 dt} = \text{effective duration.}$$

This is exact for a rectangular shot. In other words, the expressions have been chosen in such a way that an idealized rectangular shot of height α and duration T has the same real and effective amplitude and duration.

The quantities T and α are useful because they may be related to laboratory observables:

$$\text{mean voltage} = \bar{V} = \lambda T\alpha;$$

$$\text{variance} = \overline{(V - \bar{V})^2} = \lambda T\alpha^2,$$

where λ is the mean rate of the point process.

The mean and variance of the voltage may be obtained in the laboratory. The effective duration may be obtained from the experimental frequency response, and these equations may be solved simultaneously for the amplitude alpha, and for the mean rate lambda of the underlying point process. The result is shown in Fig. 16. The horizontal scale is the logarithm of the light intensity, which ranges over a factor of one hundred thousand. The one curve shows the mean rate of the point process increasing with light, while the amplitude of the quantal event decreases. This change in quantal amplitude (which was not present in the neuromuscular junction, as we saw) permits the visual receptor to function effectively over an enormous range of light intensities.

In the visual receptor of the horseshoe crab, the voltage signal which arose from one point process in turn gives birth to another point process, as we see in Fig. 17. The cell in which these measurements were

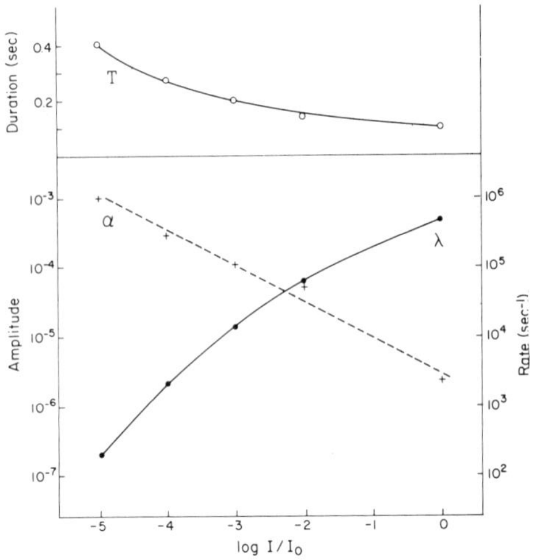

Figure 16. Deduced duration, amplitude, and occurrence rate of quantal potentials, vs. light intensity (Dodge, Knight & Toyoda, 1968).

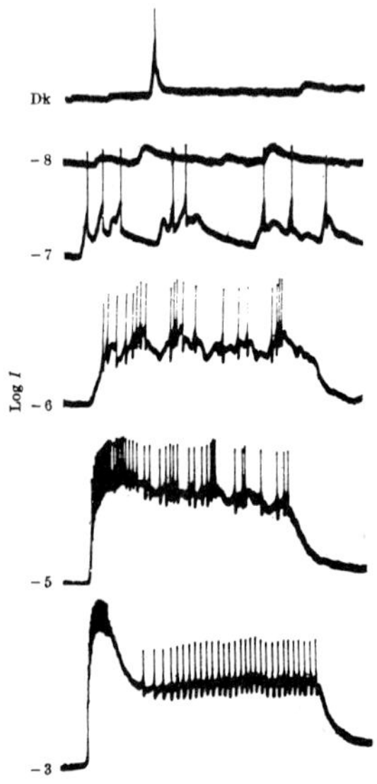

Figure 17. Intracellular potential and resulting nerve impulses (Dowling, 1968).

made is also a nerve cell, and here we see nerve impulses superimposed on the continuous voltage records. (In the records we saw earlier the nerve impulses were eliminated by an artifice.)

The discharge of nerve impulses is evidently related to the voltage, with higher voltage giving rise to more frequent impulses. Also, more nerve impulses occur on a voltage that is rising than on one that is falling. As we go from the top record to the bottom the light intensity is increased, and we notice that the stochastic features of the point process become less prominent. (These records were taken by John Dowling.)

For some purposes it is convenient to treat the point process as if some of its associated variables were continuous. The firing period Δt, for example, exists only at discrete points in time. However, for theoretical purposes, we can imagine a whole ensemble of identical neurons, one of which is firing at any particular moment. In this sense the firing period may be discussed as a continuous variable.

A better measure of nervous activity is the reciprocal firing period, or firing rate

$$N(t) = \frac{1}{\Delta t} .$$

Before we see how the point process of nerve impulse firing is related to a stochastic intracellular voltage, we should see how it is related to a determinative voltage. The idea behind the experiment is shown in Fig. 18. Light gives rise to voltage in the region marked "1". Voltage gives rise to impulses in the region marked "2". (I will not discuss the region marked "3".) The experiments discussed so far involved measurement of voltages through the microelectrode. It is also feasible to run electric current through the microelectrode and thereby establish an intracellular voltage of our own choice. Such a record is shown in Fig. 19.

The top record shows the intracellular voltage in response to sinussoidally modulated light. The middle record shows the firing rate in response to modulated intracellular voltage, controlled by the

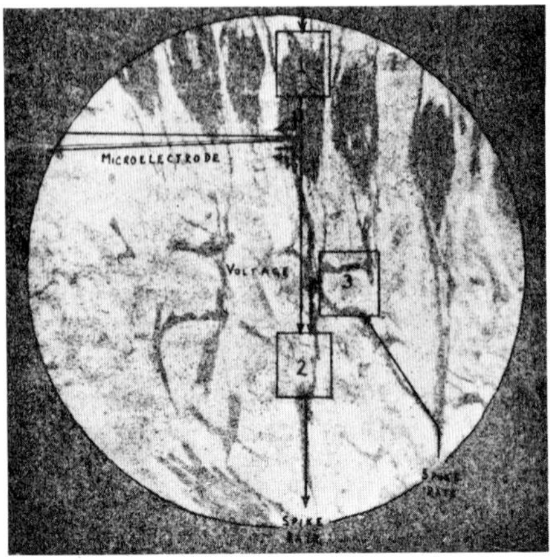

Figure 18. Schematic of *Limulus* visual organization.

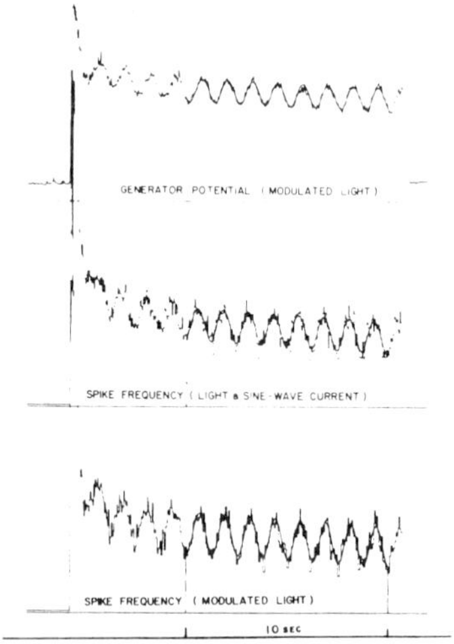

Figure 19. Raw data from the three kinds of experimental runs.

experimenter. First, we check that these component processes correctly predict the third record, which shows the firing rate in response to light. This is shown in Fig. 20.

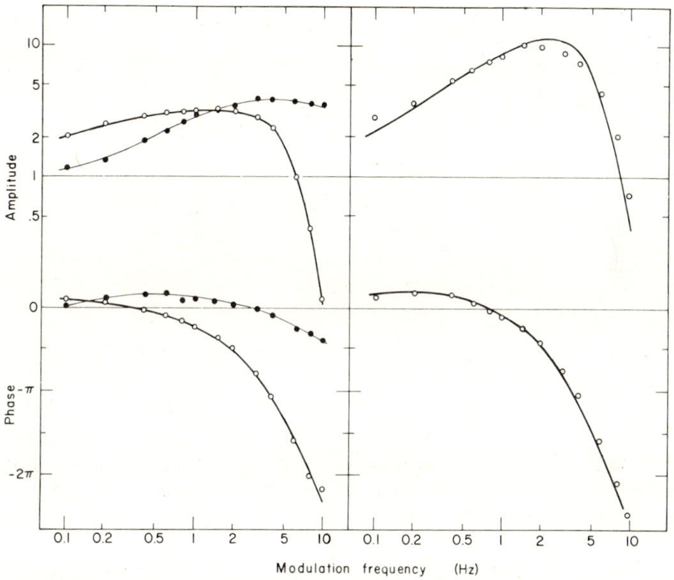

Figure 20. The frequency responses (Knight, Toyoda & Dodge, 1970).

These are frequency responses. In the left frame the open dots show voltage response to light, and the solid dots show firing rate response to voltage. From these, we predict the firing rate response to light, shown by the solid curves in the right hand frame. The points in the right hand frame were obtained by direct measurement at the indicated frequencies.

Figure 21 shows the result of a theoretical calculation. Plausible electrophysiology leads us to propose a model for the way input voltage leads to the generation of nerve impulses. The model leads to the theoretical frequency response given at the top of the figure. The formula contains two adjustable parameters. (The points shown were obtained from an electronic analog of the model.)

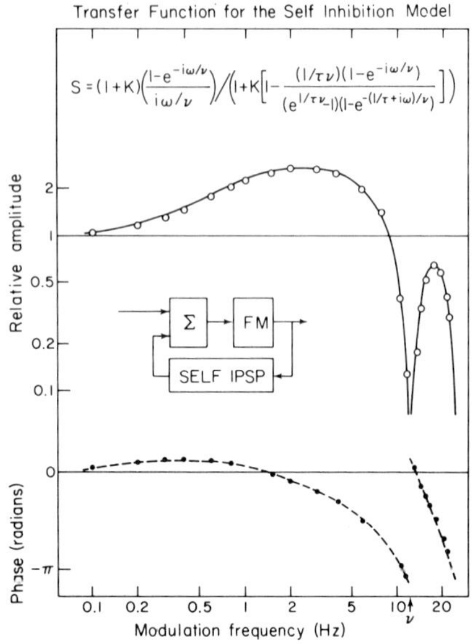

Figure 21. Model for nerve impulse encoding, with formula and curves for frequency response. The parameter ν is impulse rate in absence of modulating (Dodge, Shapley & Knight, 1970).

The theoretical result has been fitted to real horseshoe crab data in Fig. 22. (The data were taken by Robert Shapley.). We notice that the sensitivity of firing rate to intracellular voltage modulation increases with frequency and then decreases. We notice also that the response shows a phase lead up to about three cycles per second. This means that firing rate will reach a maximum while the voltage is rising, in agreement with the earlier slide Fig. 17.

We proceed now from intracellular voltages which are determinative to those which are stochastic. It will be convenient to deal with a stochastic voltage in terms of its power spectrum (or variance spectrum) $\phi(\omega)$ which is defined as

$$\phi(\omega) = \frac{1}{2\pi}\int_{-\infty}^{\infty} d\tau\ C(\tau)e^{i\omega\tau},$$

with inverse relationship,

$$C(\tau) = \int_{-\infty}^{\infty} d\omega\ \phi(\omega) e^{i\omega\tau} .$$

In brief, the power spectrum and autocorrelation form a Fourier transform pair.

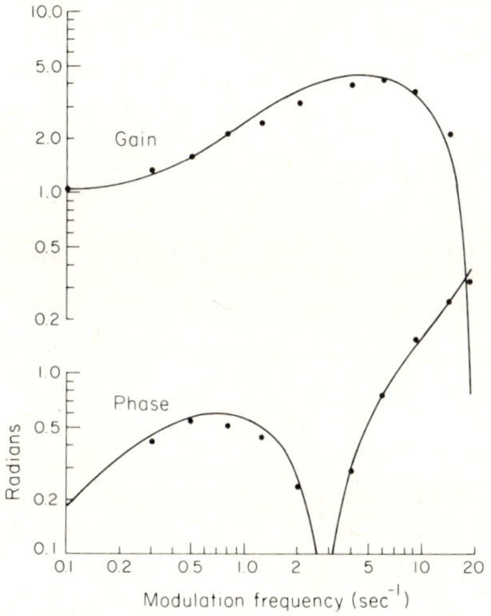

Figure 22. Measured power spectrum of intracellular potential, and predicted power spectrum of firing rate (Shapley, 1971).

Now we have just characterized the cellular machinery that translates voltage to firing rate, as a transducer with a known frequency response. A transducer acts on the power spectrum of the signal transduced in a well defined way.

If subscript N, e.g., $\phi_N(\omega)$, denotes output and subscript G denotes input, e.g., $\phi_G(\omega)$, then

$$\phi_N(\omega) = |S(\omega)|^2 \phi_G(\omega) ,$$

whence

$$C_N(\tau) = \int_{-\infty}^{\infty} d\omega |S(\omega)|^2 \phi_G(\omega) e^{i\omega\tau} .$$

The power spectrum is amplified by the square of the frequency response.

In the top frame of Fig. 23 we see the experimentally measured power spectrum of the intercellular voltage in a visual cell. In the lower

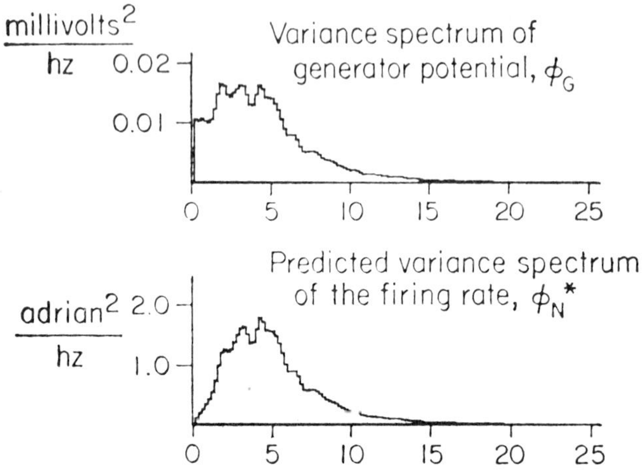

Figure 23. Measured power spectrum of intracellular potential, and predicted power spectrum of firing rate (Shapley, 1971).

frame that power spectrum has been multiplied by the squared voltage-to-firing rate frequency response measured in the same cell. The Fourier transform should yield the autocorrelation of the firing rate and is shown by the solid line in Fig. 24. The points were obtained directly by measuring irregularities in the nerve-impulse train of the same cell under steady light.

In this way, Shapley (1971) has given a complete accounting of the correlation structure of the impulse firing pattern of a sensory neuron.

Now, if we were to follow the nerve signal down the optic nerve to the brain, we would find that at the end of the neuron the signal is converted back to a voltage and then back once more into another impulse train. Following it through the creature's central nervous system, we would find that over and over the message is passed between continuous process and point process, until finally we find ourselves following a motor neuron, which carries us back to the neuromuscular junction.

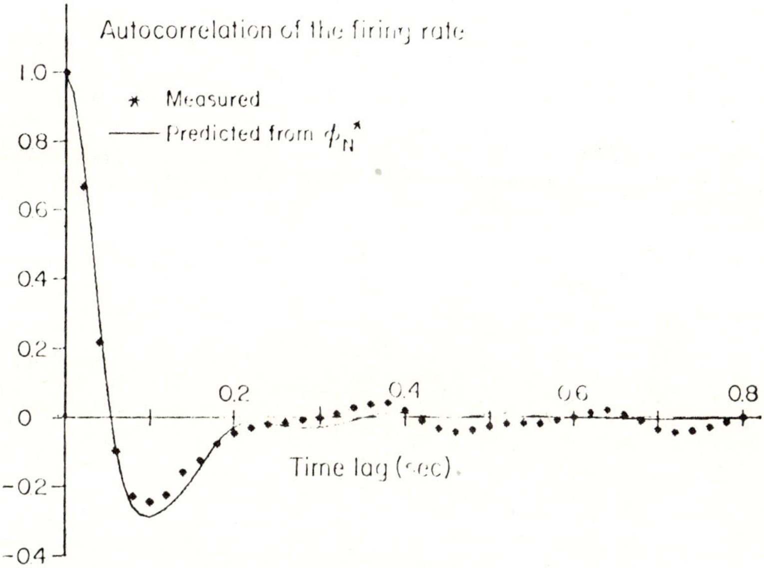

Figure 24. Autocorrelation of the firing rate, predicted and measured (Shapley, 1971).

I would like to say in closing that we are beginning to understand nervous function in quantitative terms, and point processes play an important role in that understanding.

REFERENCES

Adolph, A. (1963). A stochastic model of neuron function and its application to subthreshold processes in the *Limulus* photoreceptor and frog neuromuscular junction. Ph.D. Thesis, Rockefeller University, New York.

Castillo, J. del & Katz, B. (1954). Quantal components of the end plate potential. J. Physiol. 124, 560-73.

Dodge, F. A., Knight, B. W. & Toyoda, J. I. (1968). Voltage Noise in *Limulus* Visual Cells. Science 160, 88-90.

Dodge, F. A., Shapley, R. M. & Knight, B. W. (1970). Linear systems analysis of *Limulus* retina. Behavioral Science 15, 24-36.

Dowling, J. (1968). Discrete potentials in the dark-adapted eye of *Limulus*. Nature 217, 23-31.

Fatt, P. & Katz, B. (1952). Spontaneous subthreshold activity at motor nerve endings. J. Physiol. 117, 109-28.

Hartline, H. K., Wagner, H. G. & Ratliff, F. (1956). Inhibition in the eye of *Limulus*. J. Gen. Physiol. 39, 651-73.

Katz, B. (1966). *Nerve, Muscle, and Synapse*. New York: McGraw-Hill.

Knight, B. W., Toyoda, J. I. & Dodge, F. A. (1970). A quantitative description of the dynamics of excitation and inhibition in the eye of *Limulus*. J. Gen. Physiol. 56, 421-37.

Shapley, R. (1971). Fluctuation in impulse rate in *Limulus* eccentric cells. J. Gen. Physiol. 57, 539-56.

POINT PROCESSES IN EPIDEMIOLOGY

J. Gani
University of Sheffield, England

SUMMARY

This paper surveys recent developments in point processes which arise in the theory of epidemics. After a brief introduction, some new work on chain binomial models is described; this is followed by a report of current research on carrier-borne epidemics. The use of perturbation techniques in the solution of the general stochastic epidemic is then sketched, and some asymptotic results given for it in the case where initial numbers of susceptibles and infectives are very large. Recent results on the cost of epidemics are outlined; finally, a brief account is given of space-time interaction and clustering in the spread of an epidemic.

1. INTRODUCTION

In his basic paper on stochastic point processes, Bartlett (1954) presented some general methods for Markov processes $\{X(t)\}$, applicable to the theory of epidemics. These had previously been introduced in his contribution to the 1949 Symposium on Stochastic Processes held by the Royal Statistical Society in London (Bartlett, 1949).

These methods consisted essentially in the derivation of a symbolic equation

$$\frac{\partial f}{\partial t} = \phi(i\theta, \frac{\partial}{\partial i\theta}, t) f \tag{1}$$

for the characteristic function $f(\theta) = E\{e^{i\theta X(t)}\}$ of the process $X(t)$, with $\frac{\partial}{\partial i\theta}$ operating only on f. The function $\phi(i\theta, x, t)$ is defined as

$$\lim_{\delta t \to 0} E \left\{ \frac{\exp[i\theta\{X(t+\delta t)-X(t)\}]-1}{\delta t} \right\} \tag{1.2}$$

given that X(t) = x. An analogous equation for the probability generating function $P(z) = f(-i \ln z)$ of the process can also be derived which corresponds to the results from the forward Kolmogorov equation. The method can be extended to vector valued Markov processes $\{\underline{X}(t)\}$.

In the context of epidemics, a point process in time t will have associated with it some random vector $\underline{X}(t)$ whose components may be one or more of the following random variables: susceptibles R(t), infectives S(t), carriers Y(t), their locations $\{L(t)\}$ in the plane, and other related random variables such as the cost C(t) of the epidemic. For convenience, epidemic processes are usually assumed to be Markovian.

It is the purpose of this review to outline some of the recent work carried out in mathematical epidemiology. In selecting the material to be discussed, several valuable developments have necessarily had to be omitted. Readers interested in a comprehensive survey up to 1967 should consult Dietz (1967); a broad sketch of current trends in epidemic theory is provided by the 17 communications in the WHO Symposium in Quantitative Epidemiology, Moscow, 23-27 November 1970. (See Adv. Appl. Prob. 3 (1971) 193-228).

In the present paper I have, while attempting to cover a wide range of topics, been guided in my choice largely by personal interests. My hope is that I shall succeed in making those recent developments which have attracted my attention of as much interest to my readers as myself.

The paper consists of six sections: the first is devoted to chain binomial methods and their use in the statistical analysis of measles and hepatitis data. A second considers time dependent results for carrier-borne epidemics, and the use of matrix methods in computing probabilities of their final size. The third surveys the application of perturbation techniques to the general stochastic epidemic, and the estimation of infection and removal parameters in this model on the basis of smallpox data. The fourth section summarizes asymptotic results for the general stochastic epidemic when the initial populations of susceptibles and infectives are both very large. In the fifth, some recent results are outlined on the costs of epidemics; these depend on the stochastic path integral under the infective curve. Finally, a brief account is given of the analysis of space-time interactions in epidemic processes. We now proceed to develop these main themes.

2. CHAIN BINOMIAL METHODS

A simple advance in the discrete time treatment of small scale epidemics has been the reformulation of Greenwood and Reed-Frost chain binomial models as Markov chains. A brief outline of this was first presented by Gani (1969) and later developed in detail by Gani and Jerwood (1971). Consider a discrete time epidemic process in which the latent period of the infection is taken as the unit of time; let the random variable S_t denote the number of infected individuals just prior to time $t = 0,1,\ldots,$ who become infectious at t, while R_t is the remaining number of susceptibles. Clearly $R_t = R_{t+1} + S_{t+1}$.

If $0 < p = 1-q < 1$ is the probability of contact between any 2 individuals, then in the Greenwood model where the number of infectives at time t, when these are non-zero, is assumed not to influence the probability of infection during (t,t+1), we have

$$\text{prob}(R_{t+1}=r_{t+1} \mid R_t=r_t) = \frac{r_t!}{r_{t+1}!(r_t-r_{t+1})!} p^{r_t-r_{t+1}} q^{r_{t+1}} \qquad (t=0,1,\ldots,). \tag{2.1}$$

This clearly indicates the Markovian nature of $\{R_t\}$. For $r_0 = k$, say, the transition probability matrix of the process can be written as

$$r_t \begin{array}{c} \nearrow \\ 0 \\ 1 \\ 2 \\ \vdots \\ k \end{array} \begin{array}{c} \\ \\ \\ \\ \\ \\ \end{array} \overset{\displaystyle r_{t+1}}{\begin{array}{ccccc} 0 & 1 & 2 & \cdots & k \\ \end{array}} \left(\begin{array}{ccccc} 1 & 0 & 0 & \cdots & 0 \\ p & q & 0 & \cdots & 0 \\ p^2 & 2pq & q^2 & \cdots & 0 \\ \cdots & \cdots & \cdots & & \cdots \\ p^k & kp^{k-1}q & \binom{k}{2}p^{k-2}q^2 & & q^k \end{array}\right) = \underline{P} + \underline{Q} \tag{2.2}$$

where $\underset{\sim}{Q} = \text{diag}(1,q,\ldots,q^k)$, and $\underline{P}$ is the remaining matrix with zeros in the diagonal.

The epidemic process is assumed to stop at time $T = t$ when $r_{t-1} = r_t \geq 0$. The probability of this event is given by

$$\text{prob}(T=t) = \underset{\sim}{A}_k' \underline{P}^{t-1} \underset{\sim}{Q}\underline{E}, \tag{2.3}$$

where $\underset{\sim}{A}_k' = (0,0,\ldots,0,1)$ is a $(k+1)$ row vector, and $\underline{E}$ is the $(k+1)$ column vector of unit elements. From (2.3), the p.g.f. of time T to termination of the epidemic is seen to be

$$\underset{\sim}{A}_k' (\underline{I} - \theta\underline{P})^{-1} \theta\underset{\sim}{Q}\underline{E} \qquad (0 \leq \theta \leq 1). \tag{2.4}$$

A similar expression can readily be obtained for the joint p.g.f. of T and the final number of infected cases at the end of the epidemic.

In the case of the Reed-Frost model, the number of infected cases s_t at time t affects infection during the interval $(t,t+1)$, and

$$\text{prob}(S_{t+1} = s_{t+1},\ R_{t+1} = r_t - s_{t+1} \mid S_t = s_t, R_t = r_t)$$

$$= \frac{r_t!}{s_{t+1}!(r_t - s_{t+1})!} (1-q^{s_t})^{s_{t+1}} q^{s_t(r_t - s_{t+1})}. \tag{2.5}$$

From this, it is clear that $\{S_t, R_t\}$ form a bivariate Markov chain. Techniques used in the manipulation of this chain are similar to those outlined earlier, though the matrices are now larger and more complicated.

The Greenwood model may be considered as a Markov chain imbedded in a continuous time pure death process. Though there are some similarities in their characteristics, the Reed-Frost model does not, however, correspond to the Markov chain imbedded in a simple stochastic epidemic. But the reformulation of these chain binomial models as Markov chains enables us to overcome the restrictiveness of the Greenwood and Reed-Frost infection schemes. For example, the Markov chain imbedded in the simple stochastic epidemic (for a single latent period), or in the general stochastic epidemic may serve equally well as suitable models. We may also, in the Greenwood and Reed-Frost type models, use non-homogeneous

Markov chains to simulate changes in the probability p of infectious contact, possibly due to inoculation. An example of this is considered in detail in the final section of Gani and Jerwood's (1971) paper.

At a more practical level, Bailey and Alff-Steinberger (1970) have used information from basic chain binomial models to estimate parameters from an associated continuous time Markov process with infection parameter λ. In this, the instant of infection is followed by a latent period ξ, which is normally distributed $N(\mu,\sigma^2)$, and is succeeded by an infectious period of constant length α. On the basis of Hope Simpson's measles data, assuming either a Greenwood or Reed-Frost model in households of two or three people, Bailey and Alff-Steinberger found that μ,α were of the order of 8 and 7 days respectively.

A similar analysis of Dr. K. Peterson's data on infectious hepatitis assuming a Reed-Frost model yielded values of 16 and 22 days for μ,α. It should be pointed out that in this case, differences in the cut-off point of the data for serial intervals between infections resulted in sizeable differences in the estimates, but a perfectly satisfactory fit is obtained from the model whether the cut-off point is 9-10 or 12-13 days.

3. TIME-DEPENDENT RESULTS FOR CARRIER-BORNE EPIDEMICS

An interesting development in the continuous time stochastic theory of epidemics is Gillian Denton's (1971) time-dependent solution for the carrier-borne infection previously discussed by Downton (1968). Consider a closed population initially consisting of $n \geq 1$ susceptibles and $a \geq 1$ carriers at time $t = 0$. If at time $t \geq 0$ there are $0 \leq r \leq n$ susceptibles and $0 \leq s \leq n + a - r$ carriers, with the remaining $n + a - s - r$ individuals removed from the population, then the transition probabilities for the process in the interval $(t, t+\delta t)$ are given by

$\text{prob}\{(r,s) \to (r-1, s+1)\}$ when a susceptible is infected, and becomes an undetected carrier;

$= \pi rs\, \delta t + o(\delta t)$

$\text{prob}\{(r,s) \to (r-1,s)\}$ when a susceptible is infected, detected and removed;

$= (1-\pi)\, rs\, \delta t + o(\delta t)$

$\text{prob}\{(r,s) \to (r,s-1)\}$ when a carrier is detected and removed.

$= \rho s\, \delta t + o(\delta t)$

Here the infection rate is taken as 1, and the relative carrier removal rate as ρ, while the probability that an infective becomes an undetected carrier is $0 < \pi < 1$.

The probability generating function $P(z,w,t) = \sum_{r,s} p_{rs}(t) z^r w^s$, where $p_{rs}(t)$ is the probability of r susceptibles and s carriers at time $t > 0$, given that these are respectively n and a at $t = 0$, satisfies the second order partial differential equation

$$\frac{\partial P}{\partial t} = w\{w\pi+1-z-\pi\} \frac{\partial^2 P}{\partial z \partial w} + (1-w)\rho \frac{\partial P}{\partial w}, \tag{3.1}$$

subject to the initial condition $P(z,w,0) = z^n w^a$. If we write $P(z,w,t) = \sum_{r=0}^{n} z^r f_r(w,t)$ as in Gani (1967), and take Laplace transforms

$$F_r(w,\theta) = \int_0^\infty e^{-\theta t} f_r(w,t)\,dt, \qquad \text{Re}\ \theta > 0,$$

with respect to time t, (3.1) can be reduced to a set of first order partial differential equations.

These can be written in matrix form as

$$\underline{A}(w) \frac{\partial \underline{F}}{\partial w} + \theta \underline{F} = w^a \underline{E}, \tag{3.2}$$

where $\underline{F}' = \underline{F}'(w,\theta) = \{F_n(w,\theta), \ldots, F_0(w,\theta)\}$, $\underline{E}' = \{1,0,\ldots,0\}$ and $A(w)$ is given by the $(n+1) \times (n+1)$ matrix

$$A(w) = \begin{pmatrix} w(n+\rho)-\rho & & & & \\ -nw(\pi w+1-\pi) & w(n-1+\rho)-\rho & & & \\ & -(n-1)\ w(\pi w+1-\pi) & w(n-2+\rho)-\rho & & \\ & \cdots & \cdots & \cdots & \\ & & & -w(\pi w+1-\pi) & w\rho-\rho \end{pmatrix}$$

The solution to (3.2) is then obtained using Gani's (1967) method; this is

$$\underline{F}(w,\theta) = \sum_{i=0}^{n+a+1} \frac{w^i}{i!} \left(\prod_{j=0}^{i=1} \underline{B}_j\right)_{n+1} \underline{F}(0,\theta) - \sum_{i=a+1}^{n+a+1} \frac{w^i}{i!} \frac{a!}{\rho} \left(\prod_{j=a+1}^{i-1} \underline{B}_j\right)_{n+1} \underline{E}. \tag{3.3}$$

Here the suffix n+1 indicates $(n+1) \times (n+1)$ northwest truncation of a matrix, and $\underline{B}_j$ is the $2(n+1) \times 2(n+1)$ matrix

$$\underline{B}_j = \begin{pmatrix} \{j\underline{A}^{(1)}(0) + \theta\underline{I}\}/\rho & j(j+1)\ \underline{A}^{(2)}(0)/2\rho \\ \underline{I} & 0 \end{pmatrix} \tag{3.4}$$

with $\underline{A}^{(k)}(0) = \partial^k \underline{A}(w)/\partial w^k|_{w=0}$, k=1,2. Moreover $\underline{F}(0,\theta)$ is given by

$$\underline{F}(0,\theta) = \left(\prod_{j=0}^{n+a} \underline{B}_j\right)^{-1}_{n+1} \left\{\frac{a!}{\rho} \left(\prod_{j=a+1}^{n+a} \underline{B}_j\right)_{n+1} \underline{E}\right\}. \tag{3.5}$$

It should be noted that all products $\prod \underline{B}_j$ are taken from left to right in strict decreasing order of j, and $\prod_{k}^{k-1} \underline{B}_j = \underline{I}$ by definition.

If the probability of an epidemic of total size n-r, not counting the original a carriers, is denoted by P_{n-r} $(0 \leq r \leq n)$, then it follows that

$$P_{n-r} = \lim_{t\to\infty} p_{r0}(t) = \lim_{\theta\to 0} \theta F_r(0,\theta) = \lim_{\theta\to 0} \theta\{\underline{F}(0,\theta)\}_{n-r},$$

a result readily obtained from (3.5). In carrying out the necessary calculations, Gillian Denton discovered that if $\underline{P}(n,a)$ denotes the

$$\underline{P}(n,a) = \lim_{\theta\to 0} \frac{a!\theta}{\rho} \left(\prod_{j=0}^{n+a} \underline{B}_j\right)^{-1}_{n+1} \left(\prod_{j=a+1}^{n+a} \underline{B}_j\right)_{n+1},$$

which, with the exception of the final vector $\underline{E}$ is $\lim_{\theta\to 0} \theta \underline{F}(0,\theta)$, then

$$\underline{P}(n,a) = \{\underline{P}(n,1)\}^a. \tag{3.6}$$

Further, the first, second,..., (n+1)th columns of $\underline{P}(n,a)$ correspond to the vectors of probabilities of the total size of the epidemic for a initial carriers and n, n-1,...,0 initial susceptibles respectively.

As is pointed out in the paper, these results lend considerable power to the matrix method used, particularly if one is concerned with computing values of the probabilities of the epidemic size for increasing values of the initial susceptible population up to n. These can be obtained straightforwardly from a knowledge of the matrices $\underline{B}_j$ in (3.4). The result (3.6) also holds, with the obvious minor modifications in the $\underline{B}_j$, for the case of the general epidemic considered by Gani (1967).

4. THE GENERAL STOCHASTIC EPIDEMIC

Current studies of the general stochastic epidemic have been concerned with theoretical developments involving the use of perturbation techniques, as well as more practical methods for the estimation of parameters relying on electronic computation.

An earlier application of perturbation techniques was made by Bailey (1968) to the simple stochastic epidemic involving only susceptibles and infectives. This gave asymptotically valid approximations for a large population of size N. After a change of time variable T = Nt, the m.g.f. $M_d(\theta,T) = e^{\theta\xi(T)}$ of the proportion $\xi(T)$ of susceptibles at time T in the deterministic case is known to satisfy the partial differential equation

$$\frac{\partial M_d}{\partial T} = \theta\left(\frac{\partial^2 M_d}{\partial\theta^2} - \frac{\partial M_d}{\partial\theta}\right), \tag{4.1}$$

the infection rate being taken as unity. The initial condition is

$M_d(\theta,0) = e^{\theta/(1+\gamma)}$, where $\gamma = (N-n)/n$ is the initial ratio of infectives to susceptibles.

In the stochastic model the equivalent equation for the m.g.f. is

$$\frac{\partial M}{\partial T} = N(1-e^{-\theta/N})\left(\frac{\partial^2 M}{\partial \theta^2} - \frac{\partial M}{\partial \theta}\right).$$

Approximating to the first order in N^{-1}, this reduces to

$$\frac{\partial M}{\partial T} = \theta\left(1-\frac{\theta}{2N}\right)\left(\frac{\partial^2 M}{\partial \theta^2} - \frac{\partial M}{\partial \theta}\right), \tag{4.2}$$

with the same initial condition $M(\theta,0) = e^{\theta/(1+\gamma)} = e^{n\theta/N}$ as before. Using only the first order perturbation on the corresponding deterministic process, Bailey was able through an eigenfunction approach to obtain useful approximations for the mean, variance and epidemic curve of the simple stochastic epidemic.

Recently Weiss (1971) has suggested an alternative perturbation method depending on the moments $\nu_r = E(\xi^r)$ of the proportion ξ of susceptibles. These satisfy the differential difference equations

$$\frac{d\nu_r}{dT} = r(\nu_{r+1}-\nu_r) + N^{-1}\sum_{j=0}^{r-2}(-1)^j\binom{r}{j+2}N^{-j}(\nu_{r-j-1}-\nu_{r-j}). \tag{4.3}$$

If ν_r is expanded in powers of N^{-1} as

$$\nu_r = \nu_r^{(0)} + \nu_r^{(1)}N^{-1} + \nu_r^{(2)}N^{-2} + \ldots, \tag{4.4}$$

and the first two terms substituted in (4.3), a simple set of differential difference equations is derived which can readily be solved. An identical technique is applicable to the general stochastic epidemic, where the population now consists of susceptibles, infectives and removals, i.e., individuals who become immune or die. In Daniels' (1971) application of perturbation techniques to the general stochastic epidemic, a slightly different approach through the cumulant generating function is used. It is shown that the first approximations to the three second order cumulants of the process satisfy the set of differential equations

$$\frac{d}{dt}\begin{pmatrix} K_{20}^{(1)} \\ 2K_{11}^{(1)} \\ K_{02}^{(1)} \end{pmatrix} = \begin{pmatrix} xy \\ -2xy \\ (x+\rho)y \end{pmatrix} + 2\begin{pmatrix} -y & -x & 0 \\ y & x-y-\rho & x \\ 0 & y & x-\rho \end{pmatrix}\begin{pmatrix} K_{20}^{(1)} \\ 2K_{11}^{(1)} \\ K_{02}^{(1)} \end{pmatrix} \tag{4.5}$$

Here, $x = K_{10}^{(1)}$, $y = K_{01}^{(1)}$ are the first order cumulants, which are the solutions of the deterministic equations for the susceptibles and infectives in the process, ρ being the relative removal rate. While (4.5) cannot be solved explicitly, numerical approximations are obtainable.

The validity of the general stochastic epidemic model has sometimes been called into question. Thus, an analysis of data and estimation of parameters for such a model would prove extremely valuable. In a recent paper, Bailey and Thomas (1971) have made extensive use of an IBM system 360 (Model 40) computer to analyze data from a smallpox epidemic of 30 cases in a community of 120 in SE Nigeria. After a detailed discussion of the likelihood functions based on periods between successive removals, and the number of such removals, the ML estimators of the infection and removal rates were calculated. These are respectively

$$\hat{\beta} = .00168 \pm .00047, \qquad \hat{\gamma} = .162 \pm .050, \tag{4.6}$$

so that the relative removal rate was found to be $\hat{\rho} = 97 \pm 22$.

To cut down the rather lengthy time taken for an exact determination of the likelihood function, a gamma type approximation was used for the distribution of periods between the (u-1)th and u-th removals, when there were s remaining infectives, given the past history of the process. Computation using this approximate method was then speeded up by a factor of 18, without seriously affecting the accuracy of the estimates. Further work is to be done to establish the value of the approximation for larger bodies of data.

5. ASYMPTOTIC RESULTS FOR THE GENERAL EPIDEMIC

In some recent work, Nagaev and Startsev (1970) have obtained some interesting new asymptotic results for the general stochastic epidemic.

Let n and m respectively be the initial susceptibles and infectives in such a model. Then, the transition probabilities of the associated imbedded Markov chain are known to be

$$\text{prob}\{(r,s) \to (r+1,s)\} = \frac{n-r}{\rho+n-r} = p_r ,$$
$$\text{prob}\{(r,s) \to (r,s+1)\} = \frac{\rho}{\rho+n-r} = q_r , \tag{5.1}$$

where r,s are infectives and removals respectively, ρ being the relative removal rate. The absorbing boundary for this Markov chain is the straight line s = m+r.

If ν_n and T_n are the final size and duration of the epidemic starting with n initial susceptibles, it is difficult to obtain explicit results for their probabilities directly from the Markov chain formulation. Clearly, if we define $\{X_j\}$ as independent random variables, geometrically distributed with probabilities $\{p_j\}$ of success, and $S_k = \sum_{j=0}^{k} X_j$, then

$$\text{prob}\{\nu_n > k\} = \text{prob}\{S_0 < m, S_1 < m+1, \ldots, S_k < m+k\}. \tag{5.2}$$

But though (5.2) is a well understood probability, its explicit expression is not easily obtained. Nagaev and Startsev have, however, succeeded in deriving several asymptotic results for it when n and m are large; these results depend on the relation of m and n as they both tend to infinity. Typical of these are the following:

Theorem 1. The probability of the final size of a general stochastic epidemic is given asymptotically, for large n and m, such that m = o(n) where a δ exists for which $0 < \delta \le \frac{\rho}{n} \le 1-\delta$ or $1+\delta \le \frac{\rho}{n} = o(m)$, by

$$\text{prob}(\nu_n > a_n - xb_n) = \Phi(x)\{1+o(1)\}, \tag{5.3}$$

where

$$a_n = m/(\tfrac{\rho}{n} - 1), \qquad b_n = \{m(\tfrac{\rho}{n} + \tfrac{\rho^2}{n^2})/(\tfrac{\rho}{n} - 1)^3\}^{1/2} \qquad \text{if } \tfrac{\rho}{n} > 1,$$

and

$$a_n = n\alpha , \qquad b_n = \frac{\{n\alpha(1+\rho^2/n^2)/(1-\alpha)\}^{1/2}}{\frac{\rho}{n}(1-\alpha)-1} \qquad \text{if } \tfrac{\rho}{n} < 1.$$

Further, in (5.3) α denotes the solution of $\alpha + \frac{\rho}{n}\ \ell n(1-\alpha) = 0$, and $\Phi(x)$ is the standard normal distribution function.

Theorem 2. Let $|\beta| = |\ m(1-\frac{\rho}{n})\ | \to \infty$, where $\beta = o(m)$ and $n/m^3 = o(\beta^{-2})$, and put $a_n = (2nm)^{\frac{1}{2}}$, $b_n = (2n^3/m)^{\frac{1}{4}}$. Then in this case the probability of the final size of the epidemic is given asymptotically by

$$\text{prob}(\nu_n \geq a_n - xb_m) = \Phi(x)\{1+o(1)\}. \tag{5.4}$$

Theorem 3. Let $\frac{\rho}{mn} \geq \delta > 0$, then the probability of the final size of the epidemic is asymptotically of the Poisson form

$$\text{prob}(\nu_n = k) = \{\frac{(nm/\rho)^k}{k!} \exp(-nm/\rho)\} \ \{1+o(1)\}. \tag{5.5}$$

These results, quite apart from their analytic interest, are clearly useful in characterizing the behavior of the general stochastic epidemic in large closed populations with big initial numbers of both susceptibles and infectives.

6. THE COST OF EPIDEMICS

A different problem initially broached by Becker (1970) and developed by Jerwood (1970) in a recent note and his Ph.D. thesis (Jerwood, 1971), is the cost of epidemics. Jerwood specifies the cost of a simple stochastic epidemic in a total population of N individuals with i initial infectives, as

$$C_i = aW_i + bT_i, \tag{6.1}$$

where $a,b > 0$ are constants. In this, T_i is the duration of the epidemic and $W_i = \int_0^{T_i} S(t)dt$ the area under the stochastic path between 0 and T_i traced by the number of infectives $S(t)$ at time t. The definition (6.1), a linear combination of the duration of the epidemic and the total man-hours lost by infectives, seems realistic though possibly too simple. Using the method due to McNeil (1970), the Laplace-Stieltjes transform of the cost is found to be

$$C_i^*(\theta) = \prod_{j=1}^{N-i} [1 + \frac{\theta\{a(N-j)+b\}}{j(N-j)}]^{-1}, \qquad \text{Re } \theta \geq 0, \tag{6.2}$$

where the infection rate is taken as 1 for simplicity.

For the general stochastic epidemic, also starting from i initial infectives and s susceptibles in a total population N, the cost may be written as

$$C_{is} = \int_0^{T_{is}} \{aS(t) + b\}dt = aW_{is} + bT_{is}, \tag{6.3}$$

where a,b > 0 are again constants, S(t) represents the number of infectives at time t, and T_{is}, W_{is} are respectively the duration time of the epidemic and the area $\int_0^{T_{is}} S(t)dt$ under the stochastic infective path between 0 and T_{is}.

If we define the joint Laplace-Stieltjes transform $\phi_{is}(\theta_1,\theta_2)$ of $\{W_{is},T_{is}\}$ as

$$\phi_{is}(\theta_1,\theta_2) = E(e^{-\theta_1 W_{is} - \theta_2 T_{is}}), \qquad \text{Re } \theta_1,\theta_2 \geq 0, \tag{6.4}$$

then we can, again using McNeil's (1970) technique, show that these satisfy the bivariate difference equation

$$\alpha_{is}\phi_{is} = \rho\phi_{i-1,s} + s\,\phi_{i+1,s-1}. \tag{6.5}$$

Here ρ is the relative removal rate, and $\alpha_{is}(\theta_1,\theta_2) = (\theta_1 + \theta_2 i^{-1} + s + \rho)$. We obtain formally that

$$\phi_{is} = \prod_{j=1}^{i} \frac{\rho}{\alpha_{is}} + \frac{s}{\rho} \sum_{k=1}^{i} \phi_{k+1,s-1}\left(\prod_{j=k}^{i} \frac{\rho}{\alpha_{js}}\right) \tag{6.6}$$

$$(i = 1,\ldots, N-s;\ s = 0, 1,\ldots,N-1).$$

Jerwood has examined in some detail the distributions of $\{W_{is},T_{is}\}$, their expectations $M_{is} = E(W_{is})$, $N_{is} = E(T_{is})$, and hence the expectation $E(C_{is})$ of the cost. He has found that $\{M_{is},N_{is}\}$ satisfy the respective difference equations

$$(s+\rho)M_{is} = 1+\rho M_{i-1,s} + sM_{i+1,s-1}, \tag{6.7}$$

$$(s+\rho)N_{is} = i^{-1} + \rho N_{i-1,s} + sN_{i+1,s-1}.$$

The first equation can be solved to give results of the form

$$M_{is} = \frac{i+s}{\rho} - \sum_{j=1}^{s} \binom{s}{j} \left(\frac{\rho}{\rho+j}\right)^{i+s-j} \beta_j, \tag{6.8}$$

where the β_j are themselves positive solutions of the equations

$$\sum_{j=1}^{s} \binom{s}{j} \left(\frac{\rho}{\rho+j}\right)^{s-j} \beta_j = \frac{s}{\rho} \qquad (s = 1,2,\ldots,N).$$

The $\{M_{is}\}$ are shown to have the simple bounds

$$\frac{i}{\rho} \leq M_{is} \leq \frac{i+s}{\rho}, \tag{6.9}$$

and graphical information computed for the behavior of $\{M_{is}\}$ when $i = 1, 4, 7;\ s = 3$, has demonstrated that these bounds are reasonably good.

The equation for the $\{N_{is}\}$ yields

$$N_{is} = \frac{1}{s+\rho} \sum_{k=1}^{i} \left(\frac{\rho}{s+\rho}\right)^{i-k} \left(\frac{1}{k} + sN_{k+1,s-1}\right), \tag{6.10}$$

where $N_{io} = \sum_{k=1}^{i} k^{-1}/\rho$. Hence they can be evaluated recursively without difficulty.

A large number of interesting problems remain unsolved in this area; several of these are currently under investigation.

7. SPACE-TIME INTERACTIONS IN EPIDEMICS

A problem of great importance in epidemiology has been the identification of clustering among reported cases of a disease. Knox (1964 a,b) investigated methods for deciding whether leukaemia cases occurring at small distances from each other also occurred close in time; such space-time clustering would support the theory that the disease was contagious. His method was later improved upon by Barton and David (1966); a further extension by Pike and Smith (1968) allowed for limited periods of susceptibility and infectiousness.

Barton and David specified Knox's procedure in graph-theoretic terms.

Let X be the number of pairs of cases from among the n reported which are found to be

(i) living within a fixed distance δ from each other

(ii) infected within a fixed time period τ of each other.

Adjacency matrices S and T of size $n \times n$ may be written such that $S_{ij}=1$ if case j is at a distance no greater than δ from case i and is 0 otherwise, while $T_{ij} = 1$ if case j occurs within a period τ of i and is 0 otherwise. By definition, $S_{jj} = T_{jj} = 0$. The measure X of space-time clustering is then given by

$$X = \frac{1}{2} \sum_i \sum_j S_{ij} T_{ij}. \tag{7.1}$$

This integer is found to be distributed approximately as a Poisson variable and can consequently be used as a test statistic. Details of other graph-theoretic structures of use in epidemiology may be found in Tautu (1971).

In Knox's method the space and time measures related to cases i and j are either 0 or 1. But S_{ij}, T_{ij} need not necessarily be restricted to these values; any reasonable measure of contact in space and time may be used. The variable X will then remain an indicator of space-time clustering; its first two moments in this general case have been obtained by Barton and David (1966). Mantel (1967) later developed a generalized regression approach to the clustering problem of which the Knox and Barton and David results are special cases. Here a statistic

$$Z = \sum_i \sum_j S_{ij} R_{ij} \tag{7.2}$$

is defined where S_{ij}, R_{ij} are respectively functions of the space and time coordinates of cases i, j. Testing of the statistic is carried out by Monte Carlo methods, or in some cases where it appears justifiable, by the use of an approximate normal distribution for $U = \{Z-E(Z)\}/\sigma_Z$.

In a recent paper Klauber (1971) considers an extension of Mantel's method to test clustering between two sets of points. His methods are to be applied to the diagnosis of lymphatic leukaemia cases in both man and pet cats. Let $\{X_i, Y_i, T_i\}$ $(i = 1,2,\ldots,I)$ and $\{X_j^*, Y_j^*, T_j^*\}$ $(j = 1,2,\ldots,J)$ represent the space and time coordinates of two sets of observations, say on men and cats respectively. Then we may define the space and time distances between case i among men and case j among cats by

$$S_{ij} = S[\{(X_i - X_j^*)^2 + (Y_i - Y_j^*)^2\}^{\frac{1}{2}}],$$
$$R_{ij} = R(|T_i - T_j^*|), \tag{7.3}$$

where S, R are given functions, for example the reciprocals of the distances between cases, plus some adjustable constant.

The statistic (7.2) may still be used; it can be compared to its randomization distribution by assuming one set of points fixed, and the coordinates of the other randomly permuted, or both sets may be taken as random. The mean and variance of Z can be found for both randomization models, and in some instances the approximate standard normal variable $U = \{Z - E(Z)\}/\sigma_Z$ may again be used as the test statistic. As might be expected, this approximation improves for larger samples. Two examples are given of tests based on empirical data; one of them is concerned with lymphoma cases for 117 cats and 93 dogs in California, where no statistical significance at a reasonable level was found for clustering.

It will be apparent from this brief sketch of current work in epidemiology that the field is in a state of rapid development. The intrinsic interest of the problems considered, and the liveliness of research in this area will, I hope, encourage more probabilists and statisticians to contribute to it in future.

REFERENCES

Bailey, N. T. J. (1968). A perturbation approximation to the simple stochastic epidemic in a large population. Biometrika 55, 199-209.

Bailey, N. T. J. & Alff-Steinberger, C. (1970). Improvements in the estimation of the latent and infectious periods of a contagious disease. Biometrika 57, 141-53.

Bailey, N. T. J. & Thomas, A. S. (1971). The estimation of parameters from population data on the general stochastic epidemic. Theoret. Pop. Biol. 2, 257-70.

Bartlett, M. S. (1949). Some evolutionary stochastic processes. J. R. Statist. Soc. B 11, 211-29.

Bartlett, M. S. (1954). Processus stochastiques ponctuels. Ann. Inst. H. Poincaré 14, 35-60.

Barton, D. E. & David, F. N. (1966). The random intersection of two graphs. In *Research Papers in Statistics*, David, F. N. ed. New York: Wiley.

Becker, N. (1970). Mathematical models in epidemiology and related fields. Ph.D. thesis, Sheffield University.

Daniels, H. E. (1971). A note on perturbation techniques for epidemics. WHO symposium on Quantitative Epidemiology, Moscow, 23-27 November 1970. Adv. Appl. Prob. 3, 214-8.

Denton, Gillian M. (1971). On the time-dependent solution of Downton's carrier-borne epidemic. Manchester-Sheffield School of Probability and Statistics Research Report, July 1971.

Dietz, K. (1967). Epidemics and rumours — A Survey. J. R. Statist. Soc. A 130, 505-28.

Downton, F. (1968). The ultimate size of carrier-borne epidemics. Biometrika 55, 277-89.

Gani, J. (1967). On the general stochastic epidemic. Proc. 5th Berkeley Symp. 4, 271-9.

Gani, J. (1969). A chain binomial study of inoculation in epidemics. Bull. I.S.I. 43 (2), 203-4.

Gani, J. & Jerwood, D. (1971). Markov chain methods in chain binomial epidemic models. Biometrics 27, 591-603.

Jerwood, D. (1970). A note on the cost of the simple epidemic. J. Appl. Prob. 7, 440-3.

Jerwood, D. (1971). Cost of epidemics. Ph.D. thesis, Sheffield University.

Klauber, M. R. (1971). Two-sample randomization tests for space-time clustering. Biometrics 27, 129-42.

Knox, G. (1964a). Epidemiology of childhood leukaemia in Northumberland and Durham. Brit. J. Prev. Med. 18, 17-24.

Knox, G. (1964b). The detection of space-time interactions. Appl. Statist. 13, 25-9.

Mantel, N. (1967). The detection of disease clustering and a generalized regression approach. Cancer Res. 27, 209-20.

McNeil, D. R. (1970). Integral functionals of birth and death processes and related limiting distributions. Ann. Math. Statist. 41, 480-5.

Nagaev, A. N. and Startsev, A. N. (1970). Asymptotic analysis of a stochastic epidemic model. Teor. Veroyatnost.i Primenen. 15, 97-105.

Pike, M. C. & Smith, P.G. (1968). Disease clustering: a generalization of Knox's approach to the detection of space-time interactions. Biometrics 24, 541-56.

Tautu, P. (1971). Structural models in epidemiology: an introductory investigation. WHO Symposium on Quantitative Epidemiology, Moscow, 23-27 November 1970. Adv. Appl. Prob. 3, 196-8.

Weiss, G. H. (1971). On a perturbation method for the theory of epidemics. WHO Symposium on Quantitative Epidemiology, Moscow, 23-27 November 1970. Adv. Appl. Probl. 3, 218-20.

POINT PROCESS PROBLEMS IN RELIABILITY

D. P. Gaver
Naval Postgraduate School, Monterey, California

SUMMARY

This paper has three parts. First, there is a brief comment of certain "classical" reliability investigations that are particularly related to point stochastic processes. Second, a class of realiability-motivated models that yield families of correlated skew distributions are discussed. These "accumulation models" are suggested by the observation that times to complex system failure and the subsequent repair times should often be correlated, and not independent as has usually been assumed. Mixture arguments are then shown to generate negatively correlated variables; in particular, negatively correlated exponentials. Third, we briefly mention certain reliability problems that arise in an urban context: examples are models of traffic jams and of electric power generation availability.

1. INTRODUCTION

The classical Theory of Reliability developed in response to the revelation that complex systems of revolutionary design, giving promise of extraordinary capability, are likely to be frustratingly fallible in practice. What is more, the occurrence of failures is often highly unpredictable, sometimes even - pessimistically - related in an inverse fashion to preventive maintenance actions. Likewise, the duration of subsequent failure diagnosis and repair varies for a multiplicity of reasons, among them being availability of skilled personnel and the necessary tools and spare parts. Since failure occurrence, and repair termination, are often thought of as point events, it is not at all unnatural that reliabilists should have early embraced point process-like models, and contributed largely to the development of the theory of point stochastic processes. Some examples of point process models in reliability now follow.

2. CLASSICAL RELIABILITY

The classical theory of reliability deals largely with entities, e.g. machines, that may be in one of two states with respect to capability of carrying out an assigned task. These states may be termed the "up" or available state and the "down" or failed state. Classical reliability mathematics deal - and deals - extensively with the following topics, to mention but a few.

2.1. Distributions of Times to Failure

Barlow and Proschan (1965) and in later publications, devote thorough attention to this topic, particularly in the univariate case. See also Marshall and Olkin (1967) and Downton (1970), for multivariate models. Esary, Marshall and other of the "Seattle School" elaborate upon implications of assumptions about the failure or hazard rate: i.e., that it possess the IFR (increasing failure rate), DFR (decreasing failure rate), or other properties of plausible physical origin. The original papers should be consulted for details (see Barlow and Proschan, 1965, and Esary, Marshall and Proschan, 1970, for references).

2.2 Wear Processes, Shock Models and Fatigue

Attempts to relate the time to system failure to randomly varying environmental conditions, e.g. shocks or blows timed by a point process stochastic process, have been made by several authors: Epstein (1958), Mercer (1961), Gaver (1963), Esary, Marshall and Proschan (1970), Reynolds and Savage (1970) and Downton (1970), to mention a few. Fatigue models have been treated by Birnbaum and Saunders (1969). A more complicated situation in computer failures is treated by Lewis (1969).

2.3. Redundancy and Repairman Problems

Many actual systems feature some redundancy, i.e., have extra components or elements that replace others that have failed. If no repairs to the spares are possible, and the latter fail sequentially then Barlow and Proschan (1965) give bounds for the probability of system time to failure, assuming IFR properties for the time to failure distribution. Reliability under similar conditions, but with a random environment influencing failure rates, has been treated by Gaver (1963). Practical questions, largely untreated, relate to the allocation of a fixed budget (space, money, or both) among spares or redundant elements for competing applications.

If repair or anticipatory replacement is possible various new problems concerning queueing and decision theory arise; again see Barlow and Proschan (1965) and also Flehinger (1958), Cox (1962), Jorgenson, McCall and Radner (1967), and others. In the main this work is probabilistic rather than statistical, in that relevant distributions are assumed known. Often the convenient "independently and identically distributed" assumptions are inappropriate, particularly where reliability growth may occur; see Barlow and Scheuer (1966). With respect to queueing for repair, the long-run availability of simple systems can sometimes be estimated by use of classical methods. Frequently, however, long-run results are of less interest than are first-passage results: supposing that two (or more) components are available to carry out the same basic function, and when one fails the other is switched in while the repair is in process, then what is the time until both (or all) components are simultaneously on repair? Under reasonable conditions on individual failure times, the time to system (all component) failures will be approximately exponentially distributed, as discussed by Gaver (1963), Harris (1968) and Gnedenko et al. (1969). Mazumdar (1970) has considered this problem type with the added twist of an inspection.

3. SYSTEM RELIABILITY MODELS AND THE GENESIS OF INTERDEPENDENT FAILURE AND REPAIR TIMES

As has been mentioned, many of the usual models for failure and repair assume the independence of times to failure, and the subsequent times to repair. Situations for which such assumptions are *not* plausible can be identified; some of these will now be reviewed in summary form; see Gaver and Luckew (1971) for details.

3.1. Accumulation Models: The Basic Setup

Imagine that a complex system contains many failure-prone or degradation-prone subsystems which are not always truly essential to system operation. Supposing that the system is initially operative, let it continue to operate until a vital subsystem fails, necessitating immediate repair. We shall term this latter failure the *killing* or *emergency* event, and denote it by E. Before E occurs there may take place one or more secondary, non-killing (possibly degrading) events, i.e., failures of subsystems that are not currently required. When E eventually takes place, and repair must begin, then not only is the killing failure repaired, but so are all of the accumulated non-killing failures: hence the name *accumulation models*. If chance long postpones E, many repairs will clearly accumulate, so time to system failure and the subsequent system repair time tend to be positively related. One is motivated to make the above, and similar, assumptions by everyday experience with automobiles, or possibly with medical or dental situations.

3.2. Particular Models

Let $\underline{U}$ denote the duration of time from the moment at which a failure-free system begins operation until failure (event E) occurs; $\underline{R}$ is then the subsequent repair or down time. The following models yield correlated $(\underline{U},\underline{R})$ pairs of simple form.

Model 1. Exponential $\underline{U}$, Renewal-generated Secondary Failures, Additive Repairs. Consider the situation in which

$$\text{prob}\{\underline{U} > x\} = e^{-\lambda x} \qquad (\lambda > 0,\ x \geq 0). \tag{3.1}$$

Initially the system is failure-free, and a renewal process generating secondary events begins, to be terminated at time $\underline{U}$. If the independent inter-event times have distribution (d.f.) $G(\cdot)$, then, conditional upon $\underline{U}$, the number of secondary failures N has d.f. given by

$$\text{prob}\{N(\underline{U}) = n|\underline{U}\} = G^{n*}(\underline{U}) - G^{(n+1)*}(\underline{U}), \tag{3.2}$$

where n* represents n-fold Stieltjes convolution of distributions, and $G^{0*}(x)$ is the unit step at the origin. If the time to repair, $\underline{R}$, is the *sum* of independently and identically distributed (iid) repair times $R_n, n = 1,2,\ldots,$ with d.f. $L(\cdot)$, then

$$\underline{R} = R_1 + R_2 + \ldots + R_{N(\underline{U})+1}, \tag{3.3}$$

since the number of repairs to be made include the catastrophic failure. Then

$$\text{prob}\{\underline{R} \leq y, N(\underline{U}) = n|\underline{U}\} = L^{(n+1)*}(y)\{G^{n*}(\underline{U}) - G^{(n+1)*}(\underline{U})\} \tag{3.4}$$

and hence

$$\text{prob}\{\underline{U} \leq x, N(\underline{U}) = n, \underline{R} \leq y\}$$

$$= \int_0^x e^{-\lambda z} L^{(n+1)*}(y)\{G^{n*}(z) - G^{(n+1)*}(z)\}dz. \tag{3.5}$$

Introduction of Laplace-Stieltjes transforms, denoted by ^ , and the generating function on N furnishes joint transforms:

$$E\{e^{-\xi\underline{U}}e^{-\eta\underline{R}}u^{N(\underline{U})}\} = \frac{\lambda}{\lambda+\xi}\left[\frac{\{1-\hat{G}(\xi+\lambda)\}\hat{L}(\eta)}{1-u\hat{G}(\xi+\lambda)\hat{L}(\eta)}\right]; \tag{3.6}$$

the above converges at least for $\xi,\eta \geq 0$, and $0 \leq u \leq 1$. Although this transform is rather uninformative as it stands, it may be differentiated to provide moments, or even inverted numerically. For example, denoting differentiation by prime,

$$E(\underline{U}) = \frac{1}{\lambda}, \quad E(\underline{R}) = \frac{E(R_1)}{1-\hat{G}(\lambda)};$$

$$\text{var}(\underline{U}) = \frac{1}{\lambda^2}, \ \text{var}(\underline{R}) = \text{var}(R_1)\frac{1}{1-\hat{G}(\lambda)} = \{E(R_1)\}^2 \frac{\hat{G}(\lambda)}{\{1-\hat{G}(\lambda)\}^2}; \tag{3.7}$$

$$\text{cov}(\underline{R},\underline{U}) = -\frac{E(R_1)\hat{G}'(\lambda)}{\{1-\hat{G}(\lambda)\}^2} > 0.$$

Conditional moments are also of interest; for instance conventional renewal theory shows that

$$E(\underline{R}|\underline{U}) = E(R_1)\sum_{n=0}^{\infty} G^{n*}(\underline{U}), \tag{3.8}$$

since

$$E\{N(\underline{U})|\underline{U}\} = \sum_{n=0}^{\infty} G^{n*}(\underline{U}). \tag{3.9}$$

The transform

$$\int_0^{\infty} e^{-\xi z} E(\underline{R}|\underline{U}=z)\,dz = E(R_1)\sum_{n=0}^{\infty}\int_0^{\infty} e^{-\xi z} G^{n*}(z)\,dz$$

$$= \frac{E(R_1)}{\xi\{1-\hat{G}(\xi)\}} \tag{3.10}$$

reveals conditions under which the regression function (3.8) of $\underline{R}$ on $\underline{U}$ is linear. The left-hand side of (3.8) is linear iff it has transform $C_1/\xi^2 + C_2/\xi$; if this is equated to the right-hand side of (3.10) it may be shown that

$$\hat{G}(\xi) = \frac{1}{1+\xi}, \tag{3.11}$$

provided the secondary process inter-event time mean is adjusted to be unity; constants C_1 and C_2 are identified with the aid of Abelian theorems. Consequently it appears that *the regression of* $\underline{R}$ *on* $\underline{U}$ *is linear* iff G *is the exponential d.f.*, so that N is a Poisson process. In general, asymptotic renewal theory shows that this regression is nearly linear for large $\underline{U}$, provided G has a finite expectation. For the present model no simple expression for $E(\underline{U}|\underline{R})$ can be derived.

Model 1-A: Exponential $\underline{U}$, *Poisson Secondary Failures, Additive Exponential Repairs.* Simplification of (3.6) under the above conditions yields

$$E(e^{-\xi\underline{U}}e^{-\eta\underline{R}}) = \frac{-\lambda\mu}{(1+\lambda+\xi)(\mu+\eta)-\mu} . \tag{3.12}$$

Setting $\xi = 0$ reveals that

$$E(e^{-\eta\underline{R}}) = \frac{1}{1 + \frac{1+\lambda}{\lambda\mu}\eta} , \tag{3.13}$$

so $\underline{R}$ itself has an exponential distribution, as is clear from the fact that it is a geometrically compounded sum of exponential random variables. Hence $\underline{U}$ and $\underline{R}$ form a pair of random variables with a joint (or bivariate) exponential distribution. Expressions (3.12) describes a family of bivariate exponential distributions quite different from those treated previously by Marshall and Olkin (1967). It coincides with the distribution derived by Downton (1970), who was also motivated by (somewhat different) reliability considerations.

The density of the bivariate exponential described by (3.12) can be expressed in closed form. We have

$$f_{\underline{U},\underline{R}}(x,y) = \sum_{n=0}^{\infty} e^{-\lambda x}e^{-x}\frac{x^n}{n!}e^{-\mu y}(\frac{\mu y}{n!})^n \mu$$

$$= \exp[-\{(\lambda+1)x+\mu y\}]I_0\{2(\mu xy)^{\frac{1}{2}}\}\lambda\mu . \tag{3.14}$$

$I_0(\cdot)$ is the Bessel function of imaginary argument. A similar sort of mixing argument is utilized in Feller (1966) to generate a univariate d.f.. The above distribution has been previously written down by several investigators.

Differentiation of the transform now delivers

$$\text{cov}(\underline{U},\underline{R}) = \frac{1}{\lambda^2\mu} , \text{ and } \quad \text{corr}(\underline{U},\underline{R}) = \frac{1}{1+\lambda} , \tag{3.15}$$

both positive. Although the conditional distribution of $\underline{R}$, given $\underline{U}$, still involves the Bessel function the regression is linear:

$$E(\underline{R}|\underline{U}) = (1+\underline{U}] \frac{1}{\mu} . \tag{3.16}$$

For this model, the regression of $\underline{U}$ on $\underline{R}$ is also linear, for

$$E(\underline{U}|\underline{R}) = \frac{1}{1+\lambda} (1 + \frac{\mu R}{1+\lambda}) . \tag{3.17}$$

Model 2. Arbitrary $\underline{U}$, Poisson Secondary Failures, Additive Repairs. If $\underline{U}$ is given an arbitrary distribution F, while G is made exponential (and hence N is Poisson), relatively simple expressions again appear, and are recorded below.

$$E(e^{-\xi\underline{U}} e^{-\eta\underline{R}}) = \int_0^\infty \int_0^\infty \sum_{n=0}^\infty e^{-\xi z} F\{dz\} e^{-z} \frac{z^n}{n!} e^{-\eta y} L^{(n+1)*}(dy)$$

$$= \hat{L}(\eta)\hat{F}\{\xi + 1 - \hat{L}(\eta)\}. \tag{3.18}$$

The moments of $\underline{U}$ are those of F, while

$$E(\underline{R}) = E(R_1)\{1+E(\underline{U})\},$$

$$E(\underline{R}|\underline{U}) = E(R_1)(1+\underline{U}), \tag{3.19}$$

$$\text{var}(\underline{R}) = \text{var}(\underline{U})\{E(R_1)\}^2 + E(\underline{U})E(R_1^2) + \text{var}(R_1),$$

$$\text{cov}(\underline{R},\underline{U}) = E(R_1)\text{var}(U).$$

The previous models have dealt with a situation in which secondary failures are not repaired until the emergency event, E, occurs. In certain circumstances, however, repairs may occur while a mission is in an active status. Possibly the repair facility is unable to keep pace with demand (failures), in which case a backlog of secondary failures will await treatment when E eventually signals the necessity for mission termination and intensive overhaul. As an example in this category consider the following model.

Model 3: *Exponential $\underline{U}$, Diffusion-approximated Secondary Failure Accumulation, Additive Repairs*. Let $N(t)$ denote the number of secondary failures awaiting repair at time t after the system has been overhauled and rendered failures-free. Often $N(t)$ may be interpreted to be the state variable of a complex queueing process. We shall, however, approximate it by a diffusion process, cf., Cox and Miller (1966 Ch. 5) with infinitesimal mean and variance μ and σ^2. Remark 1 of Gaver (1968) states that if $N(0) = 0$, then $N(\underline{U})$, the backlog of unrepaired secondary failures at the exponential time $\underline{U}$, has itself an approximate exponential density

$$f_{N(\underline{U})}(x) = ce^{cx}, \tag{3.20}$$

where

$$c = -\frac{\mu}{\sigma^2}\{1 \pm (1 + 2\lambda\sigma^2/\mu^2)^{\frac{1}{2}}\},$$

the plus (+) sign being appropriate if $\mu < 0$, while the negative sign holds true if $\mu > 0$. The case $\mu > 0$ corresponds to a saturated queueing system, i.e., one for which demand rate for service (secondary failure rate) exceeds repair rate, while $\mu < 0$ describes an unsaturated process that eventually displays an equilibrium distribution, itself exponential.

For the simplest repair model, simply assume that the $N(\underline{U})$ items are repaired at *constant* rate $r = \frac{1}{E(R_1)}$, so

$$\underline{R} = \frac{1}{r} E\{N(\underline{U})\} \approx \frac{1}{rc}, \tag{3.21}$$

the approximation deriving from the replacement of $N(\underline{U})$ by a diffusion and the neglect of the time to repair the catastrophic failure *E*. Since diffusion is most appropriate under heavy traffic conditions in which $\mu > 0$, or $\mu < 0$ is close to zero, a single repair time is negligible. The joint density may easily be characterized by transform; from Gaver (1968)(with minor notation change):

$$E\{e^{-\zeta N(\underline{U})}e^{-\xi(\underline{U})}\} = \frac{\lambda}{\lambda+\xi}\,\frac{c(\xi)}{\zeta+c(\xi)},$$

where

$$c(\xi) = -\frac{\mu}{\sigma^2}[1 \pm \{1+\frac{2\sigma^2}{\mu^2}(\lambda+\xi)\}^{\frac{1}{2}}]. \tag{3.22}$$

Again the plus (minus sign) and $\mu < 0$ (>0) associate. A simple scale change gives the joint transform of $\underline{U}$ and $\underline{R}$:

$$E(e^{-\eta\underline{R}} e^{-\xi\underline{U}}) = \frac{\lambda}{\lambda+\xi}\frac{rc(\xi)}{\eta+rc(\xi)}; \tag{3.23}$$

the latter is seen to be the joint transform of a bivariate exponential distribution. Moments are easily found:

$$E(\underline{U}) = \lambda^{-1}, \qquad E(\underline{R}) = \frac{1}{rc(0)};$$

$$var(\underline{U}) = \lambda^{-2}, \qquad var(\underline{R}) = \frac{1}{r^2\{c(0)\}};$$

$$corr(\underline{R},\underline{U}) = \lambda\frac{c'(0)}{c(0)} \tag{3.24}$$

$$= \pm\frac{\sigma^2}{\mu^2}\lambda\left[(1+\frac{2\sigma^2}{\mu^2})^{\frac{1}{2}} \pm (1+\frac{2\sigma^2}{\mu^2}\lambda)\right]^{-1},$$

the appropriate sign to be chosen in the manner described before. In particular if $\lambda \to 0$, which corresponds to a rare occurrance of the emergency event E, then

$$\text{if } \mu < 0;\ corr(\underline{R},\underline{U}) \sim \frac{\sigma^2}{\mu^2}\lambda;$$

$$\text{if } \mu > 0;\ corr(\underline{R},\underline{U}) \sim \frac{1}{1+\frac{1}{2}\frac{\sigma^2}{\mu^2}\lambda}. \tag{3.25}$$

The above model may be elaborated upon, for instance by permitting a random number of interruptions to occur in the repair time $\underline{R}$. In this case $\underline{R}$ is converted into the completion time random variable $\underline{C}$ useful in the study of priority queues. If interruption durations are independent, identically distributed and exponential, and if the conditional number of interruptions in time $\underline{R}$ is Poisson, then the total *interruption* time experienced during repair will have an exponential d.f., and be correlated positively with $\underline{U}$. The mixture argument also permits us to see that if the same emergency event *E* terminates several independent diffusions, then the resulting repair times $\underline{R}_1, \underline{R}_2, \ldots, \underline{R}_k$ will enjoy a joint exponential distribution. If *E* represents a "shock," then a new type of shock-engendered joint distribution is before us.

Model 4: *Exponential* $\underline{U}$, *Renewal-generated Secondary Failures, Parallel Repairs*. Still another plausible model type results if, instead of taking total repair time $\underline{R}$ to be the sum, as in (3.3), it is assumed to be of the form

$$\underline{R} = \text{Max}\{R_1, R_2, \ldots, R_{N(\bar{U})+1}\}. \tag{3.26}$$

If (R_i) is a sequence of iid random variables, with d.f. G, then

$$\text{prob}\{\underline{R} \leq x | N(U)\} = \{L(x)\}^{N(\underline{U}+1)}. \tag{3.27}$$

Clearly (3.2) describes the situation in which individual repairs may proceed simultaneously and independently, and repair is finally completed when the longest subsystem repair is finished.

The joint probability distribution of $\underline{U}$ and $\underline{R}$ is

$$\text{prob}\{\underline{U} \leq y, \underline{R} \leq x\} = \sum_{n=0}^{\infty} \int_0^y \lambda e^{-\lambda z} dz\{G^{n*}(z) - G^{(n+1)*}(z)\} L^{n+1}(x). \tag{3.28}$$

Tractable expressions for the joint transform will not exist. However, putting

$$I_{\underline{R}}(z) \begin{cases} = 1 & \text{if} \quad \underline{R} \leq x, \\ = 0 & \text{if} \quad \underline{R} > x, \end{cases} \tag{3.29}$$

and, from (3.28),

$$E\{e^{-\xi \underline{U}} I_{\underline{R}}(x)\} = \frac{\lambda}{\lambda+\xi} \; \frac{L(x)\{1-\hat{G}(\lambda+\xi)\}}{1-L(x)\hat{G}(\lambda+\xi)}. \tag{3.30}$$

From this expression of the marginal distribution of $\underline{R}$ is

$$\text{prob}\{\underline{R} \leq x\} = \frac{L(x)\{1-\hat{G}(\lambda)\}}{1-L(x)\hat{G}(\lambda)}. \tag{3.31}$$

It cannot be anticipated that any simple expression for $E[\underline{R}]$ will reveal itself in terms of moments of G.

4. TEMPORAL DEVELOPMENT OF ACCUMULATION MODELS

The questions that are likely to concern the designer or user of such a system as we have described relate to its behavior as time passes, and up and down periods alternate.

4a. The Probability that the System is Up

Let $\phi(t)$ denote the probability that the system is up at time t, conditional upon its having just entered the up state at t = 0. That is

$$t\varepsilon\{(0,\underline{U}_1]\cup(\underline{U}_1+\underline{R}_1,\underline{U}_1+\underline{R}_1+\underline{U}_2]\cup\ldots(\sum_{i=1}^{n}(\underline{U}_i+\underline{R}_i),\sum_{i=1}^{n}(\underline{U}_i+R_i)+\underline{U}_{n+1}]\ldots\} \tag{4.1}$$

or, for short, $t\varepsilon(\underline{U})$. Then for the sum of an up time $\underline{U}$ and its immediately succeeding repair time $\underline{R}$ put

$$H(x) = \text{prob}\{\underline{U} + \underline{R} \leq x\}. \tag{4.2}$$

The system is up at time T if either, (i) it has not yet failed, or, (ii) it has failed, been repaired, and then effectively started over from its original condition. Therefore $\phi(t)$ satisfies the renewal equation

$$\phi(t) = 1 - F(t) + \int_0^t \phi(t-z)H\{dx\}. \tag{4.3}$$

From the key renewal theorem, or a direct long-run argument, it follows that

$$\lim_{t\to\infty}\phi(t) = \frac{1}{E(U+R)}\int_0^\infty\{1 - F(t)\}dt = \frac{E(\underline{U})}{E(\underline{U})+E(\underline{R})} \quad . \tag{4.4}$$

The Laplace transform of $\phi(t)$ is

$$\hat{\phi}(s) = \int_0^\infty e^{-st}\phi(t)dt = \frac{1-\hat{F}(s)}{s} \times \frac{1}{1-\hat{H}(s)} \; . \tag{4.5}$$

Since

$$\hat{H}(s) = E(e^{-s\underline{U}} e^{-s\underline{R}}), \tag{4.6}$$

an expression for the latter can be found for several of the specific models treated earlier: by setting $\xi = \eta = s$, $u = 0$ in (3.6), or $\xi = \eta = s$ in (3.12), (3.18) or (3.23). One can then study rate of approach to steady state, (4.4), by inversion of (4.5).

Example 4.1. Refer to Model 1-A, for which

$$\hat{F}(s) = \lambda(\lambda+s)^{-1}, \tag{4.7}$$

$$\hat{H}(s) = \frac{\lambda\mu}{(1+\lambda+s)(\mu+s)-\mu} .$$

Then

$$\hat{\phi}(s) = \{\frac{1}{\lambda+\sigma}\}\{\frac{s^2+s(\lambda+\mu+1)+\lambda\mu}{s(s+\lambda+\mu+1)}\} \tag{4.8}$$

and inversion gives

$$\phi(t) = \frac{\mu}{1+\lambda+\mu} + \frac{e^{-\lambda t}}{1+\mu} + \frac{e^{-(1+\lambda+\mu)t}}{(1+\mu)(1+\lambda+\mu)} . \tag{4.9}$$

Let us contrast the transient behavior of the latter to that of a system with *independent* exponential up times and repair times, both having the same rate parameters. If $\phi_i(t)$ represents the probability that such a system is up at time t, then it may easily be shown that

$$\phi_i(t) = \frac{\mu}{1+\lambda+\mu} + \frac{1+\lambda}{1+\lambda+\mu} \exp\{-\lambda t(\frac{1+\lambda+\mu}{1+\lambda})\}. \tag{4.10}$$

A comparison of the exponential terms in (4.9) and (4.10) reveals that, at least for the present example, the ultimate rate of approach to the steady-state value $\mu(1+\lambda+\mu)^{-1}$ is slower when positive dependence exists between $(\underline{U};\underline{R})$ pairs than when pair elements are independent, but have respectively identical marginal distributions.

4b. Repair Period Forward Recurrence Time

Suppose the system is observed at time t. We wish to calculate the joint probility that the system is down for repair, and it will remain down for a time $\leq$ w. Let

$$\psi(w,t) = \text{prob}\{t\varepsilon(\underline{R}), V_t \leq w\}, \tag{4.11}$$

where V_t is the forward recurrence time at t—the time until all repairs currently underway are completely finished. Let

$$\psi_1(w,t) = \text{prob}\{t\varepsilon(\underline{R}_1), V_t \leq w\}. \tag{4.12}$$

Understanding that the system was newly brought into the up state at t = 0, we have, again by use of the regeneration occurring at the instant $\underline{U}_1 + \underline{R}_1$, that

$$\psi(w,t) = \psi_1(w,t) + \int_0^t \psi(w,t-x)H\{dx\}. \tag{4.13}$$

Consideration of the events in $(o, \underline{U}_1 + \underline{R}_1]$ shows that

$$\psi_1(w,t) = \sum_{n=0}^{\infty} \int_0^t F(dz)\{G^{n*}(z) - G^{(n+1)*}(z)\}\{L_{n+1}(t+w-z) - L_{n+1}(t-z)\}, \tag{4.14}$$

where L_n, the subscripted L, represents the distribution of time to carry out n repairs. In the case of the additive repair model, (3.3), $L_n \equiv L^{n*}$, while in the parallel repair case, (3.26), $L_n \equiv L^n$. Now

$$\int_0^{\infty} \psi_1(w,t)dt \leq \int_0^{\infty} \{1-H(t)\}dt = E(U) + E(R) < \infty, \tag{4.15}$$

so the key renewal theorem shows that a long-run or steady-state value exists for the joint probability:

$$\psi(w) = \lim_{t\to\infty} \psi(w,t) = \frac{1}{E(U)+E(R)} \int_0^{\infty} \psi_1(w,t)dt. \tag{4.16}$$

In order to push this investigation in terms of specific models it is convenient to introduce transforms. For illustration consider the following example.

Example B.1. Forward Recurrence Time for Model 1. In Model 1 the d.f.

F is exponential (λ), and $L_n = L^{n*}$. If now the joint Laplace-Stieltjes transform is extracted, matters simplify. Shortcutting the justifiable formal manipulations, we get

$$\tilde{\psi}_1(\zeta,s) = \int_0^\infty e^{-\zeta w}\hat{\psi}_1(dw,s) = \int_0^\infty e^{-\zeta w}\int_0^\infty e^{-st}\psi_1(sw,t)dt$$

$$= \frac{\lambda}{\lambda+s}\{1-\hat{G}(s+\lambda)\}\frac{1}{(\zeta-s)}\left\{\frac{\hat{L}(s)}{1-\hat{L}(s)\hat{G}(s+\lambda)} - \frac{L(\zeta)}{1-\hat{L}(\zeta)\hat{G}(s+\lambda)}\right\}, \quad (4.17)$$

at least for $\zeta \geq 0$, $s \geq 0$. By Abel's theorem for transforms we have

$$\tilde{\psi}(\zeta) = \lim_{t\to\infty}\int_0^\infty e^{-\zeta w}\psi(dt,t)$$

$$= \lim_{s\to 0}\tilde{\psi}_1(\zeta,s)\ \frac{1}{E(\underline{U})+E(\underline{R})} \quad (4.18)$$

and in the present model

$$\tilde{\psi}(\zeta) = \frac{1}{E(\underline{U})+E(\underline{R})}\ \frac{1-E(e^{-\zeta\underline{R}})}{\xi}\ . \quad (4.19)$$

The latter transform is inverted by inspection, and

$$\lim_{t\to\infty}\psi(w,t) = \frac{\int_0^w \{1-F_{\underline{R}}(x)\}dx}{E(\underline{U})+E(\underline{R})}\ , \quad (4.20)$$

$F_{\underline{R}}$ denoting the distribution of the total repair time, $\underline{R}$. The latter result can easily be obtained by heuristic arguments, and comes as no surprise. Of greater interest is the possibility of obtaining transient response information from the transform of (4.17),

$$\tilde{\psi}(\zeta,s) = \frac{\tilde{\psi}_1(\zeta,s)}{1-\hat{H}(s)}\ . \quad (4.21)$$

For example, differentiation of ζ generates the first moment of the measure $\hat{\psi}(w,s)$:

$$\int_0^\infty e^{-st} \int_0^\infty w\psi(dw,t)\,dt =$$

$$\frac{\lambda}{s^2[s+\lambda-\hat{L}(s)\{\lambda+s\hat{G}(s+\lambda)\}]}\left[\frac{\{\hat{L}(s)-1\}\{1-\hat{G}(s+\lambda)\}+sE(R)\{1-\hat{L}(s)\hat{G}(s+\lambda)\}}{1-\hat{G}(s+\lambda)}\right]. \tag{4.22}$$

While the transform (4.22) may be inverted, either numerically or analytically, it may be instructive to interpret it directly. Suppose that the system is *observed*, or has its services demanded, at a random time D following a moment at which the system is newly up. If D has the exponential d.f. with rate parameter $s > 0$ (i.e., $E(D) = s^{-1}$), then

$$\hat{\psi}(w,s)s = \int_0^\infty e^{-st}\psi(w,t)s\,dt = P\{D\varepsilon(\underline{R}),\ V_D \leq w\}. \tag{4.23}$$

If system unreadiness for a mission or task is rated in part on the basis of the average of scores over n observations, where the ith score is S_i, so

$$\bar{S}_n = \frac{S_1+S_2+\ldots+S_N}{n} \tag{4.24}$$

and

$$S_i \begin{cases} = V_D(i) & \text{if the system is on repair and remains so for remaining time } V_D(i), \\ = 0 & \text{if the system is up at D,} \end{cases} \tag{4.25}$$

then (4.22) is the expected unreadiness score.

We shall omit detailed calculations for Model 2 because of its similarity to Model 1.

Example B.2. *Forward Recurrencence Time for Model 3.* In this model $\underline{U}$ is exponential, and N(t), the number of secondary failures present at t, developes according to a diffusion. We may obtain the transform of $\psi_1(w,t)$ by first conditioning on $N(\underline{U})$ and $\underline{U}$ and then removing the condition. The result is

$$\int_0^\infty e^{-st} E\{e^{-\zeta V_t(1)} | N(\underline{U})\} dt = \frac{e^{-sN(\underline{U})r^{-1}} - e^{-\zeta N(\underline{U})r^{-1}}}{\zeta - s} . \tag{4.26}$$

Now application of the convolution theorem and an appeal to the transform (3.22) brings out

$$\tilde{\psi}_1(\zeta,s) = \frac{\lambda}{\lambda+s} \frac{rd(s)}{\{s+rc(s)\}\{\zeta+rc(s)\}} . \tag{4.27}$$

Abel's Theorem then delivers the expression

$$\tilde{\psi}(\zeta) = \lim_{t\to\infty} \int_0^\infty e^{-\zeta w} \psi(dw,t) = \frac{\lambda}{\lambda+rc(0)} \cdot \frac{rc(0)}{\lambda+rc(0)} \tag{4.28}$$

and so inversion by inspection shows that the limiting forward recurrence time distribution is exponential:

$$\lim_{t\to\infty} (w,t) = \frac{\lambda}{\lambda+rc(0)} \{1-e^{-rc(0)w}\}, \tag{4.29}$$

where $c(0)$ is given by (3.20).

The time-dependent behavior of V_t is implicitly described by the Laplace transform obtainable from (4.2). We find

$$\tilde{\psi}(\zeta,s)s = \int_0^\infty e^{-st} E\{e^{-\zeta V_t}\} s dt = \frac{\lambda}{\lambda+s+rc(s)} \cdot \frac{rc(s)}{\lambda+rc(s)} . \tag{4.30}$$

Inversion gives

$$\int_0^\infty e^{-st} P\{V_t \leq w\} s dt = \frac{\lambda}{\lambda+s+rc(s)} \{1-e^{-rc(s)w}\} \qquad (w \geq 0) \tag{4.31}$$

The interpretation is that the above expression is the probability that a system randomly observed at exponential time D, where $E(D) = 1/s$, is in the repair state and will remain so for a time V_D not exceeding w. Apparently V_D has again an exponential distribution; the joint distribution of D and V_D may also be obtained and will be of bivariate exponential form as well, as will other joint distributions of possible interest. Further details will not be given here.

Example B.3 *Forward Recurrence Time for Model 4.* The parallel repair model may be studied by specializing (4.13), as was mentioned earlier.

Calculations analogous to those of Example B.1 may be made:

$$\tilde{\psi}_1(w.s) = \int_0^\infty e^{-st}\psi_1(w,t)dt$$
$$= \frac{\lambda}{\lambda+s}\{1-\hat{G}(s+\lambda)\}\int_0^\infty e^{-st}\{\frac{L(t+w)}{1-\hat{G}(s+\lambda)L(t+w)} - \frac{L(t)}{1-\hat{G}(s+\lambda)L(t)}\}dt. \tag{4.32}$$

Rarely will it be feasible to explicitly evaluate the transform. However Abel's theorem again permits the limit evaluation:

$$\lim_{s\to\infty} \tilde{\psi}_1(w,s) = \{1-\hat{G}(\lambda)\}\int_0^w \frac{L(z)}{1-\hat{G}(\lambda)L(z)} dz. \tag{4.33}$$

5. DESIGN DEFECT MODELS AND THE GENERATION OF NEGATIVELY CORRELATED BIVARIATES

When complex systems are manufactured, or even being adjusted or repaired following use, the introduction of one or more *removable design defects* or *maladjustments* may be anticipated. Such a defect is apt to make its presence felt rather early. It is even likely that, once one defect appears, others may become evident upon closer inspection. Such a parable now will be used to motivate the development of a class of bivariate exponentials (and other skew distributional families) having negative dependence.

Model 5. *Design defects geometrically distributed.* In the course of system manufacture a particular system copy receives N defective elements; here N is a random variable with the distribution given by $\{p_n\}$, $n = 1,2,\ldots$. We suppose that at least one such defect occurs, for immortality is inadmissable in the real world. Suppose that the time to first appearance (e.g. physical evidence = failure) for each defect is T, having d.f. F(t), and that $T_1,T_2,\ldots,T_N$ are independent. Our simple scenario now runs as follows: at time

$$\underline{T} = \min(T_1,T_2,\ldots,T_n) \tag{5.1}$$

following the beginning of system usage one defect makes itself evident (the ith subsystem fails). Diagnosis is initiated, and *all* defective elements are discovered. Finally, repair is carried out; the time required, $\underline{R}$, is distributed as $G^{n*}(x)$, provided N = n. Plainly the joint d.f. of $\underline{T}$ and $\underline{R}$ is obtainable from

$$\text{prob}\{\underline{T} > t,\ \underline{R} \leq x\} = \sum_{n=1}^{\infty} \{1-F(t)\}^n p_n G^{n*}(x). \tag{5.2}$$

The following results may be obtained; we omit details for the present. We shall suppose, for purposes of illustration, that N is geometric. Then

$$p_n = (1-\alpha)\alpha^{n-1}, \qquad 0 < \alpha < 1, \quad n = 1,2,\dots . \tag{5.3}$$

First, the conditional expectation of total repair time is

$$E(\underline{R}|\underline{T} > t) = \frac{E(R)}{1-\alpha\{1-F(t)\}}, \tag{5.4}$$

where E(R) is the expected value of an elementary repair time. Moreover,

$$\begin{aligned} E\{\underline{R}|\underline{T} = t\} &= E(R)\left[\frac{1+\alpha\{1-F(t)\}}{1-\alpha\{1-F(t)\}}\right] \\ &= E(R)\left[1 + \frac{2\alpha}{1-\alpha}\{1-\phi(t)\}\right], \end{aligned} \tag{5.5}$$

where ϕ is the d.f. of $\underline{T}$:

$$\phi(t) = \text{prob}(\underline{T} \leq t) = \frac{F(t)}{1-\alpha\{1-F(t)\}}. \tag{5.6}$$

The regression is seen to be highly non-linear, and given that $\underline{T}$ is long we can anticipate only one repair, while if $\underline{T}$ is short the number increases to $\frac{1+\alpha}{1-\alpha} > 1$; the tendency towards negative relationship is clearly present.

Going further, one can calculate

$$\text{cov}(\underline{R},\underline{T}) = \alpha E(R)E(T)\left[1 - \frac{E\{\underline{T}_{(1)}\}}{E(T)}\right] < 0, \tag{5.7}$$

where $\underline{T}_{(1)}$ has the distribution of the smallest of a sample of two from

$\phi(t)$, the d.f. of $\underline{T}$.

Model 5-A. *A Bivariate Exponential with Negative Correlation.* It will now be shown that one can select $F(t) = 1 - F(t)$ in such a way as to induce exponentially distributed time to *system* failure, $\underline{T}$. Since we already know that exponential G, geometrically compounded, yields exponential $\underline{R}$, the outcome will be a $(\underline{T},\underline{R})$ pair with exponential marginals and *negative* correlation.

If $F(t) = 1 - \bar{F}(t)$ satisfies

$$\sum_{n=1}^{\infty} [\bar{F}(t)]^n \alpha^{n-1}(1-\alpha) = e^{-\lambda t}, \tag{5.8}$$

then $\underline{T}$ is exponential with mean λ^{-1}. Summation of the series and solution for F yields

$$\bar{F}(t) = \frac{1}{\alpha+(1-\alpha)e^{\lambda t}} \qquad (t \geq 0), \tag{5.9}$$

which is a *logistic* distribution, left-truncated at t=0. If $\lambda = 1$, then $\underline{T}$ is exponential with unit mean, while $E\{\underline{T}_{(1)}\} = 1/2$, so

$$\text{cov}(\underline{R},\underline{T}) = -\frac{\alpha}{2} E(\underline{R})E(\underline{T}). \tag{5.10}$$

If G is chosen so as to make $\underline{R}$ exponential then it may be shown that

$$\text{corr}(\underline{R},\underline{T}) = -\frac{\alpha}{2} = -\frac{1}{2}\{1-\frac{1}{E(N)}\}. \tag{5.11}$$

Consequently, a greatest lower bound for the correlation in this model is $-\frac{1}{2}$. It is known that the "antithetic variates" scheme of Monte Carlo

$$X = -\ell n\ R, \qquad Y = -\ell n(1-R), \tag{5.12}$$

where R is uniform over [0,1], yields

$$\text{corr}(X,Y) = 1 - \frac{\pi^2}{6} \approx 0.6649, \tag{5.13}$$

the maximum attainable negative correlation (Moran, 1967). It may only be conjectured as to whether an alternatively structured setup following the present pattern can reach the latter value. One in which

$$\underline{R} = \max[R_1, R_2, \ldots, R_N] \tag{5.14}$$

probably comes closer to doing the trick.

Further investigation of the above models and mechanisms should be of interest. Professor A. Hawkes has informed this writer that he has obtained quite similar results by alternative methods.

6. SOME NOVEL RELIABILITY PROBLEMS

Important and concrete reliability problems occur in great profusion in the modern world. The following examples may stimulate others to identify further problems of this sort.

6.1 Failure Induced Traffic Jams

An all-too-common experience for the motorist is to find himself entrapped in a nearly immobile line of automobiles, awaiting the removal of an accident or other highway flow stoppage. The social cost of such incidents may perhaps be measured in terms of the total delay suffered by all passengers held up, although the length of delay to certain individuals (e.g. a surgeon on emergency call) is probably of more importance than to others. These delays, and their unpredictable nature, are often the result of the failure of a driver-vehicle system, and hence originate in unreliability.

A preliminary attempt to model a traffic jam has been made by Gaver (1969), and extended by Luckew (1971). Briefly, a model of the following type has been considered. Vehicles proceed down a freeway according to some system of space-time trajectories. An accident occurs (e.g. tire blowout) at a moment thereafter taken to be the origin. Vehicles now are forced to queue behind the accident location because of the flow restriction at that point. The subsequent arrivals of vehicles may be taken to be a point stochastic process. After the obstruction is removed, flow out of the area resumes, at first at a slow rate because of the density of vehicles present. Eventually the entire jam dissipates.

It is tempting to model the above situation in a standard queue-theoretic manner, assuming point process arrivals, and letting "service" occur at the congested area near the accident. However, reasonably adequate, and much simpler, results follow if one adopts a "pointless" or fluid approximation approach: the total number of vehicles that have arrived by time t is A(t), where A is taken to have a *continuous* non-decreasing function taking on all positive real (not merely integer) values. For further refinement, we allow post-accident queue build up to be a diffusion process, with positive drift and a specified variance. Before complete obstruction removal the jam builds in a manner that is governed by a diffusion with reflecting barrier at zero (the obstruction is still present, and the queue will quickly re-form if it happens to vanish). Following removal, the queue vanishes with negative drift, so it may be modeled by a diffusion with absorbing barrier. The expected total vehicle-hours of delay *following* accident removal is derived in Gaver (1969). Suppose w represents the number of vehicles entrapped at the moment of removal; then

$$\bar{w} = \frac{w^2}{2(1-\rho)} + \frac{w\rho}{(1-\rho)^2} \frac{E(B^2)}{E(B)}, \tag{6.1}$$

where we have selected a time scale so that the rate of departure from the jam area is unity, so that $\rho - 1$ is the expected number of vehicles arriving at the jam per unit time, and B represents the random size of a platoon of vehicles. It turns out that a simple deterministic approximation simply omits the second term in the above expression. The latter can be appreciable if the jam dissipates slowly. The variance of total waiting time may also be calculated and is proportional to $(1-\rho)^{-5}$; work of Daley (1969), Daley and Jacobs (1964) and McNeill (1970) is relevant to the total delay problem.

The cost of such delays can be reduced by (i) increasing the reliability of the driver-vehicle-road system (ii) improving the procedures for detection and removal of obstructions when they occur, (iii) rerouting traffic at indications of trouble ahead. Other possibilities also exist. Various ramifications of the problem are under further consideration.

5.2 Electric Power System Reliability

For many years the national demand for electric power has grown steadily. So dependable is the supply that only recently has the public become aware of the fact that reliability questions may be an issue in the planning of system expansion. In fact, the presence of many generators, of widely varying capacities and availabilities, in confrontation with demands that fluctuate both systematically (according to time of day, week and year) and randomly (weather-sensitive loads are on the increase, because of air conditioning) challenge the reliabilist with some difficult problems in probability and statistics. We mention several of these.

(a) The statistical estimation of long-run availability for particular generating equipments, using operational data. Since generators fail rather infrequently, there is a strong impetus to pool data from various sources. A rational approach involves use of prior knowledge of the variability of failure rates and repair process parameters for different copies of the same machine, the latter to be blended with operational data as it becomes available. A modified Bayesian approach seems suitable, particularly when new machines are added to the system.

(b) The convenient estimation of the probability that demand will exceed total available capacity. Let C_i denote the actual available capacity of the ith generator: simply,

$$C_i \begin{cases} = \text{installed capacity, with probability } p_i, \\ = 0 \qquad\qquad\qquad\quad \text{, with probability } 1 - p_i \end{cases} \tag{6.2}$$

p_i being the long-run availability of machine i. Then

$$\underline{C} = C_1 + C_2 + \dots C_n, \tag{6.3}$$

where n represents the number of machines actually on line on a day. It is now customary to compare $\underline{C}$ to $\underline{L}$, the peak load on a day, and to seek to make the probability that $\underline{L} > \underline{C}$ acceptably small. Usually the probability calculation (whose answer depends upon that of (a) above, through p_i) is made by straightforward binomial expansion. However, analytical approximations based on Edgeworth asymptotic expansions and the method of large deviations have been found useful, particularly when large systems are under discussion.

(c) The comparison of $\underline{L}$ to $\underline{C}$ above is crude in that it takes no account of the *degree* of the deficit $\underline{L} - \underline{C}$ given that the latter occurs. Neither does it account for the times between successive "upcrosses," i.e., instants t_i such that demand and capacity satisfy

$$\begin{aligned} L(t_i-) &< \underline{C}(t_i-), \\ L(t_i+) &> \underline{C}(t_i+). \end{aligned} \tag{6.4}$$

Nor does it account for the time *duration* of the outage. The latter quantities in particular may be attacked by adaptation of stationary process methods (see Cramér and Leadbetter, 1966) but much remains to be done.

7. CONCLUSION

In this paper I have attempted to provide an overview of accomplishments and directions in the reliability theoretic area, particularly as the latter relates to point stochastic processes. Lack of time prevents a more thorough discussion and more encyclopedic acknowledgment of the contributions of the many accomplished researchers in this interesting and continually evolving field.

ACKNOWLEDGMENT

Research conducted with the partial support of the Office of Naval Research and the National Science Foundation.

REFERENCES

Antleman, G. & Savage, I. R. (1965). Surveillance problems: Wiener process. Naval Research Logistics Quarterly 12, 35-55.

Antleman, G. & Savage, I. R. (1965). Characteristic functions of stochastic integrals and characteristic functions. Naval Research Logistics Quarterly 12, 199-222.

Barlow, R. & Proschan, F. (1965). *Mathematical Theory of Reliability*. New York: Wiley.

Barlow R. & Scheuer, E. (1966). Reliability growth during a development testing program. Technometrics 8, 53-60. (See also remark by R. R. Read, Technometrics 13, 199-200.)

Birnbaum, Z. W. & Saunders, S. C. (1969). A new family of life distributions. J. Appl. Prob. 6, 319-27.

Cox, D. R. (1962). *Renewal Theory*. London: Methuen.

Cox, D. R. & Lewis, P. A. W. (1966). *The Statistical Analysis of Series of Events*. London: Methuen & New York: Barnes and Noble.

Cox, D. R. & Miller, H. D. (1966). *The Theory of Stochastic Processes*. London: Methuen.

Cramér, H. & Leadbetter, M. R. (1966). *Stationary and Related Stochastic Processes*. New York: Wiley.

Daley, D. J. (1969). The total waiting time in a busy period of a stable single-server queue, I. J. Appl. Prob. 6, 550-64.

Daley, D. J. & Jacobs, D. R. Jr. (1969). The total waiting time in a busy period of a stable single-server queue, II. J. Appl. Prob. 6, 565-72.

Downton, F. (1970). Bivariate distributions of reliability theory. J. R. Statist. Soc. B 32, 408-17.

Epstein, B. (1958). The exponential distribution and its role in life-testing. Ind. Quality Control 15, 2-7.

Esary, J., Marshall, A. W. & Proschan, F. (1970). Shock models and wear processes. FSU Statistics Report M194, The Florida State University, Department of Statistics, Tallahassee, Florida.

Feller, W. (1966). *An Introduction to Probability Theory and its Applications, Vol. II*. New York: Wiley.

Flehinger, B. J. (1958). Reliability improvements through redundancy at various system levels. IBM Journal of Research and Development 2, 148-58.

Gaver, D. P. (1963). Random hazard in reliability problems. Technometrics 5, 211-26.

Gaver, D. P. (1963). Time to failure and availability of paralleled systems with repair. IEEE Trans. on Reliability R-11, 30-8.

Gaver, D. P. (1968). Diffusion approximations and models for certain congestion problems. J. Appl. Prob. 5, 607-23.

Gaver, D. P. (1969). Highway delays resulting from flow-stopping incidents. J. Appl. Prob. 6, 137-53.

Gaver, D. P. & Luckew, R. S. (1971). System reliability and bivariate skew distribution. Research Report, Naval Postgraduate School, Monterey, California.

Gnedenko, B. V., Belyayev, Yu & Solovyev, A. D. (1969). *Mathematical Methods of Reliability Theory*. New York: Academic Press.

Harris, R. (1968). Reliability applications of a bivariate exponential distribution. Operations Research 16, 18-27.

Jorgenson, D. W., McCall, J. & Radner, R. (1967). *Optimal Replacement Policy*. Chicago, Illinois: Rand McNally.

Lewis, P. A. W. (1964). A branching Poisson process model for the analysis of computer failure patterns. J. R. Statist. Soc. B 26, 398-456.

Luckew, R. S. (1971). *Generalized Alternating Renewal Process Models*. Thesis, Department of Statistics, Carnegie-Mellon University.

Marshall, A. W. & Olkin, I. (1967). A multivariate exponential distribution. J. Am. Statist. Ass. 62, 30-44.

Mazumdar, M. (1970). Reliability of two-unit redundant repairable systems when failures are revealed by inspections. SIAM J. of Applied Math. 19, 637-47.

McNeill, D. (1970). Integral functionals of birth and death processes and related limiting distributions. Ann. Math. Statist. 41, 480-85.

Mercer, A. (1961). Some simple wear-dependent renewal processes. J. R. Statist. Soc. B 23, 368-76.

Moran, P. A. P. (1967). Testing for serial correlation with exponentially distributed variates. Biometrika 54, 395-401.

Moran, P. A. P. (1969). Statistical inference with bivariate gamma distributions. Biometrika 56, 627-34.

Morey, R. C. (1966). Some stochastic properties of a compound renewal damage model. Operations Research 14, 902-08.

Reynolds, D. S. & Savage, I. R. (1970). Random wear models in reliability theory. Research Report, Statistics Department, Florida State University

POINT PROCESSES IN FORESTRY

W. G. Warren
Canadian Forestry Service, Vancouver, British Columbia

SUMMARY

An attempt is made to indicate the role of point processes in conventional forestry, i.e., the estimation and prediction of stand volume. The development is traced through the assessment of natural regeneration to static models of spatial pattern of trees and thence to models of stand development and the incorporation of growth laws. It is suggested that the current trend towards simulation could be obviated, in part, by cooperation between the forester and mathematician in the development of appropriate multivariate point process models.

1. INTRODUCTION

This paper will begin with a short personally oriented historical introduction. This will lead into an outline of the major problem areas as I see them today. The approach will be general and excursion into formal mathematics will be minimal.

There are, of course, problems which are common to forestry and to other fields, especially plant ecology, for which a clear-cut division from forestry cannot be made. To keep the subject within bounds, reference to related work in "non-forest" ecology will also be minimal. Further, most applications of point process theory have dealt with time series; for example, time series related to the occurrence of accidents, genetic mutations, the building-up of a waiting line, the growth and possible extinction of a population of parasites. The corresponding studies of point processes are, of course, of interest to forestry, but since not peculiar to forestry will not be considered here.

2. HISTORY

The author's introduction to point processes in forestry came with questions regarding the assessment of natural regeneration in cut-over areas; an outcome of this was a review paper (Warren, 1964) of limited distribution. Relevant segments of this paper are therefore reproduced here.

The obvious method of assessing regeneration is to establish a set of sample plots and count the number of stems on each. The earliest departure from this seems to be due to Lowdermilk (1921) who described restocking in cut-over areas not in total number of trees per acre, but in percentage of milacres (units 6.6 ft. square) stocked, i.e., containing one or more stems. The basis for this was the assumption that, if a given area is divided into squares of such a size that one established seedling or tree per square will fully stock the square at maturity, then the percentage of units so stocked, regardless of the total number of seedlings per acre, will indicate the proportion of land being utilized by tree growth. Cowlin (1932) rightly observed that the success of the method depended on the selection of the proper sized square.

Lowdermilk gave reasons for selecting one-thousandth of an acre but Haig (1931) observed that from the start there had been considerable discussion about the size of unit to be employed, and that the milacre had been adopted because of its convenience in field use and office compilation, but many foresters regarded it as too small a unit. Haig recommended the use of the four-milacre, or 13.2 ft. square, for stocked quadrat assessment of the type of forest in which he was interested; that is, full stocking would correspond to 250 stems per acre.

In the years which followed, the stocked-quadrat technique received considerable use. Gill (1950) reported that quadrats of various sizes, ranging all the way from 1/4 milacre to 8 milacres had been used in North America. Lynch and Schumacher (1941) noted that discussion regarding the problem of what size of quadrat square contains the most information concerning the condition of restocking of cut-over lands had at times been heated; practically all quadrat sizes between one and four milacres had been proposed, strongly attacked, and just as strongly defended.

In other words, the definition of full stocking determined the quadrat size, but there was considerable difficulty in getting agreement between foresters on what constituted full stocking. Thus, the question arose that, if an assessment were made with a certain quadrat size, what would be the result in terms of a different size?

The first attempt at empirical conversion from one quadrat size to another seems to have been made by Wellner (1940). Gill (1950) recognized the problem caused by the nature of the dispersion — regular, random (homogeneous) or heterogeneous — and developed a heterogeneity index by which conversion of stocking from one quadrat size to another could be performed without difficulty, apart from the initial determination of the index for a particular species or area. On the evidence presented Gill's method has one great virtue — it seems to work.

More attention seems to have been given, however, to the theoretical conversion of Grant (1951) which according to Sahay (1955) permitted the complete separation of the ideal size of quadrat for expressing stocking figures from the question of the most convenient size of quadrat to use in the field.

Grant's formula is based, however, on the assumption of randomness of dispersion [in the sense of a planar Poisson point process — see Matérn (1971) for a critique of the terminology]. Although Grant was very conscious of the problem of heterogeneity and devoted much of his paper to it, this aspect seems to have been ignored, or neglected, by many foresters when they used the formula. Forestry literature, however, abounds with examples of non-random dispersions, the general consensus being that natural regeneration is aggregated but, as a consequence of mortality through competition, the pattern moves towards regularity in a mature stand.

Spatial pattern is therefore of central importance in forestry, and not only with respect to the assessment of regeneration. It seems obvious that a single tree must be the foundation stone of any model of forest growth or structure which attempts to be realistic. Accordingly, forest researchers have been prominent amongst those who have investigated the various measures of dispersion which have from time to time appeared in the literature. The more recent contributions include Bary-Lenger (1967), Mawson (1968) and Payandeh (1970b). The last mentioned author investigated the quadrat method (Greig-Smith, 1952), the nearest-neighbour method (Clark and Evans, 1954), the point-to-plant distance method (Pielou, 1959), the coefficient of aggregation (Hopkins, 1954) and the ratio and correlation tests of Hoglate (1965) on five forest types and four computer-redistributed populations (random, semi-random, uniform, semi-uniform) generated from each type.

A different approach, which has yet to be properly evaluated, was recently suggested by Fraser and van den Driessche (1971), with the application of the Voroni and Delaunay tessellations. All these methods are greatly aided by the advances made in the techniques of aerial photography of forest stands.

3. PROBLEM AREAS

3.1 Static Models of the Spatial Pattern

What has so far been discussed can be placed under the heading of static models of spatial pattern of trees in a stand. Much of the work has been empirical in nature, based on actual or computer-generated stand maps and, with some notable exceptions, limited to the study of the properties of a single or few techniques.

As a consequence, many of the questions remain unresolved. Deduction of general properties after examination, no matter how detailed, of but a few actual forest patterns clearly should not be regarded as conclusive. Sampling simulation is a valuable tool — the author has used it extensively in recent years — but stability seems to be achieved only after a large number of realizations. In addition, the advent of computers has stimulated interest in models of an advanced type, but although the models may be described in stochastic-process terminology, they have often not been conceived as such. Instead the possibility to produce realizations in a computer has been the guiding principle. With these trends there is a danger of resorting to simulation even when the results can be obtained by means of ordinary probability calculus.

The idea behind the author's work (Warren 1962, 1971) was to take a hopefully physically meaningful model — the center-satellite or clustering process — and develop the mathematics required for theoretical study of plot and distance methods on the same population. Persson (1964, 1965, 1971) has made theoretical studies on regular, random and simply clustered populations. Matérn (1971b and pers. comm.) has expressed a view that the doubly stochastic Poisson process seems to be a not too bad approach to take account of variations in fertility, microclimate and similar environmental factors, but that such a model should be supplemented by a mechanism accounting for the effect of competition between neighbouring individuals; the theory was treated by Matérn in his paper (1971b) and in his earlier monograph (1960). Stiteler and Patil (1969) theoretically examined the variance-to-mean ratio for lattice patterns and for individuals and tightly grouped clumps randomly distributed in territories defined by adjacent congruent rectangles, there being the same number of individuals (or clumps) in each. Such models may not be unrealistic for man-made forests.

Such work can be criticized on the grounds that forests do not possess such idealistic mathematical structures. The writer's reply to this (Warren, 1971b) is that although it seems unlikely that any of the models will provide an exact representation of some actual population they may well provide reasonable first-order approximations to reality, so that their parameters are not entirely without a biophysical meaning. It seems important, therefore, to gain as much knowledge as possible of the behaviour of the various statistics for as wide a range of models as possible. Certainly the behaviour of any one statistic will not tell us what the model should be, but the combined behaviour of several statistics, the index of dispersion, the interplot correlation and the distance to nearest neighbour, say, may direct us to a relatively small set of possibilities.

At this point it should also be noted that a forester's interest often extends no further than the error variances of, in particular, estimates obtained by "indirect" mensurational methods (see below). In this circumstance, one need only make the "second-order" assumptions about the process and not specify its further properties.

As implied above, static models of spatial pattern have a role in other than regeneration assessment. The assessment of the volume of merchantable crop is one of the two primary tasks in forestry — the other is the prediction of yield. The argument of systematic versus random sampling has long raged and empirical studies abound; those by Hasel (1938), Osbourne (1942) and Finney (1949, 1950, 1953) must be regarded as classics in forestry literature. The effect of spatial distribution on the relative efficiency of systematic sampling has been recognized. Theoretical aspects were treated in depth by Matérn (1960). Palley and O'Regan (1961) drew attention to the possibility of extending empirical studies to computer-generated forests, and a recent study of systematic sampling of this nature has been made by Payandeh (1970a). Likewise computer simulation has been used in a series of papers by Arvanitis and O'Regan (1969a, 1969b, 1971) to examine the cost effectiveness of various sampling schemes.

Forest research also requires the laying out of field experiments, but point-process concepts as such do not appear to have played any role, as yet, in experimental design consideration, the random field concepts of agrigulture, with isotropic correlation (see, for example, Matérn, 1971a) seemingly being felt adequate.

The plot, be it circular, square, rectangular, small or large, is of course, the traditional tool of forest mensuration. However, the elegant angle-count method, first described by Bitterlich (1948), and its derivatives have also received universal acceptance. Theoretical aspects of this method have been studied by Holgate (1967) for random and square-lattice dispersions, other cases being regarded as too intractable.

Distance methods, as noted above, have been used for certain purposes, generally of an ecological nature, perhaps as a consequence of their originators being ecologists more often than foresters. There also seem to be problems arising from the change of scale; techniques reported as highly successful in, say, grass or shrublands have not always been found to work so well in forests. For example, the author has reported (Warren, 1971a) difficulty with the rather attractive wandering-quarter method of Cantana (1963), but theoretical study, or extensive simulation, based on models which attempt to reproduce forest conditions was never completed.

In general, foresters require an areal basis for information. The angle-count method, although an "indirect method", automatically gives this [basal area in square feet (metres) per acre (hectare)]. But the indirect methods generally require some form of conversion which often turns out to be a function of the (unknown) spatial pattern, which, in turn, emphasizes the need for robust methods. There is yet another method which lends itself to studies where the trees fall into one of several categories, for example species A and species B, or diseased and healthy. Consider a belt transect sufficiently narrow that there is no difficulty in ordering the trees along its length. In this way, runs of one kind relative to the other can be generated. Pielou (1962, 1963, 1965) has shown how to use this information to (i) test whether the individuals of one kind are dispersed at random with respect to individuals of the other kind and (ii) estimate the proportion of area taken up by patches of disease and the intensity of the disease within patches.

The belt transect is often attractive from the operational point of view. Traditionally, plot boundaries have to be marked out and the plot searched, a difficult task in a forest with a heavy understory. Further, plots may be well separated from each other, and, if the terrain is rugged, the measuring crew may wonder why, since they have to cover the distance, they cannot be taking measurements en route. With the belt transect, the marking of boundaries is unnecessary, walking a compass line is sometimes all that is required, whether or not a tree should be included is easily checked, and since measurement is continuous the crew does not have the frustrated feeling that they are wasting time.

Pielou's statistics are based on run length and do not incorporate any measure of distance. Suppose that the individuals of species A are randomly dispersed (planar Poisson process). Then the distance between successive sampled individuals measured along the direction of the transect will be distributed as a negative exponential with parameter, λ, which depends on the density of the population and the width of the transect. Likewise, if the individuals of species B are independently and randomly dispersed, the distances between successive individuals will be independently distributed as a negative exponential with parameter μ. This enables various properties to be deduced. For example, it is possible to derive the distribution of the number of runs generated over a transect of length T (Warren, unpubl.); the mean is

$$[2\lambda\mu(\lambda+\mu)T + (\lambda^2+\mu^2)\{1-e^{-(\lambda+\mu)T}\}]/(\lambda+\mu)^2$$

and a similar but more complicated expression exists for the variance.

What, however, are the results if, for example, species A is randomly dispersed while species B is aggregated? Further work on this type of problem would seem to be useful.

3.2 Models of Stand Development

Trees are, of course, not points. The diameter of a tree, conventionally measured at breast height (4 ft. 6 in. or 1.3 m) is generally not negligible with respect to the distance between trees. Further, competition between trees ultimately produces an area about each tree within which no other tree can exist. These restrictions are often conveniently neglected in theoretical studies (notably the writer's); exceptions are the theoretical derivation of Matern (1960, p. 47) and the simulation studies of Newnham (1968) and Newnham and Maloley (1970).

As indicated in the previous section, a forester is rarely interested in the spatial distribution per se. Trees have a size and a forester's traditional interest is in the estimation and prediction of volume, and more recently with how that volume is represented in terms of logged output, that is the distribution of log dimensions (Warren, 1959) and the quality of the logs — the proportion of the various grades of sawn timber and the relative amount of pulpwood. Much of this thinking may have to be revised if, as there is some indication, future interest lies solely in fibre production, the idea being the harvesting of fibre and its reconstitution into a variety of products with controlled properties.

As of now, however, yield table construction has formed the backbone of practical forestry, sometimes from data collected on permanent plots over a long time period, sometimes from data collected on plots in stands of different ages, and combinations thereof. In recent years, there has been an increasing interest in theoretical growth models.

The approach has been largely deterministic. Interpolations and extrapolations have been made from smooth curves fitted to empirical data with little concern for the variability exhibited. With few exceptions growth models have contained no probabilistic elements.

A stochastic approach with Markov chain concepts has been taken by Beichelt (1969), Rudra(1968), Wagener and Stephens (1970) and Suzuki (1971), but this work seems to lie outside the main stream of current thinking. More common, and growing rapidly, is the attempt to simulate forest growth on a computer, primarily to investigate the effects of different management policies, such as initial spacing or species mix, timing and quantity of thinnings, timing of pruning (if any) and time of final felling.

Methods have been varied. Earlier work tended to treat a stand as a whole, but most work today attempts to simulate the growth of individual trees and integrate these into a model of a stand. Two recent papers representative of the diverse methods being employed are those of Opie (1971) and Strand (1971). Many others could be cited, Strand (pers. comm., 1968) has well described the situation. "Most attacks have up to now been based on the first approach. One tries to study the growth in terms of number of stems per acre, basal area, mean height, etc. This gives information that is easy to handle and to use for prediction purposes. But we always look at stands as a whole, and do not try to study what is going on *inside* the stand.

"The second approach seems to be more sophisticated and is certainly more difficult. But this approach may perhaps give us more information about the interaction or competition between trees. At the same time, it should be easier to tell the effect of changes in the structure of stands. By the first approach, we get a combined effect of changes in structure and changes due to the growth of individual trees.

"The most difficult point seems to be to combine the individual trees into a stand. Here we run into difficulties in connection with the interaction between trees. But once we find a good solution to this problem we should be able to get information of much wider use than the 'stand' approach."

The simulations attempted have commonly included stochastic elements in the construction of the spatial pattern, but the growth and size of trees has been treated as if deterministic. There have been exceptions; for example Newnham (1964) related the death of a tree to its magnitude relative to its neighbours, while Newnham and Maloley (1970) related the diameter at breast height to the area occupied by a tree (the Area Potentially Available of Brown, 1965) modified by the addition of a random variable. Mitchell (1969), in one of the more sophisticated models as yet attempted, used an expected height-age relationship to determine crown radius, but also incorporated a random variable to simulate unexplained sources of variation such as heredity.

The concept of simulation has even gone past the growing of the stand to include the felling of it, with models to study the performance of logging machines (Almquist, 1969, 1970; Newnham, 1966, 1967; Newnham and Sjunnesson, 1969, Sjunnesson, 1970). Clearly the spatial distribution of trees coupled with the distribution of their dimensions must be a major element of such models.

The theory of point processes deals, of course, not only with the arrangement of points in space and time, but also treats the connection between this arrangement and the properties of the individuals. The yield problem of forestry is thus an example of multivariate point process, that is, by the definition of Cox and Lewis (1970), a point process in the plane in which a real-valued quantity is associated with each point. But the problem is clearly very complex in nature. For example, in spite of such studies as Nishizawa (1968), the question of the most applicable measure of competition between trees remains unresolved.

4. SUMMATION

Can the questions asked by foresters be answered by a theoretical approach rather than by simulation? In other words, can a tractable multivariate point process theory be developed which reflects the physical nature of a forest? There may be papers in this volume which provide at least partial answers to these questions. However, the ordinary forest researcher's concern is with the application of models of the stochastic process type and he is interested only in expositions of the theory which are not too overwhelmingly technical. For example, there is a recent interesting, possibly excellent, paper on the homogeneous planar Poisson point process (Miles, 1970) which would appear to be relevant to the ideas of Fraser and van den Driessche referred to above. But who is going to bridge the gap between such works?

There are few people with both the interest in forestry and the necessary mathematical background to make such applications. It must be remembered that most forest biometricians have had their primary training in forestry and that their knowledge of statistics was obtained from later experience, possibly with some formal instruction, although commonly restricted to one or two methods courses. Some indeed entered forestry in order to avoid mathematics.

Thus, there appears to be the need of a greater communication between the mathematical statistician and the forester, to ensure that the theory truly reflects foresters' problems and the structure of forests and also that the theoretical results get beyond the pages of some abstract journal and are put to use in the forests.

APOLOGIES AND ACKNOWLEDGMENTS

This paper is by no means a complete review of the subject. For example, questions of forest disease have been but briefly mentioned and forest insects, whose dispersion can surely be modelled by a point process, not mentioned at all. Reference to the literature is acknowledged as incomplete, and I must apologize to those whom I have chosen to omit, but who rightfully feel their work worthy of inclusion. In particular, my coverage of the non-English literature has been minimal, not because the work is lacking in quantity or merit, but because my inadequate command of other languages has prevented my examination of it in the breadth and depth that I would like.

I am greatly indebted to Prof. B. Matérn, who was originally invited to give this paper, for his notes on the subject. On the other hand, although he will recognize his hand in several places, Prof. Matérn should not be held accountable for the manner in which his material has been used, or perhaps misused, or for the additions made to it.

REFERENCES

Almquist, A. (1969). Simulation of knuckle-boom-equipped feller-bunchers. Rep. Logging Res. Fdn. Sweden 14.

Almquist, A. (1970). The effect of stem distribution on the production of logging machines in final felling operations. Rep. Logging Res. Fdn. Sweden 6.

Arvanitis, L. G. & O'Regan, W. G. (1969a). Sampling simulation computer programs. I. Fixed and variable plots. Res. Rep. Coll. Agr. Life Sci., Univ. Wis. 49.

Arvanitis, L. G. & O'Regan, W. G. (1969b). Sampling simulation computer programs. II. Cost-effectiveness and sampling efficiency. Res. Rep. Coll. Agr. Life Sci., Univ. Wis. 50.

Arvanitis, L. G. & O'Regan, W. G. (1971). Cluster or satellite sampling in forestry: a Monte Carlo simulation study. *Proc.* 3rd *Conf. Adv. Group of Forest Statist., Jouy-en-Josas, France* (in press).

Bary-Lenger, A. (1967). Étude statistique de la dispersion spatiale des arbres en forêt. Biométr.-praxim. 8, 115-48.

Beichelt, F. (1969). Die Entwicklung der Stammzahl im Bestand als stochastischer Prozess. Wis. Z. Tech. Univ., Dresden 18, 1051-3.

Bitterlich, W. (1948). Die Winkelzahlprobe. Allg. forst-u. holzw. Ztg. 59, 4-5.

Brown, G. S. (1965). Point density in stems per acre. N. Z. Forest Res. Notes, 38.

Catana, A. J. Jr. (1963). The wandering quarter method of estimating population density. Ecol. 44, 349-60.

Clark, P. J. & Evans, F. C. (1954). Distance to nearest neighbour as a measure of spatial relationships in populations. Ecol. 35, 445-53.

Cowlin, R. W. (1932). Sampling Douglas-fir reproduction stands by the stocked-quadrat method. J. Forest. 30, 437-9.

Cox, D. R. & Lewis, P. A. W. (1972). Multivariate point processes. *Proc. 6th Berkeley Symp.* (in press).

Finney, D. J. (1949). Random and systematic sampling in timber surveys. Forest. 22, 1-36.

Finney, D. J. (1950). An example of periodic variation in forest sampling. Forest. 23, 96-111.

Finney, D. J. (1953). The estimation of error in the systematic sampling of forests. J. Indian Soc. Agr. Statist. 5, 6-16.

Fraser, A. R. & van den Driessche, P. (1971). Triangles, density and pattern in point populations. *Proc. 3rd Conf. Adv. Group Forest Statist., Jouy-en-Josas, France* (in press).

Gill, C. B. (1950). A study of the relationship between the number of trees per acre and dispersion. Forest.Chron. 26, 186-96.

Grant, J. A. C. (1951). The relationship between stocking and size of quadrat. Forest. Bull. Univ. Toronto.

Greig-Smith, P. (1952). The use of random and contiguous quadrats in the study of structure of plant communities. Ann. Bot. 16, 293-316.

Haig, I. T. (1931). The stocked-quadrat method of sampling reproduction stands. J. Forest. 29, 747-9.

Hasel, A. A. (1938). Sampling errors in timber surveys. J. Agr. Res. 57, 713-36.

Holgate, P. (1965). Some new tests of randomness. J. Ecol. 53, 261-6.

Holgate, P. (1967). The angle-count method. Biometrika 54, 615-23.

Hopkins, B. (1954). A new method of determining the type of distribution of plant individuals. Ann. Bot. 18, 213-27.

Lowdermilk, W. C. (1921). A unit of area as a unit of restocking. Appl. Forest. Notes, North. Rocky Mtn. Forest Exp. Sta. 17, 1-3.

Lynch. D. W. & Schumacher, F. X. (1941). Concerning the distribution of natural regeneration. J. Forest. 39, 49-51.

Matérn, B. (1960). Spatial variation. Stochastic models and their application to some problems in forest surveys and other sampling investigations. Meddn. St. Skogsforskninst. 44, 1-144.

Matérn, B. (1971a). Performance of various designs of field experiments when applied in random fields. *Proc. 3rd Conf. Adv. Group Forest. Statist., Jouy-en-Josas, France* (in press).

Matérn, B. (1971b). Doubly stochastic Poisson process in the plane. In *Statistical Ecology, Vol. 1.* Patil, Pielou & Waters, eds. University Park: Penn. State Univ. Press, 195-214.

Mawson, J. C. (1968). A Monte Carlo study of distance measures in sampling for spatial distribution in forest stands. Forest. Sci. 14, 127-39.

Miles, R. E. (1970). On the homogeneous planar Poisson point process. Math. Biosci. 6, 85-127.

Mitchell, K. J. (1969). Simulation of the growth of even-aged stands of white spruce. Bull. Sch. Forest., Yale Univ. 75.

Newnham, R. M. (1964). The development of a stand model for Douglas-fir. Ph.D. Thesis, Fac. Forest., Univ. Brit. Columbia.

Newnham, R. M. (1966). A simulation model for studying the effect of stand structure on harvesting pattern. Forest. Chron. 42, 39-44.

Newnham, R. M. (1967). A progress report on the simulation model for pulpwood harvesting machines. Inf. Rep., Forest Manage. Inst., Ottawa, FMR-X-6.

Newnham, R. M. (1968). The generation of artificial populations of points (spatial patterns) in a plane. Inf. Rep. Forest. Manage. Inst., Ottawa, FMR-X-10.

Newnham, R. M. & Maloley, G. T. (1970). The generation of hypothetical forest stands for use in simulation studies. Inf. Rep. Forest Manage. Inst., Ottawa, FMR-X-26.

Newnham, R. M. & Sjunnesson, S. (1969). A FORTRAN program to simulate harvesting machines for mechanized thinning. Inf. Rep. Forest. Manage. Inst., Ottawa, FMR-X-23.

Nishizawa, M. (1968). Measures of competition and stand density for individual trees of *Pinus radiata*. Manage. Rep., Forest Res. Inst. Rotorua, N. Z. 14.

Opie, J. E. (1971). STANDSIM - A general model for simulating the growth of even-aged stands. *Proc. 3rd. Conf. Adv. Group Forest Statist., Jouy-en-Josas, France* (in press).

Osbourne, J. C. (1942). Sampling errors of systematic and random surveys of cover type areas. J. Amer. Statist. Assn. 37, 256-64.

Palley, M. V. and O'Regan, W. G. (1961). A computer technique for the study of forest sampling methods. I. Point sampling compared with line sampling. Forest Sci. 7, 282-94.

Payandeh, B. (1970a). Relative efficiency of two-dimensional systematic sampling. Forest Sci. 16, 271-6.

Payandeh, B. (1970b). Comparison of methods for assessing spatial distribution of trees. Forest Sci. 16, 312-7.

Persson, O. (1964). *Distance Methods. The Use of Distance Measurements in the Estimation of Seedling Density and Open Space Frequency.* Studia Forestalia Suecica, 15, Stockholm: Skogshogskolan.

Persson, O. (1965). *Distance methods II. Distributions of distance from a randomly located sample point to plants in an equilateral triangular lattice.* Institutionen for Skoglig Matematisk Statistik, Research notes, 6, Stockholm.

Persson, O. (1971). The robustness of estimating density by distance measurements. In *Statistical Ecology,* Vol. 2, Patil, Pielou & Waters, eds., University Park: Penn. State Univ. Press, 175-90.

Pielou, E. C. (1959). The use of point-to-plant distances in the study of the pattern of plant populations. J. Ecol. 47, 607-13.

Pielou, E. C. (1962). Runs of one species with respect to another in transects through plant populations. Biometrics 18, 579-93.

Pielou, E. C. (1963). Runs of healthy and diseased trees in transects through an infected forest. Biometrics 19, 603-14.

Pielou, E. C. (1965). The spread of disease in patchily infected forest stands. Forest Sci. 11, 18-26.

Rudra, A. B. (1968). A stochastic model for the prediction of diameter distribution of even-aged stands. Opsearch 5, 59-73.

Sahay, B. K. (1955). Sampling of natural regeneration. Indian Forest. 81, 742-52.

Sjunnesson, S. (1970). An example of simulation as a tool for analysis of man-machine systems for thinnings. Res. Notes, Dep. Oper. Efficiency, Roy. Coll. Forest., Stockholm, 42.

Stiteler, W. M. & Patil, G. P. (1971). Variance-to-mean ratio and Morista's index as measures of spatial patterns in ecological populations. In *Statistical Ecology,* Vol. 1, Patil, Pielou and Waters, eds., University Park: Penn. State Univ. Press, 423-53.

Strand, L. (1971). A model for stand growth. *Proc. 3rd Conf. Adv. Group Forest Statist., Jouy-en-Josas, France* (in press).

Suzuki, T. (1971). Forest transition as a stochastic process. Mitt. forstl. BundVersAnst, Wien 91, 69-88.

Wagener, P. E. & Stephens, G. R. (1970). Transition probabilities for a forest. Nature 225, 1160-1.

Warren, W. G. (1959). The distribution of stand merchantable volume by log diameter classes. N. Z. Forest. Res. Notes, 15.

Warren, W. G. (1962). Contributions to the study of spatial point processes. Ph.D. Thesis, Mimeogr. Ser., Inst. of Statist., Univ. North Carolina, No. 337.

Warren, W. G. (1964). Assessment of regeneration stocking in exotics. The nature of the problem and sampling methods. *Proc. Symp. Assessments Methods in New Zealand and Australia.* Wellington: R. E. Owen, Govt. Printer.

Warren, W. G. (1971a). The centre-satellite concept as a basis for ecological sampling. In *Statistical Ecology,* Vol. 2, Patil, Pielou, & Waters, eds., University Park: Penn. State Univ. Press, 87-116.

Warren, W. G. (1971b). Discussion on the papers by Stiteler and Patil and by Persson. In *Statistical Ecology,* Patil, Pielou & Waters, eds., University Park: Penn. State Univ. Press, Vol. 1, 453-6 and Vol. 2 187-80.

Wellner, C. B. (1940). Relationships between three measures of stocking in natural reproduction of the western white pine type. J. Forest. 38, 636-8.

SOME SPATIAL AND TEMPORAL POINT PROCESSES IN PHOTOGRAPHIC SCIENCE

J. F. Hamilton
Eastman Kodak Research Laboratories, Rochester, New York

W. H. Lawton
Eastman Kodak Research Laboratories, Rochester, New York

E. A. Trabka
Eastman Kodak Research Laboratories, Rochester, New York

SUMMARY

The photographic process is a rich source of interesting stochastic processes, yet so far little of the statistical literature has been motivated by problems in photographic science. We review some of the temporal and spatial point processes associated with the speed/grain ratio of photographic films in the hope that these problems will be swept into the mainstream of research in statistics and probability.

1. INTRODUCTION

We begin with a brief description of the photographic process in Section 2, examining the physical characteristics of the recorded image, moving on to mathematical description and physical measurement, and finally indicating briefly how these concepts lead to useful measures of merit.

Section 3 is devoted to the two-dimensional, spatial point processes associated with granularity, the noise in the photographic image. We will trace the development of models of increasing complexity, showing in particular why dishonest models continue to survive. Our own work on crowded emulsions, in which grain locations are significantly constrained because of the impenetrability of matter, leads us to formulate some theoretical questions concerning extensions of the renewal model to two dimensions, transects of isotropic random fields, and interacting point processes.

Section 4 emphasizes the one-dimensional temporal point processes associated with formation of the latent image that is developed into the photographic signal. Interest here centers around the failure of reciprocity between intensity and time of exposure. From detailed models based on solid-state physics, which so far appear to be amenable only to Monte Carlo solution, we abstract simpler models that may yield to analysis and from which significant explicit answers may still be obtained. Two phenomena that may be illuminated by proceeding in this manner are the saturation (roll-over) of low-intensity reciprocity failure and the role of competition for growth of latent image centers in high-intensity reciprocity failure.

We warn the reader that this is not a comprehensive review. The subject is too vast to be covered in a single paper. Instead, we have tried to touch upon some of the more important aspects of understanding and evaluating signal and noise in the photographic process as we see it in the course of our own work. In surveying the literature, we have made a modest attempt at historical accuracy. But we have not hesitated to present old ideas in new ways and in more useful forms.

2. THE PHOTOGRAPHIC PROCESS

2.1. Recorded Image

Most of us have used photographic film and know that it can record scenes such as that of the young lady in Fig. 1. It is also likely that most of us do not know much about how this image reproduction is accomplished. For this reason our point of departure will be a detailed, if somewhat idealized, description of the photographic process (Mees & James, 1966).

Imagine, then, that the girl in Fig. 1 is alive and here waiting for us to record this scene. What we perceive is a spatial variation in reflected light intensity, and this we convert by appropriate exposure and development to a spatial variation in transmittance or optical density of film. How this is actually accomplished will be revealed shortly, but in any case lightness becomes darkness so that the resulting

Figure 1. A continuous tone positive print.

Figure 2. The negative record.

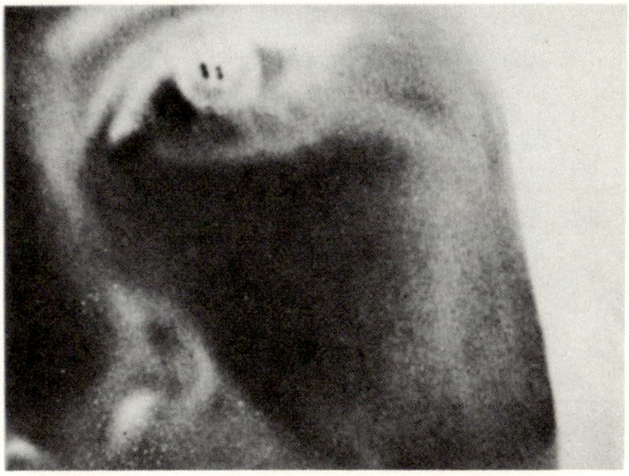

Figure 3. A 16X enlargement.

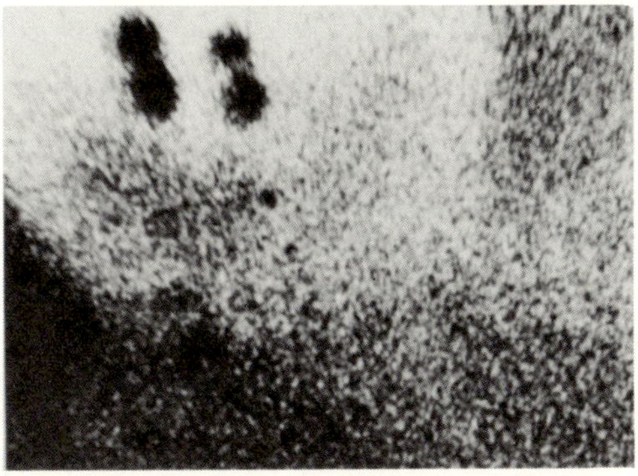

Figure 4. A 100X enlargement

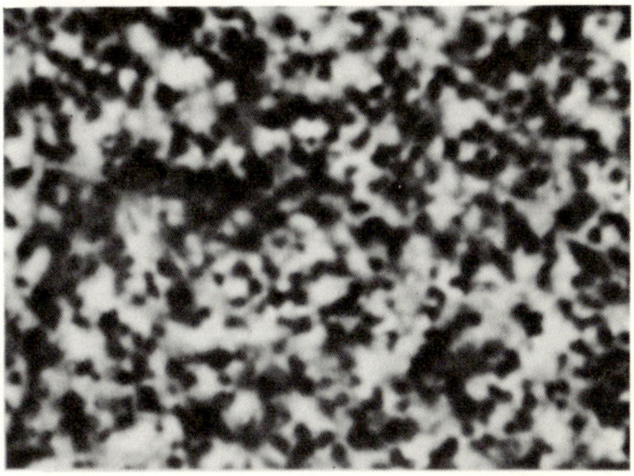

Figure 5. A 1000X enlargement.

negative shown in Fig. 2, appears to be composed of locally uniform shades or tones of darkness that vary with position. These continuous tones, or changes in them (depending on your point of view), are the signals of interest.

Let us look at a small region in order to isolate a single tone. In the 16X enlargement shown in Fig. 3 several features are still visible but the flesh tone now has a slightly grainy appearance. Enlarging by a factor of 100, we obtain the very grainy image of Fig. 4. In the 1000X enlargement shown in Fig. 5 we see discrete clumps of silver in a transparent medium.

This shows that the developed image is really a *spatial point process* of random events of different types (silver grains of varied size and shape). The number, size and configuration of grains in a region determines the amount of light transmitted to our eye from that region. This accounts for the grainy appearance at those lower magnifications 16X and 100X) where the eye could not resolve the actual grains. Without any magnification, the image appeared to consist of continuous, uninterrupted tones.

Those fluctuations in the number, size and configuration of grains from one small subregion to the next, which cause corresponding fluctuations in the transmittance (or optical density) of uniformly exposed regions, constitute the *noise* of our photographic system. When we measure this noise objectively we call it *granularity*, whereas when we refer to its subjective impression, we call it *graininess*.

But how was this record obtained? If we had looked at a microscopic area before development, we probably would have seen something like what is shown magnified 25,000X at (a) in Fig. 6. Originally, the film placed in the camera consisted of very small, light sensitive, nearly transparent silver halide crystals dispersed in transparent gelatin and coated on a transparent support; but exposure of these crystals to light caused the formation of tiny silver specks (aggregates of silver atoms) within the crystals. The opaque silver specks shown here have been chemically enlarged slightly so that they are visible to the electron microscope. When the exposed film is placed in a photographic developer,

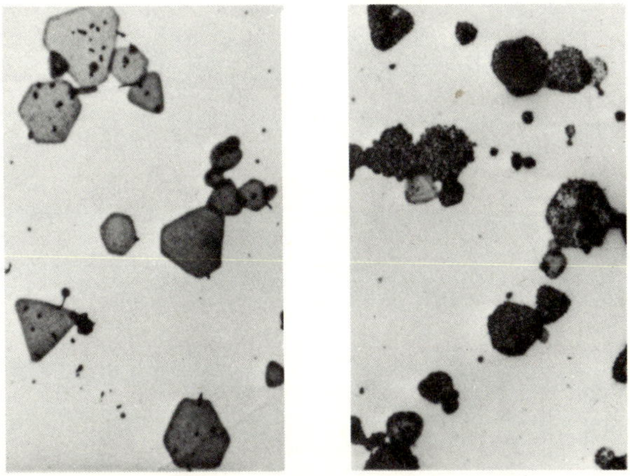

(a) (b)

Figure 6. Electron micrographs for the same exposure showing development centers consisting of aggregates of silver atoms on silver bromide emulsion grains. (a) Low-intensity. (b) High-intensity.

these silver halide crystals are converted to opaque metallic silver. The rate, or probability, that they are converted is, however, dependent on the size and number of these silver centers.

The micrograph at (b) in Fig. 6 is for the same total exposure as that at (a), but of higher intensity and shorter time. Here we use exposure in the technical sense as the product of light intensity (more precisely, irradiance) and exposure time. The strikingly different statistical appearance of these micrographs is a visual dramatization of *reciprocity failure* in the photographic process (see Section 4.2), which is related to another interesting spatial point process going on inside the silver halide crystal itself. Often we prefer to consider a silver halide crystal merely as a kind of black Pandora's box that requires a certain temporal pattern of photon absorptions for developability. Consequently, much of the photographic literature is concerned with the *temporal point process* of photon absorption.

2.2. Mathematical description and physical measurement

The photographic negative is, however, a physical object.

Mathematical analysis is carried out on a mathematical description of such objects. A complete mathematical description of the developed negative is given by the fraction t(x,y) of incident light passed by the negative at the point (x,y). This so-called *point transmittance* is equal to 0 if the negative is opaque at (x,y) and is equal to 1 if it is perfectly transparent.

Because of the random nature of the distribution of silver halide crystals and of the exposure mechanism, it is useful to consider the photographic negative for a *uniform exposure* as the realization of some two-dimensional, homogeneous, isotropic random field $\underline{t}(x,y)$. (The underlining is used here to denote a random variable, not a vector.) Of particular interest are the mean $E\{\underline{t}(x,y)\} = \mu_t(x,y) = \mu_t$ and $\mathrm{var}\{\underline{t}(x,y)\} = \sigma_t^2(x,y) = \sigma_t^2$; since the process is homogeneous, these quantitites will be independent of (x,y). (We will consistently use μ and σ to denote respectively the (ensemble) mean and standard deviation of the random variable indicated by the attached subscript.).

Another quantity of interest, for it is easily measured, is the average *area (aperture) transmittance*

$$\underline{T}_A = \frac{1}{A} \int\int_A \underline{t}(x,y)\,dxdy \tag{1}$$

of a finite area A of the negative. This too will be a random variable unless A is very large. Its expected value will be equal to

$$E(\underline{T}_A) = \mu_T = \mu_t, \tag{2}$$

which is independent of the area chosen. Although the subscript A will often be suppressed to simplify expressions, we will always use the capital $\underline{T}$ to denote area (aperture) transmittance in contrast to the lower case $\underline{t}$, which will be reserved for point transmittance.

Initially, photographic scientists concerned themselves only with the analysis of the continuous tone aspects of photography. Of principal interest was the relationship between the uniform exposure E received by a very large area A of the film, and the resulting average transmittance

$$T_\infty = \lim_{A\to\infty} \underline{T}_A \tag{3}$$

in the developed negative. (Strictly speaking, $\underline{T}_A$ is dependent on the irradiance I to which the film is subjected as well as the time of exposure τ. Over a limited range, the so-called reciprocity law may hold and $\underline{T}_A$ will depend solely on the exposure $E = I\tau$. See Mees & James (1966)). Equations (1) and (3) indicate that T_∞ is the *spatial* average of the point transmittance. Provided the point transmittance is ergodic, we have that

$$T_\infty = \mu_t.$$

In particular, $\mathrm{var}(T_\infty) = 0$, a behavior that is confirmed experimentally, and T_∞ is deterministic.

Although T_∞ was then the fundamental observable, the most meaningful quantity to many photographic scientists was the *optical density* defined by

$$D = -\log_{10} T_\infty = -\log_{10} \mu_t. \tag{5}$$

This preference is based on the additive nature of macroscopic densities as opposed to the multiplicative nature of transmittance. (The psychophysical evidence embodied in the Weber-Fechner law is sometimes also given as a justification for this preference. This law implies that just noticeable differences correspond to equal small differences in density. See Brown & Mueller (1965)).

This study of the relationship of exposure and density led to the classical Hurter and Driffield (H&D) curve in which the large-area density D is plotted against the logarithm of exposure, log E. Such a curve is shown in Fig. 7. One of the features of an H and D curve that is of special interest is the *photographic speed* as measured by

$$S = 1/E_0, \tag{6}$$

where E_0 is the exposure required to produce some specified density D_0. The higher the speed the more sensitive the film is to light.

Later interest turned from the average transmittance or density of the effectively infinite, uniformly exposed region to an examination of local nonuniformity within such a region. These studies were carried out experimentally by drastically reducing the area A over which

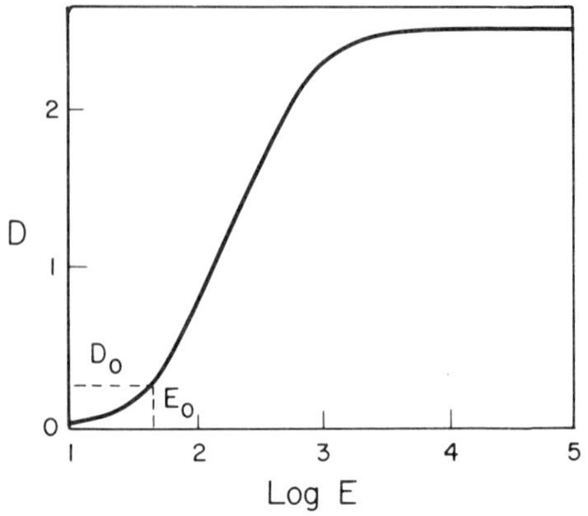

Figure 7. Classical H&D curve of large-area density D vs log exposure showing determination of speed $S = 1/E_0$.

the aperture transmittance $\underline{T}_A$ was determined. Numerous measurements (observations) of $\underline{T}_A$ were made within large, uniformly exposed regions. Its sample variance is a good estimator of $\text{var}(\underline{T}_A) = \sigma_T^2$. Note that for a small area A, $\underline{T}_A$ no longer behaves like a deterministic variable and its variance $\text{var}(\underline{T}_A) = \sigma_T^2$ provides a measure of the non-uniformity, or *granularity*, of the region.

By this time, however, large-area density as defined by (5) had become firmly established, and so granularity work was cast in terms of density. Not surprisingly, the deterministic relation (5) was used to define small-area density, or microdensity. A new random variable $\underline{D}$ was defined by

$$\underline{D} = -\log_{10}\underline{T}, \tag{7}$$

where $\underline{T} = \underline{T}_A$.

A formal rigorous relationship between the mean and variance of $\underline{T}$ and the mean and variance of $\underline{D}$ is not possible without making some additional assumptions (Trabka, 1969). Usually, it is assumed that the averaging area (aperture) A is large enough so that σ_T^2 is reasonably small. It then follows that the means and variances of these random variables are approximately related by

$$\mu_D \approx -\log_{10}\mu_T, \tag{8}$$

$$\sigma_D^2 \approx k^2 \sigma_T^2/\mu_T^2, \tag{9}$$

where $k = \log_{10}e$. In many instances (but not all), data denoted as μ_D and σ_D^2 are actually obtained from measurements on $\underline{T}_A$ and *calculated* according to the right-hand sides of (8) and (9), the transformation indicated by (7) never having been performed. In those instances, and for our purposes here, we may look upon (8) and (9) as *definitions*. (Often the event $\underline{T} = 0$, that is, $\underline{D} = +\infty$, happens with a positive probability and then $\mu_D = +\infty$ while σ_D is undefined. However, the physical observable in (7) is $\underline{T}$ and analog computation of $\underline{D}$ (in a microdensitemeter) can only approximate (7) over some range $0 < T_{min} \leq \underline{T} \leq 1$.)

An important characteristic of photographic noise is its dependence on the signal level μ_D. Plots of normalized σ_D vs μ_D are shown in Fig. 8. The curves shown there are theoretical predictions based on mathematical models that will be described later. The black dots correspond to measurements on a typical commercial emulsion. The quantity $N^{\frac{1}{2}}/k$ has been used as a convenient normalization for the theoretical predictions where $k = \log_{10}e$ and $N = A/a$ is the ratio of aperture area A to grain area a. The experimental data has been scaled (multiplied by a constant) to fit the theoretical curves for low mean densities. As the dashed curve in this log-log plot is a straight line with slope $\frac{1}{2}$, we see that the experimental data may be characterized as approximately a straight line with slope slightly less than $\frac{1}{2}$.

Since it is sometimes convenient to remove the known theoretical dependence of σ_D on the area A of the scanning aperture, the normalized quantity

$$G = (2A)^{\frac{1}{2}}\sigma_D, \tag{10}$$

known as the Selwyn granularity after Selwyn (1935), is often used.

In many instances the light transmitted by film is passed through the eye or some other optical system, which over certain ranges can be approximated as a linear system with a nonuniform spatial

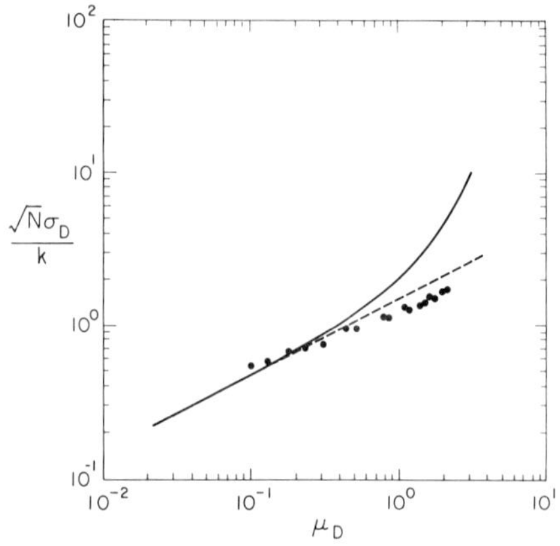

Figure 8. Normalized rms density fluctuations $N^{\frac{1}{2}}\sigma_D/k$ vs mean density μ_D. Black dots are data obtained from measurements on a commercial emulsion. Curves are theoretical predictions for opaque grains by a dishonest (dashed) and honest (solid) method.

frequency response. Consequently, it is important to know the spectrum of the transmittance fluctuations. Homogeneity and isotropy are reasonable assumptions so that a single spatial frequency variable ν in lines per millimeter suffices as the independent variable. Typical spectra are shown in Fig. 9, where the spectrum of the print has been predicted (Doerner, 1962) from that of the negative.

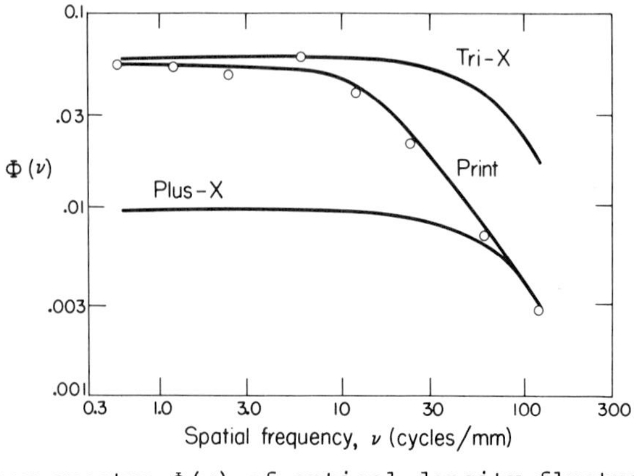

Figure 9. Wiener spectra $\Phi(\nu)$ of optical density fluctuations vs spatial frequency ν. The negative on Kodak Tri-X film is printed on Kodak Plus-X film and the spectrum of the print is shown both from measurements (circles) and theory (solid curve)(Doerner, 1962).

2.3. Measures of merit

We have noted so far two classes of problems in photographic science: signal analysis (μ_D) and noise analysis (σ_D, spectrum). Moreover, although both spatial and temporal point processes play roles in each, up to now, temporal point processes of photon absorption have been central to signal analyses, and spatial point processes have been central to noise analyses. Before reviewing these areas, we must call attention to some measures of merit that have been found useful in the evaluation of photographic systems. See also Higgins (1971). These are some *numbers* of interest to the photographic scientist and it is to them that abstract theory ought to relate if it is to be helpful.

Since high speed and low granularity are substantially incompatible, photographic scientists often seek to maximize the speed-granularity ratio (more commonly called speed-grain ratio)

$$\frac{\text{speed}}{\text{grain}} = \frac{S}{G} . \tag{11}$$

It is also possible to regard photographic film as a member of a wider class of radiation detectors (Jones, 1959; Zweig, 1964). In this context, the input signal is usually considered to be a change $\Delta\mu_E$ in exposure, to be detected by a corresponding change $\Delta\mu_D$ in the output signal. The detective quantum efficiency Q, defined as

$$Q = \frac{(\text{signal/noise})^2_{\text{out}}}{(\text{signal/noise})^2_{\text{in}}} \tag{12}$$

is found to be a useful figure of merit. It can be expressed as the product

$$Q = D \cdot D_c \tag{13}$$

of two other figures of merit, viz., the detectivity

$$D = (A\sigma_D)^{-1} \frac{d\mu_D}{d\mu_E} , \tag{14}$$

which is closely related to the speed-grain ratio in (11), and the contrast detectivity

$$D_c = \sigma_D^{-1} \frac{d\mu_D}{d(\log_e \mu_E)} = (A\mu_E)D, \tag{15}$$

which is closely related to the ability of photographic film to detect contrast *changes* in the original scene (detectable signal vs noise in input).

The capacity of film to store information is taken into account by using the area of the film point spread (response to a point exposure) in (14) and measuring σ_D with an aperture of this size. In this connection the detectivity D is also known as informational sensitivity, which is not to be confused with information capacity in the Shannon sense. The latter has also been calculated for photographic films (Jones, 1961).

3. GRANULARITY AND SPATIAL POINT PROCESSES

3.1. Historical Development of Models

We have seen that uniformly exposed and processed film has a grainy appearance (when viewed under magnification) because its transmittance is determined by small grains of silver distributed with some degree of chaos throughout the emulsion. Some of the important factors are:

1. The number of grains;
2. Their overlap configuration;
3. Emulsion thickness;
4. Light scatter.

Realistic models of granularity incorporating the effect of light scatter appear intractable at present, although emulsion point spreads can be calculated using Monte Carlo techniques (Wolfe, Marchand & DePalma, 1968). All models for granularity in the literature have assumed that silver grains do not scatter light but only absorb it. Moreover, they have only dealt with collimated incident light so that simple geometrical shadowing is used to determine the total projected area of the grains, which in turn determines the transmittance of the aperture.

3.1.1. First generation models - the random dot model and a dishonest method

As we have noted, granularity is signal-dependent so that it is important to know how μ_D depends on the factors listed above. In the very first communication from the Kodak Research Laboratories, P.G. Nutting (1913) obtained the formulae

$$\mu_T = \left(1 - \frac{\bar{n}a}{m}\right)^m, \quad \mu_D = -m \log_{10}\left(1 - \frac{\bar{n}a}{m}\right), \tag{16}$$

where $\bar{n}$ is the concentration or average number of grains per unit projected area distributed evenly among m layers, all grains having the same area a and overlapping prohibited within a single layer. Note that $\mu_D = \infty$ when $m = \bar{n}a$ and, as $m \to \infty$ (infinite thickness), we have the asymptotic expressions

$$\mu_T = e^{-\bar{n}a}, \quad \mu_D = k\bar{n}a. \tag{17}$$

Later Silberstein and Trivelli (1938) ignored emulsion thickness and considered opaque flat disks whose centers are distributed completely at random (i.e., without constraints) over a scanning aperture. These disks represent the projected area of grains distributed throughout the volume of the emulsion. Note that since the absence of constraints would hold for an infinitely thick emulsion, Silberstein and Trivelli, in effect, gave another derivation of (17). Their model was later to be christened *the random-dot model* (Marchant & Dillon, 1961). By this time (17) had become to photographic scientists what F = ma is to the physicist.

The numbers factor appears in both (16) and (17) via $\bar{n}$, whereas emulsion thickness m appears only in the original Nutting formula. Both the Nutting and Silberstein-Trivelli formulas for mean density μ_D allow overlap of the grain projections. For without overlap m = 1 and we would have

$$\mu_T = 1 - \bar{n}a, \quad \mu_D = -\log(1-\bar{n}a). \tag{18}$$

However, average transmittance in (17) is an exponential rather than a linear function of $\bar{n}$ exactly because we lose coverage due to overlap.

Early workers did not know how to take the overlap configuration factor into account in a rigorous way when computing variances. Heuristic attempts yielded such strange results (Selwyn, 1935) that the following "dishonest" method became established: one assumes that (17), a relation between means, is also true for the corresponding random variables, viz.,

$$\underline{D} = k\,\underline{n}\,a. \tag{19}$$

Here, $\underline{n} = \underline{s}/A$ where $\underline{s}$ denotes the number of silver grains whose centers project into a given area A. The assumption of no constraints implies that $\underline{s}$ has a Poisson distribution with $\sigma_s^2 = \mu_s = A\bar{n}$. Taking the variance of both sides of (19) yields

$$\sigma_D = (k\,\frac{a}{A}\,\mu_D)^{\frac{1}{2}}. \tag{20}$$

Equation (20), which states that rms density is proportional to the square root of the mean density, together with (17), succinctly summarizes the signal and noise characteristics of many photographic negatives. The assumptions made in their derivation are inconsistent, with variances in overlap configuration contributing to μ_D, but not to σ_D except indirectly through the dependence of σ_D on μ_D. Yet, note that (20) plotted on a log-log scale is a stright line with slope $\frac{1}{2}$, shown as the dashed curve in Fig. 8, and that the data for a typical emulsion plotted there show much the same thing, except that the slope is slightly less than $\frac{1}{2}$. The predictions of this dishonest method are reasonably good.

The period from 1935 to 1955 added little to granularity theory. Emphasis was on improving measurement techniques and establishing the psychophysical relation between granularity and graininess. There were some extensions of the square-root law, which we can readily obtain by generalizing (19) to

$$\underline{D} = \frac{k}{A}\sum_{i=1}^{\underline{s}} \underline{a}_i, \tag{21}$$

allowing the grain area *a* to be a random variable $\underline{a}$, so that the density is expressed as the sum of a random number of $\underline{s}$ of identically distributed random variables $\underline{a}_i$. We can immediately write

$$\mu_D = \frac{k}{A}\mu_s\mu_a \tag{22}$$

and

$$\sigma_D^2 = \frac{k^2}{A^2}(\mu_s\sigma_a^2 + \mu_a^2\sigma_s^2), \tag{23}$$

which can be combined to yield

$$\sigma_D = \{\frac{k}{A}\mu_a\mu_D(\frac{\sigma_s^2}{\mu_s} + \frac{\sigma_a^2}{\mu_a^2})\}^{\frac{1}{2}}. \tag{24}$$

The formula for the Poisson case when $\sigma_s^2/\mu_s = 1$ was given earlier by Siedentopf (1937), whereas the more general form given by (24) was first derived by Ingarden and Mikusinski (1953) using a method different from that given here.

3.1.2 Second generation models - focus on point transmittance and honest methods

The second phase of development of the random-dot model was initiated by B. Picinbono (1955a,b); see O'Neill (1963). Whereas phase one generally did not treat the film and measuring apparatus separately, but was concerned mostly with aperture transmittance, Picinbono focused his attention on the intrinsic point transmittance $\underline{t}(P)$ of the two-dimensional film. Still using opaque circular grains, he took $\underline{t}(P)$ equal to 0 or 1 depending on whether or not the point P was covered by a circular grain. The centers are Poisson-distributed and the grain radius is allowed to be a random variable. For this case, Picinbono derived a rigorous formula for the mean, covariance and spectrum of point transmittance fluctuations.

The results of Picinbono were subsequently generalized by Savelli (1957, 1958), who allowed the grain and film-base transmittance to be parameters but who considered only circular grains of a single size (monodisperse emulsions). He obtained more information about the random-dot model than Picinbono by deriving also an explicit formula for the characteristic function of the joint distribution of the numbers $\underline{s}(P_1)$ and $\underline{s}(P_2)$ of grains covering the points P_1 and P_2, respectively.

Recently, Benton and Kronauer (1971) generalized Picinbono's formula for the covariance function of point transmittance fluctuations for the random-dot model so that it is valid for an isotropic (invariant under rotation) ensemble of grains of arbitrary size, shape and absorptance profile. The assumption of isotropy, however, is unnecessary as the following development (Kemperman, 1971) shows. It includes all the earlier special cases, in particular a generalization of Savelli's results.

Introducing the transform

$$\psi(u_1,u_2) = E[\{\underline{t}(P_1)\}^{u_1}\{\underline{t}(P_2)\}^{u_2}], \tag{25}$$

the mean μ_t and covariance $C(P_1-P_2)$ of $\underline{t}(P_1)$ and $\underline{t}(P_2)$ can be expressed as

$$\mu_t = \psi(1,0) = \psi(0,1), \tag{26}$$

$$C(P_1-P_2) = \psi(1,1) - \mu_t^2.$$

The grains are taken to be partially transparent objects of arbitrary shape. Each grain is characterized by a reference point P_0 (usually somewhere near the center of the grain) and a point absorptance function $A(P)$, which can be any measurable function satisfying $0 \leq A(P) \leq 1$. We assume that at the point $P' = P_0 + P$ the grain absorbs a fraction $A(P) = A(P'-P_0)$ of the incident light. Normally, $A(P) = 0$ for large values $|P|$.

Now consider a random collection Ω of absorptance functions. More precisely, let Ω be a probability space and suppose that to each point $W \epsilon \Omega$ there corresponds a function $A(W,P)$ satisfying $0 \leq A(W,P) \leq 1$. Selecting a random absorptance function from this collection amounts to selecting a random point $\underline{W}$ from the probability space Ω, subject to $\text{prob}(\underline{W}\ \epsilon B) = P(B)$ for measurable subsets $B \subset \Omega$.

Select a random field of grains as follows: In the first place, the reference points $\underline{P}_1$, $\underline{P}_2$,... of these grains are chosen as a two-dimensional Poisson field with density $\bar{n}$. Further, we associate with each reference point $\underline{P}_i$ and absorptance function chosen at random from the above collection, the different absorptance functions being independent of each other and independent of the reference points $\underline{P}_i$. The corresponding random grain absorbs at the point P a fraction $\underline{A}_i(P-\underline{P}_i) = A(\underline{W}_i, P-\underline{P}_i)$ of the incident light and we allow the different grains to overlap each other. Thus the resulting random field of grains will be such that at the point P there is a random transmittance $\underline{t}(P)$ given by

$$\underline{t}(P) = t_b \cdot \prod_{i=1}^{\infty} \{1-A(\underline{W}_i, P-\underline{P}_i)\}, \tag{28}$$

where the film base transmittance t_b will generally not be equal to 1.

Using this definition of $\underline{t}(P)$ in (25), one can show (Kemperman, 1971) that

$$\psi(u_1,u_2) = t_b^{u_1+u_2} e^{-\bar{n}\Phi(u_1,u_2)}. \tag{29}$$

Here, $\bar{n}$ is the average number of grain centers (reference points) per unit area while $\Phi(u_1,u_2)$ is given by the planar integral

$$\Phi(u_1,u_2) = \int\{+1-E_W[\prod_{j=1}^{2} \{1-A(\underline{W},P_j-P)\}^{u_j}]\}dP. \tag{30}$$

The only random contribution left in (30) is from $\underline{W}$ (the ensemble of grains), and E_W denotes expectation with respect to $\underline{W}$, (that is, integration relative to the probability measure on Ω). The points P_1 and P_2 in (32) are not reference points of grains but are merely those points for which we desire to evaluate the transform defined by (25).

In order to calculate the mean and variance, it is convenient to introduce the quantities

$$\langle a\rangle = \int E_W\{A(\underline{W},P)\}dP, \tag{31}$$

$$\Gamma(P_1-P_2) = \langle a\rangle^{-1}\int E_W\{A(\underline{W},P)A(\underline{W},P+P_1-P_2)\}dP, \tag{32}$$

which are often easily computable. Both integrals extend over the full plane. Clearly $0 \leq \Gamma(P_1-P_2) \leq 1$; further $\Gamma(P_1-P_2) \leq \Gamma(0) \leq 1$ by Cauchy's inequality. The quantity $\langle a\rangle$ is an effective average absorptive area and $\Gamma(P_1-P_2)$ is a normalized average autocorrelation function of grain absorptance profiles. Combining (26), (27), (29) and (30) we obtain the relations

$$\mu_t = t_b e^{-\bar{n}\langle a\rangle}, \tag{33}$$

$$C(P_1-P_2) = \mu_t^2\{(t_b/\mu_t)^{\Gamma(P_1-P_2)} - 1\}. \tag{34}$$

Note that in view of (2), (33) is a generalization of Silberstein's formula (17); further $C(P_1-P_2) \geq 0$. If we assume isotropy and write $\rho = |P_1-P_2|$, we have $C(P_1-P_2) = C^*(\rho)$ and $\Gamma(P_1-P_2) = \Gamma^*(\rho)$ corresponding to the result of Benton and Kronauer (1971). The results are also applicable to color granularity where the grains correspond to dye clouds whose absorptance is nonuniform.

Let us now specialize our results to circular grains having a uniform absorptance g, i.e.,

$$A(\underline{r},P) \begin{cases} = g & (|P| < \underline{r}), \\ = 0 & (|P| > \underline{r}), \end{cases} \tag{35}$$

where the random grain radius $\underline{r}$ has a given distribution function $\text{prob}\{\underline{r} < r\} = F(r)$ and $0 < g \leq 1$ is an absolute constant. In this case the point transmittance can be expressed as

$$\underline{t}(P) = t_b(1-g)^{\underline{s}(P)}, \tag{36}$$

where $\underline{s}(P)$ denotes the number of grains covering the point P. Furthermore, the transform (25) can be regarded as the characteristic function of the joint distribution of the numbers $\underline{s}(P_1)$ and $\underline{s}(P_2)$. More precisely,

$$E[e^{i\{v_1\underline{s}(P_1)+v_2\underline{s}(P_2)\}}] = t_b^{-u_1-u_2}\,\psi(u_1,u_2), \tag{37}$$

provided we choose $u_j = iv_j/\log_e(1-g)$ for $j = 1,2$; here $i = (-1)^{\frac{1}{2}}$. The left-hand side of (37) was evaluated by Savelli (1971), but only for monodisperse emulsions. The effective absorptive area in the present case is given by

$$\langle a \rangle = g\mu_a, \tag{38}$$

where μ_a is the average grain area.

In obtaining results for the covariance, we need not restrict ourselves to two dimensions as the above analysis holds for a one-dimensional random-dot model also. (More precisely, the plane can be replaced by any additive group with underlying σ-infinite invariant measure dP.) We emphasize this because the one-dimensional result will be useful for comparison with one-dimensional models of crowded emulsions considered later. Introducing the subscript j = 1 or 2 to indicate dimensionality, the quantities of interest become

$$\mu_t = t_b e^{-\bar{n}ga_j}, \tag{39}$$

$$C_j^*(\rho) = \mu_t^2\{(t_b/\mu_t)^{g\gamma_j(z)} - 1\}, \tag{40}$$

$$\sigma_t^2 = \mu_t^2\{(t_b/\mu_t)^g - 1\}, \tag{41}$$

$$z = \rho/\rho_j. \tag{42}$$

Here, $\rho = |P_1 - P_2|$ is variable while ρ_j is a convenient normalization constant. Moreover, a_j is the average j-dimensional grain area and

$$\gamma_j(z) = a_j^{-1}\int_{\frac{1}{2}\rho}^{\infty} I_j(r,\rho)\,dF(r). \tag{43}$$

The distribution function of grain radii is F(r), and $I_j(r,\rho)$ denotes the j-dimensional area of the *intersection* of two j-dimensional circles of diameter 2r having their centers a distance ρ apart. Clearly, $I_j(r,\rho) = 0$ if $\rho > 2r$. Further, if $\rho < 2r$, then

$$I_1(r,\rho) = 2r - \rho, \tag{44}$$

whereas

$$I_2(r,\rho) = 2r^2 \arccos(\rho/2r) - \rho\{r^2 - (\rho/2)^2\}^{\frac{1}{2}}. \tag{45}$$

In the special case that all grains have the same radius r_0, we have $\gamma_j(z) = a_j^{-1} I_j(r_0, \rho)$. Taking $\rho_j = 2r_0$, we have for one dimension that

$$a_1 = \rho_1 = \ell \text{ (grain length)}, \tag{46}$$

$$\gamma_1(z) \begin{cases} = 1-z & (0 \le z \le 1), \\ = 0 & (z > 1), \end{cases} \tag{47}$$

whereas for two dimensions, we have

$$a_2 = \pi r_0^2 = a \text{ (grain area)}, \tag{48}$$

$$\rho_2 = 2r_0, \tag{49}$$

$$\gamma_2(z) \begin{cases} = \frac{2}{\pi}\{\arccos z - z(1-z^2)^{\frac{1}{2}}\} & (0 \le z \le 1), \\ = 0 & (z > 1). \end{cases} \tag{50}$$

The point-transmittance spectrum for the two-dimensional random-dot model, namely,

$$P(\nu) = 2\pi \int_0^{\infty} C_2^*(\rho) J_0(2\pi\nu\rho)\rho d\rho, \tag{51}$$

is shown in Fig. 10 as calculated by Savelli (1957). The normalized spectrum $P(\nu)/8a\mu_t^2$ is plotted vs normalized spatial frequency νr_0 for various mean transmittances at (a) and the value of the spectrum at zero frequency is plotted vs mean transmittance at (b). Note the rapid build-up of low frequency power as the mean transmittance decreases.

The variance σ_T^2 of the aperture transmittance is given by

$$\sigma_T^2 = \int_{-\infty}^{+\infty} \phi_j(P) C_j(P) dP, \tag{52}$$

in view of Parseval's theorem. The integral is either single or double depending on j, $\phi_j(P)$ is the self-convolution of the aperture transmittance distribution and $C_j(P)$ is the covariance of point transmittance fluctuations. For a transparent circular aperature of radius R_0 and area $A = R_0^2$, we have

$$\phi_2(P) = A^{-1}\gamma_2\{|P|/2R_0)\}, \tag{53}$$

whereas for a rectangular aperture of length L, we have

$$\phi_1(P) = L^{-1}\gamma_1(|P|/L). \tag{54}$$

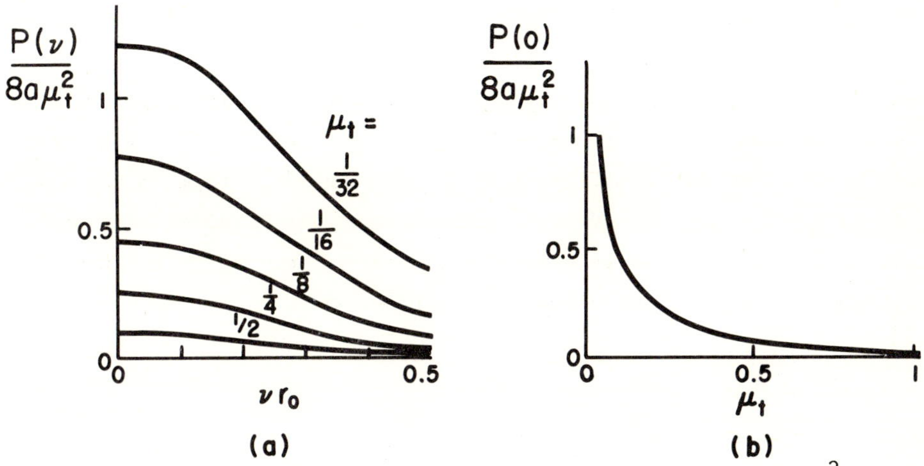

Figure 10. Normalized point transmittance fluctuations $P(\nu)8a\mu_t^2$. (a) Wiener spectrum vs normalized spatial frequency νr_0. (b) Zero-frequency intercept vs mean transmittance (Savelli, 1958).

The covariances $C_j(P)$ are given by (40), while $\gamma_1(z)$ is given by (47) and $\gamma_2(z)$ by (50). In the special case of a constant grain radius r_0, recalling Definition (9) and Identity (2), we can write

$$\frac{N_j^{\frac{1}{2}}\sigma_D}{k} = [2^{2j-1}\int_0^{\min(\alpha_j,1)} \gamma_j(\frac{u}{\alpha_j})\{(t_b/\mu_T)^{g\gamma_j(u)} - 1\}u^{j-1}\,du]^{\frac{1}{2}}, \tag{55}$$

where $N_1 = L/\ell$, $N_2 = A/a$, and $\alpha_j = N_j^{1/j}$. Since aperture transmittance is usually the measured quantity, we take (8) as the definition of μ_D. Thus (55) allows us to plot normalized σ_D vs μ_D as the solid curve in Fig. 8 for g = 1. The full consequence of the honest random-dot model, incorporating the configuration overlap term in both μ_D and σ_D, turns out to be a *greater* discrepancy with experimental results than that given by the simpler dishonest method. Although this result follows almost immediately from the work of Picinbono or Savelli, photographic scientists did not learn of the disastrous effect of the term they had long wished to incorporate into the random-dot model until much later when Bayer (1964) gave an *ad hoc* derivation of this same result and pointed out its consequences. He also noted that in practice the incident light is usually not collimated and there is considerable light scatter in the emulsion, all of which could account for the overestimation of granularity by the honest random-dot model.

There are some approximate closed-form expressions for σ_D that are useful for large apertures, i.e., when $\alpha_j >> 1$. The upper limit of integration in (55) becomes 1, and the first term in the integrand is approximately 1 over the range where the second term is different from zero. If we approximate γ_2 by γ_1, then the integrations are performed readily in both dimensions and we get

$$\frac{(N_j)^{\frac{1}{2}}\sigma_D}{k} = \{2^{2j-1}Q_j(Z)\}^{\frac{1}{2}}, \tag{56}$$

where

$$Q_1(Z) = \frac{1}{Z}(e^Z-1-Z), \tag{57}$$

$$Q_2(Z) = \frac{1}{Z^2}(e^Z-1-Z-\frac{1}{2}Z^2), \tag{58}$$

and

$$Z = \frac{g}{k}(\mu_D-D_b), \tag{59}$$

with $D_b = -\log_{10}t_b$ being the density of the base. For $j = 1$,(57) is an asymptotic expression as $\alpha_1 \to \infty$. For $j = 2$, there is a slight bias error, as shown in Fig. 11, although the shape of the curve over the region of interest has negligible distortion. Making a comparison of (56) and (58) with (20) for small Z, we find that our approximation is larger by a factor of $2/\sqrt{3}$, and this factor may be readily incorporated into the forumula (56) for $j = 2$. Equations (56) and (57) for small Z yield (20), as we expected.

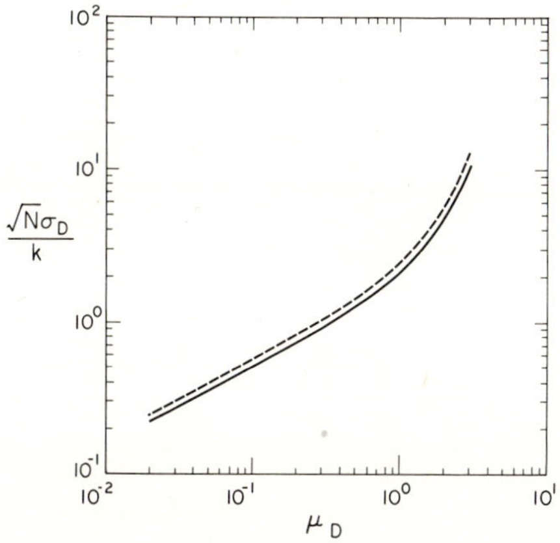

Figure 11. Normalized rms density fluctuations $N^{\frac{1}{2}}\sigma_D/k$ vs mean density μ_D for the honest, two-dimensional, random-dot model with opaque grains. The lower (solid) curve is exact whereas the upper (dashed) curve comes from the approximation described in the text.

These closed form expressions for σ_D are very useful. We have seen that by reducing to (20) for small Z they predict the straight line with slope $\frac{1}{2}$ on a log-log plot for small Z, while the exponential terms in the $Q_j(Z)$ yield the rapid buildup of σ_D for large Z. We can also see from the way that the grain absorptance g enters the expression for Z that the straight line portion of the curve extends over a larger range of μ_D for small g. This is illustrated in Fig. 12, which shows curves for various values of g. When g = 0.1 the curve is a straight line with slope $\frac{1}{2}$ up to a density of 3. This suggests the possibility of interpreting granularity measurements on real emulsions that scatter light (and for which the incident light is not completely collimated) by an honest random-dot model with an effective absorptance appreciably less than the true absorptance. This is analogous to the current practice of using an optical area appreciably greater than the geometrical area for a phenomenological interpretation (Romer & Morawski, 1963) of mean density data based on (17).

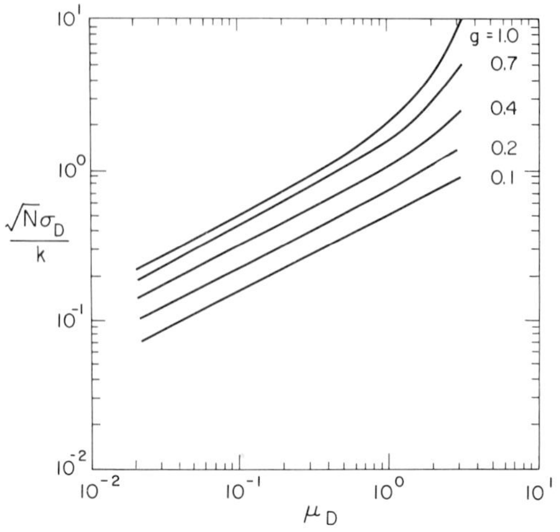

Figure 12. Normalized rms density fluctuations $N^{\frac{1}{2}}\sigma_D/k$ vs mean density μ_D for the honest, two dimensional, random-dot model and for grains of different absorptances g.

Meanwhile, the dishonest random-dot model survives. Haugh (1963) predicted linear log-log plots with slopes less than $\frac{1}{2}$ that were typical of experimental data. He estimated that the slope was less than $\frac{1}{2}$ by an amount proportional to

$$\delta = (q-p)\sigma^2_{\ell na}, \tag{60}$$

where $\sigma^2_{\ell na}$ is the variance of the log normal distribution *assumed* for the grain areas of undeveloped grains, the average number of quanta absorbed by a grain during exposure was assumed proportional to a^q, and the average number of quanta required for developability was assumed proportional to a^p.

The salient feature of this analysis was inclusion of the dependence of mean developed grain size μ_a on exposure. Up to this time, developed silver grains were represented in the random-dot model by opaque disks distributed at random, whereas in reality silver halide crystals are distributed at random, and exposure to light randomly selects which of these will be developed. If we start with a completely random distribution and each crystal has the same probability of being selected, we will end up with a completely random (Poisson) distribution of developed grains but at a lower concentration. However, with a distribution of unexposed grain sizes, the larger crystals will reach threshold first so that we might expect the average size of the developed silver grains to be inversely proportional to the optical density. If we heuristically set $a \propto \mu_D^{-b}$ in (20) with $b > 0$, we get $\sigma_D \propto \mu_D^{(1-b)/2}$ and slopes less than $\frac{1}{2}$ on a log-log plot. Incorporating exposure and development into the honest random-dot model should yield similar results.

3.1.3. Third generation models - focus on crowded emulsions and emulsion thickness

Photographic grains cannot always behave like Poisson-distributed disks as the random-dot model assumes. Because real grains cannot occupy the same space at the same time, they crowd one another out in thin emulsions at those high concentrations that yield high optical densities. This imposes constraints on their locations, which, consequently, cannot be completely random.

The dishonest random-dot model first allowed for this in (24) through the term σ_s^2/μ_s (the variance to mean ratio of the number of silver grains projecting into an area A), which is less than 1 in the presence of crowding constraints. Later, Langner (1963) introduced the dependence of this ratio on exposure. Starting with a similar, more fundamental ratio σ_h^2/μ_h, for coated silver-halide crystals before exposure and development ($\sigma_s^2/\mu_s \to \sigma_h^2/\mu_h$ as the development probability $\to 1$), he predicted that σ_D^2 would decrease as μ_D approached the maximum possible density at complete development if $\sigma_h^2/\mu_h \ll 1$. Yet we do not know how to relate this ratio to the thickness and concentration $\bar{n}$ of the emulsion.

At the Kodak Research Laboratories, we have started research that we hope will lead to a third generation of models focusing on crowded emulsions with emulsion thickness playing an explicit role. In the thinnest possible emulsion, one that is not thicker than a single grain, grains cannot overlap. If the effect of this constraint on granularity at high concentrations in such a monolayer could be understood, it ought to be possible to extend the results to crowded multilayers by an analysis of stacked, crowded monolayers.

The equations describing the mean and average aperture transmittance for a monolayer (Trabka, 1971) are

$$\mu_T = T_{min} + (1-p)(1-T_{min}), \tag{61}$$

and

$$N\sigma_T^2 = (1-\mu_T)(1+\varepsilon p). \tag{62}$$

Although these equations are extremely simple, they yield an interesting phenomenon, as we shall see shortly. The four parameters N, p, T_{min} and ε are defined as follows:

$$N = A/a \tag{63}$$

$$p = \text{development probability}, \tag{64}$$

$$T_{min} = 1 - \bar{n}a, \tag{65}$$

$$\varepsilon = \varepsilon(A) = \frac{\sigma_h^2}{\mu_n} - 1. \tag{66}$$

The dependence of development probability on exposure need not concern us now.

When p = 1, the average transmittance is equal to the minimum possible transmittance T_{min}, which in turn depends on the concentration $\bar{n}$ of silver halide crystals. The parameters $\bar{n}$ and ε describe statistically how silver halide crystals were coated on the film base. The random variable h denotes the number of silver halide crystals coated in an aperture of area A. (Note that $\bar{n} = \mu_h/A$.) Without constraints, h is Poisson-distributed, $\sigma_h^2 = \mu_h$ and $\varepsilon = 0$. For this reason we refer to ε as the Poisson excess. Note that ε can both increase or decrease σ_T^2 depending on whether it is positive or negative. The effect of crowding in a monolayer is to decrease σ_h^2 so that $\varepsilon < 0$ in this case. Figure 13 shows plots of normalized σ_D vs μ_D using Definitions (8) and (9) for various values of ε. Note that on this log-log plot $\varepsilon = -0.8$ yields a straight line, whereas the granularity decreases for μ_D near D_{max} when $\varepsilon = -0.91$. This effect is more pronounced for higher D_{max}, being as much as a factor of five for $D_{max} = 2$.

In order to understand the reason for this difference

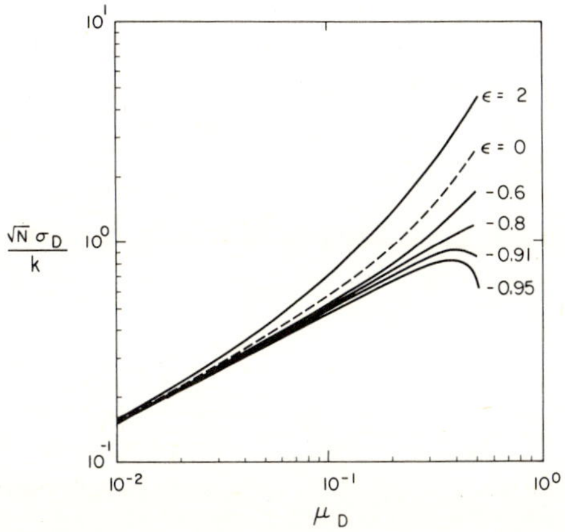

Figure 13. Normalized rms density fluctuations $N^{\frac{1}{2}}\sigma_D/k$ vs mean density μ_D of a monolayer for various values of the Poisson excess ε when $T_{min}^D = 0.3$ ($D_{max} = 0.52$). The dashed curve ε = 0 corresponds to a completely random distribution of grains, while the curves for ε < 0 correspond to constrained randomness due to grains of finite size crowding one another (Trabka, 1971).

in behavior, we need to relate ε, which is a macroscopic parameter, to the microstructure of the grains. Although (61) and (62) follow from a two-dimensional analysis, we turn to a one-dimensional model based on renewal theory for quick results.

A one-dimensional grain is considered to cover a line segment of fixed length ℓ. The events whose statistics we shall be interested in are the *leading, left-hand edges of grains* spread out along a line. The number of such events (edges) in an interval of length L is the one-dimensional analog of the random variable $\underline{h}$ whose mean and variance determine the Poisson excess ε. Let $\underline{x}$ be the random variable representing the gaps between successive left-hand edges (*not* the space between successive grains). A fundamental result of renewal theory states that, for large L,

$$\frac{\sigma_h^2}{\mu_h} = C^2 = \frac{\sigma_x^2}{\mu_x^2}\,, \tag{67}$$

i.e., the variance-to-mean ratio for the number is equal to the square of the coefficient of variation of the gaps. This gives

$$\varepsilon = C^2 - 1, \tag{68}$$

and gross variations as measured by ε are thus related to local random structure as measured by C. In order to determine ε we need a probability density function for the gaps.

For events occurring completely at random, the gap distribution F(x) is exponential. We modify this slightly and write

$$F(x) = p_0\delta(x-\ell) + (1-p_0)U(x-\ell)\lambda e^{-\lambda(x-\ell)}, \tag{69}$$

which can be considered as a mixture of packing and parking. The delta function $\delta(x)$ of strength $0 \leq p_0 \leq 1$ is capable of providing a nonzero probability p_0 that $x = \ell$, i.e., that successive grains will touch one another. We shall refer to p_0 simply as a packing factor. The parameter λ is the old Poisson rate parameter which determines the average gap when $p_0 = \ell = 0$. The second term may be looked upon as a parking factor. The Heaviside step U(x) guarantees that the gap between successive leading edges of grains is not smaller than the grain length, yet the probability that two grains will touch because of this term is zero, which would be good parking parctice if grains were cars.

The coefficient of variation in this case is

$$C^2 = \frac{1+p_0}{1-p_0} T_{min}{}^2, \tag{70}$$

where

$$T_{min} = \frac{1-p_0}{1+\lambda\ell-p_0} . \tag{71}$$

Note that $C \geq T_{min}$ with equality for $p_0 = 0$. Figure 14 shows two samples of one-dimensional grains for $T_{min} = 0.32$ ($D_{max} = 0.5$). In Fig. 14(a) the packing factor p_0 is zero and the pattern is quite regular. In Fig. 14(b) the packing factor is $p_0 \approx 0.4$ and the pattern is more irregular. Large gaps are more in evidence, being necessary to compensate for clumps introduced by packing so that the average transmittance is the same. Note that the irregular pattern with packing corresponds to a value of ε for

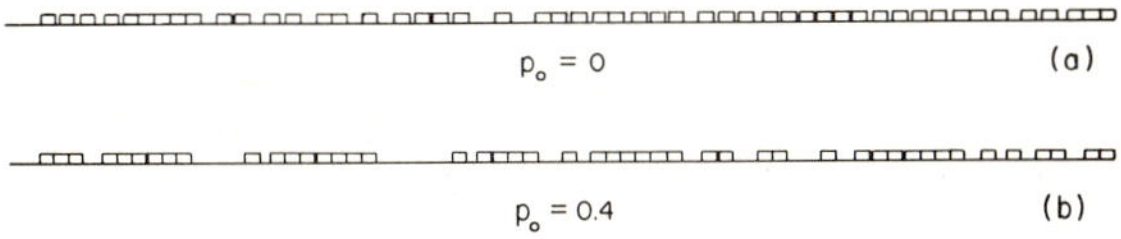

Figure 14. Absorptance profiles for crowded monolayers with mean transmittance $\mu_t = T_{min} = 0.32$ ($D_{max} = 0.5$). (a) Packing factor $p_0 = 0$. (b) Packing factor $p_0 = 0.4$ (Trabka, 1971).

which the total large-aperture granularity did decrease near D_{max}. This suggests that we look at the spectrum of the granularity.

The spectrum is the same for fluctuations of transmittance $\underline{t}$ or absorptance $\underline{a}$. For, since $\underline{t} = 1-\underline{a}$, we have

$$C(\rho) = \text{cov}\{\underline{t}(u), \underline{t}(u+\rho)\} = \text{cov}\{\underline{a}(u), \underline{a}(u+\rho)\}. \tag{72}$$

Let the absorptance profile of a single grain with left-hand edge located at u = 0 be given by the pulselike function A(u), where A(u) = 0 for $u < 0$ and $u > \ell$ but is otherwise only restricted to values between 0 and 1. Then the absorptance at any point of a semi-infinite monolayer can be expressed as

$$\underline{a}(u) = \sum_{n=1}^{\infty} \underline{y}_n A(u-\underline{S}_n), \tag{73}$$

where

$$\underline{S}_n = \sum_{i=1}^{n} \underline{x}_i \tag{74}$$

is the cumulative sum of the gap random variable $\underline{x}_i$, and $\underline{y}_n$ is a binary valued random variable, independent of the $\underline{S}_n$, which indicates whether or not the nth grain received a developable exposure, i.e., prob$\{\underline{y}_i = 1\} = p$ and prob$\{\underline{y}_i = 0\} = 1-p$. Using the expressions given by Smith (1958, p. 267), the single-sided Wiener spectrum can be shown to consist of the product of two terms

$$P(\nu) = 2S(\nu) \cdot g(\nu), \tag{75}$$

where $S(\nu)$ is the "energy" density spectrum of the average absorptance profile and $g(\nu)$ is the spectrum of the point processes associated with the leading left-hand edges of grains. (The "power" density spectrum of a random process is known as the Wiener spectrum in the optical literature to distinguish it from other kinds of spectra that are also important.)

Spectra for grain absorptance profiles that are rectangular, i.e.,

$$A(u) \begin{cases} = 1 & (0 \leq u \leq \ell) \\ = 0 & (u < 0,\ u > \ell), \end{cases} \tag{76}$$

are shown in Fig. 15(a) and (b) plotted vs ν in cycles per unit length. The spectrum goes to zero at $\nu = 0.25$, which is the reciprocal of the grain length $\ell = 4$ used in the computation. The second hump after cutoff is the first side lobe in the $(\sin^2\omega/\omega^2)$ spectrum of the grain absorptance profile. The peak just before cutoff in Fig. 15(a) for a packing factor of zero is the periodicity introduced by the spectrum of the point process associated with grain locations and is double the zero frequency asymptote. In order to achieve low transmittance without packing, the

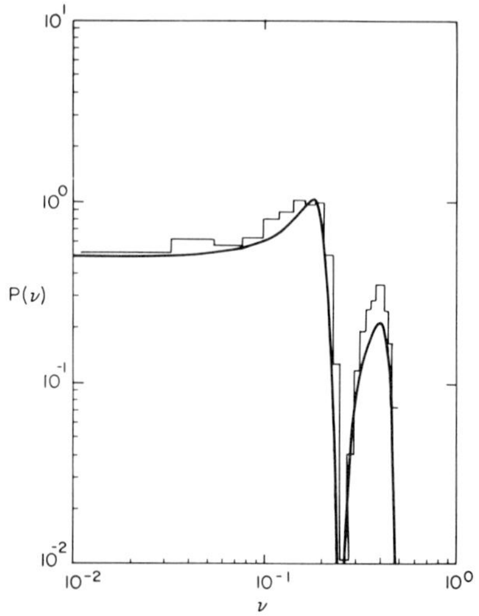

Figure 15a. Wiener spectrum $P(\nu)$ of point-transmittance fluctuations vs ν in cycles per unit length for a completely exposed, crowded monolayer with no packing. Here $T_{min} = 0.3$, $p_\rho = 0$, $C^2 = 0.09$, and $\varepsilon = -0.91$. The continuous curve is theoretical, whereas the curve with jumps depicts spectral estimates from Monte Carlo simulation whose resolution is given by the length of the line segment between jumps (Trabka, 1971).

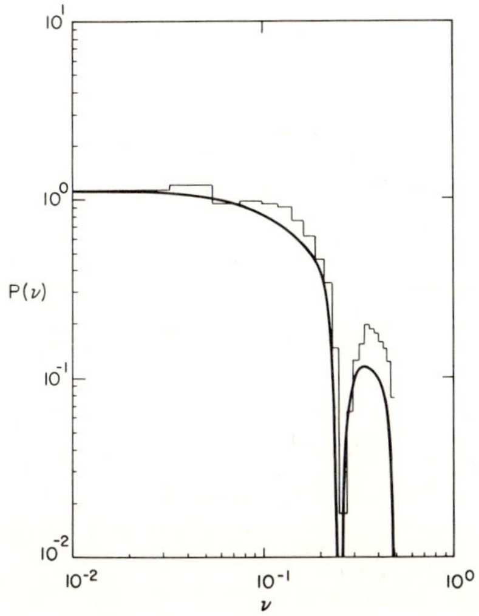

Figure 15b. Wiener spectrum $P(\nu)$ of point-transmittance fluctuations vs ν in cycles per unit length for a completely exposed, crowded monolayer with packing. Here $T_{min} = 0.3$, $P_0 = 0.38$, $C^2 = 0.2$ and $\varepsilon = -0.8$. The continuous curve is theoretical, whereas the curve with jumps depicts spectral estimates from Monte Carlo simulation whose resolution is given by the length of the line segment between jumps (Trabka, 1971).

grains have to be closely and nearly regularly spaced, accounting for the periodic component near $\nu \approx 1/\ell$. Note that $p_0 = 0$ yields $\varepsilon = -0.91$. This is the value for which the total large aperture granularity in Fig. 13 decreases near D_{max}. When the packing factor increases to $p_0 = 0.38$, the periodic component in the spectrum disappears, as shown in Fig. 15(b) because, as we have seen, the pattern becomes more irregular. Note that $p_0 = 0.38$ corresponds to $\varepsilon = -0.8$, the value for which the total large-aperture granularity in Fig. 13 did not decrease near D_{max}. The solid curves in Figs. 15(a) and (b) are theoretical, whereas the staircase-like functions correspond to spectral estimates based on Monte Carlo simulation.

We can now see why $\varepsilon = -0.91$ yields lower large-aperture granularity than $\varepsilon = -0.8$. According to (41), the total area under the point transmittance spectrum is fixed once the mean transmittance μ_t is fixed. For aperture lengths $L >> \ell$, the frequency response of the aperture cuts off at $1/L << 1/\ell$. With packing ($\varepsilon = -0.8$) we have relatively more noise power at those low frequencies that are passed by the aperture.

3.2. Some Theoretical Questions

We have seen how renewal theory has been useful in explaining the granularity to be expected in crowded monolayers. However, many interesting questions remain.

3.2.1. Two-dimensional analysis

To what extent can the renewal theory of point processes on the line be extended to point processes in the plane? Consider square grains, all the same size, distributed without overlap in a plane. What is the analog of (67) which allowed us to relate the Poisson excess ε to the microstructure of the grains as measured by the coefficient of variation of the gaps? In two dimensions, what is a gap?

Because of the complete ordering possible with points in one dimension we could take the no-overlap constraint into account merely by noting the position of the last grain and suitably restricting the size of the next gap, which is independent of the preceding gaps (Markovian). In two dimensions, because of lack of complete ordering, the position of any grain is restricted by the location of all the others, so that a more general point process, rather than a renewal process, would seem to be required for a realistic model. Yet we might still expect some relation similar to (67) between the macrostructure and microstructure. (Related questions have been studied by Blum, 1965.)

How do we construct realizations in two dimensions? Throwing down grains completely at random and removing any that overlap is time consuming and will not yield low transmittances. Recall that from the one-dimensional parking problem we get a mean fractional coverage of 0.748 or a mean transmittance of 0.252 whereas, by our introduction of the packing factor p_0 in (69), we can achieve zero transmittance. A low mean transmittance is even harder to achieve by parking alone in two dimensions. The Palasti (1960) conjecture for small congruent squares distributed at random without overlap over a larger square, with sides parallel to the larger square, predicts a mean fractional coverage in two dimensions of $(0.748)^2 = 0.56$ or a mean transmittance of 0.44. What kinds of packing algorithms are practical in two dimensions to achieve any specified mean transmittance and how does the Poisson excess depend on the parameters of the algorithm (See Eqs. (68), (70), and (71) for the one-dimensional case)?

3.2.2. Transects of two-dimensional random fields

Because the granularity processes of interest are a subclass of *t*wo-dimensional, *h*omogeneous, *i*sotropic, *r*andom *f*ields (2HIRF). they are characterized by the statistical properties along any transect (linear section or scan). Working with one-dimensional models may then be justified provided they actually are transects of real granularity patterns. Here we will be interested in only the second-order properties of these processes as given by their covariances or, equivalently, by their spectra. Since these properties generally do not determine the processes uniquely, we will phrase our questions in terms of classes of covariance functions rather than classes of processes. Before restricting our attention to granularity patterns, however, it will be helpful to recall the relationships between 2HIRF and their transects.

The covariance function $C_2(\rho_1,\rho_2)$ and spectrum $S_2(\omega_1,\omega_2)$ of a 2HIRF, being isotropic, can be characterized by sections, $C_2^*\{(\rho_1^2+\rho_2^2)^{\frac{1}{2}}\} = C_2(\rho_1,\rho_2)$ and $S_2^*\{(\omega_1^2+\omega_2^2)^{\frac{1}{2}}\} = S_2(\omega_1,\omega_2)$, which are Fourier-Bessel transform pairs, viz.,

$$S_2^*(\omega) = 2\pi\int_0^\infty \rho C_2^*(\rho)J_0(\rho\omega)d\rho, \tag{78}$$

$$C_2^*(\rho) = 2\pi\int_0^\infty \omega S_2^*(\omega)J_0(\rho\omega)d\omega. \tag{79}$$

The covariance function $C_1(\rho)$ of a transect of this 2HIRF is given by

$$C_1(\rho) = C_2(\rho,0) = C_2^*(\rho) \tag{80}$$

and has a one-dimensional spectrum given by the Fourier transform

$$S_1(\omega) = \int_{-\infty}^{+\infty} C_1(\rho)e^{-i\omega\rho}d\rho, \tag{81}$$

which is also the marginal of $S_2(\omega_1,\omega_2)$, i.e.,

$$S_1(\omega) = \int_{-\infty}^{+\infty} S_2(\omega,\omega_2)d\omega_2 = \int_{-\infty}^{\infty} S_2^*\{(\omega^2+\omega_2^2)^{\frac{1}{2}}\}d\omega_2. \tag{82}$$

Now let J be the class of all covariance functions $C(\rho)$ of transects from 2HIRF. We know how to characterize J; see for example Blanc-Lapierre and Faure (1965). Given a one-dimensional covariance function $C(\rho)$, $C\varepsilon J$ if and only if it has the representation

$$C(\rho) = 2\pi\int_0^{\infty} \omega S(\omega)J_0(\rho\omega)d\omega, \tag{83}$$

where $S(\omega) \geq 0$.

For us a random function is binary valued if it takes on only the values 0 or 1. Characterization of the class $J_b \subset J$ of all covariance functions $C(\rho)$ of transects from binary-valued 2HIRF (B2HIRF), which has applications in ecology, is closer to our needs, although not quite sufficient. In this context we note the interesting result of Switzer (1965) that showed how to construct a B2HIRF with correlation coefficient

$$C(\rho) = e^{-2\rho\lambda} \tag{84}$$

and corresponding spectrum

$$S(\omega) = \frac{8\pi}{(4\lambda^2+\omega^2)^{3/2}}. \tag{85}$$

At present, we are most interested in the class $J_g \subset J_b \subset J$ of all covariance functions $C(\rho)$ of transects from grain-generated B2HIRF (GB2HIRF). A GB2HIRF (granularity process) has a point transmittance $\underline{t}(x,y)$, which is 1 if no grain covers (x,y) and 0 otherwise. Realizations are constructed by stochastically distributing opaque grains in a plane. This requires the specification of (a) an ensemble of grain shapes, and (b) a distribution of grain locations. See Section 3.1.3.

For the subclass $J_{gp} \subset J_g$ obtained by distributing grains according to a *Poisson* distribution, we conjecture that $C(\rho)\,\varepsilon\, J_{gp}$ if and only if $C(\rho)$ has the representation given by (34) where $\Gamma(x)$ is positive semidefinite since it is to be interpreted as an average correlation function. In our study of crowded monolayers we would like necessary and sufficient conditions that $C(\rho)\varepsilon J_{gno}$ where the subclass $J_{gno} \subset J_g$ is obtained by constraining grain locations so that grains do *not overlap*.

Note that Switzer's (1965) example with covariance (84) is probably in J_g also. But what characterizes the set J_g? If L is the class of all covariance functions $C(\rho)$ of linear (one-dimensional) stationary random processes, then clearly $J_g \subset L$. Note that the transect of any granularity process as described above must appear as shown in Fig. 16, where intervals of random length on which the film absorptance is 1 (pulses) are separated by intervals (gaps) on which the film absorptance is 0. Focusing attention on the discontinuities (edges of the pulses), we may consider the transect of any granularity process as an *alternating bivariate point process* consisting of two types of events denoted by $\uparrow$ and $\downarrow$ respectively. Using this fact, can we find come characterization of J_g analogous to (83)? Matérn (1960, P. 42) uses this characterization to show that (downward) convexity of $C(\rho)$ to the right of the origin is a necessary condition for membership in J_g, that is

$$C(\rho) > C(0) + \rho C'(0). \tag{86}$$

Benton and Kronauer (1971) have argued that a necessary condition is that $S(\omega)$ be asymptotic to ω^{-3}. Note that this condition is also satisfied by $C(\rho)$ given by (84), belonging to J_g.

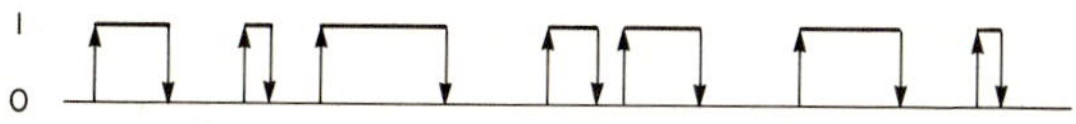

Figure 16. A one-dimensional binary-valued random process defined by an alternating, bivariate point process of two types of events, the up and down arrows.

The characterization of J_g in terms of the covariances of an alternating, bivariate, point processes calls attention to the simpler *alternating renewal processes* (Cox, 1962) as models for transects of granularity processes. Let $\underline{X}_i$ be the lengths of intervals on which film absorptance is +1 with distribution function $F_x(X) = \text{prob}(\underline{x} \leq x)$

and Y_i be the lengths of intervals on which film absorptance is 0 with distribution function $F_Y(y) = \text{prob}(\underline{Y} \le y)$. P. A. W. Lewis (1971) has shown that if $R_X(x) = 1 - F_X(x)$ denotes the survivor function of $\underline{X}_i$, and if

$$\bar{R}(\tau) = \frac{\int_\tau^\infty R_X(u)\,du}{\mu_X + \mu_Y}, \tag{87}$$

then the covariance function is

$$C(\rho) = \bar{R}(\rho) + \int_0^\rho R_X(\rho-y)\,d\bar{G}(y) - \left(\frac{\mu_X}{\mu_X+\mu_Y}\right)^2, \tag{88}$$

where $\bar{G}(y)$ is the infinite sum of convolutions

$$\bar{G}(y) = \bar{F}(y)*F_Y(y)*\sum_{i=0}^\infty \{F_Y(y)*F_X(y)\}^{i*}, \tag{89}$$

with $\bar{F}(y) = 1 - \bar{R}(y)$.

Let L_{ar} be the class of all covariance functions satisfying (88). Can we represent a transect of some GB2HIRF by an alternating renewal process or is the set $J_g \cap L_{ar}$ empty? Are all transects alternating renewal processes, i.e., is $J_g \subset L_{ar}$; and if not, is there some modification of the alternating renewal process that will include all of J_g?

Note that not all alternating renewal processes are transects, i.e., $L_{ar} \mid J_g$. For, as we have seen, a necessary condition that a covariance $C(\rho) \varepsilon J$ is that it have the representation (83) with non-negative spectrum. Now, if we take the alternating renewal model of Section 3.1.3 with fixed grain length i.e., $dF_X/dx = \delta(x-\ell)$, and gap probability density function dF_y/dy given by (69), compute its covariance function $C(\rho)$ as shown in Fig. 17 for $p_0 = 0$ and $T_{min} = 0.4$, and substitute into (78), we obtain a section $S_2^*(\omega)$ with negative values as shown in Fig. 18. Hence, the covariance function of Fig. 17 is not in J and so not in J_g. By construction it is in L_{ar}. Therefore, $L_{ar} \not\subset J_g$ as we inquired above.

However, if we take the spectrum in Fig. 18, set all negative values to zero, and then obtain $C(\rho)$ from (79), we obtain the

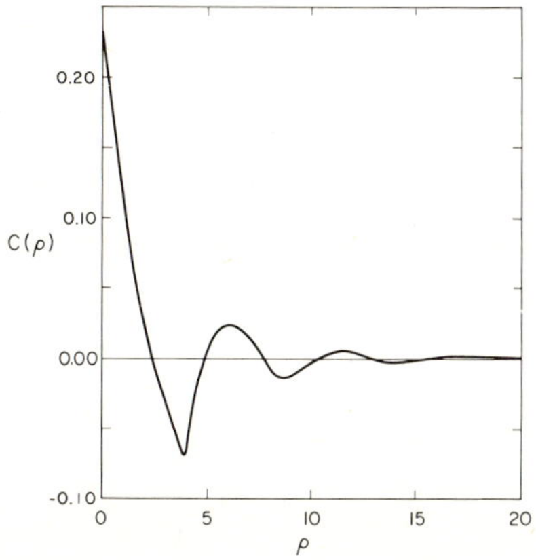

Figure 17. Covariance function of a one-dimensional granularity pattern consisting of single-size grains without packing for $T_{min} = 0.4$.

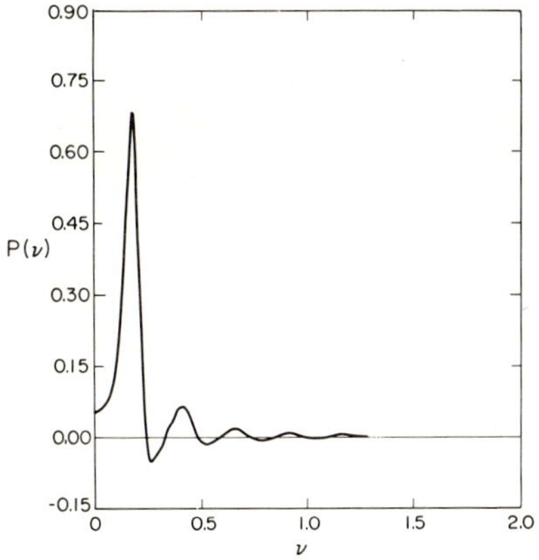

Figure 18. Fictitious spectrum of an isotropic two-dimensional process having the covariance function of Fig. 17.

covariance function, shown normalized in Fig. 19 (dashed curve), which differs very little from the original normalized covariance function (solid curve). In what sense, if any, is this new process an approximation to the original alternating renewal process? Given any alternating renewal process, how do we find the "closest" process in J_b (or J_g)? More precisely, given $C(\rho)\varepsilon J_g$, which grain distribution F_X and gap distribution F_Y will yield an alternating renewal covariance function $C_{ar}(\rho)$ "closest" to $C(\rho)$? Or given a gap distribution F_Y, what grain distribution F_X will yield $C_{ar}(\rho)\varepsilon J_g$?

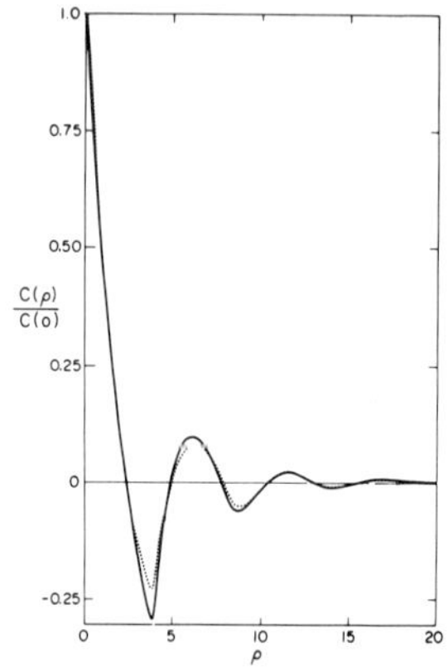

Figure 19. Normalized covariance function (solid curve) corresponds to the spectrum of Fig. 18 and dashed curve corresponds to the spectrum (Figure 9) when negative values are set to zero.

3.2.3. Interacting point processes

In the enlargement printing of a negative, we have the spatial distribution of effectively large grains selecting, by shadowing, which grains in the print material will not develop. A one-dimensional model of this process is depicted in Fig. 20. The dark grains in the print material developed because the exposure was not blocked by grains in the negative. If we postulate an alternating renewal model for the negative and undeveloped print, how do we describe the developed print? In particular, what is the relationship between the spectrum of the print and the grain length and gap distributions of the negative? The clustering of developed grains in the print due to gaps in the negative ought to produce a buildup of noise power at low frequencies similar to that predicted by Bartlett (1964) for a clustering model in plant ecology. How does one extend Bartlett's model to take into account the crowding effect of finite-sized grains (or plants) in the print material?

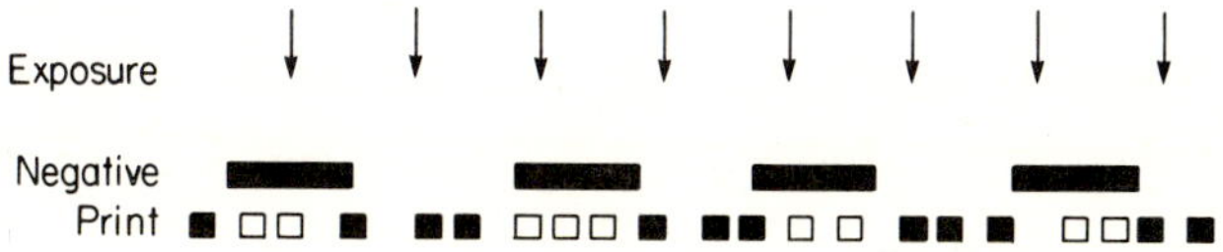

Figure 20. Clustering of fine grains (dark) in print material copies from a coarse-grained negative.

4. SIGNAL (SENSITIVITY) AND TEMPORAL POINT PROCESSES

Photographic film is not a pure energy detector. The mean optical density μ_D, which is the photographic signal, depends upon the specific time sequence of incident photons during exposure and not just their total number. This phenomenon is known as *reciprocity failure* and the understanding of the inefficiencies that cause it is essential to progress in photographic science. Modern high-speed digital computers have now provided adequate Monte Carlo solutions (Bayer & Hamilton, 1965). But the formation of the latent image (Hamilton, 1968, 1970) is a rich and fascinating type of Markov population process (Kingman, 1969) that has not yet found its way into the statistical literature.

4.1 Latent Image Formation

The silver specks in the silver halide crystals shown in Fig. 6 constitute the latent image. The size of a speck, the number of silver atoms in the speck, determines the rate of development of the crystal. For instance, four atom specks are a good, nominal size for simplified go-no go detector models of the photographic process. A schematic diagram of latent image formation is shown in Fig. 21. The physical effect of photon absorption is the creation of an electron hole pair that interacts with other colonies of trapping centers and point ionic defects leading eventually to nucleation and growth of the latent image. Nucleation involves the combination of two electrons from the cycle with two mobile silver ions to form a stable two-silver-atom nucleus. Additional electrons trapped by this nucleus attract other silver ions, and the number of silver atoms in the nucleus grows. Here we have the spatial-temporal point process of photon absorptions interacting with competing spatial-temporal nucleation and growth processes. It is not the simple picture that it was in the beginning. Let us review some of the earlier purely mathematical work attempting to relate temporal point processes to latent image formation.

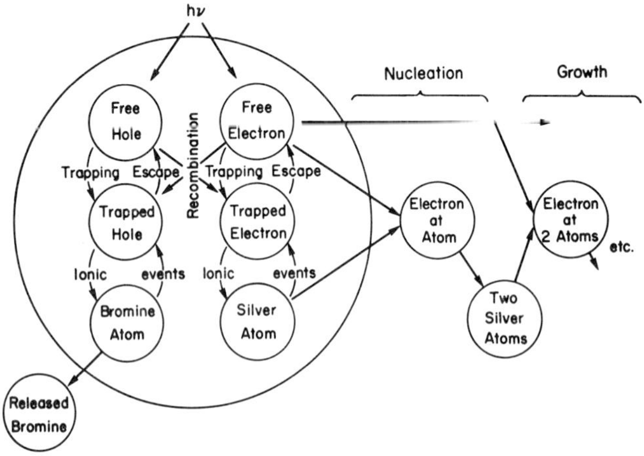

Figure 21. Schematic diagram of latent image formation. Both electrons and holes are in equilibrium among the three states within the large circle. Stable products are formed by the interactions that take place outside that circle (Hamilton, 1968).

4.2 Reciprocity Failure

We can modify (17) to take into account exposure and development by writing

$$\mu_D = k(\bar{n}p)a, \tag{90}$$

where now $\bar{n}$ is the concentration of silver halide crystals and p is their development probability. In general,

$$p = p(I,\tau) \tag{91}$$

is a function of both irradiance I and exposure time τ. Note that if p depended only on the exposure $E = I\tau$ and exposure was constant, then

$$\log E = \log I + \log \tau = \text{constant} \tag{92}$$

would give a constant mean density μ_D independent of I and τ. A plot of logE vs logI for constant mean density would then be a horizontal straight line. In general, a plot of logE vs logI for a specified mean density is called a reciprocity curve, and deviations from a horizontal straight line indicate departures from pure energy detection or failure of reciprocity between time and irradiance (intensity). A typical reciprocity curve is shown in Fig. 22, where lines of constant exposure time correspond to 45° diagonals if the same logarithmic scales are used on both axes. The value of intensity at the minimum of the reciprocity curve is called the optimum intensity because it yields the minimum exposure. Any increase in exposure (decrease in sensitivity) at intensities less than the optimum is called low-intensity reciprocity failure, whereas at intensities higher than the optimum it is called high-intensity reciprocity failure.

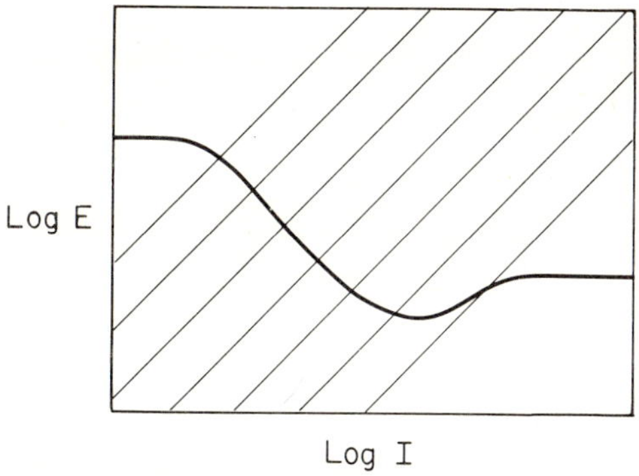

Figure 22. A typical reciprocity failure curve of log exposure vs log irradiance for constant optical density. The 45° lines correspond to constant exposure times.

A classical problem in queueing theory, the probability that service will stagnate in the interval $(0,\tau)$ (i.e., that K customers or more will arrive within the service time τ_0), was studied quite early in in the history of photography (Silberstein, 1939) in connection with low-intensity reciprocity failure although it is most often associated (ten Hoopen and Reuver, 1965; Meyer, 1967) with visual perception in phychophysics.

4.2.1. Low-intensity reciprocity failure

In photography, the K customers of queueing theory correspond to the K photons that must be absorbed within an interval τ_0 that is the decay time of the nucleus, analogous to the service time. For K = 2 and photons being absorbed at a rate λ, Silberstein (1939) gave

$$p = 1 - e^{-E} \sum_{i=0}^{m+1} \frac{(E)^i}{i!} \{1 - (i-1)\frac{E_o}{E}\}^i \tag{93}$$

as the exact development probability, where $E = \lambda\tau$ is the number of photons absorbed during the exposure time, $E_0 = \lambda\tau_0$ is the number of photons absorbed during the decay time, and $m = \tau/\tau_0$ is the largest integer less than or equal to τ/τ_0. Later Silberstein (1945) reformulated the problem in geometrical language, treating the problem of a random distribution of n points along a line of unit length, and calculated the expected number $Q_K(n)$ of occurrences of K of these points falling within a given fractional element (a K-aggregate). Identifying the n points with E and the fraction element with τ_0/τ, he gave

$$Q_K(E) = K(\frac{E_0}{E})^{K-1} \frac{E!}{(E-K)!K!} [1-(\frac{E_0}{E})\{\frac{4+C}{K+1} (E-K) - \frac{K-1}{K}\}], \tag{94}$$

where

$$C = K - 1 + (\frac{1}{2} + \frac{1}{3} + \ldots + \frac{1}{K}) \tag{95}$$

for $E_0/E << 1$ and $E_0 < 1/2$.

Similar results have evolved from the problem of visual excitation, where K is the number of photons that must be absorbed within a summation period τ_0 for seeing. Exact results for the case K = 3 have been obtained by Van Elteren and Gerrits (1961) whereas ten Hoopen and Reuver (1965a, 1965b) have presented approximate methods of calculation for exponentially distributed service times τ_0 (silver-atom lifetimes).

Later Webb (1950) used (94) and wrote

$$p = 1 - e^{-Q_K(E)} \tag{96}$$

for the development probability, or probability of occurrence of one or more K-aggregate photon absorptions. Moreover, he noted that for extremely low intensities λ and long exposure times

$$Q_K(E) \approx \frac{E(E_0)^{K-1}}{(K-1)!} \tag{97}$$

so that the reciprocity curve defined by $Q_K(E)$ = constant yields

$$\log E = -(K-1)\log I + \text{constant}, \tag{98}$$

where we have let $I = \lambda$. Low-intensity asymptotes for the reciprocity curve are lines with slope $-(K-1)$, where K is the number of silver atoms required to form a stable nucleus (K-1 is the number of servers in the queueing problem).

The most general result motivated by problems of visual excitation has been established by Ikeda (1965). For events occurring with Poisson rate λ having random lifetimes τ_i with finite first moment $\alpha = E\{\tau_i\}$, the probability that K events will be alive simultaneously exactly m times in $[0,\tau]$ as $\tau \to \infty$ and $\lambda^K \tau \to \mu$ is given by

$$p_m^K(\tau) \sim e^{-\xi} \frac{\xi^m}{m!}, \tag{99}$$

where

$$\xi = \frac{\mu\alpha^{K-1}}{(K-1)!}. \tag{100}$$

Consequently the probability of at least one K^{th}-order response in $[0,\tau]$ is given by

$$p = 1 - e^{-\xi}, \tag{101}$$

but for $\alpha = \tau_0$, we have

$$\xi = \frac{(\lambda^K \tau)\tau_0^{K-1}}{(K-1)!} = \frac{E(E_0)^{K-1}}{(K-1)!}, \tag{102}$$

and (101) is identical with (96) used by Webb.

The relevance of Silberstein's work to the probability of a Kth-order visual response seems to have been overlooked, although his work did not go unnoticed (Mack, 1948, 1950; Naus, 1965, 1966). It is strange that mathematical interest in this problem is fostered by mechanisms of vision that are relatively little understood compared with those of latent image formation.

4.2.2 High-intensity reciprocity failure

One of the earliest mechanisms proposed for high-intensity reciprocity failure is one in which the ionic time required for a silver ion to join an electron trapped at a nucleus acts as a dead time τ^* during which additional electrons that arrive are ineffective. Instead of considering the initial Poisson processes of photon absorptions with rate λ, it is then sufficient to consider the output of a type I counter with dead time τ^* and the initial Poisson process as input (Zweig, 1965). The average time between output events is then $\frac{1}{\lambda} + \tau^*$. At high intensities the probability of forming the stable nucleus is one, and we need only consider the probability of adding M more silver atoms to the nucleus. The average time between M consecutive counter output events is $M(\frac{1+\lambda\tau^*}{\lambda})$, so that the mean number in a large interval τ is

$$\mu = \frac{\lambda\tau}{M(1+\lambda\tau^*)}, \tag{103}$$

By analogy with the heuristic procedure that allowed us to write (96), we can obtain the development probability from

$$p = 1 - e^{-\mu}. \tag{104}$$

The reciprocity curve for μ = constant and large λ becomes

$$\log E = \log I + \log M = \text{constant}, \tag{105}$$

which would predict an asymptote with slope +1 at high intensity.

4.3 Some Applied Problems

The theoretical low- and high-intensity reciprocity failure asymptotes are seldom achieved in real emulsions as other factors enter which cause these curves to roll over and become horizontal at both ends. Knowing the locations and causes of the roll-over points is important to photographic science. Analysis of the complete model of spatial and temporal interacting point processes, schematized in Fig. 21, is the desired end. Yet, in the words of Carrier (1962), "In particular, the applied mathematician must be very skillful at finding that question (or family of questions) such that the answer will fill the scientific need while the extraction of the answer and its interpretation are not prohibitively expensive." We do not wish to deprive anyone of this interesting and rewarding activity in the area of latent image formation, but what follows will merely be a few suggestions to stimulate interest.

4.3.1. Low-intensity roll-over

We have seen that the formation of a K-aggregate of silver atoms for a stable nucleus ought to have a low-intensity asymptote with slope $-(K-1)$. A very plausible mechanism for low-intensity roll-over is hole removal caused by the released bromine depicted in Fig. 21. This can be interpreted as providing a finite probability that the lifetime of a single silver atom is infinite so that it becomes important to know how Ikeda's result embodied in (96) and (97) is modified when $\alpha = E\{\tau_0\} = \infty$.

4.3.2. Competition for growth

The dead time required for a silver ion to join a trapped electron (the case of the slow ion in Section 4.2.2) is thought to be the primary mechanism of high-intensity reciprocity failure only in relatively large silver halide crystals. For relatively small crystals, the growth of a nucleus to developable size in competition with the formation of new nuclei, with growth of other existing nuclei, and with recombination, is thought to be the primary mechanism of high-intensity reciprocity failure.

Consider the limiting case in which the exposure is so short that all the photons are absorbed simultaneously. The photoelectrons generated will very quickly reach equilibrium among the free, trapped, and silver-atom states depicted inside the large circle in

Fig. 21. Neglecting the trapped state of an electron, it will be in the free state with probability ϕ and in the atom state with probability α. Let $\vec{a}(t) = (a_1, a_2, \ldots, a_N)$ be the state vector, where a_K gives the number of K-atom aggregates. Initially we have $\vec{a}(0) = (C, 0, \ldots, 0)$ where C is the number of one-atom aggregates equal to the number of photoelectrons. The probability of a nucleation event per unit time is

$$\lambda_{12} = (a_1\phi)\{(a_1-1)\alpha\}\rho, \tag{106}$$

where $a_1\phi$ is the number of free electrons, $(a_1-1)\alpha$ is the number of single atoms available for nucleation (the silver atom from which the free electron came is not eligible), and ρ is a kind of cross section parameter typical for this event. The probability of a growth event per unit time is

$$\lambda_{1K} = (a_1\phi)a_K\,\rho, \tag{107}$$

where $K \neq 1,2$ and we have assumed the same cross section for growth, independent of size. A nucleus can grow only by the addition of a single atom, so that $\lambda_{iK} = 0$ for $i \neq 1$.

For this simplified model we are not so much interested in the exact time history of the process as in the distribution of final equilibrium states. Assuming $\alpha \approx 1$, Equations (106) and (107) yield the ratio

$$\frac{\text{chances for nucleation}}{\text{chances for a K growth}} = \frac{a_1-1}{a_K},$$

which can be used to compute the transition probabilities for the resulting Markov chain.

Figure 23 illustrates the competition between nucleation and growth starting with $a_1 = 7$ silver atoms. Single atoms are unstable, so the first nucleation occurs with probability one. The chances now become 4 to 1 for another nucleation, i.e., $p_{12} = 0.8$ and $p_{13} = 0.2$. Continuing in this manner until the single atoms are exhausted, we obtain the possible final states enclosed in double circles. The question is: What is the probability of obtaining an m-aggregate or larger of silver atoms given N single atoms. A solution obtained for m = 5 using

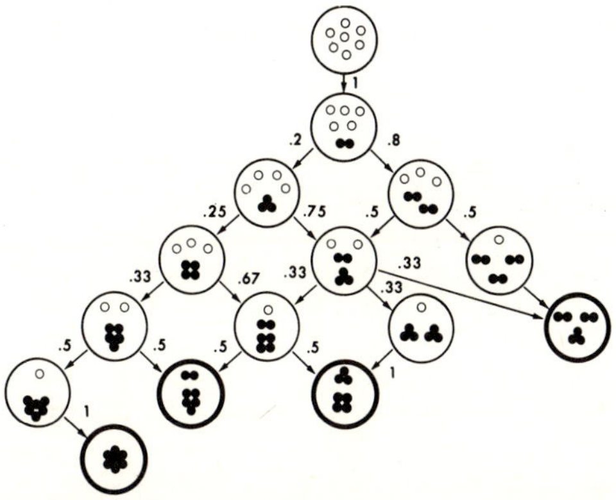

Figure 23. Possible stages of nucleation and growth starting with 7 silver atoms. Numbers along paths are transition probabilities and the final states are enclosed by double circles.

a computer algorithm is shown in Fig. 24, but formal analytical expressions for these probabilities would be useful.

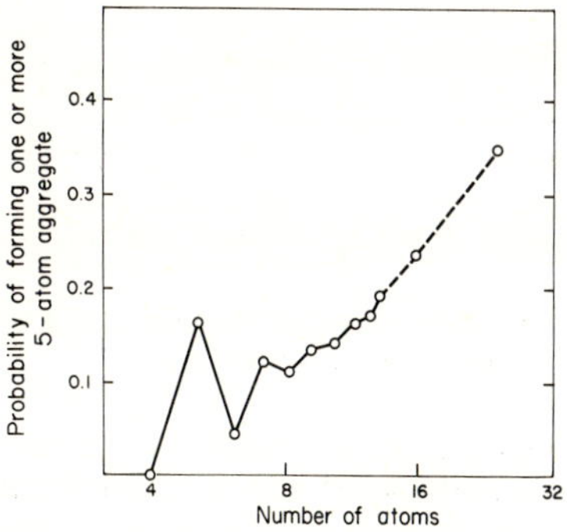

Figure 24. Probability of forming a 5-atom or larger aggregate as a function of the total number of silver atoms per grain, assuming no change in cross section with aggregate size (Hamilton, 1970).

ACKNOWLEDGMENTS

The authors are indebted to Professor J. H. B. Kemperman for many stimulating discussions, suggestions, and careful reading of the manuscript, to Mr. Edward C.Doerner for preparing the first five figures, and to Miss Toby C. Moidel for her development of the reliable computer programs that provided numerical results for illustration.

REFERENCES

Bartlett, M. S. (1964). The spectral analysis of two-dimensional point processes. Biometrika 51, 299-311.

Bayer, B. E. (1964). Relation between granularity and density for a random-dot model. J. Opt. Soc. Amer. 54, 1485-90.

Bayer, B. E. & Hamilton, J. F. (1965). Computer investigation of a latent-image model. J. Opt. Soc. Amer. 55, 439-52.

Benton, S. A. & Kronauer, R. E. (1971). Properties of granularity Wiener spectra. J. Opt. Soc. Amer. 61, 524-9.

Blanc-Lapierre, A. & Faure, P. (1965). Stationary and isotropic random functions. In Bernoulli, Bayes, Laplace Anniversary Volume, J. Neyman & L. LeCam, ed. New York: Springer-Verlag.

Blum, E. H. (1965). Statistical geometric approach to the random packing of spheres. Ph.D. thesis submitted to the Department of Mathematics, Princeton University.

Brown, J. L. & Mueller, C. G. (1965). Brightness discrimination and brightness contrast. In *Vision and Visual Perception*. C. H. Graham, ed. New York: Wiley.

Carrier, G. F. et al (1962). Applied mathematics: what is needed in research and education. SIAM Review 4, 297-320.

Cox, D. R. (1962). *Renewal Theory*. New York: Wiley.

Doerner, E. C. (1962). Wiener-spectrum analysis of photographic granularity. J. Opt. Soc. Amer. 52, 669-72.

Elteren, P. van & Gerrits, H. J. M. (1961). Eeen wachtprobleem voorkomende bij drempelwaardemetingen aan het oog. Statistica Neerlandica 15, 385-401.

Hamilton, J. F. & Bayer, B. E. (1965). Investigation of a latent image model: recombination of trapped electrons and free holes. J. Opt. Soc. Amer. 55, 528-33.

Hamilton, J. F. (1968). Latent-image formation as a probabilistic process. Photogr. Sci. Eng. 12, 143-5.

Hamilton, J. F. (1970). Investigation of a latent image model by an analytical approximation. Photogr. Sci. Eng. 14, 102-11.

Hamilton, J. F. (1972). The photographic grain. Appl. Opt. 11, 13-21.

Haugh, E. F. (1963). A structural theory for the Selwyn granularity coefficient. J. Photogr. Sci. 11, 65-8.

Higgins, G. C. (1971). Methods for anlyzing the photographic system, including the effects of nonlinearity and spatial frequency response. Photogr. Sci. Eng. 15, 106-18.

Hoopen, M. ten & Reuver, H. A. (1965a). On a waiting time problem in physiology. Statistica Neerlandica 19, 27-34.

Hoopen, M. ten & Reuver, H. A. (1965b). An n-fold coincidence problem in physiology. J. Theor. Biol. 9, 117-23.

Ikeda, S. (1965). On Bouman-Velden-Yamamoto's asymptotic evaluation formula for the probability of visual response in a certain experimental research in quantum-biophysics of vision. Ann. Inst. Stat. Math. 17, 295-310.

Ingarden, R. A. & Mikusinski, J. G. (1953). Considérations théoriques sur la granulation. Sci. Ind. Photogr. 24, 280-1.

Jones, R. C. (1959). Quantum efficiency of detectors for visible and infrared radiation. In *Advances in Electronics and Electron Physics*. L. Marton, ed. New York: Academic Press.

Jones, R. C. (1961). Information capacity of photographic films. J. Opt. Soc. Amer. 51, 1159-71.

Kemperman, J. H. B. (1971). Private communication.

Kingman, J. F. C. (1969). Markov population processes. J. Appl. Prob. 6, 1-18.

Langner, G. (1963). Zur Berechnung der Kornigkeit and der Kontrastschwelle photographischer Schichten. Photogr. Korr. 99, 177-85.

Lewis, P. A. W. (1971). Private communication.

Mack, C. (1948). An exact formula for $Q_k(n)$, the probable number of k-aggregates in a random distribution of n points. Phil. Mag. (Seventh series) 39, 778-90.

Marchant, J. C. & Dillon, P. L. P. (1961). Correlation between random-dot samples and the photographic emulsion. J. Opt. Soc. Amer. 51, 641-4.

Matérn, B. (1960). Spatial variation. Stochastic models and their application to some problems in forest surveys and other sampling investigations. Medd. fron Statens Skogsforskningsinstitut. 49, 1-144.

Mees, C. E. & James, T. H., eds. (1966). *The Theory of the Photographic Process*. Third Ed. New York: Macmillan.

Meyer, R. M. (1967). Some Poisson-type limit theorems for sequences of dependent rare events, with application. Ph.D. Thesis submitted to The Department of Statistics, University of North Carolina.

Naus, J. I. (1965). The distribution of the size of the maximum cluster of points on a lines. J. Am. Statist. Assoc. 60, 532-8.

Naus, J. I. (1966). Some probabilities, expectations, and variances for the size of largest clusters and smallest intervals. J. Am. Statist. Assoc. 61, 1191-9.

Nutting, P. G. (1913). On the absorption of light in heterogeneous media. Phil. Mag. (6) **26**, 423-6.

O'Neill, E. L. (1963). *Introduction to Statistical Optics*, Chapter 7. Reading, Mass.: Addison-Wesley.

Palasti, I. (1960). On some random space filling problems. Publ. Math. Inst. Hung. Acad. Sci. 5, 353-60.

Picinbono, B. (1955a). Modele statistique suggéré par la distribution de grains d'argent dans les films photographiques. Comptes Rendus 240, 2206-08.

Picinbono, B. (1955b). Fluctuations de la transparence locale d'un film photographique uniformément impressioné. Comptes Rendus 240, 2296-8.

Romer, W. & Morawski, T. (1963). The absorption cross section for light of the silver grains in photographic deposits. J. Photogr. Sci. 11, 132-35.

Savelli, M. (1957). Étude d'un modèle suggéré par des recherches sur les propriétés statistiques de la transparence des émulsions photographiques. Comptes Rendus 244, 1710-2.

Savelli, M. (1958). Résultats pratiques de l'étude d'un modèle à trois parametres pour la representation des propriétés statistiques de la granularité des films photographiques et notamment des propriétés spectrales. Comptes Rendus 246, 3605-8.

Selwyn, E. W. H. (1935). A theory of graininess. Photogr. J. 75, 571-80.

Siedentopf, H. (1937). Uber Körnigkeit, Dichteschwankungen und Vergrösserungsfahigkeit photographischer Negative. Physik. Z. 38, 454-9.

Silberstein, L. & Trivelli, A. P. H. (1938). Relations between the sensitometric and the size-frequency characteristics of photographic emulsions. J. Opt. Soc. Amer. 28, 441-59.

Silberstein, L. (1939). A theoretical treatment of the two-quanta hypothesis as applied to the photographic reciprocity law failure. J. Opt. Soc. Amer. 29, 432-47.

Silberstein, L. (1945). The probable number of aggregates in random distributions of points. Phil. Mag. (7) 36, 319-36.

Smith, W. L. (1958). Renewal theory and its ramifications. J. R. Statist. Soc. B 20, 243-302.

Switzer, P. (1965). A random set process in the plane with a Markovian property. Ann. Math. Statist. 36, 1859-63.

Trabka, E. A. (1969). Relationship between rms density and transmittance fluctuations of photographic film. J. Opt. Soc. Amer. 59, 662-3.

Trabka, E. A. (1971). Crowded emulsions: granularity theory for monolayers. J. Opt. Soc. Amer. 61, 800-10.

Webb, J. H. (1950). Low intensity reciprocity-law failure in photographic exposure: energy depth of electron traps in latent-image formation; number of quanta required to form the stable sublatent image. J. Opt. Soc. Amer. 40, 3-13.

Wolfe, R. N., Marchand, E. W. & Depalma, J. J. (1968). Determination of the modulation transfer function of photographic emulsions from physical measurements. J. Opt. Soc. Amer. 58, 1245-56.

Zweig, H. J. (1964). Performance criteria for photodetectors - concepts in evolution. Photogr. Sci. Eng. 8, 305-11.

Zweig, H. J. & Gaver, D. P. (1965). Coincidence counter models with applications to photographic detection theory. IBM J. Res. & Dev. 9, 100-7.

POTENTIAL CONTRIBUTIONS OF THE THEORY OF STOCHASTIC POINT PROCESSES TO THE SOCIAL SCIENCE STUDY OF MAP DISTRIBUTIONS

M. F. Dacey
Department of Geography, Northwestern University, Evanston, Illinois

SUMMARY

The social scientist's experiences with the application of two-dimensional stochastic point processes for study of spatial distributions formed by social and cultural phenomena include many unresolved problems and unfulfilled expectations but few success stories. For this reason, the conventional summary of published results is replaced by a highly generalized summary of the objectives and procedures of this research area with the objective of identifying some of the problems that confront social science research and do not seem tractable with the analytic tools and statistical methods currently provided by the theory of stochastic point processes. The objective of this summary is to identify some of the mathematical and statistical results that could contribute to the social science study of map representations of spatial distributions.

1. INTRODUCTION

Maps are frequently used to record observations on physical reality. Maps that show physical phenomena constitute the substance of the field of geomorphology. Maps that show locations of plants and animals are of interest to ecology. The maps that show cultural and social phenomena are part of the substance of the social sciences. The purpose of this paper is the evaluation of the pertinence of these social science data to stochastic point processes in two-dimensional space. This restricts interest to phenomena that are or can be represented as small dots located on a point map so as to indicate occurrences of a phenomenon in a region of the earth's surface. Because the point maps can become prohibitively complex when the points are related to identifying attributes that may number five, ten, twenty or more, many of the point maps actually analyzed by social scientists are frequently displayed on tables instead of maps. It is nevertheless useful to refer to these data sets as map distributions.

There are three major ways that map distributions become involved in social science research. At one level, they show the locations of social and cultural events that result from the location decisions and actions of individuals and groups. Map distributions are empirical evidence of the results of human behavior and the evaluation of many models involves the comparison of predicted properties with measurements obtainable from map distributions. At another level, map distributions show the locations of social and cultural events that influence man's movement in and use of his terrestrial environment. Because of this dependence, many aspects of human behavior need to be studied with respect to the spatial distributions of the events that structure and direct man's use of space. A model of human behavior that effectively contributes to the making and evaluation of policy that affects land use planning, transportation and network design, or economic and social development, must usually take into account that observed behavior in space reflects many components of the spatial environment in which the behavior occurs. At a different level there are many studies in the social sciences that have no direct reference to behavioral connotations but only require the identification and measurement of attributes of map distributions. Typical topics of these studies are the description and comparison of levels of racial or economic segregation in residential patterns or spatial concentration of health care services.

There are several ways to summarize the pertinent literature. The conventional approach is to identify who did what and how and the results that were obtained. I doubt that this type of review properly conveys the current status and priority problems of this research area to a group whose primary interest is the mathematical and statistical aspects of stochastic point process. An alternative approach is the display of typical data sets. Here again it is difficult to articulate why the data are interesting and the types of results that are pertinent to social science objectives. Instead, I have attempted a highly generalized summary of the objectives and procedures of this research area and its relation to the tools and statistical methods currently provided by the theory of stochastic point processes.

This type of summary may proceed from one of two basic premises. It may be accepted that the currently available probability theory and statistical methods are adequate to handle the problems and questions that arise in social science research. In this case, it is hard to avoid the conclusion that the social sciences have not made particularly effective use of the structure that is available and, moreover, have frequently misapplied methods and techniques. The criticisms of the uses of inferential procedures and stochastic model-building are well-known. In the past, the critics may have been dominated by non-social scientists, but the currently increasing disillusionment with science and its methods has been accompanied by an increase in the frequency and intensity of criticism by social scientists, many of whom were trained in the tradition of inferential research. Despite the ease of defending this premise and the resulting indictment of social science research, the failures of this argument should not be overlooked. It has not been effective in redirecting or restructuring the methodological approach of social science research. Also, the criticism is seldom constructive and accompanied by suggestions of more appropriate procedures for design, implementation or evaluation. Since frequent criticism has had little impact, except possibly the negative effect that many social scientists are reluctant to interact with probabilists and statisticians, there may be merit to the consideration of the alternative premise that the limitation is that the available probabilistic and statistical structure is not sufficiently rich and powerful to encompass the motivation and objectives of social science research.

The ensuing remarks suggest that this failure may be attributable to two over-riding obstacles - the scale of the spatial distributions that interest social scientists and the personal involvement in the phenomena that form these spatial distributions. It is difficult to approach the spatial implications of human behavior with the same level of detachment as for the study of spatial distributions formed by locations of atoms, lunar craters, diseased trees and prairie dogs. In addition, a spatial distribution defined at the scale of a city, state or nation is not amenable to experimentation and has an inherent uniqueness that is difficult to replicate at other regions of the earth's surface.

With this premise in mind, I have tried to construct a perspective or meta model that encompasses the objectives and structure of this research area. To achieve some concreteness, I attempted to identify a model that will "fit" about half a dozen selected studies that seem representative of the best model building and statistical analysis in the social science study of map distributions. The topics of these studies include intraurban migration with emphasis on the old and new homes of movers (Kaiser and Weiss, 1969); the location of cities and towns in rural areas (King, 1961); diffusion of an innovation with emphasis on the location of adopters (Hägerstrand, 1967); the locations of retail establishments in an urban area (Rogers, 1969); intraurban travel patterns with emphasis on shopping trips (Marble, 1967, and Nystuen, 1967); and social contact studies that emphasize social distance (Morrill and Pitts, 1967). These studies include statistical analysis of map distributions or statements of models that generate theoretical map distributions.

Unfortunately, I have not been able to complete the construction of the desired meta model. Possibly, though, I can indicate components of such a model that may illuminate some of the major methodological and technical obstacles that must be resolved before the social science study of map distributions can be effectively related to the currently available theory of stochastic point processes.

First, it is worth mentioning the evidence that fosters a priori acceptance that integration of social science studies with the theory and methods of point processes is an achievable objective. The genesis of this faith is the knowledge that spatial distributions are studied in many fields - physics, astronomy, geology, plant and animal ecology, etc. - which to the outsider appear to have successfully adapted the theory of stochastic point processes to the study of spatial distributions formed by a wide variety of phenomena. These spatial distributions have the same structural and geometric properties as many of the map distributions that occur in social science study. Moreover, these studies use probabilistic models, analytic tools and statistical methods which appear adaptable to the properties and relations that are of concern to social science research. These similarities underlie the conjecture that the theory and methods of stochastic point processes can be integrated with social science studies of map distributions. The adaptability may be illusionary because, thus far, the evidence is not encouraging.

2. POINT MAP DISTRIBUTIONS

One part of this meta model concerns the relation of map distributions to point processes. The concern here is any type of point process and not restricted to those that contain a chance or stochastic component. There are two basic ways of establishing this relation. One approach uses point map distributions that indicate the occurrences of discrete cultural objects that occupy fixed locations for, usually, a lengthy period of time. These map distributions are representations of spatial distributions that have evolved, by a specific time, from man's decisions and actions within the terrestrial confines of his physical, social and economic environments. This evolutionary development comprises a location process that for phenomena that are represented on a map by point symbols may reasonably be treated as a point process. Settlement patterns (such as cities and towns) and locations of economic activities (such as retail establishments) are typical examples of spatial distributions that frequently are considered to be generated by point location processes. A social science study may be directed to identification of properties of this type of point location process.

The other approach is for point map distributions that indicate locations that are indicative of man's behavior in space. These map distributions are typically representations of behavior that is structured by existing spatial distributions, such as the influence during a brief time period of one location event upon the spatial behavior of many individuals, or the influence over a long period of many location events upon a single individual or household. An example of the former is the locations of residences of shoppers at a supermarket during a day, and an example of the latter is the locations of retail establishments visited by members of a family during a month. When these map distributions use point symbols to represent the spatial behavior that results from man's interaction with existing spatial distributions, the underlying decisions and actions comprise a behavioral process that may be reasonably treated as a point process. The behavioral processes that are most commonly studied with respect to point map distributions include travel, shopping, migration and social contacts.

3. MAP DISTRIBUTIONS AND STOCHASTIC PROCESSES

The relationships between map distributions and point processes are more readily obtained than is the identification of the point processes that are of the stochastic type. Because it is not possible to prove or disprove the assertion that a particular spatial distribution is generated by a location or behavioral process that includes a chance element, there is considerable freedom in designating the processes that will be treated as stochastic processes. A decisive factor is the philosophical implications to the specification of the role of chance in human decisions and behavior. For most social scientists, the decision is to reject the probabilistic formulations and statistical analyses of all human and institutional behavior. For the social science studies that are to be integrated into the meta model there are several possible approaches to behavior.

One is to accept that location decisions and behavior in space of individuals and groups are governed by probabilistic laws. Though this view potentially provides for numerous stochastic processes, it is an anathema to most social scientists because it suggests (Curry, 1966, p. 40) "regarding the earth's surface as governed by the mechanism of a roulette wheel and its development as a permanent floating crap game."

A second approach is to accept that the observable results of the location decisions and actions of individuals and groups appear to an observer as though they obey probabilistic laws. While it is philosophically more palatable to accept that the results of behavior only appear as though they were governed by probabilistic laws, this approach contradicts the emphasis of social science research on the explanation of behavior rather than the description or prediction of observable results of location and behavioral processes. To admit models that treat observed behavior as though it obeyed probabilistic laws is, to most social scientists, prima facie evidence that the models fail to explain the constituents and determinants of human behavior. This defect is critical because many social scientists make a sharp distinction between models that explain behavior as opposed to those that only replicate the spatial and other consequences of behavior. Possibly this attitude is adequately reflected in the criticisms that have been directed to the use of stochastic models.

"...These models are verifiable, but they tell us little or nothing about processes. We must recognize these two forms of model - the process model and the predictive model - and realize that process and verifiability are almost mutually contradictory...apart from their normative role, (process models) exist to help us to understand reality, not to predict that reality." (Webber, 1971, p. 27)
Similarly, (Guelke, 1971, p. 42) "the patterns they simulate and describe are frequently beyond their competence to explain."

It is my impression that chance enters studies of behavior in two different ways: by serving as a surrogate for the aspects of behavior that need to be, but are not yet, incorporated within the known and specified deterministic laws, and by encompassing behavior that accompanies inability either to discriminate the effects of alternative decisions or to express a preference for the effects of alternative decisions. As an example, a model for selection of residential sites by households may relate the household's selection of an urban neighborhood to the household's attributes (income, life cycle, etc.) and reaction to the environment (distance to job, perceived amenities, etc.), but may treat the selection of a specific house in this neighborhood as a random selection from the houses that are available within a specified price range. Further, when the model indicates that the household is unable to discriminate among several neighborhoods, the selection of a specific neighborhood is treated as a random selection from the indistinguishable neighborhoods.

This use of probability notions primarily as a convenient representation for ignorance of real world behavior and of behavior that is arbitrary or capricious permits the social scientist to incorporate probabilistic structures for the study of behavior without a deep commitment to a particular probabilistic interpretation of human behavior. The suspicion of probabilistic interpretations of location and behavioral processes in part reflects that many of the models adaptable to map distributions can be derived from not only alternative assumptions but also from contradictory assumptions. As a consequence, it frequently is not clear what interpretation or significance can be attached to the finding that a model fits a map distribution.

There are several important corollaries to this approach.

First, to accept that a map distribution has a random point pattern or is a realization of a Poisson point process is to many social scientists an indication of complete indifference or totally arbitrary decisions, so that there is no interesting location or behavior process to study. As a consequence, few social scientists share the mathematicians' and statisticians' enthusiasm for the Poisson process.

Second, the restriction to aspects of behavior that are not known or cannot be known means that the chance component is typically the least interesting part of a model that purports to explain location decisions and behavior and, accordingly, receives little or no emphasis in the specification of a model. To many social scientists, probability serves only as a fudge factor that accounts for the discrepancies between observation and theory.

Third, the objective of many studies is primarily to explain a specific type of human behavior or, at a more preliminary stage, only to increase the level of understanding of the behavior. These objectives seem to defy incorporation within an acceptable system of hypotheses. At the least, it is difficult to compare the level of understanding that is associated with competing models.

Fourth, a study may be deemed successful if it only succeeds in achieving a description or classification of the pattern formed by a map distribution.

4. THE ELUSIVE UNIVERSE OF MAP DISTRIBUTIONS

Probability theory enters into many social science studies at the operational level by asserting that a set of observations constitutes a random sample from some larger universe of possible observations. However, the specification of this universe is, frequently, embarrassingly difficult to achieve. While the one or more point map distributions that are used for analysis are treated as map representations of a sample from a universe composed of spatial distributions that could be analyzed, the typical situation is considerable ambiguity about the universe that is actually being studied. This ambiguity precludes selection of a random sample and casts suspicion upon the frequent claim that the spatial distributions used for analysis are typical members of the universe. Part of the difficulty reflects that the study of map distributions seldom involves processes that can be studied in the context of experiments. Moreover, because terrestrial space is finite, man occupies and consumes space in a way that distinguishes it from time. Observations can frequently be replicated at other times but, because man's behavior in space is studied only with respect to existing spatial distributions that occupy definite regions, it typically is not possible to replicate the observations at other regions. The finiteness of terrestrial space and the essential uniqueness of the spatial distributions in each region combine to frustrate identification of a universe of map distributions that can be subjected to statistical analysis. Two examples illustrate some of the critical constraints.

The first example is typified by the study of a point map distribution that is defined by the locations of objects within a specified region, such as the cities and towns in a state or the retail establishments in a city. There are many social science studies that formulate a stochastic point process for the locations of these types of objects or that use statistical methods for analysis of properties of the point map distribution that indicates the observed locations. It is difficult to specify the universe that contains as a typical member the map distributions indicating, for examples, the locations of the cities and towns in Iowa or retail establishments in Stockholm. The universe cannot be the same region at different times because the locations observed at a succession of times show only minor changes; these spatial distributions typically have evolved prior to careful observation so that their historic development cannot be constructed with adequate fidelity and, once developed, the locations are sticky so that additions and deletions of objects occur relatively infrequently. The universe cannot include different regions because it is difficult to produce other regions that are comparable in all pertinent respects to Iowa or Stockholm. The alternative to specifying that the observed point map distribution is a sample either from a universe defined by the region at various time points or from a universe defined by comparable regions is that it is a sample from a hypothetical universe that contains all possible Iowas or all possible Stockholms. While the speculation of a super-universe that is not observable is the typical justification for treating a spatial distribution as a sample, this defense conflicts with the view of probability as relative frequency.

The second example is typical of point map distributions that are analyzed with use of social survey data. The map distribution shows the residences of users of the Chicago beaches on a selected summer day. Although such data indicate that the density of users tends to decrease with increasing distance from a beach, distance alone is insufficient either to predict or to explain use of the beach. So, a study of beach use by inhabitants of Chicago would typically include a formulation that relates beach use to distance of residence from the beach, along with factors such as family income, age, distance to alternative recreation facilities, and other attributes that are considered to affect use of public beaches. Typically, statistical methods are used in the analysis of these data. Although the sample day may not be randomly selected, these observations will be treated, probably without strong objections, as a representative sample selected from a universe of summer days in Chicago. In addition, many social scientists would like to treat these observations as representative of beach use in the numerous urban areas that have public beaches. In this case, the failure to randomly select the sample will be overlooked on the grounds that it is no worse to have convenience dictate the selection of the city of study than the day of study. Even granting these concessions, an investigator is reluctant to identify the cities that constitute an appropriate universe because prior to analysis it is not clear for which, if any, cities it is reasonable to assume that comparable observations will have the same statistical properties as the Chicago data. The uncertainty about the existence and composition of a universe of cities conflicts with the view of probability as relative frequency.

A simple example may help to explain how the social scientist can conceive of a location or behavioral process that is pervasive but does not produce a universe of regions that are realizations of a common stochastic location or behavioral process. This example consists of a location process that may be formulated in terms of locating poker chips on a table. The table may represent a city that has established corporate limits. The chips represent public facilities, such as schools or fire stations, that provide a service to the surrounding area. These chips are located on the table one at a time. The location of each chip is defined by an aim point and a random vector that has its origin centered on the aim point, and its length and direction defined by a bivariate probability law, subject to the constraint that the terminal point of the vector is located on the table. Each aim point is defined in terms of the locations of previously located chips and the boundary of the table. A simple model that may have some relevance to the location of public facilities is that the probability that a chip is located at a particular point on the table is proportionate to the sum of distances from that point to the nearest chip and to the nearest boundary point. This location process is readily formulated as a Monte Carlo simulation of a location game.

The game may be played many times at the same table. Alternatively, the game can be played under identical location rules but with each game played at tables that differ in size and shape. The social scientist's problem, in the context of location processes, is analogous to that of given the results of the games played at a variety of tables, determine if the location rule was constant from table to table. The magnitude of the problem is appreciably increased in the context of behavioral processes for which the focus of the study is analogous to the use of space that is structured by the locations of the chips on the various tables. The reason that the social scientist plays the game under rules that do not allow repeated play at any one table is that once objects of a particular type are located within a region it usually is not conceivable that the objects be removed and another location process initiated. It is, however, conceivable that the same type of objects will be located in accordance with the same location process in other regions, but the typical circumstance is that it is not possible to identify two or more regions that have the same physical and geometric attributes. Thus, the social scientist is denied the possibility of observing the same location process operating under identical conditions in the same or a different regional setting.

To return to the game, the locations of poker chips are subject to the same pervasive rules, distance to nearest chip and nearest boundary point, and these basic principles are unaffected by the size or shape of the table on which the game is played. For a study that is motivated by explication of the location process, it is not a matter of great concern that the actual locations are affected by geometric attributes of the table. The attempt to identify the process that produces a spatial distribution for any table may be frustrated by need to take into account attributes of the table but, for the objective of isolating the underlying location process, to the social scientist it is as effective to study the results of a game played at one table as at any other table. The fact that the measurements may vary from table to table may complicate the inferential problem, but it is not an especially relevant issue to a study based on the supposition that the location process is consistent from table to table and the task is to explain the nature of the process.

Though this is a highly simplistic example of the situation that typically arises in social science research, I think it displays some of the basic elements of situations in which social scientists attempt to use stochastic models and statistical methods for the analysis of spatial distributions. It contains a location rule that operates from region to region. This rule contains a chance component. While the chance component varies from region to region in a way that defines unique realizations of different stochastic processes, to the view of the social scientist, it may be the least critical component of a model that purports to explain location decisions and actions. As part of the study of location decisions and actions, it is reasonable to evaluate the hypothesis that the spatial distributions of chips on the various tables are generated by a common location process. Sometimes this process is suggested by a location theory or other considerations, but the more typical circumstance is that the location rule is unknown. It is the anticipation of many social scientists that probability theory and statistical methodology can contribute to the evaluation of the hypothesis. At least, the social science literature is replete with studies that attempt to adapt this theory and method to their needs. While the reason for failure may be obvious, the remedy is not.

A glib explanation is that there is need to "normalize" measurements for the size and shape of table, but it is not at all clear what this operation involves or how it is accomplished. One alternative is evidently for the researcher to treat each table or point map distribution as a unique realization of a stochastic point process, but this approach makes it difficult to justify the use of statistical methods. Alternatively, if he does not adopt statistical methods, then it is not at all clear how to make statements regarding the degree of conformity of his observations to a postulated process.

This is as much of the model as I have been able to construct. It is admittedly a gross simplification of a large literature that is directed to a class of difficult and, evidently, important questions concerning man's use of his terrestrial environment. Possibly, though, the model may help to explain the interests of social science studies that attempt to make use of the theory and statistical methods of point processes. My subjective impression is that the objectives and methods of this research area are reasonable but the tools of analysis are quite inadequate to the task.

5. MISCELLANEOUS PROBLEMS

There are a number of technical problems that emanate from the fact that the tools usually used for analysis of point processes - the cell count method and the method of order distance - are not readily adapted to many of the spatial distributions that arise in empirical analysis. Because many of these problems are widely recognized, reference is made only to the complications that frequently arise in the study of map distributions of social and cultural phenomena.

Though mathematical work has largely concentrated on the Poisson point process, there are few spatial distributions of social and cultural phenomena that have properties of this process. In contrast, a small number of probability laws, such as the negative binomial and Neyman type A, fit cell counts for a wide variety of phenomena. The models that generate these and other discrete probability laws that are usually fit to cell count data seldom are formulations of a process that satisfies the social scientist's intuitive notion of a location process. Though the literature on these probability laws is extensive, it appears to provide only scant information about the location processes that underlie these probability laws or the interpretations that can be attached to the finding that observed cell counts are approximated by a particular probability law.

Though the results of a cell count analysis appear to be affected by the choice of cells, there are few guidelines on appropriate specification of size and shape of cell. Also, use of several sizes of cells *seem* to increase the amount of information available about a spatial distribution, but there is little ability to manipulate the additional data sets.

Similar types of problems confront the use of order distance and other spacing measures. Only one who has worked with many data sets can appreciate how many regions of empirical interest are not convex, and it is not clear how to apply the various spacing measures to such regions. Similarly, it is an empirically verifiable property that most objects will locate close to the boundaries of regions so that most schemes for removing boundary effects also remove most of the observations.

When order distance is used, it is not clear how many orders are to be used or how to interpret results obtained from several orders.

Measurements between points of a point map distribution must reflect different attributes of a spatial distribution than distance to these points from random sample points. The attributes that are being measured are not clear and, except with reference to the Poisson process, it is not clear how to integrate and interpret results obtained from these two kinds of measurements.

A subjective impression is that methods of order distance are not particularly useful in discriminating various types of location processes. Distributions that visually appear quite different frequently yield similar values for the means and variances of the lower order distances. It appears that even when there exists a well-defined universe of map distributions, spacing measures are frequently of little assistance in evaluating alternative hypotheses.

Except for the Poisson process and a few general purpose terms, such as "clustered" and "more regular than random," the social scientist is ill-equipped to study point map distributions. One of the most critical defects is that there is little knowledge of the ways in which measures obtained from point map distributions are indicative of varying kinds and attributes of point distributions. The following example is typical of a conceptual scheme that is frequently used to illustrate the types of attributes that many social scientists want to incorporate in the description or analysis of point map distributions. A point map is conceived as having three independent attributes: density, pattern and

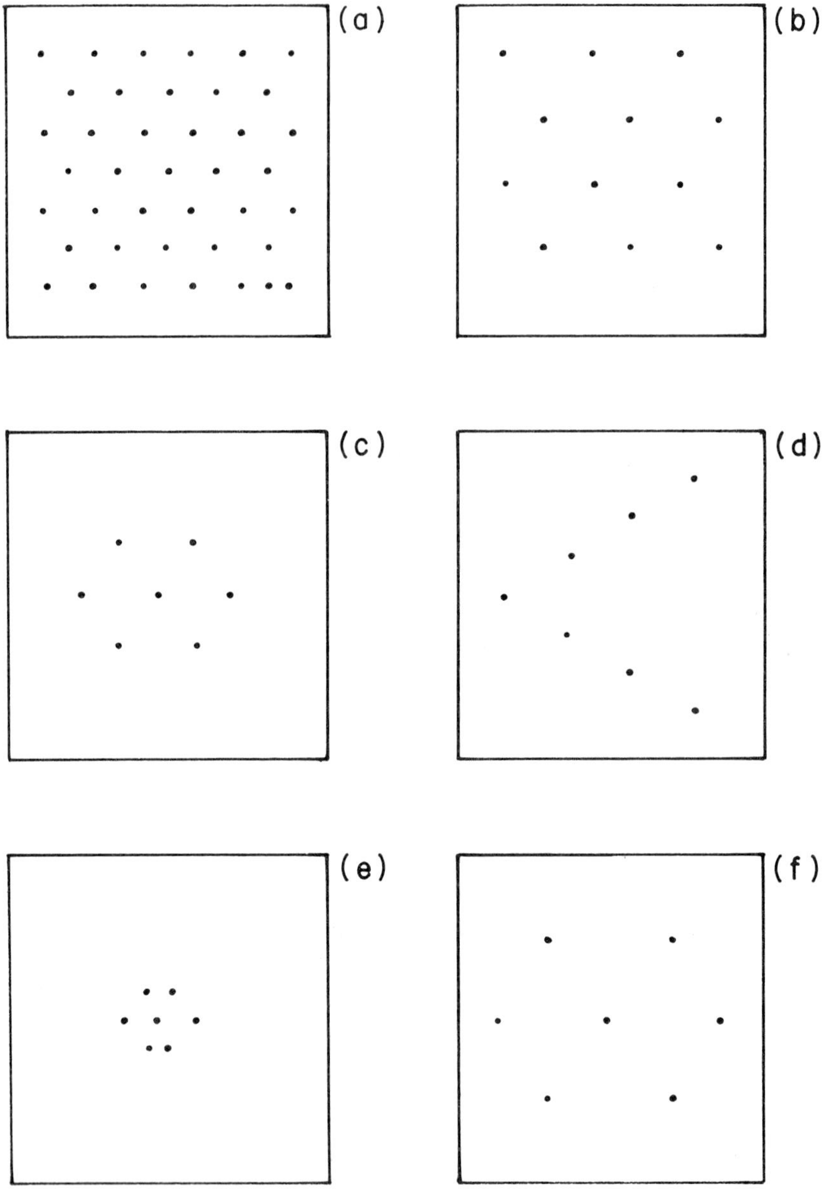

Figure 1. These diagrams are typical of the diagrams used to illustrate independent attributes of point map distributions. Diagrams a and b have the same patterns and dispersions but differ in density. Diagrams c and d have the same densities and dispersions but differ in pattern. Diagrams e and f have the same densities and patterns but differ in disperson.

dispersion. These attributes are explicated by reference to highly stylized diagrams, such as Fig. 1, and to definitions of the following type, which are adapted from Thomas (1965).

The *density* of a spatial distribution is the overall frequency of occurrence of points within a study area *relative* to the size of the study area.

The *pattern* of a spatial distribution is the areal arrangement of points without regard to the size of the study area.

The *dispersion* of a spatial distribution is the extent of the spread of the points within a study area relative to the size of the study area.

Diagrams (Fig. 1) are used to illustrate these attributes and to demonstrate that they may be independently varied.

A type of study of great interest to many social scientists would express these concepts in terms of measurable properties, relate these measures to varying types of patterns and dispersions, identify the pattern and dispersion characteristics of alternative stochastic point processes and develop an experimental design that provides for evaluation of hypotheses concerning the pattern and dispersion characteristics of observed point map distributions.

6. CONCLUSIONS

A posed but as yet unanswered question concerns the prevalence of point map distributions of social and cultural phenomena that may be treated as realizations of stochastic point processes. The answer depends upon the selected frame of reference. For many social scientists, there are none. For others they are innumerable and pervade an extensive variety of social and cultural phenomena, but it cannot be said that this group of social scientists have successfully adapted the theory and methods of stochastic point processes to the study of these map distributions. Hopefully, this survey of the methodological premises that motivate and underline the geographic study of map distributions has identified some of the mathematical and statistical results that could potentially contribute to the social science study of map representations of spatial distributions.

ACKNOWLEDGMENTS

The support of the National Science Foundation, Grant GS-2967, is acknowledged.

REFERENCES

Curry, L. (1966). Chance and landscape. *Northern Geographical Essays in Honour of G. H. J. Daysh*. J. W. House, ed. Newcastle-upon Tyne: Oriel Press.

Guelke, L. (1971). Problems of scientific explanation in geography. *Canadian Geographer* 15, 38-53.

Hagerstrand, T. (1967). *Innovation Diffusion as a Spatial Process*. Chicago: University of Chicago Press.

Kaiser, E. J. & Weiss, S. F. (1969). Decision agent models of the residential development process - a review of recent research. Traffic Quarterly 23, 597-631.

King, L. J. (1961). A multivariate analysis of the spacing of urban settlements in the United States. *Annals of the Association of American Geographers* 51, 222-33.

Marble, D. F. (1967). A theoretical exploration of individual travel behavior. *Quantitative Geography, Part I-Economic & Cultural Topics*. D. F. Marble & W. L. Garrison, eds. Evanston: Northwestern University.

Morrill, R. L. & Pitts, F. R. (1967). Marriage, migration, and the mean information field: a study in uniqueness and generality. *Annals of the Association of American Geographers* 57, 401-22.

Nystuen, J. D. (1967). A theory and simulation of intraurban travel. *Quantitative Geography, Part I-Economic & Cultural Topics*. D. F. Marble & W. L. Garrison, eds. Evanston: Northwestern University.

Rogers, A. (1969). Quadrat analysis of urban dispersion: 1. Theoretical techniques 2. Case studies of urban retail systems. Environment & Planning 1, 47-80 and 155-71.

Thomas, E. N. (1965). *A Structure of Geography*. Boulder: University of Colorado.

Webber, M. J. (1971). Empirical verifiability of classical central place theory. Geographical Analysis 3, 15-28.

Author Index

Abell, G.O., 666
Abbe, E., 468
Adolph, A., 739, 740
Adrian, E.D., 713
Agnew, R. A.,558
Akilov, G. P., 483
Albus, J. S. 707, 708
Alff-Steinberger, C., 760
Almquist, A., 810
Almqvist, E., 189
Ambartzumian, R. W., 399, 520, 551, 555, 597, 598, 635, 636
Arvanitis, L. G., 806
Asmus, F., 134

Bailey, N. T. J., 292, 505, 675, 760, 765
Balagangadharan, K., 338, 406, 409, 412
Bar-David, I., 29
Barlow, R., 775, 776
Barndorff-Nielsen, O., 218, 226, 227
Barr, M. L.,704
Bartlett, M. S., 2, 7, 18, 19, 24, 28, 41, 42, 44, 48, 61, 136, 137, 138, 142, 146, 208, 218, 220, 221, 225, 234, 235, 243, 259, 261, 313, 315, 316, 331, 332, 343, 406, 453, 468, 472, 473, 479, 480, 485, 496, 505, 537, 538, 651, 675, 676, 686, 690, 756, 854
Barton, D. E.,769
Bartoszyński, R., 677, 678
Barr, R., 3, 49
Bary-Lenger, A., 804
Bates, G. E., 652
Bauersachs, E., 126
Bayer, B. E., 855
Bayly, E. J., 709, 721
Beall, G., 505
Becker, N., 767
Bédard, G., 29, 218, 226, 227
Beichelt, F., 809
Bellman, R. E., 537
Belyaev, Yu. K., 300, 346, 348, 349, 350, 352, 374, 438, 446, 449, 450, 452, 457, 458, 459, 460, 461
Belyayev, Yu., 776
Bendat, J. S., 721
Benton, S. A., 832, 834
Berman, S. M., 460, 641
Beutler, F. J., 344, 359
Bhabba, H. J., 472, 473, 534, 541
Bickel, P. J., 23
Billingsley, P., 306, 327, 367, 479
Birnbaum, Z. W., 775
Bishop, P. O., 714
Bitterlich, W., 806
Blackith, R. E., 132
Blanc-LaPierre, A., 850
Blum, E. H.,848
Blumenthal, S., 27
Bochner, S., 317, 330, 331, 332, 337, 338, 344
Boswell, M. T., 33, 285, 295
Bowman, R. J., 722
Breiman, L., 67, 167, 362, 365
Brillinger, D. R., 27, 42, 322, 341
Bromley, D. A., 182, 184, 188, 189 197
Bronowski, J., 650
Brown, G. S., 810
Brown, J. L., 823
Brown, M., 48, 68, 69, 72, 73, 81, 87, 373
Brown, S., 131
Brunk, H. D., 33
Buckley, D. J., 167
Bühler, W. J.,678
Bulinskya, E. V.,437, 441
Bush, B. M. H., 708

Calvin, W. H., 709
Cannings, C., 285
Carslaw, H. S., 706
Cartwright, D. E.,463

Carrier, G. F., 861
Castillo, J. del, 736, 737
Catana, A. J.,132, 807
Chakrabarty, S. K., 541
Chasman, C., 182
Chernov, Yu. A., 691
Chua, L., 188
Chung, K. L., 550, 557
Cinlar, E., 372, 558, 572, 574, 617
Clark, A. J., 132
Clark, J. R., 2, 29, 218, 216, 227, 228
Clark, L. F., 703, 722
Clark, P. J., 804
Cohen, J. E., 295
Cormack, R. M., 469
Cottam, G., 126, 128
Cooley, J. W., 7, 718
Cowan, J. D.,703, 711
Cowlin, R. W., 802
Cox, D. R., 1, 2, 3, 4, 6, 7, 8, 9, 10, 11, 13, 17, 18, 22, 23, 24, 25, 27, 28, 30, 31, 34, 36, 39, 41, 43, 45, 48, 55, 60, 61, 63, 65, 66, 69, 83, 84, 136, 137, 201, 203, 204, 206, 208, 212, 214, 215, 218, 219, 220, 222, 225, 226, 248, 249, 259, 273, 274, 275, 277, 278, 280, 282, 300, 310, 311, 315, 332, 364, 368, 406, 415, 479, 496, 551, 562, 563, 598, 709, 715, 776, 782, 810, 851
Craig, C. E., 132
Cramér, H., 201, 300, 330, 364, 436, 437, 439, 440, 445, 448, 449, 452, 454, 460, 461
Curry, L., 873
Curtis, J. T., 126, 128

Dacey, M. F., 869
Daley, D. J., 252, 323, 332, 334, 335, 343, 346, 347, 349, 351, 364, 368, 384, 408, 425, 429, 468, 479, 617, 651, 691, 795
Daniels, H. E., 675, 764
Darian-Smith, I., 716, 722
Darling, D. A., 218, 225, 226
Darwin, J. H., 286
David, F. N., 769
Deltheil, R., 132
Denton, Gillian M., 760, 762
DePalma, J. J., 828
DeWitt, C. T.,134
Dice, L. R.,133
Dillon, P. L. P., 829
Dimitriev, N. A., 534
Dietz, K., 757
Dodge, F. A.,722, 732, 740, 744, 745, 747, 750, 751
Doerner, E. C., 826
Doksum, K. S., 23
Doob, J. L., 313, 318, 332, 337, 338, 373, 402, 562
Dowling, J., 747, 748
Downton, F., 212, 760, 775, 780
Davidson, R., 516, 519, 527, 531
Davies, R. D., 7, 343
Debes, H., 372, 402
Dobrushin, R. L., 344, 372, 373, 402
Dudley, R. M., 339
Durand, D., 38, 39
Durbin, J., 18, 22

Eadie, G. S., 468
Eberhardt, L. L., 133
Eccles, J. C., 708, 712
Eysel, U. T., 711
Effron, B., 503
Eiby, G. A., 7
Eikler, A. R., 469
Ekholm, A., 208, 272, 274, 277
Elteren, P. Van, 858
Epstein, B., 775
Erlander, S., 107
Esary, J., 775
Essed, F. E., 126
Evans, R. D., 176, 177, 194
Evans, F. C., 132, 804

Fatt, P., 755
Faure, P., 850
Feller, W., 75, 369, 370, 498, 505, 691, 780
Feynman, R., 544
Fieger, W., 348, 351, 356, 359, 443, 447
Fielder, G., 684, 690
Finney, D. J., 806

Firk, F. W., 181
Fisher, L., 313, 503
Flehinger, B. J.,776
Fisher, R. A., 36
Foote, R. S., 151
Forrest, J. S., 34
Franken, P., 240, 241, 342, 392, 551, 563, 566, 574, 581, 582, 584, 587, 589, 590, 591, 594, 595
Frankenhaeuser, B., 706
Fraser, A. R., 804
French, A. S., 708, 710, 716, 717, 718, 719, 720, 722, 725
Frisch, H. L., 469
Fuortes, M.G.F., 711
Furry, W. H., 533

Gafarian, A. J., 167
W. R. Gaffey, 653, 654, 655
Gani, J., 296, 469, 675, 758, 760, 761, 762, 763
Gaver, D. P. 9, 28, 212, 217, 218, 228, 229, 232, 775, 776, 777, 782, 794
Gazis, D. C., 151, 152, 154, 157, 158
Geffroy, J., 503
Geisler, C. D., 711
Geman, D., 452
Gerrits, H.J.M., 858
Gelfand, I. M., 303, 318, 330, 331, 332, 334, 338, 406, 409, 479, 408, 481
Gerstein, G. L., 6, 7, 709, 710
Ghosh, B., 133
Gilbert, E. N., 469, 504
Gill, C. B., 802, 803
Gilles, D. C., 26, 243
Gnanadesikan, R., 7
Gnedenko, B. V., 300, 371, 567, 607, 608, 610, 776
Gobbi, A., 188
Goldberg, J. M., 711
Goldman, D. E., 707, 708
Goldman, J. R., 67, 372, 373, 374, 499, 501, 563, 568, 617
Goldman, L., 707, 708
Goodall, D. W., 133
Goodman, A. S., 11, 12, 13, 17, 19, 49
Govier, L. J., 26, 237, 249
Grandell, J., 28, 30, 90, 92, 96, 99, 109, 117, 218, 221, 235
Grant, J.A.C., 803
Greenwood, M., 651
Greenwood, J. A., 27, 38, 39
Grenander, V., 91, 105, 106, 107
Grieg-Smith, P., 133, 142, 469, 491, 504, 804
Griffiths, R. C., 9
Grigelionis, B., 371, 503, 563, 566, 574, 582, 587, 589, 590, 590, 607, 617
Guelke, L., 874
Gumbel, E. J., 38, 39
Gurland, J., 344, 505
Gustavson, J., 107

Hagerstrand, T., 871
Hagger, R. A., 704
Haig, I. T., 802
Hale, B. W., 128
Hamilton, J. F., 855, 856, 863
Hammersley, J. M., 133, 469
Hannan, E. J., 41, 114, 115, 116
Harris, R., 776
Harris, T. E., 300, 304, 373, 468, 485, 505, 537, 672
Hasel, A. A., 806
Haugh, E. F., 840
Hausberg, H., 133
Hawkes, A. G., 261, 267, 270, 345, 479, 794
Hebb, D. O., 703
Heitler, W., 534
Hellstrom, C. W., 2
Henneman, E., 703
Herbach, L., 27
Hertz, P., 123
Higgins, G. C., 827
Hille, E., 483
Hinz, P., 506
Hodgkin, A. L., 705, 707
Holden, A. V., 710, 716, 717, 719, 720, 725
Holgate, P., 124, 129, 130, 468, 503, 504, 804, 806
Hopkins, B., 129, 130
Hoversten, E., 29,
Hoyle, F., 672
Hubbard, J. I., 712

Hughes, G. W., 722
Humason, M. L., 669
Hutchinson, N. A., 707
Huxley, A. F., 706, 707
Hyvärinen, J., 208, 272, 273, 278 729

Ikeda, S., 859
Ingarden, R. A., 831
Ito, K., 318, 332, 406, 409, 412, 437
Ivanov, V. A., 437, 441

Jack, J. J. B., 712
Jackson, O. A. Y., 23
Jacobs, D. R., Jr., 795
Jaeger, J. C., 706
Jagers, P., 374, 503, 563, 569
Johannesma, D. I. M., 709, 711
Jakeman, E., 29
James, T. H., 818, 823
Janossy, L., 537, 538
Jerwood, D., 758, 760, 767
Jirina, M., 300, 306, 331, 374
Jones, R. C., 827, 828
Jorgenson, D. W., 776
Joshi, S. W., 469
Jowett, J., 338

Kac, M., 225, 226, 227, 228, 437, 439, 440, 441
Kaiser, E. J., 871
Kalman, R. E., 155
Kandel, E. R., 712
Kantorovich, L. V., 483
Kaplan, E. L., 354
Karlin, S., 7,311
Karp, S., 2, 218, 226, 227, 228
Katcher, A. M., 7
Katz, B., 732, 736, 737, 738, 739
Kemperman, J. H. B., 832
Kendall, D. G., 298, 331, 485, 533, 534, 537, 538, 675, 676
Kenall, M. G., 125, 503, 520
Kennedy, D. P., 67, 374
Kerstan, J., 258, 360, 369, 370, 384, 399, 402, 574, 583, 617, 635
Kenton, B., 725
Keuls, M., 134

Khintchine, A. Y., 202, 224, 236, 300, 307, 315, 348, 350, 354, 358, 371, 372, 529, 563, 570, 607, 617, 635, 636
Kiang, N. Y., 714, 722
Kibble, W. F., 212, 217
King, L. J., 871
Kingman, J. F. C., 221, 233, 307, 520, 855
Kirschfield, K., 722
Klauber, M. R., 770
Knapp, C. H., 152, 154, 157, 158
Knight, B. W., 709, 710, 716, 722, 732, 740, 744, 745, 747, 750, 751
Knox, C. K., 709
Knox, G., 769, 770
Koles, Z. J., 707, 722
Kolmogorov, A. N., 331, 332, 480, 534, 567, 608, 610
König, D., 315
Koppel, D. E., 29
Kornhuber, H. H., 716, 722
Kovalenko, I. N., 300, 371, 577
Krickeberg, K., 515, 519
Kronauer, R. E., 832, 834
Kruger, L., 725
Kuehner, J. A., 189
Kummer, G., 398, 528
Kurth, H., 812

Lampard, D. G., 8, 9, 212, 215, 216
Lamperti, J., 498
Langner, G., 841
Lárusson, E., 107
Lawrance, A. J., 28, 242, 345, 424
Lawrence, R., 167
Leadbetter, M. R., 201, 300, 321, 330, 346, 348, 349, 350, 352, 356, 358, 364, 430, 437, 439, 440, 441, 445, 446, 448, 449, 452, 454, 456, 457, 461, 501, 797
LeCam, L., 313, 331, 344, 480, 581, 582
Lee, P. A., 215, 217
Lee, P. J., 258, 368, 369, 370, 394, 396, 608, 610
Lehman, E. L., 81, 82
Leneman, O. A. Z., 344, 359
Levick, W. R., 714
Levitan, H., 7, 715

Lewis, P. A. W., 1, 2, 3, 4, 6, 7, 11, 12, 13, 17, 18, 19, 22, 23, 24, 25, 26, 27, 28, 30, 31, 34, 36, 37, 38, 39, 41, 42, 44, 45, 46, 47, 55, 60, 61, 66, 69, 83, 136, 137, 197, 201, 203, 206, 208, 218, 225, 235, 236, 240, 241, 242, 243, 247, 248, 249, 250, 251, 252, 259, 273, 274, 275, 277, 278, 280, 282, 300, 310, 311, 313, 315, 332, 343, 364, 368, 406, 415, 479, 496, 506, 551, 575, 577, 598, 651, 715, 775, 810, 852
Lewis, T., 237, 248, 249, 364
Liebherr, H. G., 134
Liemant, A., 342, 372, 392, 401, 402, 574
Lindgren, G., 439, 463, 464
Llinás, R., 712
Longuet-Higgins, M. S., 438, 449, 452, 453, 454, 458, 463
Lord, R. D., 134
Loś, J., 678
Lowdermilk, W. C., 802
Luckew, R. S., 777, 794
Lundberg, O., 285, 286
Lynch, D. W., 802

Macchi, O., 218, 220, 227
Mack, C., 860
Maffei, L., 722
Maher, J. V., 184
Malevich, T. L., 463
Maloley, G. T., 808, 810
Mandelbrot, B., 709, 710
Mannard, A., 722, 725
Mantegazzini, F., 711
Mantel, N., 770
Marble, D. F., 871
Marcinkiewicz, J., 562
Marchand, E. W., 828
Marchant, J. C. 829
Marcus, A. H., 684, 686, 689, 690, 691, 695, 696, 697
Marshall, A. W., 775, 780
Matern, B., 134, 300, 469, 496, 498, 503, 504, 692, 693, 851, 805, 806, 808, 812
Matthes, K., 258, 261, 315, 354, 356, 360, 362, 365, 369, 370, 372, 384, 392, 393, 394, 398, 399, 402, 456, 498, 499, 528, 574, 617, 635

Mathews, P. M., 541
Matthews, P. B. C., 714, 716, 722, 723
Maurenzig, P. A., 188
Mawson, J. C., 134, 804
May, A. D. 167
Mayall, N. V. 669
Mazumdar, M., 776
McCall, J., 776
McCulloch, W. S., 703, 709
McFadden, J. A., 201, 202, 235, 243, 364, 553, 554
McNeil, D. R., 767, 768
McNeill, D., 795
McNeill, I. B., 2
Mecke, J., 360, 370, 374, 384, 398, 399, 555
Mees, C. E., 818, 823
Melvill Jones, G., 722
Mercer, A., 775
Meyer, R. M., 858
Meyer, P. A., 530, 531
Mikusinski, J. G., 831
Miles, R. E., 316, 469, 520, 811
Miller, H. D., 222, 226, 709, 782
Milne, R. K., 251, 252, 257, 259, 314, 315, 329, 369, 469, 501, 502
Milsum, G. H., 722
Mitchell, K. J., 810
Mönch, G., 385
Moore, G. P., 6, 7
Moore, P. G., 123, 124, 129, 130, 714, 715
Moran, P. A. P., 9, 10, 11, 13, 125, 209, 211, 503, 520, 793
Morawski, T., 839
Morisita, M., 134
Morrill, R. L., 871
Mode, C. J., 286
Mountcastle, V. B., 710, 716, 722
Moyal, J. E., 304, 324, 343, 468, 469, 472, 481, 484, 485, 505
Mueller, C. G., 823
Murali Rao, K., 522, 529, 531, 532

Murthy, V. K., 167
Mountford, M. D., 129

Nagaev, A. N., 765
Naus, J. I., 504, 860
Nawrotzki, K., 304, 374
Nelson, R. B., 237, 248, 249, 250
Neveu, J., 72, 354, 360
Newell, G. F., 166
Newman, D. S., 251, 252, 257, 258, 329
Newnham, R. M., 808, 810
Neyman, J., 236, 312, 342, 343, 468, 492, 493, 495, 501, 504, 575, 643, 648, 650, 652, 657, 661, 666, 667, 668, 669, 673, 676
Nicholson, C., 712
Nishizawa, M., 810
Noble, D., 705
Nosko, V. P., 461
Nutting, P. G., 829
Nystuen, J. D., 871

Olkin, I., 775, 780
O'Neill, E. L., 831
Oosting, H., 718
Opie, J. E., 809
O'Regan, W. G., 806
Orey, S., 462
Osbourne, J. C., 806
Ososkov, G. A., 371, 563, 570, 607, 617
Over, H. J., 134

Palasti, I., 848
Palm, C., 354, 371, 562, 563, 617
Paloheimo, J. E., 484, 506
Park, T., 674
Parthasarathy, K. R., 331, 368
Papengelou, F., 354, 360, 516, 522, 525, 526, 527, 528, 529
Parker, P. D., 188
Parzen, E., 95, 96
Patil, G. P., 285, 295, 469, 805
Payandeh, B., 804, 806
Penfound, W. T., 128
Perkel, D. H., 6, 7, 18, 25, 714, 715
Persson, O., 125, 131, 503, 504, 805
Phillips, R. S., 483
Picinbono, B., 831, 838
Pickands, J., 462
Pielou, E. C., 129, 130, 142, 143, 469, 804, 807
Piersol, A. G., 721
Pike, M. C., 769
Piterbarg, V. I., 452
Pitts, F. R., 871
Pitts, W. H., 703, 709
Poppele, R. E., 716, 722
Prekopa, A., 315
Pringle, J. W. S., 720
Prohorov, Yu. V., 331, 480, 582
Proschan, F., 23, 775, 776
Purple, R. L., 716
Pusey, P. N., 29
Pyke, E. R., 29
Pyke, R., 22, 23, 45, 623

Quastel, D. M. J., 712
Quenouille, M. H., 17
Qureishi, A. S., 6

Radha, T. K., 539
Radner, R., 776
Ramakrishnan, A., 322, 473, 534, 537, 538, 539, 540, 541, 543, 544, 546
Rall, W., 712
Ranson, S. W., 703
Ranganathan, N. R., 541, 544
Rao, C. R., 82, 505
Redman, S. J., 712
Reichardt, W., 722
Reiffer, B., 29, 30
Renyi, A., 374, 503, 559
Rescia, R. R., 505
Rescigno, A., 730
Reuver, H. A., 237, 858
Reynolds, D. S., 775
Rice, F. L., 128
Rice, S. O., 437, 438, 453
Richter, G., 240, 241, 342
Rizzolatti, G., 722
Roberts, A., 708
Roberts, F. D. K., 469, 504
Roberts, W. J., 722
Rogers, A., 871
Romer, W., 839

Rosenblatt, M., 105, 106, 107, 366
Rosenthal, N. P., 722
Ross, S. M., 27
Roy, B. K., 711
Rozanov, Yu. A., 338, 406, 409, 412, 459, 460
Rudra, A. B., 809
Rushton, W. A. H., 705, 706, 708
Ryll-Nardzewski, C. 354, 360, 362, 522, 526, 527, 528, 529

Sachs, M. W., 182, 184, 188
Sage, A. P., 155
Sahay, B. K., 803
Sakata, H., 722
Sandage, A. R., 653, 669
Santalo, L. A.,644
Sapagouas, J., 622, 623
Saunders, S. C., 775
Savage, I. R., 775
Savelli, M., 831, 832, 835, 836 837, 838
Schaufele, R., 623
Scheuer, E., 776
Schumacher, F. X., 802
Schwartz, L., 409
Scott, A. J., 469
Scott, E. L., 236, 312, 343, 468, 492, 493, 495, 501, 504, 575, 643, 652, 657, 666, 667, 668, 669, 673, 676
Scott, W. T., 534
Segundo, J. P., 7, 714, 715
Selwyn, E. W. H., 825, 830
Sereling, R. J., 364
Serfozo, R. F., 218, 221, 233, 235
Shane, C. E., 657
Shank, R. E., 128
Shannon, C. E., 725
Shapira, D., 188
Shapley, R.M., 732, 751, 753, 754
Sherman, H., 29, 30
Shiryaev, A.M., 331
Shlien, S., 7, 26
Shorrock, R., 67
Sie, S., 184
Siedentopf, H., 831
Siegert, A.J.F., 218, 225, 226, 227, 228
Siemssen, R. H., 184
Silberstein, L., 829, 858, 860
Sjunnesson, S., 810
Skellam, J. G., 123, 129, 130, 501, 505
Skorohod, A. V., 619
Smith, D. R., 711
Smith, R. S.,707
Smith, W. L., 27, 310, 429, 562, 563, 845
Slivnyak, I. M., 351, 354, 360, 362, 400
Smith, P. G., 769
Sneath, P. H., 469
Snyder, D. L., 30, 434
Sobel, M. J., 574
Sokal, R. R., 469
Solovyev, A. D., 776
Sparks, G. A., 166
Spekreijse, H., 718
Spencer, W. A.,712
Spitzer, F., 373
Srinivasan, S. K., 300, 322, 539 540, 541
Srivastava, R. C., 296
Stark, L., 721
Startsev, A. N., 765
Stein, R. B., 705, 708, 709, 710, 711, 712, 713, 714, 716, 717, 718, 722, 723, 725
Stephens, G. R., 809
Stephens, M. A., 39
Stevens, C. F., 709
Stevens, W. L., 475
Steuart, G. N., 170
Stiteler, W. M., 805
Stoffels, A., 135
Stone, C., 67, 619, 372, 374, 402
Störmer, H., 561
Straf, M., 374
Strand, L., 135, 809
Student, 468
Sulanke, R., 503
Suzuki, T., 809
Swain, S., 29
Swanson, M.D., 657
Switzer, P., 851
Symons, M. J., 469
Szász, D., 608
Szeto, M. W., 151

Takacs, L., 691, 692
Talbot, W. H., 716, 722
Tautu, P., 770
Ten-Hoopen, M., 237, 713, 858
Terzuolo, C. A., 722
Thedeen, T., 67, 344, 372
Thomas, A. S., 765
Thomas, E. C., 714, 722
Thomas, E. N., 884
Thompson, H. R., 123, 504
Thompson, S., 131
Toksöz, M. N., 7, 27
Tortrat, A., 398, 399
Toyoda, J. I., 722, 732, 740, 744, 745, 747, 750
Trabka, E. A., 824, 842, 843, 845, 846, 847
Trivelli, A. P. H., 829
Tsaregradsky, I. P., 585
Tukey, J. W., 36, 718
Tulya-Muhika, S., 499, 501
Turner, M. E., 468
Turnovsky, S., 7

Uhlenbeck, G. E., 534

Van den Driessche, P., 804
Vasil'ev, P. I., 346
Vere-Jones, D., 7, 9, 24, 25, 26, 212, 217, 235, 237, 246, 247, 251, 252, 257, 259, 261, 322, 324, 326, 334, 337, 338, 343, 384, 408, 414, 425, 429, 430, 434, 468, 485, 496, 617

Viktorov, S. V., 135
Vilenkin, N. Ya., 303, 318, 330, 331, 334, 338, 406, 409, 479, 480, 481
Vinogradov, O. P., 520
Volkonskii, V. A., 347, 459, 460
Von Neumann, J., 703

Wagener, P. E., 809
Walloe, L., 8, 9, 13, 27
Warren, W. G., 131, 504, 802, 805, 807, 808
Watanabe, T., 714, 722
Watson, G. S., 38, 39
Webb, J. H., 859
Webber, M. J., 847
Weiss, A. H., 7
Weiss, G. H., 764
Weiss, L, 87
Weiss, S. F., 871
Weissblum, W., 554
Weidinger, A., 184
Welch, P.D., 7
Wellner, C. B., 803
Werner, G., 703, 704
Westcott, M., 218, 220, 221, 251, 252, 257, 258, 324, 326, 328, 329, 342, 366, 367, 369, 370, 429
White, I., 34
Whitt, W., 374
Whittle, P., 675
Williams, E. J., 38, 39
Williams, T., 237, 248, 249, 250
Williams, W. O., 714
Wilson, V. J., 720
Wold, H., 8, 202, 212, 214, 215
Wolfe, R. N., 828
Woyczyński, W. A., 608
Wright, G., 140, 145
Wycech-Loś, M., 678

Yaglom, A. M., 338, 406, 418
Yang, G. L., 13
Yemm, R., 722, 725
Yeo, G. F., 218, 226, 227
Ylvisaker, N. D., 437, 439, 441
Yule, G., 651

Zaregradskii, I. P., 520
Ziezold, H., 520
Zitek, F., 350
Zonn, W., 669
Zweig, H. J., 827, 860
Zwickey, F., 656

Applied Probability and Statistics (Continued)

WOLD and JUREEN · Demand Analysis
WONNACOTT and WONNACOTT · Introduction to Econometric Methods
YOUDEN · Statistical Methods for Chemists
ZELLNER · An Introduction to Bayesian Inference in Econometrics

Tracts on Probability and Statistics

BILLINGSLEY · Ergodic Theory and Information
BILLINGSLEY · Convergence of Probability Measures
CRAMÉR and LEADBETTER · Stationary and Related Stochastic Processes
JARDINE and SIBSON · Mathematical Taxonomy
KINGMAN · Regenerative Phenomena
RIORDAN · Combinatorial Identities
TAKÁCS · Combinatorial Methods in the Theory of Stochastic Processes